Mitochondria

METHODS IN MOLECULAR BIOLOGY™

John M. Walker, SERIES EDITOR

METHODS IN MOLECULAR BIOLOGY™

Mitochondria

Practical Protocols

Edited by

Dario Leister

Department für Biologie I, Botanik
Ludwig-Maximilians-Universität München
München, Germany

Johannes M. Herrmann

Zellbiologie
Universität Kaiserslautern
Kaiserslautern, Germany

HUMANA PRESS ✳ TOTOWA, NEW JERSEY

999 Riverview Drive, Suite 208
Totowa, New Jersey 07512

www.humanapress.com

This publication is printed on acid-free paper. ∞
ANSI Z39.48-1984 (American Standards Institute) Permanence of Paper for Printed Library Materials.

Cover design by Nancy K. Fallatt.

Cover illustrations: *(Background)* Images of GFP fluorescence after transfection of mouse embryonic fibroblasts. *See* Chapter 29 for details. *(Inset boxes)* Intracellular localization of mRNA molecules: Transcripts of different genes appear not to be dispersed equally in the cell but show a gene-specific distribution. This localization of mRNAs might improve the targeting of certain proteins to mitochondria. For details *see* Chapter 35.

For additional copies, pricing for bulk purchases, and/or information about other Humana titles, contact Humana at the above address or at any of the following numbers: Tel.: 973-256-1699; Fax: 973-256-8341; E-mail: orders@humanapr.com; or visit our Website: www.humanapress.com

Printed in the United States of America. 10 9 8 7 6 5 4 3 2 1

eISBN 978-1-59745-365-3

ISSN 1064-3745

Library of Congress Cataloging-in-Publication Data

Mitochondria: practical protocols / edited by Dario Leister, Johannes Herrmann.
p. ; cm. -- (Methods in molecular biology, ISSN 1064-3745 ; v. 372)
Includes bibliographical references and index.
ISBN 1-58829-667-9 (alk. paper)
1. Mitochondria--Laboratory manuals. I. Leister, Dario.
II. Herrmann, Johannes (Johannes M.) III. Series: Methods in molecular biology (Clifton, N.J.) ; v. 372.
[DNLM: 1. Mitochondria--physiology--Laboratory Manuals. 2. Mitochondria--genetics--Laboratory Manuals. 3. Mitochondrial Proteins --Laboratory Manuals. 4. Models, Biological--Laboratory Manuals.
W1 ME9616J v.372 2007 / QU 25 M6838 2007]
QH603.M5M58 2007
571.6'57--dc22

2006014176

Preface

Among life scientists, the mitochondrial community originally formed a rather separate and isolated subgroup studying a peculiar organelle. The relevance of mitochondrial research was thought to be limited to the field of bioenergetics. This view changed considerably during the last 10 to 15 years when it became more and more evident that mitochondrial processes are closely intertwined with many other cellular processes. Prominent examples are apoptosis or cellular aging and the observation that several human diseases are related to mitochondrial dysfunction, including Parkinson's and Alzheimer's diseases. As a result, mitochondrial processes are studied by an increasing number of scientists, ranging from molecular biologists, biochemists, structural biologists, and cell biologists to pharmacists and medical researchers.

Our intention in compiling this book is to offer readers a wide range of protocols and reproducible step-by-step procedures for the study of mitochondrial functions. The wide spectrum of methods in a single book will be a valuable resource both for beginners and for experienced scientists. All chapters are written by specialists who have solid experience in the procedures described and who provide the readers with valuable tips and tricks in a special "Notes" section integrated into each chapter.

In Part I of *Mitochondria: Practical Protocols,* a series of chapters introduces the major experimental model systems and discusses their specific advantages and limitations for the analysis of mitochondrial functions. In addition to a short overview on the general properties of these systems, detailed protocols are provided for cultivation of these organisms and on basic experiments in "mitochondriology."

Parts II through VI of *Mitochondria: Practical Protocols* comprise an ample collection of detailed protocols for studying different molecular aspects of mitochondrial functions: genetics and molecular biology (Part II), biochemistry (Part III), physiology (Part IV), dynamics and morphology (Part V), and functional genomics (Part VI). The chapters not only cover the classical methodology of mitochondrial research but also focus on more recent aspects, such as the examination of apoptotic effects, fusion and fission of mitochondria, and proteome and transcriptome analyses.

We are grateful to all contributing authors for the time they devoted to sharing their knowledge and experience.

Johannes M. Herrmann
Dario Leister

Contents

III Biochemistry

IV Physiology

Contributors

Sharon H. Ackerman • *Department of Biochemistry and Molecular Biology, Wayne State University School of Medicine, Detroit, MI*
Cristina Adán • *Departamento de Bioquímica, Instituto de Investigaciones Biomédicas "Alberto Sols" CSIC-UAM, Facultad de Medicina, Universidad Autónoma de Madrid, Madrid, Spain*
Uwe Ahting • *Institute of Human Genetics, GSF National Research Center for Environment and Health, Neuherberg, Germany*
Katrin Altmann • *Institut für Zellbiologie, Universität Bayreuth, Bayreuth, Germany*
Elizabeth A. Amiott • *Department of Biochemistry, University of Utah, Salt Lake City, UT*
Karine Auré • *Institut de Myologie, INSERM U582; Université Pierre et Marie Curie, Groupe Hospitalier Pitié-Salpêtrière, Paris Cedex, France*
Janneke Balk • *Institut für Zytobiologie und Zytopathologie, Philipps-Universität Marburg, Marburg, Germany; Department of Plant Sciences, Cambridge University, Cambridge, United Kingdom*
Pascale Belenguer • *Laboratoire Biologie Cellulaire et Moléculaire du Contrôle de la Prolifération, UMR 5088, Université Paul Sabatier, Toulouse Cedex, France*
Shashi Bhushan • *Department of Biochemistry and Biophysics, Arrhenius Laboratories for Natural Science, Stockholm University, Stockholm, Sweden*
Annette Biederbick • *Institut für Zytobiologie und Zytopathologie, Philipps-Universität Marburg, Marburg, Germany*
Stefan Binder • *Molekulare Botanik, Universität Ulm, Ulm, Germany*
Alexander M. van der Bliek • *Department of Biological Chemistry, David Geffen School of Medicine at UCLA, Los Angeles, CA*
Nina Bolle • *Genetische Botanik und Molekularbiologie, Botanisches Institut und Botanischer Garten, Christian-Albrechts-Universität, Kiel, Germany*
Nathalie Bonnefoy • *Centre de Génétique Moléculaire, UPR 2167, Gif-sur-Yvette Cedex, France*
Nabile Bouzaidi-Tiali • *Departmant of Biology, University of Fribourg, Fribourg, Switzerland*

IRINA BRANDINA • *FRE 2375 CNRS "MLPH," IPCB, Strasbourg, France; Molecular Biology Department, Biology Faculty of Moscow State University, Moscow, Russia*

HANS-PETER BRAUN • *Abt. Angewandte Genetik, Naturwissenschaftliche Fakultät, Universität Hannover, Hannover, Germany*

LAURENCE BULLIARD • *Department of Biology, University of Fribourg, Fribourg, Switzerland*

ANNE-LAURE CHANEZ • *Department of Biology, University of Fribourg, Fribourg, Switzerland*

FABIEN CHARRIÈRE • *Department of Biology, University of Fribourg, Fribourg, Switzerland*

STÉPHANE CHIRON • *Centre de Génétique Moléculaire, UPR 2167, Gif-sur-Yvette Cedex, France*

SARA CIPOLAT • *Dulbecco-Telethon Institute, Venetian Institute of Molecular Medicine, Italy*

G. DESMOND CLARK-WALKER • *Molecular Genetics and Evolution Group, Research School of Biological Sciences, Australian National University, Canberra, Australian Capital Territory, Australia*

RACHEL CLIFTON • *ARC Centre of Excellence in Plant Energy Biology, M316, University of Western Australia, Crawley, Australia*

XAVIER DARZACQ • *Laboratoire de Génétique Moléculaire CNRS-ENS, Paris Cedex, France; Department of Anatomy and Structural Biology, Albert Einstein College of Medicine, Bronx, NY*

SEAN M. DAVIDSON • *The Hatter Institute and Centre for Cardiology, University College London, London, United Kingdom*

FREDERIC DEVAUX • *Laboratoire de Génétique Moléculaire CNRS-ENS, Paris Cedex, France*

MICHAEL R. DUCHEN • *Department of Physiology, University College London, London, United Kingdom*

MARK DÜRR • *Institut für Zellbiologie, Universität Bayreuth, Bayreuth, Germany*

NINA ENTELIS • *FRE 2375 CNRS, "MLPH" IPCB, Strasbourg, France*

HOLGER EUBEL • *Abt. Angewandte Genetik, Naturwissenschaftliche Fakultät, Universität Hannover, Hannover, Germany*

CAROL L. FARR • *Department of Biochemistry and Molecular Biology, Michigan State University, East Lansing, MI*

MIGUEL ANGEL FERNÁNDEZ-MORENO • *Departamento de Bioquímica, Instituto de Investigaciones Biomédicas "Alberto Sols" CSIC-UAM, Facultad de Medicina, Universidad Autónoma de Madrid, Madrid, Spain*

JOACHIM FORNER • *Molekulare Botanik, Universität Ulm, Ulm, Germany*

FRANÇOISE FOURY • *Unité de Biochimie Physiologique, Louvain-la-Neuve, Belgium*
THOMAS D. FOX • *Department of Molecular Biology and Genetics, Cornell University, Ithaca, NY*
LARS-GUNNAR FRANZÉN • *School of Business and Engineering, Naturrum, University of Halmstad, Halmstad, Sweden*
CHRISTIAN FREZZA • *Dulbecco-Telethon Institute, Venetian Institute of Molecular Medicine, Padova, Italy*
SOLEDAD FUNES • *Institut für Physiologische Chemie, Ludwig-Maximilians-Universität München, München, Germany*
MAURICETTE GAISNE • *Centre de Génétique Moléculaire, UPR 2167, Gif-sur-Yvette Cedex, France*
SHILPA GANDRE • *Department of Biological Chemistry, David Geffen School of Medicine at UCLA, Los Angeles, CA*
MATHILDE GARCIA • *Laboratoire de Génétique Moléculaire CNRS-ENS, Paris Cedex, France*
RAFAEL GARESSE • *Departamento de Bioquímica, Instituto de Investigaciones Biomédicas "Alberto Sols" CSIC-UAM, Facultad de Medicina, Universidad Autónoma de Madrid, Madrid, Spain*
ANNA C. GAY • *Department of Anatomy and Cell Biology, College of Physicians and Surgeons, Columbia University, New York, NY*
ELZBIETA GLASER • *Department of Biochemistry and Biophysics, Arrhenius Laboratories for Natural Science, Stockholm University, Stockholm, Sweden*
STEFFI GOFFART • *FinMIT Centre of Excellence; Institute of Medical Technology and Tampere University Hospital, University of Tampere, Tampere, Finland*
DIEGO GONZÁLEZ-HALPHEN • *Departamento de Genética Molecular, Instituto de Fisiología Celular, Universidad Nacional Autónoma de México, D.F., México*
LESLIE I. GRAD • *Membrane Protein Research Group, Department of Biochemistry, University of Alberta, Edmonton, Alberta, Canada*
EMMANUELLE GUILLOU • *Laboratoire Biologie Cellulaire et Moléculaire du Contrôle de la Prolifération, UMR 5088, Université Paul Sabatier, Toulouse Cedex, France*
JOSHUA L. HEAZLEWOOD • *ARC Centre of Excellence in Plant Energy Biology and School of Biomedical and Chemical Sciences, The University of Western Australia, Crawley, Australia*
JOHANNES M. HERRMANN • *Zellbiologie, Universität Kaiserslautern Kaiserslautern, Germany*

IAN J. HOLT • *MRC Dunn Human Nutrition Unit, Cambridge, United Kingdom*
ELKE K. HORN • *Department of Biology, University of Fribourg, Fribourg, Switzerland*
CLAUDE JACQ • *Laboratoire de Génétique Moléculaire CNRS-ENS, Paris Cedex, France*
JUDITH A. JAEHNING • *Department of Biochemistry and Molecular Genetics, University of Colorado at Denver and Health Sciences Center, Aurora, CO*
LAURIE S. KAGUNI • *Department of Biochemistry and Molecular Biology, Michigan State University, East Lansing, MI*
PETR KAMENSKI • *FRE 2375 CNRS "MLPH," IPCB, Strasbourg, France; Molecular Biology Department, Biology Faculty of Moscow State University, Moscow, Russia*
FRANK KEMPKEN • *Genetische Botanik und Molekularbiologie, Botanisches Institut und Botanischer Garten, Christian-Albrechts-Universität, Kiel, Germany*
IGOR A. KRASHENINNIKOV • *Molecular Biology Department, Biology Faculty of Moscow State University, Moscow, Russia*
THOMAS LANGER • *Institut für Genetik, Universität zu Köln, Köln, Germany*
CHRISTOPHER J. LEAVER • *Department of Plant Sciences, University of Oxford, Oxford, United Kingdom*
BERNARD D. LEMIRE • *Membrane Protein Research Group, Department of Biochemistry, University of Alberta, Edmonton, Alberta, Canada*
ROLAND LILL • *Institut für Zytobiologie und Zytopathologie, Philipps-Universität Marburg, Marburg, Germany*
ANNE-MARIE MAGER-HECKEL • *FRE 2375 CNRS "MLPH," IPCB, Strasbourg, France*
FLORENCE MALKA • *INSERM U582 - Institut de Myologie; Université Pierre et Marie Curie, IFR14, Groupe Hospitalier Pitié-Salpêtrière, Paris Cedex, France*
ROBERT P. MARTIN • *FRE 2375 CNRS "MLPH," IPCB, Strasbourg, France*
PETER MARTINSSON • *FinMIT Centre of Excellence; Institute of Medical Technology and Tampere University Hospital, University of Tampere, Tampere, Finland*
YUICHI MATSUSHIMA • *Department of Biochemistry and Molecular Biology, Michigan State University, East Lansing, MI*
J. MICHAEL MCCAFFERY • *Integrated Imaging Center, Department of Biology, The Johns Hopkins University, Baltimore, MD*
SHELLY L. MEEUSEN • *Department of Molecular and Cellular Biology, University of California Davis, Davis, CA*

A. Harvey Millar • *ARC Centre of Excellence in Plant Energy Biology and School of Biomedical and Chemical Sciences, The University of Western Australia, Perth, Australia*

Dejana Mokranjac • *Institut für Physiologische Chemie, Ludwig-Maximilians-Universität München, München, Germany*

Ulrich Mühlenhoff • *Institut für Zytobiologie und Zytopathologie, Philipps-Universität Marburg, Marburg, Germany*

Frank E. Nargang • *Department of Biological Sciences, University of Alberta, Edmonton, Alberta, Canada*

Walter Neupert • *Institut für Physiologische Chemie, Ludwig-Maximilians-Universität München, München, Germany*

Jodi Nunnari • *Department of Molecular and Cellular Biology, University of California Davis, Davis, CA*

Pavel F. Pavlov • *Department of Biochemistry and Biophysics, Arrhenius Laboratories for Natural Science, Stockholm University, Stockholm, Sweden*

Edward M. Perkins • *Integrated Imaging Center, Department of Biology, The Johns Hopkins University, Baltimore, MD*

Liza A. Pon • *Department of Anatomy and Cell Biology, College of Physicians and Surgeons, Columbia University, New York, NY*

Holger Prokisch • *Institute of Human Genetics, Technical University of Munich and GSF National Research Center for Environment and Health, Neuherberg, Germany*

Mascha Pusnik • *Department of Biology, University of Fribourg, Fribourg, Switzerland*

Doron Rapaport • *Institut für Physiologische Chemie der Universität München, München, Germany*

Jörg Reinders • *Protein Mass Spectrometry and Functional Proteomics Group, Rudolf-Virchow-Center for Experimental Biomedicine, Julius-Maximilians-University of Würzburg, Würzburg, Germany*

Aurelio Reyes • *MRC Dunn Human Nutrition Unit, Wellcome Trust/MRC Building, Cambridge, United Kingdom*

Manuel Rojo • *INSERM U582, Institut de Myologie, Groupe Hospitalier Pitié-Salpêtrière, Université Pierre et Marie Curie, Paris Cedex, France*

Charlotta Rudhe • *Department of Biochemistry and Biophysics, Arrhenius Laboratories for Natural Science, Stockholm University, Stockholm, Sweden*

Leanne C. Sayles • *Membrane Protein Research Group, Department of Biochemistry, University of Alberta, Edmonton, Alberta, Canada*

André Schneider • *Department of Biology, University of Fribourg, Fribourg, Switzerland*

LUCA SCORRANO • *Dulbecco-Telethon Institute, Venetian Institute of Molecular Medicine, Padova, Italy*
ALBERT SICKMANN • *Protein Mass Spectrometry and Functional Proteomics Group, Rudolf-Virchow-Center for Experimental Biomedicine, Julius-Maximilians-University of Würzburg, Würzburg, Germany*
ROBERT H. SINGER • *Department of Anatomy and Structural Biology, Albert Einstein College of Medicine, Bronx, NY*
PAUL M. SMITH • *Institut für Zytobiologie und Zytopathologie, Philipps-Universität Marburg, Marburg, Germany*
JOHANNES N. SPELBRINK • *FinMIT Centre of Excellence, Institute of Medical Technology and Tampere University Hospital, University of Tampere, Tampere, Finland*
LINDA L. SPREMULLI • *Department of Chemistry, University of North Carolina, Chapel Hill, NC*
OLIVER STEHLING • *Institut für Zytobiologie und Zytopathologie, Philipps-Universität Marburg, Marburg, Germany*
STEPHANIE SUNDERHAUS • *Abt. Angewandte Genetik, Naturwissenschaftliche Fakultät, Universität Hannover, Hannover, Germany*
THERESA C. SWAYNE • *Department of Anatomy and Cell Biology, College of Physicians and Surgeons, Columbia University, New York, NY*
LEE J. SWEETLOVE • *Department of Plant Sciences, University of Oxford, Oxford, United Kingdom*
IVAN TARASSOV • *FRE 2375 CNRS "MLPH," IPCB, Strasbourg, France*
TAKASHI TATSUTA • *Institut für Genetik,Universität zu Köln, Köln, Germany*
NICOLAS L. TAYLOR • *Department of Plant Sciences, University of Oxford, Oxford, United Kingdom*
ALEXANDER TZAGOLOFF • *Department of Biological Sciences, Columbia University, New York, NY*
THOMAS WAIZENEGGER • *Institut für Physiologische Chemie der Ludwig-Maximilians-Universität München, München, München, Germany*
BENEDIKT WESTERMANN • *Institut für Zellbiologie, Universität Bayreuth, Bayreuth, Germany*
JAMES WHELAN • *Plant Molecular Biology Group, Biochemistry Department, School of Life and Biomedical Sciences, University of Western Australia, Crawley, Australia*
TAKEHIRO YASUKAWA • *MRC Dunn Human Nutrition Unit, Cambridge, United Kingdom*
DEREK YELLON • *The Hatter Institute and Centre for Cardiology, University College London, London, United Kingdom*

I

Model Systems to Study Mitochondrial Functions

1

The Mitochondria of Cultured Mammalian Cells

I: Analysis by Immunofluorescence Microscopy, Histochemistry, Subcellular Fractionation, and Cell Fusion

Florence Malka, Karine Auré, Steffi Goffart, Johannes N. Spelbrink, and Manuel Rojo

Summary

Mitochondria form a dynamic network in which continuous movement, fusion, and division ensure the distribution and exchange of proteins and deoxyribonucleic acid (DNA). The recent past has seen the identification and characterization of the first proteins governing the organization, function, and dynamics of mitochondria and mitochondrial DNA, and it is predictable that numerous new proteins will require localization and functional characterization in the future. In this chapter, we describe methods for the visualization of mitochondria and mitochondrial activity in cultured mammalian cells to establish the localization or distribution of its components and to study mitochondrial fusion.

Key Words: Cytochrome-*c* oxidase; mitochondrial fusion; mitochondrial morphology; succinate dehydrogenase.

1. Introduction

Mitochondria are involved in numerous essential cellular processes: they produce adenosine triphosphate (ATP) by oxidative phosphorylation, participate in various metabolic pathways, contribute to calcium homeostasis and signaling, and play a key role in apoptosis. Mitochondria represent a single cellular compartment where deoxyribonucleic acid (DNA) and proteins are exchanged through continuous fusion and fission reactions ***(1,2)***. Mitochondrial dynamics are governed by specific proteins, a majority of which have been described and characterized in the recent past ***(3,4)***. Mutations in genes encoding such proteins

From: *Methods in Molecular Biology, vol. 372: Mitochondria: Practical Protocols*
Edited by: D. Leister and J. M. Herrmann © Humana Press Inc., Totowa, NJ

are responsible for severe human diseases *(5,6)*, and their knockout hampers development of viable mice *(7)*, demonstrating the relevance of mitochondrial dynamics for the development and function of entire organisms. Dynamics ensure the distribution and exchange of mitochondrial DNA, which is organized in hundreds of punctate nucleoids that distribute throughout mitochondria *(2,8)*. The mitochondrial dysfunctions provoked by defects in mitochondrial DNA maintenance provoke severe diseases in humans *(9)* and in mice *(10)*.

Mammalian cells in culture represent valuable models to study diverse aspects of mitochondrial function, to characterize the organization and dynamics of mitochondria, and to determine the subcellular localization and intramitochondrial distribution of proteins and DNA. In this chapter, we describe protocols for (1) the localization of molecules by immunofluorescence microscopy and subcellular fractionation, (2) the visualization of mitochondria and of mitochondrial respiratory activities, and (3) the study of mitochondrial fusion. Together with Chapter 2, we provide an overview of available methods that allow characterization of mitochondria, identification and localization of new (mitochondrial) molecules, and study of their functions.

2. Materials

2.1. Cell Culture

1. Tissue culture dishes, flasks, and multiwell dishes for culture of adherent cells: the cells described in this study grow directly on plastic or glass, but all protocols can be adapted to coated material (gelatin, poly-L-lysin, etc.).
2. Standard culture media for cell culture, Dulbecco's phosphate-buffered saline (D-PBS) without $CaCl_2$, $MgCl_2$, and trypsin-EDTA (ethylenediaminetetraacetic acid) solution, are stored as indicated by the manufacturer. For culture of human fibroblasts, human 143B, green monkey COS-7, and mouse NIH3T3 cells, we use Dulbecco's modified Eagle's medium (DMEM) containing 4.5 g/L glucose, glutaMAX™ I, and pyruvate. For culture of human HeLa cells, we use minimum essential medium (MEM) with Earle's salts and glutaMAX I. All media are supplemented with 10% fetal bovine serum (FBS), 50 IU/mL penicillin, and 50 µg/mL streptomycin. Media for cells with respiratory deficits (rho-zero cells devoid of mitochondrial DNA and patient-derived cells) are supplemented with 0.2 m*M* uridine (*see* **Notes 1** and **2**). Supplemented media are stored at 4°C for up to 2 mo.
3. 5-Bromo-2-deoxy-uridine (BrdU) stock solution: 10–15 m*M* in H_2O.

2.2. Immunofluorescence Microscopy

1. Microscope slides (76 × 26 × 1 mm) and cover slips (12- to 14-mm diameter for standard immunofluorescence microscopy and 25-mm diameter for cell fusion experiments). Cover slips are autoclaved in large numbers and stored until use.
2. Methanol preequilibrated to –20°C.

3. Solution of 3% (w/v) paraformaldehyde (PFA) in PBS. PFA is dissolved in PBS prewarmed to 80°C and stored frozen in single-use aliquots (10 mL) at –20°C. Frozen aliquots can be stored for months.
4. Quenching solution: 50 m*M* NH_4Cl in PBS.
5. Permeabilization or denaturation solutions: 0.1% (w/v) Triton X-100 in PBS, 0.1% (v/v) or (w/v) sodium dodecyl sulfate (SDS) in PBS (stable at room temperature). Solutions of 8 *M* urea or 2*N* HCl in water are stored at room temperature.
6. Antibody dilution buffer prepared directly before use: 10% (w/v) fetal bovine serum (FBS) in PBS. The addition of 0.04% (w/v) NaN_3 to an aliquot of 100% FBS stored at 4°C prevents microbial growth.
7. To mount cover slips on microscope slide, you can use Mowiol mounting medium (as described in this item) or commercially available mounting medium (e.g., Vectashield, Vector Laboratories). Mowiol mounting medium: add 2.4 g Mowiol 4-88 (= polyvinyl alcohol 4-88, Fluka), 6 g glycerol, 1.6 mL 1.5 *M* Tris-HCl at pH 8.8, and 10.4 mL H_2O to a 50 mL Falcon tube. Agitate at 37°C for several hours to overnight, centrifuge, and freeze supernatant in 1 mL aliquots. After thawing, add 50 ng/mL DAPI (4,6-diamidino-2-phenylindole) for nuclear staining (if desired).
8. Primary antibodies: store as indicated by the manufacturer. In the absence of precise indications and to avoid freeze-thawing, antibodies and antisera are stored at –20°C after addition of glycerol to a final concentration of 50% (v/v).
9. Secondary antibodies specific for mouse, rabbit, or rat immunoglobulin Gs: for colocalization studies with mixtures of primary antibodies from different species, it is essential to use preabsorbed secondary antibodies that do not cross-react with immunoglobulin Gs from other species. We routinely use secondary antibodies labeled with Alexa Fluor 350, 488, or 568 (Molecular Probes). Small aliquots (50 µL) are stored at –20°C, and thawed aliquots are stored at 4°C for up to 2 mo.

2.3. Histochemistry

It is advisable to reserve dedicated vessels or cover slip holders for each reaction as the solutions can irreversibly color the glassware.

2.3.1. Cytochrome-c Oxidase Histochemistry

1. PBS as described in **Subheading 2.3**.
2. Cytochrome-*c* oxidase (COX) preincubation buffer: 50 m*M* Tris-HCl at pH 7.6. Store at room temperature.
3. COX preincubation medium (CPIM): dissolve 28 mg $CoCl_2.6H_2O$ and 10 g sucrose in 100 mL COX preincubation buffer. Add 50 µL dimethyl sulfoxide (DMSO) to 10 mL CPIM immediately before use. CPIM without DMSO can be stored at 4°C for up to 1 mo (*see* **Note 3**).
4. COX rinse medium (CRM): dissolve 10% sucrose (w/v) in 50 m*M* $NaHPO_4$, pH 7.3. 50 m*M* $NaHPO_4$ is stored at room temperature. CRM can be stored at 4°C for up to 1 mo (*see* **Note 3**).
5. Catalase solution: dissolve at 2 mg/mL in water and freeze in 100 µL aliquots for single use.

6. COX incubation medium (CIM): dissolve 10 mg cytochrome-*c* from horse heart prepared without using trichloroacetic acid (TCA) (previously called type IV) and 10 mg hydrochloride diaminobenzidine (DAB; it is recommended to use 10 mg tablets) in 10 mL CRM. DAB is highly toxic and must be disposed properly. Add 100 µL catalase solution and 25 µL DMSO. Pass through a 0.22-µm syringe filter (*see* **Note 4**).
7. Gelatin mounting solution: add 15 g gelatin to 80 mL glycerol and fill up to 100 mL with distilled water. Dissolve by heating and stirring and store in aliquots at 4°C.

2.3.2. Succinate Dehydrogenase Histochemistry

1. PBS as described in **Subheading 2.3**.
2. Prepare a 1 *M* solution of potassium cyanide the day of use (65 mg KCN/mL).
3. Phenazine methosulfate (*N*-methyldibenzopyrazine methyl sulfate salt). Prepare solution at 20 m*M* in water and freeze in single-use (100 µL) aliquots at –20°C.
4. Succinate dehydrogenase (SDH) incubation buffer (SIB): 0.2 *M* $NaHPO_4$, pH 7.6.
5. SDH incubation medium (SIM): 50 m*M* succinic acid, 1.5 m*M* nitroblue tetrazolium (NBT), 5 m*M* EDTA, and 1 m*M* KCN in SIB. Add 59 mg succinic acid and 12.26 mg NBT to 10 mL SIB. Add 0.25 mL of 0.2 *M* EDTA and 10 µL of 1 *M* KCN. Filter with a 0.22 µm filter and add 100 µL of 20 m*M* phenazine methosulfate. SIM without KCN and phenazine methosulfate can be stocked in aliquots at –20°C.
6. Gelatin mounting solution (*see* **Subheading 2.3.1, step 7**).

2.4. Subcellular Fractionation

1. To collect cells, you can use cell scrapers or trypsin-EDTA solutions. Syringes (1–5 mL) are equipped with 22-gage (0.7 × 50 mm) needles.
2. PBS: prepare a 10X stock with 1.37 *M* NaCl, 27 m*M* KCl, 15 m*M* KH_2PO_4, and 81 m*M* Na_2PO_4. Store 10X stock at room temperature. Prepare working solution (1X) by dilution of one part with nine parts water. Store 1X working solution at 4°C.
3. HEPES sucrose (HS): 10 m*M* HEPES at pH 7.5 and 250 m*M* sucrose. Store at 4°C (*see* **Note 3**).
4. Solutions of protease inhibitors. Phenylmethylsulfonylfluoride (PMSF) (200 m*M* in isopropanol) is stored at room temperature. Protease inhibitor cocktail tablets (e.g., Roche Applied Science) are handled and stored as indicated by the manufacturer.
5. For protein quantification, you can use bicinchoninic acid protein assay kit (Pierce) or any other method.

2.5. Polyethylene Glycol-Mediated Cell Fusion

Polyethylene glycol (PEG) 1500 (granulate) from BDH or other supplier: prepare 50% (w/v) PEG in medium without FBS directly before fusion (you will need around 500 µL per 35 mm well or 25 mm cover slip). Add 1 mL medium per gram of PEG and incubate at 37°C for 5–15 min (you can shake during incubation). When solution is transparent, vortex so the PEG from the bottom of the tube is mixed well (*see* **Note 5**).

3. Methods

Mitochondrial distribution, organization, morphology, and dynamics can vary significantly between cell lines (**Fig. 1**) and even between the subclones maintained by different laboratories. Mitochondrial morphology also varies with culture conditions, with the passage number, and so on. It is therefore important to use similar culture conditions in all experiments and always to perform control experiments in parallel.

3.1. Cell Culture

The volumes given here for passage and maintenance of 100 mm dishes (~92 mm diameter, ~57 cm^2 area) must be adapted when using dishes or flasks of other sizes (areas). Immortalized cells are routinely passed twice a week when approaching 80–100% confluence. HeLa, COS-7, or NIH3T3 cells diluted 1:5 to 1:10 and 143B cells diluted 1:10 to 1:20 (surface ratio) become confluent within 3–4 d. In contrast to immortalized cell lines, human primary fibroblasts stop growing and become quiescent when confluent. Human primary fibroblasts diluted 1:2 to 1:3 are usually confluent 1 wk after plating. They can be kept for several weeks without passage, provided that medium is exchanged regularly.

1. Remove medium and wash cells with 10–15 mL PBS. Add 2 mL trypsin/EDTA solution and incubate at 37°C until cells detach from the bottom of the dish (5–15 min).
2. Shake cells gently and add 8 mL fresh medium. Homogenize all cells by gentle up-and-down pipeting (four to six times) throughout the entire dish surface.
3. For fusion experiments only: mix different cell populations by pipeting.
4. Introduce the adequate amount of cells in a new dish/well containing fresh medium and cover slips if desired (*see* **Note 6**). The amount of cells depends on the cell type, its growth rate, the predicted culture time, and the desired degree of confluence (*see* **Note 7**). Distribute cells homogeneously by shaking the dish (not in a circle, but reproducing a figure eight).
5. For labeling of replicating mitochondrial DNA with BrdU: add BrdU (final concentration 10–15 μ*M*) into culture medium 2–24 h before fixation (*see* **Note 2**). The amount of incorporated BrdU will increase with the length of the pulse.

3.2. Immunofluorescence Microscopy

The subcellular localization of molecules is a prerequisite for all studies of their functional and molecular characterization. In this chapter, we describe various fixation and permeabilization protocols with specific advantages and drawbacks. We have found that some antigens are labeled with all fixation or permeabilization conditions, and other antigen-antibody pairs only work with a specific protocol (*see* **Table 1**). If possible, we prefer protocols based on fixation with PFA because mitochondrial structure is preserved (**Fig. 1A–E**).

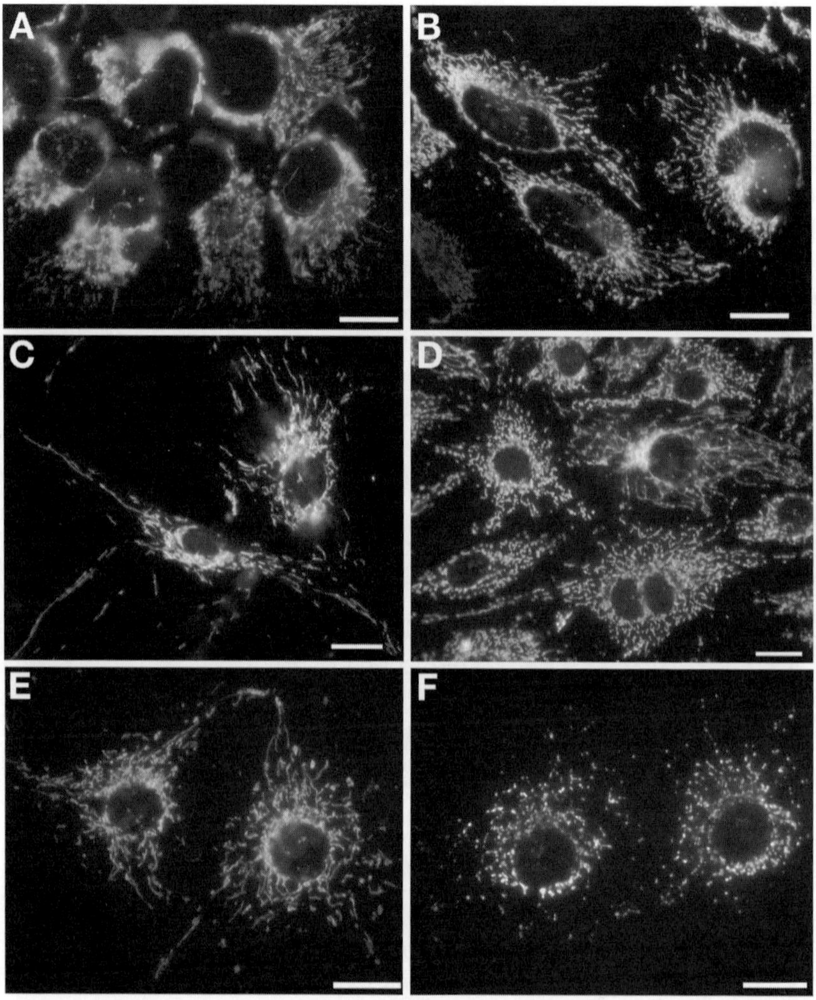

Fig. 1. Morphology and distribution of mitochondria in cultured cells. (**A**) Stably transfected 143B cells expressing GFP targeted to the mitochondrial matrix. (**B**) Stably transfected HeLa cells expressing red fluorescent protein DsRed targeted to the mitochondrial matrix. (**C**) Human skin fibroblasts labeled with antibodies against COX2. (**D**) NIH3T3 cells labeled with antibodies against VDAC. (**E**), (**F**) COS-7 cells labeled with antibodies against cytochrome-*c* (**E**) or Hsp60 (**F**). Cells were fixed with paraformaldehyde (**A–E**) or methanol (**F**), permeabilized with Triton X-100 (**C–E**), and treated with urea (**D**). Bars: 20 μm.

Nevertheless, some antigens are only visualized after fixation and permeabilization with cold methanol, which distorts and fragments mitochondrial filaments (*see* **Table 1** and **Fig. 1F**). An intermediate protocol with short PFA fixation and

Table 1
Antigen/Antibody Pairs Illustrating the Diversity of Adequate Fixation and Permeabilization Conditions

Fixation/ permeabilization	M	P/M	P/TX	P/TX/U	P/TX/HCl	P/SDS
COX2 *(**13**)*	–	–	++	++	++	n.t.
COX4 *(**13**)*	++	±	–	n.t.	n.t.	n.t.
Mfn2 *(**14**)*	+	–	–	–	–	–
mtTFA *(**2**)*	++	++	++	++	+	n.t.
Bromodeoxyuridine (Abcam)	–	–	–	–	++	n.t.
DNA (Progen GmbH)	++	++	++	++	++	++
Cytochrome-*c* (BD Pharmingen)	–	±	++	n.t.	n.t.	n.t.
Hsp60 (Sigma)	++	+	±	n.t.	n.t.	++
VDAC/porin (BD Calbiochem)	±	±	–	++	+	n.t.

M, methanol; P/M, PFA/methanol; P/TX, PFA/Trition X-100; P/TX/U, P/TX followed by urea; P/TX/HCl, P/TX followed by HCl; P/SDS, PFA/SDS; ++, strong signal and low background; ±, low signal or high background; –, no signal; n.t., not tested.

subsequent methanol permeabilization represents a compromise that can work in some cases. In addition, it is also possible to use protocols that use formaldehyde (instead of PFA) for fixation. They are described in Chapter 2.

The circular mitochondrial DNA of mammals was discovered in the mid-1960s, but its localization and dynamics were not reported until recently. Among the reasons for this delay was certainly the difficulty of visualizing it with stains (like DAPI or Hoechst) that allow easy detection of nuclear DNA. In this subheading, we describe the two strategies that achieve efficient visualization of mitochondrial DNA: (1) the visualization of incorporated BrdU with BrdU-specific antibodies *(**8,11**)* and (2) the direct detection of DNA molecules with DNA-specific antibodies (**Fig. 2**) *(**2,12**)*.

Colocalization of new molecules can be performed by using antibodies against known mitochondrial proteins (as discussed in this chapter), by labeling cells with specific mitochondrial dyes, or by using cells expressing fluorescent proteins targeted to mitochondria (*see* Chapter 2).

Cells are plated on sterile cover slips 1–2 d before fixation. Unless for cell fusion experiments (for which high confluence is mandatory), we prefer subconfluent cultures with cells spread flat. Unless otherwise indicated, all steps are performed at room temperature. PFA and methanol are toxic chemicals that require adequate disposal.

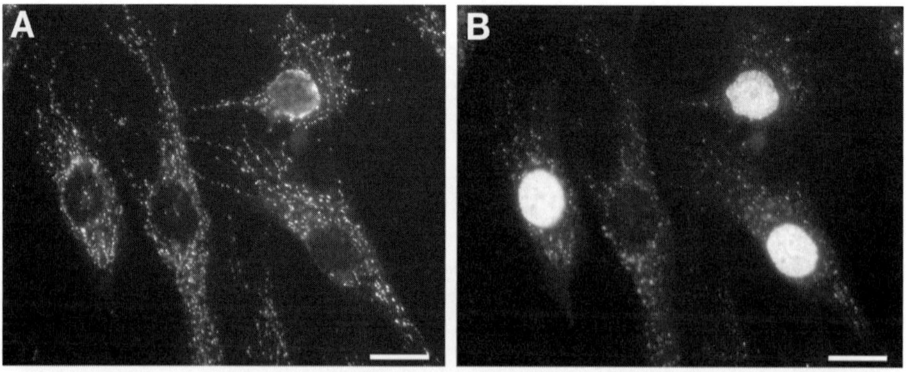

Fig. 2. Visualization of DNA and incorporated bromodeoxyuridine. Human fibroblasts were incubated with BrdU for 16 h, fixed with PFA, permeabilized with Triton X-100, treated with HCl, and labeled with (**A**) mouse antibodies against DNA and (**B**) rat antibodies against BrdU. Bars: 20 μm.

3.2.1. Fixation With Methanol

1. Remove medium and wash once with PBS equilibrated to room temperature.
2. Transfer cover slips to vessels or dishes containing methanol preequilibrated at –20°C. Alternatively, add a large volume of cold methanol to the dish. Incubate in a freezer at –20°C for 5 min.
3. Wash the cells once or twice with PBS. Label with antibodies within the next 1–3 d.

3.2.2. Fixation With Paraformaldehyde

1. Eliminate medium and wash once with PBS equilibrated to room temperature.
2. Transfer cover slips to vessels or dishes containing 3% (w/v) PFA in PBS. Alternatively, add PFA to dish. Incubate at room temperature for at least 20 min (*see* **Notes 8** and **9**).
3. Wash three times with PBS and incubate with quenching solution for 10 min.
4. Permeabilize for 5 min with 0.1% (w/v) Triton X-100 or 0.1% SDS in PBS.
5. Wash three times with PBS and label with antibodies or treat as described in **steps 6** or **7**.
6. For visualization of voltage-dependent anion channel (VDAC): treat with 8 *M* urea for 15 min (*see* **Note 10**).
7. For visualization of BrdU: denature DNA with 2 *N* HCl for 15 min (*see* **Note 11**).
8. Wash extensively with water to remove urea or HCl. Wash twice with PBS and label with antibodies within the next 1–7 d.

3.2.3. Decoration With Antibodies

1. Dilute primary and secondary antibodies in antibody dilution buffer. Prepare 50 μL per 12 mm cover slip and 150 μL per 25 mm cover slip. Short centrifugation of secondary antibodies eliminates precipitates and reduces background.

2. Dispose parafilm on an appropriate flat surface (e.g., glass plate), pipet the antibody solution on the hydrophobic parafilm, and place the cover slip onto the droplet. Incubate for 30 min.
3. Wash three times with PBS in 300–500 μL drops of PBS on parafilm or in the wells of 6- or 12-well dishes.
4. Incubate with secondary antibodies for 30 min and wash three times with PBS as described in **step 3**.
5. Place 5–20 μL mounting medium onto microscope slide. Rinse cover slip with distilled water, adsorb excess water by quickly touching a filter paper, and place cover slip onto mounting medium. Slides can be viewed after 30–60 min, when remaining water is evaporated and mounting medium is hardened. They can be stored at 4°C in the dark for several weeks or months.

3.3. Histochemistry

Histochemical analysis allows visualization of the activity of mitochondrial complexes *in situ*. The protocols presented here for activities of COX (complex IV of the respiratory chain; **Fig. 3**) and SDH (complex II of the respiratory chain) are adapted from those used in muscle histochemistry. Muscle tissue has more mitochondria and higher enzymatic activities than cultured cells.

1. Before histochemistry, cells are cultured on square or round cover slips for at least 24 h.
2. Wash cover slip once with PBS for 2 min.
3. Fix cells by drying in the air for 15–20 min. Carefully note cell side. The cover slip can be used immediately or stored at –80°C wrapped in aluminum foil.

3.3.1. Cytochrome-c Oxidase Histochemistry

Analysis of COX activity is based on precipitation of DAB, an electron donor for cytochrome-*c*. Oxidation of DAB by COX reveals the structure of the mitochondrial network and points to eventual deficits of COX activity (**Fig. 3**). The incubations are performed in specialized vessels that can hold cover slips or in vessels that can accommodate a cover slip holder. We use 10 mL of each solution for vessels and holders with capacity for up to eight cover slips.

1. Incubate cover slips with freshly prepared CPIM for 15 min at room temperature.
2. Rinse once with CRM.
3. Incubate with freshly prepared CIM for 4.5 h at 37°C (*see* **Note 12**).
4. Wash once for 5 min with CRM.
5. Wash once for 5 min with PBS.
6. Wash once for 5 min with distilled water.
7. Place a 10- to 40-μL drop of heated gelatin mounting solution medium onto microscope slide.
8. Immediately place cover slip onto mounting solution.

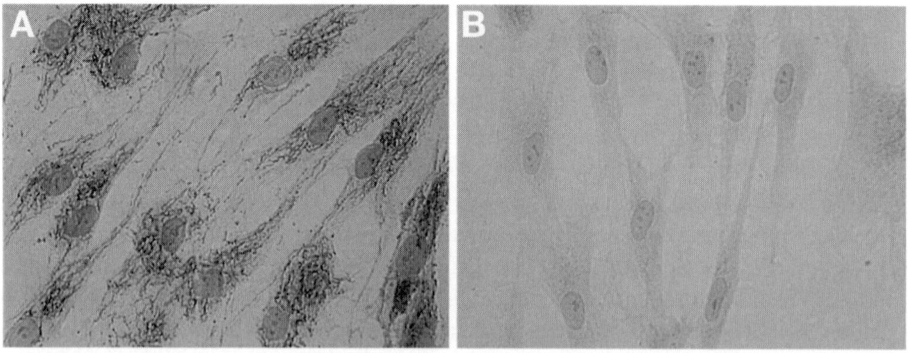

Fig. 3. Histochemistry of cytochrome-*c* oxidase (COX) in human fibroblasts: (**A**) control cells; (**B**) COX-deficient cells.

3.3.2. Succinate Dehydrogenase Histochemistry

The SDH catalyzes the conversion of succinate to fumarate. SDH activity demonstration in microscopy is based on the use of a tetrazolium salt (NBT) as an electron acceptor with phenazine methosulfate.

1. Incubate cover slips with freshly prepared SIM for 4 h at 37°C (*see* **Note 12**).
2. Wash three times in PBS for 5 min.
3. Wash once for 5 min with distilled water.
4. Place a 10- to 40-μL drop of heated gelatin mounting solution medium onto microscope slide.
5. Immediately place cover slip onto mounting solution.

3.4. Subcellular Fractionation

The enrichment of mitochondria by subcellular fractionation and the analysis of fractions by Western blot represent a valuable alternative to investigate or confirm the subcellular distribution of proteins. This is especially true when available antibodies fail to label a given protein in fixed cells (*see* **Subheading 3.2.**) but label the denatured protein after SDS polyacrylamide gel electrophoresis and Western blot.

The volumes given here for 120 dishes that are 100 mm (~92 mm diameter, ~57 cm^2 area) must be adapted when using dishes or flasks of other sizes (areas). Cell washing and collection are performed at room temperature. Homogenization and fractionation must be done at 4°C, and all fractions must be kept on ice. Low-speed centrifugations are performed with standard tabletop centrifuges for 1.5 mL tubes. High-speed centrifugations are performed with the TLA 45 rotor in a Beckman TL 100 Ultracentrifuge.

1. Wash cells twice with PBS (10 mL/dish), add 4 mL PBS per dish, and scrape cells with a moistened scraper. Collect cells in PBS with a moistened plastic Pasteur

pipet, transfer into a 15- or 50-mL Falcon tube, and centrifuge for 5 min at approx 450*g*. Alternatively, collect cells by trypsinization (*see* **Subheading 3.1.**) and wash twice with PBS by centrifugation (5 min at ~450*g*).

2. Wash the cell pellet in HS (2 mL/dish) and centrifuge 5 min at 450*g*. This washing step removes salts (of PBS and media) and is critical for successful homogenization. Gently resuspend cells in HS to a final volume of 1 mL (if cells are collected from a single dish) or in 200–500 µL per dish (if cells are collected from several dishes). Add protease inhibitors (e.g., 0.5 m*M* PMSF) at this step, if required.
3. Aspirate cells through a 22-gage needle into a syringe of a volume similar to that of the cell suspension. Aspirate the cell suspension slowly and continuously in the syringe, then eject strongly against the tube side (avoid air bubbles). The number of passages required to break up cells (without breaking up the nuclei) depends on the cell type (commonly between 5 and 15 passages). The efficacy of cell disruption is established by phase contrast microscopy. After optimal homogenization, you should observe a majority of free nuclei, lots of small particles (intracellular organelles), and very few unbroken cells (*see* **Note 13**). Passages through the syringe must be interrupted when all nuclei are free or when broken or aggregated nuclei begin to be observed.
4. Centrifuge 5 min at 500*g* to separate nuclear pellets and postnuclear supernatant. This step can be repeated if necessary (*see* **Note 14**).
5. Centrifuge the supernatant for 10 min at 6000*g* to separate mitochondrial pellet and postmitochondrial supernatant. A yellow-brown mitochondrial pellet should be recognizable (*see* **Note 15**).
6. If required, the postmitochondrial supernatant can be separated into a microsomal and a cytosolic fraction by centrifugation for 60 min at 80,000*g* (e.g., Beckman TLA 45, 40,000 rpm).
7. All pellets are resuspended in an adequate volume of HS. Fractions can be stored at –20 or –80°C. To minimize protein degradation, it is advisable to determine the protein concentration and add the reagents for SDS polyacrylamide gel electrophoresis immediately.

3.5. Polyethylene Glycol-Mediated Cell Fusion

The overall appearance of the mitochondrial compartment is determined by the equilibrium between antagonistic fusion and fission reactions and thus indirectly reflects mitochondrial dynamics. Nevertheless, the fusion of cells containing different or differently labeled mitochondria allows the study of mitochondrial fusion (and fusion-mediated intermitochondrial exchanges) directly ***(1,2)***. The protocol described here works for HeLa cells, 143B cells, and human skin fibroblasts but can probably be adapted to other types of cells. All solutions and media are prewarmed to 37°C, and fusion experiments are done at room temperature under a sterile hood.

1. Plate cells 1 or 2 d before fusion, preferentially onto 25 mm cover slips. Cells should be 70–100% confluent the day of fusion. If confluence is too low, then the

probability of fusion is low, and cells tend to be killed by PEG treatment. If confluence is too high, then the observation and analysis of fusion events by fluorescence microscopy is difficult.

2. Wash cells three times using medium without FBS. Add freshly prepared 50% PEG solution dropwise and gently onto cover slips (3–4 drops per cover slip; do not fill the well with PEG). Do not shake. After 40–50 s, tilt plate and aspirate PEG. Immediately add medium with FBS to the side, not on the top, of the cells.
3. Tilt slowly two to three times and aspirate medium. Wash cells three times as described and put cells back into the incubator for 10 min. Wash cells again three times every 10 min. Washes are critical for cell survival and must be done very gently. Dark droplets of unwashed PEG can be identified by phase contrast microscopy. Polykaryons can be identified 30–60 min after PEG treatment. Fusion-mediated exchanges achieve diffusion of mitochondrial components throughout the mitochondrial compartment of polykaryons within 8–16 h. Fused cells survive for at least 24 h.

4. Notes

1. Respiratory-deficient cells produce all ATP by glycolysis and die after consumption of all glucose in the medium. It is important to survey medium acidification by lactate and to change medium when required.
2. High concentrations of uridine interfere with the incorporation of BrdU into replicating DNA. Uridine must be removed from media during the duration of the bromodeoxyuridine pulse.
3. It is imperative to avoid microbial contamination for storage of sucrose-containing solutions.
4. Even freshly prepared CIM will be reddish and contain numerous precipitates. It must be prepared and filtered directly before use.
5. The solution should be "pinkish to orange," indicating neutral pH. If the solution is yellowish, as can happen with old batches of PEG, then do not reestablish pH with NaOH (such solutions tend to kill cells).
6. Cover slips tend to float early after immersion into medium. It is therefore convenient to immerse them into culture medium 1–2 h before addition of cells. We use 25-mm diameter cover slips for fusion experiments and 12- to 14-mm diameter cover slips for standard immunofluorescence experiments.
7. Cells attach and start to spread on substrate within hours, but most cells need 1–2 d to reacquire normal morphology. Some cells need more time to reacquire a normal morphology on glass cover slips.
8. For intermediate conditions, interrupt PFA fixation after 5 min and permeabilize cells with cold methanol as described in **Subheading 3.4.1**.
9. After fixation for 20 min, cells can be stored in PFA solution at 4°C for 1 to several weeks. After quenching, permeabilization or denaturation, cells can be stored in PBS at 4°C for 1–7 d.

10. Denaturation with urea enables detection of VDAC. Urea treatment does not alter mitochondrial appearance and does not affect green fluorescent protein (GFP) and DsRed fluorescence.
11. DNA denaturation with HCl is required for detection of incorporated bromodeoxyuridine. HCl treatment eliminates GFP and DsRed fluorescence and lowers detection of some antigens. Alternatively, it is possible to detect incorporated bromodeoxyuridine with kits containing nucleases (e.g., Bromo-2′-deoxy-uridine Labeling and Detection Kit, Roche Applied Science).
12. The optimal incubation time for a high signal-to-noise ratio must be established for each cell type. Given the variability between experiments, samples must be incubated in parallel for comparison.
13. The observation of cells before and after homogenization will help discriminate unbroken cells and "free nuclei."
14. The nuclear pellet is difficult to see in some cell types.
15. Depending on the cell line, it is common to observe yellowish mitochondrial pellets contaminated (to variable extents) with white nuclei.

Acknowledgments

M. R. is an investigator for CNRS. Work in the laboratory of M. R. is supported by INSERM and by grants from AFM and from Ministère Délégué à la Recherche (A.C.I. B.C.M.S.). We thank Anne Lombès for support, advice, and stimulating discussions. S. G. and H. S. are supported by the Academy of Finland, the Medical Fund of Tampere University Hospital, and the European Community's sixth Framework Programme for Research, Priority 1 "Life Sciences, Genomics and Biotechnology for Health," contract LSHM-CT-2004–503116.

References

1. Legros, F., Lombes, A., Frachon, P., and Rojo, M. (2002) Mitochondrial fusion in human cells is efficient, requires the inner membrane potential and is mediated by mitofusins. *Mol. Biol. Cell.* **13,** 4343–4354.
2. Legros, F., Malka, F., Frachon, P., Lombès, A., and Rojo, M. (2004) Organization and dynamics of human mitochondrial DNA. *J. Cell Sci.* **117,** 2653–2662.
3. Shaw, J. M., and Nunnari, J. (2002) Mitochondrial dynamics and division in budding yeast. *Trends Cell Biol.* **12,** 178–184.
4. Westermann, B. (2003) Mitochondrial membrane fusion. *Biochim. Biophys. Acta* **1641,** 195–202.
5. Delettre, C., Lenaers, G., Griffoin, J. M., et al. (2000) Nuclear gene *OPA1*, encoding a mitochondrial dynamin-related protein, is mutated in dominant optic atrophy. *Nat. Genet.* **26,** 207–210.
6. Zuchner, S., Mersiyanova, I. V., Muglia, M., et al. (2004) Mutations in the mitochondrial GTPase mitofusin 2 cause Charcot-Marie-Tooth neuropathy type 2A. *Nat. Genet.* **36,** 449–451.

7. Chen, H., Detmer, S. A., Ewald, A. J., Griffin, E. E., Fraser, S. E., and Chan, D. C. (2003) Mitofusins Mfn1 and Mfn2 coordinately regulate mitochondrial fusion and are essential for embryonic development. *J. Cell Biol.* **160,** 189–200.
8. Garrido, N., Griparic, L., Jokitalo, E., Wartiovaara, J., Van Der Bliek, A. M., and Spelbrink, J. N. (2003) Composition and dynamics of human mitochondrial nucleoids. *Mol. Biol. Cell* **14,** 1583–1596.
9. Spelbrink, J. N., Li, F. Y., Tiranti, V., et al. (2001) Human mitochondrial DNA deletions associated with mutations in the gene encoding Twinkle, a phage T7 gene 4-like protein localized in mitochondria. *Nat. Genet.* **28,** 223–231.
10. Trifunovic, A., Wredenberg, A., Falkenberg, M., et al. (2004) Premature ageing in mice expressing defective mitochondrial DNA polymerase. *Nature* **429,** 417–423.
11. Magnusson, J., Orth, M., Lestienne, P., and Taanman, J. W. (2003) Replication of mitochondrial DNA occurs throughout the mitochondria of cultured human cells. *Exp. Cell Res.* **289,** 133–142.
12. Iborra, F. J., Kimura, H., and Cook, P. R. (2004) The functional organization of mitochondrial genomes in human cells. *BMC Biol.* **2,** 9.
13. Bakker, A., Barthelemy, C., Frachon, P., et al. (2000) Functional mitochondrial heterogeneity in heteroplasmic cells carrying the mitochondrial DNA mutation associated with the MELAS syndrome (mitochondrial encephalopathy, lactic acidosis, and strokelike episodes). *Pediatr. Res.* **48,** 143–150.
14. Rojo, M., Legros, F., Chateau, D., and Lombes, A. (2002) Membrane topology and mitochondrial targeting of mitofusins, ubiquitous mammalian homologs of the transmembrane GTPase Fzo. *J. Cell Sci.* **115,** 1663–1674.

2

The Mitochondria of Cultured Mammalian Cells

II: Expression and Visualization of Exogenous Proteins in Fixed and Live Cells

Steffi Goffart, Peter Martinsson, Florence Malka, Manuel Rojo, and Johannes N. Spelbrink

Summary

Mitochondria are almost ubiquitous organelles in Eukaryota. They are highly dynamic and often complex structures in the cell. The mammalian mitochondrial proteome is predicted to comprise as many as 2000–2500 different proteins. Determination of the subcellular localization of any newly identified protein is one of the first steps toward unraveling its biological function. For most mitochondrial proteins, this can now be done relatively easily by cloning a complementary deoxyribonucleic acid of interest in frame with an additional sequence for a fluorescent or nonfluorescent protein tag. Transfection and subsequent visualization, either by direct fluorescence microscopy or by indirect immunofluorescence microscopy, will give the first clue to mitochondrial localization. In combination with a fluorescent "marker" dye, the mitochondrial localization can be confirmed. This chapter describes some of the methods used in determining mitochondrial protein localization, which can also be used to study dynamics of mitochondria or individual mitochondrial proteins or protein complexes.

Key Words: DsRed; fluorescent microscopy; GFP; mammalian cell culture; MitoTracker; PicoGreen; transfection.

1. Introduction

Mitochondria are highly dynamic organelles that continuously fuse and divide *(1)*. Advances in fluorescent protein tagging, as well as advances in live cell imaging, have made it possible to study these processes in detail. With good image resolution, it is also possible to detect details in mitochondrial structure. It has been shown that some of the mitochondrial enzymatic processes occur at discrete foci within the mitochondrial network, such as shown for pyruvate

From: *Methods in Molecular Biology, vol. 372: Mitochondrial: Practical Protocols*
Edited by: D. Leister and J. M. Herrmann © Humana Press Inc., Totowa, NJ

dehydrogenase *(2)*. In addition, we and others have shown that mammalian mitochondrial deoxyribonucleic acid (mtDNA) is organized in discrete protein–DNA complexes *(3,4)* called nucleoids *(5)*, with DNA synthesis *(3,4,6)* as well as transcription *(6)* localized within or near these structures.

We describe some of the methods that allow us to visualize mitochondria and nucleoids, either by live cell fluorescent imaging or imaging after cell fixation, using fluorescent protein tagging, transfection, and application of fluorescent dyes. These methods can be applied for the study of the dynamics of mitochondria or mitochondrial protein complexes and are generally applicable to establish the mitochondrial localization of a protein of interest. In combination with Chapter 1, a comprehensive overview of available methods is provided to establish mitochondrial protein localization in mammalian cells.

2. Materials

2.1. Fluorescent and Nonfluorescent Protein Tagging

1. Mammalian expression vectors: (usually commercial) vectors for fluorescent tagging such as enhanced green fluorescent protein (GFP), cyan fluorescent protein (CFP), DsRed2, and the like (e.g., Clontech, Palo Alto, CA) or for nonfluorescent tagging. Small tags also can be directly introduced by polymerase chain reaction.
2. Complementary DNA (cDNA) for proteins of interest.
3. Common reagents for molecular biology.

2.2. Cell Culture and Transfection

2.2.1. Transfection of 143B Osteosarcoma Cells With TransFectin™ for Use in Fluorescent Microscopy

1. Regular media for cell culture are essentially as described in Chapter 1. Typically, we use Dulbecco's modified Eagle's medium containing 4.5 g/L glucose, 10% (v/v) fetal calf serum, 2 m*M* L-glutamine, and 1 m*M* Na-pyruvate. Medium is stored at 4°C and always prewarmed to 37°C before use.
2. 10-cm ∅ Plastic cell culture plates.
3. 143B Osteosarcoma cells (*see* **Note 1**).
4. 10 m*M* EDTA (ethylenediaminetetraacetic acid) in H_2O: filter sterilize by passing through a 0.2-μm filter. Alternatively, use a commercially available trypsin-EDTA solution.
5. Six-well culture plates or similar.
6. Glass coverslips of convenient size, either square or round. Alternatively, use glass-bottom culture dishes (MatTek, Ashland, MA) or chambered coverglass (e.g., Nalge Nunc, Naperville, IL). These are particularly useful for time-lapse life cell imaging with inverted microscopes.
7. Expression vector for (tagged) protein of interest at approx 0.5 mg/mL of sufficient (midi- or maxiprep) purity.
8. Mammalian β-galactosidase expression vector (e.g., pcDNA3.1(-)Myc-His/LacZ, Invitrogen, Carlsbad, CA).

9. OptiMEM® I (Gibco-Invitrogen) at room temperature before use.
10. TransFectin (Bio-Rad, Hercules, CA).

2.2.2. Ca-Phosphate Transfection of Mouse C2C12 Myoblasts

In addition to the common cell culture reagents, plasmids, and plasticware described in **Subheading 2.2.1.**, this method requires the following:

1. HBS (HEPES-buffered saline): 50 m*M* HEPES, 280 m*M* NaCl, 10 m*M* KCl, 1.5 m*M* Na_2HPO_4, 12 m*M* α-D-glucose; adjust with NaOH to pH 7.05 and store in 200-µL aliquots at –20°C (*see* **Note 2**).
2. 500 m*M* $CaCl_2$, cell culture grade in water, filter sterilize and store in 200-µL aliquots at –20°C.
3. Sterile water: MilliQ or cell culture grade.
4. A carrier plasmid (*see* **Note 3**).
5. Sterile 1X phosphate-buffered saline (PBS): 140 m*M* NaCl, 27 m*M* KCl, 6.5 m*M* Na_2HPO_4, 1.5 m*M* KH_2PO_4 (we usually prepare a 10X stock in MilliQ water, adjust to pH 7.4, and sterilize by autoclaving; this is diluted in autoclaved MilliQ to obtain 1X working solutions).

2.3. Posttransfection: Use of Dyes to Visualize Mitochondria and Mitochondrial DNA

2.3.1. MitoTracker Red and MitoTracker Green Staining of Mitochondria

1. Regular cell culture medium prewarmed to 37°C.
2. MitoTracker CMXRos (MitoTracker Red)/MitoTracker Green FM (Molecular Probes, Eugene, OR).

2.3.2. PicoGreen Staining of Mitochondrial Nucleoids

1. Regular cell culture medium prewarmed to 37°C.
2. PicoGreen (Molecular Probes).

2.4. Posttransfection: Cell Fixation, Mounting, and β-Galactosidase Staining

2.4.1. Cell Fixation and Mounting for Imaging and Storage

1. 1X PBS (*see* **Subheading 2.2.2.**), prewarmed to 37°C.
2. Fixative in PBS at 37°C, such as 1X PBS diluted from a 10X stock solution, 3.7% (v/v) formaldehyde, 5% (w/v) sucrose (formaldehyde in this case is from a 37% liquid stock solution, which typically also contains 10–15% methanol for stabilization). Alternatively, the fixatives described in Chapter 1 can be used. In our experience, the presence of sucrose in the fixative results in better preservation of mitochondrial ultrastructure as one would normally see without fixation.
3. Glass slides.
4. Scalpel.
5. Blotting paper.

6. Nonfluorescent mounting medium (e.g., Vector Laboratories) and colorless nail polish or mounting medium that solidifies (Mowiol, Vector Laboratories Hard Set Medium). Mounting medium is stored at 4°C but should best be equilibrated at room temperature before use.

2.4.2. β-Galactosidase Staining Procedure

1. 1X PBS (*see* **Subheading 2.2.2.**), prewarmed to 37°C.
2. Fixative in 1X PBS: 1X PBS, 2% (v/v) formaldehyde, 0.2% (v/v) glutaraldehyde.
3. X-Gal (= 5-bromo-4-chloro-3-indolyl-β-D-galactopyranoside): 40 mg/mL in dimethyl sulfoxide stored at –20°C.
4. 0.1 *M* K-Ferricyanide; store at room temperature protected from light.
5. 0.1 *M* K-Ferrocyanide; store at room temperature protected from light.
6. 1.0 *M* $MgCl_2$.
7. X-Gal staining solution in 1X PBS: 5 m*M* K-ferricyanide, 5 m*M* K-ferrocyanide, 2 m*M* $MgCl_2$, 1 m*M* X-Gal (fresh dilution from the above stock solutions); protect from light.

3. Methods

There are various reasons for wanting to introduce exogenous proteins in mammalian cells and to tag those proteins. Fluorescently tagged proteins are commonly used to study the subcellular localization of the proteins in question. This is usually fast and requires little more than basic molecular biology skills and equipment, cell culture facilities, a fluorescent microscope, and common sense. A second use for introduced fluorescent proteins that is becoming more important and feasible in modern cell biology is the possibility to study the dynamics of the protein itself, the complex in which the protein normally resides, or whole organelles in live cells using live cell imaging. Live cell imaging requires an expensive fluorescent microscopy setup typically with an intensity-adjustable light source; a highly sensitive, cooled charge-coupled device (CCD) camera; fast shutter speeds; fast filter changers for multicolor imaging; carefully chosen combinations of excitation and emission filters; a vertically motorized stage or piezo-motorized objectives; and a cell culture incubator encasing part of the microscope. In addition, it requires powerful computers capable of handling and storing large data sets; the analysis often requires additional specialized image analysis software, such as deconvolution, tracking, and possibly three-dimensional rendering software. However, if interested in a process on a large time-scale, from hours to days, then very often it suffices to use cell fixation at various times during an experiment. This of course will not allow tracking of single molecules, complexes, or organelles over time but often provides enough information to understand at least some of the dynamics (e.g., when studying mitochondrial fusion) (*see* ref. *7*).

A variety of commercial fluorescent and nonfluorescent protein tagging vectors is available. These include not only regular fluorescent proteins such

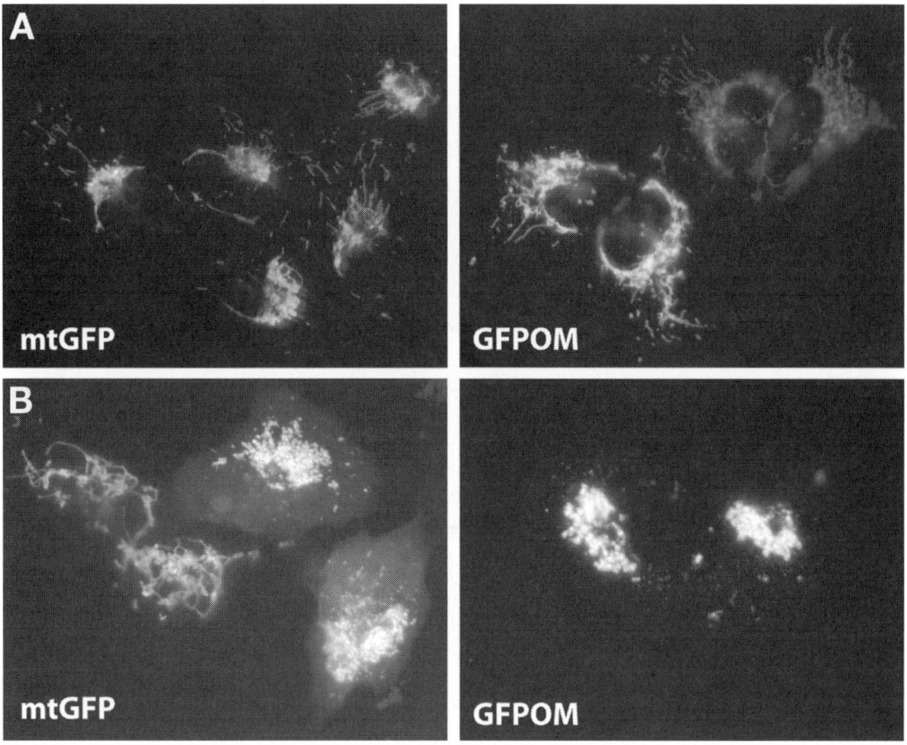

Fig. 1. Visualization of mitochondria with fluorescent proteins on transient transfection of HeLa cells. (**A**) GFP molecules targeted to the mitochondrial matrix (mtGFP) or to the mitochondrial outer membrane (GFPOM) allow visualization of mitochondrial filaments. (**B**) Mitochondrial morphology is severely and unspecifically modified at very high expression levels.

as GFP and its variants, such as CFP and yellow fluorescent protein, DsRed (a red fluorescent protein derived from the coral *Discosoma* sp.) but also more specialized proteins such as photoactivatable GFP ***(8)*** and photoswitchable fluorescent proteins (*see* ref. ***9***). Commercially available DsRed1 and 2 are obligate tetramers ***(10)***, which can result in mistargeting and aggregation (*see*, e.g., ref. ***11*** and our own unpublished data). Variants are now available that form either dimers or monomers (*see* ref. ***12*** and references therein). Monomeric DsRed is now also commercially available from Clontech. Fluorescently tagged mitochondrial proteins are also particularly useful in the study of organelle dynamics as they can selectively label one of the mitochondrial compartments like the mitochondrial matrix or the mitochondrial outer membrane (as shown in **Fig. 1A**). Fluorescent proteins can be targeted to the mitochondrial matrix by the addition of an N-terminal targeting sequence, as with mitochondrial GFP (mtGFP) *(7)* or to the outer membrane (OM) with a C-terminal transmembrane domain, as with GFP-OM *(4)*. Again, one must be aware that transient expression can achieve very high expression levels, and that this can lead to mistargeting of fluorescent proteins or to artificial modification of mitochondrial morphology and dynamics (**Fig. 1B**). This problem can be circumvented by using stable transfection, which generally achieves lower expression levels (not shown).

To determine the localization of an endogenous protein, first one needs a specific antibody that works in immunofluorescence (*see* Chapter 1) for that particular protein; in addition, expression levels of the endogenous protein need to be sufficiently high to allow detection at a single-cell level. Although specific antibodies can also be used to determine subcellular localization biochemically, the results of these methods are often less straightforward to interpret but can provide additional information that will not be obtained easily by fluorescent protein tagging (e.g., using submitochondrial fractionation). Nonfluorescent tags are usually small peptide tags such as polyhistidine tags that can be used for purification. Some of the most commonly used epitope tags are the so-called c-myc, FLAG, HA, VSVG, and V5 tags. These are well defined small peptide sequences that are specifically recognized by monoclonal antibodies and allow the study of *in situ* protein localization using immunofluorescence and, biochemically, subcellular fractionation and Western blot analysis. For both biochemical subfractionation methods and *in situ* methods using tagged proteins, one should always be conscious of overexpression artifacts.

Most noninducible expression vectors use viral promoters that commonly result in higher than normal expression levels. This can often result in at least partial mislocalization or aggregation of the protein of interest and even in cell death. In some cases, it could prove useful to establish a stable expressing cell line. These often show lower expression levels than transient expression. However, even low expression levels of both large fluorescent protein and small epitope tags can result in mistargeting and toxicity. So, experiments need to be interpreted with care and, if possible, confirmed using alternative methods.

A comparison of the targeting of mitochondrial transcription (and DNA packaging) factor A (TFAM) using either immunofluorescence for the endogenous protein or using different tags is given in **Figs. 2** and **3**. These pictures demonstrate some of the artifacts observed with tagging and overexpression. The majority of the proteins of the mitochondrial matrix, inner membrane and intermembrane space possess cleavable N-terminal targeting sequences (for review, *see* ref. ***13***). Therefore, internal or C-terminal tags should be used.

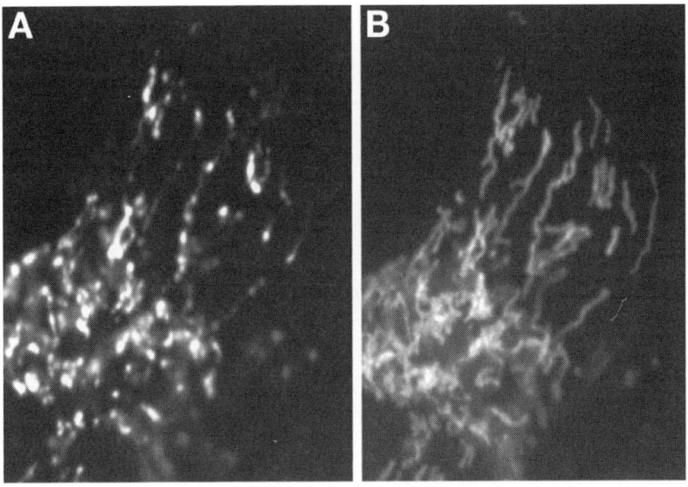

Fig. 2. Immunofluorescence detection of endogenous TFAM in 143B cells. (**A**) Endogenous TFAM was detected using a rabbit polyclonal antibody against human TFAM (kind gift of Dr. R. Wiesner). A secondary fluorescein-labeled antirabbit antibody was used for detection of TFAM-containing foci. (**B**) The same cells were counterstained prior to fixation and lysis using MitoTracker RED as described in this chapter. Immunofluorescence was done essentially as described in ref. ***3*** and Chapter 1. By comparison of panels **A** and **B**, it can be observed that endogenous TFAM is concentrated in foci within the mitochondrial network (note also that the panels show only part of a single 143B cell with the nucleus situated in the lower left corner). These foci have been shown to contain mtDNA *(3,4)* and are therefore by definition nucleoids.

3.1. Fluorescent and Nonfluorescent Protein Tagging

Cloning of a cDNA of interest in any of the above-mentioned tagging vectors uses common molecular biological techniques beyond the scope of this chapter (*see* **Note 4** for a few pointers).

3.2. Cell Culture and Transfection

A wide variety of cell culture conditions and transfection techniques is used, very often depending on the cell types studied. We present two transfection methods frequently used in our laboratories; one uses a commercial transfection reagent (TransFectin, Bio-Rad) because this reagent in our hands gives the best results for the widest range of cell types, and one is a noncommercial (Ca-phosphate) method. It is beyond the scope of this chapter to discuss all possible methods and cell lines in great detail, and even if one chooses to use the cell lines and methods we describe, we recommend first to try to optimize the method with a well-established reporter vector such as GFP or β-galactosidase

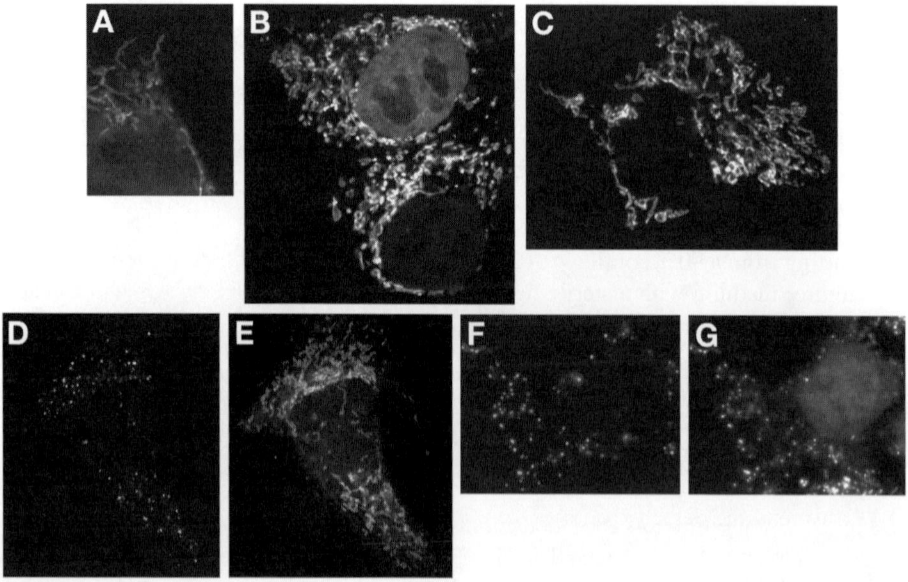

Fig. 3. A comparison of the intramitochondrial localization of TFAM recombinant proteins with various epitope tags and expressed in 143B cells. (**A**) TFAM-EGFP (enhanced GFP) at low expression levels shows faint, mostly uniform mitochondrial fluorescence (shown is a detail of a single cell). (**B**) High level of TFAM-EGFP expression results in mostly uniform mitochondrial fluorescence, often showing abnormal mitochondrial morphology. In addition, in this case we also frequently observe nuclear green fluorescence, presumably an overexpression artifact enhanced by the nonspecific DNA-binding ability of TFAM. (**C**) Immunofluorescence detection of TFAM containing a small c-myc epitope tag. Although in this case punctate foci can be observed, fluorescence is still mostly uniform. (**D**), (**E**) TFAM-DsRed2: (**D**) transfected cells show clear intramitochondrial foci, as judged by comparison of the mitochondrial morphology seen with MitoTracker Green staining (**E**). Although this is more or less what we observe for endogenous TFAM (as shown in **Fig. 1**), TFAM-DsRED2 transfected cells also frequently show aberrant nucleoid morphology because of apparent protein aggregation and especially when expression levels are high (not shown) (**F**), (**G**) TFAM-DsRed2 (**F**) colocalizes with mtDNA as seen by PicoGreen costaining (**G**) (note that PicoGreen also stains nuclear DNA but not at very high intensity). All procedures to obtain these fluorescent images (i.e., transfection; MitoTracker and PicoGreen staining; fixation for panels **A–E**; and mounting) were as described in this chapter. Panels **A–E** were obtained by confocal microscopy; panels **F** and **G** are epifluorescent live cell microscopy images.

because small differences in routine cell culture procedures and reagents could result in differences in transfection efficiencies between laboratories (*see* **Note 5**).

3.2.1. Transfection of 143B Osteosarcoma Cells With TransFectin for Use in Fluorescent Microscopy

The protocol applies to a wide variety of cells. The amount of TransFectin and DNA used might have to be optimized for each cell type.

1. Starting from a 90% confluent 10-cm ∅ plate of adherent 143B cells, remove (cell culture) medium with a sterile tip or Pasteur pipet by vacuum suction and spread 450 µL of prewarmed 10 m*M* EDTA on top of the cells.
2. Incubate the cells for 5 min at room temperature.
3. In the meantime, prepare as many 6-well plates as needed; add 2 mL fresh medium to each well and add a sterilized (e.g., a 24 × 24 mm^2) coverslip (*see* **Notes 6–8**).
4. Following the 5-min incubation, gently tap the side of the 10-cm ∅ plate to the palm of your hand to loosen the cells; if necessary, check by microscope.
5. Resuspend the detached cells in 10 mL medium.
6. Add 1/10 volume of cells to each well (i.e., 200 µL) and gently mix the cells to homogeneity.
7. Return the cells to the cell culture incubator. Using this dilution of cells, the cells will now be ready for transfection the second day after dilution (*see* **Notes 9** and **10**).
8. For transfection, mix 1 µg of DNA with 250 µL OptiMEM I for each well to be transfected (*see* **Notes 11** and **12**); in a second tube, for each well pipet 4 µL TransFectin in 250 µL OptiMEM and mix gently.
9. Add the TransFectin mixture to the DNA mixture; mix gently by flicking or inverting the tube and leave at room temperature for 15 min.
10. Dropwise add approx 500 µL TransFectin-DNA mixture to each well and swirl gently (the cell culture medium should *not* be removed prior to adding the mixture).
11. Incubate cells for 4 h in the cell culture incubator.
12. Remove the mixture, gently wash once with cell culture medium, add 2 mL medium, and incubate for 1 or 2 d additional (*see* **Note 13**).

3.2.2. Ca-Phosphate Transfection of Mouse C2C12 Myoblasts

1. Grow cells in normal cell culture medium to a confluency of 50–80%.
2. At 1 h before transfection, refresh medium.
3. Mix DNA and $CaCl_2$ at a final concentration of 250 m*M* $CaCl_2$ and 0.1 µg/µL DNA total (*see* following table and **Note 14**).

	500 m*M* $CaCl_2$	Vector	Carrier DNA	Final volume
12-Well plate (1 well)	12.5 µL	0.1–1 µg	2 µg	25 µL
35-mm dish, 6-well plate	25 µL	0.2–2 µg	4 µg	50 µL
60-mm dish	75 µL	0.5–5 µg	10 µg	150 µL
100-mm dish	250 µL	1.3–13 µg	25 µg	500 µL

4. Vortex well and incubate at room temperature for 20–30 min.
5. Add the DNA/$CaCl_2$ mixture slowly to an equal volume of HBS buffer (*see* **Note 15**).
6. Incubate mixture for 20–30 min at room temperature.
7. Vortex briefly to resuspend the precipitate.
8. Add to the cell culture dish and swirl gently to spread equally (*see* **Note 16**).
9. Incubate overnight (*see* **Note 17**).
10. On the next day, wash cells several times with PBS to remove precipitate and add fresh medium (*see* **Note 18**).

3.3. Posttransfection: Use of Dyes to Visualize Mitochondria and Mitochondrial DNA

Mitochondria can be specifically labeled not only with targeted fluorescent proteins, but also with various dyes. Several of these dyes can be used as highly specific counterstains to confirm the mitochondrial localization of a tagged fluorescent protein. The most versatile and frequently used of these dyes is the rhodamine derivative chloromethyl-X-rhosamine (CMXRos or MitoTracker Red™). In contrast to Rhodamine 123, MitoTracker Red retains a strong fluorescence and mitochondrion-specific localization following fixation and permeabilization, making it suitable for use also in immunofluorescence ***(14)***. Another frequently used dye is MitoTracker Green, which can be used for live cell imaging and can be fixed but is lost on cell permeabilization. Last, a study has shown that the DNA intercalating reagent PicoGreen very effectively stains mtDNA in living cells and can be used for visualization following paraformaldehyde fixation, at least in some cell lines ***(15)***. The accumulation of MitoTracker Red inside mitochondria, in contrast to MitoTracker Green, is very much dependent on the mitochondrial membrane potential ***(16)***. The next protocols discuss staining by MitoTracker dyes and PicoGreen.

3.3.1. MitoTracker RED and MitoTracker Green Staining of Mitochondria

1. Both dyes are typically dissolved at 100 µM or 1 m*M* concentrations in dimethyl sulfoxide and can be aliquoted for rapid thawing; store at –20°C (*see* **Notes 19** and **20**).
2. For staining cells grown in 6-well plates, dilute either dye 1:1000 or 1:10,000 (depending on the stock concentration) in 2 mL medium per well to give a final concentration of 100 n*M* (*see* **Note 21**).
3. Remove medium of cells by vacuum suctioning and carefully add 2 mL staining solution.
4. Place cells back in the incubator and incubate for 10–15 min.
5. Remove staining solution and carefully wash cells twice with 2 mL medium. Add 2 mL medium and incubate cells for an additional 2 h (*see* **Note 22**).
6. Cells are now ready for direct live cell imaging (*see* **Note 23**) or for further processing by fixation and possibly immunofluorescence (the latter only with MitoTracker RED; *see* introduction to this subheading).

3.3.2. PicoGreen Staining of Mitochondrial Nucleoids

In mammalian cells, mtDNA can be visualized as discrete foci within the mitochondrial network ***(5)***. These foci, called *nucleoids*, contain mtDNA as well as various proteins ***(3,4)***. Several approaches have been used to visualize the DNA in these structures directly or indirectly, including staining with ethidium bromide (EtBr) ***(6,17)***, immunofluorescence to detect the incorporated nucleotide analog 5-bromo-2′-deoxy-uridine (BrdU) ***(3,4,6)*** (*see also* Chapter 1) or by using an anti-DNA antibody ***(4,6)***, and by fluorescent *in situ* hybridization ***(18,19)***. One of the advantages of PicoGreen staining over fluorescent *in situ* hybridization, BrdU, and antibody detection of mtDNA is that it does not require cell fixation and can be easily used in combination with coexpression of a fluorescent protein. Compared to EtBr, it gives less background because it does not stain mitochondrial RNA. Like EtBr, PicoGreen rapidly bleaches.

1. Add 2–5 µL of PicoGreen stock solution directly to the cell culture medium and mix gently (*see* **Note 24**).
2. Place cells back in the incubator for 15 min or more (incubation has been done for up to 5 h without problems; ***15*** and our unpublished observations).
3. Wash cells twice with fresh medium and proceed with imaging or cell fixation (*see* **Note 25**).

3.4. Posttransfection: Cell Fixation, Mounting, and β-Galactosidase Staining

To store samples for later inspection or to prepare them for immunofluorescence, cells grown on coverslips need to be fixed. We give a simple procedure that works nicely for cells transfected with a GFP-tagged protein and stained with MitoTracker Red. For a more detailed overview of various fixation methods, *see* Chapter 1. In addition, we provide an *in situ* β-galactosidase detection method, principally as described in ref. ***20***.

3.4.1. Cell Fixation and Mounting for Imaging and Storage

Most of the following cell-handling steps we routinely do at ambient light conditions.

1. Prepare fixative fresh before use and preheat to 37°C.
2. Following transfection and MitoTracker Red staining, for instance, remove medium by vacuum suction.
3. Gently wash cells once with PBS.
4. Carefully add fixation solution and incubate cells for an additional 20 min at 37°C.
5. Remove fixative and wash three times with PBS.
6. Following the last wash step, leave the cells in PBS while marking microscope slides for mounting.
7. To each slide, add 18–20 µL mounting medium (*see* **Note 26**).

8. Use a pointed device (e.g., a scalpel) to lift one side of the coverslip from the well and carefully take the coverslip with a pair of tweezers; confirm the side on which the cells are attached.
9. Drain excess liquid and gently lower the side of the coverslip, with cell side facing down, against the drop of mounting medium.
10. Gently lower the rest of the coverslip, trapping the mounting medium underneath.
11. Drain excess liquid, sometimes still present on the sides, with a small piece of blotting paper but be careful not to drain the liquid from underneath the coverslip.
12. Seal and fix the coverslip with colorless nail polish and leave to dry for 5–10 min (*see* **Note 27**). Store in the dark at 4°C.

3.4.2. β-Galactosidase Staining Procedure

1. At 1 d following transfection of a β-galactosidase expression vector in a 6-well plate, wash cells once with PBS.
2. Fix cells for 15 min with 2 mL 2% formaldehyde and 0.2% glutaraldehyde in PBS at 37°C.
3. In the meantime, prepare a *fresh* staining solution in PBS by dilution from the various stock solutions.
4. Wash cells three times with PBS.
5. Add 2 mL staining solution per well.
6. Incubate the cells until they become noticeably blue and estimate efficiency of transfection by examining cells under the microscope (*see* **Note 28**).
7. The plate can be stored for a considerable period at room temperature or 4°C by adding a 10% glycerol solution in PBS and spreading the solution over the cells.

4. Notes

1. A variety of adherent human and mouse cell lines can be used for transfection, such as human HeLa, 143B osteosarcoma and derivatives thereof, A549 lung-carcinoma and derivatives, human embryonal kidney (HEK) 293 and derivatives, or mouse 3T3 fibroblasts and C2C12 myoblasts. Transfection efficiency with different reagents and cell lines, however, does need careful comparison and optimization (*see also* additional **Notes 3, 5, 9,** and **10**).
2. The HBS buffer is the crucial component for calcium–DNA complex formation and therefore for transfection efficiency and low cytotoxicity. Adjust pH carefully. To achieve best results, compare several batches of HBS buffer and use only the one with the highest transfection efficiency. A quick performance test can be made by slowly adding 1 volume 0.25*M* $CaCl_2$ to 1 volume HBS. Within a few minutes, $CaPO_4$ precipitates and can be observed under a microscope. Fine, sandy precipitate will be easily taken up by cells and is optimal; big crystals lead to low transfection efficiency and cytotoxicity.
3. To increase the DNA amount and facilitate calcium–DNA precipitation, a non-specific plasmid is used. Suitable is any vector without impact on the transfected cells, such as commonly used bacterial expression or cloning vectors.

4. First, as mentioned, a majority of mitochondrial proteins have N-terminal-targeting peptides. It is routine practice, when first testing whether a protein is mitochondrial, not only to clone the full-length protein in front of a C-terminal tag, but also as a negative control to generate a variant that lacks the putative N-terminal-targeting peptide. The latter usually requires the engineering of an alternative start codon. Alternatively, N-terminal segments of the protein can be put in front of GFP to test their ability to target GFP to the mitochondrial compartment. Second, some proteins are naturally expressed at very low levels because their genes have poor Kozak sequences for translation initiation. To increase expression levels, a consensus Kozak sequence (5′-C-C-A/G-C-C-A-T-G-G-3′) can be engineered at the 5′ end of the cDNA. Last, if several differentially "colored" fluorescent proteins are to be used, then it is advisable to purchase all vectors from the same company as this will often allow quick swapping of cDNAs from one vector to the other.
5. When trying transfection for the first time, start with an easy-to-transfect cell line such as a HEK293 line or HeLa because these should generally transfect at approx 50% or higher efficiency with any of the most common transfection reagents.
6. Coverslips can be sterilized immediately prior to use by direct flaming or by dipping in 70% ethanol followed by flaming. Alternatively, they can be batch sterilized for 1 h at 150°C (e.g., in a glass beaker).
7. Transfections can always be up- or downscaled to other culture plate formats by maintaining equal volume and concentration to surface area ratios, similar to the example given for Ca-phosphate.
8. Because quite a few commercial transfection reagents do not tolerate the presence of antibiotics, we generally grow "transfection" cells without them.
9. Although many commercial suppliers of transfection reagents recommend transfection 1 d following cell dilution, we prefer 2 d with some of the cell types we use, such as HEK293, because the cells are not yet properly attached to the glass slides after 1 d. For cells that show fast attachment and spreading, transfection can be done after 1 d or even during the same day but make sure to have high enough cell density by decreasing the dilution approx 2.5-fold on d 0 (i.e., dilute ~ 1 in 4). Although for TransFectin we recommend a cell density of at least 50% at the time of transfection, the Ca-phosphate method can be used at lower density for cells that are easy to transfect, such as HeLa cells. This might in fact be desirable if cells are to be used for fluorescent microscopy because the cells will show nice spreading for imaging.
10. Despite promises by manufacturers, transfection reagents can be quite toxic to cells. In the case of TransFectin, we notice considerable toxicity, especially when cell density is low (i.e. below 50% or so) at the day of transfection.
11. Vector DNA should be highly pure and reasonably concentrated (i.e., >0.5 μg/μL). We have used various commercial midi- or maxiprep kits to purify DNAs with satisfactory results. For consistent results, aliquot the purified DNA samples, dissolved in 10 m*M* Tris-HCl at pH 7.4 –/+ 0.1 m*M* EDTA and store at –20°C to avoid too many freeze-thaw cycles. Miniprep DNA samples should be avoided.

12. It is possible to cotransfect two or more vectors simultaneously (e.g., for GFP- and DsRed-tagged proteins for use in multicolor imaging). In this case, we maintain a final DNA concentration of 1 μg using 0.5 μg of each in case of cotransfection of two vectors.
13. In our experience, the additional wash step following transfection significantly reduces cell death for several cell lines, including 143B, although it is not specifically recommended by Bio-Rad.
14. Although the total DNA amount should not be varied greatly, the amounts of vector and carrier DNA can be modulated according to desired expression levels.
15. Slow mixing is important for a fine precipitate and can best be achieved by pipeting the DNA-$CaCl_2$ solution in 10-μL steps into the HBS buffer. Do not vortex at this point.
16. With cells showing little cytotoxicity during normal transfection, the efficiency can be increased by removing most of the medium before adding the Ca-HBS mixture and thus reaching a higher $CaPO_4$ concentration. Medium should be readded after 1–2 h.
17. Increasing the CO_2 concentration in the cell culture incubator to 6% for transfection and overnight incubation lowers the pH of the cell culture medium and can help to increase the transfection rate.
18. Check cells under a microscope. Extensive cell death might be a sign of impurities of the DNA and chemicals used. Expression of the transfected gene normally starts within 10 h; however, full expression is visible after 1–2 d.
19. Although both MitoTracker dyes are obviously light sensitive, no extreme measures need to be taken to avoid light exposure. It usually suffices to avoid exposure to intense light.
20. MitoTracker Red is a stable, highly fluorescent dye with an excitation maximum at 594 nm and an emission maximum at 608 nm in cells ***(14)***. This makes it the dye of choice for use in combination with GFP- or CFP-tagged proteins.
21. As a general guideline for MitoTracker Red, we recommend concentrations between 50 and 100 n*M* and 10–30 min labeling followed by a 10 min to 2 h "wash/chase." If needed, labeling can be checked on an inverted fluorescence microscope before MitoTracker removal.
22. The intensity of fluorescence of MitoTracker Red is such that even a small amount of nonmitochondrial background fluorescence can give a blurred result. Washing and especially the additional chase help prevent this problem. If cells are lysed after fixation (for further immunofluorescence), then a short wash/chase usually suffices for MitoTracker Red. MitoTracker Green generally gives a much less intense fluorescence and can do without the additional 2-h chase. In addition, for some cell lines it might prove beneficial to use a somewhat higher concentration (200 n*M*) and longer incubation times to obtain a stronger MitoTracker Green signal.
23. For long time-lapse experiments, we recommend the use of a fluorescently tagged protein as a mitochondrial marker because it is not excluded that MitoTracker Red could photosensitize at least some cell lines, resulting in mitochondrial damage and cell death (***21*** and our own unpublished data).

24. The best results need to be determined for each cell line by varying the amount of PicoGreen used and times of staining (***15*** and our own unpublished observations). The given protocol gives good results with 143B osteosarcoma cells.
25. In our hands, paraformaldehyde fixation results in loss of signal and specificity, so it is not recommended without careful testing.
26. Mounting medium not only allows long-term storage of samples, but also prevents rapid bleaching during fluorescent microscopy. We generally use mounting medium that also contains DAPI (4,6-diamidino-2-phenylindole), allowing for visualization of the nucleus.
27. Mounting medium is now also available as hard set, which means it will harden during drying. When hard-set mounting medium is used, there is no need to seal and fix the coverslip with nail polish.
28. Depending on the cell line under study, transfection efficiency, and the vector promoter, incubation times can vary greatly, from 15 min to several hours. To allow for longer incubation, use a humidified incubator; otherwise, the plate will dry out.

Acknowledgments

S. G., P. M., and H. S. are supported by the Academy of Finland, the Medical Research Fund of Tampere University Hospital, and the European Community's sixth Framework Programme for Research, Priority 1 "Life Sciences, Genomics and Biotechnology for Health," contract LSHM-CT-2004–503116. M. R. is an investigator of the CNRS. Work in the group of M. R. is supported by INSERM and by grants from AFM and from Ministère Délégué à la Recherche (A. C. I.).

References

1. Scott, S. V., Cassidy-Stone, A., Meeusen, S. L., and Nunnari, J. (2003) Staying in aerobic shape: how the structural integrity of mitochondria and mitochondrial DNA is maintained. *Curr. Opin. Cell Biol.* **15,** 482–488.
2. Margineantu, D. H., Brown, R. M., Brown, G. K., Marcus, A. H., and Capaldi, R. A. (2002) Heterogeneous distribution of pyruvate dehydrogenase in the matrix of mitochondria. *Mitochondrion* **1,** 327–338.
3. Garrido, N., Griparic, L., Jokitalo, E., Wartiovaara, J., Van Der Bliek, A. M., and Spelbrink, J. N. (2003) Composition and dynamics of human mitochondrial nucleoids. *Mol. Biol. Cell* **14,** 1583–1596.
4. Legros, F., Malka, F., Frachon, P., Lombes, A., and Rojo, M. (2004) Organization and dynamics of human mitochondrial DNA. *J. Cell Sci.* **117,** 2653–2662.
5. Satoh, M., and Kuroiwa, T. (1991) Organization of multiple nucleoids and DNA molecules in mitochondria of a human cell. *Exp. Cell Res.* **196,** 137–140.
6. Iborra, F. J., Kimura, H., and Cook, P. R. (2004) The functional organization of mitochondrial genomes in human cells. *BMC Biol.* **2,** 9.
7. Legros, F., Lombes, A., Frachon, P., and Rojo, M. (2002) Mitochondrial fusion in human cells is efficient, requires the inner membrane potential, and is mediated by mitofusins. *Mol. Biol. Cell* **13,** 4343–4354.

8. Patterson, G. H., and Lippincott-Schwartz, J. (2002) A photoactivatable GFP for selective photolabeling of proteins and cells. *Science* **297,** 1873–1877.
9. Chudakov, D. M., Verkhusha, V. V., Staroverov, D. B., Souslova, E. A., Lukyanov, S., and Lukyanov, K. A. (2004) Photoswitchable cyan fluorescent protein for protein tracking *Nat. Biotechnol.* **22,** 1435–1439.
10. Baird, G. S., Zacharias, D. A., and Tsien, R. Y. (2000) Biochemistry, mutagenesis, and oligomerization of DsRed, a red fluorescent protein from coral. *Proc. Natl. Acad. Sci. U. S. A.* **97,** 11,984–11,989.
11. Lauf, U., Lopez, P., and Falk, M. M. (2001) Expression of fluorescently tagged connexins: a novel approach to rescue function of oligomeric DsRed-tagged proteins. *FEBS Lett.* **498,** 11–15.
12. Campbell, R. E., Tour, O., Palmer, A. E., et al. (2002) A monomeric red fluorescent protein. *Proc. Natl. Acad. Sci. U. S. A.* **99,** 7877–7882.
13. Koehler, C. M. (2004) New developments in mitochondrial assembly. *Annu. Rev. Cell Dev. Biol.* **20,** 309–335.
14. Poot, M., Zhang, Y., Kramer, J., et al. (1996) Analysis of mitochondrial morphology and function with novel fixable fluorescent stains *J. Histochem. Cytochem.* **44,** 1363–1372.
15. Ashley, N., Harris, D., and Poulton, J. (2005) Detection of mitochondrial DNA depletion in living human cells using PicoGreen staining. *Exp. Cell Res.* **303,** 432–446.
16. Pendergrass, W., Wolf, N., and Poot, M. (2004) Efficacy of MitoTracker Green and CMXrosamine to measure changes in mitochondrial membrane potentials in living cells and tissues. *Cytometry A* **61,** 162–169.
17. Spelbrink, J. N., Li, F. Y., Tiranti, V., et al. (2001) Human mitochondrial DNA deletions associated with mutations in the gene encoding Twinkle, a phage T7 gene 4-like protein localized in mitochondria. *Nat. Genet.* **28,** 223–231.
18. van de Corput, M. P., van den Ouweland, J. M., Dirks, R. W., et al. (1997) Detection of mitochondrial DNA deletions in human skin fibroblasts of patients with Pearson's syndrome by two-color fluorescence *in situ* hybridization. *J. Histochem. Cytochem.* **45,** 55–61.
19. Margineantu, D. H., Cox, W. G., Sundell, L., et al. (2002) Cell cycle dependent morphology changes and associated mitochondrial DNA redistribution in mitochondria of human cell lines. *Mitochondrion* **1,** 425–435.
20. Sanes, J. R., Rubenstein, J. L., and Nicolas, J. F. (1986) Use of a recombinant retrovirus to study post-implantation cell lineage in mouse embryos. *EMBO J.* **5,** 3133–3142.
21. Minamikawa, T., Sriratana, A., Williams, D. A., Bowser, D. N., Hill, J. S., and Nagley, P. (1999) Chloromethyl-X-rosamine (MitoTracker Red) photosensitizes mitochondria and induces apoptosis in intact human cells. *J. Cell Sci.* **112,** 2419–2430.

3

Drosophila melanogaster as a Model System to Study Mitochondrial Biology

Miguel Angel Fernández-Moreno, Carol L. Farr, Laurie S. Kaguni, and Rafael Garesse

Summary

Mitochondria play an essential role in cellular homeostasis. Although in the last few decades our knowledge of mitochondria has increased substantially, the mechanisms involved in the control of mitochondrial biogenesis remain largely unknown. The powerful genetics of *Drosophila* combined with a wealth of available cell and molecular biology techniques, make this organism an excellent system to study mitochondria. In this chapter we will review briefly the opportunities that *Drosophila* offers as a model system and describe in detail how to purify mitochondria from *Drosophila* and to perform the analysis of developmental gene expression using *in situ* hybridization.

Key Words: *Drosophila*; gene expression; molecular localization.

1. Introduction

The fruit fly *Drosophila melanogaster*, a tiny insect about 3 mm long, was used extensively as an animal model in biology throughout the last century. In the famous Fly Room at Columbia University, T. H. Morgan and his students A. H. Sturtevant, C. B. Bridges, and H. J. Muller carried out a series of genetic analyses of *Drosophila* that led them to formulate the chromosome theory of heredity. This important achievement led to Morgan's 1933 Nobel Prize.

Between 1913 and 1930, several essential techniques required for genetic analysis were introduced. These include (1) the use of balancers, which are chromosomes with multiple inversions that cannot recombine with their homologs, thus allowing the maintenance of lethal mutations in heterozygotes without further selection; (2) the discovery of polytene chromosomes, which allow the physical mapping of genes; and (3) the introduction of x-rays as a

From: *Methods in Molecular Biology, vol. 372: Mitochondria: Practical Protocols*
Edited by: D. Leister and J. M. Herrmann © Humana Press Inc., Totowa, NJ

mutagenic agent, a finding that led to Muller's 1946 Nobel Prize. Most of the techniques developed at that time are still used in genetic work and make *Drosophila* the most genetically manipulable metazoan.

In the 1970s, many powerful biochemical, molecular, and cellular techniques were developed that allowed the use of *Drosophila* as a model system to study many complex biological phenomena. Paradigmatic examples of the feasibility of *Drosophila* in biological research were the identification and cloning of the bithorax complex by E. Lewis, D. Hogness, and their colleagues, and the genomewide mutational screen carried out by C. Nüsslein-Volhard and E. Wieschaus in 1981 that led to the discovery of dozens of genes involved directly in regulating embryonic development. Lewis, Nüsslein-Volhard, and Wieschaus shared the Nobel Prize in 1995. Another breakthrough in *Drosophila* research was the development in 1981 by A. Spradling and G. M. Rubin of efficient techniques based on P-transposons to generate transgenic flies.

During the last two decades of the 20th century, an arsenal of cellular and molecular tools have also been developed in *Drosophila* or adapted to work with this organism. The complete genome sequence was first reported in 2000, and its analysis is proceeding rapidly. The possibility to combine the power of classical genetics with a wide variety of cellular and molecular techniques has attracted more and more scientists to work with *Drosophila* in the context of many different fields, including regulation of gene expression, cell biology, neurobiology, behavior, development, aging, and more recently the physiopathology of human diseases.

However, in spite of the many advantages, *Drosophila* has not achieved priority status as an animal model in the mitochondrial field, in which scientists traditionally have been more focused on yeast and mammals. In this chapter, we present a brief introduction to the system, emphasizing some aspects that may be useful for laboratories interested in using *Drosophila* to study mitochondrial biogenesis and function. The reader is redirected to some excellent and extremely useful bench books and World Wide Web utilities that explain the genetics, biology, and manipulability of *Drosophila* in detail that is beyond the scope of this chapter. An introduction to *Drosophila* research may be found at http://flybase.bio.indiana.edu/allied-data/introductory.html.

1.1. The Drosophila *Life Cycle*

The *Drosophila* life cycle is short, and therefore it is easy to raise a large number of individuals for genetic, biochemical, and molecular analyses. In the laboratory, *Drosophila melanogaster* is usually cultured at 25 or 18°C (the latter mainly for maintaining stocks); we provide all the timing for 25°C, except where specifically indicated. The generation time is roughly 10 d from fertilized egg to eclosed adult, and the maximum life span ranges from 60 to 80 d depending on

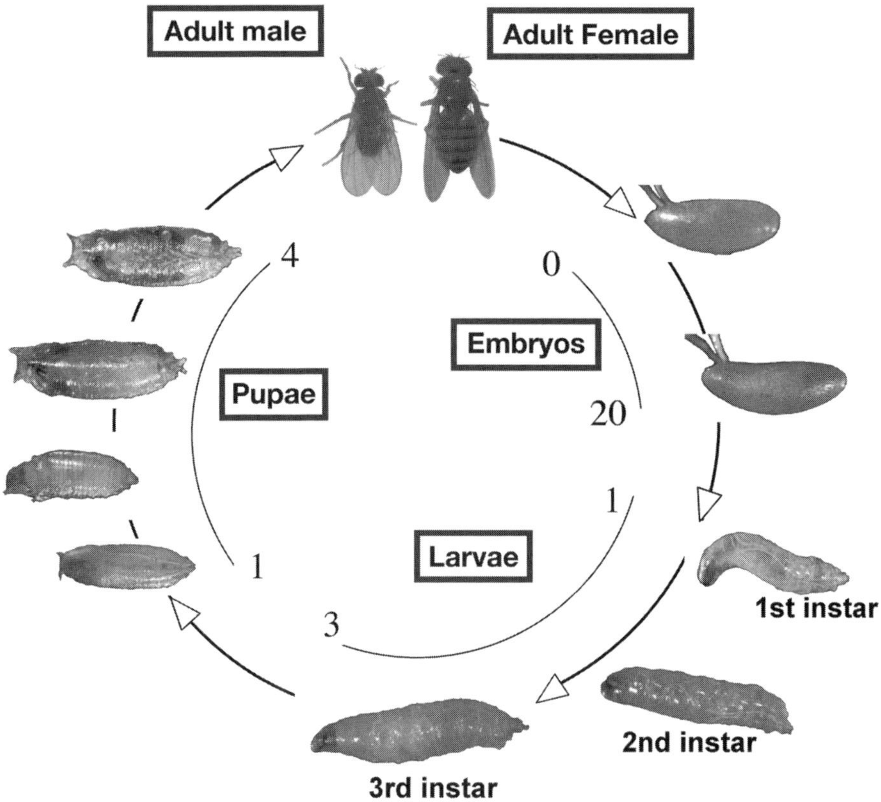

Fig. 1. The *Drosophila* life cycle is divided into four stages: embryo, larva, pupa, and adult. The time length of the stages is approximate and is shown in hours for embryos and days for larvae and pupae.

the culture conditions. *Drosophila* is a holometabolous insect, and its life cycle can be divided into four stages: embryo, larva, pupa, and adult (**Fig. 1**).

Females lay roughly 100 embryos per day, and embryogenesis lasts only 24 h (for a detailed description of embryonic stages, *see* http://www.sdbonline.org/fly/aimain/2stages.htm or http://flymove.uni-muenster.de/). The first instar larva begins to feed immediately on the surface of the medium and passes through two molts (**Fig. 2**). Second instar larvae burrow into the medium, and when the third instar larva is mature, it leaves the culture medium and wanders up the walls of the flask, searching for a place to pupariate for 24–48 h. During pupariation, a complete body metamorphosis from larva to adult takes place; most larval tissues are degraded, and adult organs develop from an undifferentiated sac of cells, the imaginal disks. In *Drosophila*, there are 10 pairs of imaginal disks, which reconstruct the entire adult body except the abdomen, and a genital disk, which forms

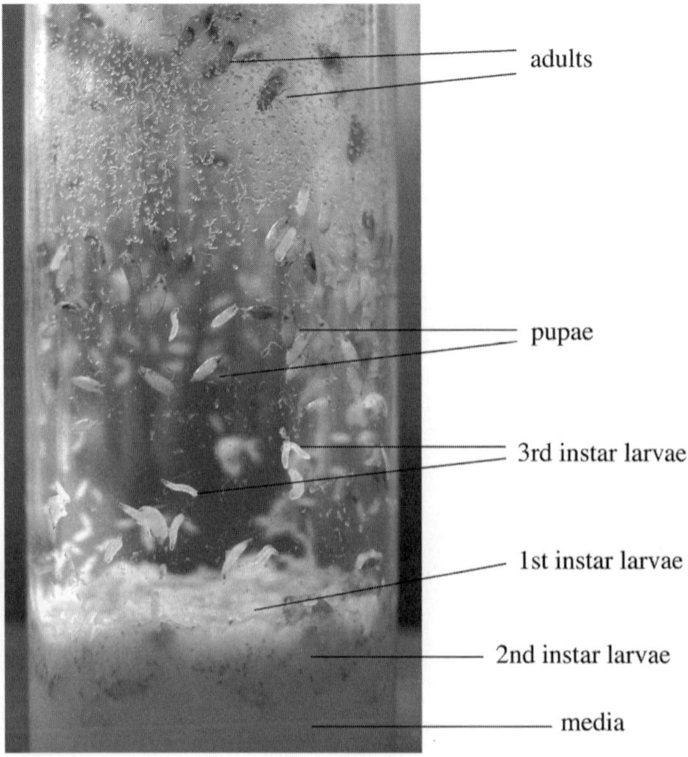

Fig. 2. The different stages of *Drosophila* life cycle growing in a bottle. First instar larvae feed on the surface of the medium. Second instar larvae burrow into the medium to feed (small black dots are the jaws of second instar larvae). Third instar larvae wander up the walls of the bottle, where they will pupariate. Adults are at the top of the bottle.

the reproductive organs. The abdominal epidermis forms from histoblasts, a group of specialized imaginal cells. The imaginal disks constitute cellular territories that have been extensively used to unravel the role of many genes involved directly in the morphogenesis of adult structures. Finally, the adult emerges between 9 and 10 d after egg fertilization (at 18°C, development takes twice as long).

1.2. The Drosophila *Genome*

Drosophila has four pairs of chromosomes: X/Y, II, III, and IV, with most of the gene content located on chromosomes X, II, and III. The first annotated sequence, release 1, was published in March 2000 ***(1)***. The haploid genome size is estimated to be 175 Mb by flow cytometry of propidium-stained nuclei, a value very similar to that obtained in the release 3.2 genome sequence (176 Mb). The number of protein-coding genes based on *in silico* methods of gene prediction is roughly 15,000, approx half of those predicted in the human genome. Release 4.0 was made public in April 2004 (the, last update was on March 03, 2006 [http://flybase.net/annot/release.html]), differing from release 3.2 with very few new annotations. Release 5.0 of genomic sequences was available on March 29, 2006 (http://www.fruitfly.org/sequence/release5genomic.shtml).

The mitochondrial genome of *Drosophila* shows the general features of other animal mitochondrial deoxyribonucleic acids (mtDNAs) regarding gene order, density, structure, and a genetic code that differs from the universal code, although some genes are rearranged compared to the mammalian mitochondrial genome *(2)*. A striking difference lies in the noncoding region, which contains 90–96% deoxyadenylate and deoxythymidylate residues (the A+T-rich region) and ranges in size from 1 to 5 kb in different *Drosophila* subgroups *(3)*. In *D. melanogaster*, the total length of the mtDNA molecule is 19,517 bp.

1.3. Drosophila *Mitochondrial Proteins*

As in other eukaryotic systems, the *Drosophila* mitochondrial genome encodes only a very small fraction of mitochondrial proteins that share a very high degree of evolutionary conservation. Many of the nuclear-encoded mitochondrial proteins are also very well conserved. An excellent analysis of the latter is presented in the MitoDrome database (http://www.ba.itb.cnr.it/BIG/Mito DromeOLD/), where one can find the *Drosophila* nuclear genes encoding mitochondrial proteins, their human counterparts, functions, and ontology. MitoDrome2 (http://www2.ba.itb.cnr.it/MitoDrome/index.php) is an enhanced version in which the authors identify, characterize, and show tools for analyzing genes encoding proteins that constitute the five large respiratory chain complexes in *D. melanogaster, D. pseudoobscura*, and *Anopheles gambiae* *(4)*.

Although analysis of the mitochondrial proteome is well under way in several organisms (e.g., *Arabidopsis*, rice, yeast, mouse, and human) *(5)*, to date no similar studies have been initiated in *Drosophila*.

1.4. Working With Drosophila

Working with *Drosophila* in the laboratory is relatively easy and requires neither special technical skill nor sophisticated infrastructure. Flies are generally grown in plastic vials and bottles containing medium (fly food) (**Fig. 3**). It is also possible to culture them in mass using special containers if you need to work with large numbers of flies (e.g., to carry out protein purification. Several media have been described, all based on simple components such as agar, yeast (not yeast extract), sucrose, and propionic acid. Culture medium can be prepared in a simple kitchen or with more complex and automated facilities depending on the number of stocks and specific needs. An excellent Web page to learn in detail how to prepare a complete series of media for *Drosophila* culturing, either animals or cells,

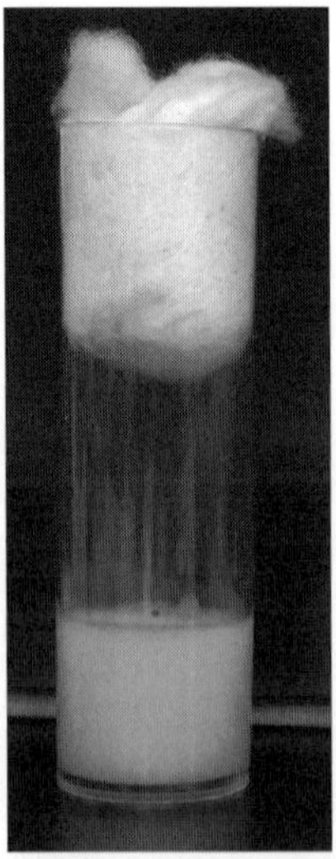

Fig. 3. Plastic vial containing medium for growing flies. The vial is covered with hydrophobic cotton to avoid condensation of humidity that could interfere with air supply.

under different conditions or for different purposes is http://www.protocol-online.org/prot/Model_Organisms/Drosophila/Drosophila_Culture_Handling/.

Stocks are usually maintained in vials at 18°C with four to five generation cycles before transfer. Because fly stocks can only be maintained by live culturing, it is crucial to keep two to four different cultures for each individual stock, with alternate generations separated by 1–2 wk if it is possible. Flies in experimental use are maintained routinely at 25°C.

To carry out crosses, you must start with virgin females. Female flies do not mate within the first 8–12 h after emergence as adults from the pupae. Thus, using this window of time, flies can be collected, and females can be separated from the males and kept separately until needed. Males can be collected at any time, with the best efficiency of mating when they are between 3 and 10 d old. The number of flies needed to start a new culture varies, mainly depending on the genotype and the specific requirements of the experiment. In general terms, 4–8 virgins and a smaller number of males are required for vials, and 10–20 flies are needed for bottles of small and medium size. To collect virgins, examine phenotypic markers, and manipulate *Drosophila* stocks, CO_2 is generally used to anesthetize flies instead of the traditional ether because is safer, easier, and avoids overanesthetization of the animals.

It is important to note the striking conservation of biological processes from flies to mammals. When a *Drosophila* homolog of an essential but poorly understood mammalian gene is identified, as happens with a large number of mitochondrial genes, powerful genetic and molecular techniques available in *Drosophila* can be applied to its characterization. These techniques include those discussed next.

1.4.1. Loss of Function Phenotype

1.4.1.1. Use of Mutagenic Agents

Mutagenic agents are used, followed by analysis of the different phenotypes produced and characterization of those caused by the mutation in the gene of interest.

1.4.1.2. Gene Disruption Mediated by P-Elements

This method is based on the insertion of a DNA flanked by transposase target sequences (the so-called P-element). The DNA inserts randomly into the genome. There are thousands of *Drosophila* lines available with P-elements inserted in different locations. In addition, a project to disrupt each gene in the *D. melanogaster* genome is under way (P-Element Screen/Gene Disruption Project; ***6***). Excellent information can be found at http://flypush.imgen.bcm.tmc.edu/pscreen/.

The power of these techniques can be increased using deletion mutants. Detailed information about the *Drosophila* Deletion Project such as construction, maps, and available stocks can be found at http://www.drosdel.org.uk/.

1.4.1.3. Ribonucleic Acid Interference (*see* Chapter 15)

Knockdown of *Drosophila* genes by ribonucleic acid interference (RNAi) either in cells or in animals is described in detail in Chapter 15. An excellent Web page to visit in relation to ribonucleic acid interference is http://flyrnai.org/.

1.4.1.4. Homologous Recombination

Although historically it was thought that *Drosophila* lacks the homologous recombination process, the method developed by Golic and collaborators demonstrated that it does occur ***(7)***. This technique allows precise substitution of a specific DNA region of the *Drosophila* genome by another homologous, although not identical, sequence.

1.4.2. Overexpression Phenotype (see *Chapter 15)*

By a relatively easy transgenesis, one can introduce extra copies of a complementary DNA or a gene under the control of a selected promoter. In addition, the UAS/GAL4 system is a powerful and extensive transgenesis-based method described in detail in Chapter 15. There are collections of transgenic flies available for this technique, such as those found at http://flystocks.bio.indiana.edu/Browse/misc-browse/gal4.htm.

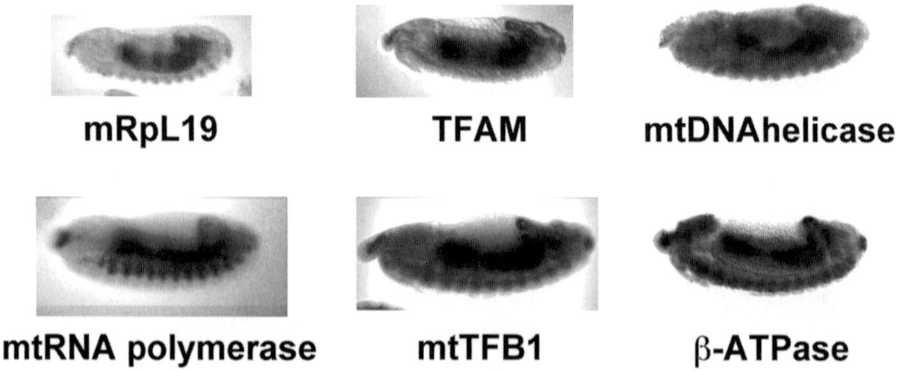

Fig. 4. Pattern of transcription of some nuclear-encoded mitochondrial genes during *Drosophila melanogaster* embryogenesis. All encode proteins involved in mtDNA metabolism: mRpL19, mitochondrial large ribosomal subunit protein 19; TFAM, mitochondrial transcription factor A; mtDNA helicase; mtRNA polymerase; mtTFB1, mitochondrial transcription factor B1; β-ATPase, β subunit of mitochondrial adenosine triphosphate synthase.

1.4.3. Developmental Pattern of Expression

Conservation of the developmental pattern of expression of a gene in different organisms may be an initial indicator of a similar function of the gene or of the process in which it is involved. An excellent Web site to access this method as a first approach is http://www.ceolas.org/VL/fly/protocols.html. In addition, we describe here a protocol for visualizing gene expression during *Drosophila* embryogenesis that we use in our laboratory and suggest a visit to http://www.fruitfly.org/cgi-bin/ex/insitu.pl. During *Drosophila* embryonic development, many nuclear-encoded mitochondrial genes involved in mtDNA replication, mtDNA maintenance, transcription, and translation share the midgut as a common territory of transcription (**Fig. 4**). Transcription of the genes encoding the mitochondrial ribosomal proteins mRpS17 and mRpL22 and the mtDNA maintenance factor TFAM (Mitochondrial Transcription Factor A) is also active in the midgut, as one can see at http://www.fruitfly.org/cgi-bin/ex/bquery.pl?qpage=entry&qtype=summary.

1.4.4. Phylogenetic Footprinting

Myriad computer programs have been developed to assist in the analysis of sequence data. Availability of genome sequences from other *Drosophila* species such as *Drosophila yakuba*, *Drosophila simulans*, or *Drosophila pseudoobscura* and other insects such as *Anopheles* or *Apis* open the possibility of using phylogenetic footprinting for identification of common regulatory elements that might suggest functional relationships among genes or groups of genes.

One notable example of this approach is the work of Caggese and collaborators *(8)*, which revealed the presence of a putative regulatory element termed NRG (nuclear respiratory gene) in 100% of respiratory chain genes and in many other nuclear-encoded mitochondrial genes in *D. melanogaster.* These are 90% conserved in *D. pseudoobscura* and in the respiratory chain complex V genes of *A. gambiae*. These authors also identified and annotated the *D. melanogaster*, *D. pseudoobscura*, and *A. gambiae* orthologs of 78 nuclear genes encoding mitochondrial proteins involved in oxidative phosphorylation by comparative analysis of their genomic sequences and organization.

2. Materials

2.1. Partially Purified Mitochondria

The ionic strength of buffers is determined using a radiometer conductivity meter.

1. Phenylmethylsulfonyl fluoride (PMSF) is prepared as a 0.2 *M* stock solution in isopropyl alcohol. Store aliquots at –20°C.
2. Sodium metabisulfite: 1.0 *M* stock solution at pH 7.5. Store aliquots at –20°C.
3. Leupeptin (Peptide Institute, Inc., code 4041): 1 mg/mL stock solution in 50 m*M* Tris-HCl, pH 7.5, 2 m*M* EDTA (ethylenediaminetetraacetic acid). Store aliquots at –20°C.
4. 0.5 *M* EDTA, pH 8.0.
5. 1 *M* Sucrose, ultrapure.
6. 1 *M* HEPES-KOH, pH 8.0; store at 4°C.
7. 1 *M* Dithiothreitol (DTT). Store aliquots at –20°C.
8. 3 *M* Potassium chloride.
9. 1 *M* Calcium chloride.
10. 10% (v/v) Triton X-100.
11. Homogenization buffer: 15 m*M* HEPES-KOH, pH 8.0, 5 m*M* KCl, 2 m*M* $CaCl_2$, 0.5 m*M* EDTA, 0.5 m*M* DTT, 0.28 *M* ultrapure sucrose, 1 m*M* PMSF, 10 m*M* sodium metabisulfite, 2 µg/mL leupeptin.
12. Dounce tissue grinder (homogenizer) (Wheaton), 7 mL, with tight and loose pestles.
13. Oak Ridge centrifuge tubes (screw-capped polypropylene copolymer tubes), 50 mL and 10 mL (Nalge Nunc International).
14. 25-mL Glass graduated cylinder.
15. Small plastic funnel.
16. Camel hair or similar brush with bristles clipped to approx 5 mm long.
17. 75-µm Nitex screen (Sefar America, Inc.), four 15-cm squares.

2.2. Mitochondrial Extraction

1. 20% (w/v) Sodium cholate. Cholic acid is dissolved in hot ethanol, filtered through Norit A (J. T. Baker Chemical Co.), and recrystallized twice before titration to pH 7.4 with sodium hydroxide.

2. Extraction buffer: 25 m*M* HEPES-KOH, pH 8.0, 10% (v/v) glycerol, 0.3 *M* NaCl, 1 m*M* EDTA, 1 m*M* DTT, 1 m*M* PMSF, 10 m*M* sodium metabisulfite, 2 μg/mL leupeptin.
3. Stabilization buffer: 25 m*M* HEPES-KOH, pH 8.0, 2 m*M* EDTA, 80% (v/v) glycerol.
4. 5 *M* Sodium chloride.
5. 1.5-mL Microcentrifuge tubes.
6. Other materials are as in **Subheading 2.1**.

2.3. *Visualizing Mitochondrial Messenger RNAs in* Drosophila *Embryos*

2.3.1. Preparation of the Probe

1. For transcription, we use the in vitro labeling kit from Roche (DIG RNA Labelling Kit SP6/T7; cat. no. 1175025). Although not included, T3 RNA polymerase can be used with this kit.
2. Phenol/chloroform (1:1).
3. Carbonate buffer: 120 m*M* Na_2CO_3, 80 m*M* $NaHCO_3$, pH 10.2. Store at –20°C.
4. Degradation stop solution: 0.2 *M* NaAc, pH 6.0.
5. 4 *M* LiCl.
6. Transfer RNA (tRNA) from baker's yeast (Sigma, cat. no. R5636).
7. 3 *M* Sodium acetate.
8. 100% Ethanol.
9. 70% (v/v) Ethanol in water.
10. Hybridization solution: 50% (v/v) deionized formamide, 5X SSC (Saline–Sodium Citrate), 50 μg/mL heparin, 100 μg/mL tRNA, and 0.1% (v/v) Tween-20.
11. Heparin sodium salt (Sigma- Aldrich, **ref.** H-3393).
12. 20X SSC: 3 *M* NaCl, 0.3 *M* sodium citrate, pH 7.

2.3.2. Preparation of the Embryos

1. 2.25% (w/v) Sodium hypochlorite.
2. A small spatula, a soft brush, and a filter to retain embryos.
3. Fixing solution: 1.3 mL 37% (v/v) formaldehyde, 5 mL heptane, 0.5 mL 10X phosphate-buffered saline (PBS), and 3.2 mL water.
4. PBS: 136 m*M* NaCl, 2 m*M* KCl, 8 m*M* Na_2HPO_4, 2 m*M* KH_2PO_4, pH 7.4.
5. 100% methanol.

2.3.3. Hybridization, Developing, and Visualization

1. PBT: PBS, 0.01% (v/v) Tween-20.
2. Rotator mixer.
3. 70, 50, and 30% (v/v) Methanol in PBT.
4. 4% (v/v) formaldehyde in PBT.
5. Hybridization solution/PBT (8:2 and 1:1) (*see* **item 10** in **Subheading 2.3.1.**).
6. Antidigoxigenin antibody from Roche (cat. no. 1093274).
7. Developing solution: 4 *M* NaCl, 50 m*M* $MgCl_2$, 100 m*M* Tris-HCl, pH 9.0, 0.1% (v/v) Tween-20.

8. *p*-Nitroblue tetrazolium chloride from Roche (cat. no. 1383213).
9. 5-Bromo-4-chloro-3-indolyl phosphate from Roche (cat. no. 1383221).
10. 70, 50, and 30% (v/v) Ethanol in PBT.
11. Xylene.
12. Permount SP15-500 (Fisher Chemicals).
13. Glass slides and coverslips.
14. Clear nail polish.
15. Other materials as in **Subheading 2.3.2**.

2.4. Visualizing Mitochondrial Proteins With Fluorescent Antibodies in Drosophila *Embryos*

1. 1.25% (w/v) Sodium hypochlorite.
2. AbFixing solution (fixing solution for using antibody): 0.6 mL 37% (v/v) formaldehyde, 8 mL heptane, 2.8 mL water, and 0.6 mL 5X buffer B (50 m*M* potassium phosphate buffer, pH 6.8, 225 m*M* KCl, 75 m*M* NaCl, 65 m*M* $MgCl_2$).
3. 10% (w/v) bovine serum albumin (BSA) in PBT.
4. Vectashield H-1000 (Vector Laboratories, Inc., Burlingame, CA).
5. Other materials are as in **Subheading 2.3.2**.

3. Methods

3.1. Partially Purified Mitochondria

1. Collect *D. melanogaster* (Oregon R) embryos (average age 9 h) immediately before use by rinsing from agar collection plates using 0.1% (v/v) Triton X-100 and 0.7% (w/v) NaCl, brushing with a camel hair brush, and collecting onto a 75-µm Nitex screen *(9)*.
2. Dechorionate embryos by incubation in 2.25% (w/v) sodium hypochlorite for 2 min with stirring, then rinse embryos thoroughly using Triton-NaCl solution *(9)*.
3. Settle embryos for 15 min in 20 mL Triton-NaCl solution in a 25 mL graduated cylinder to remove remaining chorions, yeast, and fly fragments; aspirate supernatant; and repeat settling twice (*see* **Note 1**) *(9)*.
4. Collect dechorionated settled embryos onto a tared 75-µm Nitex screen and blot until damp between layers of paper towels; then, weigh embryos.
5. Suspend processed embryos at a ratio of 4 mL/g (*see* **Note 2**), wet weight, in homogenization buffer containing 15 m*M* HEPES-KOH, pH 8.0, 5 m*M* KCl, 2 m*M* $CaCl_2$, 0.5 m*M* EDTA, 0.5 m*M* DTT, 0.28 *M* ultrapure sucrose, 1 m*M* PMSF, 10 m*M* sodium metabisulfite, and 2 µg/mL leupeptin; homogenize in approx 7-mL portions in a standard (7-mL) Dounce homogenizer using six strokes of the loose pestle followed by six strokes of the tight pestle (*see* **Note 3**).
6. Filter the homogenate through a 75-µm Nitex screen into a 50 mL Oak Ridge centrifuge tube.
7. Rehomogenize the sample retained on the Nitex screen as in **step 5** using the same buffer (1 mL/g), filter as in **step 6**, and combine with the original filtrate (*see* **Note 4**).
8. Centrifuge the combined filtrate at 1000*g* for 7 min at 3°C to pellet nuclei and cellular debris (*see* **Note 5**).

9. Remove the supernatant using a 10-mL pipet, transfer to a fresh centrifuge tube, and repeat centrifugation as in **step 8** (*see* **Note 5**).
10. Repeat **step 9** once.
11. Pellet mitochondria by centrifugation at 7400*g* for 10 min at 3°C (*see* **Note 6**). Aspirate supernatant and discard.
12. Resuspend the mitochondrial pellet at a ratio of 2 mL of homogenization buffer per gram of starting embryos, transfer suspension into a 10-mL Oak Ridge centrifuge tube, centrifuge at 8000*g* for 15 min at 3°C, and aspirate supernatant and discard.
13. Repeat **step 12** once.
14. Resuspend the third pellet at a ratio of 0.5 mL homogenization buffer per gram, combine sample into one 10-mL tube or distribute into two 1.5-mL microcentrifuge tubes, and centrifuge as in **step 12**.
15. Freeze the final mitochondrial pellet in liquid nitrogen and store at –80°C.

3.2. Mitochondrial Extraction

1. Thaw frozen, partially purified mitochondria from freshly harvested and dechorionated *Drosophila* embryos (5 g) on ice for at least 30 min.
2. Resuspend mitochondria at a ratio of 0.5 mL/g of starting embryos (*see* **Note 7**) in extraction buffer containing 25 m*M* HEPES-KOH, pH 8.0, 10% (v/v) glycerol, 0.3 *M* NaCl, 1 m*M* EDTA, 1 m*M* DTT, 1 m*M* PMSF, 10 m*M* sodium metabisulfite, and 2 μg/mL leupeptin.
3. Add sodium cholate to a final concentration of 2% (v/v) (*see* **Note 8**) and incubate the suspension on ice for 30 min with gentle mixing by inversion at 5-min intervals.
4. Centrifuge the resulting extract at 96,000*g* for 30 min at 3°C.
5. Recover the supernatant fluid (*see* **Note 9**) and add an equal volume of stabilization buffer containing 25 m*M* HEPES-KOH, pH 8.0, 2 m*M* EDTA, and 80% (v/v) glycerol.
6. Store the mitochondrial extract (fraction I) at –20°C.

3.3. Visualizing Mitochondrial Messenger RNAs in Drosophila Embryos

3.3.1. Preparation of the Probe

Smaller probes penetrate the embryo more readily. Thus, after transcription, the probe is usually degraded by alkali treatment and purified (*see* **Note 10**).

1. The fragment to be labeled must be previously cloned by standard methods in an *Escherichia coli* vector (i.e., pBluescript) flanked by a T7, T3, or SP6 RNA polymerase promoter (*see* **Note 11**).
2. Digest 2–3 μg of the plasmid with a suitable restriction enzyme. This must produce a linear fragment containing the promoter and the gene or fragment of the gene (*see* **Note 12**).
3. Check the digestion by agarose gel electrophoresis. If it is complete, then treat it three times with phenol/chloroform (*see* **Note 13**).
4. Precipitate the DNA with sodium acetate and ethanol by standard procedures and resuspend it to yield a 1-mg/mL concentration in water.

5. Make the riboprobe using the appropriate RNA polymerase (e.g., use an in vitro labeling kit following manufacturer's recommendations). A standard reaction includes 1 µg DNA; digU-NTP mix (labeling mix containing the font rNTPs plus digoxigenin-UTP); ribonuclease inhibitor; buffer; and RNA polymerase in a final volume of 10 µL.
6. After 2 h at 37°C, check 1 µL of the transcription reaction on a 1% agarose gel. You should see a single band of the expected size, although more diffuse than a DNA band (*see* **Note 14**).
7. To degrade the probe, add first 15 µL water (*see* **Note 10**).
8. Add 25 µL carbonate buffer and keep at 65°C for 40 min.
9. Add 50 µL degradation stop solution.
10. To purify the already degraded riboprobe, precipitate the RNA by adding 10 µL 4 *M* LiCl, 5 µL tRNA (20 µg/µL), and 300 µL ethanol.
11. Incubate 30 min at –20°C.
12. Centrifuge at 12,000*g* for 20 min at 4°C.
13. Wash twice with 70% (v/v) ethanol in water and resuspend in 100 µL of hybridization solution. Check 5 µL by agarose gel electrophoresis. Degradation must be observed.
14. Store at –20°C for days or up to several months.

3.3.2. Preparation of Embryos

1. Harvest embryos from 8 h collection using a soft brush, water, and a small filter that permits liquid to pass through but retains the embryos (*see* **Note 15**).
2. Submerge the filter in 2.25% (w/v) sodium hypochlorite for 2 min to remove the chorion (*see* **Note 16**).
3. Rinse exhaustively with water (*see* **Note 17**).
4. Fix the embryos by taking them with a spatula and submerging in 10 mL fixing solution (*see* **Note 18**).
5. Mix vigorously for 20 min (i.e., 300 rpm in a shaker).
6. Remove the formaldehyde phase (the lower phase).
7. Add 10 mL methanol and mix vigorously by hand for 60 s. This removes the embryonic vitelin membrane.
8. Embryos sediment in a few seconds. Those with vitelin membranes remain suspended.
9. Remove everything except embryos at the bottom.
10. Add 5 mL methanol, mix gently, and remove it. Repeat twice.
11. Add 1 mL methanol, transfer embryos carefully with a cut pipetor tip to a 1.5-mL tube, and store at 4°C or –20°C (*see* **Note 19**).

3.3.3. Hybridization, Developing, and Visualization

Although embryos are already fixed, we recommend fixing the embryos again after storage. This requires a previous hydration, which is made as follows (*see* **Note 20**):

1. Use approx 50 µL of embryos in a 1.5-mL tube.
2. Remove the methanol and add 1 mL 70% methanol in PBT. Mix gently for 10 s.

3. Remove the methanol and add 1 mL 50% methanol in PBT. Mix gently for 10 s.
4. Remove the methanol and add 1 mL 30% methanol in PBT. Mix gently for 10 s.
5. Remove the methanol and add 1 mL PBT. Mix gently for 2 min. Repeat.
6. To refix the already hydrated embryos, remove the PBT and add 1 mL PBT/4% (v/v) formaldehyde.
7. Mix on a rotator mixer for 20 min at room temperature.
8. Remove the PBT/formaldehyde solution and wash with 1 mL PBT on a rotator mixer for 5 min. Repeat five times.
9. Wash with 1 mL PBT/hybridization solution (1:1).
10. For hybridization, remove the PBT/hybridization solution and prehybridize by adding 1 mL hybridization solution.
11. Incubate at 55°C for 60 min (no rotator mixer required) (*see* **Note 21**).
12. Prepare the probe: 1 µL of probe is added to 50 µL of hybridization solution and heated to 80°C for 10 min. Place on ice for 5 min (*see* **Note 22**).
13. Remove the prehybridization solution from the embryo tube and add the probe.
14. Incubate at 56°C overnight.
15. For washing, remove (and store) the probe containing hybridization solution and add 1 mL 55°C preheated hybridization solution. Incubate 20 min at 55°C. Repeat twice (*see* **Note 23**).
16. Remove solution and wash with hybridization solution/PBT (8:2). Rotator mix for 1 min.
17. Remove solution and wash with hybridization solution/PBT (1:1). Rotator mix for 1 min.
18. Remove solution and wash with PBT. Rotator mix for 20 min. Repeat four times.
19. For developing, remove PBT and add 400 µL PBT containing 10 µL pretreated antidigoxigenin antibody (*see* **Note 24**).
20. Incubate 60 min in a rotator mixer at room temperature.
21. Remove antibody solution and wash 5 min with 1 mL PBT in a rotator mixer. Repeat four times.
22. Remove PBT and wash twice with 1 mL freshly prepared developing solution (*see* **Note 25**).
23. Remove and add 1 mL developing solution containing 9 µL p-nitroblue tetrazolium chloride and 7 µL 5-bromo-4-chloro-3-indolyl phosphate.
24. When embryos are colored, stop reaction by washing with PBT (*see* **Note 26**).
25. Finally, to prepare embryos for the microscope, we dehydrate them by washing with 30% (v/v) ethanol in PBT and leave 2 min on the bench.
26. Wash with 50% (v/v) ethanol in PBT. Leave 2 min on the bench.
27. Wash with 70% (v/v) ethanol in PBT. Leave 2 min on the bench.
28. Wash with 100% ethanol. Leave 2 min on the bench. Repeat twice (*see* **Note 27**).
29. Remove ethanol and add 1 mL xylene, which removes all possible traces of water (*see* **Note 28**).
30. Remove xylene and add 200 µL Permount.
31. Remove the embryos carefully with a cut pipetor tip and place on a glass slide; try to separate individual embryos. Add a glass coverslip and seal with clear nail polish.

32. Take good pictures under the microscope.

3.4. Visualizing Mitochondrial Proteins With Fluorescent Antibodies in *Drosophila* Embryos

The preparation of the embryos is the same as for riboprobes (*see* **Subheading 2.3.2.**) except for the following:

The chorion is removed with 1.25% (v/v) sodium hypochlorite treatment.
Substitute fixing solution for AbFixing solution.
Do not store embryos. Depending on the antibody, fresh embryos are crucial.

Thus, after embryo hydration, we incubate with primary antibody as follows:

1. Incubate embryos with 10% (w/v) BSA in PBT. Incubate 60 min in a rotator mixer at room temperature (*see* **Note 29**).
2. Remove the solution and wash with 1 mL PBT in rotator mixer for 10 min. Repeat three times.
3. Add primary antibody in PBT (*see* **Note 30**).
4. Incubate overnight at 4°C in a rotator mixer.
5. To incubate with the secondary antibody, we remove primary antibody solution and wash four times for 10 min with PBT in a rotator mixer at room temperature.
6. Add 200 μL of secondary fluorescent antibody 1:200 in PBT. Incubate at least 60 min in a rotator mixer at room temperature.
7. Wash 5 min with PBT. Repeat three times.
8. Remove PBT and add 3 drops Vectashield.
9. Remove the embryos carefully with a cut Pipetman tip and put on a glass slide. Place a glass coverslip and seal with clear nail polish.
10. Take good pictures. (*Avoid immersion oil contacts with Vectashield.*)
11. Store in dark at 4°C. Embryos will remain fluorescent for approx 1 mo.

4. Notes

1. All operations are performed at 0–4°C.
2. The procedures from **step 5** through **step 15** are designed for 5 g starting material and may be adjusted proportionally.
3. Push the pestle slowly through the sample. To prevent sample loss, try wrapping parafilm around the top of the homogenizer and the pestle.
4. The sample retained on the Nitex screen is rehomogenized to break any remaining intact embryos.
5. The nuclear and cellular debris pellet is pale yellow and somewhat loose; try to remove supernatant without disturbing the pellet.
6. The mitochondrial pellet is beige and forms a tighter pellet than that observed in **step 8**.
7. Resuspend the mitochondrial pellet in one-third of the total extraction buffer volume, using the remaining two-thirds in two aliquots to wash out residual mitochondria and combine.

8. Before adding sodium cholate, make certain that the mitochondria are completely suspended. Immediately following the addition of sodium cholate, the sample should become slightly viscous.
9. The top of the supernatant will have a white lipid layer, and the top of the pellet is somewhat loose. Remove the supernatant carefully, avoiding the lipid layer and leaving behind approx 5% of the supernatant near the pellet.
10. This is for probes bigger than 300 bp.
11. It is not necessary to clone a complete complementary DNA; a fragment of approx 300–400 bp is sufficient.
12. Only 1 μg of the digestion is used for the transcription reaction, and the remainder may be stored. The promoter and the insert must be arranged in order to transcribe antisense molecules to hybridize with the mRNA of interest.
13. From this point, take care to protect the RNA (use gloves, aqueous solutions treated with diethyl pyrocarbonate, and sterilized materials).
14. The reaction may be stopped by deoxyribonuclease I treatment. For *in situ* experiments, this is not strictly required because of the high number of transcribed RNA vs DNA template molecules.
15. Because *Drosophila* can store embryos in the abdomen several hours, the time of laying is delayed. Thus, two or three consecutive layings are discarded, and the fourth laying period is harvested with few old embryos. For visualizing RNA, embryos can be stored at –20°C so that a large number can be harvested for several experiments.
16. Some researchers prefer 1.25% (w/v) sodium hypochlorite for 4 min or other combinations.
17. Some consider inserting a final wash with 0.7% (w/v) NaCl/0.02% (v/v) Triton X-100.
18. We use small glass vials that are sold for scintillation counting. Embryos go to the interface. It is important that they form a monolayer. If there are too many embryos, then they aggregate and do not fix properly.
19. Some prefer to store embryos in ethanol. This requires extra washes with ethanol before storage.
20. Incubation of non-refixed embryos with hybridization solution may break many structures in the embryo. For refixing, previous rehydration is required.
21. Prehybridization saturates the nonspecific nucleic acid-binding elements in embryos.
22. This treatment denatures the RNA so that hybridization is facilitated.
23. After hybridization, probes can be used several times, although this is often unnecessary because of the excess of unused probe stored at –20°C. From here, steps are at room temperature.
24. To avoid unspecific interactions with embryos, a 1:50 dilution in PBT of the antidigoxigenin antibody is incubated overnight at 4°C with rehydrated embryos. Thus, final dilution of the antibody is 1/2000.
25. Sometimes, this solution crystallizes at room temperature. Discard in this case.
26. After the developing reaction, the staining will be partially removed, so it is better to overdevelop than underdevelop. This choice is usually difficult to make because one tends worry about overdevelopment. Embryos not stained in 2 h will remain as such.

27. Sometimes, keeping overnight at 4°C increases the contrast by eliminating nonspecific staining.
28. If embryos collapse or aggregate, then repeat the 100% ethanol washes.
29. This is for BSA to block the nonspecific protein–protein interaction points in the embryos.
30. Make a dilution of serum at 1:100 to 1:500. If the antiserum is particularly poor, then use a 1:50 dilution; if excellent, then use 1:1000. For monoclonal antibodies, use a 1:10 dilution.

Acknowledgments

The work in our laboratories was supported by Ministerio de Ciencia y Tecnología, Spain (grant BFU2004-04591) and Instituto de Salud Carlos III, Redes de centros RCMN (C03/08) and Temáticas (G03/011) to R. G.; Fondo de Investigaciones Sanitarias, (PI041001) to M. A. F.-M.; and National Institutes of Health grant GM45295 to L. S. K.

References

1. Adams, M. D., Celniker, S. E., Holt, R. A., et al. (2000) The genome sequence of *Drosophila melanogaster. Science* **287,** 2185–2195.
2. Garesse, R. (1988) *Drosophila melanogaster* mitochondrial DNA: gene organization and evolutionary considerations. *Genetics* **118,** 649–663.
3. Lewis, D. L., Farr, C. L., and Kaguni, L. S. (1995) *Drosophila melanogaster* mitochondrial DNA: completion of the nucleotide sequence and evolutionary comparisons. *Insect Mol. Biol.* **4,** 263–267.
4. Tripoli, G., D'Elia, D., Barsanti, P., and Caggese, C. (2005) Comparison of the oxidative phosphorylation (OXPHOS) nuclear genes in the genomes of *Drosophila melanogaster, Drosophila pseudoobscura and Anopheles gambiae. Genome Biol.* **6,** R11
5. Gabaldon, T. and Huynen, M. A. (2004) Shaping the mitochondrial proteome. *Biochim. Biophys. Acta* **1659,** 212–220.
6. Bellen, H. J., Levis, R. W., Liao, G., et al. (2004) The BDGP Gene Disruption Project: single transposon insertions associated with 40% of *Drosophila* genes. *Genetics* **167,** 761–781.
7. Rong, Y. S., and Golic, K. G. (2000) Gene targeting by homologous recombination in *Drosophila. Science* **288,** 2013–2018.
8. Sardiello, M., Tripoli, G., Romito, A., et al. (2005) Energy biogenesis: one key for coordinating two genomes. *Trends Genet.* **21,** 12–16.
9. Brakel, C. L. and Blumenthal, A. B. (1977) Multiple forms of *Drosophila* embryo DNA polymerase: evidence for proteolytic conversion. *Biochemistry* **16,** 3137–3143.

4

Isolation and Functional Analysis of Mitochondria From the Nematode *Caenorhabditis elegans*

Leslie I. Grad, Leanne C. Sayles, and Bernard D. Lemire

Summary

Mitochondria are essential organelles with central roles in diverse cellular processes such as apoptosis, energy production via oxidative phosphorylation, ion homeostasis, and the synthesis of heme, lipid, amino acids, and iron-sulfur clusters. Defects in the mitochondrial respiratory chain lead to or are associated with a wide variety of diseases in humans. The nematode *Caenorhabditis elegans* provides a powerful genetic and developmental model system for reproducing deleterious mutations causing mitochondrial dysfunction and investigating their metabolic consequences and their mechanisms of pathology. In this chapter, we describe the growth of *C. elegans* in liquid culture, the isolation of crude and purified mitochondria, and polarographic and histochemical approaches for measuring mitochondrial respiratory chain function.

Key Words: Clark-type electrode; cuticle permeabilization; cytochrome-*c* oxidase; fixation; histochemistry; NADH-ubiquinone oxidoreductase; polarography; respiration; rotenone; succinate dehydrogenase.

1. Introduction

Our knowledge of *Caenorhabditis elegans* genetics and biology is both refined and extensive *(1)*. *Caenorhabditis elegans* is a small, free-living nematode worm that lives in the soil of the temperate regions of the world. In the mid-1960s, Sydney Brenner selected it as a model animal system for his investigations into the development of the nervous system *(2)*. The nematode has a short but complex life cycle, taking about 3 d to complete at 25°C. *Caenorhabditis elegans* exists primarily as a self-fertilizing hermaphrodite capable of producing approx 300 progeny per generation. Fertilized embryos develop into first-stage larvae, hatch, and proceed through four distinct larval stages (L1–L4) prior to becoming reproductive adults. Under conditions of food

From: *Methods in Molecular Biology, vol. 372: Mitochondria: Practical Protocols*
Edited by: D. Leister and J. M. Herrmann © Humana Press Inc., Totowa, NJ

deprivation or stress, an alternative developmental stage, called the dauer larva, can occur *(3)*. Dauer larvae can survive for months without feeding, but when favorable conditions return, they can exit this stage and resume maturation to the adult.

The number of somatic cells in the hermaphrodite adult (959 total; 302 neuronal cells) is invariant, and the organism's complete cell lineage has been described *(4)*. Although a simple animal, *C. elegans* possesses differentiated tissue systems, including a nervous system, pharyngeal and body muscles, intestine, epidermis, and a reproductive system.

Caenorhabditis elegans has become an organism of choice for studying biological processes linked to human disease. It has been estimated that *C. elegans* has orthologs for approx 50% of human disease genes *(5)*; many such genes have been investigated, yielding important insights into their functions. The availability of the complete *C. elegans* mitochondrial deoxyribonucleic acid (mtDNA) and nuclear DNA sequences has greatly stimulated research in the nematode system and facilitated the use of both forward and reverse genetic approaches *(6,7)*.

The structure, metabolism, and bioenergetics of nematode mitochondria share many similarities with their mammalian counterparts *(8,9)*. The nematode mtDNA is slightly smaller than the human mitochondrial genome, possessing 12 of the 13 protein-coding genes; it is missing the ATP8 gene, encoding a subunit of the adenosine triphosphate synthase *(6)*. Pathways of intermediary metabolism such as the Krebs cycle and oxidative phosphorylation are highly conserved in *C. elegans* *(10,11)*. In particular, both the mtDNA-and nuclear DNA-encoded subunits that constitute the functional cores of the mitochondrial respiratory chain (MRC) complexes are highly conserved *(9)*. It is worth noting that the *C. elegans* complex I (NADH-ubiquinone oxidoreductase), which consists of at least 36 subunits, resembles the complex I of higher eukaryotes and is sensitive to the inhibitor rotenone.

The close structural and functional conservation of mitochondrial protein complexes combined with the general ease of genetic manipulation make *C. elegans* an attractive system for studying the effects of mitochondrial function and dysfunction on organismal metabolism, development, and aging. In the following subheadings, we present some of the methodologies we utilize in our investigations into nematode mitochondrial energy metabolism, beginning with the growth of worms in liquid culture. We describe our procedures for isolating and purifying mitochondria and for measuring complex I-dependent respiration. We also present protocols for measuring respiration in live animals and for the histochemical analysis of succinate dehydrogenase (SDH) and cytochrome-*c* oxidase (COX) activities.

2. Materials

2.1. Growth of Bacteria for Worm Liquid Cultures

1. Superbroth: 12 g Bacto™ tryptone, 24 g yeast extract, 4 mL 100% (v/v) glycerol, 900 mL H_2O. Autoclave, then add 100 mL 0.17 *M* KH_2PO_4, 0.72 *M* K_2HPO_4 (23.1 g KH_2PO_4, 125.4 g K_2HPO_4 dissolved in water to a final volume of 1 L and autoclaved) (*see* **Note 1**).
2. 10,000X Streptomycin: 200 mg/mL streptomycin sulfate dissolved in water. Store at –20°C.
3. 1000X Nystatin: 40 mg/mL suspension in water. Store at –20°C.

2.2. Growth of C. elegans *in Liquid Culture*

1. Nematode growth medium plates: mix 3 g NaCl, 5 g Bacto peptone, 17 g agar with 975 mL H_2O in a 2-L Erlenmeyer flask. Cover the mouth of the flask with aluminum foil and autoclave for 40 min. Cool the flask to approx 55°C and add 1 mL 1 *M* $CaCl_2$, 1 mL 1 *M* $MgSO_4$, 25 mL 1 *M* potassium phosphate, pH 6.0, and 1 mL of 5 mg/mL cholesterol dissolved in ethanol, mixing after each addition.
2. 1 *M* Potassium phosphate, pH 6.0: 136.1 g KH_2PO_4 adjusted to pH 6.0 with KOH in a final volume of 1 L. Sterilize by autoclaving.
3. Cholesterol: 5 mg/mL in ethanol. Do not autoclave (*see* **Note 2**).
4. 1 *M* $CaCl_2$: 147.0 g $CaCl_2 \cdot 2H_2O$ in 1 L H_2O. Sterilize by autoclaving.
5. 1 *M* $MgSO_4$: 246.5 g $MgSO_4 \cdot 7H_2O$ in 1 L H_2O. Sterilize by autoclaving.
6. M9 buffer: 3 g KH_2PO_4, 6 g Na_2HPO_4, 5 g NaCl, water to 1 L. After autoclaving, add 1 mL 1 *M* $MgSO_4$.
7. S basal complete: 5.85 g NaCl, 1 g K_2HPO_4, 6 g KH_2PO_4, add water to 1 L. After autoclaving, add 1 mL cholesterol (5 mg/mL in ethanol), 10 mL 1 *M* sodium citrate, pH 6.0, 10 mL trace metals solution, 3 mL 1 *M* $CaCl_2$, and 3 mL 1 *M* $MgSO_4$.
8. 1 *M* Sodium citrate, pH 6.0: 210.1 g citric acid monohydrate ($C_6H_8O_7 \cdot H_2O$) adjusted to pH 6.0 with NaOH in a final volume of 1 L. Sterilize by autoclaving.
9. Trace metals solution: 1.86 g disodium ethylenediaminetetraacetic acid (EDTA) (5 m*M*), 0.69 g $FeSO_4 \cdot 7H_2O$ (2.5 m*M*), 0.2 g $MnCl_2 \cdot 4H_2O$ (1 m*M*), 0.29 g $ZnSO_4 \cdot 7H_2O$ (1 m*M*), 0.025 g $CuSO_4 \cdot 5H_2O$ (0.1 m*M*); add water to a final volume of 1 L. Sterilize by autoclaving and store in the dark.

2.3. Harvesting C. elegans *Cultures*

To harvest *C. elegans* cultures, use M9 buffer: 3 g KH_2PO_4, 6 g Na_2HPO_4, 5 g NaCl; add water to a final volume of 1 L. After autoclaving, add 1 mL 1 *M* $MgSO_4$.

2.4. Cleaning C. elegans *by Sucrose Flotation*

1. 0.1 *M* NaCl.
2. 60% (w/v) Sucrose.
3. Worm lysis buffer: 0.8 *M* sucrose, 1 m*M* EDTA, 10 m*M* Tris-HCl, pH 7.4 (*see* **Note 3**).

2.5. Isolation of Crude Mitochondria

1. Complete mini protease inhibitor cocktail tablets (Roche 1836153).
2. Acid-washed glass beads, 100-μm diameter. New beads are soaked overnight in concentrated HCl. Using a Buchner funnel, the beads are washed once with water and resuspended and stirred in 2 volumes 0.5 *M* HCl. The beads are washed with water until the pH is approx 4.0, washed with 2 volumes of 0.5 *M* NaOH followed by water until the pH is approx 8.0. They are dried overnight in a drying oven.

2.6. Isolation of Purified Mitochondria

1. 1 *M* Sucrose: 1 *M* sucrose, 10 m*M* Tris-HCl, pH 7.4, 1 m*M* EDTA.
2. 2 *M* Sucrose: 2 *M* sucrose, 10 m*M* Tris-HCl, pH 7.4, 1 m*M* EDTA.

2.7. Polarographic Analysis: NADH-Dependent, Rotenone-Sensitive Respiration of Isolated Mitochondria

1. Sodium sulfite.
2. MSE: 0.2 *M* Mannitol, 0.07 *M* sucrose, 0.1 *M* EDTA, pH 7.4.
3. 0.1 *M* 4-(2-Hydroxyethyl)-1-piperazineethanesulfonic acid (HEPES; adjusted to pH 7.4 with NaOH).
4. 10 m*M* β-Nicotinamide adenine dinucleotide, reduced form (NADH). This solution is light sensitive and should be made fresh daily.
5. 1 m*M* Rotenone in absolute ethanol. Store at –20°C.
6. 1 *M* KCN (made fresh daily in a fume hood). KCN solutions should be kept alkaline as highly toxic HCN is produced under acidic conditions.

2.8. Polarographic Analysis: Whole Animal Respiration

For polarographic analysis of whole animal respiration, use M9 buffer: 3 g KH_2PO_4, 6 g Na_2HPO_4, 5 g NaCl; add water to 1 L final volume. After autoclaving, add 1 mL 1 *M* $MgSO_4$.

2.9. Determination of Protein Content After Whole Animal Respiration Measurements

1. 1.85 *M* NaOH, 7.4% (v/v) β-mercaptoethanol.
2. 50% (w/v) Trichloroacetic acid (TCA).
3. Acetone.
4. 5% (w/v) Sodium dodecyl sulfate (SDS), 62.5 m*M* Tris-HCl, pH 6.8.

2.10. Histochemical Analysis of Mitochondrial Function: Fixation and Permeabilization of Worms

1. 2X MRWB: 160 m*M* KCl, 40 m*M* NaCl, 20 m*M* ethyleneglycol-bis-(β-aminoethyl ether) *N,N,N′,N′*-tetraacetic acid (EGTA), 10 m*M* spermidine HCl, 30 m*M* piperazine-*N,N′*-bis[2-ethanesulfonic acid] (PIPES), pH 7.4 (PIPES free acid adjusted to pH 7.4 with NaOH), 50% (v/v) methanol.

2. Tris Triton buffer (TTB): 100 m*M* Tris-HCl, pH 7.4, 1% (v/v) Triton X-100, 1 m*M* EDTA.
3. 40X Borate buffer: 1 *M* H_3BO_3, 0.5 *M* NaOH.
4. Phosphate-buffered saline (PBS): 136 m*M* NaCl, 2.6 m*M* KCl, 10 m*M* Na_2HPO_4, 1.7 m*M* KH_2PO_4, pH 7.4.
5. 20% (w/v) Paraformaldehyde: Add 200 mg paraformaldehyde to 0.9 mL 5 m*M* NaOH. Incubate at 65°C for 15 min. Prepare fresh.
6. 1 *M* Dithiothreitol (DTT).
7. β-Mercaptoethanol.

2.11. Histochemical Analysis of Mitochondrial Function: SDH Activity Stain

1. 0.5 *M* EDTA, pH 7.4.
2. 1 *M* KCN (made fresh daily).
3. 100 m*M* Phenazine methosulfate (PMS).
4. 1 *M* Sodium succinate, pH 7.4.
5. 1 *M* Sodium malonate, pH 7.4.
6. 25 m*M* Nitroblue tetrazolium (NBT).
7. 2% (w/v) Agarose.

2.12. Histochemical Analysis of Mitochondrial Function: COX Activity Stain

1. 3,3′-Diaminobenzidine.
2. Cytochrome-*c* (from horse heart; Sigma).
3. Catalase.

3. Methods

The growth of *C. elegans* in liquid culture requires that sufficient bacteria be available as food for the entire duration of the liquid culture growth. If the culture starves or the aeration is insufficient, then the worms will enter the dauer stage, an alternative nonfeeding, nonreproductive stage. Contamination of the liquid culture can severely reduce the yield of worms. Therefore, it is important to use aseptic technique and uncontaminated inocula. The presence of antibiotics is highly recommended.

Worms must be essentially free of bacteria before measuring respiration or proceeding with the isolation of mitochondria. The worms are isolated by centrifugation and washed several times to remove bacteria. The harvested worms are suspended in buffer and placed on a shaker platform to allow the digestion of *E. coli* remaining in the gut. Worms can be further cleaned by a sucrose flotation step. This step is optional but will remove much of the debris that accumulates during liquid culture. It will reduce the yield and result in the loss of some juvenile worms because of osmotic shrinking. The worms must be kept cold

(0–4°C) for proper flotation of the worms. It is important to work quickly because prolonged exposure to high sucrose concentrations is lethal.

Grinding with glass beads is used to break open the worms, and crude mitochondria are isolated by differential centrifugation. It is important to use protease inhibitors during the lysis. The crude mitochondria can be used for Western blot analysis or enzyme assays or further purified by sedimentation on a sucrose gradient.

Polarographic analysis provides a quick and reproducible means of measuring the rate of oxygen consumption by isolated mitochondria or by whole nematodes. A variety of respiratory substrates, cofactors, and inhibitors can be used and be instrumental in localizing an enzymatic defect to a specific portion of the respiratory chain. The majority of mutations that cause mitochondrial disease occur within genes encoding subunits of the MRC.

Beginning with the oxidation of NADH by complex I, electrons are passed through the MRC via a series of electron carriers to complex IV, where molecular oxygen is reduced to water. The rate of oxygen consumption can be measured directly using a Clark-type electrode, which consists of a probe with an exposed platinum cathode and a silver anode. When the anode and cathode are polarized (typically at–0.6 V), the current produced is directly proportional to the partial pressure of oxygen. The current is generated by the following reactions:

$$4Ag^+ + 4Cl^- \rightarrow 4AgCl + 4e^-$$

$$O_2 + 4H^+ + 4e^- \rightarrow 2H_2O$$

The reactions are connected via an electrolyte solution, such as KCl. The cathode is typically covered by an oxygen-permeable membrane, such as a polypropylene membrane, to exclude contaminating species, ions, or sample that might interfere with the reaction.

All polarographic assays are conducted using a Mitocell (MT200) respiration chamber with magnetic stirrer and 1302 Clark-type microcathode oxygen electrode attached to a 782 oxygen meter (Strathkelvin Instruments, Glasgow, UK, or Warner Instruments, Hamden, CT, USA). The MT200 is designed for the measurement of respiration rates using very small (50 or 100 µL) sample volumes, an important asset when limited amounts of material are available for analysis. For example, reproducible respiration rates can be achieved using as few as 150 adult nematodes. This is an important advantage when studying mutant strains that are difficult to culture because of the severity of the mutant phenotype or when manually isolating a specific genotype or developmental stage from a heterogeneous population.

In **Subheading 3.7.**, we first describe the measurement of complex I-dependent respiration in isolated mitochondria. The enzyme transfers electrons from NADH to ubiquinone and pumps protons across the mitochondrial inner membrane into the intermembrane space. In the method described, isolated mitochondria are broken open because, in intact mitochondria, the active site of complex I faces the matrix, and NADH cannot be transported across the inner membrane. Rotenone is a highly specific inhibitor of complex I, and that portion of NADH oxidation that is inhibitable by rotenone is considered to be mediated by complex I.

Histochemical staining can be a useful method for examining *C. elegans* mitochondrial mutants and for assessing whether mitochondrial deficiencies are tissue specific. In addition, staining may be used to screen for new mutants among populations of mutagenized animals. The assays require the permeabilization of the *C. elegans* cuticle; the permeabilization does not kill the embryos within the adult gravid hermaphrodite, allowing for the recovery of progeny from interesting animals ***(12)***. Cuticle proteins are solubilized with reducing agents such as β-mercaptoethanol with little or no effect on enzyme activity and on the viability of the embryos. In **Subheadings 3.11.** and **3.12.**, we describe histochemical staining protocols for SDH and for COX activity.

SDH (complex II) is a membrane-bound enzyme that catalyzes the oxidation of succinate to fumarate in the TCA cycle and reduces ubiquinone to ubiquinol. In most organisms, all SDH subunits are encoded by nuclear genes. Thus, SDH activity is not decreased by mutation of the mtDNA or by conditions that affect mtDNA expression. Histochemical detection of SDH activity is based on the reduction of a tetrazolium salt (NBT) and the formation of a blue precipitate. PMS serves as an intermediate electron carrier. Tissues with high SDH activity, such as the pharynx, intestine, and gonad will normally appear dark blue. Tissues with lower levels of activity such as the body wall muscle will appear as a lighter shade of blue. To test for staining specificity, samples can be incubated with the SDH-specific inhibitor sodium malonate (10 m*M*). Although a complete absence of SDH activity should result in colorless tissue, we often detect residual light blue staining. Our protocol has been optimized to minimize this background staining.

COX (complex IV) catalyzes the oxidation of cytochrome-*c* and the reduction of molecular oxygen. In *C. elegans*, complex IV consists of 13 subunits; the 3 largest subunits (COI, COII, COIII) form the catalytic core and are encoded by the mtDNA. Reduced COX activity may therefore indicate the presence of mutations or conditions that affect mtDNA expression. COX requires myriad nuclear-encoded factors for proper assembly and function. Misassembly of the COX complex usually results in its degradation and complete absence of the enzyme. Therefore, there is a strong correlation between the steady-state levels of complex IV polypeptides and COX activity. Histochemical staining of COX activity is a traditional and reliable assay used extensively in the diagnosis of human pathology. It is based on the oxidation of 3,3′-diaminobenzidine tetrahydrochloride and the formation of a brown precipitate; the intensity and distribution of the precipitate correspond directly with COX activity. The specificity of the COX activity staining is determined by the addition of the COX-specific inhibitor KCN. KCN inhibition is complete, resulting in a colorless specimen.

3.1. Growth of Bacteria for Worm Liquid Cultures

1. We grow HB101 (*supE44 hsdS20(r_B-m_B-) recA13 ara-14 proA2 lacYI galK2 rpsL20 xyl-5 mtl-1*) rather than OP50, a uracil auxotroph commonly used as worm food on plates because of higher yields. The use of a fermenter allows for the easy production of the large quantities of *E. coli* needed.
2. Prepare a 50-mL culture of HB101, grown overnight in superbroth to stationary phase as inoculum.
3. Assemble the 5-L Biostat fermenter vessel, fill with superbroth, and autoclave. Allow the vessel to cool. Add the superbroth salts, streptomycin to 20 µg/mL, and inoculate. Connect the vessel to the processor, set the temperature at 37°C, and begin stirring and aeration. Grow overnight (*see* **Note 4**).
4. Once the growth is finished, aseptically siphon the culture to sterile 1-L centrifuge bottles and centrifuge at 4000*g* for 15 min. Wash the pellet once with S basal complete. Resuspend the pellets in 2 volumes of S basal complete and transfer to 50-mL polypropylene tubes for storage at –20°C.

3.2. Growth of C. elegans *in Liquid Culture*

1. *C. elegans* is grown in baffled flasks at 20°C for 5–12 d depending on the strain used and the size of the inoculum. Healthy liquid cultures require adequate aeration and are best started with a large inoculum. This protocol can be scaled up for even larger quantities of worms.
2. Wash off the worms from one or two uncontaminated 6-cm worm plates just cleared of bacteria with M9 buffer into a 1-L baffled flask containing no more than about 150 mL S basal complete. Add about 25 mL of HB101 suspension, streptomycin to 20 µg/mL, and nystatin to 40 µg/mL.
3. Shake cultures at 210 rpm at 20°C.
4. Check the culture daily by aseptically removing an aliquot for microscopic examination. If the culture is not turbid with bacteria (usually about 3 d), then add another 25 mL of HB101. The culture will require more frequent feeding as the number of worms increases (*see* **Note 5**).
5. The culture is harvested when the worms have reached a high density. All developmental stages will be present (*see* **Note 6**).

3.3. Harvesting C. elegans *Cultures*

1. The culture is harvested by centrifugation in 50-mL polypropylene tubes in a swinging bucket rotor at 1100*g* for 5 min (*see* **Note 7**).
2. The supernatant is either carefully poured off or removed by aspiration; the worm pellet is soft. The worm pellets are pooled and washed several times in M9 buffer until the supernatant is clear.

3. The final worm pellets are resuspended in M9 and allowed to shake on an orbital shaker for 30 min. The worms are centrifuged, and the supernatant is removed. The yield ranges from 10 to 17 mL of soft packed worms per 150 mL of culture.
4. If the worms are to be used to isolate mitochondria, then it is best to continue without freezing them. Otherwise, the worm pellets can be frozen at –20°C until needed.

3.4. Cleaning C. elegans *by Sucrose Flotation*

1. Wash worm pellets once in ice-cold 0.1 *M* NaCl and resuspend in 100 mL of 0.1 *M* NaCl. Aliquot 25 mL into four 50-mL polypropylene tubes and place on ice for several minutes to chill.
2. Add an equal volume of ice-cold 60% sucrose and invert several times. Centrifuge the worms for 5 min at 1100*g*. It is important to work quickly because the high osmolarity of the sucrose will kill the worms if exposed for too long.
3. The worms will float to the top, and the debris will pellet to the bottom of the tube. Quickly remove the worms using a glass Pasteur pipet and dilute at least fourfold in 0.1 *M* NaCl. Wash worms twice in 0.1 *M* NaCl.
4. Resuspend the final worm pellet in 2 volumes of worm lysis buffer with added protease inhibitor cocktail. At this stage, the worms can be frozen in liquid N_2 or, preferably, used directly for mitochondrial isolation (*see* **Note 3**).

3.5. Isolation of Crude Mitochondria

1. A Bead-Beater (Biospec Products) is assembled, and the chamber is filled one-half to two-thirds full with acid-washed glass beads. The worms (in worm lysis buffer with protease inhibitor cocktail) are added to the chamber, and the chamber is filled to the top with cold worm lysis buffer. The rotor assembly is lowered into the chamber, displacing a small amount of liquid. It is important to exclude all air during the operation of the Bead-Beater (*see* **Note 8**).
2. The assembled chamber is surrounded with ice. Grinding proceeds with three pulses of 1 min each interspersed with 1-min intervals to allow for heat dissipation. A small aliquot of the supernatant is examined to assess the extent of breakage.
3. The supernatant is recovered and homogenized by hand in a glass-Teflon homogenizer for 30 s. Recovery is increased by rinsing the glass beads in worm lysis buffer and pooling the supernatants.
4. Wash the beads several times with water (until the water is clear) between samples. After all the samples are processed, soak the beads in lab detergent overnight and rinse thoroughly with water. Dry the beads overnight in an oven.
5. Centrifuge the lysate at 2500*g* for 10 min at 4°C to pellet debris.
6. Centrifuge the supernatant at 15,000*g* for 10 min at 4°C. Resuspend the pellet in cold worm lysis buffer and centrifuge again at 15,000*g* for 30 min at 4°C.
7. Resuspend the pellet in a small volume of worm lysis buffer and briefly homogenize in a glass-Teflon homogenizer.
8. Aliquot the crude mitochondria into microcentrifuge tubes, freeze in liquid N_2, and store at –80°C.

3.6. Isolation of Purified Mitochondria

1. Pour a 10-mL, 1 *M* to 2 *M* sucrose gradient in a 15-mL tube for a swinging bucket rotor such as the Beckman SW27. Up to 4 mL of crude mitochondria in worm lysis buffer can be layered onto the gradient.
2. Centrifuge at 80,000*g* for 90 min at 4°C. Intact mitochondria will be found in the brown band in the middle of the gradient.
3. Remove the mitochondria with a glass Pasteur pipet and dilute with 3 volumes of cold worm lysis buffer. Centrifuge at 30,000*g* for 30 min at 4°C to pellet the mitochondria.
4. Gently resuspend the pellet in a small volume of worm lysis buffer and homogenize with a glass-Teflon homogenizer. Aliquot the purified mitochondria into microcentrifuge tubes, freeze in liquid N_2, and store at –80°C.

3.7. Polarographic Analysis: NADH-Dependent, Rotenone-Sensitive Respiration of Isolated Mitochondria

1. Set the Mitocell chamber volume to 50 µL. Insert the plunger, which contains a capillary tube through which substrates and inhibitors are introduced, to seal the chamber. Substrates and inhibitors are introduced into the Mitocell chamber using modified 1-µL and 5-µL Mikroliterspritze syringes (Innovative Labor Systeme, Stützerbach, Germany) included with the Mitocell system. Experiments can be performed at room temperature or be temperature controlled by circulating water through the MT200 chamber jacket. It is important that the temperatures of all samples and controls be identical as the solubility of oxygen in water varies considerably with temperature. The electrode is calibrated with oxygen-free and oxygen-saturated water. Oxygen-free water is prepared by adding a pinch of sodium sulfite to distilled water and mixing (*see* **Note 9**). Oxygen-saturated water (267 µmol/L) is prepared by bubbling air through distilled water for approx 15 min. Crude or sucrose-purified mitochondria can be used in these assays; they can be prepared fresh or thawed no more than once. Respiration rates decrease in mitochondria that have undergone multiple freeze-thaw cycles.
2. Crude isolated mitochondria initially suspended in worm lysis buffer or in MSE buffer are diluted 10-fold in 0.1 *M* HEPES, pH 7.4, and sonicated for 2–3 min in a Branson 1200 bath sonicator (Branson Ultrasonics Corp., Danbury, CT) in ice water (*see* **Note 10**).
3. Centrifuge the sonicated mitochondria at 18,500*g* for 30 min at 4°C. Remove the supernatant and resuspend in 0.1 *M* HEPES, pH 7.4.
4. 20 µg mitochondrial protein is diluted to a volume of 60 µL with 0.1 *M* HEPES, pH 7.4, and introduced into the Mitocell chamber. The plunger is carefully replaced to avoid the introduction of air bubbles into the chamber. The magnetic stirrer is turned on to keep contents of the chamber homogeneous. Data are recorded with a computer or directly on a chart recorder.
5. The basal level of oxygen consumption is recorded for 2–4 min and should be linear.

6. NADH is injected into the chamber to a final concentration of 600 μ*M*. Oxygen consumption is recorded for 2–4 min or until linear.
7. Rotenone is injected to a final concentration of 100 n*M*, and oxygen consumption is monitored for up to 10 min or until linear (*see* **Note 11**). An example of results is shown in **Fig. 1**.
8. In a similar manner, malonate-sensitive, succinate-dependent respiration can be determined by adding sodium succinate and then sodium malonate to final concentrations of 5 m*M*. Succinate is the substrate for complex II (SDH or succinate-ubiquinone oxidoreductase), and malonate is a potent competitive inhibitor.

3.8. Polarographic Analysis: Whole Animal Respiration

1. L4 to early adult animals are transferred into 1.5-mL microcentrifuge tubes using M9 buffer. The animals are centrifuged at 350*g* for 3 min, and the supernatant is carefully removed. The wash is repeated twice with 1 mL M9 buffer.
2. After the third wash, 1 mL M9 buffer is added, and the animals are incubated at room temperature with constant agitation for 30 min. This allows the contents of the digestive system to empty and reduces the amount of contaminating bacteria. The worms are centrifuged as above, and the supernatant is removed.
3. Animals are resuspended in fresh M9 buffer at approx 10,000/mL. A 60-μL aliquot of resuspended washed animals is introduced into the Mitocell chamber, the plunger is replaced, and stirring is initiated.
4. Oxygen consumption is recorded for a minimum of 10 min or until linear.
5. The sensitivity of the respiration to cyanide, a specific inhibitor of complex IV, is determined by adding 1 μL 1 *M* KCN into the chamber. All respiration should be abolished.
6. The animals are removed from the chamber into a fresh microcentrifuge tube for the determination of protein content.

3.9. Determination of Protein Content After Whole Animal Respiration Measurements

1. The nematode sample from the Mitocell chamber is diluted to 1 mL in distilled water in a microcentrifuge tube. Then 0.15 mL of 1.85 *M* NaOH, 7.4% (v/v), β-mercaptoethanol is added. The tube is mixed by inversion and incubated on ice for 10 min to disrupt the cuticle and cell membranes (*see* **Note 12**).
2. Add 0.15 mL of 50% (w/v) TCA. The tube is mixed by inversion and incubated on ice for 10 min to precipitate the protein.
3. The samples are centrifuged at 14,000*g* for 12 min, and the supernatants carefully removed.
4. The pellets are washed with 1 mL ice-cold acetone to remove excess lipid, recentrifuged, and allowed to air dry for 5 min.
5. The pellets are solubilized in 50 μL of 5% (w/v) sodium dodecyl sulfate, 62.5 m*M* Tris-HCl, pH 6.8, by thoroughly vortexing (10–20 min on an automated mixer).

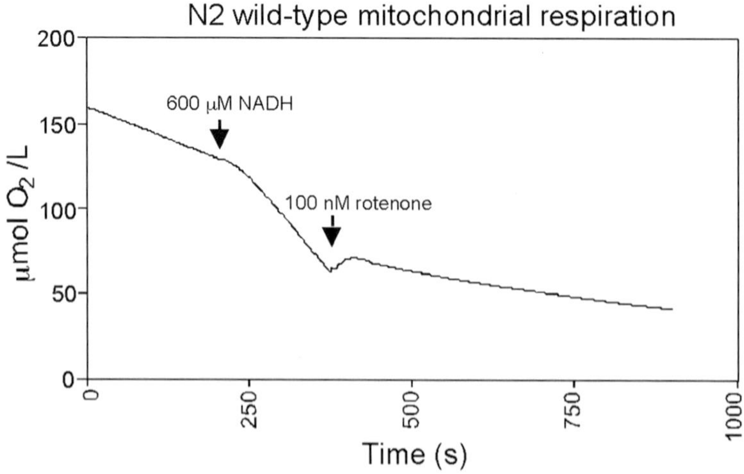

Fig. 1. Complex I-dependent respiration of crude isolated mitochondria. Crude isolated mitochondria from the wild-type *C. elegans* strain N2 were lysed by dilution and sonication. Introduction of 20 μg of mitochondrial protein was made into the Mitocell chamber, and respiration was measured. After measuring the basal level of oxygen consumption, NADH and rotenone were added sequentially into the chamber to final concentrations of 600 μ*M* and 100 n*M*, respectively.

6. The protein concentrations are determined using a detergent-compatible protein assay such as the Bio-Rad D_C Protein Assay (Bio-Rad Laboratories, Hercules, CA) following the manufacturer's instructions.

3.10. Histochemical Analysis of Mitochondrial Function: Fixation and Permeabilization of Worms

1. Approximately 1000 synchronized animals are washed free of bacteria as described above and resuspended in 1.5 mL of M9 buffer in a microcentrifuge tube.
2. The worms are pelleted by centrifugation at 350*g* for 3 min, and the rotor is allowed to coast to a stop without a brake (younger animals may require spins of up to 5 min to pellet).
3. The supernatant is carefully removed, and the pellet is washed once with 1 mL H_2O.
4. The worms are pelleted, the supernatant is removed, and the pellet is placed on ice.
5. The following are added: 500 μL ice-cold 2X MRWB, 400 μL H_2O, and 100 μL 20% (w/v) paraformaldehyde. The tubes are mixed by inversion and incubated for 35 min at 4°C with constant rotation (1–2 rpm).
6. The fixed worms are pelleted and washed twice with 1 mL TTB.
7. The final pellet is resuspended in 1 mL TTB with 1% (v/v) β-mercaptoethanol (10 μL β-mercaptoethanol in 990 μL TTB) and incubated for 15 min at room temperature with constant rotation.

8. The worms are pelleted and washed once in 1.5 mL 1X borate buffer.
9. The worms are pelleted, resuspended in 1 mL 0.9X borate buffer containing 10 m*M* DTT (10 µL DTT, 22.5 µL 40X borate buffer in 967.5 µL H_2O), and incubated at room temperature for 15 min with constant rotation.
10. The worms are pelleted and washed twice with 1.5 mL PBS. After the final centrifugation, all but approx 200 µL of supernatant are removed before proceeding to histochemical staining.

3.11. Histochemical Analysis of Mitochondrial Function: SDH Activity Stain

1. The assay mixture is light sensitive and should be stored in the dark when not in use. The following recipe prepares enough assay solution for 10 assays: 5 m*M* EDTA, 1 m*M* KCN, 0.2 m*M* PMS, 50 m*M* sodium succinate, 0.25 m*M* NBT in 10 mL PBS.
2. Add 1 mL of the assay solution per tube of fixed and permeabilized nematodes. For a negative control, 20 µL 1 *M* sodium malonate are added (final concentration of 20 m*M* sodium malonate). The tubes are mixed gently, but thoroughly, by inversion.
3. The tubes are incubated for 50 min at 37°C in the dark with constant rotation to ensure uniform exposure to the reagents.
4. Following the staining reaction, the worms are centrifuged at 350*g* for 3 min, and the rotor is allowed to coast to a stop (for younger animals, the centrifugation may take longer). The supernatant is carefully removed.
5. The worms are washed three times for 5 min each with 1.5 mL H_2O to remove excess stain.
6. After the final wash, the supernatant is removed, leaving a volume equal to that of the nematode pellet.
7. Stained nematodes are mounted onto 2% (w/v) agarose pads on glass slides and examined by light microscopy. SDH activity appears blue, with dark-blue stain identifying tissue with high levels of SDH activity. The absence of SDH activity, as in the negative control, will appear as a very light blue. If background staining is problematic, then the incubation time of staining (**step 3**) can be decreased. An example of results is shown in **Fig. 2**.

3.12. Histochemical Analysis of Mitochondrial Function: COX Activity Stain

1. The assay mixture is light sensitive and should be stored in the dark when not in use. The following recipe prepares enough assay solution for 10 assays: 10 mg 0.1% (w/v) 3,3′-diaminobenzidine, 10 mg 0.1% (w/v) cytochrome-*c*, 2 mg 0.02% (w/v) catalase in 10 mL PBS. Not all of the reagents will dissolve completely. Decant as much of the supernatant as possible into a fresh tube, leaving undissolved material behind.
2. Add 1 mL of the assay mixture per tube of fixed and permeabilized nematodes. As a negative control, 10 µL of 1 *M* KCN are added (final concentration of 10 m*M* KCN). The tubes are mixed gently, but thoroughly, by inversion.

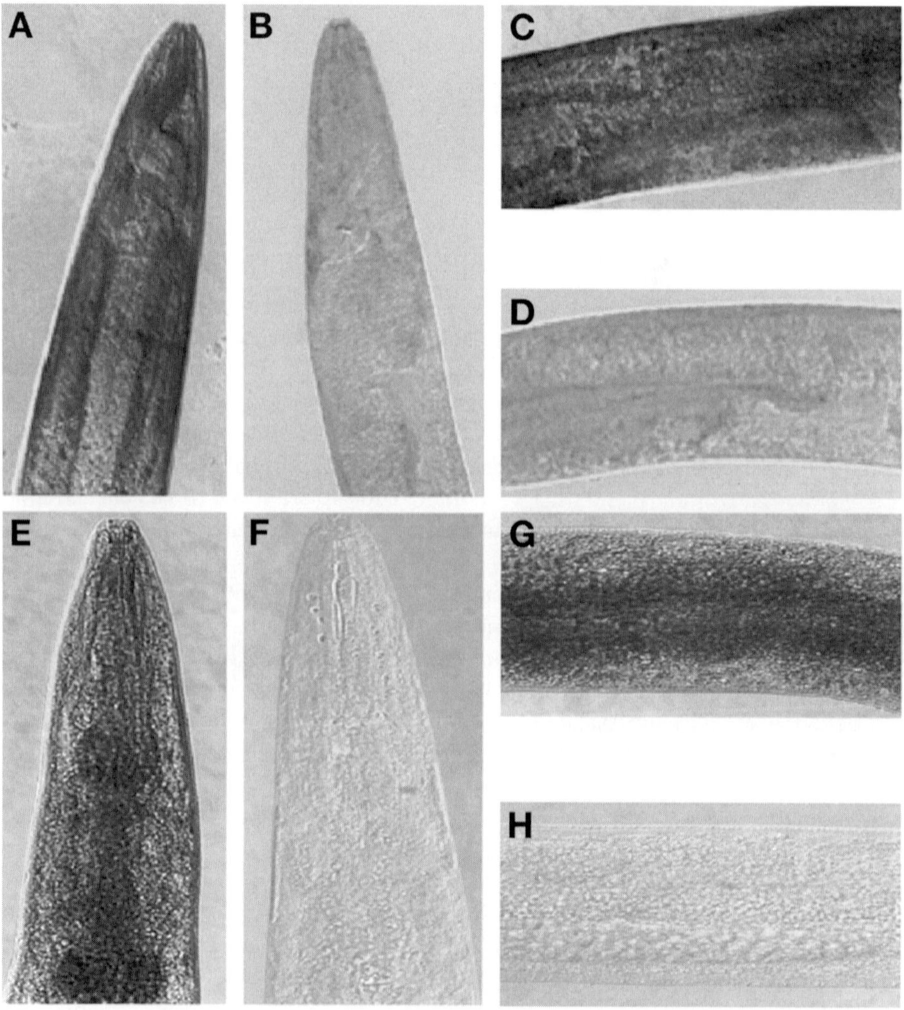

Fig. 2. Histochemical staining for mitochondrial enzyme function. Wild-type *C. elegans* hermaphrodites (strain N2) were fixed, permeabilized, and stained for SDH activity (**A**)–(**D**) or cytochrome-*c* oxidase activity (**E**)–(**H**). Representative examples of the head (**A**), (**B**), (**E**), (**F**) or body regions (**C**), (**D**), (**G**), (**H**) were photographed. Control reactions contained the SDH-specific inhibitor, sodium malonate (**B**), (**D**) or the COX-specific inhibitor KCN (**F**), (**H**).

3. The tubes are incubated for 75 min at 37°C in the dark with constant rotation.
4. Following the staining reaction, the worms are centrifuged at 350*g* for 3 min, and the rotor is allowed to coast to a stop (younger animals may require longer centrifugation). The supernatant is carefully removed.

5. The worms are washed three times for 5 min each with 1.5 mL H_2O to remove excess stain. After the final wash, the supernatant is removed, leaving a volume equal to that of the nematode pellet.
6. Stained nematodes are mounted onto 2% (w/v) agarose pads on glass slides and examined by light microscopy. COX activity produces a brown stain. Lack of color production indicates the absence of activity. This assay is less susceptible to problems with high background and is sensitive enough to discriminate between relatively subtle differences in enzyme activity. An example of results is shown in **Fig. 2**.

4. Notes

1. All solutions should be prepared in water that has been deionized and filtered or distilled to remove organic contaminants.
2. *C. elegans* is unable to synthesize cholesterol and requires the addition of exogenous sterol for growth ***(13)***.
3. As an alternative to worm lysis buffer, worm pellets can be resuspended in MSE (*see* **Subheading 2.7.** for recipe). Protease inhibitor cocktail can be replaced with 1 m*M* phenylmethyl sulfonyl fluoride (0.25 *M* stock in ethanol). The phenylmethyl sulfonyl fluoride has a short half-life in aqueous solutions and should be added immediately prior to use.
4. HB101 is resistant to streptomycin. If streptomycin-sensitive strains of *E. coli* are used, then the antibiotic should be omitted from the medium.
5. Worms grown in liquid culture are longer and thinner than worms grown on agar plates. Large clumps of eggs may form in liquid culture.
6. It is important not to wait too long to harvest the liquid culture because very high worm densities will result in them entering the dauer stage as dauer pheromone accumulates.
7. If larger cultures are grown, then the worms can be allowed to settle in a 1-L glass graduated cylinder for 1 h. Most of the medium can be removed, and the remainder containing the worms is transferred to 50-mL polypropylene tubes for centrifugation.
8. Chambers of different sizes (15, 50, and 350 mL) are available for the Bead-Beater depending on the volume of sample to be lysed. Other forms of mechanical shearing such as a motorized homogenizer can be used to break the worm cuticle.
9. In older literature, the use of sodium dithionite is recommended for producing oxygen-free water. This is not recommended because the breakdown products can impair electrode function. All traces of sulfite must be rinsed from the electrode after calibration of the zero point.
10. Sonication disrupts both the outer and inner mitochondrial membranes and releases soluble endogenous substrates. NADH is impermeable to the inner membrane, and the complex I active site faces the mitochondrial matrix. In some organisms, outwardly facing NADH dehydrogenases can catalyze NADH-dependent respiration in intact mitochondria; these dehydrogenases are not rotenone sensitive.
11. Inhibition by rotenone is slow, and complete inhibition is not immediate. NADH-dependent respiration is determined by subtracting the basal rate of oxygen consump-

tion from the rate in the presence of NADH. Rotenone-sensitive NADH-dependent respiration is determined by subtracting the rate of respiration in the presence of rotenone from the NADH-dependent rate. It is important to wash the Mitocell chamber thoroughly several times with ethanol and water to remove all traces of inhibitors such as rotenone or malonate before beginning measurements on a new sample.

12. By determining the total protein content of each sample, the respiration rates can be normalized. Although rates can also be expressed per number of animals, the counting of nematodes is laborious and prone to error, especially when numerous samples are to be analyzed.

References

1. Harris, T. W., Chen, N., Cunningham, F., et al. (2004) WormBase: a multi-species resource for nematode biology and genomics. *Nucleic Acids Res.* **32,** D411–D417.
2. Riddle, D. L., Blumenthal, T., Meyer, B. J., and Priess, J. R. (1997) Introduction to *C. elegans*, in *C. elegans II* (Priess, J., ed.), Cold Spring Harbor Laboratory Press, New York, pp. 1–22.
3. Riddle, D. L., and Albert, P. S. (1997) Genetic and environmental regulation of dauer larva development, in *C. elegans II* (Priess, J., ed.), Cold Spring Harbor Laboratory Press, New York, pp. 739–768.
4. Sulston, J. (1988) Cell lineage, in *The Nematode* Caenorhabditis elegans (Wood, W. B., ed.), Cold Spring Harbor Laboratory Press, New York, pp. 123–155.
5. Culetto, E., and Sattelle, D. B. (2000) A role for *Caenorhabditis elegans* in understanding the function and interactions of human disease genes. *Hum. Mol. Genet.* **9,** 869–877.
6. Okimoto, R., Macfarlane, J. L., Clary, D. O., and Wolstenholme, D. R. (1992) The mitochondrial genomes of two nematodes, *Caenorhabditis elegans* and *Ascaris suum. Genetics* **130,** 471–498.
7. *C. elegans* Sequencing Consortium. (1998) Genome sequence of the nematode *C. elegans*: A platform for investigating biology. *Science* **282,** 2012–2018.
8. Murfitt, R. R., Vogel, K., and Sanadi, D. R. (1976) Characterization of the mitochondria of the free-living nematode, *Caenorhabditis elegans. Comp. Biochem. Physiol.* **53B,** 423–430.
9. Tsang, W. Y., and Lemire, B. D. (2003) The role of mitochondria in the life of the nematode, *Caenorhabditis elegans. Biochim. Biophys. Acta* **1638,** 91–105.
10. O'Riordan, V. B., and Burnell, A. M. (1990) Intermediary metabolism in the dauer larva of the nematode *Caenorhabditis elegans*–II. The glyoxylate cycle and fatty-acid oxidation. *Comp. Biochem. Physiol.* **95B,** 125–130.
11. O'Riordan, V. B., and Burnell, A. M. (1989) Intermediary metabolism in the dauer larva of the nematode *Caenorhabditis elegans*–I. Glycolysis, gluconeogenesis, oxidative phosphorylation and the tricarboxylic acid cycle. *Comp. Biochem. Physiol.* **92B,** 233–238.
12. Xie, G., Jia, Y., and Aamodt, E. (1995) A *C. elegans* mutant screen based on antibody or histochemical staining. *Genet. Anal. Biomol. Eng.* **12,** 95–100.
13. Kurzchalia, T. V., and Ward, S. (2003) Why do worms need cholesterol? *Nat. Cell Biol.* **5,** 684–688.

5

Isolation of Mitochondria From Procyclic *Trypanosoma brucei*

André Schneider, Fabien Charrière, Mascha Pusnik, and Elke K. Horn

Summary

The mitochondrion of the parasitic protozoon *Trypanosoma brucei* shows a number of unique features, many of which represent highly interesting research topics. Studies of these subjects require the purification of mitochondrial fractions. Here, we describe and discuss the two most commonly used methods to isolate mitochondria from insect stage *T. brucei*. In the first protocol, the cells are lysed under hypotonic conditions, and mitoplast vesicles are isolated on Percoll gradients; in the second method, lysis occurs isotonically by N_2 cavitation, and the mitochondrial vesicles are isolated by Nycodenz gradient centrifugation.

Key Words: N_2 cavitation; Nycodenz; Percoll; subcellular fractionation; trypanosome; *Trypanosoma brucei*.

1. Introduction

The parasitic protozoon *Trypanosoma brucei* is not only an important pathogen but also has proven to be an excellent model for basic science in general. Two main reasons for this are that (1) it is amenable to a wide range of molecular genetic, cell biological, and biochemical techniques and (2) it has unique biology ***(1,2)***. The first point is illustrated by the fact that transfection of *T. brucei* by homologous recombination was achieved in 1990 ***(3)***. Furthermore, in 1995 transfection of *T. brucei* was used to establish a highly inducible gene expression system ***(4)*** and thus greatly expanded the repertoire of molecular genetic techniques. Most important, shortly after the discovery of RNA interference (RNAi) in *Caenorhabditis elegans* in 1998 ***(5)***, it was shown that the process is also operational in *T. brucei* ***(6)***. In 2000, finally, inducible gene expression was combined with RNAi ***(7,8)***. The resulting system has

From: *Methods in Molecular Biology, vol. 372: Mitochondria: Practical Protocols*
Edited by: D. Leister and J. M. Herrmann © Humana Press Inc., Totowa, NJ

since revolutionized research on all aspects of *T. brucei* biology as it allows stable and inducible downregulation of any desired messenger ribonucleic acid (mRNA).

The unique biology of trypanosomes has attracted and still is attracting much interest. That this interest is justified is illustrated by the fact that processes such as transsplicing, glycosyl phosphatidylinositol anchoring of membrane proteins, and RNA editing were originally discovered in *T. brucei* and only later shown to occur in essentially all eukaryotic cells.

Phylogenetic analyses based on ribosomal RNA sequences suggest that *T. brucei* belongs to the deepest branching eukaryotes having bona fide mitochondria involved in oxidative phosphorylation ***(9)***. Thus, it might not be a surprise that the *T. brucei* mitochondrion shows many unique features that represent highly interesting research topics ***(10)***. Unlike most other eukaryotes, *T. brucei* has a single mitochondrion. Its genome is not distributed all over the matrix but localized to a specific region inside the organelle opposite the basal body of the flagellum ***(11)***.

The genome itself is also very unusual: it consists of a large structure of two highly concatenated genetic elements: the maxi- and the minicircles. Thus, the replication of the mitochondrial genome and how it is segregated during cell division represents a fascinating problem ***(12)***. The maxicircle encodes typical mitochondrial genes. However, many of these represent cryptogenes, meaning that their primary transcripts need to be processed by RNA editing to become functional mRNAs.

The intriguing process of RNA editing has been the focus of much research and is still actively investigated ***(13)***. It has been known for many years that the trypanosomal mitochondrial deoxyribonucleic acid (DNA) does not encode any transfer RNA (tRNA) genes, indicating that, unlike most other eukaryotes, all mitochondrial tRNAs have to be imported from the cytosol ***(14)***. Synthesis of mitochondrial-encoded proteins in *T. brucei* shows interesting deviations to other translation systems. Not only does it require edited mRNAs (at least in some cases) and imported tRNAs, but also it uses mitochondrial ribosomes that have among the shortest known ribosomal RNAs ***(15)***. Finally, it is known that the trypanosomal mitochondrion has some unusual metabolic pathways, such as a plantlike alternative oxidase and the adenosine triphosphate-producing acetyl:succinate coenzyme A-transferase cycle, which normally is only found in hydrogenosomes ***(16)***.

These examples represent some of the topics of mitochondrial biology that are actively investigated in *T. brucei* and serve to illustrate the rich and unusual biology of the *T. brucei* mitochondrion. It is clear that research on any of these problems requires at some point the purification of mitochondria. It is the aim of this review to summarize and discuss the two main methods used for this purpose. The main difference between the two protocols is that in one the cells are lysed under hypotonic conditions, whereas in the other lysis occurs in an isotonic buffer. The hypotonic protocol, which is based on **refs. *17*** and ***18***, is the method of choice if purity of the preparation is the main concern, whereas if functionality is the main issue, then it is recommended to use to the isotonic protocol ***(19)***.

2. Materials

2.1. Isolation of Mitochondria: Hypotonic Procedure

2.1.1. Growth and Harvesting of Cells

1. Procyclic *T. brucei* cells (*see* **Note 1**).
2. SDM-79 medium supplemented with 5% heat-inactivated fetal bovine serum ***(20)***.
3. Centrifuge; fixed-angle rotor, 6 × 500 mL capacity; six centrifuge bottles.
4. Disposable counting chamber: KOVA Glasstic Slide 10 with grid chamber (cat. no. 22-270141; Hycor Biomedical).
5. Wash buffer: 20 m*M* sodium phosphate buffer, pH 7.9, 20 m*M* glucose, 0.15 *M* NaCl. Prepare as 4X stock (*see* **Note 2**).

2.1.2. Hypotonic Cell Breakage and Deoxyribonuclease Digestion

1. Hypotonic lysis buffer: 1 m*M* Tris-HCl, pH 8.0, 1 m*M* ethylenediaminetetraacetic acid (EDTA), pH 8.0. Prepare as 10X stock.
2. 40-mL Dounce tissue homogenizer with a large clearance pestle.
3. 5-L Pressure vessel (cat. no. XX6700P05; Millipore).
4. Luer-Lok syringe and hypodermic needle no. 26 (brown) and no. 25 (orange).
5. Sucrose stock: 1.75 *M*.
6. Centrifuge; fixed-angle rotor, 8 × 50 mL capacity; eight 50-mL centrifuge tubes.
7. STM buffer: 20 m*M* Tris-HCl, pH 8.0, 0.25 *M* sucrose, 5 m*M* $MgCl_2$. Prepare as 4X stock (sterilize by filtration).
8. Deoxyribonuclease (DNase) I, from bovine pancreas, grade II (cat. no. 104159; Roche).
9. STE buffer: 20 m*M* Tris-HCl, pH 8.0, 0.25 *M* sucrose, 2 m*M* EDTA, pH 8.0. Prepare as 4X stock.
10. EDTA stock: 0.5 *M*, pH 8.0.

2.1.3. Percoll Step Gradients

1. Ultracentrifuge; large swinging bucket ultracentrifuge rotor, 6 × 38.5 mL capacity; six Ultra-Clear centrifuge tubes 38.5 mL (25 × 89 mm) (cat. no. 344058; Beckmann). Tubes can be washed and reused.
2. Percoll 100% (cat. no. P-1644; Sigma), keep at 4°C.
3. STE buffer containing 20, 25, 30, 35, and 75% (v/v) of Percoll each; keep at 4°C.
4. 40-mL Dounce tissue homogenizer with the small clearance pestle.
5. 10-mL Syringe with attached 100-μL glass capillary (*see* **Note 3**).

2.1.4. Removal of Percoll and Storage

1. Centrifuge; fixed-angle rotor, 8 × 50 mL capacity; eight 50-mL centrifuge tubes.
2. STE buffer.
3. BCA protein assay kit (cat. no. 23227; Pierce).
4. Fatty acid-free bovine serum albumine (BSA) (cat. no. A-6003; Sigma), prepare 100-mg/mL stock.

2.2. Isolation of Mitochondria: Isotonic Procedure

2.2.1. Growth and Harvesting of Cells

For growth and harvesting of cells, *see* **Subheading 2.1.1**.

2.2.2. Isotonic Cell Breakage, DNase Digestion, and Low-Speed Spins

1. SoTE buffer: 20 m*M* Tris-HCl, pH 7.5, 0.6 *M* sorbitol, 2 m*M* EDTA, pH 7.5. Prepare as 2X stock.
2. Cell disruption bomb for N_2 cavitation, capacity 920 mL (cat. no. 4635; Parr Instrument Co.).
3. Centrifuge; fixed-angle rotor, 8 × 50 mL capacity; 50-mL centrifuge tubes.
4. SoTM buffer: 20 m*M* Tris-HCl, pH 8.0, 0.6 *M* sorbitol, 5 m*M* $MgCl_2$. Prepare as 4X stock (sterilize by filtration).
5. Luer-Lok syringe and hypodermic needle no. 25 (orange).
6. DNase I, from bovine pancreas, grade II (cat. no. 104159; Roche).
7. EDTA stock: 0.5 *M*, pH 7.5.

2.2.3. Nycodenz Step Gradients

1. Ultracentrifuge; large swinging bucket ultracentrifuge rotor, 6 × 38.5 mL capacity; 38.5-mL Ultra-Clear centrifuge tubes (25 × 89 mm) (cat. no. 344058; Beckmann). Tubes can be washed and reused.
2. Nycodenz powder (cat. no. 1002424; Nycomed): prepare 80% (w/v) stock (*see* **Note 4**).
3. SoTE buffer containing 18, 21, 25, 28, and 50% (w/v) Nycodenz each.
4. 40-mL Dounce tissue homogenizer with the small clearance pestle.
5. 10-mL Syringe with attached 100-µL glass capillary (*see* **Note 3**).

2.2.4. Removal of Nycodenz and Storage

1. Centrifuge; fixed-angle rotor, 8 × 50 mL capacity; eight 50-mL centrifuge tubes.
2. SoTE buffer.
3. BCA protein assay kit (cat. no. 23227; Pierce).
4. Fatty acid-free BSA (cat. no. A-6003; Sigma), prepare 100-mg/mL stock.

3. Methods

Trypanosoma brucei contains a single large mitochondrion that cannot be isolated as an intact structure. Thus, independent of the chosen isolation method the mitochondrion will fragment into vesicles. However, although the native mitochondrial morphology is disrupted during purification, the isolated vesicles retain many mitochondrial functions.

3.1. Isolation of Mitochondria: Hypotonic Procedure

The hypotonic procedure represents a modified version of the one described in **ref.** ***18***. It relies on initial cell lysis under hypotonic conditions by passage through a narrow hypodermic needle. Subsequently, the extract is treated with DNAse and separated on Percoll gradients. Hypotonic cell lysis is very efficient; thus, the hypotonic procedure is the method of choice to obtain biochemically pure mitochondria (e.g., for isolating mtRNAs). However, during lysis not only the cell membrane but also the mitochondrial outer membrane becomes ruptured; thus, the purified mitochondrial vesicles represent mitoplasts (*see* **Note 5**) *(21)*. These mitoplasts show little if any activity when assayed for mitochondrial protein import. Thus, for studying mitochondrial functions, it is better to use mitochondrial vesicles purified by the isotonic method (*see* **Subheading 3.2.**).

3.1.1. Growth and Harvesting of Cells

1. Procyclic *T. brucei* are grown in suspension at 27°C in a total volume of 1–5 L SDM-79 medium containing 5% FCS. Cells are harvested at a density of 2.5–5.0 $\times 10^7$ cells/mL, corresponding to late log phase (*see* **Note 6**).
2. Noted that all further steps are on ice.
3. Spin cells in 500-mL centrifuge bottles in a fixed-angle rotor at 4°C for 10 min at 11,000 $\times g$ (*see* **Note 7**).
4. During centrifugation, determine cell concentration by microscopic counting using a disposable counting chamber.
5. After centrifugation, immediately remove the medium (*see* **Note 8**). Add fresh cell culture to the same set of centrifuge bottles and repeat **step 3**.
6. Resuspend pellets in a small volume of wash buffer, combine pellets in one bottle, add wash buffer to approx 450 mL, and spin as in **step 3** (*see* **Note 9**).

3.1.2. Hypotonic Cell Breakage

1. Resuspend cell pellet in hypotonic lysis buffer at 1.2×10^9 cells/mL (*see* **Note 10**). Add only a little hypotonic lysis buffer at first and homogenize in a 40-mL glass Dounce tissue homogenizer using the large clearance pestle to prevent clogging of the hypodermic needle. Check under the microscope whether, as expected, under hypotonic conditions the cells have rounded up.
2. Pour suspension into the pressure vessel sitting in an ice bath and containing a magnetic stirrer. Close the vessel, lock the outlet valve, and apply 5 bars of pressure using N_2 (*see* **Note 11**).
3. Attach a hypodermic needle (no. 26, brown) to the outlet of the pressure vessel.

4. Open the outlet valve, collect suspension in a glass beaker on ice, and measure the volume.
5. Add one-sixth volume of 1.75 *M* sucrose stock. Mix well to reestablish isotonic conditions (*see* **Note 12**).
6. Examine lysis microscopically (*see* **Note 13**).

3.1.3. DNase Digestion

1. Spin lysate in 50-mL centrifuge tubes in a fixed-angle rotor at 4°C for 10 min at 17,500*g*.
2. Pour out supernatants (*see* **Note 14**). The cloudy supernatants represent the cytosolic fraction and, depending on the experiment, may be kept.
3. Resuspend pellets by vortexing in one-sixth volume of STM buffer. Pool in a single tube and estimate total volume.
4. Add solid DNase I to 0.1 mg/mL final concentration (*see* **Note 15**).
5. Push the extract through a hypodermic needle (no. 25, orange) using a Luer-Lok syringe.
6. Incubate for 45 min at 4°C. During the incubation, repeat **step 5**. The viscosity should drop during DNase digestion.
7. Add an equal volume of STE buffer.
8. Add 1/125 of the total volume of EDTA stock (0.5 *M*) (*see* **Note 16**).
9. Spin in 50-mL centrifuge tubes at 4°C for 10 min at 17,500*g*. The resulting pellet will be very soft. Thus, the cloudy supernatant should be removed with a pipet.

3.1.4. Percoll Step Gradients

1. Precool large swinging bucket ultracentrifuge rotor.
2. Determine the number (two, four, or six) of gradients you will need. Each gradient should be loaded with lysate corresponding to 2.0–3.5 × 10^{10} cell-equivalents (*see* **Note 17**). To simplify balancing, an even number of tubes should be used.
3. During the DNase digestion, prepare Percoll step gradients: pipet 8 mL cold 35% Percoll containing STE buffer into each of the 38.5-mL ultracentrifuge tubes. Carefully overlay 8 mL each of cold 30/25/20% Percoll containing STE buffer (the gradients can also be prepared the day before and should in this case be kept at 4°C).
4. Resuspend pellets (*see* **Subheading 3.1.3., step 9**) in a total volume of 3–6 mL 75% Percoll containing STE buffer per gradient, pool them, and homogenize in a 40-mL glass Dounce tissue homogenizer using the small clearance pestle (*see* **Note 18**).
5. Place equal volumes of the samples (3–6 mL) at the bottom of each gradient. Practically, this is done using the 10-mL hypodermic syringe with the attached glass capillary (*see* **Note 19**).
6. Balance gradients with the 20% Percoll containing STE buffer on an electronic balance.
7. Spin gradients in the large swinging bucket ultracentrifuge rotor at 4°C for 45 min at 100,000*g*.
8. After centrifugation, the gradients should show three bands. The middle one, at the 25/30% Percoll interphase, is the most diffuse and may account for up to a third of the total gradient volume. Microscopic examination shows this band is most enriched for mitoplast vesicles (*see* **Note 20**).

9. Collect 10–15 mL of this zone (25/30% Percoll interphase) using the same 10-mL hypodermic syringe with the attached glass capillary used for loading.

3.1.5. Removal of Percoll and Storage

1. Distribute the collected mitoplast fractions of all six gradients into eight 50-mL centrifuge tubes and dilute to 50 mL each with STE buffer.
2. Cap tubes with two layers of parafilm and mix vigorously by inversion (*see* **Note 21**).
3. Spin in a fixed-angle rotor at 4°C for 15 min at 33,000*g*. The resulting pellet will be very soft; thus, discard supernatant with a pipet, leaving approx 3 mL STE in the tube.
4. Combine two pellets into a single tube each. Dilute the combined pellets in the resulting four tubes to 50 mL with STE buffer and repeat **steps 2** and **3**. The obtained pellets will be tighter now, and essentially all STE can be removed.
5. Combine all four pellets into one tube, dilute to 50 mL with STE buffer, and repeat **steps 2** and **3**.
6. Resuspend the mitoplast pellet in a small volume of STE buffer and examine it in the microscope. The fraction should look as shown in **Fig. 1A**.
7. Take a small aliquot and determine the protein concentration by the BCA protein assay kit (*see* **Note 22**).
8. Aliquots of mitoplasts can directly be flash frozen in liquid N_2 and stored at –70°C. However, the best way to preserve the membrane integrity of the mitoplast is to add one-ninth volume of 100 mg/mL fatty acid-free BSA before freezing (*see* **Note 23**) ***(22)***.

3.2. Isolation of Mitochondria: Isotonic Procedure

The isotonic procedure is based on the one described in **ref. *19***; the low-speed spins were modified from **ref. *23***. In this method, the cells are lysed in an isotonic buffer by N_2 cavitation. Subsequently, the extract is treated with DNase, intact cells are removed by low-speed spins, and organellar vesicles are separated on Nycodenz gradients. Cell breakage by N_2 cavitation is less efficient than hypotonic lysis. Thus, the obtained mitochondrial fraction is often less pure than the one obtained by the hypotonic lysis procedure. However, isotonically isolated mitochondrial vesicles have an intact outer membrane (*see* **Note 4**) ***(21)***. Furthermore, it was shown that, for many functional studies, such as investigating mitochondrial protein import, the isotonic procedure is the method of choice ***(19)***.

3.2.1. Growth and Harvesting of Cells

For growth and harvesting of cells, *see* **Subheading 3.1.1**.

3.2.2. Isotonic Cell Breakage

1. Resuspend pellet in SoTE buffer at 2×10^9 cells/mL. Take a small sample as a control for the microscopic examination of the extent of the cell lysis (*see* **step 7**).
2. Homogenize in a 40-mL glass Dounce tissue homogenizer using the large clearance pestle.

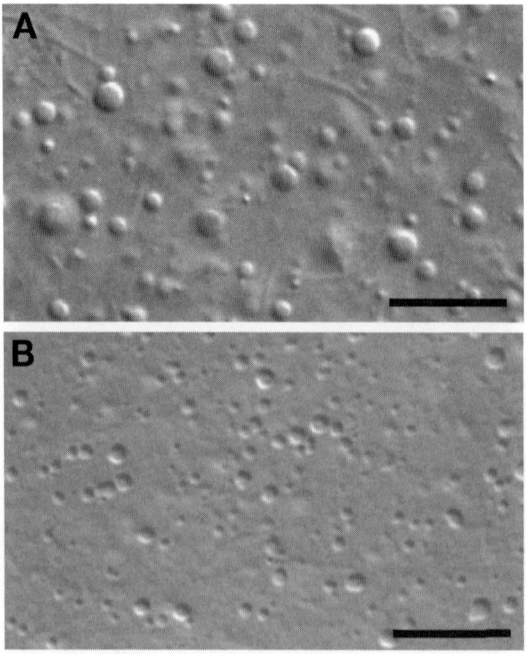

Fig. 1. Nomarski microscopy of purified mitochondrial fractions. **(A)** Final mitoplast vesicle fraction isolated by the hypotonic purification procedure. **(B)** Final mitochondrial vesicle fraction using the isotonic purification method. Vesicles isolated by the hypotonic procedure are larger in size than the ones purified by the isotonic method. Bar = 20 μm.

3. Put cell suspension into the cell disruption bomb sitting in an ice bath. Close bomb; lock the outlet valve.
4. Apply 55 bars using N_2 and close the inlet valve. Incubate for 30 min under constant stirring while bomb sits in an ice bath (*see* **Note 24**).
5. Depressurize the cell disruption bomb and collect the foamy suspension.
6. Let the foam settle for few minutes.
7. Examine lysis microscopically by comparing samples before and after lysis (*see* **Note 25**).
8. Spin lysate in 50-mL centrifuge tubes in a fixed-angle rotor at 4°C for 10 min at 24,000*g*. Pour out supernatants (*see* **Note 14**). The supernatants represent the cytosolic fraction and, depending on the experiment, may be kept.

3.2.3. DNase Digestion

1. Resuspend pellet in equal volume of SoTM buffer.
2. Add solid DNase I to 0.1 mg/mL final concentration (*see* **Note 15**).
3. Push the extract through a hypodermic needle (no. 25, orange) using a Luer-Lok syringe.

4. Incubate for 45 min at 4°C. During incubation, repeat **step 3**. The viscosity should drop during DNase digestion.
5. Add an equal volume of STE buffer.
6. Add 1/125 volume of EDTA stock (0.5 *M*) (*see* **Note 16**).

3.2.4. Low-Speed Spins

1. Spin lysate in 50-mL centrifuge tubes in a fixed-angle rotor at 4°C for 10 min at 490*g*. Fill tubes to the top; if necessary, add SoTE buffer.
2. Transfer supernatant to a beaker and keep on ice. The pellet will be very soft; thus, leave 1–2 mL of the supernatant in the tube.
3. Resuspend each pellet in approx 10–15 mL SoTE by homogenizing in a 40-mL glass Dounce tissue homogenizer using the large clearance pestle. Pool supernatants.
4. Spin in 50-mL centrifuge tubes in a fixed-angle rotor at 4°C for 10 min at 375*g*. Fill tubes to the top; if necessary, add SoTE buffer (*see* **Note 26**).
5. Pool supernatants with the previous ones (*see* **step 2**).
6. Distribute the pooled supernatants to 50-mL centrifuge tubes and spin in a fixed-angle rotor at 4°C for 10 min at 24,000*g*. Discard supernatants.

3.2.5. Nycodenz Step Gradients

1. Precool large swinging bucket ultracentrifuge rotor.
2. Determine the number (two, four, or six) of gradients you will need. Each gradient should be loaded with lysate corresponding to 3.5–6.5 × 10^{10} cell-equivalents. To simplify the balancing, an even number of tubes should be used.
3. During the DNase digestion, prepare Nycodenz step gradients: pipet 8 mL cold 28% Nycodenz containing SoTE buffer into each 38.5-mL ultracentrifuge tube. Carefully overlay 8 mL each of cold 25/21/18% Nycodenz containing SoTE buffer.
4. Resuspend pellets (*see* **Subheading 3.2.6., step 6**) in a total volume of 3–6 mL 50% Nycodenz containing SoTE buffer per gradient, pool them, and homogenize in a 40-mL glass Dounce tissue homogenizer using the small clearance pestle (*see* **Note 18**).
5. Place equal volumes of the samples (3–6 mL) at the bottom of each gradient. Practically, this is done by using the 10-mL hypodermic syringe with the attached glass capillary (*see* **Note 19**).
6. Balance gradients with the 18% Nycodenz containing SoTE buffer on an electronic balance.
7. Spin gradients in the large swinging bucket ultracentrifuge rotor at 4°C for 45 min at 100,000*g*.
8. After centrifugation, the gradients should show three bands. Microscopic examination shows that the middle band at the 25/28% Nycodenz interphase, which generally is the most prominent one, is enriched for mitochondrial vesicles (*see* **Note 27**).
9. Collect approx 5 mL of this band (25/28% Nycodenz interphase) using the same 10-mL hypodermic syringe with the attached glass capillary used for loading.

3.2.6. Removal of Nycodenz and Storage

1. Distribute the collected 25/28% Nycodenz interphase mitochondrial fractions of all gradients into 50-mL centrifuge tubes and dilute at least fivefold with SoTE buffer.
2. Cap tubes with two layers of parafilm and mix by inversion.
3. Spin in a fixed-angle rotor at 4°C for 15 min at 33,000*g* and discard supernatants.
4. Resupend pellets in approx 1 mL per gradient of SoTE buffer and pool. Distribute the resulting suspension into 1.5-mL Eppendorf tubes.
5. Spin in a Eppendorf centrifuge at approx 10,000*g* and discard as much supernatant as possible.
6. Resuspend the mitochondrial pellet in a small volume of SoTE buffer and examine by microscope. The fraction should look as shown in **Fig. 1B**.
7. Take a small aliquot and determine the protein concentration by the BCA protein assay kit (*see* **Note 28**).
8. Aliquots of mitochondrial vesicles can directly be flash frozen in liquid N_2 and stored at –70°C. However, the best way to preserve the membrane integrity of the mitochondrial fraction is to add one-ninth volume of 100-mg/mL fatty acid-free BSA before freezing (*see* **Note 23**) *(22)*.

4. Notes

1. The procedures appear to work for any *T. brucei* cell line. We have used it for the *T. brucei* 427 and 29-13 strains, as well as for many transgenic cell lines, including induced RNAi strains.
2. If not indicated otherwise, then the buffers and solutions described in this chapter were not sterilized. EDTA-containing solutions are not prone to microbiological contaminations, and thus for short-term storage were kept at 4°C; long-term storage was done at –20°C. All other solutions were kept at –20°C and thawed overnight before use.
3. The glass capillary can easily be attached to the syringe using parafilm.
4. Weigh 80 g of Nycodenz, add aliquots to approx 45–50 mL H_2O mixed by a magnetic stirrer. 80% Nycodenz will take a long time to solubilize; to accelerate the process, the solution can be warmed to approx 37°C. After complete solubilization, add waterto 100 mL.
5. This has been demonstrated by direct comparison of the mitochondrial adenosine triphosphate production pathways in organellar vesicles isolated by either the hypotonic or the isotonic purification protocols *(21)*.
6. Both the hypotonic and the isotonic procedures work best on a large scale (5 L of culture) and are not recommended for less than 1 L of well-grown cells. The indicated cell densities are for *T. brucei* 427 grown in SDM-79 and may be different for other cell lines or media. Large-scale cultures are grown at 27°C in 2-L Erlenmeyer flasks containing 1–1.3 L culture each on a shaking incubator set at 115 rpm. Some transgenic strains may be more fragile than wild-type cells; in this case, shaking needs to be reduced.
7. Indicated *g* forces always refer to g_{max} at the bottom of the tube.

8. *Trypanosoma brucei* cells are highly motile, which results in soft pellets. Thus, the medium needs to be poured off immediately after centrifugation. This should always be done with the pellets facing downward.
9. 1X PBS can be used instead of wash buffer. The total cell number can be estimated by weighing and be compared to the one obtained by microscopic counting: 5.8 g wet weight of cells correspond to approx 10^{11} cells.
10. This is the highest recommended concentration. Breakage of cells using the pressure vessel works best for volumes 100 mL or larger. For smaller preparations, the cells can be diluted two- to fourfold more.
11. If no pressure vessel is available, then the cells can be lysed manually using a 20- to 40-mL Luer-Lok syringe and pushing them once or twice with as much force as possible through a no. 26 hypodermic needle.
12. The time the lysed cells remain in the hypotonic lysis buffer before the sucrose is added is critical and should be minimized since otherwise the mitochondrial vesicles will lyse as well.
13. Lysis is expected to be complete. Thus, cell fragments, flagella, and floating vesicles but no live cells are observed.
14. A white floating layer will appear on the solution. This layer probably represents broken membranes and is indicative of efficient cell lysis; it can most easily be removed using a paper tissue.
15. DNase digestion is essential to allow efficient separation on either Percoll or Nycodenz gradients.
16. Addition of EDTA complexes the magnesium and thus stops the DNase digestion. Furthermore, addition of EDTA also serves to prevent aggregation of mitochondrial preparations, which is observed in the presence of magnesium.
17. We have never loaded more the 3.5×10^{10} cell equivalents; however, we expect the gradients to tolerate higher loadings.
18. The dilution of the pellet with 75% Percoll containing STE buffer in the hypotonic procedure or with the 50% Nycodenz containing SoTE buffer in the isotonic preparation needs to be sufficient to allow the suspension to sink beneath the lowest layers of the step gradients.
19. It is best to insert the capillary into the gradient along the tube wall and to keep it there until the whole sample has been applied. First, load a small volume of the sample only and wait few seconds to make sure it remains at the bottom of the tube. If it floats up, then remove the syringe, add more of the 75% Percoll containing STE buffer for the hypotonic procedure or 50% Nycodenz containing SoTE for the isotonic preparation, and try again.
20. Microscopic examination shows that the top band (20/25% Percoll interphase), which is the most intense one, mainly contains flagella and some cell fragments, whereas the lowest band (35/75% Percoll interphase), which normally is the least intense, contains a uniform population of vesicular structures of unknown origin that are much smaller than the ones observed in the mitoplast fraction. Large mitoplast vesicles are seen in all three fractions but are most enriched in the central part of the gradient (25/30% Percoll interphase) (**Fig. 1A**). The main contaminants of the

central mitoplast fraction are flagella. Depending on the cell line and the efficiency of the cell lysis, there can be variations in the relative intensities of the three zones. Furthermore, if only little material is loaded on the gradient, then it is possible that no clear accumulation of material is seen in the central zone. Although the yield will be low in this case, it is still worthwhile to collect the zone and to proceed with the purification as a mitoplast pellet may only become visible after the washes.

21. Percoll does not mix easily so vigorous mixing is important. The more the Percoll is diluted, the tighter the pellets will be that are obtained after centrifugation.
22. The obtained yields can be quite variable but are typically in the range of 20 mg protein or 250–300 μg RNA per 10^{11} cells.
23. Although the integrity of the mitoplasts or the mitochondrial fractions during storage at –70°C is best maintained in the presence of BSA, it should be considered that the presence of BSA might interfere with some downstream applications, such as Western blot analysis.
24. Pressure will slightly drop during incubation; therefore, readjust to 55 bar after the first 5 min.
25. Besides cell ghosts and floating vesicles, a significant number of intact live cells are observed. Thus, cell lysis is less efficient—typically around 80–90%—than with the hypotonic protocol.
26. The resulting pellet should be smaller than the one obtained in the first low-speed spin.
27. Microscopic examination shows that the top band (18/21% Nycodenz interphase) mainly contains intact cells, whereas the lowest band (28/50% Nycodenz interphase) is enriched for flagella. Mitochondrial vesicles are seen in all three fractions but are most enriched in the central part of the gradient (25/28% Nycodenz interphase) (**Fig. 1B**). The main contaminants of the central mitochondrial vesicle fraction are flagella. Occasionally, few intact cells may be observed as well. As for the hypotonic procedure, depending on the cell line and the efficiency of the cell lysis, there can be variations in the relative intensities of the three zones.
28. The obtained yields can be quite variable but are typically in the range of 15–30 mg protein per 10^{11} cells.

Acknowledgments

This study was supported by grant 31-067906.02 from the Swiss National Foundation and by a grant from the Novartis Foundation.

References

1. Gull, K. (2001) The biology of kinetoplastid parasites: insights and challenges from genomics and post-genomics. *Int. J. Parasitol.* **31,** 443–452.
2. Cross, G. A. (2001) African trypanosomes in the 21st century: what is their future in science and in health? *Int. J. Parasitol.* **31,** 427–433.
3. tenAsbroek, A. L. M. A., Ouellette, M., and Borst, P. (1990) Targeted insertion of the neomycin phosphotransferase gene into the tubulin gene cluster of *Trypanosoma brucei. Nature* **348,** 174–175.

4. Wirtz, E., and Clayton, C. (1995) Inducible gene expression in trypanosomes mediated by a prokaryotic repressor. *Science* **268,** 1179–1183.
5. Fire, A., Xu, S., Montgomery, M. K., Kostas, S. A., Driver, S. E., and Mello, C. C. (1998) Potent and specific genetic interference by double-stranded RNA in *Caenorhabditis elegans. Nature* **391,** 806–811.
6. Ngo, I., Tschudi, C., Gull, K., and Ullu, E. (1998) Double-stranded RNA induces mRNA degradation in *Trypanosoma brucei. Proc. Natl. Acad. Sci. USA* **95,** 14,687–14,692.
7. Shi, H., Djikeng, A., Mark, T., Wirtz, E., Tschudi, C., and Ullu, E. (2000) Genetic interference in *Trypanosoma brucei* by heritable and inducible double-stranded RNA. *RNA* **6,** 1069–1076.
8. Wang, Z., Morris, J. C., Drew, M. E., and Englund, P. T. (2000) Inhibition of *Trypanosoma brucei* gene expression by RNA interference. *J. Biol. Chem.* **275,** 40,174–40,179.
9. Sogin, M. L., Elwood, H. J., and Gunderson, J. H. (1986) Evolutionary diversity of eukaryotic small-subunit rRNA genes. *Proc. Natl. Acad. Sci. USA* **83,** 1383–1387.
10. Schneider, A. (2001) Unique aspects of mitochondrial biogenesis in trypanosomatids. *Int. J. Parasitol.* **31,** 1403–1415.
11. Ogbadoyi, E. O., Robinson, D. R., and Gull, K. (2003) A high-order trans-membrane structural linkage is responsible for mitochondrial genome positioning and segregation by flagellar basal bodies in trypanosomes. *Mol. Biol. Cell.* **14,** 1769–1779.
12. Morris, J. C., Drew, M. E., Klingbeil, M. M., et al. (2001) Replication of kinetoplast DNA: an update for the new millennium. *Int. J. Parasitol.* **31,** 453–458.
13. Koslowsky, D. J. (2004) A historical perspective on RNA editing: how the peculiar and bizarre became mainstream. *Methods Mol. Biol.* **265,** 161–197.
14. Schneider, A., and Marechal-Drouard, L. (2000) Mitochondrial tRNA import: are there distinct mechanisms? *Trends Cell Biol.* **10,** 509–513.
15. Horvath, A., Nebohacova, M., Lukes, J., and Maslov, D. A. (2002) Unusual polypeptide synthesis in the kinetoplast-mitochondria from *Leishmania tarentolae.* Identification of individual *de novo* translation products. *J. Biol. Chem.* **277,** 7222–7230.
16. vanHellemond, J. J., Opperdoes, F. R., and Tielens, A. G. M. (1998) Trypanosomatides produce acetate via a mitochondrial acetate:succinate CoA transferase. *Proc. Natl. Acad. Sci. USA* **95,** 3036–3041.
17. Braly, P., Simpson, L., and Kretzer, F. (1974) Isolation of kinetoplast-mitochondrial complexes from *Leishmania tarentolae. J. Protozool.* **21,** 782–790.
18. Harris, M. E., Moore, D. R., and Hajduk, S. L. (1990) Addition of uridines to edited RNAs in trypanosome mitochondria occurs independently of transcription. *J. Biol. Chem.* **265,** 11,368–11,376.
19. Hauser, R., Pypaert, M., Häusler, T., Horn, E. K., and Schneider, A. (1996) In vitro import of proteins into mitochondria of *Trypanosoma brucei* and *Leishmania tarentolae. J. Cell Sci.* **109,** 517–523.
20. Brun, R., and Schönenberger, M. (1979) Cultivation an in vitro cloning of procyclic culture forms of *Trypansoma brucei* in a semi-defined medium. *Acta Tropica* **36,** 289–292.

21. Allemann, N., and Schneider, A. (2000) ATP production in isolated mitochondria of procyclic *Trypanosoma brucei*. *Mol. Biochem. Parasitol.* **111,** 87–94.
22. Kozlowski, M., and Zagorski, W. (1988) Stable preparation of yeast mitochondria and mitoplasts synthesizing specific polypeptides. *Anal. Biochem.* **172,** 382–391.
23. Priest, J. W., and Hajduk, S. L. (1996) In vitro import of the rieske iron-sulfur protein by trypanosome mitochondria. *J. Biol. Chem.* **271,** 20,060–20,069.

6

Saccharomyces cerevisiae as a Model Organism to Study Mitochondrial Biology

General Considerations and Basic Procedures

Katrin Altmann, Mark Dürr, and Benedikt Westermann

Summary

Budding yeast *Saccharomyces cerevisiae* is widely used to study mitochondrial biogenesis and function. We review some basic properties that make yeast an ideal model organism to investigate various aspects of mitochondrial biology. We discuss genetic features of commonly used yeast strains that are important for mitochondrial studies. Furthermore, this chapter provides protocols describing yeast culture conditions and procedures for isolation and purification of mitochondria.

Key Words: Mitochondria isolation; mitochondrial biogenesis; model organism; *Saccharomyces cerevisiae*; yeast strains.

1. Introduction

Budding yeast *Saccharomyces cerevisiae* has proven to be an excellent model organism to study a great variety of basic cellular functions that are conserved in eukaryotic cells. Several properties make yeast particularly suitable for genetic, biochemical, and cell biological studies. For instance, *S. cerevisiae* can be cultured in an economic manner and has a short generation time (under optimal conditions, less than 2 h). This allows the isolation of biological material in amounts sufficient for further biochemical studies. Maybe most important, genetic engineering is highly efficient in yeast. *Saccharomyces cerevisiae* is viable with numerous markers, a large selection of different kinds of plasmids and gene fusion cassettes is available, homologous recombination is very efficient, and laboratory yeast strains are stable as both haploid and diploid strains *(1)*. The *S. cerevisiae* genome sequence has been completely known since 1996. The genome has a size of approx 12 million bp and harbors about 6000 genes; only about 4% of nuclear genes have introns *(2)*.

From: *Methods in Molecular Biology, vol. 372: Mitochondria: Practical Protocols*
Edited by: D. Leister and J. M. Herrmann © Humana Press Inc., Totowa, NJ

In recent years, several comprehensive genomewide gene deletion and protein fusion libraries have been constructed. These are now available to the scientific community as a great resource for systematic studies ***(3–7)***.

The following databases provide good starting points to retrieve information about yeast genetics and biology: (1) Saccharomyces Genome Database ***(8)***, http://www.yeastgenome.org/; (2) Yeast Virtual Library, http://www.yeastgenome.org/VL-yeast.html; (3) Comprehensive Yeast Genome Database ***(9)***, http://mips.gsf.de/genre/proj/yeast/.

Saccharomyces cerevisiae is a facultative anaerobic yeast capable of satisfying its energy requirements with adenosine triphosphate (ATP) generated by fermentation. Thus, only relatively few mitochondrial proteins are essential for cell viability. These include a handful of factors essential for import and assembly of nuclear-encoded precursor proteins, iron/sulfur cluster assembly, and flavin mononucleotide synthesis. The fact that many mitochondrial functions can be studied using viable knockout mutants makes budding yeast an ideal organism for dissecting the molecular processes required for biogenesis of respiratory-competent mitochondria.

Saccharomyces cerevisiae can live on a variety of carbon sources, but glucose and fructose are the preferred ones. On these carbon sources, most of the cellular ATP is generated in the cytosol by fermentation, and the expression of enzymes required for the utilization of other carbon sources is strongly reduced. This phenomenon is known as *glucose repression* or *catabolite repression* ***(10)***. Glucose repression affects the expression of many mitochondrial factors, including enzymes of the citric acid cycle and respiratory chain complexes. As synthesis of ATP by oxidative phosphorylation is a major function of mitochondria, mitochondrial size, volume, and structure are adapted to the carbon source of the growth medium ***(11–14)***. Typically, wild-type yeast cells growing logarithmically on glucose-containing medium display a relatively simple mitochondrial network that consists of few branched organelles (**Fig. 1**, top). In contrast, on nonfermentable carbon sources such as glycerol, mitochondria are much more numerous and form a highly branched interconnected network (**Fig. 1**, bottom).

The mitochondrial genome of *S. cerevisiae* is roughly 80,000 bp in size and encodes eight major proteins, which are all essential for oxidative phosphorylation. These are cytochrome-*b* (a subunit of the ubiquinol-cytochrome-*c* oxidoreductase); Cox1, Cox2, and Cox3 (subunits of the cytochrome-*c* oxidase); Atp6, Atp8 and Atp9 (subunits of the F_o part of ATP synthase); and Var1 (a component of the small subunit of the mitochondrial ribosome) ***(15)***. Many domesticated yeast strains produce high frequencies of mutants lacking intact mitochondrial genomes at rates of 2% or more ***(1)***. Strains harboring an intact mitochondrial genome are designated ρ^+ (rho$^+$). Respiratory-deficient strains harboring a defective mitochondrial genome are ρ^- (rho$^-$), and strains completely lacking mitochondrial DNA are ρ^0 (rho^0). Despite the capacity of mitochondria

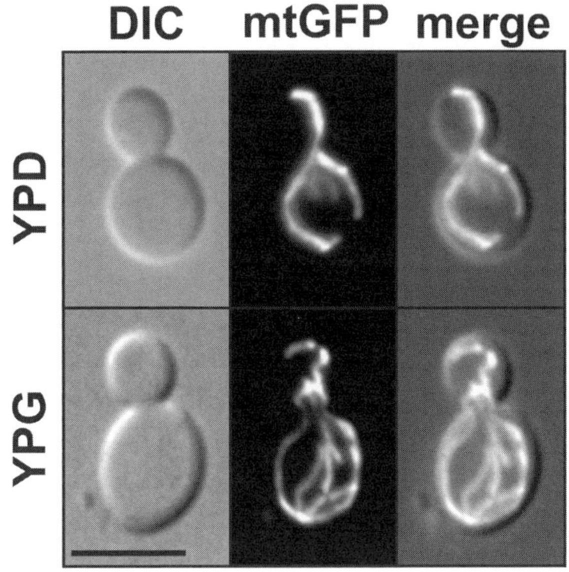

Fig. 1. Mitochondrial structure and volume depend on the carbon source of the medium. Wild-type yeast cells (BY4741) expressing mitochondria-targeted green fluorescent protein (mtGFP) ***(32)*** were grown to logarithmic growth phase in glucose-containing medium (YPD; top) or glycerol-containing medium (YPG; bottom) and analyzed by differential interference contrast microscopy (DIC) and fluorescence microscopy (mtGFP). Bar = 5 μm.

to encode and synthesize proteins, more than 300 genes located in the nucleus are required for respiratory competence ***(16)***. Mutants in these genes are commonly referred to as nuclear *petite* or *pet* mutants ***(17)***.

The mitochondrial proteome of *S. cerevisiae* has been extensively characterized by mass spectrometric analysis of purified mitochondria ***(18,19)***. These approaches have led to the identification of most of the estimated 700–800 proteins that constitute the yeast mitochondrial proteome ***(19,20)***.

Mitochondrial research using *S. cerevisiae* as a model organism has been instrumental in elucidating the biogenesis and biological function of this organelle ***(21)***. It is safe to predict that the knowledge of the mitochondrial proteome combined with the availability of comprehensive mutant collections will give mitochondrial research with yeast another boost in the coming years.

2. Materials

2.1. Commonly Used Yeast Strains

A selection of commonly used yeast strains together with their genotypes and sources where they can be obtained is presented in **Table 1**.

Table 1
Commonly Used Yeast Strains

Strain	Genotype	Source
BY4741	*MATa his3Δ1 leu2Δ0 met15Δ0 ura3Δ0*	EUROSCARF:Y00000
BY4742	*MATα his3Δ1 leu2Δ0 lys2Δ0 ura3Δ0*	EUROSCARF:Y10000
BY4743	*MATa/MATα his3Δ1/his3Δ1 leu2Δ0/leu2Δ0 met15Δ0/MET15 LYS2/lys2Δ0 ura3Δ0/ura3Δ0*	EUROSCARF:Y20000
D273-10B	*MATα mal*	ATCC:24657
S288C	*MATα SUC2 mal mel gal2 CUP1 flo1 flo8-1 hap1*	ATCC:204508
W303 MATa	*MATa ura3-52 trp1Δ2 leu2-3_112 his3-11 ade2-1 can1-100*	EUROSCARF:20000A
YPH499	*MATa ura3-52 lys2-801_amber ade2-101_ochre trp1-Δ63 his3-Δ200 leu2-Δ1*	ATCC:204679
YPH500	*MATα ura3-52 lys2-801_amber ade2-101_ochre trp1-Δ63 his3-Δ200 leu2-Δ1*	ATCC:204680
YPH501	*MATa/MATα ura3-52/ura3-52 lys2-801_amber/lys2-801_amber ade2-101_ochre/ade2-101_ochre trp1-Δ63/trp1-a63 his3-Δ200/his3-Δ200 leu2-Δ1/leu2-Δ1*	ATCC:204681

Strains can be obtained from the Yeast Genetics Stock Culture Center of the American Type Culture Collection, ATCC (http://www.atcc.org/common/catalog/yeastGeneticStock/yeastGeneticStockIndex.cfm) or EUROSCARF (http://www.uni-frankfurt.de/fb15/mikro/euroscarf/col_index.html).

2.2. Yeast Culture

Amounts are per liter medium; for liquid medium, omit agar (*see* **Note 1**).

1. YPD (yeast extract/peptone/dextrose) medium: 10 g Bacto™ yeast extract, 20 g Bacto peptone, 20 g glucose (100 mL of a 20% stock solution), 20 g Bacto agar, and distilled water to bring to 1000 mL.
2. YPG (yeast extract/peptone/glycerol) medium: 10 g Bacto yeast extract, 20 g Bacto peptone, 3% (v/v) glycerol (100 mL of a 30% stock solution), 20 g Bacto agar, and distilled water to bring to 1000 mL.
3. Yeast extract/peptone/galactose medium: 10 g Bacto yeast extract, 20 g Bacto peptone, 20 g galactose (100 mL of a 20% w/v stock solution), 20 g Bacto agar, and distilled water to bring to 1000 mL.
4. Yeast extract/peptone/raffinose medium: 10 g Bacto yeast extract, 20 g Bacto peptone, 20 g raffinose (100 mL of a 20% w/v stock solution), 20 g Bacto agar, and distilled water to bring to 1000 mL.
5. SD (synthetic minimal medium, dextrose): 6.7 g yeast nitrogen base (including ammonium sulfate), 2% glucose (100 mL of a 20% w/v stock solution), 20 g Bacto agar, distilled water to bring to 1000 mL. Depending on the auxotrophic markers of

the yeast strain, supplements may have to be added (the amounts are as indicated for SC medium, **item 6**).

6. SC (synthetic complete medium, dextrose): SD medium supplemented with adenine sulfate (20 mg/L), uracil (20 mg/L), L-tryptophan (20 mg/L), L-histidine-HCl (20 mg/L), L-arginine-HCl (20 mg/L), L-methionine (20 mg/L), L-tyrosine (30 mg/L), L-leucine (30 mg/L), L-isoleucine (30 mg/L), L-lysine-HCl (30 mg/L), L-phenylalanine (50 mg/L), L-glutamic acid (100 mg/L), L-aspartic acid (100 mg/L), L-valine (150 mg/L), L-threonine (200 mg/L), and L-serine (400 mg/L). For selection on auxotrophic markers, relevant supplements are omitted.
7. Lactate medium: 3 g Bacto yeast extract, 1 g KH_2PO_4, 1 g NH_4Cl, 0.5 g $CaCl_2 \cdot 2H_2O$, 0.5 g NaCl, 0.6 g $MgSO_4 \cdot H_2O$, 3 mg $FeCl_3$, 2% (v/v) lactate, and distilled water to bring to 1000 mL. Adjust to pH 5.5 (~7.5 g/L NaOH pellets).

2.3. Isolation of Mitochondria by Differential Centrifugation

Amounts are for 2 L yeast culture at OD_{600} 1–2 (~10 g wet weight of cells).

1. Tris-SO_4 buffer (30 mL): 100 m*M* Tris-SO_4, pH 9.4, 10 m*M* dithiothreitol (add just before use from freshly prepared 1 *M* stock solution).
2. Sorbitol buffer (80 mL): 1.2 *M* sorbitol, 20 m*M* phosphate buffer, pH 7.4 (chill 40 mL on ice).
3. Zymolyase 20T (~30 mg).
4. Buffer A (60 mL): 0.6 *M* sorbitol, 20 m*M* HEPES-KOH, pH 7.4, 1 m*M* phenylmethylsulfonyl fluoride (PMSF); chill buffer on ice before use (*see* **Note 2**).
5. Buffer B (40 mL): 0.6 *M* sorbitol, 20 m*M* HEPES-KOH, pH 7.4; chill buffer on ice before use.
6. Bovine serum albumin (BSA) solution (50 μL): 75 mg/mL fatty acid-free BSA, 0.6 *M* sorbitol.
7. Liquid nitrogen.

2.4. Purification of Mitochondria by Sucrose Gradient Purification

1. SEM buffer: 250 m*M* sucrose, 1 m*M* ethylenediaminetetraacetic acid, 10 m*M* 4-Morpholinepropanesulfonic acid–KOH (MOPS-KOH), pH 7.2.
2. Sucrose step gradient: 20, 30, 40, 50, and 60% sucrose (w/w) in 100 m*M* KCl, 1 m*M* ethylenediaminetetraacetic acid, 1 m*M* PMSF, 10 m*M* MOPS-KOH, pH 7.2 (*see* **Note 2**).

3. Methods

3.1. Choice of Suitable Yeast Strains

Commonly used laboratory yeast strains are not truly wild-type *S. cerevisiae* strains. Many laboratory stocks have been inbred with related *Saccharomyces* species. Thus, genetic backgrounds and growth properties of different yeast strains might differ considerably *(1)*. Therefore, care should be taken in choosing strains for research on mitochondria (*see* **Note 3**).

S288C *(22)* is often used as a normal standard because it was used to sequence the yeast genome *(2)*. Furthermore, BY wild-type strains BY4741, BY4742, and BY4743 *(23)*, which were used in the worldwide gene deletion project *(3)*, are derived from S288C. Unfortunately, S288C and its derivatives carry an insertion of a defective Ty1 retrotransposon element in the coding region of the *HAP1* gene *(24)*. *HAP1* encodes a transcriptional regulator that is involved in the regulation of a variety of genes involved in electron transfer reactions, sterol metabolism, and protein synthesis. As Hap1 function is compromised in S288C and its derivatives, these are not the best strains for mitochondrial research.

W303 is widely used for studies of mitochondrial biology. This strain contains the *ybp1-1* mutation, which makes it more sensitive to oxidative stress *(25)*. W303 also contains a *bud4* mutation that causes defects in the budding pattern of haploid cells. In addition, W303 strains contain the *rad5-535* allele (Saccharomyces Genome Database).

D273-10B has been used for mitochondrial studies in numerous laboratories. It has normal cytochrome content and respiration, shows a low frequency of spontaneous ρ^- generation, and is relatively resistant to glucose repression *(26)*.

Strains YPH499, YPH500, and YPH501 contain six nonrevertible auxotrophic mutations that can be conveniently used for selection of vectors *(27)*.

3.2. Yeast Culture

Yeasts are grown either on agar plates or in Erlenmeyer flasks in liquid cultures under constant agitation (~140 rpm). The optimal growth temperature for *S. cerevisiae* is 30°C. For long-term preservation of strains, yeast cells are resuspended in 15% (v/v) glycerol and stored at –80°C *(1)*.

YPD is a glucose-containing rich medium for routine growth. It supports growth of respiratory-deficient mutants. However, mitochondrial functions may be reduced because of glucose repression. YPG is a complex medium containing a nonfermentable carbon source (glycerol). It does not allow growth of respiratory-deficient (ρ^- or *petite*) mutants. Mitochondrial functions are induced on glycerol (compare **Fig. 1**). Yeast extract/peptone/galactose contains galactose as a carbon source. This medium is often used to induce genes that have been placed under control of the *GAL* promoter. Galactose is a fermentable carbon source that allows growth of ρ^- or *petite* mutants and does not induce glucose repression. Yeast extract/peptone/raffinose similarly supports growth of respiratory-deficient mutants without causing glucose repression, but it does not induce the *GAL* promoter. It should be noted that the *GAL* promoter is repressed in the presence of even minor amounts of glucose.

SD is used for selection on auxotrophic markers. It is a synthetic minimal medium containing salts, trace elements, vitamins, a nitrogen source, and glucose.

Depending on the auxotrophic markers, certain supplements may have to be added. SC contains all possible supplements, except those omitted to select on auxotrophic markers. Depending on the type of experiment, glucose can be replaced by any other carbon source in both types of minimal media.

Lactate medium is a semisynthetic medium that is often used to grow yeast cultures for preparation of mitochondria ***(28)***. To have an optimal induction of mitochondrial functions, yeasts are precultured for several generations in lactate medium (*see* **Note 4**).

3.3. Isolation of Mitochondria by Differential Centrifugation

Yeast mitochondria can be conveniently isolated by differential centrifugation ***(28–30)***. The following protocol is outlined for 2 L of culture with an OD_{600} of 1–2, corresponding to approx 10 g wet weight of cells (*see* **Note 5**).

1. Collect the cells at 2000*g* for 5 min and determine wet weight.
2. Resuspend the pellet in 100 mL distilled water and centrifuge at 2000*g* for 5 min (*see* **Note 6**).
3. Resuspend the cells in 30 mL Tris-SO_4 buffer and incubate the suspension for 10 min at 30°C under agitation (~140 rpm).
4. Collect the cells by centrifugation at 2000*g* for 5 min and resuspend them in 40 mL sorbitol buffer. Add Zymolyase 20T (2 mg/g cells) to the suspension and incubate under gentle agitation for 20–40 min at 30°C until spheroplasts have formed (*see* **Note 7**).
5. All the following steps will be performed on ice using ice-cold buffers, and centrifugation steps are performed at 4°C. Harvest the spheroplasts by centrifugation at 2000*g* for 5 min.
6. Resuspend the pellet in 40 mL sorbitol buffer (gently shaking or stirring with a pipet) and spin at 2000*g* for 5 min.
7. Resuspend the spheroplasts carefully in 30 mL buffer A, transfer the suspension to a 50-mL Dounce homogenizer (tight-fitting glass pistil) and homogenize with 15 strokes.
8. Centrifuge the homogenate at 2000*g* for 5 min and keep the supernatant.
9. Resuspend the pellet carefully in 30 mL buffer A, homogenize with 15 strokes, and spin at 2000*g* for 5 min.
10. Combine the supernatants and centrifuge at 12,000*g* for 10 min.
11. Resuspend the pellets in 1 mL buffer B using cut pipet tips and fill up to 30 mL with buffer B.
12. Spin down any remaining cell debris at 2000*g* for 5 min, transfer the supernatant to a fresh tube and centrifuge at 12,000*g* for 10 min.
13. Resuspend the pellet in 0.5 mL buffer B.
14. Take an aliquot to determine protein concentration.
15. Add 35 µL BSA solution (7% v/v) (*see* **Note 8**).
16. Make aliquots of 30–50 µL; immediately snap-freeze in liquid nitrogen. Store at –80°C (*see* **Note 9**).

3.4. Purification of Mitochondria by Sucrose Gradient Purification

Mitochondria isolated by differential centrifugation can be further purified by centrifugation on a sucrose step gradient ***(31)***.

1. Load mitochondrial suspension on top of a gradient consisting of 20, 30, 40, 50, and 60% sucrose buffer (w/w) in a Beckman SW41 centrifuge tube (*see* **Note 10**).
2. Centrifuge 15 min at 240,000*g* at 4°C.
3. Collect mitochondria from the band between the 40 and 50% sucrose layers.
4. Concentrate mitochondria by centrifugation in a microfuge tube at 12,000*g* at 4°C.
5. Wash pellet with SEM and resuspend purified mitochondria in SEM.

4. Notes

1. A liter of medium is sufficient for approx 30 plates. To avoid hydrolysis of the agar, caramelized glucose, and mushy plates, it is recommended to autoclave the components of the medium separately (e.g., 20 g agar in 500 mL H_2O, carbon source in 100 mL H_2O, and the other components in 400 mL H_2O). Combine the solutions directly after autoclaving (15 min at 120°C, 1 atm), mix thoroughly, and pour plates. *For safety reasons, melting solidified agar in a microwave oven must be avoided.*
2. PMSF is dissolved at 200 m*M* in ethanol. Prepare freshly before use. PMSF is a protease inhibitor. It works fine for many applications. However, some assays using isolated mitochondria might be inhibited by the presence of PMSF during mitochondria isolation. In this case, other protease inhibitor cocktails might be tried.
3. Several laboratory yeast strains (e.g., W303, YPH499, YPH500, YPH501) carry the *ade2* marker. When adenine in the growth medium becomes limiting, these strains accumulate a pink pigment as an intermediate during the purine nucleotide biosynthetic pathway and form red colonies. As formation of the pigment is dependent on oxidative metabolism, the presence of the *ade2* marker is often useful to judge respiratory activity of mutants with defective oxidative phosphorylation.
4. A small amount of glucose (0.05%) may be added to the first culture to help the cells adapt to the culture conditions. However, the addition of glucose to later cultures should be avoided to prevent glucose repression. The incubation time depends markedly on the yeast strain. Therefore, it is recommended to measure the growth rate of a preculture before inoculating the big culture that will be used for mitochondria isolation. Precultures should be always kept in the logarithmic growth phase (i.e., $OD_{600} < 2.0$).
5. To obtain a high mitochondria yield, cultures are usually grown in lactate medium (*see* **Subheading 3.2.** and **Note 4**). Alternatively, YPG or minimal media may be used. For respiratory-deficient mutants, fermentable nonrepressing carbon sources such as galactose or raffinose are recommended.
6. When working with large amounts of culture, it is convenient to pool the cells.
7. To test for spheroplast formation, add 50 μL cells to 2 mL H_2O and sorbitol buffer. When the suspension in water clears, stop the incubation and proceed with the next step.
8. The addition of BSA may stabilize the mitochondria but is not required to maintain them in a functional condition during storage.

9. Frozen mitochondria are good for many applications. However, some in vitro assays might require that mitochondria are prepared freshly.
10. Use 14 × 89 mm ultraclear centrifuge tubes (e.g., Beckman no. 344059) for a Beckman SW41 ultracentrifuge swing-out rotor or equivalent equipment. The gradient should have a total volume of 10 mL and is overlaid with 1 mL mitochondria suspension containing not more than 50–100 mg mitochondrial protein in SEM.

References

1. Sherman, F. (1991) Getting started with yeast. *Methods Enzymol.* **194,** 3–21.
2. Goffeau, A., Barrell, B. G., Bussey, H., et al. (1996) Life with 6000 genes. *Science* **274,** 546–552.
3. Giaever, G., Chu, A. M., Ni, L., et al. (2002) Functional profiling of the *Saccharomyces cerevisiae* genome. *Nature* **418,** 387–391.
4. Huh, W. K., Falvo, J. V., Gerke, L. C., et al. (2003) Global analysis of protein localization in budding yeast. *Nature* **425,** 686–691.
5. Ghaemmaghami, S., Huh, W. K., Bower, K., et al. (2003) Global analysis of protein expression in yeast. *Nature* **425,** 737–741.
6. Mnaimneh, S., Davierwala, A. P., Haynes, J., et al. (2004) Exploration of essential gene functions via titratable promoter alleles. *Cell* **118,** 31–44.
7. Martin, A. C. and Drubin, D. G. (2003) Impact of genome-wide functional analyses on cell biology research. *Curr. Opin. Cell Biol.* **15,** 6–13.
8. Christie, K. R., Weng, S., Balakrishnan, R., et al. (2004) Saccharomyces Genome Database (SGD) provides tools to identify and analyze sequences from *Saccharomyces cerevisiae* and related sequences from other organisms. *Nucleic Acids Res.* **32,** D311–D314.
9. Güldener, U., Münsterkötter, M., Kastenmüller, G., et al. (2005) CYGD: the Comprehensive Yeast Genome Database. *Nucleic Acids Res.* **33,** D364–D368.
10. Gancedo, J. M. (1998) Yeast carbon catabolite repression. *Microbiol. Mol. Biol. Rev.* **62,** 334–361.
11. Stevens, B. (1981) Mitochondrial structure, in *The Molecular Biology of the Yeast Saccharomyces: Life Cycle and Inheritance* (Strathern, E. W., Jones, E. W., and Broach, J. R., eds.), Cold Spring Harbor Press, Cold Spring Harbor, NY, pp. 471–504.
12. Pon, L. and Schatz, G. (1991) Biogenesis of yeast mitochondria, in *The Molecular Biology of the Yeast Saccharomyces: Genome Dynamics, Protein Synthesis, and Energetics* (Broach, J. R., Pringle, J. R., and Jones, E. W., eds.), Cold Spring Harbor Press, Cold Spring Harbor, NY, pp. 333–406.
13. Egner, A., Jakobs, S., and Hell, S. W. (2002) Fast 100 nm resolution 3D-microscope reveals structural plasticity of mitochondria in live yeast. *Proc. Natl. Acad. Sci. U. S. A.* **99,** 3370–3375.
14. Visser, W., van Spronsen, E. A., Nanninga, N., Pronk, J. T., Gijs Kuenen, J., and van Dijken, J. P. (1995) Effects of growth conditions on mitochondrial morphology in *Saccharomyces cerevisiae. Antonie Van Leeuwenhoek* **67,** 243–253.

15. de Zamaroczy, M. and Bernardi, G. (1985) Sequence organization of the mitochondrial genome of yeast—a review. *Gene* **37,** 1–17.
16. Dimmer, K. S., Fritz, S., Fuchs, F., et al. (2002) Genetic basis of mitochondrial function and morphology in *Saccharomyces cerevisiae. Mol. Biol. Cell* **13,** 847–853.
17. Tzagoloff, A. and Dieckmann, C. L. (1990) *PET* genes of *Saccharomyces cerevisiae. Microbiol. Rev.* **54,** 211–225.
18. Sickmann, A., Reinders, J., Wagner, Y., et al. (2003) The proteome of *Saccharomyces cerevisiae* mitochondria. *Proc. Natl. Acad. Sci. U. S. A.* **100,** 13,207–13,212.
19. Prokisch, H., Scharfe, C., Camp, D. G., 2nd, et al. (2004) Integrative analysis of the mitochondrial proteome in yeast. *PLoS Biol.* **2,** e160.
20. Reichert, A. S. and Neupert, W. (2004) Mitochondriomics or what makes us breathe. *Trends Genet.* **20,** 555–562.
21. Scheffler, I. E. (2000) A century of mitochondrial research: achievements and perspectives. *Mitochondrion* **1,** 3–31.
22. Mortimer, R. K. and Johnston, J. R. (1986) Genealogy of principal strains of the yeast genetic stock center. *Genetics* **113,** 35–43.
23. Brachmann, C. B., Davies, A., Cost, G. J., et al. D. (1998) Designer deletion strains derived from *Saccharomyces cerevisiae* S288C: a useful set of strains and plasmids for PCR-mediated gene disruption and other applications. *Yeast* **14,** 115–132.
24. Gaisne, M., Becam, A. M., Verdiere, J., and Herbert, C. J. (1999) A "natural" mutation in *Saccharomyces cerevisiae* strains derived from S288c affects the complex regulatory gene *HAP1* (*CYP1*). *Curr. Genet.* **36,** 195–200.
25. Veal, E. A., Ross, S. J., Malakasi, P., Peacock, E., and Morgan, B. A. (2003) Ybp1 is required for the hydrogen peroxide-induced oxidation of the Yap1 transcription factor. *J. Biol. Chem.* **278,** 30896–30904.
26. Sherman, F. (1963) Respiration-deficient mutants of yeast. I. Genetics. *Genetics* **48,** 375–385.
27. Sikorski, R. S. and Hieter, P. (1989) A system of shuttle vectors and host strains designed for efficient manipulation of DNA in *Saccharomyces cerevisiae. Genetics* **122,** 19–27.
28. Daum, G., Böhni, P. C., and Schatz, G. (1982) Import of proteins into mitochondria: cytochrome b_2 and cytochrome *c* peroxidase are located in the intermembrane space of yeast mitochondria. *J. Biol. Chem.* **257,** 13,028–13,033.
29. Diekert, K., de Kroon, A. I. P. M., Kispal, G., and Lill, R. (2001) Isolation and subfractionation of mitochondria from the yeast *Saccharomyces cerevisiae. Methods Cell Biol.* **65,** 37–51.
30. Glick, B. S. and Pon, L. A. (1995) Isolation of highly purified mitochondria from *Saccharomyces cerevisiae. Methods Enzymol.* **260,** 213–223.
31. Rowley, N., Prip-Buus, C., Westermann, B., et al. (1994) Mdj1p, a novel chaperone of the DnaJ family, is involved in mitochondrial biogenesis and protein folding. *Cell* **77,** 249–259.
32. Westermann, B. and Neupert, W. (2000) Mitochondria-targeted green fluorescent proteins: convenient tools for the study of organelle biogenesis in *Saccharomyces cerevisiae. Yeast* **16,** 1421–1427.

7

Studying Mitochondria in an Attractive Model: *Schizosaccharomyces pombe*

Stéphane Chiron, Mauricette Gaisne, Emmanuelle Guillou, Pascale Belenguer, G. Desmond Clark-Walker, and Nathalie Bonnefoy

Summary

The fission yeast *Schizosaccharomyces pombe*, widely used for studies of cell cycle control and differentiation, provides an alternative and complementary model to the budding yeast *Saccharomyces cerevisiae* for studies of nucleo-mitochondrial interactions. There are striking similarities between *S. pombe* and mammalian cells, in both their respiratory physiology and their mitochondrial genome structure. This technical review briefly lists the general and specific properties that are helpful to know when starting to use fission yeast as a model system for mitochondrial studies. In addition, advice is given for cell growth and genetic techniques, tips for disruption of genes involved in respiration are presented, and a basic differential centrifugation protocol is provided for the isolation of purified mitochondria that are suitable for diverse applications such as subfractionation and in vitro import.

Key Words: Cytochrome spectra; fractionation; gene disruption; mitochondria preparation; petite-negative yeast; respiratory mutants; *Schizosaccharomyces pombe*; *ura4*.

1. Introduction

The fission yeast *Schizosaccharomyces pombe* was the sixth model eukaryotic organism to be completely sequenced. Among the nearly 4900 *S. pombe* genes distributed on three chromosomes, 14% are unique to *S. pombe*, and 3% are common to *Caenorhabditis elegans* but absent in *Saccharomyces cerevisiae* ***(1)***. *S. pombe* and *S. cerevisiae* are actually considered as divergent from each other as either yeast is from higher eukaryotes. The evolutionary position of *S. pombe*

From: *Methods in Molecular Biology, vol. 372: Mitochondria: Practical Protocols*
Edited by: D. Leister and J. M. Herrmann © Humana Press Inc., Totowa, NJ

is discussed in **refs. *2*** and ***3***, and general properties of fission yeast are presented in **refs. *4–7***.

Interest for fission yeast is increasing in the scientific community because of several attractive features and despite some disadvantages compared to the widely used budding yeast. Among the general advantages of *S. pombe* as a model system are the uniformity of the genetic background (because all strains are derived from the same original isolate), the low redundancy level of the genes, and, above all, the occurrence of higher eukaryote-related genes or functions that are absent or strongly diverged in *S. cerevisiae*. Likewise, specific benefits for mitochondriologists come from the similarity to higher eukaryotes' mitochondrial physiology: mitochondrial inheritance is mediated by microtubules like in higher eukaryotes ***(8)***, *S. pombe* mitochondrial deoxyribonucleic acid (mtDNA) is reminiscent of mammalian mtDNA in its small size and structure, and *S. pombe* cells are dependent on respiration for survival (they are *petite* negative; i.e., they cannot tolerate the loss of mtDNA). Petite negativity can be beneficial when working with functions for which deficiency would cause in *S. cerevisiae* a drastic accumulation of deletions or loss of mtDNA.

The general drawbacks of *S. pombe* compared to *S. cerevisiae* are slower growth rate, tedious manipulation of diploids, lack of centromeric plasmids, and low availability of marker genes. In addition, specific technical difficulties for mitochondrial studies include the reduced viability of respiratory mutants (or even lethality in some case), the limited possibility to manipulate the mitochondrial genome, the resistance of *S. pombe* mitochondria to disruption, and the small number of mitochondria-specific antibodies available. Generally, the mitochondrial field is much less documented than for *S. cerevisiae*, which means that specific tools, protocols, and knowledge are often missing and must be inferred from the data available in other systems.

1.1. Schizosaccharomyces pombe *Mitochondrial Genome and Respiratory Physiology*

S. pombe mtDNA structure was thoroughly reviewed in **ref. *9***: it is very compact (19 kb) with a low intron content and encodes cytochrome-*b*; subunits 1–3 of cytochrome oxidase (Cox1–Cox3); subunits 6, 8, and 9 of adenosine triphosphate (ATP) synthase (Atp6, Atp8, Atp9); and a ribosomal protein that might have an additional function in mtDNA binding ***(10)***. Mitochondrial genes for complex I are not present; instead, two nuclear genes coding predicted single-peptide NADH dehydrogenases are found, like in *S. cerevisiae* (similarly, *S. pombe* lacks an alternative oxidase).

In addition, the mtDNA encodes the two large ribosomal ribonucleic acids (RNAs) and a complete set of transfer RNAs, which separate protein-coding genes. Recent messenger RNA (mRNA) mapping studies have revealed that the removal of transfer RNAs from larger transcripts participates in mRNA processing as in human cells (for review, *see* **ref. *11***). The resulting mitochondrial mRNAs contain rather short untranslated regions of 38–220 nt on the 5′ side and 0–15 nt on the 3′ side ***(12)***. These short flanking sequences are reminiscent of the compact human mitochondrial mRNAs, suggesting that the translational mechanisms and control might be rather similar in both organisms. Conforming to this idea is that at least two mitochondrial translation factors are conserved in fission yeast and human but absent in budding yeast ***(13)***.

Another similarity between *S. pombe* and cells from higher eukaryotes is the inability to tolerate the loss of mtDNA, which classifies *S. pombe* as a *petite*-negative yeast ***(14)***. The two-component hypothesis ***(15)*** proposes that the lack of mtDNA, which causes a simultaneous loss of electron transport complexes that translocate protons and of the F_0 part of F_1F_0-ATP synthase, compromises viability because a membrane potential cannot be generated. In this hypothesis, the remaining F_1 part of *S. pombe* ATP synthase lacks ability to hydrolyze ATP to generate adenosine 5′-diphosphate feeding the electrogenic adenosine 5′-diphosphate/ATP exchange translocator ***(15)***. Accordingly, loss of either the electron transporters or the ATP synthase is compatible with survival by allowing the production of a membrane potential, either through direct proton transport by the respiratory complexes or by reversal of the F_1F_0-ATP synthase, resulting in ATP hydrolysis and the export of protons through this complex.

Mutations simultaneously affecting both systems but still allowing the production of a minimal membrane potential are viable (e.g., *Δtsf1*; ***13***) to poorly viable (e.g., *Δoxa1Sp2*; ***16***). Lethality occurs when the two systems generating the membrane potential are too drastically affected or completely lacking, such as in a double *oxa1Sp1-oxa1Sp2* deletion mutant ***(16)***, in a complete mitochondrial translation block (e.g., *Δtuf1*; ***13***), or when the mtDNA is lost (*rho⁰* mutants). However, two unlinked nuclear mutants of *S. pombe*, called *ptp1* and *ptp2*, for *petite* positive ***(17)***, can permit viability of *rho⁰*, *Δtuf1*, and some *mrp1* (RNA component of the ribonuclease MRP) mutants ***(17,18)***. The nature of the *ptp* genes is still unknown. Moreover, at least four other unlinked mutations are able to make *S. pombe* become *petite* positive, suggesting that the genetic basis of petite-negativity might be more complex in *S. pombe* than in *Kluyveromyces lactis* (G. D. Clark-Walker and N. Bonnefoy, 2004, unpublished results). The occurrence of large mtDNA deletions has not been described so far in *S. pombe*; however, a strong depletion of an otherwise wild-type genome has been observed in a mitochondrial translation mutant in a *ptp* background ***(13)***.

Mutations in nuclear genes coding for mitochondrial functions display rather different phenotypes whether they affect the respiratory chain integrity or not. We found, when creating over a dozen respiratory mutants, that respiratory chain defects always prevented the growth not only on glycerol but also on galactose as the main carbon source. This contrasts with the respiratory mutants that are [Gal$^+$] in *S. cerevisiae* and is reminiscent of the [Gal$^-$] phenotype of respiratory-deficient hamster or human cells. As in higher eukaryotes ***(19)***, a possibility is that the Leloir pathway, which permits the utilization of galactose, is not efficient enough in *S. pombe* to draw sufficient energy from galactose for growth in the absence of glucose. Finally, the adenine auxotrophic mutation *ade6*, like *ade2* for *S. cerevisiae*, can be used to screen respiratory-deficient mutants because, under limiting adenine conditions, *S. pombe ade6* mutants accumulate a red pigment only if strains are respiratory competent.

1.2. Schizosaccharomyces pombe *Genetics*

S. pombe is naturally a homothallic organism, but heterothallic haploid cells with stabilized mating types (h– or h+) are available for genetic studies. Mating of haploids is activated under starvation conditions, like sporulation of diploids. Thus, diploid cells tend to sporulate readily after the cross. To prevent sporulation, diploid cells must be selected or rapidly transferred onto rich medium devoid of peptone. Because mitotic recombination occurs at a high rate, selection of diploids is based on the complementation of intragenic mutations, like the *ade6-M210* and *ade6-M216* mutations. Stable h+/h+ or h–/h– diploids are also available and can sporulate at a low rate. The selection of diploid cells prior to sporulation can become important when determining the mitochondrial or nuclear nature of a mutation by genetic segregation because the mixing of mitochondria in *S. pombe* is a slow process compared to sporulation. Thus, if microdissection of asci is conducted directly from a cross without selection of the diploid cells, then mitochondrial mutations will often display a surprising 2:2 segregation. This can also represent a way to propagate a mitochondrial mutation in different nuclear backgrounds. Finally, tetrad dissections are performed on plates containing 5% glucose, which improves the germination of respiratory-deficient spores significantly, and without prior digestion with enzymes, because the ascus cell wall is spontaneously lysed on incubation at 30–37°C. To complete this summary, all the basic techniques for *S. pombe* genetic manipulation can be found in **ref. *19***, and advice and protocols can be found at http://www-rcf.usc.edu/~forsburg/pombeweb.html. It should be noted that, because of the instability of diploids, all gene symbols are written in lowercase because recessivity and dominance cannot be assessed easily.

Mitochondrial genetics is less advanced, and currently cytoductions and transformation of *S. pombe* mitochondria are not available, but setting these up will be facilitated by the small size of the mtDNA and by further knowledge on the molecular basis of petite-negativity. However, an interesting tool is provided by the *mutator* strain, which carries a mutation in the dual-function mitochondrial ribosomal protein gene *rps3/urfa*, carried on the mtDNA. This mutation

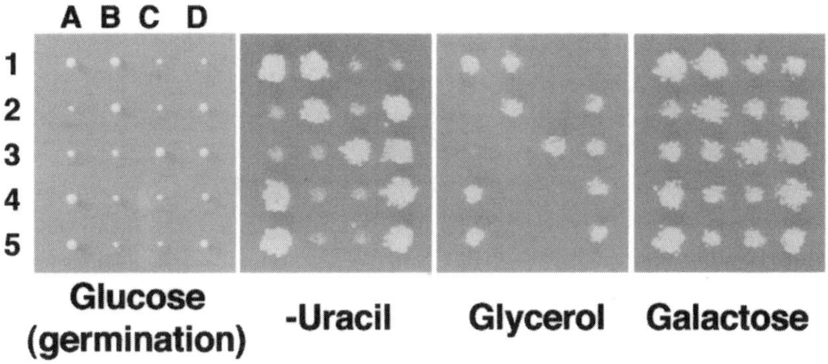

Fig. 1. Effect of the widely used *ura4* mutation on glycerol growth. [Ura^+] *S. pombe* cells were crossed to *ura4-D1.8* cells, which are auxotroph for uracil. Tetrads were dissected directly from the sporulation mixture and germinated on complete medium containing 5% (w/v) glucose to improve germination of the spores. Colonies were replica plated on minimal medium lacking uracil, as well as complete glycerol or galactose media (*see* **Subheading 2.1.**).

has allowed the isolation of numerous deletion and point mutations in *S. pombe* mtDNA ***(20)***.

1.3. Respiratory Gene Disruption in Schizosaccharomyces pombe

Homologous gene disruption is available in *S. pombe* ***(21)*** but must be carried out with some caution when generating respiratory-deficient mutants because of their low viability. First, only a small number of marker genes are available. A single copy of *S. cerevisiae LEU2* can (slowly) complement the *leu1-32* mutation, whereas high copy, but not single copy, of *S. cerevisiae URA3* can complement the *ura4* mutation. In addition, the commonly used *ura4* mutation should be used parsimoniously for mitochondria-related studies despite its convenient ability to confer 5′-fluoro-orotic acid resistance because it leads to drastic growth decrease on glycerol (but not on galactose; *see* **Fig. 1**). Thus, bona fide *S. pombe* markers like *ade6*, *his3*, *his7*, and *arg3* are recommended for disruption, or even better, antibiotic resistance genes like the Kan^R gene. Kan^R is a marker of choice for the disruption of *S. pombe* respiratory genes because selection is performed on complete glucose medium supplemented with the antibiotic G418. Respiratory-deficient [Gal^-] disruptants that are highly counterselected on minimal medium are significantly more easily recovered on complete medium. Other antibiotics are used for *S. pombe*, also on complete medium (http://www.biotwiki.org/twiki/bin/view/Main/NewMarkers).

Second, the usual high-efficiency transformation protocol for gene disruption is a lithium acetate-based technique as described in **ref.** ***13)***. We recommend growing the recipient strain in complete glucose medium containing adenine before harvesting the cells for transformation, even though some *S. pombe* protocols suggest growing the cells in minimal medium with low glucose to improve the transformation efficiency. We found that growth in stringent minimal medium yields more transformants but decreases the ratio of [Gal$^-$] disruptants up to six- to sevenfold. Dimethyl sulfoxide is also used in some protocols, but in our hands did not significantly improve the transformation or disruption efficiency. However, single-stranded carrier DNA, often lacking in common *S. pombe* transformation protocols, is highly beneficial in terms of efficiency. Last, an easy way to substantially improve the recovery of [Gal$^-$] clones is to plate the transformation mixture on medium containing 5% glucose instead of the usual 2% (*see* **Fig. 2**).

S. pombe is prone to ectopic recombination. Thus, fragments with large homology regions are best, but polymerase chain reaction fragments generated with oligonucleotides containing 75–80 bases of homology with the recipient *locus* on both sides give a quick and very reasonable result, especially if a phenotype facilitates the screening (**Fig. 2**). In addition, the candidate clones must be tested on both sides using primers external to the transforming fragment, and it is also wise to verify whether there are insertions additional to the bona fide disruption. Genomic DNA of the transformants is prepared according to **ref.** ***22***.

1.4. Molecular, Cytological, and Biochemical Tools

For molecular biology studies, a large range of *S. pombe* high-copy plasmids allows expression of genes, shuffling experiments, and gap repair ***(23)***. However, no single-copy plasmids are available because of the large size of the centromeres. Expression plasmids with both constitutive and regulatable promoters (e.g., *nmt1* repressible by thiamine) must generally be used for cross-complementation studies because the fission yeast transcriptional machinery does not always recognize the transcriptional signals from other species and vice versa. For expression of *S. pombe* genes in other species, complementary DNA must often be used because 43% of the *S. pombe* genes carry (generally short) introns. Transformations of single plasmids can be performed according to the one-step method designed for *S. cerevisiae* ***(24)*** using a more acidic pH 4.9 lithium acetate solution.

Plasmids containing CoxIV-green fluorescent protein (GFP) fusions have been devised ***(25)*** for mitochondrial staining, and cytological studies can generally be carried out in *S. pombe* (**Fig. 3**) using common *S. cerevisiae* protocols ***(26)*** (*see also* Chapter 31), with a few precautions. When inducing a gene under the control of the *nmt1* promoter (e.g., a GFP fusion), cells should not be grown for more than 18 h at 25°C in minimal medium devoid of thiamine to avoid alterations of the mitochondrial network. Similarly, cells should be fixed with

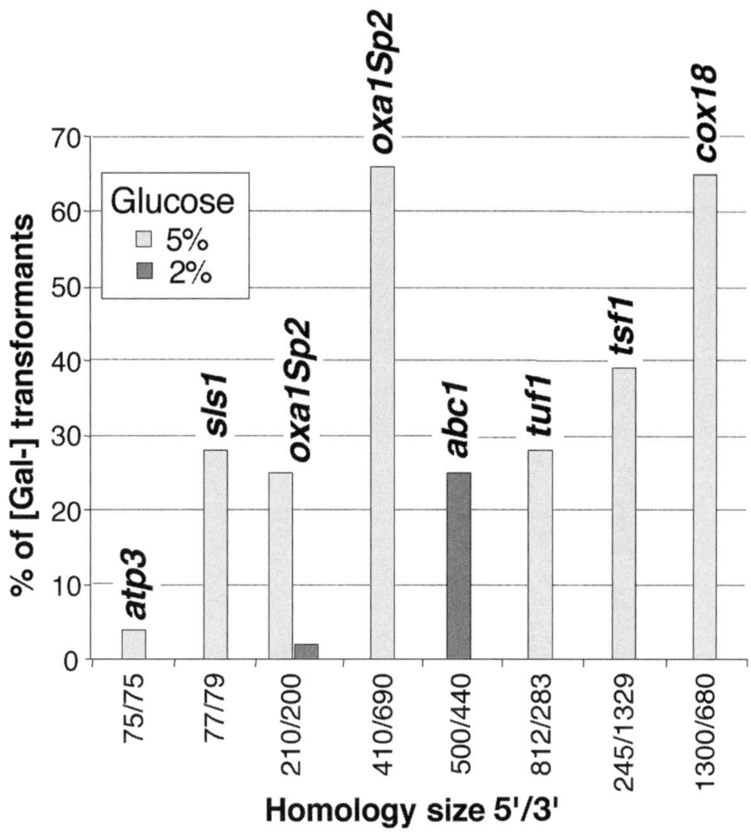

Fig. 2. Levels of disruption for respiratory genes depending on the homology size available for recombination. The levels are the percentage of [Gal⁻] clones among the transformants tested (generally 50–200). In each case, several [Gal⁻] clones were confirmed to contain the expected deletion. The homology sizes are indicated in basepairs for both sides (5′/3′) of each of the seven target genes presented.

formaldehyde for a maximum of 10 min. In addition, MitoTracker Red can be used for transient visualization of mitochondria in living cells (**Fig. 3B**), but fixation of the cells generates, in our hands, a high red background that prevents observation. Extended excitation of the dye modifies the mitochondrial network.

S. pombe is also amenable to mitochondrial biochemical studies. Low-temperature cytochrome spectra can be carried out on whole cells frozen in liquid nitrogen. Spectra differ clearly from those recorded for *S. cerevisiae* because the cytochrome-*c* peak shows a shoulder at 544 nm. However, there is only one cytochrome-*c* gene in *S. pombe*, and both the main peak and its shoulder disappear in a cytochrome-*c* mutant (N. Bonnefoy, 2001, unpublished results;

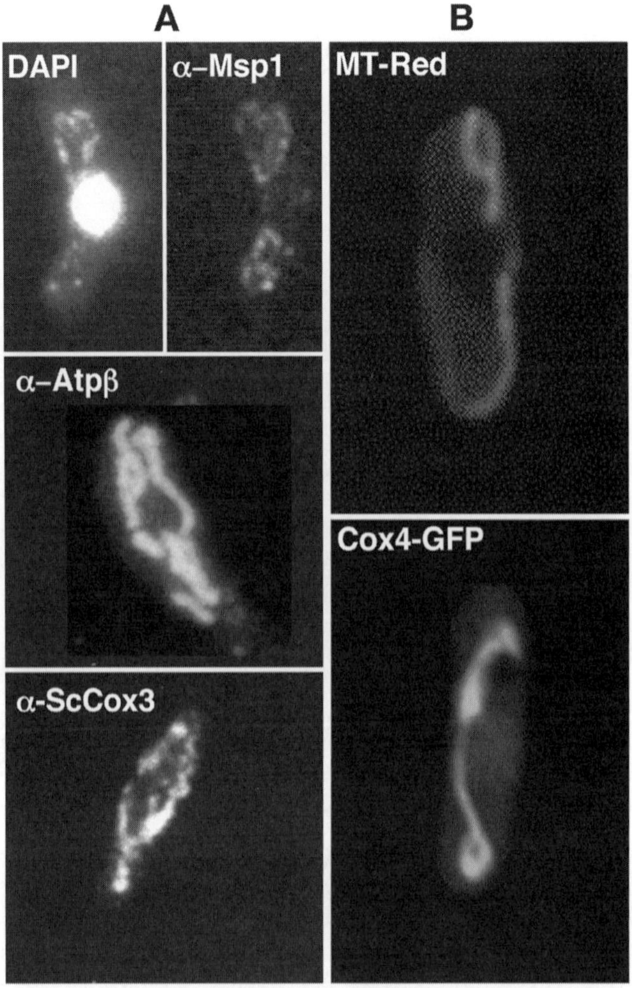

Fig. 3. Visualization of the *S. pombe* mitochondrial network. (**A**) DNA labeling with DAPI and indirect immunofluorescence using antibodies recognizing Msp1p *(28)*, Atpβ (a gift from M. Boutry), or the *S. cerevisiae* Cox3p (Molecular Probes). Note that this monoclonal anti-*S. cerevisiae* Cox3p antibody does not recognize the *S. pombe* Cox3p in Western blots. (**B**) Fluorescence microscopy of *S. pombe* cells stained with MitoTracker Red CMXRos (MT-Red, Molecular Probes) or containing a GFP targeted into mitochondria (Cox4-GFP; *25*).

see **Fig. 4**). In addition, mitochondria can be purified (*see* **Subheading 3., Fig. 5**) and used for fractionation, in vitro import, enzyme activity, and oxygen consumption measurements.

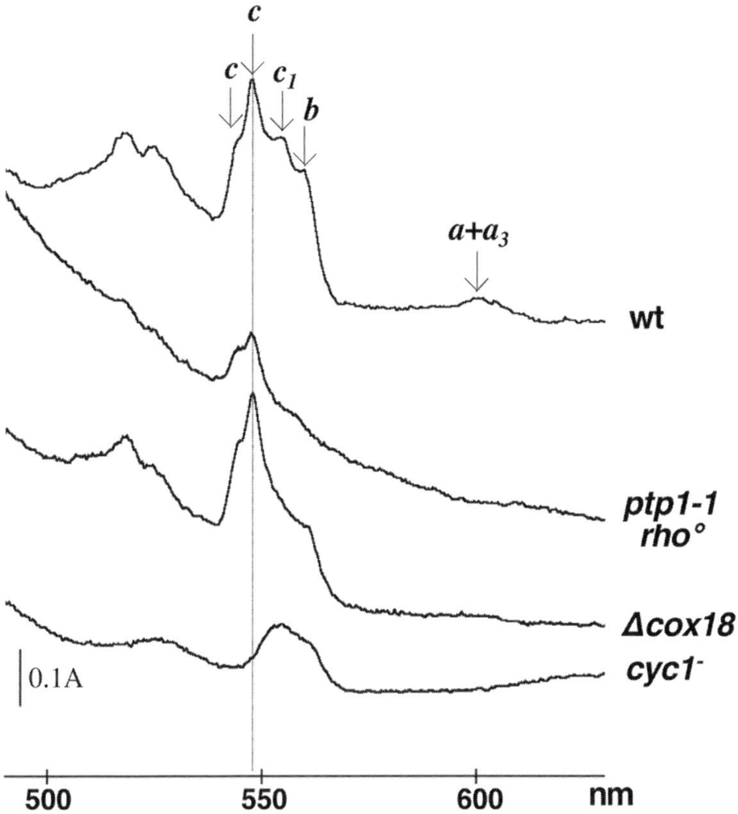

Fig. 4. Whole cell cytochrome spectra of various *S. pombe* wild-type or mutant strains. *S. pombe* cells were grown on complete glucose plates, harvested, and dried between two paper filters. These cell pastes were frozen in liquid nitrogen after addition of sodium dithionite, which fully reduces the cytochromes, and directly used to record cytochrome spectra from 630 to 490 nm using a Cary400 spectrophotometer as described in **ref. *30***. Peak maxima for *S. pombe* are 603, 560, 554, and 548/544 nm for cytochrome-*aa3*, *b*, *c1*, and *c*, respectively.

2. Materials

2.1. Cell Growth and Purification of Mitochondria

2.1.1. Growth Media for Schizosaccharomyces pombe

1. Glucose medium (YPGA): 1% (w/v) yeast extract, 2% (w/v) peptone, 2% (w/v) glucose, 50 µg/mL adenine. If peptone is omitted (e.g., for diploids), then supplements must be at 200 µg/mL.
2. Nonfermentable medium is YPGalA (*see* **Subheading 1.1.**): 1% (w/v) yeast extract, 2% (w/v) peptone, 2% (w/v) galactose, 0.1% (w/v) glucose, 50 µg/mL adenine. Glycerol can also replace galactose.

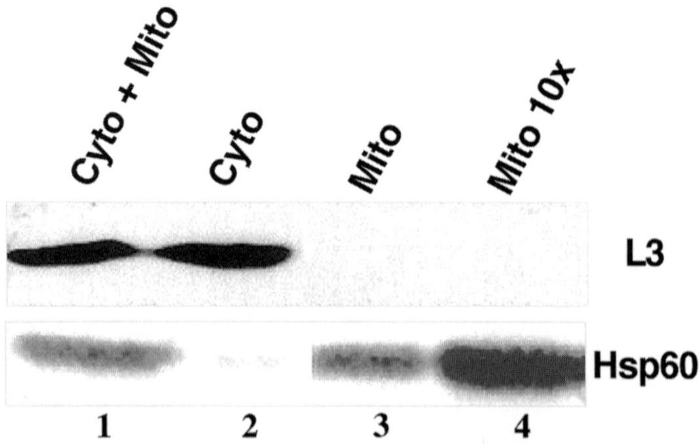

Fig. 5. Purification of *S. pombe* mitochondria. Postnuclear (Cyto + Mito, **step 14**), cytosolic (Cyto, **step 16**), or mitochondrial (Mito, **step 21**) fractions were analyzed using antibodies recognizing the human Hsp60 (Sigma) or the *S. pombe* cytosolic ribosomal protein L3 (a gift from J. R. Warner). The postnuclear fraction in lane 1 contains 100 μg protein, lanes 2 and 3 correspond to an equivalent number of cells, and lane 4 corresponds to 10 times more cells.

3. Minimal medium is synthetic defined (SD) (yeast nitrogen base without amino acids, 2% (w/v) glucose with appropriate supplements at 20–50 μg/mL) or Edinburgh Minimal Medium (EMM) (Qbiogen; *see* http://www-rcf.usc.edu/~forsburg/ pombeweb.html).
4. Mating/sporulation medium is ME: 3% (w/v) Bacto™ malt extract with appropriate supplements at 10 μg/mL, pH 5.5.

2.1.2. Buffers for Preparation and Fractionation of Mitochondria

1. Distilled water.
2. 10 m*M* EDTA (ethylenediaminetetraacetic acid).
3. Digestion buffer: 1.2 *M* sorbitol, 10 m*M* sodium citrate, pH 5.8, 0.2 m*M* EDTA.
4. 98% β-Mercaptoethanol (Merck).
5. Zymolyase 100T (Seikagaku Co.).
6. Lytic enzyme from *Trichoderma harzianum* (Sigma).
7. Lysis buffer: 0.6 *M* sorbitol, 10 m*M* imidazol-HCl, pH 6.4, 2 m*M* EDTA (*see* **Note 1**).
8. BSA (bovine serum albumin), <0.02% fatty acid (Sigma A7030).
9. PMSF (phenylmethylsulfonyl fluoride), freshly made 0.1 *M* stock solution in ethanol.
10. Protease inhibitor tablets (Roche 11873580001).
11. Homogenizer with tight glass pestle (Wheaton).

2.2. Subfractionation of Mitochondria

1. Swelling buffer: 10 m*M* imidazol-HCl, pH 6.4 (*see* **Note 1**).
2. Sonication buffer: 0.06 *M* sorbitol, 10 m*M* imidazol-HCl, pH 6.4, 2 m*M* EDTA (*see* **Note 1**).

3. Methods

3.1. Cell Growth and Purification of Mitochondria

1. Grow the cells in 1 L YPGA medium up to 2.5×10^7 cells/mL (*see* **Note 2**).
2. Collect the cells by a 10-min centrifugation at 2000*g* at room temperature.
3. Discard the supernatant and wash the cell pellet with 200 mL distilled water. Centrifuge 10 min at 2000*g* at room temperature.
4. Resuspend the pellet in 50 mL 10 m*M* EDTA, transfer in smaller preweighted tubes, and centrifuge 10 min at 2000*g* at room temperature.
5. Discard the supernatant, weigh the tube to determine the wet weight of the cell pellet, and resuspend at 3 mL/g of cells (or 2×10^9 cells/mL) in digestion buffer, freshly supplemented with 0.3% (v/v) β-mercaptoethanol.
6. Add 1 mg/mL Zymolyase 100T and 1 mg/mL *T. harzianum* lytic enzyme (*see* **Note 3**).
7. Incubate 30 min at 37°C under gentle shaking to generate protoplasts (*see* **Note 4**).
8. Transfer the tubes at 4°C to stop the digestion (*see* **Note 5**). All subsequent steps must be conducted on ice, and ice-cold buffers must be used.
9. Centrifuge 15 min at 2000*g* and 4°C. Discard the supernatant (*see* **Note 6**).
10. Resuspend the pellet in 15–20 mL lysis buffer (*see* **Note 7**). Break the protoplasts by pipeting 10 times with a 10-mL pipet or with 10 strokes of a glass homogenizer.
11. Incubate 15 min at 4°C.
12. Centrifuge 15 min at 2500*g* and 4°C.
13. Transfer the supernatant to a new tube; discard the pellet that contains unbroken cells (*see* **Note 8**), cell debris, and nuclei; and spin again 5 min at 2500*g* and 4°C.
14. Transfer the supernatant to a fresh tube and discard the pellet. Repeat **steps 13** and **14** if the pellet was disturbed when decanting the supernatant.
15. Centrifuge the supernatant 15 min at 12,000*g* and 4°C (*see* **Note 9**).
16. Keep a sample from the supernatant, which corresponds to the postmitochondrial fraction (*see* **Note 10**).
17. Resuspend the mitochondrial pellet with 2 mL lysis buffer supplemented with 0.5% (w/v) BSA (*see* **Note 11**).
18. Transfer in an Eppendorf tube and spin 2 min at 800*g* and 4°C.
19. Transfer the supernatant to a fresh tube and spin 15 min at 12,000*g* and 4°C. Discard the supernatant and eliminate the floating lipids, if any (a paper tissue can be used to clean the tube).
20. Repeat **steps 17–19** twice (*see* **Note 11**).
21. Resuspend the mitochondrial pellet in 100 μL lysis buffer supplemented with 0.5% (w/v) BSA. The pellet has a light pinkish color because of cytochromes. Measure the protein concentration (the typical yield is around 300 μL at 20 mg/mL for 1 L of culture).
22. Transfer aliquots in tubes, freeze in liquid nitrogen, and store at –70°C. Alternatively, 5- or 10-μL drops of mitochondria can be frozen directly in liquid nitrogen and pooled in a tube stored at –70°C.

3.2. Subfractionation of Mitochondria

1. Add 450 μL swelling buffer to 50 μL mitochondria adjusted to 10 μg/μL.
2. Incubate 30 min at 4°C, vortexing every 5 min.
3. Centrifuge 10 min at 12,000*g* and 4°C; the supernatant corresponds to the intermembrane space fraction.
4. Resuspend the pellet, corresponding to mitoplasts and unlysed mitochondria, in 400 μL sonication buffer.
5. Sonicate the mitochondrial extract (for a Bioblock 130-W sonicator 75186, perform four 10-s pulses at 1-s intervals and 70% amplitude, i.e., 23 J delivered). The lysate should become clear.
6. Spin 10 min at 12,000*g* at 4°C and discard the pellet, which contains unlysed mitochondria.
7. Centrifuge the supernatant 20 min at 100,000*g* and 4°C; the supernatant corresponds to the matrix fraction, and the pellet corresponds to internal and external membranes (*see* **Notes 12** and **13**).

4. Notes

1. Imidazol-HCl can be replaced by other buffering systems.
2. Because mutants affecting the respiratory chain are [Gal$^-$], mitochondria are extracted from glucose-grown cells, but preparation can also be done from galactose-grown respiring cells or from transformants grown in minimal medium to select for a plasmid. Because *S. pombe* cells carry out only a few divisions under anaerobic conditions, the cultures must be sufficiently oxygenated (i.e., 1 L of culture can be grown in a 5-L flask). In addition, take care to harvest cells at the beginning of the exponential phase because overgrown cells will be less efficiently digested to give protoplasts, which will reduce the final yield of mitochondria. Because respiratory mutants and wild type display very different doubling times and cell maxima, it is best to record the growth curve of a given strain before inoculating for the preparation of mitochondria, knowing that upscaling a culture will generally increase its doubling time. Typically, inoculate around 3 mL freshly grown culture (1–2 × 10^8 cells/mL) of wild-type *S. pombe* 972h$^-$ in 1 L medium and grow overnight for 14 h. Use a larger volume to inoculate a wild-type strain carrying auxotrophies (around 10 mL/L) or a respiratory mutant (up to 50–80 mL/L) and grow the cells up to 20 h.
3. For cost reasons, Zymolyase can be omitted to perform the digestion only with lytic enzymes. The incubation time must be increased, up to 90 min.
4. The production of protoplasts can be controlled during the incubation by transferring 10-μL aliquots to two tubes containing 1 mL of either digestion buffer or water. Generally, a lysis of 90% is obtained in water within a few minutes, as determined by measuring the OD_{600} of both samples. Lysis can also be assessed in 0.3 *M* sorbitol and typically reaches 75%. The protoplast formation and breakup can also be analyzed using a microscope.
5. We generally omit the usual washes of the pellet protoplast with 1.2 *M* sorbitol because they often cause premature breaking of the protoplast. The enzymes are washed away during subsequent cycles of centrifugation.

6. Optional: if the supernatant obtained after centrifugation of the digestion mixture is very cloudy, then it may contain a large amount of mitochondria. Keep it on ice and pool with the supernatant obtained after lysis at **step 12**.
7. If working with proteins sensitive to degradation, then the protoplast pellet should be resuspended in lysis buffer supplemented with 1 m*M* PMSF and protease inhibitor tablets (Roche) before lysis.
8. Repeating **steps 10–12** and pooling both supernatants not only almost doubles the yield, but also can damage the mitochondrial membrane.
9. Repeat **step 15** to recover more mitochondria and increase the yield.
10. In addition, aliquots corresponding to total protoplasts and to the mitochondria-containing cytoplasm can be taken at **steps 10** and **14**, respectively, and adjusted to a constant cell number.
11. We usually perform three cycles of differential centrifugation washes, steps that have proved in *S. cerevisiae* to yield high-purity mitochondria as analyzed by electron microscopy *(27)*. However, these washes also reduce the yield significantly, and **steps 21–24** can be replaced by a single cycle of differential centrifugation washes under more diluted conditions (30 mL). When using the mitochondria strictly for localization of a protein (and not for any functional test), one of the washes can be performed using lysis buffer supplemented with 0.2 *M* KCl to remove cytoplasmic proteins sticking to the mitochondrial surface.
12. The *S. pombe* mitochondria are difficult to break; thus, the purity of each fraction needs to be analyzed thoroughly with control antibodies. Crossreaction with antibodies recognizing mitochondrial proteins from other species is unpredictable but generally low, and a limited number of antisera have been raised against *S. pombe* proteins. Available *S. pombe* antibodies in our labs include an anti-Cox2p polyclonal antipeptide *(13)* and an anti-Msp1p polyclonal antibody *(28)* for the inner membrane and an anti-Tuf1p antibody that recognizes the mtEF-Tu translation elongation factor in the matrix *(13)*. We also use the anti-Arg8p serum *(29)* and the anti-human hsp60 monoclonal antibody (clone LK2, Sigma), which recognize the *S. pombe* equivalents in the matrix. In addition, c-Myc-tagged version of the etch virus protease fused to presequences of *S. cerevisiae* cytochrome-b_2 or *Neurospora crassa* Atp9 can be used directly as compartment markers or to direct digestion of a target protein (containing the protease cleavage site) in a specific compartment *(25)*. Finally, generation of antipeptides recognizing Dnm1p, Fzo1p, cytochrome-*c*, Anc1p, and Atpβ is under way in our laboratories.
13. Protease protection experiments to localize protease-sensitive proteins can be conducted as in other systems by treatment of total, swelled, or sonicated mitochondria for 30 min on ice with 100 μg/mL proteinase K compared to an untreated control. Stop the digestion with 2 m*M* of freshly made PMSF.

Acknowledgments

We thank L. Pelloquin for invaluable contribution to the development of mitochondrial studies on *S. pombe* in P. B.'s laboratory; M. Boutry and J. R. Warner for antibodies; and G. Dujardin for interesting discussions. S. C. and

E. G. were fellows from the Ministère de la Recherche et la Technologie and from the Association pour la Recherche sur le Cancer, respectively. This work has been supported by grants from the Association Française contre les Myopathies to P. B. and N. B.

References

1. Wood, V. (2005) *Schizosaccharomyces pombe* comparative genomics, from sequence to systems, in *Comparative Genomics; Using fungi as models* (Sunnerhagen, P., and Piskur, J., eds.), Springer-Verlag, Heidelberg, pp. 233–235.
2. Sipiczki, M. (2000) Where does fission yeast sit on the tree of life? *Genome Biol.* **1,** REVIEWS1011.
3. Lenaers, G., Pelloquin, L., Olichon, A., et al. (2002) What similarity between human and fission yeast proteins is required for orthology? *Yeast* **19,** 1125–1126.
4. Zhao, Y., and Lieberman, H. B. (1995) *Schizosaccharomyces pombe*: a model for molecular studies of eukaryotic genes. *DNA Cell Biol.* **14,** 359–371.
5. Forsburg, S. L. (1999) The best yeast? *Trends Genet.* **15,** 340–344.
6. Gomez, E. B., Bailis, J. M., and Forsburg, S. L. (2002) Fission yeast enters a joyful new era. *Genome Biol.* **3,** REPORTS4017.
7. Yanagida, M. (2002) The model unicellular eukaryote, *Schizosaccharomyces pombe*. *Genome Biol.* **3**, COMMENT2003.
8. Weir, B. A., and Yaffe, M. P. (2004) Mmd1p, a novel, conserved protein essential for normal mitochondrial morphology and distribution in the fission yeast *Schizosaccharomyces pombe*. *Mol. Biol. Cell* **15,** 1656–1665.
9. Schäfer, B. (2003) Genetic conservation vs variability in mitochondria: the architecture of the mitochondrial genome in the petite-negative yeast *Schizosaccharomyces pombe*. *Curr. Genet.* **43,** 311–326.
10. Neu, R., Goffart, S., Wolf, K., and Schafer, B. (1998) Relocation of *urf a* from the mitochondrion to the nucleus cures the mitochondrial mutator phenotype in the fission yeast *Schizosaccharomyces pombe*. *Mol. Gen. Genet.* **258,** 389–396.
11. Schäfer, B. (2005) RNA maturation in mitochondria of *S. cerevisiae* and *S. pombe*. *Gene* **354,** 80–85.
12. Schäfer, B., Hansen, M., and Lang, B. F. (2005) Transcription and RNA-processing in fission yeast mitochondria. *RNA* **11,** 785–795.
13. Chiron, S., Suleau, A., and Bonnefoy, N. (2005) Mitochondrial translation: elongation factor tu is essential in fission yeast and depends on an exchange factor conserved in humans but not in budding yeast. *Genetics* **169,** 1891–1901.
14. Bulder, C. J. (1964) Induction of *petite* mutation and inhibition of synthesis of respiratory enzymes in various yeasts. *Antonie Van Leeuwenhoek* **30,** 1–9.
15. Clark-Walker, G. D., and Chen, X. J. (2001) Dual mutations reveal interactions between components of oxidative phosphorylation in *Kluyveromyces lactis*. *Genetics* **159,** 929–938.
16. Bonnefoy, N., Kermorgant, M., Groudinsky, O., and Dujardin, G. (2000) The respiratory gene *OXA1* has two fission yeast orthologues which together encode a function essential for cellular viability. *Mol. Microbiol.* **35,** 1135–1145.

17. Haffter, P., and Fox, T. D. (1992) Nuclear mutations in the petite-negative yeast *Schizosaccharomyces pombe* allow growth of cells lacking mitochondrial DNA. *Genetics* **131,** 255–260.
18. Paluh, J. L., and Clayton, D. A. (1996) A functional dominant mutation in *Schizosaccharomyces pombe* RNase MRP RNA affects nuclear RNA processing and requires the mitochondrial-associated nuclear mutation *ptp1-1* for viability. *EMBO J.* **15,** 4723–4733.
19. Seo, B. B., Kitajima-Ihara, T., Chan, E. K., Scheffler, I. E., Matsuno-Yagi, A., and Yagi, T. (1998) Molecular remedy of complex I defects: rotenone-insensitive internal NADH-quinone oxidoreductase of *Saccharomyces cerevisiae* mitochondria restores the NADH oxidase activity of complex I-deficient mammalian cells. *Proc. Natl. Acad. Sci. USA* **95,** 9167–9171.
20. Ahne, F., Merlos-Lange, A. M., Lang, B. F., and Wolf, K. (1984) The mitochondrial genome of the fission yeast *Schizosaccharomyces pombe* 5. Characterization of mitochondrial deletion mutants. *Curr. Genet.* **8,** 517–524.
21. Klinner, U., and Schäfer, B. (2004) Genetic aspects of targeted insertion mutagenesis in yeasts. *FEMS Microbiol. Rev.* **28,** 201–223.
22. Hoffman, C. S. and Winston, F. (1987) A 10-min DNA preparation from yeast efficiently releases autonomous plasmids for transformation of *Escherichia coli*. *Gene* **57,** 267–272.
23. Siam, R., Dolan, W. P., and Forsburg, S. L. (2004) Choosing and using *Schizosaccharomyces pombe* plasmids. *Methods* **33,** 189–198.
24. Chen, D. C., Yang, B. C., and Kuo, T. T. (1992) One-step transformation of yeast in stationary phase. *Curr. Genet.* **21,** 83–84.
25. Guillou, E., Bousquet, C., Daloyau, M., Emorine, L. J., and Belenguer, P. (2005) Msp1p is an intermembrane space dynamin-related protein that mediates mitochondrial fusion in a Dnm1p-dependent manner in *S. pombe*. *FEBS Lett.* **579,** 1109–1116.
26. Pelloquin, L., Belenguer, P., Menon, Y., Gas, N., and Ducommun, B. (1999) Fission yeast Msp1 is a mitochondrial dynamin-related protein. *J. Cell Sci.* **112** (Pt 22), 4151–4161.
27. Pflieger, D., Le Caer, J. P., Lemaire, C., Bernard, B. A., Dujardin, G., and Rossier, J. (2002) Systematic identification of mitochondrial proteins by LC-MS/MS. *Anal. Chem.* **74,** 2400–2406.
28. Pelloquin, L., Belenguer, P., Menon, Y., and Ducommun, B. (1998) Identification of a fission yeast dynamin-related protein involved in mitochondrial DNA maintenance. *Biochem. Biophys. Res. Commun.* **251,** 720–726.
29. Steele, D. F., Butler, C. A., and Fox, T. D. (1996) Expression of a recoded nuclear gene inserted into yeast mitochondrial DNA is limited by mRNA-specific translational activation. *Proc. Natl. Acad. Sci. USA* **93,** 5253–5257.
30. Claisse, M. L., Pere-Aubert, G. A., Clavilier, L. P., and Slonimski, P. P. (1970) Method for the determination of cytochrome concentrations in whole yeast cells. *Eur. J. Biochem.* **16,** 430–438.

8

Neurospora crassa as a Model Organism for Mitochondrial Biogenesis

Frank E. Nargang and Doron Rapaport

Summary

Neurospora crassa has proven to be an excellent organism for studying various aspects of the biology of mitochondria by biochemical and genetic approaches. As *N. crassa* is an obligate aerobe and contains complex I, its mitochondria are more similar to mammalian mitochondria than those of yeast. The recent sequencing of the genome of *N. crassa* and a gene knockout project that is under way make the organism even more attractive. We describe some of the advantages of *N. crassa* as a model organism and present methods for isolation of mitochondria, fractionation of these organelles, and disruption of essential genes in this organism.

Key Words: Digitonin fractionation; *Neurospora crassa*; sheltered disruption.

1. Introduction

Neurospora crassa is a filamentous ascomycete that was originally described in 1843 as a contaminant that infected French bakeries. The fungus was developed as a laboratory organism in the 1920s and gained considerable fame as the organism used by Beadle and Tatum in their studies on the relationship between genes and proteins that resulted in a Nobel Prize *(1)*. Past reviews are available with detailed information for the growth and handling of the organism, descriptions of the many mutations that have been isolated, and the usefulness of the organism in past and present-day science *(2–5)*. Working with *N. crassa* is facilitated by resources at the Fungal Genetics Stock Center (FGSC), which houses and supplies a large number of mutant strains, many natural isolates of various *Neurospora* species, as well as molecular tools such as complementary deoxyribonucleic acid (cDNA) and cosmid libraries. The FGSC Web site at http://www.fgsc.net/ and the Neurospora home page at http://www.fgsc.net/Neurospora/index.htmL contain information on

From: *Methods in Molecular Biology, vol. 372: Mitochondria: Practical Protocols*
Edited by: D. Leister and J. M. Herrmann © Humana Press Inc., Totowa, NJ

techniques and links to genome sequences. For these reasons and others described below, the organism continues to be a useful model for studies in genetics, biochemistry, cell biology, development, and population biology.

Neurospora crassa can be easily and cheaply grown in the laboratory as it requires only carbon and nitrogen sources, other simple salts, trace elements, and the vitamin biotin. Vegetative cultures consist of branched filaments called hyphae, which grow from their tips and frequently fuse with each other. Collectively, the hyphal system is referred to as the *mycelium*. Hyphae contain many haploid nuclei and are subdivided into compartments by crosswalls called *septa*. These compartments are not truly analogous to cells because the septa are not complete. Each compartment may contain multiple nuclei, and movement of organelles occurs between compartments through pores in the septa. Strains containing genetically identical nuclei are called *homokaryons*. Genetically non-identical strains can fuse to form a heterokaryon in which the different haploid nuclei are maintained in a shared cytoplasm. Formation of heterokaryons requires that strains be of the same mating type and that they contain similar alleles at heterokaryon incompatibility loci. Because there is no vegetative diploid phase in the life cycle of the organism, heterokaryons are used for complementation studies and for sheltering mutations in essential genes that can be introduced into one nucleus if there is a wild-type copy in the other (*see* **Subheading 3.6.**).

Subculturing is accomplished by transfer of asexual spores, called *conidia* or *conidiaspores*, to fresh medium. Large liquid cultures for experimental purposes, such as isolation of mitochondria, are also started by inoculation with conidia. Conidia are produced in abundance on aerial hyphae that arise from mycelium growing on the surface of solid medium. Long-term storage of strains can be accomplished by simply freezing agar-containing slants on which conidia have been produced or by storing conidia in vials containing desiccated silica gel granules. Genetic crosses are easily done using strains of opposite mating type (*A* and *a*). The generation time is short, and ascospores, the products of crosses, are large and easily isolated. Transformation can be accomplished by a number of techniques, and many resistance markers for selection of transformants are available.

The genome of *N. crassa* has been sequenced ***(6,7)***, and a gene knockout project is now under way (http://www.dartmouth.edu/~neurosporagenome/). The genome has been estimated to contain about 10,000 protein-coding genes, which is considerably more than the number predicted for *Saccharomyces cerevisiae* or *Schizosaccharomyces pombe* and approaches the 14,000 predicted for *Drosophila melanogaster* ***(6)***. Thus, the organism may be a better model for more complex organisms than the yeasts.

The limitations of the organism are usually expressed in terms of what is lacking in comparison to yeast. For example, the lack of a vector that replicates in the nucleus of the organism means that all stable transformants must be the result of integration into the genome, and identification of multicopy suppressors from cloned libraries cannot be achieved. For many years, the inability to generate knockouts easily at target loci was probably the major difficulty. However, two developments have virtually eliminated this problem. First, using a split resistance marker approach, a high frequency of knockouts was obtained if the flanking sequences on each component of the split marker were about 3 kb long ***(8)***. Second, transformation of strains carrying mutations in the *mus-51* or *mus-52* genes, which encode proteins that function in nonhomologous end-joining of double-stranded DNA breaks, resulted in virtually 100% homologous replacements when the flanking sequences were about 1 kb long ***(9)***. Time and effort can also be saved by having the recombination machinery in *S. cerevisiae* assemble the knockout cassettes rather than having to clone the appropriate fragments in vitro ***(8)***.

Harnessing the natural phenomenon of repeat induced point mutation (RIP) is another means of generating essentially null mutants of a target gene ***(10)*** and has been used to make several mutants in genes encoding mitochondrial proteins ***(11–13)***. RIP depends on the presence of an artificially created duplication of the target gene in one nucleus of an *N. crassa* strain. When taken through a sexual cross, both members of the duplication are subject to RIP, which results in frequent GC-to-AT transitions, methylation, and transcriptional silencing ***(10,14)***. The popularity of this technique will undoubtedly fade with the development of the knockout procedures.

The mitochondrial genome of *Neurospora* is a 64,840-bp circle. Many strains isolated from the wild have also been found to contain mitochondrial plasmids ***(15)***. The mitochondrial DNA (mtDNA) contains two ribosomal ribonucleic acids (RNAs), at least 27 transfer RNAs, and 26 open reading frames ***(15)***. Of these open reading frames, 14 encode subunits of the inner membrane respiratory complexes. Initial proteome studies with *N. crassa* mitochondria have identified a set of 251 proteins as mitochondrial ***(15a)***. The number of proteins predicted to make up *N. crassa* mitochondria has been estimated to be as high as 2000–2200 based on analysis of the genome sequence ***(7)***.

Neurospora crassa is an obligate aerobe so that no mtDNA mutants corresponding to the rho^- or rho^0 mutants of yeast have been isolated. Similarly, point mutations in mitochondrial or nuclear genes that result in complete loss of respiration would be expected to be inviable. Nonetheless, several mutants affecting respiratory components encoded on both mtDNA and in the nucleus have been isolated in *N. crassa* ***(15)*** and are assumed to be mutants that retain at least some function. Targeted mutations in *N. crassa* mtDNA have not been made.

The mitochondria of *N. crassa* are more similar to mammalian mitochondria than those of yeast regarding the presence of complex I, which catalyses electron transfer from NADH plus H^+ (reduced nicotinamide adenine dinucleotide) to ubiquinone in the mitochondrial inner membrane. The characterization of mutants affecting specific subunit proteins of complex I in *N. crassa* has proven useful in understanding the elaborate biogenesis, structure, and function of this oligomeric enzyme ***(16)***. *Neurospora* also possesses alternative oxidase. This enzyme is present in many other fungi, all green plants, some protists, some bacteria, and some animal species ***(17–21)***. In many systems, alternative oxidase is regulated by complex mechanisms that respond to stress, developmental signals, or the presence of reactive oxygen species. In *N. crassa*, the enzyme is only present when electron transport via the standard electron transport chain is interrupted ***(22)***. Thus, *N. crassa* serves as an excellent model system for studying the mechanisms of expression of the protein ***(23,24)***.

Neurospora crassa has proven to be an excellent organism for studying various aspects of mitochondrial biology. These studies are aided by the fact that relatively large amounts of mitochondria can be readily obtained from easily harvested mycelium. One of the most important roles for *N. crassa* as an experimental organism continues to be in the field of protein import into mitochondria. Historically, *N. crassa* was at the forefront of many of the milestone events in this field. Some examples include the demonstration of posttranslational import ***(25)***, identification of TOM (translocase of the outer mitochondrial membrane) components ***(26,27)***, and the functional reconstitution and structural analysis of the TOM complex ***(28–30)***. We describe the methods for isolation of mitochondria from *N. crassa* and a detergent-based method that is helpful in determining the localization of proteins in the different mitochondrial sub-compartments.

In addition, because many of the proteins required for mitochondrial protein import are essential for viability, it is important to have methods for generating mutants in the genes encoding these proteins that allow the mutants to be maintained in a viable background. Therefore, we describe a procedure for disrupting essential genes in the organism (sheltered disruption). This technique produces a heterokaryotic strain that can be manipulated to deplete the target gene product so that the effects on mitochondrial biogenesis can be studied ***(31)***.

2. Materials

2.1. Production and Harvesting of Conidia

1. The wild-type strain 74-OR23-1A (74*A*) is generally used for biochemical studies (FGSC 987).
2. Biotin solution: dissolve 20 mg biotin in 100 mL H_2O plus 100 mL 95% ethanol.
3. Trace element stock solution (amounts dissolved in 1 L H_2O): 50 g citric acid, 50 g $ZnSO_4$, 10 g $Fe[(NH_4)_2\ SO_4]$, 2.5 g $CuSO_4$, 0.5 g $MnSO_4{\cdot}H_2O$, 0.5 g water-free H_3BO_3, 0.5 g Na_2MoO_4.

4. Autoclaved 50% (w/v) sucrose stock solution.
5. Vogel's stock solution (amounts for 1 L of 50X stock solution): 150 g Na_3-citrate·$2H_2O$, 250 g KH_2PO_4, 100 g NH_4NO_3, 10 g $MgSO_4$, 5 g $CaCl_2$, 2.5 mL biotin solution, and 5 mL stock solution of trace elements. Autoclave the solution.
6. Vogel's growth medium: 1 L contains 20 mL Vogel's stock solution, 40 mL 50% (w/v) sucrose, and 940 mL demineralized water. This is referred to as Vogel's minimal medium. Auxotrophic strains require additional nutrients.
7. Agar.
8. Erlenmeyer flasks (250 mL) with cotton or sponge plugs.
9. Sterile distilled water and sterile bottle or flask.
10. A Buchner funnel (about 10-cm diameter), the surface of which has been covered with three to four layers of cheesecloth. The funnel is then wrapped in aluminum foil and autoclaved.
11. A hemacytometer for counting the harvested conidia.

2.2. *Isolation of Mitochondria From* Neurospora crassa *Mycelium*

1. Vogel's medium as in **Subheading 2.1**.
2. SEM buffer: 250 m*M* sucrose, 1 m*M* ethylenediaminetetraacetic acid (EDTA), 10 m*M* 3-(*N*-morpholino)propanesulfonic acid (MOPS)-KOH, pH 7.2.
3. Phenylmethylsulfonyl fluoride (PMSF) stock solution (200 m*M*). Prepared fresh daily by dissolving PMSF in 95% ethanol (35 mg PMSF per milliliter of ethanol).

2.3. *Digitonin Fractionation of* Neurospora crassa *Mitochondria*

1. Digitonin stock solution of 1% (e.g., 10 mg digitonin in 1 mL SEM buffer).
2. Proteinase K stock solution at 2 mg per milliliter of SEM buffer.
3. PMSF stock solution (*see* **Subheading 2.2.**).

2.4. *Transformation of* Neurospora crassa *Conidia by Electroporation*

1. Materials for growth, harvesting, and counting of conidia as in **Subheading 2.1**.
2. Sterile 1 *M* sorbitol.
3. Linearized DNA to be transformed. The DNA should be dissolved in sterile water to give 2 μg per 5 μL. A construct developed for disruption will generally contain an antibiotic resistance gene that replaces, or is inserted into, the coding sequence of the target gene. This provides a selectable marker for transformants. The hygromycin resistance gene is often used for this purpose in *N. crassa* ***(32)***.
4. 10X Sugars (amounts for 1 L): 200 g L-sorbose, 5 g fructose, 5 g glucose, 2 g myoinositol. Dissolve in distilled water, then autoclave (*see* **Note 1**).
5. Top agar (amounts for 1 L): 20 mL 5X Vogel's stock solution (*see* **Subheading 2.1.**), 182 g sorbitol, 15 g agar, and 850 mL H_2O. Auxotrophic strains will require additional supplements. The top agar should be autoclaved and then mixed with 100 mL sterile 10X sugars. The top agar should be cooled to 45°C before use.
6. Transformation plates: the components of these plates are identical to the top agar except that they do not contain sorbitol.

7. When required for selection of transformants, hygromycin B is added to both the top agar and the transformation plates (after autoclaving) to a final concentration of 175 μg/mL.

2.5. Isolation of Neurospora crassa Genomic DNA

1. Harvested conidia (*see* **Subheading 2.1.**) and Vogel's medium (*see* **Subheading 2.1.**).
2. DNA isolation buffer: 100 m*M* Tris-HCl, pH 8.0, 10 m*M* EDTA, 1% sodium dodecyl sulfate.
3. 8 *M* Potassium acetate (pH 4.3).
4. Solutions of 70 and 95% ethanol.
5. Solution of 1 m*M* EDTA, pH 8.0.
6. High-salt buffer: 25 m*M* Tris-HCl, pH 7.4, 100 m*M* NaCl, 2 m*M* EDTA.
7. Ribonuclease A (10 mg/mL): the solution should be boiled for 15 min to inactivate contaminating deoxyribonucleases.

2.6. Generation of Mutations in Essential Genes by Sheltered Disruption

1. The DNA construct for knocking out the target gene. The target gene must be cloned with at least 3 kb of flanking sequence on both sides to increase the chances of homologous replacement. The target gene-coding sequence is modified as in **Subheading 2.4., step 3** to allow for disruption of the gene and selection of transformants.
2. The HP1 heterokaryotic strain (available by request from the Nargang lab). HP1 (**Fig. 1**) is constructed from strains 71-18 (*pan-2 BmlR a*) and 76-26 (*his-3 mtrR a*) and is maintained on Vogel's medium lacking both histidine and pantothenate. This forces the maintenance of the heterokaryon because the component nuclei must complement each other's auxotrophies. Once conidia form in the slant, use them to inoculate 250-mL Erlenmeyer flasks containing 50 mL agar-solidified minimal Vogel's medium. Once abundant conidia have formed in the flasks, harvest (*see* **Subheading 3.1.**) and prepare them for electroporation (*see* **Subheading 3.4.**).
3. Slants containing minimal Vogel's medium and slants containing minimal Vogel's medium plus 0.5X the concentration of the antibiotic used for selection of transformants (e.g., hygromycin).
4. Plates for growth of *N. crassa* colonies. The basic composition of the plates is identical to the transformation plates described in **Subheading 2.4**. Plates may also contain drugs and additional nutrients for shifting the ratio of nuclei in the heterokaryon (*see* **Subheading 3.6.6.**). When required, histidine is added to a final concentration of 200 μg/mL and pantothenate to 10 μg/mL. Benomyl is made up as a stock solution at 250 μg/mL in ethanol and stored at –20°C. The final concentration in medium is 0.5 μg/mL. Benomyl is added after the medium has been autoclaved. p-Fluorophenylalanine (FPA) is added directly to the flask containing the medium prior to autoclaving because it is difficult to dissolve at room temperature.

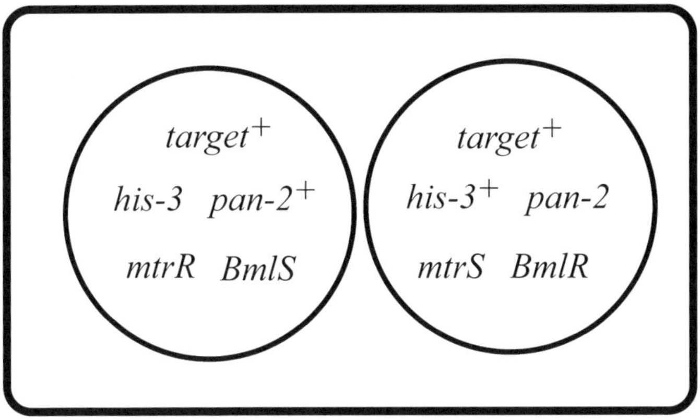

Fig. 1. Heterokaryon HP1. The component nuclei of the heterokaryon are represented by circles within the rounded box that represents the heterokaryon. Genetic markers in each nucleus are shown. *target*$^{+}$ represents any essential gene of interest that may be targeted for disruption. *his*, histidine; *pan*, pantothenate; *mtrR*, 4-methyltryptophan or p-fluorophenylalanine (FPA) resistant; *mtrS*, 4-methyltryptophan or FPA sensitive; *BmlR*, benomyl resistant; *BmlS*, benomyl sensitive.

The final concentration should be 400 µ*M*. When required, histidine, pantothenate, benomyl, and FPA are also added to liquid medium at the same concentrations (*see* **Subheading 3.6.9.**).

3. Methods

3.1. Production of Conidia

1. Prepare 500 mL Vogel's growth medium. Aliquot 50 mL to each of 10 Erlenmeyer flasks (250-mL size) (*see* **Note 2**).
2. Add 1 g agar to each flask and plug with a cotton or sponge plug.
3. Autoclave and allow the agar to solidify.
4. Inoculate each flask with conidia of the desired strain from a slant.
5. Grow at 30°C until mycelium covers the surface of the agar and climbs 2–3 cm up the sides of the flask. This should require 2–3 d for wild-type strains (*see* **Note 3**).
6. Remove the flasks from the incubator and place them in a well-lit room at about 22° C for 3–7 d to allow formation of conidia.
7. Harvest conidia by adding about 50 mL sterile water to each flask and vigorously swirling with the plug in place. After the "dust" of conidia has settled in the flask, pour the suspension through a sterile Buchner funnel covered with three or four layers of cheesecloth and collect the flowthrough in a sterile flask or bottle.
8. Count the conidia using a hemacytometer. Dilutions of 1:10 to 1:1000 may be necessary. Expect about 10^{10} conidia from each flask (*see* **Note 3**).

9. Store the harvested conidia in a refrigerator at a concentration of 10^8 conidia per milliliter or higher.
10. The suspension can be concentrated by centrifugation in a clinical centrifuge and resuspension in fresh distilled water or by allowing the conidia to settle overnight and carefully pouring off as much of the liquid as desired.
11. Use conidia within 10 d of harvesting to avoid loss of viability.

3.2. Isolation of Mitochondria From *Neurospora crassa* Mycelium

1. On the evening prior to the day of the experiment, start a 2-L culture (which is sufficient for import experiments). Mix 1900 mL sterile water, 80 mL sterile 50% sucrose stock solution, and 40 mL sterile Vogel's 50X stock.
2. Add conidial stock suspension to give a final concentration of 1 to 1.5×10^6 conidia/mL. Connect the flask to a forced-air supply and aerate at 1–1.5 bar. The air should pass through a sterile filtering apparatus prior to reaching the flask.
3. Place the aerated flask into a water bath at 25°C overnight under bright illumination.
4. Mitochondria are of optimal quality if the hyphae are harvested 14–16 h after inoculation.
5. The following morning, cool a mortar and pestle and prepare the PMSF stock solution.
6. Harvest the mycelium by filtration, wash with water, and let the "pancake" become very dry. The washing step is especially important if the medium contains inhibitors.
7. Weigh the harvested mycelium on a filter paper. Do not touch with bare hands or contaminating proteins may be transferred to the mycelium.
8. For each 1 g mycelium measure 1.5 g quartz sand and 2 mL SEMP (SEM buffer plus 5 μL PMSF stock solution per milliliter of SEM buffer).
9. From this point, perform all operations at 2°C or on ice.
10. Rip the mycelium into small pieces and place it in the cooled mortar with 1X its wet weight of SEMP. Add 1.5X the mycelium weight of quartz sand. Use the cooled-pestle to homogenize the mixture gently to a consistency resembling a liquid dough.
11. Add another 1X the original mycelium weight of SEMP and grind the cells with the pestle for 30–45 s (*use timer*) using a circular motion.
12. Pour the slurry into 40- to 50-mL centrifugation tubes. Carefully clean the sand from outside the tubes to avoid damaging the centrifuge rotor.
13. Process by differential centrifugations at 2°C:
    a. 5 min at 2000*g*. Pour the supernatant into fresh tubes and discard the pellets.
    b. 12 min at 17,400*g*. Discard the supernatant and resuspend the pellets carefully in 1 mL SEM buffer (*see* **Note 4**) using a 1-mL pipetor. Combine the suspended mitochondria into one tube and fill to 30 mL with SEM.
    c. Repeat **step a**.
    d. 12 min at 17,400*g*. Discard the supernatant.
14. Gently resuspend the pellets with about 0.5 mL SEM using a 1-mL pipetor. Store the mitochondria in a 1.5-mL centrifuge tube on ice.
15. Determine the protein concentration in the mitochondrial sample (*see* **Note 5**).
16. For use in import experiments, dilute the mitochondrial suspension with SEM buffer to 5 mg/mL (*see* **Note 6**).

17. It is possible to add bovine serum albumin to a concentration of 5 mg/mL to stabilize the mitochondria.

3.3. Digitonin Fractionation of Neurospora crassa *Mitochondria*

Many studies require the localization of proteins or import intermediates to a distinct submitochondrial compartment. A basic step in such fractionation is often the selective rupturing of the outer membrane of isolated mitochondria. In some organisms, this can be achieved by osmotic shock. However, this procedure does not work well for *N. crassa* mitochondria, and fractionation using the mild detergent digitonin can be used in its place. The use of digitonin for this purpose is based on the observation that a concentration of the detergent can be found where the outer membrane will be ruptured while the inner membrane remains intact. Eventually, further increases in the digitonin concentration will result in opening of the inner membrane.

Generally, the outer membrane is ruptured with a digitonin concentration of about 0.15% (w/v), whereas concentrations of 0.2% and higher are required to affect the inner membrane. As digitonin is a natural product with quality that varies from batch to batch, it is recommended that the effectiveness of the digitonin be titrated for each lot purchased instead of using predetermined concentrations. The titration should include concentrations from 0.1 to 0.25%. The following method is described for determining if a protein is localized to the intermembrane space by Western blotting of the digitonin-treated mitochondria.

1. Use 100- to 200-μg mitochondria samples (prepared as in **Subheading 3.2.**) for each digitonin concentration to be tested.
2. Calculate the volume of digitonin stock solution and SEM buffer to obtain the final digitonin concentrations desired.
3. Resuspend mitochondria with the required amount of SEM buffer (calculated in **step 2**). The final volume of the mitochondria suspension during digitonin treatment should be 100 μL or less because standard 1.5-mL plastic centrifuge tubes are used, and it is necessary to dilute 14-fold with SEM at a later step.
4. Add the amount of digitonin that was calculated in **step 2**.
5. Mix the samples gently by flicking against the walls of the tubes.
6. Place the samples on ice for exactly 3 min.
7. Dilute the samples with 1400 μL SEM buffer.
8. Reisolate the mitochondria/mitoplasts by centrifuging for 7 min at 17,400*g* at 2°C (*see* **Note 7**).
9. Resuspend the mitochondria/mitoplasts in SEM buffer.
10. Split each sample in half.
11. Add proteinase K to one sample at a final concentration of 50 μg/mL. Mix gently.
12. Incubate all samples for 20 min on ice.
13. Add PMSF to all the samples at a final concentration of 1 m*M*. Mix gently.
14. Incubate the samples for 10 min on ice.

15. Reisolate the mitochondria/mitoplasts as in **step 8**.
16. Add Laemmli cracking (sample) buffer to the pelleted samples and analyze them by sodium dodecyl sulfate-polyacrylamide gel electrophoresis and Western blotting.
17. Immunodecorate the blots. As controls for the effectiveness of disrupting the outer membrane, use antibodies against cytochrome-*c* heme lyase and the adenosine triphosphate/adenosine 5′-diphosphate carrier (AAC). In mitoplasts, the intermembrane space protein cytochrome-*c* heme lyase will be digested by proteinase K treatment, whereas the inner membrane protein AAC will show a fragment approx 2–3 kDa smaller than the original AAC band. In intact mitochondria, both proteins remain unaffected. Use antibodies against matrix proteins like Hsp60 or the matrix processing peptidase to control for the integrity of the inner membrane.

3.4. *Electroporation of* Neurospora crassa *Conidia*

1. Harvest conidia from two to four flasks as described in **Subheading 3.1**.
2. Pellet the conidia for 2 min at 2000*g* in a benchtop centrifuge.
3. Wash the conidia three times with 50 mL sterile 1 *M* sorbitol by centrifugation as in **step 2**.
4. Resuspend the final pellet in 10–15 mL sterile 1 *M* sorbitol.
5. Count the conidia in a hemacytometer. The desired concentration is 2.0 to 2.5 × 10^9 conidia per milliliter of 1 *M* sorbitol. It may be necessary to recentrifuge and suspend in a smaller volume or to dilute with more 1 *M* sorbitol to achieve the correct concentration. The conidia can be used directly, or they can be frozen in 40-µL aliquots at –80°C for later use. When using conidia that have been frozen, it is necessary to wash each aliquot twice with 200 µL cold sterile 1 *M* sorbitol before resuspending in 40 µL cold 1 *M* sorbitol. Keep the conidia on ice at all times and centrifuge at 4°C. At least for some strains, fresh conidia give a higher transformation frequency than frozen ones.
6. Place 2 µg of the linearized DNA to be transformed in a 1.5-mL centrifuge tube on ice and add 40 µL of the conidial suspension (should be about 10^8 cells). Mix gently with a pipetor and incubate for 5 min on ice.
7. Gently add the mixture to an electroporation cuvette (2-mm gap) that has been prechilled on ice. Tap gently to get the suspension to the bottom of the cuvette.
8. Electroporate using a BTX model ECM 630 electroporator set at 2.1 kV, 475 Ω, and 25-µF capacitance. The time constant should be 10–12 ms (*see* **Note 8**).
9. Immediately add 1 mL cold sterile 1 *M* sorbitol to the cuvette and then transfer the entire contents of the cuvette to a 1.5-mL centrifuge tube. Place the tube at 30°C for 30–60 min.
10. Mix the electroporated conidial suspension gently by inversion and take 50–250 µL of the suspension to 50 mL of top agar at 45°C. Mix gently but completely. Spread equal aliquots over five transformation plates. The amount of the suspension added to the top agar will depend on the transformation frequency and the amount of background growth from untransformed cells. Background is influenced by the viability of the condia and the selection scheme for transformants (*see* **Note 9**).
11. Incubate the plates at 30°C for 2–5 d, until colonies form.

3.5. Isolation of Genomic DNA From **Neurospora crassa**

1. Grow a liquid culture overnight and harvest by filtration as described in **Subheading 3.2**. However, for genomic DNA isolation only a 50-mL culture is required. These cultures can be grown in 250-mL Erlenmeyer flasks and can be aerated with shaking rather than by using forced air.
2. Grind in a mortar and pestle with sand as in **Subheading 3.2.** except that the buffer used for grinding is DNA isolation buffer, and the grinding should be more extensive, about 1–2 min.
3. Place the ground slurry in a 40- to 50-mL centrifuge tube and bring the volume up to 10 mL with DNA isolation buffer. Mix vigorously and place the tube in a 70°C water bath for 1 h.
4. Chill the sample on ice for 10 min and add 640 μL of 8 *M* potassium acetate (pH 4.3). Mix well by inversion and store on ice for 1 h.
5. Centrifuge at 18,000*g* for 15 min at 4°C. Transfer the supernatant to a fresh centrifuge tube.
6. Add an equal volume of isopropanol and mix gently. A visible clot of DNA and a "cloud" of RNA should be visible.
7. Centrifuge at 18,000*g* for 10 min at 4°C. Pour off the supernatant, wash the pellet with 70% ethanol, and allow the pellet to air dry.
8. Suspend the pellet in 400 μL 1 m*M* EDTA (pH 8.0) and place the suspension in a 1.5-mL centrifuge tube. Add 200 μL high-salt buffer and 15 μL boiled ribonuclease A (10 mg/mL). Mix gently and incubate at 37°C for 30 min.
9. Phenol extract the sample and take the aqueous phase to a fresh tube. Add 2 volumes of 95% ethanol and 1/10 volume 3 *M* sodium acetate. After mixing gently by inversion, a visible clot of DNA should form. Remove the clot to a fresh tube with a pipet tip and rinse with 70% ethanol. Allow the clot to air dry and then suspend in 400 μL sterile distilled water.

3.6. Generation of Mutations in Essential Genes by Sheltered Disruption

Sheltered disruption is a technique that allows the loss of an essential target gene in one nucleus of a heterokaryon because the essential gene product is supplied by the second nucleus in the heterokaryon (*see* **Note 10**). Genetic markers in the component nuclei allow manipulation of the heterokaryon so that the nuclear ratios can be altered. When the ratio is heavily skewed toward the nucleus carrying the disrupted essential gene, the level of the gene product can be significantly reduced, and the resulting phenotype can be studied.

1. Electroporate (**Subheading 3.4.**) the linearized disruption DNA construct of the target gene into HP1 conidia.
2. Plate the transformed cells using top agar and transformation plates containing minimal Vogel's medium plus hygromycin. Incubate at 30°C.
3. When colonies form (2–4 d), pick about 25–50 to slants containing minimal Vogel's medium plus 0.5X hygromycin concentration used in the original plates

(*see* **Note 11**). Incubate the slants at 30°C for 2–3 d and then place them at room temperature in constant light for conidiation.

4. Streak conidia from each isolate onto a separate transformation plate containing minimal Vogel's medium plus hygromycin and place at 30°C for 2–4 d. One colony is picked from each and placed into a slant of minimal Vogel's medium without hygromycin. Grow for 2–3 d at 30°C and then place the slants at room temperature with constant light to allow conidiation (*see* **Note 12**).
5. Examine the transformants for proper integration of the disruption construct by Southern analysis (*see* **Note 13**). This is greatly aided by the availability of the *N. crassa* genome sequence, which allows the exact prediction of fragments from the endogenous, nondisrupted gene. In addition, all restriction sites in the flanking sequence of the target gene will be known, as will the sites in the hygromycin resistance gene *(32)*. Only strains showing the predicted pattern by Southern analysis should be chosen for further work. Strains showing incomplete patterns or correct patterns plus extra bands should be discarded (*see* **Note 14**).
6. Analyze the chosen strains for growth characteristics to determine the nucleus in which the integration and disruption have taken place. Harvest conidia from a minimal Vogel's slant of each strain to be tested. Use the parent strain HP1 as a control. Count conidia using a hemocytometer and adjust their concentration to 2×10^7 conidia/mL. Make serial dilutions to give 2×10^6, 2×10^5, and 2×10^4 conidia/mL. From each of the four suspensions, spot 10 μL onto plates in a horizontal row. Take care to mix the suspensions just before pipeting as the conidia will settle rapidly. Following spotting, incubate the plates at 30°C for 2–4 d. Spot the strains on transformation plates containing three different sets of supplements. The first serves as a control and contains minimal Vogel's medium. Growth of all strains should be equal on this plate. The second contains the same medium, but FPA and histidine are added; the third contains minimal Vogel's medium plus benomyl and pantothenate (*see* **Note 15**).
7. If the disruption of the target gene is shown to reduce growth of one of the sheltered heterokaryons on one of the media discussed in **step 6**, then test the strain to show that the disrupted gene is indeed essential for viability (*see* **Note 16**). Streak conidia from the sheltered heterokaryon onto transformation plates containing both histidine and pantothenate. This should allow for growth of heterokaryons and the two possible homokaryons. After incubation at 30°C for 3–4 d, pick about 100 colonies to slants containing Vogel's medium plus histidine and pantothenate. Incubate the slants at 30°C for 2–3 d and then remove to a well-lit room for conidiation. Spot conidia from each slant onto transformation plates containing minimal Vogel's medium, minimal Vogel's plus histidine and pantothenate, minimal Vogel's plus histidine, and minimal Vogel's plus pantothenate. Examine the growth on each plate from each isolated colony to determine if one class of homokaryon is not viable.
8. Rescue the disrupted nucleus to a homokaryotic state from the sheltered heterokaryon by introducing a wild-type copy of the disrupted gene. This will prove that it is the disruption of the original target gene that results in the inviability of a particular nucleus in the sheltered heterokaryon rather than an undetected small integration into an unrelated essential gene or the occurrence of a spontaneous

mutation. Transform the sheltered heterokaryon with a wild-type construct of the target gene and select transformants on specific medium. For example, if a disruption occurred in the histidine-requiring nucleus, then conidia of the strain would be transformed with a wild-type copy of the gene cloned on a plasmid carrying another resistance marker (such as bleomycin resistance; *see* **Note 17**). Select transformants on medium containing bleomycin, histidine, and FPA in both the top agar and the plates. Pick colonies and purify as in **step 4** using medium containing bleomycin, histidine, and FPA. Test the isolates for histidine requirement. Homokaryons requiring histidine should be obtained (*see* **Note 18**).

9. Mitochondria that are deficient in the essential gene product can be obtained from the sheltered heterokaryon to perform import studies or for examination of the levels of different mitochondrial proteins. For example, if the heterokaryon carries a disruption in the histidine-requiring nucleus, then grow cells in liquid cultures of Vogel's medium containing histidine and FPA (as in **Subheading 3.2.**) from conidia generated on Vogel's minimal medium (**Subheading 3.1.**). About 2–3 d of growth may be required to produce enough mycelium to isolate sufficient mitochondria. The level of FPA may be reduced from 400 to 300 μ*M* or lower if required to obtain sufficient mycelium within 3 d.

4. Notes

1. A high concentration of L-sorbose causes excessive hyphal branching and results in the formation of compact colonies.
2. The number of conidia flasks started at one time will depend on the number and type of experiments planned.
3. Mutants, particularly those affecting mitochondrial function, may take several days longer than wild-type strains to cover the surface of the agar. They may also be unable to climb the walls of the flasks and often produce fewer conidia than wild-type strains.
4. The decision regarding whether to use SEM or SEMP to suspend the mitochondria at this point is based on the purpose intended for the mitochondria. In import experiments, PMSF must not be used because it will inhibit the proteinase K that is required in a later step.
5. Typically, the procedure yields 10–20 g hyphae from 2 L medium and 0.3 mg mitochondria per gram of hyphae. If the yield of mitochondria is much higher, then grinding was too harsh. In this case, the outer membrane of a large fraction of the mitochondria may be ruptured.
6. For in vitro import experiments *N. crassa* mitochondria have to be prepared freshly each day. Storage in the freezer ruptures the outer membrane. It is best to use the mitochondria for in vitro import experiments within 1–3 h after preparation.
7. Mitoplasts are mitochondria with holes or breaks in the outer membrane. The inner membrane is intact.
8. Other electroporation machines can also be used. It will be necessary to optimize the conditions to achieve maximum transformation efficiency.

9. A "no DNA" control electroporation should also be used to verify that the colonies observed in the actual experiment are not simply at background level.
10. Conidia produced from both homokaryons and heterokaryons are typically multinucleate. Thus, conidia from a heterokaryon will frequently carry both of the component nuclei of the strain. Usually, only one nucleus in a conidium takes up transforming DNA *(33)*, so the other nuclei are unaffected. Therefore, if the disruption occurs in one nucleus of a conidium containing two genetically distinct nuclei, then the other will complement the lack of the essential product in the heterokaryotic strain.
11. The hygromycin concentration is reduced because the drug can adversely affect formation of conidia, even in resistant strains.
12. This single-colony isolation procedure purifies the hygromycin-resistant transformants from any untransformed background cells that may have been picked up along with the original colony.
13. Here, one must keep in mind that the strains analyzed are heterokaryons. Thus, the pattern seen by Southern analysis will represent two different restriction patterns: the pattern of the nondisrupted gene from the sheltering nucleus and the pattern from the nucleus where the target gene has been disrupted. Select an appropriate probe that will demonstrate the presence of both patterns effectively.
14. Because most integrations of transforming DNA in *N. crassa* are at ectopic locations in the genome, there can be great variation in the frequency of homologous replacements that give the desired product. The range can be as high as 1 in 10 to less than in 100. The frequency depends on the amount of flanking sequence and probably properties of the target locus itself. As discussed in the Introduction, a split marker approach has been developed that raises the homologous replacement frequency to as much as 50%. Although the method had not yet been published *(8)* at the time this chapter was written, the technique will be beneficial for achieving a higher frequency of sheltered disruption. On the other hand, the technique mentioned in the Introduction for achieving virtually 100% homologous replacements by transforming strains with a *mus-51* or *mus-52* genetic background is not well suited for sheltered disruption because it may be desirable to work with strains that do not carry the *mus-51* or *mus-52*. That is, attempts to cross a disrupted essential gene out of the *mus* background would be fruitless as individual haploid ascospore progeny carrying the disruption would be inviable.
15. In heterokaryons in which the disruption occurred in the histidine-requiring nucleus (**Fig. 1**), the presence of FPA will force that nucleus to predominate to allow growth in the presence of the inhibitor. However, if the disruption has occurred in an essential gene or any gene with a product that is required at high levels, then growth of the strain should be slower than the HP1 control because the numerically superior disruption-containing nucleus cannot supply the essential gene product. Similarly, any strain with a disruption of an essential gene in the pantothenate-requiring nucleus will grow more slowly on the medium containing pantothenate plus benomyl. If the disrupted gene is not essential, then the growth of the control and disruptant strains should be roughly equal on all three media.

16. This test depends on the random segregation of nuclei into conidiaspores from the heterokaryon. As the conidiaspores are typically multinucleate, the conidia arising from the heterokaryon can contain both nuclei, only the histidine-requiring nucleus, or only the pantothenate-requiring nucleus. Thus, if the disruption occurred in an essential gene in the histidine-requiring nucleus, then no viable homokaryons carrying only that nucleus should be produced from the heterokaryon. Similarly, no pantothenate-requiring homokaryons should be viable if the disruption occurred in that nucleus.
17. Bleomycin is used at a final concentration of 1.5 μg/mL. Caffeine is also added to a concentration of 0.5 mg/mL to enhance the effects of bleomycin.
18. Transformation with a tagged version of the wild-type gene can also be done. If the tagged version rescues the strain, then the result is a homokaryon expressing only the tagged version of the target gene product.

Acknowledgments

This work was supported by grants from the Natural Sciences and Engineering Research Council of Canada and the Canadian Institutes of Health Research to F. E. N. and from the Deutsche Forschungsgemeinschaft—Sonderforschungsbereich 594 to D. R.

References

1. Beadle, G. W., and Tatum, E. L. (1941) Genetic control of biochemical reactions in *Neurospora. Proc. Natl. Acad. Sci. USA* **27,** 499–506.
2. Davis, R. H., and De Serres, F. J. (1970) Genetic and microbiological research techniques for *Neurospora crassa. Methods Enzymol.* **17,** 79–143.
3. Perkins, D. D. (1992) *Neurospora*: the organism behind the molecular revolution. *Genetics* **130,** 687–701.
4. Perkins, D. D., Radford, A., and Sachs, M. S. (2001) *The* Neurospora *Compendium. Chromosomal Loci,* Academic Press, San Diego, CA.
5. Davis, R. H. (2000) Neurospora. *Contributions of a Model Organism,* Oxford University Press, Oxford, UK.
6. Galagan, J. E., Calvo, S. E., Borkovich, K. A., et al. (2003) The genome sequence of the filamentous fungus *Neurospora crassa. Nature* **422,** 859–868.
7. Borkovich, K. A., Alex, L. A., Yarden, O., et al. (2004) Lessons from the genome sequence of *Neurospora crassa*: tracing the path from genomic blueprint to multicellular organism. *Micro. Mol. Biol. Rev.* **68,** 1–108.
8. Colot, H. V., Park, G., Turner, G.E., et al. (2006) A high-throughput gene Knockout procedure for *Neurospora* reveals functions for multiple transcription factors. *Proc. Natl. Acad. Sci.* USA **103,** 10,352–10,357.
9. Ninomiya, Y., Suzuki, K., Ishii, C., and Inoue, H. (2004) Highly efficient gene replacements in *Neurospora* strains deficient for nonhomologous end-joining. *Proc. Natl. Acad. Sci. USA* **101,** 12,248–12,253.
10. Selker, E. U. (1990) Premeiotic instability of repeated sequences in *Neurospora crassa. Annu. Rev. Genet.* **24,** 579–613.

11. Harkness, T. A. A., Metzenberg, R. L., Schneider, H., Lill, R., Neupert, W., and Nargang, F. E. (1994) Inactivation of the *Neurospora crassa* gene encoding the mitochondrial protein import receptor MOM19 by the technique of "sheltered RIP." *Genetics* **136,** 107–118.
12. Grad, L., Descheneau, A., Neupert, W., Lill, R., and Nargang, F. (1999) Inactivation of the *Neurospora crassa* mitochondrial outer membrane protein TOM70 by repeat-induced point mutation (RIP) causes defects in mitochondrial protein import and morphology. *Curr. Genet.* **36,** 137–146.
13. Taylor, R., McHale, B., and Nargang, F. E. (2003) Characterization of *Neurospora crassa* Tom40-deficient mutants and effect of specific mutations on Tom40 assembly. *J. Biol. Chem.* **278,** 765–775.
14. Rountree, M. R., and Selker, E. U. (1997) DNA methylation inhibits elongation but not initiation of transcription in *Neurospora crassa. Genes Dev.* **11,** 2383–2395.
15. Kennell, J. C., Collins, R. A., Griffiths, A. J. F., and Nargang, F. E. (2004) Mitochondrial genetics of *Neurospora*, in *The Mycota II. Genetics and Biotechnology* (Kück, U., ed.), 2nd Ed., Springer-Verlag, Berlin, pp. 95–112.
15a. Schmitt, S.H., Prokisch, H., Schlunk, T., et al. (2006) Proteome analysis of mitochondrial outer membrane from *Neurospora crassa. Proteomics* **6,** 72–80.
16. Videira, A., and Duarte, M. (2002) From NADH to ubiquinone in *Neurospora* mitochondria. *Biochim. Biophys. Acta* **1555,** 187–191.
17. Chaudhuri, M., Ajayi, W., Temple, S., and Hill, G. C. (1995) Identification and partial purification of a stage specific 33 kDa mitochondrial protein as the alternative oxidase of *Trypanosoma brucei brucei* bloodstream trypanosomes. *J. Eukaryot. Microbiol.* **42,** 467–472.
18. Joseph-Horne, T., Holloman, D. W., and Wood, P. M. (2001) Fungal respiration: a fusion of standard and alternative components. *Biochim. Biophys. Acta* **1504,** 179–195.
19. Moore, A. L., Albury, M. S., Crichton, P. G., and Affourtit, C. (2002) Function of the alternative oxidase: is it still a scavenge? *Trends Plant Sci.* **7,** 478–481.
20. Stenmark, P., and Nordlund, P. (2003) A prokaryotic alternative oxidase present in the bacterium *Novosphingobium aromaticivorans. FEBS Lett.* **552,** 189–192.
21. McDonald, A. E., and Vanlerberghe, G. C. (2004) Branched mitochondrial electron transport in the animalia: presence of alternative oxidase in several animal phyla. *IUBMB Life* **56,** 333–341.
22. Li, Q., Ritzel, R. G., McLean, L. T. T., et al. (1996) Cloning and analysis of the alternative oxidase of *Neurospora crassa. Genetics* **142,** 129–140.
23. Lambowitz, A. M., Sabourin, J. R., Bertand, H., Nickels, R., and McIntosh, L. (1989) Immunological identification of the alternative oxidase of *Neurospora crassa* mitochondria. *Mol. Cell. Biol.* **9,** 1362–1364.
24. Descheneau, A. T., Cleary, I. A., and Nargang, F. E. (2005) Genetic evidence for a regulatory pathway controlling alternative oxidase production in *Neurospora crassa. Genetics* **169,** 123–135.

25. Hallermayer, G., Zimmermann, R., and Neupert, W. (1977) Kinetic studies on the transport of cytoplasmatically synthesized proteins into mitochondria in intact cells of *Neurospora crassa*. *Eur. J. Biochem.* **81,** 523–532.
26. Kiebler, M., Pfaller, R., Söllner, T., et al. (1990) Identification of a mitochondrial receptor complex required for recognition and membrane insertion of precursor proteins. *Nature* **348,** 610–616.
27. Söllner, T., Rassow, J., Wiedmann, M., et al. (1992) Mapping of the protein import machinery in the mitochondrial outer membrane by crosslinking of translocation intermediates. *Nature* **355,** 84–87.
28. Künkele, K.-P., Heins, S., Dembowski, M., et al. (1998) The preprotein translocation channel of the outer membrane of mitochondria. *Cell* **93,** 1009–1019.
29. Ahting, U., Thun, C., Hegerl, R., et al. (1999) The TOM core complex: the general protein import pore of the outer membrane of mitochondria. *J. Cell Biol.* **147,** 959–968.
30. Vasiljev, A., Ahting, U., Nargang, F. E., et al. (2004) Reconstituted TOM core complex and Tim9/Tim10 complex of mitochondria are sufficient for translocation of the ADP/ATP carrier across membranes. *Mol. Biol. Cell* **15,** 1445–1458.
31. Nargang, F. E., Künkele, K.-P., Mayer, A., Ritzel, R. G., Neupert, W., and Lill, R. (1995) "Sheltered disruption" of *Neurospora crassa* MOM22, an essential component of the mitochondrial protein import complex. *EMBO J.* **14,** 1099–1108.
32. Staben, C., Jensen, B., Singer, M., Pollock, J., and Schechtman, M. (1989) Use of bacterial hygromycin B resistance gene as a dominant selectable marker in *Neurospora crassa* transformation. *Fungal Genet. Newsl.* **36,** 79–81.
33. Grotelueschen, J., and Metzenberg, R. (1995) Some property of the nucleus determines the competence of *Neurospora crassa* for transformation. *Genetics* **139,** 1545–1551.

9

Isolation of Intact, Functional Mitochondria From the Model Plant *Arabidopsis thaliana*

Lee J. Sweetlove, Nicolas L. Taylor, and Christopher J. Leaver

Summary

The ability to isolate intact, functional mitochondria from plant tissues is a key technique in the study of the genome, proteome, and metabolic function of the plant mitochondrion. Traditionally, mitochondrial plant researchers have turned to specific plant systems and organs (such as potato tubers and pea shoots) from which mitochondria are readily isolated in large quantities. However, increasingly, research is focused on a small number of model species, and there is a need to adapt existing protocols to allow the isolation of mitochondria from these model species. Arguably, the most important of these is *Arabidopsis thaliana*, for which a formidable array of genetic resources is available. However, because of its relatively small size and the absence of large heterotrophic organs, *Arabidopsis* is a challenging plant from which to isolate mitochondria. Here, we present two methods for isolating mitochondria from *Arabidopsis*, either from heterotrophic cell suspension cultures or from hydroponic seedling cultures. We also present details of commonly used assays to assess the physical and functional integrity of the isolated organelles.

Key Words: *Arabidopsis*; cell suspension culture; leaf; mitochondria isolation; outer mitochondrial membrane integrity; respiratory control ratio.

1. Introduction

Arabidopsis thaliana was the first plant in which all three cellular genomes (nuclear, mitochondrial, plastidic) were sequenced *(1)*, and the subsequent development of a range of genetic resources has established its position as *the* model plant species. Both reverse and forward genetic mutagenesis strategies are now routine, and the former can be achieved at high throughput and with genomic saturation *(2)*. Although these genetic approaches are facilitating the discovery of gene function in *Arabidopsis*, increasingly the precise dissection of gene function requires a combination of genetic and biochemical approaches.

From: *Methods in Molecular Biology, vol. 372: Mitochondria: Practical Protocols*
Edited by: D. Leister and J. M. Herrmann © Humana Press Inc., Totowa, NJ

Among these, the ability to isolate and study the genome, proteome, and biochemistry of key organelles such as the mitochondrion is paramount *(3)*. This is particularly true of plants in which the degree of subcellular compartmentation is high and leads to a unique metabolic and biochemical complexity.

The choice of *Arabidopsis* as a model species is largely driven by genetic factors. Although the applicability of *Arabidopsis* as a genetic model is unchallenged, it is a far from ideal plant for conducting biochemical investigations. The main problem is that *Arabidopsis* is small and therefore yields much less experimental material than larger crop plants such as maize, soybean, and potato, which have traditionally been the organisms of choice for biochemists. Thus, it has been necessary to adapt existing biochemical assays to overcome the constraint of limited starting material. For example, highly sensitive amplifying assays have been developed to allow the measurement of a large number of enzymes from *Arabidopsis* leaf material *(4)*. In the context of isolating mitochondria from *Arabidopsis*, as well as the constraint of limited starting material, there is also the problem that *Arabidopsis* has little nonphotosynthetic tissue (root and seed mass are a relatively small proportion of the total plant mass). Isolation of mitochondria from photosynthetic plant tissues is inherently more difficult than from nonphotosynthetic tissue because of contamination from the thylakoid membrane system of the chloroplast, which has a similar density to mitochondria.

In this chapter, we present two methods for isolating mitochondria from *Arabidopsis* using density gradient centrifugation. The first uses a heterotrophic cell suspension culture to minimize the problem of thylakoid contamination *(5)*, and the second uses *Arabidopsis* seedlings grown in large quantities using a simple shaking hydroponic culture system and a specific gradient of Percoll and polyvinylpyrrolidone (PVP) designed to separate mitochondria from thylakoids *(6)*.

2. Materials

2.1. Growth of Heterotrophic Arabidopsis *Cell Suspension Cultures*

Culture medium (1 L): MS salts including vitamins (Ducheefa, Haarlem, The Netherlands), 3% (w/v) sucrose, 0.5 mg napthalene acetic acid (from stock of 1 mg/mL in water), 50 μg kinetin (from stock of 1 mg/mL in water). Adjust pH to 5.8 with KOH and autoclave.

2.2. Growth of Hydroponic Arabidopsis *Seedling Cultures*

1. Glass culture vessels (100 mL) with a polypropylene Magenta B-cap (Sigma-Aldrich Company Ltd., Poole, Dorset, UK). Sterilize by autoclaving.
2. Culture medium (500 mL): MS salts including vitamins (Ducheefa), 2% (w/v) sucrose, 0.2 g MES. Adjust pH to 5.8 with KOH. Transfer 60-mL aliquots to sterile glass culture vessels; add 0.06 g agar and autoclave.

3. Sterilizing solution 1: 70% (v/v) ethanol.
4. Sterilizing solution 2: 5% (v/v) sodium hypochlorite, 0.1% (v/v) Tween-20.
5. Sterile distilled water.

2.3. Preparation of Density Gradients

1. Gradient buffer 1: 0.6 *M* mannitol, 20 m*M* TES-KOH, pH 7.5, 0.2% (w/v) bovine serum albumin (BSA).
2. Gradient buffer 2: 0.6 *M* sucrose, 20 m*M* TES-KOH, pH 7.5, 0.2% (w/v) BSA.
3. Percoll (Amersham Biosciences UK Ltd., Chalfont St. Giles, UK).

2.4. Isolation of Mitochondria

2.4.1. *From Heterotrophic* Arabidopsis *Cell Suspension Cultures*

1. Miracloth (Merck Biosciences, Nottingham, UK).
2. Muslin.
3. Buchner funnel.
4. Waring blender.
5. Vacuum pump.
6. Cell extraction medium: 0.45 *M* mannitol, 50 m*M* sodium pyrophosphate ($Na_4P_2O_7$), 0.5% (w/v) BSA, 0.5% (w/v) PVP-40, 2 m*M* EGTA. Adjust pH to 8.0 with phosphoric acid. Add cysteine to 20 m*M* on day of use.
7. 250-mL Centrifuge tubes.
8. Centrifuge with fixed-angle rotor (Beckman JA-14 or equivalent).
9. Cell mitochondria wash medium: 0.3 *M* mannitol, 10 m*M* TES-KOH, pH 7.5.
10. Paintbrush.
11. 5-mL Pipet.
12. 40/23/18% Percoll step gradient (*see* **Subheading 2.3.** and **3.3.**).

2.4.2. *From Hydroponic* Arabidopsis *Seedling Cultures*

1. Buchner funnel.
2. Sterile distilled water.
3. Perspex vessel (45 × 60 × 200 mm).
4. Seedling extraction medium: 0.3 *M* sucrose, 25 m*M* $Na_4P_2O_7$, 2 m*M* EDTA (ethylenediaminetetraacetic acid; disodium salt), 10 m*M* KH_2PO_4, 1% (w/v) PVP-40, 1% (w/v) BSA. Adjust to pH 7.5 with NaOH. Add ascorbic acid to 20 m*M* on day of use.
5. Polytron® homogenizer equipped with a 20-cm long × 2-cm diameter dispersing head.
6. Miracloth (Merck Biosciences).
7. Muslin.
8. Centrifuge and centrifuge tubes as in **Subheading 2.4.1**.
9. 50-mL Centrifuge tubes.
10. Paintbrush.
11. 28% Percoll, 0–4.4% PVP-40 gradients (*see* **Subheadings 2.3.** and **3.3.**).

12. Leaf mitochondria wash medium: 0.3 *M* sucrose, 10 m*M* TES-KOH, pH 7.5.
13. 5-mL Pipet.

2.5. Determination of Mitochondrial Integrity

1. Mitochondria isolated according to **Subheading 3.4**.
2. Clark-type oxygen electrode, such as those manufactured by Hansatech (Kings Lynn, UK) or Rank Brothers (Bottisham, UK).
3. Mitochondrial reaction medium: 0.3 *M* mannitol, 10 m*M* TES-KOH, pH 7.5, 3 m*M* $MgSO_4$, 10 m*M* NaCl, 5 m*M* KH_2PO_4, 0.1% (w/v) BSA.
4. Stock solutions for cytochrome-*c* oxidase (COX) assay: 0.5 *M* Na-ascorbate; 5 m*M* cytochrome-*c* (horse); 10% (v/v) Triton X-100.

2.6. Measurement of Mitochondrial Respiratory Function

1. Mitochondria isolated according to **Subheading 3.4**.
2. Stock solutions of respiratory substrates: 100 m*M* nicotinamide adenine dinucleotide (NADH), 1 *M* succinate, 50 m*M* ATP, 500 m*M* pyruvate, 50 m*M* malate, 30 m*M* NAD^+, 10 m*M* thiamine pyrophosphate. All substrates should be made up in 500 m*M* TES-KOH, pH 7.5.
3. 10 m*M* Adenosine 5′-diphosphate (ADP) made up in 500 m*M* TES-KOH, pH 7.5.

3. Methods

Isolation of intact, functional mitochondria from plant tissues using density gradient centrifugation has been a standard procedure for many years. For a general practical guide, consult **ref.** *7*. The key aspects to the methodology are disruption of plant cells without rupture of the organelles, protection of organelles from harmful compounds released from other subcellular compartments, and separation of mitochondria from other organelles by density gradient centrifugation. We describe methods to isolate mitochondria from *Arabidopsis* tissue. In addition, methods are described to assess organelle integrity and the extent of maintenance of normal respiratory function.

3.1. Growth of Heterotrophic Arabidopsis Cell Suspension Cultures

1. Cell suspension cultures are established from *Arabidopsis* callus material *(8)*. Mature cultures are maintained under sterile conditions with 100 mL culture medium in 250-mL conical flasks sealed with aluminum foil (*see* **Note 1**). Flasks are shaken on an orbital shaker at 120 rpm and maintained at 22°C under illumination (16-h photoperiod, 110 μmol photons/m^2/s).
2. Cells are subcultured after 7 d of growth. Working in a laminar flow hood, a sterile plastic pipet is used to withdraw 10 mL culture (swirl the flask directly before doing this to prevent cells from settling). This aliquot of the culture is then pipeted into 100 mL fresh culture medium in a clean, sterile, 250-mL conical flask.
3. Stock cultures (generally five or six 250-mL flasks) are maintained under illuminated conditions. To prepare fully heterotrophic cultures, subculture from the

stock cultures as above and grow in dark conditions. As many as eight or nine culture flasks can be generated from a single stock culture. For mitochondrial isolation, 10 cultures (250 mL each) are an ideal starting point. Mitochondria are best isolated at d 6 when the cells are still in the logarithmic phase of growth.

3.2. *Growth of Hydroponic* Arabidopsis *Seedling Cultures*

1. This method is adapted from **ref. *9***.
2. Place 7 mg seeds in a 1.5-mL microtube. All subsequent steps should be performed using sterile technique in a laminar flow hood. Add 1 mL sterilization solution 1. Incubate for 2 min and remove solution by pipeting. Add 1 mL sterilization solution 2 and incubate for 15 min, mixing by inversion every 5 min. Remove solution by pipeting and wash seeds with 1 mL sterile distilled water three times. Resuspend seeds in 0.5 mL water. Using a blue pipet with the distal 5 mm cut off, gently pipet the seeds onto the surface of the culture medium in glass culture vessels (*see* **Note 2**).
3. Culture vessels are shaken on an orbital shaker at 80 rpm and maintained at 22°C under illumination (16-h photoperiod, 110 µmol photons/m^2/s). Seedlings should be harvested between 10 and 14 d later.

3.3. *Preparation of Density Gradients*

1. Appropriate concentrations of Percoll are prepared in the appropriate gradient buffer. Gradients should be prepared in advance and stored on ice until required.
2. For isolation of mitochondria from heterotrophic cell suspension cultures, two different gradients are used. The first is a step gradient consisting of 5 mL 40% (v/v) Percoll overlaid with 20 mL 23% (v/v) Percoll and then 10 mL 18% (v/v) Percoll, all made up using gradient buffer 1. The layers can be conveniently poured over one another by allowing the solution to run through a 19-gauge hypodermic needle placed against the inside of the centrifuge tube held at a 45° angle. The second gradient consists of 30 mL 28% (v/v) Percoll made up in gradient buffer 2. A sigmoidal gradient self-forms during centrifugation because of the sedimentation of the Percoll polydispersed silica colloid.
3. For isolation of mitochondria from *Arabidopsis* hydroponic seedling cultures, a single linear PVP-40 gradient in 28% (v/v) Percoll is used. Preparation of this gradient requires a gradient maker, which consists of a Perspex block with two cylindrical chambers connected by a small pipe at their base (**Fig. 1**). One of the chambers has an outflow. Both the connecting pipe and outflow are metered by taps. Two solutions are prepared: a "heavy" solution, in this case 28% (v/v) Percoll and 4.4% (v/v) PVP-40 from a 20% (w/v) stock made up in gradient buffer 2, and a "light" solution (28% v/v Percoll made up in gradient buffer 2). The light solution (15 mL) is placed in the chamber with no outflow. The tap connecting the two chambers should be briefly opened to displace air in the connecting pipe and ensure a flow. The heavy solution (15 mL) is placed into the chamber with the outflow. A magnetic stir bar is placed in the chamber with the outflow, and the solution rapidly stirred. The outflow pipe should be secured against the inside of a 40 -mL centrifuge tube held at a 45°

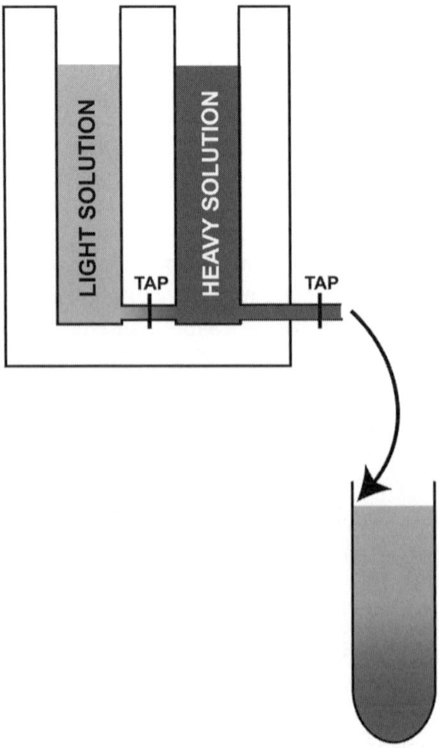

Fig. 1. Preparation of linear gradients with a gradient maker. High-density (heavy solution) and low-density (light solution) are arranged in the chambers of the gradient maker as shown. A magnetic stir bar in the chamber containing the heavy solution should rotate at a speed sufficient to ensure complete mixing. To pour the gradient, both taps should be opened simultaneously, and the outflow should be allowed to run out by gravity and down the side of a centrifuge tube held below the apparatus.

angle (above the final level to which the gradient solution will reach in the tube). It is important that the end of the outflow tube is held below the level of the gradient maker so that solution will flow by gravity (**Fig. 1**). The taps on both the connecting pipe and the outflow are then opened simultaneously. As the gradient pours, the level of solution in both chambers should drop at the same rate.

3.4. Isolation of Mitochondria

3.4.1. From Heterotrophic *Arabidopsis* Cell Suspension Cultures

1. To preserve the integrity of isolated mitochondria, it is essential that all plasticware and glassware be free from traces of detergent (*see* **Note 3**). All procedures should be done at 4°C, solutions should be prechilled, and the work should be

done in a cold room. Centrifugation should be done using a refrigerated centrifuge.

2. Separate cells from culture medium by filtration through a single layer of Miracloth using a Buchner funnel and a vacuum pump. Weigh cells.
3. Transfer up to 150 g FW of cells to a 500-mL Waring blender vessel. Add at least 2 volumes of cell extraction medium per weight of cells (*see* **Note 4**) and blend for 15 s at high speed and twice for 15 s at low speed. Leave 30-s intervals between bursts of blending to prevent excessive foaming and heating.
4. Filter extract by wringing through one layer of Miracloth and two layers of muslin to remove starch, cell debris, and unbroken cells (*see* **Note 5**).
5. Transfer filtrate to 250-mL centrifuge tubes and centrifuge at 1500*g* for 10 min. Gently pour resulting supernatant into fresh centrifuge tubes. Take care not to transfer any of the pellet, which is quite loose.
6. Centrifuge the supernatant from **step 5** at 18,000*g* for 15 min. Pour off and discard the resultant supernatant.
7. Add 1 mL cell mitochondria wash medium to the pellet and gently resuspend using a small paintbrush. Transfer all resuspended pellets to a single 50-mL centrifuge tube and add wash medium to 30 mL.
8. Repeat **steps 5** and **6**.
9. Resuspend the final pellet in 1 mL cell mitochondria wash medium using a paintbrush as before. You now have an organelle suspension. It is important to complete **steps 1–9** and move on to the density gradient centrifugation as quickly as possible. Carefully pipet the organelle suspension onto the surface of the 40/23/18% step gradient (*see* **Subheading 3.3.**) The organelle suspension from up to 10 cell cultures can be loaded onto a single gradient. Centrifuge at 40,000*g* for 30 min. It is important that the centrifuge brake be disengaged at the end of the run because rapid deceleration can disturb the contents of the gradient.
10. Mitochondria appear as a diffuse white/pale brown band at the interface of the 40/23% Percoll layers (*see* **Note 6**). Peroxisomes are just beneath but are not usually visible. Plastid membranes appear as a bright yellow band at the 23/18% Percoll interface. Aspirate the upper part of the gradient to waste. Using a 5-mL pipet, carefully transfer the mitochondria band to a fresh 50-mL centrifuge tube. Generally, 7–10 mL are sufficient to transfer all of the mitochondria. As much as possible, avoid the peroxisomal band below.
11. Add 20 mL cell mitochondria wash medium to the mitochondria and centrifuge at 18,000*g* for 10 min. Remove and discard the supernatant by aspiration. Take care not to disturb the pellet, which is very loose. It is not necessary to remove all the supernatant; leave 3–4 mL in the bottom of the tube.
12. Resuspend the mitochondrial pellet by gentle swirling and carefully pipet onto the 28% Percoll gradient. Centrifuge at 40,000*g* for 30 min with the brake off as before. Mitochondria will form a white/pale brown band in the upper part of the gradient, and peroxisomes will band toward the bottom of the gradient. Carefully transfer the mitochondria to a fresh tube using a 5-mL pipet.

13. Add 20 mL cell mitochondria wash medium to the mitochondria and centrifuge at 18,000*g* for 10 min. Remove and discard the supernatant by aspiration.
14. Repeat **step 13**. Pour off and discard the supernatant (the pellet is firm at this stage) and gently resuspend the mitochondria in 0.5 mL cell mitochondria wash medium using a small paintbrush.

3.4.2. Isolation of Mitochondria From Hydroponic Arabidopsis *Seedling Cultures*

1. To preserve the integrity of isolated mitochondria, it is essential that all plasticware and glassware be free from traces of detergent (*see* **Note 3**). All procedures should be done at 4°C, solutions should be prechilled, and the work should be done in a cold room. Centrifugation should be done using a refrigerated centrifuge.
2. Remove seedlings from six to eight culture vessels (~60 g), place in a Buchner funnel, and wash with sterile distilled water.
3. Place washed seedlings in a Perspex vessel and cover with 300 mL seedling extraction medium. Homogenize tissue with a Polytron homogenizer equipped with a 20-cm long × 2-cm diameter dispersing head using three or four 2-s bursts.
4. Filter extract by wringing through one layer of Miracloth and two layers of muslin to remove starch, cell debris, and unbroken cells.
5. Transfer filtrate to 250-mL centrifuge tubes and centrifuge at 1100*g* for 5 min. Gently pour resulting supernatant into fresh centrifuge tubes. Take care not to transfer any of the pellet, which is quite loose.
6. Centrifuge the supernatant from **step 5** at 18,000*g* for 20 min. Pour off and discard the resultant supernatant.
7. Add 1 mL leaf mitochondria wash medium to the pellets and gently resuspend using a small paintbrush. Transfer resuspended pellets to two 50-mL centrifuge tubes and add wash medium to each to 30 mL.
8. Repeat **steps 5** and **6**.
9. Resuspend the final pellets in 1 mL leaf mitochondria wash medium using a paintbrush as before. You now have an organelle suspension. It is important to complete **steps 1–8** and move on to the density gradient centrifugation as quickly as possible. Carefully pipet the two organelle suspensions onto the surface of two 28% Percoll, 0–4.4% PVP-40 gradients (*see* **Subheading 3.3.**). Centrifuge at 40,000*g* for 40 min. It is important that the centrifuge brake be disengaged at the end of the run because rapid deceleration can disturb the contents of the gradient.
10. Mitochondria will form a white/pale brown band toward the bottom of the gradient, and thylakoids are apparent as a dark green band in the upper part of the gradient (**Fig. 2**). Aspirate the upper part of the gradient to waste. Using a 5-mL pipet, carefully transfer the mitochondria band to a fresh 50-mL centrifuge tube. Generally, 7–10 mL are sufficient to transfer all of the mitochondria.
11. Add 20 mL leaf mitochondria wash medium to the mitochondria and centrifuge at 31,000*g* for 15 min. Remove and discard the supernatant by aspiration. Take care not to disturb the pellet, which is very loose. It is not necessary to remove all the supernatant; generally, leave 3–4 mL in the bottom of the tube.

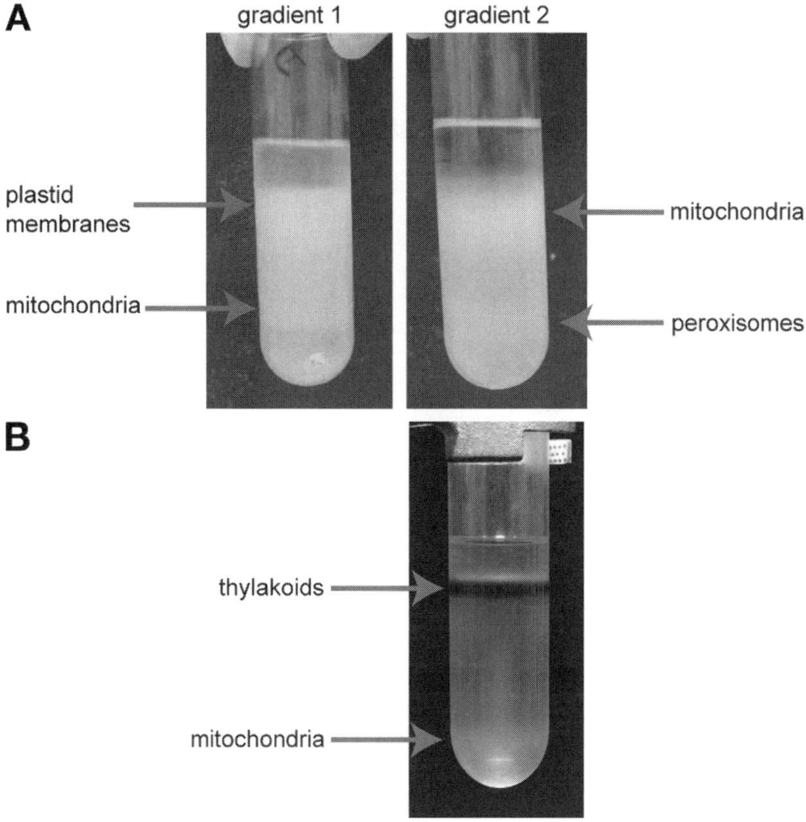

Fig. 2. Appearance of organelle bands on different density gradients: (**A**) organelles from heterotrophic *Arabidopsis* cell suspension cultures; (**B**) organelles from hydroponic seedling culture tissue.

12. Repeat **step 10** and pour off the resultant supernatant. Add 0.5 mL leaf mitochondria wash medium and resuspend pellet gently with a small paintbrush.

3.5. Determination of Mitochondrial Integrity

1. Mitochondrial integrity can be assessed by measuring the latency of the assay of COX to added cytochrome-*c*. COX is an inner membrane respiratory complex (complex IV) and requires reduced cytochrome-*c* in the intermembrane space as a substrate. Cytochrome-*c* is too large to traverse an intact outer mitochondrial membrane, and thus when intact mitochondria are incubated with reduced cytochrome-*c*, COX activity should remain at zero. Thus, by comparing COX activity in the presence and absence of Triton X-100 to rupture the outer membrane, an estimation of outer mitochondrial membrane integrity can be obtained.

2. COX is measured as the cytochrome-*c*-dependent consumption of oxygen, detected by using a Clark-type oxygen electrode. Set up the electrode according to the manufacturer's instructions using 50% saturated KCl as an electrolyte and calibrate between air-saturated water (253 nmol O_2/mL at 25°C) and zero (established by adding a few crystals of sodium hydrosulfite to the water) (*see* **Note 7**).
3. The assay should be conducted at 25°C. To 1 mL mitochondrial reaction medium add 5–20 µL mitochondrial suspension (*see* **Note 8**) and the following in sequence, waiting for a linear rate of oxygen consumption to be established each time: 20 µL 500 m*M* Na-ascorbate (*see* **Note 9**), 10 µL 5 m*M* cytochrome-*c*, and 5 µL 10% (v/v) Triton X-100. COX activity is given by [rate (c) – rate (a)]. Mitochondrial outer membrane integrity is given by:

$$1-\left[\frac{rate(b)-rate(a)}{rate(c)-rate(a)}\right]*100$$

3.6. Measurement of Mitochondrial Respiratory Function

1. Respiration of a variety of substrates can be measured as oxygen consumption using an oxygen electrode (*see* **Subheading 3.5.**). The dependence of reaction rate on added ADP can be used to calculate the respiratory control ratio that gives an indication of the extent to which the mitochondria are coupled.
2. Add 1 mL mitochondrial reaction medium and 5–20 µL of mitochondrial suspension to the reaction chamber of the oxygen electrode. Determine the background rate of oxygen consumption [rate (a)]. Add 10 µL of each of the respiratory substrates listed in **Subheading 2.6.** and determine the rate of oxygen consumption [rate (b)]. Add 10 µL 10 m*M* ADP. The rate of oxygen consumption should increase: measure the initial linear rate [rate (c)]. After a few minutes, the ADP will be depleted, and the rate of oxygen consumption will be reduced to [rate (d)]. The respiratory control ratio is given by:

$$\left[\frac{rate(d)-rate(a)}{rate(b)-rate(a)}\right]$$

4. Notes

1. To prevent contamination and poor cell growth, it is vital that the 250-mL conical flasks used for cell cultures and the 100-mL glass culture vessels used for the hydroponic seedling cultures be scrupulously clean and sterile. We recommend that a dedicated supply of flasks be established. When cleaning flasks, it is advisable to soak the flasks in detergent solution overnight before scrubbing thoroughly by hand. To remove traces of detergent, we run the flasks through a detergent-free dishwasher cycle.
2. It is important that the seeds remain close to the surface (within 1 cm of the upper surface) of the medium so that on germination seedlings can establish in the air space above the culture medium. Ensure that the medium has fully cooled

after autoclaving before attempting to transfer seeds. It may help to cool the medium to 4°C.

3. It is recommended that a dedicated stock of detergent-free glass- and plasticware is kept solely for mitochondrial isolations. Buy new glass- and plasticware for this purpose and clean using hot water only.
4. One of the common causes of poor recovery of intact mitochondria is insufficient extraction medium in relation to the amount of tissue extracted. *Arabidopsis* cells are highly vacuolated, and on rupture, the cell volume considerably dilutes the osmoticum. If insufficient extraction medium is used, then this dilution effect can lead to organelle rupture.
5. When wringing extracts through muslin/Miracloth, it is advisable to apply steady, *gentle* pressure. Too much pressure can cause the cloth to rupture suddenly.
6. A common reason for low yield of mitochondria is a failure to recognize correctly the mitochondrial band on the gradient. Refer to **Fig. 2** for guidance.
7. The performance of oxygen electrodes deteriorates over time because of electrochemical deposition of chloride and oxide salts on the silver anode. It is therefore necessary to clean the anode periodically using aluminum oxide polishing paste.
8. It is important not to add too much mitochondrial suspension; otherwise, the rate of oxygen consumption will exceed the response time of the oxygen electrode. It is therefore advisable to test several different concentrations of mitochondrial suspension and choose one in which the oxygen consumption rate in the presence of cytochrome-*c* and Triton X-100 is proportional to the amount of mitochondrial suspension used. This comment also applies to measurement of respiratory capacity.
9. The Na-ascorbate solution oxidizes rapidly and should therefore be made fresh on the day of use.

Acknowledgments

L. J. S. and C. J. L. acknowledge financial support from the Biotechnological and Biological Sciences Research Council, United Kingdom. N. L. T. is funded by a long-term EMBO fellowship. The authors would like to thank Joshua Heazlewood, University of Western Australia, for the pictures of organelles from *Arabidopsis* cell suspension cultures separated by density gradient centrifugation and Megan Morgan and John Baker for the pictures of organelles from hydroponic seedling culture tissue separated by density gradient centrifugation.

References

1. The Arabidopsis Genome Initiative. (2000) Analysis of the genome sequence of the flowering plant *Arabidopsis thaliana. Nature* **408,** 796–815.
2. Sessions, A., Burke, E., Presting, G., et al. (2002) A high-throughput *Arabidopsis* reverse genetics system. *Plant Cell* **14,** 2985–2994.
3. Millar, A. H. (2004) Location, location, location: surveying the intracellular real estate through proteomics in plants. *Funct. Plant Biol.* **31,** 563–571.

4. Gibon, Y., Blaesing, O. E., Hannemann, J., et al. (2004) A robot-based platform to measure multiple enzyme activities in *Arabidopsis* using a set of cycling assays: comparison of changes of enzyme activities and transcript levels during diurnal cycles and in prolonged darkness. *Plant Cell* **16,** 3304–3325.
5. Millar, A. H., Sweetlove, L. J., Giege, P., and Leaver, C. J. (2001) Analysis of the *Arabidopsis* mitochondrial proteome. *Plant Physiol.* **127,** 1711–1727.
6. Day, D. A., Neuberger, M., and Douce, R. (1985) Biochemical characterization of chlorophyll-free mitochondria from pea leaves. *Austr. J. Plant Physiol.* **12,** 219–228.
7. Millar, A. H. M., Liddell, A., and Leaver, C. J. (2001) Isolation and subfractionation of mitochondria from plants. *Methods Cell Biol.* **65,** 53–74.
8. May, M., and Leaver, C. (1993) Oxidative stimulation of glutathione synthesis in *Arabidopsis thaliana* suspension cultures. *Plant Physiol.* **103,** 621–627.
9. Xiang, C., and Oliver, D. J. (1998) Glutathione metabolic genes coordinately respond to heavy metals and jasmonic acid in Arabidopsis. *Plant Cell* **10,** 1539–1550.

10

Chlamydomonas reinhardtii: The Model of Choice to Study Mitochondria From Unicellular Photosynthetic Organisms

Soledad Funes, Lars-Gunnar Franzén, and Diego González-Halphen

Summary

Chlamydomonas reinhardtii is a model organism to study photosynthesis, cellular division, flagellar biogenesis, and, more recently, mitochondrial function. It has distinct advantages in comparison to higher plants because it is unicellular, haploid, and amenable to tetrad analysis, and its three genomes are subject to specific transformation. It also has the possibility to grow either photoautotrophically or heterotrophically on acetate, making the assembly of the photosynthetic machinery not essential for cell viability. Methods developed allow the isolation of *C. reinhardtii* mitochondria free of thylakoid contaminants. We review the general procedures used for the biochemical characterization of mitochondria from this green alga.

Key Words: Blue native polyacrylamide gel electrophoresis; *Chlamydomonas reinhardtii*; green alga; import-competent mitochondria; OXPHOS.

1. Introduction

Bovine heart mitochondria, which may be prepared in large quantities, have been the system of choice for the purification and structural characterization of oxidative phosphorylation (OXPHOS) complexes *(1)*. Because a large variety of molecular genetic tools may be used with *Saccharomyces cerevisiae (2)*, many respiratory mutants of this yeast have been generated and characterized. Therefore, yeast is the model organism to study the effect of mutations on OXPHOS components *(3)*, mitochondrial protein import *(4)*, and the biogenesis of mitochondrial complexes *(5)*.

Whole-genome sequencing of various organisms has attracted interest in the mitochondrial proteome of other less-studied organisms, particularly of photo-

From: *Methods in Molecular Biology, vol. 372: Mitochondria: Practical Protocols*
Edited by: D. Leister and J. M. Herrmann © Humana Press Inc., Totowa, NJ

synthetic species. Such is the case of the higher plant *Arabidopsis thaliana* ***(6–8)***. Nevertheless, for experiments involving the isolation of purified mitochondria such as in vitro protein import ***(9)***, it is difficult to obtain enough material from *Arabidopsis* plants. Instead, spinach leaves or potato tubers may be used. In general, mitochondria of photosynthetic organisms have been poorly characterized from a biochemical point of view, mainly because of the difficulties in obtaining preparations free of chloroplast contaminants. As a model system for eukaryotic photosynthetic organisms, the chlorophycean alga *Chlamydomonas reinhardtii* is suitable for many studies, and it may be regarded as a "photosynthetic yeast" ***(10)***.

Chlamydomonas reinhardtii is unicellular, haploid, and subject to tetrad analysis, and its three genomes (nuclear, chloroplastic, and mitochondrial) are amenable to specific transformation. The green alga may grow either photoautotrophically or heterotrophically on acetate, making the assembly of the photosynthetic machinery not essential for cell viability. Therefore, photosynthesis-deficient mutants may be grown heterotrophically, and respiratory-deficient mutants may be grown as obligate photoautotrophs. *Chlamydomonas reinhardtii* has become a model system for studying chloroplast biogenesis, photosynthesis, and flagellar structure and assembly. This green alga is also a suitable model for study of the interaction between chloroplasts and mitochondria in a unicellular organism. In pioneering works, its 15.8-kb linear mitochondrial genome was sequenced ***(11)***, and enriched fractions of several OXPHOS complexes were characterized ***(12)***. In addition, several nuclear genes encoding polypeptides that participate in OXPHOS were identified ***(13–21)***. In parallel, procedures to obtain pure mitochondria from *C. reinhardtii* free of thylakoid contaminants were developed ***(16,22)***, a feat considering that the cellular volume occupied by mitochondria is around 4%, whereas the cell volume occupied by the chloroplast is around 40% ***(23)***. Also, several *C. reinhardtii* mutants affecting mitochondrial components were characterized ***(24,25)***. Cell-wall-less *C. reinhardtii* mutants have allowed the isolation of chlorophyll-free mitochondrial preparations and therefore the resolution of the major mitochondrial complexes of *C. reinhardtii* with blue native polyacrylamide gel electrophoresis (BNE) ***(26,27)***. In addition, the large amount of information generated by the *C. reinhardtii* genome-sequencing project ***(28)*** has also allowed the prediction of the OXPHOS proteome of this green alga ***(29)***.

Other than the characterization of classical OXPHOS complexes, pure *C. reinhardtii* mitochondria also allow the study of the nonorthodox respiratory components peculiar of photosynthetic organisms, including several alternate nicotinamide adenine dinucleotide (NADH) dehydrogenases and at least two alternative terminal oxidases.

Summarizing, *C. reinhardtii* is the model system of choice to study mitochondria of unicellular photosynthetic organisms because it allows combining genetical analysis of mutants with a biochemical approach. Here, we concentrate on three aspects of *C. reinhardtii* mitochondria: (1) preparation of the organelles free of thylakoid membrane contamination; (2) preparation of mitochondrial samples for BNE, and (3) preparation of import-competent organelles.

2. Materials

2.1. Media and Growth Conditions

1. 20X Beijerinck's solution: 8 g/L NH_4Cl, 1 g/L $CaCl_2{\cdot}2H_2O$, 2 g/L $MgSO_4{\cdot}7H_2O$.
2. 1 *M* Potassium phosphate buffer, pH 7.0.
3. Ethylenediaminetetraacetic acid (EDTA) solution: dissolve 50 g acid-free EDTA-Na_2 in 250 mL H_2O.
4. Zinc sulfate solution: dissolve 22 g $ZnSO_4{\cdot}7\ H_2O$ in 100 mL H_2O.
5. H_3BO_3 solution: dissolve 11.3 g of H_3BO_3 in 200 mL H_2O.
6. Iron sulfate solution: dissolve 4.99 g $FeSO_4{\cdot}7\ H_2O$ in 50 mL H_2O.
7. Manganese chloride solution: dissolve 5.06 g $MnCl_2{\cdot}4H_2O$ in 50 mL H_2O.
8. Cobalt chloride solution: dissolve 1.61 g $CoCl_2{\cdot}6H_2O$ in 50 mL H_2O.
9. Cupper sulfate solution: dissolve 1.57 g $CuSO_4{\cdot}5H_2O$ in 50 mL H_2O.
10. Ammonium molybdate solution: dissolve 1.1 g $(NH_4)_6Mo_7O_{24}{\cdot}4H_2O$ in 50 mL H_2O.
11. Trace elements stock solution: mix solutions from **items 4–10**. The resulting solution should be violet. Heat the solution at 100°C until the color changes to clear green. Add the EDTA solution and let the mixture cool (the temperature should not drop below 70°C). Adjust to pH 6.5–6.8 with 20% (w/v) KOH. Dilute the solution to 1000 mL with water and let stand at room temperature in a loosely stoppered Erlenmeyer flask until the color changes to violet. Filter the solution through a 0.22-µm Millipore filter to clear the solution from a red precipitate. Store the solution in a brown bottle at room temperature.
12. Tris-acetate-phosphate medium *(23)*: 50 mL 20X Beijerinck's solution, add glacial acetic acid to pH 7.0 (~1 mL). Add 2.42 g Tris-HCl, 1 mL 1 *M* potassium phosphate buffer, pH 7.0, 1 mL trace elements stock solution. Add water to bring to 1 L final volume. Sterilize the media by autoclaving 20 min at 15 psi (1.02 atm) using a liquid cycle.

2.2. Isolation of Mitochondria Free of Thylakoid Contaminants

1. Washing buffer: 20 m*M* HEPES-KOH, pH 7.2.
2. Breaking buffer: 50 m*M* HEPES-KOH, pH 7.2, 5 m*M* EDTA, 250 m*M* sorbitol, 4 m*M* cysteine, 0.5% (w/v) polyvinylpyrrolidone-40 (PVP-40), 0.1% (w/v) bovine serum albumin (BSA).
3. Assay buffer: 250 m*M* sorbitol, 10 m*M* KCl, 10 m*M* potassium phosphate buffer, pH 7.2, 0.1% (w/v) BSA, 5 m*M* $MgCl_2$.
4. Percoll buffer: 250 m*M* sorbitol, 10 m*M* 3-(*N*-morpholino)propanesulfonic acid (MOPS)-KOH, pH 7.2, 0.5% (w/v) PVP-40, 20% (v/v) Percoll, 0.1% (w/v) BSA, 1 m*M* EDTA.
5. Percoll dilution buffer: 10 m*M* potassium phosphate buffer, pH 7.2, 250 m*M* sorbitol, 10 m*M* EDTA, 0.1% (w/v) BSA.

6. Percoll dilution buffer supplemented with protease inhibitors: supplement the Percoll dilution buffer with the following: 0.5 m*M* phenylmethylsulfonyl fluoride (PMSF), 1 m*M* benzamidine, 5 m*M* ε-amino caproic acid.

2.3. Oxymetric Measurements

1. Respiration medium: 0.2 *M* sucrose, 10 m*M* KCl, 5 m*M* $MgCl_2$, 0.1% (w/v) BSA, 10 m*M* potassium phosphate buffer, pH 7.2.
2. Stock solutions of substrates (10X): (a) 100 m*M* malate, 100 m*M* glutamate, 10 m*M* NAD^+, 10 m*M* potassium phosphate buffer, pH 7.2; or (b) 100 m*M* succinate, 10 m*M* potassium phosphate buffer, pH 7.2.
3. Stock solution of uncoupler (10X): 10 μ*M* carbonylcyanide-*p*-trifluoromethoxy-phenylhydrazone in ethanol or dimethyl sulfoxide (DMSO).
4. Stock solutions of inhibitors (10X): (a) freshly made solution of 30 m*M* rotenone in ethanol or DMSO (complex I inhibitor); (b) 1 m*M* antimycin A and 1 m*M* myxothiazol in ethanol or DMSO (complex III inhibitors); (c) freshly made solution of 10 m*M* KCN in water (complex IV inhibitor); or (d) 20 m*M* salicylhydroxamic acid in water (alternative oxidase inhibitor).
5. Adenosine 5′-diphosphate (ADP) solution: 100 m*M* ADP (disodium salt dissolved in water).

2.4. Preparation of Mitochondrial Samples for Blue Native Electrophoresis

1. Washing buffer: 0.25 *M* sorbitol, 15 m*M* BisTris, pH 7.0.
2. Sample buffer: 50 m*M* BisTris, pH 7.0, 0.75*M* ε-amino caproic acid, pH 7.0.
3. Detergent, stock solution: 10% (w/v) *n*-dodecyl maltoside.
4. Coomassie Serva Blue G, stock solution: 5% (w/v) Coomassie Serva Blue G.

2.5. Separation of OXPHOS Complexes by BNE

1. 3X (threefold concentrated) Gel buffer: 1.5 *M* ε-amino caproic acid, 150 m*M*, BisTris (or, alternatively, 75 m*M* imidazole); adjust to pH 7.0 (4°C) with HCl.
2. Cathode buffer (upper buffer): 50 m*M* Tricine,15 m*M* BisTris (or alternatively, 7.5 m*M* imidazole), 0.02% (w/v) Coomassie Blue G (Serva); adjust to pH 7.0 (4°C) with HCl.
3. Anode buffer (lower buffer): 50 m*M* BisTris (or, alternatively, 25 m*M* imidazole); adjust to pH 7.0 (4°C) with HCl.
4. AB mix (acrylamide-bisacrylamide mixture stock solution): 48 g acrylamide and 1.5 g of bisacrylamide in 100 mL water.
5. APS (10% w/v): ammonium persulfate solution 10% (w/v), freshly prepared.
6. Solution for stacking gel (7.5 mL total volume): 0.6 mL AB mix, 2.5 mL gel buffer (3X), 60 μL APS 10% (w/v), 6 μL *N,N,N,N*′-tetramethylethylenediamine (TEMED), 4.33 mL water.
7. Solution A for gradient gel (21 mL total volume): 2.5 mL AB mix, 7.0 mL gel buffer (3X), 90 μL APS 10% (w/v), 9 μL TEMED, 11.4 mL water.
8. Solution B for gradient gel (18 mL total volume): 4.7 mL AB mix, 6.0 mL gel buffer (3X), 3.6 g glycerol, 60 μL APS 10% (w/v), 6 μL TEMED, 6.7 mL water.

9. Denaturing solution for BNE gels: cathode buffer containing 1% (w/v) sodium dodecyl sulfate (SDS), 1% (v/v) β-mercaptoethanol.

2.6. Separation of OXPHOS Components in Second-Dimension Tricine-SDS Gels

1. AB mix: 48.5% (w/v) acrylamide, 1.5% (w/v) bisacrylamide.
2. 3X stock solution (threefold concentrated) Gel buffer: 3 *M* Tris-HCl, pH 8.45; 0.3% (w/v) SDS.
3. 10X (10-fold concentrated) Cathode buffer: 1 *M* Tris-HCl, 1 *M* Tricine 1% (w/v) SDS. No pH adjustment is necessary.
4. 10X (10-fold concentrated) Anode buffer: 1 *M* Tris-HCl, pH 8.9.
5. APS solution: 10% (w/v) ammonium persulfate. Freshly prepared.
6. Stacking gel solution (12 mL total volume): 1.0 mL AB mix, 4.0 mL gel buffer (3X), 100 μL APS (10%) (w/v), 25 μL TEMED, 7.0 mL water.
7. Running gel, 15% solution (36 mL total volume): 11.1 mL AB mix, 12.0 mL 3X gel buffer, 7.2 mL 80% (v/v) glycerol, 150 μL 10% (w/v) APS, 25 μL TEMED, 7.9 mL water.

2.7. Preparation of Import-Competent Mitochondria From Chlamydomonas reinhardtii

1. Preparation buffer: 0.25 *M* sucrose, 50 m*M* MOPS-KOH, pH 7.4, 5 m*M* EDTA, 10 m*M* cysteine, 0.5 m*M* PMSF, 0.6% (w/v) BSA.
2. Percoll buffer: 0.25 *M* sucrose, 10 m*M* MOPS-KOH, pH 7.2, 1 m*M* EDTA, 0.2% (w/v) BSA, 32% (v/v) Percoll (Pharmacia).
3. Wash buffer: 0.25 *M* sucrose, 25 m*M* MOPS-KOH, pH 7.2, 1 m*M* EDTA, 0.1% BSA.
4. Import buffer: 0.25 *M* mannitol, 50 m*M* KCl, 10 m*M* MOPS-KOH, pH 7.4, 2 m*M* adenosine triphosphate (ATP), 2 m*M* guanosine 5′-triphosphate, 10 μ*M* ADP, 1 m*M* methionine,1 m*M* glycine, 5 m*M* potassium phosphate, 1% (w/v) BSA.
5. Sonication buffer: 0.25 *M* sucrose, 10 m*M* MOPS, pH 7.5, 5 m*M* $MgCl_2$, 5 m*M* $MnCl_2$, 2 m*M* PMSF.
6. Yeda press.

2.8. In Vitro Expression of Complementary Deoxyribonucleic Acid Clones

1. Transcription/translation System: TNT T7 coupled reticulocyte system (Promega).
2. [^{35}S]-Methionine.

2.9. In Vitro Protein Import

1. Stock solution: 100 m*M* malate in 10 m*M* MOPS-KOH, pH 7.4.
2. Stock solution: 100 m*M* NADH in 10 m*M* MOPS-KOH, pH 7.4.
3. Stock solution: 100 μg/mL proteinase K in 10 m*M* MOPS-KOH, pH 7.4 (freshly made).
4. 100 m*M* PMSF in DMSO (freshly made).

2.10. In Vitro Protein Processing

1. Processing buffer: 15 m*M* Tris-HCl, pH 8.0, 0.5% (v/v) Triton X-100, 2 m*M* $MnCl_2$, 1 m*M* PMSF.
2. Solubilization buffer (stock solution): 5% (v/v) Triton X-100 in 2 m*M* PMSF and 15 m*M* Tris-HCl, pH 8.0.

2.11. SDS-Polyacrylamide Gel Electrophoresis and Autoradiography

For SDS-polyacrylamide gel electrophoresis (PAGE) and autoradiography, use Amplify (Amersham).

2.12. Protein and Chlorophyll Determination

1. Bio-Rad protein assay.
2. 80% (v/v) aqueous acetone.

3. Methods

3.1. Media and Growth Conditions

Chlamydomonas reinhardtii cells (*see* **Note 1** and **ref. *30***) are grown mixotrophically at 25°C under continuous light and moderate stirring or orbital agitation (*see* **Note 2**). To decrease duplication time, cultures may be bubbled with 5% (v/v) CO_2 in air.

Strains may be preserved in Tris-acetate-phosphate solid media cultures containing 1.5% (w/v) agar. Long-term stocks of the strains can also be done by freezing in the presence of 5% (v/v) DMSO, although no freezing method is known to guarantee the stability of mutant strains.

3.2. Isolation of Mitochondria Free of Thylakoid Contaminants

Eriksson et al. ***(22)*** developed a method to isolate *C. reinhardtii* mitochondria free of thylakoid contaminants. van Lis et al. ***(26)*** introduced some modifications to the procedure. The method is based on mild breakage of cells, differential centrifugations, and Percoll gradients.

1. Collect the cells in their late exponential growth phase (at a chlorophyll concentration of 10–12 μg/mL) by centrifugation at 3000*g* for 8 min.
2. Wash the cells twice in washing buffer (*see* **Note 3**).
3. Resuspend the cell pellet in breaking buffer at a final concentration of 0.5 g cells/mL.
4. Mix the cell suspension with 2 volumes of glass beads (0.5-mm diameter).
5. Vortex the mixture 1 min at low speed (~95% of the cells are broken in this step). From this point, it is important to keep solutions and samples at 4°C at all times (*see* **Note 4**).
6. Centrifuge the mixture for 5 min at 2000*g* and collect the supernatant.
7. Wash the glass beads three times with breaking buffer, centrifuging each time 5 min at 2000*g* and collecting the supernatants.
8. Mix all the supernatants and disrupt the solution with a prechilled Potter-Elvehjem homogenizer.

9. Centrifuge the homogenized solution at 1500*g*. Collect the supernatant.
10. Centrifuge for 10 min at 12,000*g*. Discard the supernatant. The pellet obtained during this step includes the crude mitochondrial extract (*see* **Note 5**).
11. Resuspend the mitochondrial pellet in 5 mL assay buffer and mix with Percoll buffer; take to a final volume of 35 mL.
12. Homogenize the sample with a Potter-Elvehjem homogenizer.
13. Centrifuge 40 min at 20,000*g*. Recover the mitochondrial fraction, which is found very close to the bottom of the tube, sometimes as a pellet.
14. Dilute the mitochondrial fraction with 40 mL Percoll dilution buffer.
15. Centrifuge 10 min at 12,000*g*. Discard the supernatant.
16. Repeat this washing step one more time.
17. Resuspend the final mitochondrial pellet in 35 mL Percoll dilution buffer supplemented with protease inhibitors. Typically, 2–3 mg of mitochondrial proteins are obtained from 1.5 L of the original culture.

3.3. Oxymetric Measurements

1. Take 300 μg mitochondrial protein extract and resuspend in 2 mL respiration buffer in the oxymeter chamber.
2. Add the substrate solution of choice (to obtain a final concentration of 10 m*M* malate, 10 m*M* glutamate, and 1 m*M* NAD^+ or of 10 m*M* succinate).
3. Follow the oxygen consumption.
4. To characterize the respiratory chain, the different complexes may be inhibited with 3 m*M* rotenone, 10 μ*M* antimycin A, 10 μ*M* myxothiazol, 1 m*M* KCN, or 2 m*M* salicylhydroxamic acid.
5. The uncoupling solution can be used in a 1:100 dilution to verify if the mitochondrial preparation exhibits respiratory control. Alternatively, ADP may be added to a final concentration of 200 μ*M* to observe state 3–state 4 transitions.

3.4. Preparation of Mitochondrial Samples for BNE

1. Mitochondria are centrifuged for 5 min at 4°C at maximal velocity in a tabletop centrifuge. From this point, it is important to keep solutions and samples at 4°C at all times.
2. Resuspend in washing buffer and recentrifuge. Remove and discard supernatant.
3. Resuspend the mitochondrial pellet in a small amount of sample buffer.
4. Mitochondria are solubilized at a final protein concentration of 5 mg/mL in sample buffer and in the presence of 1% (w/v) final concentration of *n*-dodecyl maltoside. Incubate 30 min with gentle agitation (by inversion of the tube).
5. Ultracentrifuge at 100,000*g* at 4°C for 20 min. Take supernatant.
6. Add Coomassie Serva Blue G to the supernatant (0.25 Coomassie/detergent ratio w/w) (*see* **Note 6**). Mix by inversion and load into blue native gels.

3.5. Separation of OXPHOS Complexes by BNE

The development of native electrophoretic techniques like BNE ***(31)*** has allowed the resolution and partial characterization of the *C. reinhardtii* OXPHOS complexes ***(26,27)***.

For BNE, mitochondria may be prepared as described in **Subheading 3.4**. Best results are obtained when mitochondria are prepared the same day. Once solubilized with lauryl maltoside, mitochondrial proteins may be stored on ice at 4°C for up to 1 wk. Freezing and thawing must be avoided.

BNE in gradient gels (6–13%) is run in conditions described in **refs. *26*** and ***31***. Make 6–13% gradient gel with solutions A and B in a gradient former. After polymerization, load stacking gel solution in the presence of a comb. Set gel in electrophoresis chamber and add anode and cathode buffers. Run gel at 25 mA (constant current) at 4°C.

3.6. Separation of OXPHOS Components in Second-Dimension Tricine-SDS Gels

Make 15% Tricine-SDS gel with solution for running gel. Let polymerize. Cut a lane of interest from the BNE gel and incubate for 30 min in denaturing solution for BNE gels, with gentle agitation at room temperature. Wash briefly in cathode buffer to remove excess β-mercaptoethanol. Lay on top of the freshly made 15% Tricine-SDS gel. Add stacking gel solution on top and let polymerize. Set gel in electrophoresis chamber and add anode and cathode buffers. Run at 50 V (constant voltage) until the blue front has entered the running gel and then set voltage to 100 V (approximate running time 5 h for 20 × 14 cm gels).

3.7. Preparation of Import-Competent Mitochondria From Chlamydomonas reinhardtii

In vitro import into *C. reinhardtii* mitochondria ***(32)*** is based on the method developed for higher plant mitochondria ***(33)***. Import-competent mitochondria from *C. reinhardtii* are harder to get than mitochondria exhibiting respiratory control. Familiarity with the system allows a "quick-and-cold" isolation of the organelles ***(32)***, which is essential for success of in vitro import studies. Below, we describe the original method ***(16)*** that allowed the isolation of import-competent mitochondria from the green alga. It is worth noting that both independently developed preparation methods ***(16,22)*** yield mitochondria that can be used in in vitro import experiments ***(32)***. The algal cell rupture seems to be a critical step in mitochondrial preparation: the more rapid (3–4 min) and gentle is the lysis, the more import-competent the organelles are. In addition, for both methods it is important to keep the samples at a low temperature during the isolation procedure (maximum 4°C, preferably on ice) and to perform the isolation procedure as quickly as possible.

1. *C. reinhardtii* CW15 or CW92 cells (about 7 g wet weight from 5 L of culture) are suspended in a small volume of preparation buffer at room temperature and then rapidly diluted into 75 mL ice-cold preparation buffer.
2. The cells are loaded in an ice-cold Yeda press, equilibrated at 6 bars for 2 min, and lysed quickly.

3. The homogenate is diluted with an equal volume of ice-cold preparation buffer containing 1.2% PVP-40 and centrifuged at 480*g* for 3 min.
4. The supernatant is recentrifuged at 3000*g* for 4 min (in the last two steps, unbroken cells, cell debris, and most of the chloroplasts are removed).
5. The supernatant is then centrifuged at 13,300*g* for 7 min.
6. The pellet, containing crude mitochondria, was suspended (*see* **Note 7**) in Percoll buffer and centrifuged at 37,000*g* for 50 min at 4°C (the mitochondria band forms close to the bottom of the tube). Mitochondria are collected, diluted 10 times with wash buffer, and centrifuged at 12,000*g* for 10 min.
7. The final pellet is suspended in a small volume (usually 0.5 mL) of the wash buffer and stored at –80°C until use (for import experiments, the mitochondria should be used directly, without freezing). Typically, 3–4 mg mitochondrial protein is obtained from 7 g (wet weight) of cells.
8. Mitochondria are washed once with import buffer, adjusted to a protein concentration of 20 mg/mL, and used directly for in vitro import experiments.

3.8. In Vitro Expression of Complementary Deoxyribonucleic Acid Clones

Complementary deoxyribonucleic acid (cDNA) clones of the *C. reinhardtii* genes encoding mitochondrial proteins may be expressed in vitro in the presence of [^{35}S]-methionine using the TNT T7 Coupled Reticulocyte System (Promega). This system has been used successfully with the cDNAs of *C. reinhardtii* ATP synthase α- and β-subunits and the mitochondrial Rieske FeS protein (*see* **Note 8**).

3.9. In Vitro Protein Import

The samples (100 μL, containing 100 μg mitochondrial proteins for each import reaction) are incubated at 23°C with continuous shaking for 10–20 min. Malate and NADH (2 m*M* each) are used as substrates. A protease protection assay is used to evaluate protein import. After the import reaction, a batch of mitochondria is incubated with proteinase K (8–10 μg/mL) on ice for 30 min to digest proteins that were not imported. The proteinase K was then inhibited by addition of PMSF (2 m*M*). The samples are washed once in import buffer before running gel electrophoresis.

3.10. In Vitro Protein Processing

1. Mitochondria are diluted to a protein concentration of 10 mg/mL with processing buffer and sonicated four times for 30 s at 0°C.
2. Centrifuge at 12,000*g* for 10 min at 4°C (residual intact mitochondria and large fragments are removed in this step).
3. Submitochondrial particles are collected by centrifugation at 20,000*g* for 50 min.

The membranes and matrix (pellet and supernatant fraction, respectively) are tested for processing activity. Solubilized mitochondria with 0.5% (v/v) Triton

X-100 in the presence of 2 m*M* PMSF and 15 m*M* Tris-HCl, pH 8.0) are used as control. In vitro processing of radiolabeled precursor proteins is carried out in a final volume of 50 μL processing buffer. The sample includes 50 mg mitochondrial proteins and 0.5 μL translation products. The processing reaction is carried out at 30°C for 30 min and stopped by addition of 50 μL twofold concentrated sample buffer of Laemmli ***(34)***. The samples are heated at 75°C for 5 min and loaded on the gel.

3.11. SDS-PAGE and Autoradiography

Samples are analyzed by SDS-PAGE on 12% gels using the buffer system of Laemmli ***(34)***. The gels are fixed, impregnated with Amplify (Amersham), and autoradiographed.

3.12. Protein and Chlorophyll Determination

Protein content is measured using the Bio-Rad protein assay. Chlorophyll concentration is determined by Arnon's method ***(35)***. For this purpose, a known volume of sample is extracted with aqueous acetone (bring to a 10-mL final volume with 80% v/v acetone), vortexed, and filtered through a Whatman no. 1 paper filter. Absorbance is read in a spectrophotometer at 663 and 645 nm. The chlorophyll concentration is calculated using the following equations:

$$\text{Chlorophyll b } (\mu\text{g/mL}) = (22.9 \times A_{645}) - (4.68 \times A_{663})$$

$$\text{Chlorophyll a } (\mu\text{g/mL}) = (12.7 \times A_{663}) - (2.69 \times A_{645})$$

$$\text{Total chlorophyll} = \text{Chlorophyll a} + \text{Chlorophyll b} = (20.2 \times A_{645}) + (8.02 \times A_{663})$$

4. Notes

1. The use of *C. reinhardtii* cell wall-deficient strains facilitates the isolation of intact mitochondria because milder procedures for cell disruption are required. CW15 and CW92 mutants are frequently chosen. These strains may be obtained from the *Chlamydomonas* Center at Duke University (http://www.chlamy.org/index.html). Nevertheless, the modifications introduced by van Lis et al. ***(26)*** to the procedure of Eriksson et al. ***(22)*** for the isolation of *C. reinhardtii* mitochondria now allows the use of cell wall-containing strains as starting material. To overcome the problem of chloroplast contaminants, other *C. reinhardtii* strains that have been used for mitochondria isolation procedures are mutants defective in the assembly of thylakoid membranes, the so-called yellow-in-the-dark mutants ***(30)***.
2. When using strains lacking cell walls, the addition of 1% (w/v) sorbitol is necessary to avoid osmotic shock.
3. When working with a strain that has a cell wall, resuspend in washing buffer (around 50 mg wet weight/mL), add 50 μ*M* cetyltrimethylammoniumbromide from a 10 m*M* stock solution and incubate with gentle agitation at room temperature for 5 min. Dilute fivefold and wash twice with washing buffer ***(26)***.

4. Alternatively, cells may be broken using a BioNeb® disruption system (Glas-Col, Terre Haute, IN) or in a Yeda press.
5. If an orange precipitate is present on top of the mitochondrial fraction, then remove gently by pipeting ***(26)***.
6. The ratio (w/w) of Coomassie Brilliant Blue G to detergent is critical for the resolution of OXPHOS complexes in BNE. A pellet of mitochondria (1 mg) has a volume of around 25 μL; detergent is added (20 μL of a 10% (w/v) solution to have 1% (w/v) final concentration), and 155 μL of sample buffer are added. After solubilization and ultracentrifugation, 10 μL of a 5% (w/v) solution of Coomassie Brilliant Blue G is added. This gives the desired 4:1 detergent/dye (w/w) ratio.
7. A mild-but-efficient method to suspend a pellet is to use a small paintbrush. Use the paintbrush to suspend the pellet in a small volume of buffer, then dilute into a larger volume.
8. When used in in vitro import experiments, the ATP synthase α-subunit and the Rieske FeS protein were imported with high efficiency; the ATP synthase β-subunit was imported with much lower efficiency.

Acknowledgments

We wish to acknowledge Dr. David Krogmann (Purdue University) for his comments and corrections to the manuscript and to Miriam Vázquez-Acevedo for the continuous technical assistance. Our work is supported by CONACyT (Mexico), a PAPIIT-UNAM grant (Mexico), a National Institutes of Health-Fogarty grant (United States), and a long-term fellowship from the European Molecular Biology Organization (EMBO) to S. F.

References

1. Saraste, M. (1999) Oxidative phosphorylation at the *fin de siècle*. *Science* **283,** 1488–1493.
2. Bonnefoy, N., and Fox, T. D. (2002) Genetic transformation of *Saccharomyces cerevisiae* mitochondria. *Methods Enzymol.* **350,** 97–111.
3. Fisher, N., Castleden, C. K., Bourges, I., Brasseur, G., Dujardin, G., and Meunier, B. (2004) Human disease-related mutations in cytochrome *b* studied in yeast. *J. Biol. Chem.* **279,** 12,951–12,958.
4. Herrmann, J. M., and Neupert, W. (2003) Protein insertion into the inner membrane of mitochondria. IUBMB Life **55,** 219–225.
5. Barrientos, A., Barros, M. H., Valnot, I., Rotig, A., Rustin, P., and Tzagoloff, A. (2002) Cytochrome oxidase in health and disease. *Gene* **286,** 53–63.
6. Jansch, L., Kruft, V., Schmitz, U. K., and Braun, H. P. (1996) New insights into the composition, molecular mass and stoichiometry of the protein complexes of plant mitochondria. *Plant J.* **9,** 357–368.
7. Eubel, H., Jansch, L., and Braun, H. P. (2003) New insights into the respiratory chain of plant mitochondria: supercomplexes and a unique composition of complex II. *Plant Physiol.* **133,** 274–286.

8. Heazlewood, J. L., Tonti-Filippini, J. S., Gout, A. M., Day, D. A., Whelan, J., and Millar, A. H. (2004) Experimental analysis of the *Arabidopsis* mitochondrial proteome highlights signaling and regulatory components, provides assessment of targeting prediction programs, and indicates plant-specific mitochondrial proteins. *Plant Cell* **16,** 241–256.
9. Glaser, E., Sjöling, S., Tanudji, M., and Whelan, J. (1998) Mitochondrial protein import in plants—signals, sorting, targeting, processing and regulation. *Plant Mol. Biol.* **38,** 311–338.
10. Rochaix, J. D. (1995) *Chlamydomonas reinhardtii* as the photosynthetic yeast. *Annu. Rev. Genet.* **29,** 209–230.
11. Gray, M. W., and Boer, P. H. (1988) Organization and expression of algal (*Chlamydomonas reinhardtii*) mitochondrial DNA. *Philos. Trans. R. Soc. Lond. B Biol. Sci.* **319,** 135–147.
12. Atteia, A. (1994) Identification of mitochondrial respiratory proteins from the green alga *Chlamydomonas reinhardtii. C. R. Acad. Sci. III* **317,** 11–19.
13. Amati, B. B., Goldschmidt-Clermont, M., Wallace, C. J., and Rochaix, J. D. (1988) cDNA and deduced amino acid sequences of cytochrome *c* from *Chlamydomonas reinhardtii*: unexpected functional and phylogenetic implications. *J. Mol. Evol.* **28,** 151–160.
14. Franzén, L.-G., and Falk, G. (1992) Nucleotide sequence of cDNA clones encoding the beta subunit of the mitochondrial ATP synthase from the green alga *Chlamydomonas reinhardtii*: the precursor protein encoded by the cDNA contains both an N-terminal presequence and a C-terminal extension. *Plant Mol. Biol.* **19,** 771–780.
15. Atteia, A., and Franzén, L.-G. (1996) Identification, cDNA sequence and deduced amino acid sequence of the mitochondrial Rieske iron-sulfur protein from the green alga *Chlamydomonas reinhardtii*. Implications for protein targeting and subunit interaction. *Eur. J. Biochem.* **237,** 792–799.
16. Nurani, G., and Franzén, L.-G. (1996) Isolation and characterisation of the mitochondrial ATP synthase from *Chlamydomonas reinhardtii*. cDNA sequence and deduced protein sequence of the a subunit. *Plant Mol. Biol.* **31,** 1105–1116.
17. Pérez-Martínez, X., Vázquez-Acevedo, M., Tolkunova, E., et al. (2000) Unusual location of a mitochondrial gene. Subunit III of cytochrome *c* oxidase is encoded in the nucleus of chlamydomonad algae. *J. Biol. Chem.* **275,** 30,144–30,152.
18. Pérez-Martínez, X., Antaramian, A., Vázquez-Acevedo, M., et al. (2001) Subunit II of cytochrome *c* oxidase in chlamydomonad algae is a heterodimer encoded by two independent nuclear genes. *J. Biol. Chem.* **276,** 11,302–11,309.
19. Dinant, M., Baurain, D., Coosemans, N., Joris, B., and Matagne, R. F. (2001) Characterization of two genes encoding the mitochondrial alternative oxidase in *Chlamydomonas reinhardtii. Curr. Genet.* **39,** 101–108.
20. Funes, S., Davidson, E., Claros, M. G., et al. (2002) The typically mitochondrial DNA-encoded ATP6 subunit of the F_1F_0-ATPase is encoded by a nuclear gene in *Chlamydomonas reinhardtii. J. Biol. Chem.* **277,** 6051–6058.

21. Atteia, A., van Lis, R., Wetterskog, D., et al. (2003) Structure, organization and expression of the genes encoding mitochondrial cytochrome c_1 and the Rieske iron-sulfur protein in *Chlamydomonas reinhardtii*. *Mol. Genet. Genomics* **268,** 637–644.
22. Eriksson, M., Gardeström, P., and Samuelsson, G. (1995) Isolation, purification, and characterization of mitochondria from *Chlamydomonas reinhardtii*. *Plant Physiol.* **107,** 479–483.
23. Harris, E. H. (1989) *The* Chlamydomonas *Sourcebook : A Comprehensive Guide to Biology and Laboratory Use*, Academic Press, San Diego, CA.
24. Matagne, R. F., Michel-Wolwertz, M. R., Munaut C., Duyckaerts, C., and Sluse, F. (1989) Induction and characterization of mitochondrial DNA mutants in *Chlamydomonas reinhardtii*. *J. Cell Biol.* **108,** 1221–1226.
25. Remacle, C., Duby, F., Cardol, P., and Matagne, R. F. (2001) Mutations inactivating mitochondrial genes in *Chlamydomonas reinhardtii*. *Biochem. Soc. Trans.* **29,** 442–446.
26. van Lis, R., Atteia, A., Mendoza-Hernández, G., and González-Halphen, D. (2003) Identification of novel mitochondrial protein components of *Chlamydomonas reinhardtii*. A proteomic approach. *Plant Physiol.* **132,** 318–330.
27. Cardol, P., Vanrobaeys, F., Devreese, B., Van Beeumen, J., Matagne, R., and Remacle, C. (2004) Higher plant-like subunit composition of the mitochondrial complex I from *Chlamydomonas reinhardtii*: 31 conserved components among eukaryotes. *Biochim. Biophys. Acta* **1658,** 212–214.
28. Shrager, J., Hauser, C., Chang, C. W., et al. (2003) *Chlamydomonas reinhardtii* genome project: a guide to the generation and use of the cDNA information. *Plant Physiol.* **131,** 401–408.
29. Cardol, P., González-Halphen, D., Reyes-Prieto, A., Baurain, D., Matagne, R. F., and Remacle, C. (2005) The mitochondrial oxidative phosphorylation proteome of *Chlamydomonas reinhardtii* deduced from the Genome Sequencing Project. *Plant Physiol.* **137,** 447–459.
30. Bennoun, P., Atteia, A., Pierre, Y., and Delosme, M. (1995) Etiolated cells of *Chlamydomonas reinhardtii:* a choice material to study mitochondrial respiratory complexes. *Proc. Natl. Acad. Sci. USA* **92,** 10,202–10,206.
31. Schägger, H. (1995) Native electrophoresis for isolation of mitochondrial oxidative phosphorylation protein complexes. *Methods Enzymol.* **260,** 190–203.
32. Nurani, G., Eriksson, M., Knorpp, C., Glaser, E., and Franzén, L.-G. (1997) Homologous and heterologous protein import into mitochondria isolated from the green alga *Chlamydomonas reinhardtii*. *Plant Mol. Biol.* **35,** 973–980.
33. Whelan, J., Knorpp, C., and Glaser, E. (1990) Sorting of precursor proteins between isolated leaf mitochondria and chloroplasts. *Plant Mol. Biol.* **14,** 977–982.
34. Laemmli, U. K. (1970) Cleavage of structural proteins during the assembly of the head of bacteriophage T4. *Nature* **227,** 680–685.
35. Arnon, D. I. (1949) Copper enzymes in isolated chloroplasts. Polyphenol oxidase in *Beta vulgaris*. *Plant Physiol.* **24,** 1–15.

II

Genetics and Molecular Biology

11

Directed Alteration of *Saccharomyces cerevisiae* Mitochondrial DNA by Biolistic Transformation and Homologous Recombination

Nathalie Bonnefoy and Thomas D. Fox

Summary

Saccharomyces cerevisiae is currently the only species in which genetic transformation of mitochondria can be used to generate a wide variety of defined alterations in mitochondrial deoxyribonucleic acid (mtDNA). DNA sequences can be delivered into yeast mitochondria by microprojectile bombardment (biolistic transformation) and subsequently incorporated into mtDNA by the highly active homologous recombination machinery present in the organelle. Although transformation frequencies are relatively low, the availability of strong mitochondrial selectable markers for the yeast system, both natural and synthetic, makes the isolation of transformants routine. The strategies and procedures reviewed here allow the researcher to insert defined mutations into endogenous mitochondrial genes and to insert new genes into mtDNA. These methods provide powerful in vivo tools for the study of mitochondrial biology.

Key Words: Biolistic method; gene replacement; homologous recombination; mitochondria transformation; *Saccharomyces cerevisiae.*

1. Introduction

A key feature of the yeast nuclear genetic system that has made it a preeminent tool for genetic and cell biological research is the fact that deoxyribonucleic acid (DNA) transformed into the nuclear chromosomes of *Saccharomyces cerevisiae* is incorporated into the genome only via homologous recombination. This fact allows the researcher to add, subtract, and alter genetic information in a highly controlled fashion, easily and cheaply, and essentially rewrite the yeast genome at will. In *S. cerevisiae*, and to date only in that species, similar manipulations based on homologous recombination can be carried out on the mitochondrial genome using the biolistic transformation method to deliver DNA into the organelle.

From: *Methods in Molecular Biology, vol. 372: Mitochondria: Practical Protocols*
Edited by: D. Leister and J. M. Herrmann © Humana Press Inc., Totowa, NJ

1.1. Summary of Strategies to Create Strains With Modified Mitochondrial DNA

Two properties of *S. cerevisiae* facilitate mitochondrial transformation. First, *S. cerevisiae* cells can survive when they lack part (*rho*$^-$) or all (*rho*^{0}) of the mitochondrial DNA (mtDNA). Second, *rho*$^-$ mtDNAs replicate independently of protein synthesis and show no clear requirement for a specific replication origin sequence. These two features are advantageous in creating mitochondrial transformants containing defined mtDNAs because *rho*^{0} yeast strains, entirely lacking mtDNA, can be transformed with bacterial plasmid DNAs that subsequently propagate as "synthetic" *rho*$^-$ molecules ***(1)***. The plasmids used for the transformation typically contain a mutant version of a mitochondrial gene or a foreign piece of DNA flanked by mtDNA sequences. These synthetic *rho*$^-$ strains are then used as donors of the new or modified sequence that can be integrated into a *rho*$^+$ mtDNA by homologous recombination to yield the desired new strain. Alternatively, when a mutated DNA sequence provides a function that can be selected phenotypically, direct transformation of *rho*$^+$ strains bearing deletions in the region of interest can be used to integrate mutations into mtDNA ***(2,3)***. Using this strategy, *rho*$^+$ recombinants may be obtained quicker, although the efficiency of transformation is lower than for *rho*^{0} cells (**Fig. 1**).

1.2. Bombardment

The standard device for microprojectile bombardment, which functions reproducibly for transformation of *S. cerevisiae* mitochondria, is the PDS-1000/He system. This instrument uses a helium shock wave in an evacuated chamber to accelerate microscopic metal particles coated with DNA toward a lawn of cells on a Petri plate. The shock wave is generated by rupture of a membrane at high pressure and accelerates a second membrane (the macrocarrier or flying disk), carrying the metal particles, toward the plate. Some cells on the plate are penetrated by particles and survive. DNA precipitated on the particles is thus introduced into cells and is readily taken up by the nucleus. In addition, the mitochondria of a small fraction of such transformants also take up DNA.

1.3. Generation of a Synthetic rho$^-$ Transformant

In a typical mitochondrial transformation experiment, a large number of *rho*^{0} cells are randomly bombarded by a large number of particles. In the first step, cells that have been hit and survived are allowed to make colonies on the Petri plates by selecting for a nuclear genetic marker that is included in the DNA precipitated on the particles (**Fig. 1**, top plates). Mitochondrial transformants are identified among these colonies by genetic tests for the presence of new genetic information in the mitochondrial genome (**Fig. 1**, bottom plates). This new information is typically a

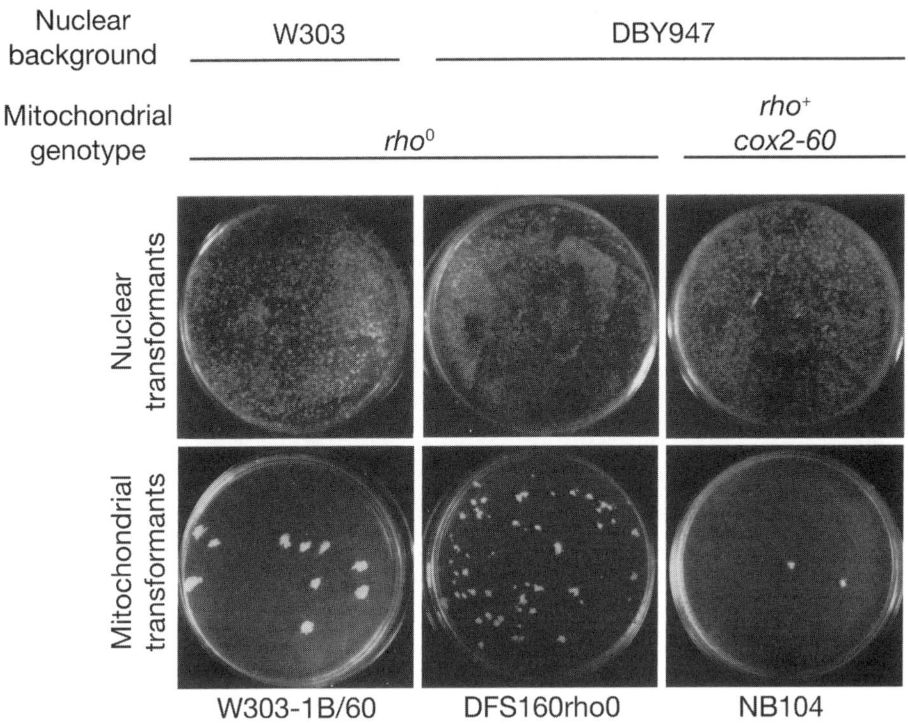

Fig. 1. Nuclear transformants and mitochondrial cotransformants obtained by bombardment of different yeast strains. The nuclear *LEU2* plasmid Yep351 ***(8)*** and the *COX2* plasmid pNB69 ***(2)*** were precipitated together onto tungsten particles and bombarded on lawns of the *rho*^{0} strains W303-1B/A/50 (*MATa ade2-1, ura3-1, his3-11,15, trp1-1, leu2-3,112, can1-100* [*rho*^{0}]) (a *rho*^{0} derivative of W303-1B; ***13***) and DFS160 (*MATα ade2-101, leu2Δ ura3-52, arg8Δ::URA3, kar1-1,* [*rho*^{0}]) ***(6)***, or on lawns of the *rho*$^+$ strain NB104. NB104 *rho*$^+$ mtDNA carries a 129-bp deletion, *cox2-60*, located around the *COX2* first codon ***(2)*** and is isonuclear to DFS160. The top plates correspond to minimal medium supplemented with sorbitol and lacking leucine. Typical plates showing about 3000 nuclear transformants for each strain have been presented. Nuclear transformants were crossed by replica plating to the nonrespiring tester strain (NB160) carrying a mutation of *COX2* initiation codon ***(2)***, and mitochondrial transformants (bottom plates) were detected by replica plating the mated cells onto nonfermentable medium. (Reprinted from **ref.** ***7*** with permission.)

portion of wild-type mitochondrial DNA (mtDNA) sequence that can rescue a known mitochondrial marker mutation by recombination after the transformants are mated to an appropriate *rho*$^+$ tester strain, resulting in recombinants with a detectable growth phenotype. The new wild-type sequence may be an unaltered region of the gene of interest, or it may be another piece of wild-type mtDNA incorporated into a vector. Such marker rescue can work with as little as 50 bp of homologous sequence flanking the site of the mutation in the tester mtDNA. The mitochondrial transformants are then purified by series of restreaking and crosses to the tester before obtaining the final synthetic *rho*$^-$ strains.

1.4. Generation of a rho$^+$ *Strain With a Modified mtDNA*

In a second step, synthetic *rho*$^-$ are mated with a wild-type *rho*$^+$ recipient strain. As a result of this second mating, mitochondria from the two strains fuse, and recombination between the two mtDNAs produces recombinant *rho*$^+$ strains in which the new mtDNA sequence is integrated by double crossover events. Pure recombinant strains are generated by subsequent mitotic segregation. Because mitochondrial DNA recombination and segregation is so frequent, this simple procedure typically yields the desired integrants at frequencies between 1 and 50% of clones derived from zygotes. Using a *rho*$^+$ recipient that carries a deletion in the region present in the synthetic *rho*$^-$ can greatly facilitate the detection of homologous recombinants if one of the strains in such a cross carries the karyogamy-defective mutation *kar1-1* *(4)* (**Fig. 2**). In this case, haploid mitochondrial mutant cytoductants can be identified after such a mating by their acquired ability to rescue a defined marker within the deleted region.

1.5. Transformation of a rho$^+$ *Strain Carrying a Deletion in the Region of Interest*

Selection for transformation by DNA fragments restoring wild-type or near wild-type function to the *rho*$^+$ recipient is straightforward by selecting first for nuclear transformants as described above in **Subheading 1.2.** and then screening for the mitochondrial phenotype by replica plating (**Fig. 3**, right) *(5)*. Respiring mitochondrial transformants can also be selected in a single step, by directly bombarding lawns of mutant *rho*$^+$ cells spread on nonfermentable plates supplemented with 0.1% glucose, to allow a brief period of outgrowth, and 1 *M* sorbitol (**Fig. 3**, left). We have also been able to select for *rho*$^+$ transformants expressing the recoded gene *ARG8*^{m} *(6)* by directly bombarding lawns spread on appropriate minimal glucose medium supplemented with sorbitol *(2,3)*. In addition, transformation can also be performed with linear DNA molecules obtained either from plasmid clones or polymerase chain reaction amplification *(7)*. Linear DNA fragments having as little as 260 bp of homologous sequence flanking each side of a deletion mutation in a *rho*$^+$ recipient were able to yield respiring transformants at frequencies similar to those obtained with circular plasmids. The ability to use fragments generated by polymerase chain reaction to transform defined mtDNA deletion recipients can substantially accelerate strain construction.

The detailed protocols used to conduct all the different steps of mitochondria transformation and integration of directed mutations or new genes in mtDNA

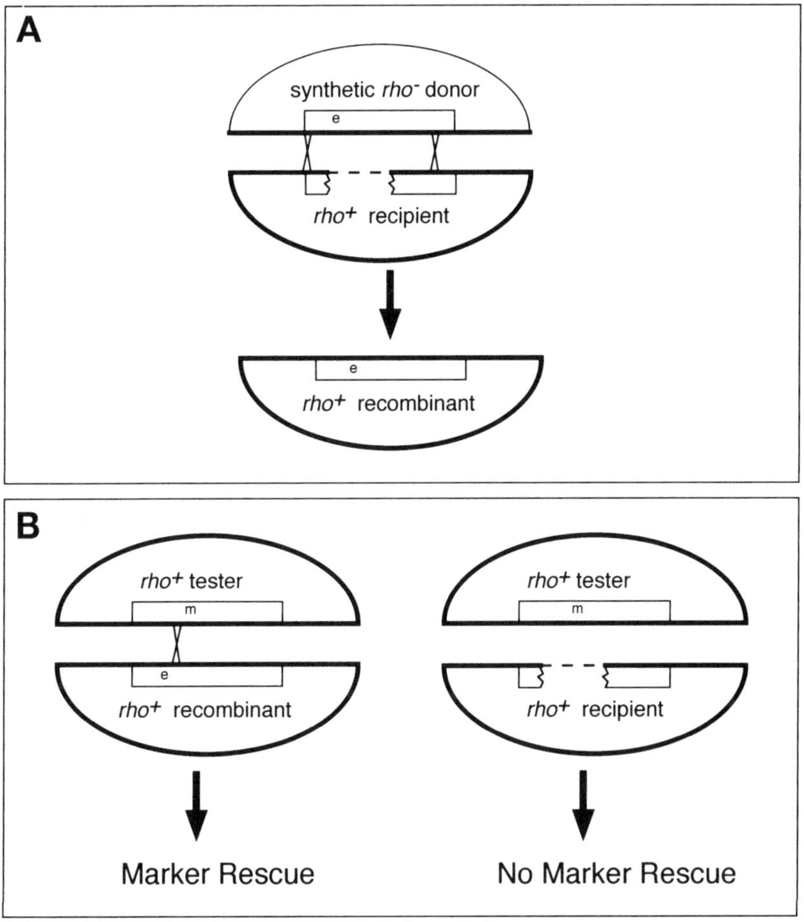

Fig. 2. Schematic diagram of recombination events that allow identification of nonrespiring recombinant cytoductants by marker rescue. Thick lines represent mtDNA sequences; thin lines represent vector DNA. The box represents a gene under study. (**A**) A karyogamy defective (*kar1-1*) synthetic *rho*$^-$ donor containing an experimentally induced mutation *e* mated to a *rho*$^+$ recipient strain with a deletion in the region of interest. Among the cells present in the mixture after mating are the desired *rho*$^+$ recombinant cytoductants. (**B**) To distinguish the desired *rho*$^+$ recombinant cytoductants from the unaltered recipient cells and other cell types present, clones derived from the mating mixture are mated to a *rho*$^+$ tester strain bearing the marker mutation *m*. The desired *rho*$^+$ recombinant cytoductants can yield respiring recombinants when mated to this tester by a crossover between *e* and *m* (and a second resolving crossover anywhere else). The ability to produce respiring recombinants identifies the desired cytoductant clones. Unaltered recipient clones, and other cells types present, cannot yield such respiring recombinants. (Reprinted from **ref. 7** with permission.)

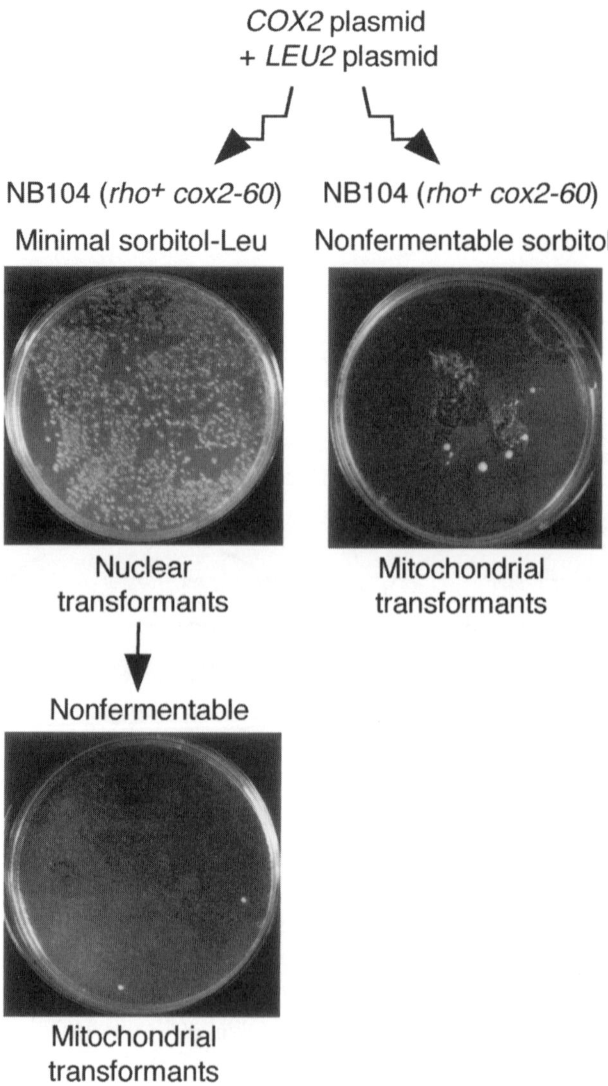

Fig. 3. Selection of *rho*⁺ mitochondrial transformants directly after bombardment. The nuclear shuttle vector Yep351 (*LEU2*) and plasmid pNB69 (*COX2*) were bombarded together onto lawns of the *rho*⁺ *cox2-60* strain NB104 (*see* **Fig. 1** legend). The bombarded lawns had been spread either on minimal medium supplemented with sorbitol but lacking leucine (top right) or on nonfermentable YPEG medium supplemented with sorbitol and 0.1% glucose (top left). Leu⁺ transformants were replica plated to nonfermentable medium (bottom plate) to select for mitochondrial transformants. (Reprinted from **ref.** *7* with permission.)

by homologous recombination are described in **Subheading 3**. For a review of the important features of *S. cerevisiae* mitochondrial genetics that underlie the methods presented, refer to **ref.** *7*.

2. Materials

2.1. Preparation of Cells

1. *rho*^{0} or *rho*$^+$ recipient strain, preferably of DBY947 genetic background (*see* **Note 1**) and *kar1-1*.
2. YPR: 1% (w/v) yeast extract, 2% (w/v) Bacto™ peptone, 2% (w/v) raffinose, 0.1% (w/v) glucose, 40 µg/mL adenine (*see* **Note 2**).
3. YPD: 1% (w/v) yeast extract, 2% (w/v) Bacto peptone, 2% (w/v) glucose, 40 µg/mL adenine.
4. Synthetic defined (SD)-sorbitol: 0.67% (w/v) yeast nitrogen base, 5% (w/v) glucose, 1 *M* sorbitol, 100 µg/mL adenine, 3% (w/v) agar, other supplements as required 40 µg/µL except for the selection marker.
5. YPEG-sorbitol is used for direct selection of respiring transformants into *rho*$^+$ cells: 1% (w/v) yeast extract, 2% (w/v) Bacto peptone, 3% (w/v) glycerol, 3% (w/v) ethanol, 0.1% (w/v) glucose, 3% (w/v) agar.

2.2. Preparation of Microprojectiles and Precipitation of DNA

1. Tungsten powder <1 µm 99.95% (metals basis; Alfa Aesar/Johnson Matthey, item 44210, CAS 7440-33-7; *see* **Note 3**).
2. Ethanol: 70 (v/v) and 100%, room temperature.
3. Sterile water.
4. Glycerol: 50% (w/v), sterile, –20°C.
5. DNA for nuclear transformation, carrying the nuclear marker and a nuclear replication origin, such as Yep351 *(8)* or pRS315 *(9)* for the *LEU2* marker. Qiagen-based midi- or maxipreps are recommended at 2 µg/µL or more.
6. DNA for mitochondrial transformation, Qiagen preps, at least 2 µg/µL.
7. Spermidine free base (Sigma): 1 *M* in water.
8. Calcium chloride: 2.5 *M* in water, –20°C, filter sterilized.
9. 100% ethanol at –20°C.

2.3. Bombardment

1. Macrocarrier holders sterilized in a Pasteur oven.
2. Macrocarriers (Bio-Rad).
3. Rupture disks: 1100 psi (*see* **Note 4**).
4. PDS-1000/He biolistic gun (Bio-Rad) with helium bottle and vacuum pump.

2.4. Identification of Mitochondrial Transformants

1. *rho*$^+$ *mit*$^-$ tester strain carrying a mitochondrial mutation in the region covered by the DNA for mitochondrial transformation and mating type opposite to the recipient *rho*^{0} strain.

2. SD plates: 0.67% (w/v) yeast nitrogen base, 2% (w/v) glucose, 2% (w/v) agar; supplements as required at 40 µg/µL.
3. YPD plates: like liquid YPD but with 2% (w/v) agar.
4. YPEG plates: 1% (w/v) yeast extract, 2% (w/v) Bacto peptone, 3% (w/v) glycerol, 3% ethanol, 2% agar.

2.5. Mating and Isolation of Recombinant Cytoductants

For mating and isolation of recombinant cytoductants, use recipient wild-type *rho*$^+$ strain or *mit*$^-$ *rho*$^+$ strain. To recover haploid recombinants, this strain must carry the *kar1-1* mutation unless the *rho*^{0} strain used for the transformation (**Subheading 2.1.**, **item 1**) is *kar1-1*.

3. Methods

Optima for the parameters of **Subheadings 3.1.** to **3.3.** are summarized in the **Table 1**.

3.1. Preparation of Cells

1. Grow the *rho*^{0} (or *rho*$^+$) strain to be bombarded in a few milliliters of liquid YPR for 1 to 2 nights at 30°C with agitation.
2. Use this fresh culture to inoculate at 1/100 a larger volume of YPR and grow for 2 or 3 d (stationary phase) at 30°C with agitation (*see* **Notes 2** and **5**). On average, 30–50 mL cultured cells are used for six shots.
3. Harvest cells and concentrate 40–100 times in liquid YPD medium to reach a cell density of 1–5 × 10^9 cells/mL.
4. Spread 0.1 mL cells onto the appropriate SD-sorbitol supplemented to provide the appropriated prototrophic selection (*see* **Notes 6** and **7**).

3.2. Preparation of Microprojectiles and Precipitation of DNA

1. Precool a microcentrifuge.
2. Weigh and sterilize in a microfuge tube 10–50 mg tungsten particles by vigorous suspension in 1.5 mL 70% ethanol (100% ethanol for gold particles; ***10***) and incubation at room temperature for 10 min. Centrifuge 15 min at room temperature at maximum speed in a microcentrifuge and carefully remove the supernatant. Wash the particles with 1.5 mL sterile water, resuspend at 60 mg/mL in frozen 50% glycerol, and keep on ice (*see* **Note 8**).
3. Precipitation of DNA onto particles is conducted on ice with ice-cold or freezer-stored reagents. For six shots, mix in a microfuge tube (*see* **Note 9**) 5 µg of plasmid for the nuclear selection (Yep351) with 15–30 µg plasmid carrying the mitochondrial DNA of interest in a total volume of 15–20 µL (*see* **Note 10**). Add 100 µL tungsten particles (*see* **Note 11**), 4 µL 1 *M* spermidine, and 100 µL 2.5 *M* $CaCl_2$ from the freezer, in that order, vortexing immediately after each addition. Incubate 10–15 min with occasional vortexing.

Table 1
Factors Influencing Mitochondrial Transformation Efficiency and Their Known Optima

Recipient strain		DNA precipitation		Biolistic parameters	
Parameter	Optimum	Parameter	Optimum	Parameter	Optimum
Genetic background:		DNA:		Rupture disks	1100 psi
		Size	5–6 kb		
		Purity	Qiagen		
Nuclear	DBY947	Concentration	>2 µg/µL	Stopping screens	None
		Volume	<15–20 µL		
Mitochondrial	*rho*^{0}	Quantity	5 µg (nuclear) 20–30 µg (mito)	Plate distance	5 cm
Carbon source	Raffinose	Particles	Aesar 44210 Tungsten powder, <1 µm	Vacuum	29–29.5
		Temperature	Ice cold		
		Precipitate	Finely dispersed		

4. Spin briefly in the precooled microcentrifuge, remove the supernatant, and add 200 µL freezer stock 100% ethanol. Take extreme care to scrape the side of the tube and to fragment aggregates of particles using the pipet tip, then resuspend the particles thoroughly. Repeat spinning and resuspension at least once until the particles resuspend easily (*see* **Note 12**).
5. Spin briefly, remove the supernatant, and add 60 µL 100% ethanol. Distribute the resulting suspension evenly at the center of six macrocarriers (flying disks) previously placed in their holders. Allow the ethanol to evaporate (*see* **Note 13**).

3.3. Bombardment

1. Thoroughly wash the chamber and the removable parts by soaking with 70% ethanol. Wipe dry the parts and the chamber meticulously because remaining ethanol will prevent efficient evacuation of the chamber (*see* **Note 14**).
2. Carefully follow the manufacturer's instructions for use of the PDS-1000/He apparatus (it employs gas under high pressure). Place the rupture disk in its retaining cap and tighten using the torque wrench (*see* **Note 15**).
3. Load the macrocarrier in its holder into the assembly system. Do not assemble the stopping screen (*see* **Note 16**).
4. Place the open Petri plate carrying the lawn of cells at 5 cm from the macrocarrier assembly (*see* **Note 17**).
5. Evacuate the vacuum chamber to reach a reading of 29–29.5 inches Hg (the higher the better) on the PDS-1000 gage (*see* **Note 18**).
6. Fire.

7. Remove any fragments of the macrocarrier disk with sterile forceps.
8. Incubate the plate at 30°C for 4–5 d until colonies appear. Expect between 1000 and 10,000 nuclear transformants per plate for S288c-related strains transformed to prototrophy or from 0 to 20 transformants per plate for direct selection of *rho*$^+$ mitochondrial transformants from a *rho*$^+$ recipient. For transformation of *rho*^{0} recipients, proceed according to **Subheading 3.4.** to identify *rho*$^-$ transformants. In the case of *rho*$^+$ transformation, nuclear prototrophic transformants are printed to selective medium to isolate mitochondrial transformants, or mitochondrial transformants can be picked directly if the bombardment was made on selective medium. Analyze them genetically as in **Fig. 2B**.

3.4. Identification of Mitochondrial rho$^-$ Transformants After Bombardment of a rho^0 Recipient Strain

1. During the incubation of the bombarded plates, set up a liquid YPD culture of an appropriate *rho*$^+$ mutant (*mit*$^-$) tester strain.
2. Replica plate the transformants onto SD medium with the appropriate supplements (to keep the transformants) and onto a lawn of the tester strain freshly spread on a YPD plate. Mark the plates precisely to facilitate the **step 5**.
3. Incubate at 30°C for 2 d to allow mating and recombination.
4. Print to YPEG medium to detect respiring diploids (or another appropriate selection medium if scoring another phenotype) (*see* **Note 19**). Incubate for 2–3 d (*see* **Note 20**).
5. Pick colonies off the bombarded plate (or its direct replica) that correspond to the position of respiring recombinants. Streak these colonies on YPD and repeat the marker rescue with the tester strain as in **step 2**. Such subcloning and retesting must usually be done three times before pure stable synthetic *rho*$^-$ clones are obtained. Cells usually lose the nuclear marker plasmid during these subcloning steps if no selection is applied for its maintenance.

3.5. Mating and Isolation of Recombinant Cytoductants

1. Grow cultures of the subcloned synthetic *rho*$^-$ strain and the recipient wild-type *rho*$^+$ strain overnight in liquid YPD. At least one of these two strains (usually the synthetic *rho*$^-$) must carry the *kar1-1* mutation (*see* **Note 21**).
2. Mix 0.5 mL of each parent (alternatively 1 mL of synthetic *rho*$^-$ and 0.2 mL wild-type *rho*$^+$; *see* **Note 22**) in a microfuge tube, spin, remove the supernatant, resuspend in residual liquid, and spread the mixture onto a YPD plate.
3. Incubate at 30°C for 4–5 h. Check zygote formation microscopically. Scrape the mating cells from the plate and use them to inoculate fresh YPD liquid medium. Incubate at 30°C with agitation for a few hours to overnight.
4. Dilute the culture and plate to obtain single colonies on minimal medium, selecting for the recipient nuclear genotype and against the donor nuclear genotype, if possible. Alternatively, plate on YPD medium. Densities of 50–200 colonies per plate should be obtained.
5. Replica plate the colonies obtained to medium that will reveal the altered phenotype of the recipient strain as a result of integration of the mutant donor sequences into

its mtDNA. For example, print to YPEG to identify clones that have acquired a mutation preventing respiratory growth.

6. Mate nonrespiring candidate clones to a known *rho*$^-$ strain that covers the region you are seeking to mutate. The desired *rho*$^+$ recombinant cytoductants will produce respiring diploids after mating to this known *rho*$^-$ strain. This step eliminates cytoductants that simply acquired the transformed synthetic *rho*$^-$ mtDNA by cytoduction.

4. Notes

1. The strain genetic background is an important factor affecting the efficiency of transformation (**Fig. 1** and **Table 1**). We have obtained the best results with *rho*^{0} strains in the S288c background, particularly those derived from DBY947 ***(11)***, such as MCC109rho0 ***(12)*** (American Type Culture Collection [ATCC] 201440); MCC123rho0, which is the identical strain with *MAT***a** (ATCC 201442); and DFS160, which is recommended for *ARG8^m* constructs ***(6)***. Strains derived from W303 ***(13)***, (ATCC 200060) give lower but satisfactory efficiencies (*see* W303-1B/A/50, **Fig. 1**); strains in the D273-10B (ATCC 24657) background are very difficult to transform. In our hands, mitochondrial transformation is 10–20 times more efficient during bombardment of a *rho*^{0} strain than of an isogenic *rho*$^+$ strain containing a small deletion in mtDNA (**Fig. 1**). This effect could be caused by either physiological differences between the strains such as the properties of the mitochondrial inner membrane or an advantage in establishing an incoming DNA molecule in the absence of endogenous mtDNA. Effects consistent with the latter notion have been observed in comparisons of mtDNA behavior after *rho*$^+$ × *rho*$^+$ matings as opposed to *rho*$^+$ × *rho*^{0} matings ***(14)***.
2. The 0.1% glucose supplement accelerates growth, and adenine seems to increase the transformation efficiency, even for Ade$^+$ prototrophs. Raffinose can be replaced by galactose (which is less expensive) with no or very limited decrease in transformation efficiency.
3. We routinely use Alfa Aesar tungsten powder <1 µm, which is inexpensive and very effective, but tungsten powder 0.4–0.7 µm is also available from Bio-Rad (catalog nos. 165–2265 or 165–2266, respectively), as is gold powder 0.6 µm (Bio-Rad 165–2262). In our hands, gold powder gave a similar yield to the Aesar tungsten, whereas tungsten powder from Bio-Rad was three times less efficient.
4. Rupture disks of 1350 psi can also be used for efficient transformation of yeast, but in our hands 1100-psi disks tend to give better results.
5. For *rho*^{0} cells in the DBY947 background, a 3-d culture is optimum, but this could differ for other backgrounds.
6. Cells can be used between 1 and 3 h after plating. Plates that are still wet can be bombarded without drastic changes in efficiency. We usually plate cells with a glass rod spreader rather than beads because the heterogeneity of plating will provide both some zones of optimum cell density for the transformation and a pattern of transformants that plays the role of useful identification marks to compare the plates when picking the mitochondrial transformants at **Subheading 3.4., step 5**.
7. YPEG-sorbitol can be used instead of SD-sorbitol when transforming a respiratory-deficient *rho*$^+$ strain and selecting directly for respiratory proficient transformants.

SD-sorbitol lacking arginine is used for selection of transformants from *rho*$^+$ *arg8* cells transformed with a construct allowing the production of functional Arg8p from the *ARG8*^{m} reporter.

8. Tungsten particles can be kept frozen in 50% glycerol for several months to several years without loss of transformation efficiency. We have not tried gold.
9. Precipitations can be upscaled by a factor of 2; a siliconized tube can be used.
10. Try to keep the DNA volume to a minimum to optimize the precipitation.
11. Vortex the tungsten suspension vigorously before pipeting because the particles tend to settle extremely rapidly.
12. This step is crucial for obtaining a finely dispersed precipitate that will efficiently transform the cells.
13. We usually work under a laminar flow hood, using sterile forceps to insert the macrocarrier disks into the holders. Holders can be conveniently kept in sterile Petri dishes. There is no need to prewash the macrocarriers or to desiccate them after coating; the ethanol will quickly evaporate.
14. For a more efficient sterilization, the chamber and parts can be soaked with 70% ethanol before starting the cell preparation and DNA precipitation, allowed to sit, and wiped dry just before the bombardment. The chamber is washed in a similar way after the bombardment.
15. Be careful not to use two rupture disks stacked together.
16. Interestingly, we have found that simply allowing the carrier disk to fly to the surface of the Petri plate, by not assembling the stopping screen, yields more transformants than obtained if the stopping screen is employed. However, in this case it is important to use sorbitol plates containing 3% agar, as advised in **Subheading 2**, to prevent severe damage to the agar surface.
17. Shorter distances result in very high colony densities in the center of the plate with few colonies at the periphery, and longer distances decrease the transformation efficiency.
18. We have found that failure to draw the greatest vacuum possible dramatically reduces the transformation efficiency. Teflon sealing tape can be used to reduce air leakage in the connections between the chamber and the pump. Cell viability is not significantly affected by a prolonged stay under these vacuum conditions.
19. When a high number of nuclear transformants are present, it may be useful also to replicate the mated cells on medium that selects for the diploids because these diploid plates may by comparison facilitate the identification of the mitochondrial transformants on the original bombarded plate.
20. Respiratory-proficient colonies typically appear earlier if the transformed DNA allows complementation of the *mit*$^-$ mutation in *trans*.
21. The introduction of a mutation causing respiratory deficiency in an otherwise *rho*$^+$ genome can be facilitated by using as *rho*$^+$ recipient a strain that carries a deletion mutation in the region of interest and that can recombine with the synthetic *rho*$^-$ sequences. Some of the colonies obtained in **Subheading 3.5., step 4** will be respiratory-deficient *rho*$^+$ haploid recombinants that have the deleted region restored but contain the desired mutation (**Fig. 2A**). Find them by mating to

freshly spread lawns of two different testers: a *rho*$^+$ tester strain that has another mutation within the recipient's deleted region distinct from the new mutation to be introduced (**Fig. 2B**) and a *rho* strain carrying wild-type information in the region deleted in the original recipient strain. The desired colonies will yield respiratory-proficient diploids after 2-d incubation at 30°C and print to YPEG. Identify the corresponding haploid *rho*$^+$ cytoductant colonies on the plates from **Subheading 3.5.**, **step 4**; restreak; and retest.

22. If the synthetic *rho*$^-$ donor and the *rho*$^+$ recipient strains share nuclear markers and therefore cannot be distinguished selectively on glucose medium, then mating mixtures should contain equal numbers of cells of both strains. If nuclear auxotrophic or drug resistance markers allow selection against the synthetic *rho*$^-$ donor strain, then the mating mixture should contain a fivefold excess of donor cells.

Acknowledgments

We thank Matthieu Caron (Mitoprod, Bordeaux, France) for the communication of results and Academic Press for permission to reprint figures and significant text from **ref. 7**. N. B. is supported by the Association Française contre les Myopathies, and T. D. F. is supported by a grant from the U.S. National Institutes of Health (GM29362).

References

1. Fox, T. D., Sanford, J. C., and McMullin, T. W. (1988) Plasmids can stably transform yeast mitochondria lacking endogenous mtDNA. *Proc. Natl. Acad. Sci. U.S.A.* **85,** 7288–7292.
2. Bonnefoy, N. and Fox, T. D. (2000) In vivo analysis of mutated initiation codons in the mitochondrial *COX2* gene of *Saccharomyces cerevisiae* fused to the reporter gene *ARG8*^{m} reveals lack of downstream reinitiation. *Mol. Gen. Genet.* **262,** 1036–1046.
3. Bonnefoy, N., Bsat, N., and Fox, T. D. (2001) Mitochondrial translation of *Saccharomyces cerevisiae COX2* mRNA is controlled by the nucleotide sequence specifying the pre-Cox2p leader peptide. *Mol. Cell Biol.* **21,** 2359–2372.
4. Conde, J. and Fink, G. R. (1976) A mutant of *S. cerevisiae* defective for nuclear fusion. *Proc. Natl. Acad. Sci. USA* **73,** 3651–3655.
5. Johnston, S. A., Anziano, P. Q., Shark, K., Sanford, J. C., and Butow, R. A. (1988) Mitochondrial transformation in yeast by bombardment with microprojectiles. *Science* **240,** 1538–1541.
6. Steele, D. F., Butler, C. A., and Fox, T. D. (1996) Expression of a recoded nuclear gene inserted into yeast mitochondrial DNA is limited by mRNA-specific translational activation. *Proc. Natl. Acad. Sci. USA* **93,** 5253–5257.
7. Bonnefoy, N. and Fox, T. D. (2001) Genetic transformation of *Saccharomyces cerevisiae* mitochondria. *Methods Cell Biol.* **65,** 381–396.
8. Hill, J. E., Myers, A. M., Koerner, T. J., and Tzagoloff, A. (1986) Yeast/*E. coli* shuttle vectors with multiple unique restriction sites. *Yeast* **2,** 163–167.

9. Sikorski, R. S. and Hieter, P. (1989) A system of shuttle vectors and yeast host strains designed for efficient manipulation of DNA in *Saccharomyces cerevisiae. Genetics* **122,** 19–27.
10. Butow, R. A., Henke, M., Moran, J. V., Belcher, S. M., and Perlman, P. S. (1996) Transformation of *Saccharomyces cerevisiae* mitochondria by the biolistic gun. *Meth. Enzymol.* **264,** 265–278.
11. Neff, N. F., Thomas, J. H., Grisafi, P., and Botstein, D. (1983) Isolation of the β-tubulin gene from yeast and demonstration of its essential function in vivo. *Cell* **33,** 211–219. 12. Costanzo, M. C. and Fox, T. D. (1993) Suppression of a defect in the 5′-untranslated leader of the mitochondrial *COX3* mRNA by a mutation affecting an mRNA-specific translational activator protein. *Mol. Cell. Biol.* **13,** 4806–4813.
13. Thomas, B. J. and Rothstein, R. (1989) Elevated recombination rates in transcriptionally active DNA. *Cell* **56,** 619–630.
14. Azpiroz, R. and Butow, R. A. (1993) Patterns of mitochondrial sorting in yeast zygotes. *Mol. Biol. Cell* **4,** 21–36.

12

Screens for Mitochondrial Mutants in Yeast

Françoise Foury

Summary

We describe methods that allow isolation, identification, and counting of mitochondrial mutants that are resistant to antibiotics (*ant*^{R}) or respiratory deficient (*rho*$^-$). (1) Analysis of diploid and meiotic progenies generated in crosses between mutants and tester strains allows distinguishing nuclear from mitochondrial mutants, for either antibiotic resistance or respiratory deficiency. (2) The mutation rate of mitochondrial deoxyribonucleic acid (mtDNA) can be estimated from the average frequency of *ant*^{R} mutants produced in a large number of independent clones. (3) The frequency of retention of mtDNA fragments in *rho*$^-$ genomes accumulating in nuclear respiratory-deficient mutants can be determined by a genetic test based on the ability of these *rho*$^-$ genomes to restore cellular growth on glycerol in crosses with selected mutants bearing punctual mutations in their mtDNA (*mit*$^-$).

Key Words: Antibiotic resistance; mtDNA instability; mutation rate; respiration deficiency.

1. Introduction

The first mitochondrial mutants were discovered in *Saccharomyces cerevisiae* more than 50 years ago *(1)*. They occurred spontaneously at high frequency (mutation rate on the order of 2×10^{-3} per cell generation) and respiratory deficient; because they formed small colonies, they were named *petite*. Later, the petite phenotype was associated with an irreversible loss of the mitochondrial genome. In fact, the petite mutants have large deletions of mitochondrial deoxyribonucleic acid (mtDNA), with reiteration of the retained DNA fragment (*rho*$^-$) *(2)*, or they may even be devoid of mtDNA (*rho*^{0}). Mitochondrial respiratory-proficient mutants exhibiting resistance to specific antibiotics (*ant*^{R}) were identified in the late 1960s *(3,4)*; in 1975, the first respiratory-deficient mutants with point mutations in mtDNA (*mit*$^-$) were

From: *Methods in Molecular Biology, vol. 372: Mitochondria: Practical Protocols*
Edited by: D. Leister and J. M. Herrmann © Humana Press Inc., Totowa, NJ

obtained *(5,6)*. These mtDNA mutants exhibit non -Mendelian inheritance and are characterized by rapid mitotic segregation of their traits *(7)*. However, mutations in the mtDNA may be secondary events to a primary mutation in a nuclear gene. Numerous nuclear genes (*PET*) play a role in stability and maintenance of the mtDNA *(8)*. Moreover, the frequency of *ant*^{R} mutations can be increased several hundredfold by defects in the proofreading activity of the Mip1 mitochondrial DNA polymerase *(9)* or in the mismatch repair activity of Msh1 *(10,11)*.

In this chapter, we first describe a procedure that allows estimation of mutation rates of mtDNA. Then, we review very simple techniques that allow identification of respiratory-deficient mutants with *pet* or *rho*$^-$ genotypes. Finally, we describe genetic methods to distinguish *rho*$^-$ from *rho*^{0} mutants.

2. Materials

2.1. Strains

1. Parental strains: W303-1B (*MATα ade2-1 his3-15 leu2-3,115 trp1-1 ura3-1 rho*$^+$), W303-1A (*MATa ade2-1 his3-15 leu2-3,115 trp1-1 ura3-1* rho$^+$), and D273-10B/A1 (*MATα met6 rho*$^+$). More information concerning the origin of these laboratory strains is given in the *Saccharomyces* Genome Database (http://www.yeastgenome.org/).
2. *rho*^{0} tester strains : IL166-6C/*rho*^{0} (*MATα ura1 rho*^{0}) and W303-1B/*rho*^{0}.
3. *rho*$^+$ tester strains: D273-10B/A1 and its isogenic counterpart NW38-4C (*MATα his1 rho*$^+$).
4. *mit*$^-$ tester strains: aM9-94-4B (*MATα ade1 rho*$^+$ *mit*$^-$ *cox2*), aM9-3-5B (*MATα ade1 rho*$^+$ *mit*$^-$ *cox3*), aM7-40-5B (*MATα ade1 rho*$^+$ *mit*$^-$ *cob*), M9-3/A3 (*MATα met rho*$^+$ *mit*$^-$ *cox3*), and M17-162 (*MATα met rho*$^+$ *mit*$^-$ *cob*).

2.2. Growth Media

1. YPD: 2% (w/v) glucose, 1% (w/v) yeast extract (Difco), 2% (w/v) Bacto™ peptone (Difco), and 40 mg/L adenine.
2. GLY: 3% (w/v) glycerol, 1% (w/v) yeast extract (Difco), 2% (w/v) Bacto peptone (Difco) adjusted to pH 5.8 with HCl, and 40 mg/L adenine.
3. OLI: 3% (w/v) glycerol, 1% (w/v) yeast extract (Difco), Bacto peptone (Difco), 25 m*M* sodium phosphate, pH 6.5, 40 mg/L adenine, and 3 mg/L oligomycin solubilized in 1 mL ethanol. Phosphate is added from a 1 *M* stock solution after autoclaving.
4. ERY: 3% (w/v) glycerol, 1% (w/v) yeast extract (Difco), Bacto peptone (Difco), 25 m*M* sodium phosphate, pH 6.5, 40 mg/L adenine, and 4 g/L erythromycin solubilized in 30 mL ethanol.
5. Glucose minimum: 2% (w/v) glucose, 0.7% (w/v) yeast nitrogen base (Difco), and the required amino acids and bases.
6. All media described above are supplemented with 2% (w/v) agar (Sigma).
7. Sporulation medium (SPO): 1% potassium acetate, 0.05% (w/v) glucose, 0.1% (w/v) yeast extract (Difco), and the required amino acids and bases. This medium is supplemented with 2% (w/v) agar (Difco).

8. Tetrazolium medium: 2% (w/v) glucose, 50 m*M* potassium phosphate, pH 7.0, 1.5% (w/v) agar, and 0.5% (w/v) 2,3,5-triphenyltetrazolium chloride. This medium is prepared immediately before use and cooled at 45°C.

3. Methods

3.1. Obtaining Mitochondrial ant^R Mutants

The most commonly used antibiotics are erythromycin and chloramphenicol, which inhibit mitochondrial protein synthesis, and oligomycin, a very specific inhibitor of the mitochondrial adenosine triphosphate synthase. In addition, a panoply of antibiotics that interact with cytochrome-*b* can be used to select *ant*^{R} mutants ***(12)***. Spontaneous *ant*^{R} mutants are easily obtained by selection on antibiotic-containing media in the presence of a respiratory carbon source such as glycerol (*see* **Notes 1–3**).

3.1.1. Obtaining a Pure ant^R Mitochondrial Mutant Clone

1. A cell patch of the wild-type parental strain is refreshed by incubation on a GLY plate for 2 d at 30°C.
2. The patch is replica plated on YPD with a velvet and incubated 1 d at 30°C.
3. The YPD cell patch is replica plated on an antibiotic-containing medium and incubated at 30°C. After 4–10 d, the *ant*^{R} colonies will develop. Mitochondrial and nuclear mutations can be distinguished as explained in **Subheading 3.1.2**. We assume here that antibiotic resistance is caused by a mitochondrial mutation. Despite the rapid mitotic segregation of the mtDNA molecules in daughter cells ***(7)***, these primary clones may contain a mixed population of wild-type and *ant*^{R} mtDNA molecules (heteroplasmy). It is therefore necessary to obtain secondary clones with pure mtDNA populations (homoplasmy) as follows.
   a. An *ant*^{R} colony is streaked for isolated colonies on a YPD plate and incubated 3 d at 30°C.
   b. Isolated colonies are gridded on a YPD plate, incubated 1 d at 30°C, and replicated on an antibiotic-containing plate.
   c. After 2–3 d at 30°C, antibiotic resistance can be observed. Cells, which are now homoplasmic for the *ant*^{R} mutation, are selected from the YPD grid (not from the antibiotic-containing plate to avoid selection of secondary *ant*^{R} mutations).

3.1.2. Discrimination Between Mitochondrial and Nuclear ant^R Mutants

Two types of crosses should be carried out, using strains of opposite mating types and complementary auxotrophies to select diploids on minimum medium. (1) *rho*$^+$ *ant*^{R} × *rho*$^+$ *ant*^{S} and (2) *rho*$^+$ *ant*^{R} × *rho*^{0}.

In cross 1, a mtDNA mutation will produce a mixture of *ant*^{S} and *ant*^{R} homoplasmic diploids, with the relative frequency of each type of diploid depending on the mtDNA output of each strain. In cross 2, all diploids will be *ant*^{R}. For a nuclear *ant*^{R} mutation, the same kind of diploids will be obtained for the two crosses. All diploids will be *ant*^{S}, except in rare cases of *ant*^{R} dominant mutation (or semidominant mutation). The same procedure described next is followed for crosses 1 and 2.

1. The *rho*$^+$ strains are refreshed on GLY plates 2 d at 30°C.
2. The *rho*$^+$ and *rho*^{0} strains grown on GLY are incubated overnight at 30°C on YPD plates.
3. The strains to be crossed are mixed carefully on a YPD plate with a flat toothpick. The cell patch is incubated 1 d at 30°C.
4. Diploids are selected by replica plating the cell patches on glucose minimum medium supplemented with the required amino acids or bases. The plates must not contain those compounds for which strains have complementary auxotrophies. The plates are incubated 2 d at 30°C.
5. Cells are spread on glucose minimum medium plates to give 100–200 diploid colonies. Plates are incubated 2–3 d at 30°C.
6. The colonies are replicated on antibiotic-containing plates and incubated 2–3 d at 30°C. Resistance to antibiotic in the two crosses is compared.

3.2. Mitochondrial DNA Mutation Rate

In an actively dividing cell population, the number of *ant*^{R} mutants in a culture at time *t* will depend on the time when the mutations occur. In other words, if the mutation is an early event occurring soon after inoculation, then the number of mutants will be high compared to a late mutation. For rare mutations, this will lead to a large variance in the number of mutants obtained in parallel cultures. It is thus important to run a substantial number of independent cultures to calculate the average number of mutants. Moreover, the mitochondrial genome is a multicopy system, so that a rare *ant*^{R} mutation occurring in a single mtDNA molecule must be propagated and fixed before it is expressed. Propagation of the *ant*^{R} mutation will depend on replication and recombination processes as well as mitotic segregation of the mtDNA molecules bearing the mutation. It will also depend on the relative growth rate of the mutants and wild-type strains.

To obtain a reliable estimate of the mtDNA mutation rate, we propose the following method inspired from the fluctuation test described for bacteria by Luria and Delbrück ***(13)***. To minimize the role of the "cell division" factor in the accumulation of *ant*^{R} mutants, the mutants are selected on solid antibiotic media on which cellular growth of antibiotic sensitive cells is severely restricted. The procedure described can be applied to other types of mitochondrial mutants (*see* **Note 4**).

1. Antibiotic-sensitive strains (wild type and mutant to be tested) are refreshed on GLY plates at 30°C for 2 d.
2. Strains are spread for single colonies (100–150 per plate) on YPD plates and incubated 3 d at 30°C. For low mutation rates, it can be assumed that there is no *ant*^{R}

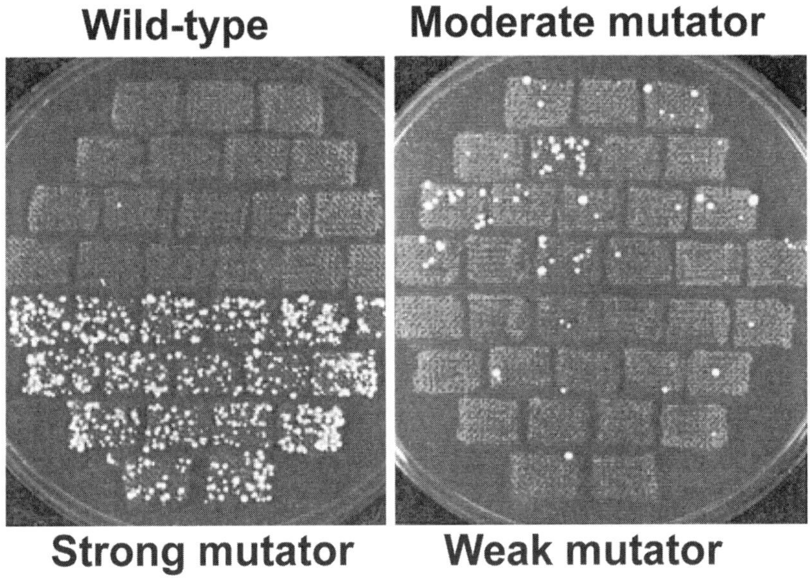

Fig. 1. Analysis of E^R mutant accumulation in wild-type and mutator strains of *Saccharomyces cerevisiae*. YPD plates containing independent clones of wild-type, strong, moderate and weak mutator strains have been replica plated on ERY. Picture was taken after 10 d at 30°C.

mutant among the 100–150 cells that have been spread. However, in rare cases a mtDNA molecule may already bear an ant^R mutation. During cell division in YPD, preexisting and new spontaneously arising ant^R mutations may be lost or fixed. The final colony on YPD plate may be composed of only ant^S cells, or a mixture of ant^S and ant^R cells.

3. YPD plates are divided into several identical small rectangles. Each rectangle is filled with the totality of a colony obtained as reported in **Subheading 2**. To compare wild-type and mutant mutation rates more accurately, half of the rectangles of the same plate are filled with wild-type colonies and the other half with mutant colonies (**Fig. 1**). For each strain, at least 25 colonies obtained as reported in **Subheading 2.** must be used. Several plates can be used. These YPD plates are incubated overnight at 30°C.
4. The cell patches are replicated in a very even manner using a good velvet on two antibiotic-containing plates. The cells deposited on one plate are washed out with 5 mL water to estimate the average number of cells contained in one rectangle. The second plate is incubated at 30°C. It is assumed that background growth is negligible. The ant^R colonies appear after several days (**Fig. 1**). The number of ant^R colonies is determined after a fixed number of days (~10 d). The number of ant^R colonies developing on each rectangle depends on whether fixation of the ant^R mutation was an early or a late event, so that for strains exhibiting low mutation rates variance is high ***(13)***.

5. An approximate mutation rate is calculated as follows. If N is the average number of cells plated in one rectangle, C is the number of rectangles (independent colonies and thus initially cells), and m is the number of ant^R mutants, then the mutation rate **a** is given by the following equation: $m = \mathbf{a}N \ln (N \times C \times \mathbf{a})$ ***(13)***. More simply, the relative frequency of ant^R mutants in a given strain may be normalized to wild type (*see* **Notes 5** and **6**).

3.3. Cellular Growth Tests for Respiratory Deficiency

3.3.1. Respiratory Carbon Source

Wild-type and mutant cells are streaked on YPD and GLY plates. Respiratory-deficient mutants do not grow on glycerol. Respiratory deficiency may be caused by mutations in nuclear (*pet*) or mitochondrial (*mit*⁻, *rho*⁻) DNA. Poor growth on glycerol may be caused by mutations that do not completely inhibit respiration. Some *pet* mutations are associated with instability of the mitochondrial genome. To estimate mtDNA instability, the mutant strain is spread on YPD plate for single colonies. After 4–5 d at 30°C, *rho*⁻ colonies can easily be distinguished from *rho*⁺ colonies by their smaller size and white color. The *rho*⁺ colonies accumulating *rho*⁻ mutants are scalloped (**Fig. 2**).

3.3.2. Low Adenine Media

An *ade1* or *ade2* strain that is respiratory competent is pink in low adenine media such as YPD prepared with Difco products. Cells that do not respire are generally white. Although often useful, this test must be taken with caution.

3.3.3. Tetrazolium Overlay

This simple test ***(14)***, based on the discrimination of respiring cells that accumulate a red pigment (formazan) in the presence of tetrazolium from respiratory-deficient cells that remain white, has previously been reported ***(15)***.

1. Strains are spread for single colonies on YPD plates and incubated at the desired temperature for 2–3 d.
2. Plates are overlaid with the tetrazolium medium as described in **Subheading 2.2.8.**
3. Plates are kept in the dark for 1–3 h. White colonies are *rho*⁻; pink colonies are *rho*⁺.

3.4. Identification of Respiratory-Deficient Mutants

Three types of respiratory-deficient mutants can be obtained: mutants with wild-type nuclear genome and altered mitochondrial genome (*rho*⁻), mutants with mutated nuclear genome (*pet*) and *rho*⁺ mitochondrial genome, and mutants with mutated nuclear genome causing alteration of the mitochondrial genome (*pet/rho*⁻).

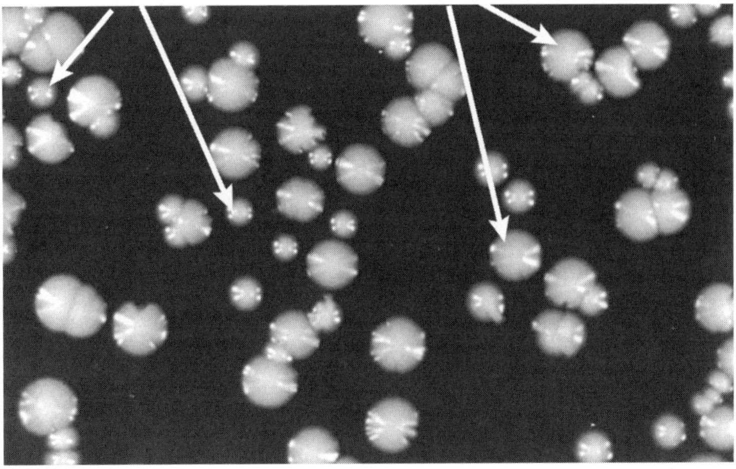

Fig. 2. Scalloped *rho*$^+$ and *rho*$^-$ colonies grown on YPD plate.

3.4.1. Discrimination Between pet and rho$^-$ Mutants

The following cross allows distinguishing a *pet* from a *rho*$^-$/*rho*^{0} mutant.

1. The respiratory-deficient mutant is crossed on a YPD plate with a *rho*^{0} tester strain of opposite mating type and complementary auxotrophies, as described in **Subheading 3.1.2**.
2. After selection of the diploids on glucose minimum medium, diploid patches are replica plated on GLY and incubated 2 d at 30°C.
3. No restoration of cellular growth of diploids on glycerol implies that the respiratory-deficient mutant has irreversibly lost its mitochondrial genome and thus is *rho*$^-$ (or *rho*^{0}).

Tetrad analysis is required to determine whether the *rho*$^-$ trait is caused by a nuclear *pet* mutation.

1. The respiratory-deficient mutant (refreshed on YPD) and wild-type strain (refreshed on GLY) are grown overnight at 30°C in 3 mL liquid YPD medium in a test tube with good shaking and aeration.
2. From each strain, 25-μL drops are spotted and mixed together on a freshly made YPD plate and incubated 1 d at 30°C.
3. The cell mixture is streaked on a freshly made YPD plate and incubated at 30°C for 8–10 h.
4. Cells are transferred to a freshly made SPO plate (*see* **Note 7**).
5. After 3–5 d at 25°C, tetrads are ready for dissection. A 2:2 segregation of the respiratory-deficiency trait (small size of the spore, no growth on glycerol) will be

observed for a *pet*/*rho*⁻ mutant. If the isolate is a simple *rho*⁻ mutant, then the respiratory deficiency trait will not be recovered in tetrads except for rare *rho*⁻ mutants, which are said to be "suppressive" ***(7)***. A non-Mendelian segregation (0:4 and 4:0) of the defect will be observed.

3.4.2. Discrimination Between rho⁻ *and* rho^0 *Mutants*

Staining with 4,6-diamidino-2-phenylindole is not a very sensitive method. A simple and reliable alternative is the genetic test described next (*see* **Note 8**). Although *rho*^{0} mutants are totally devoid of mtDNA, *rho*⁻ mutants have retained fragments of mtDNA of various sizes and different localizations (*see* **Note 9**). The test is based on the ability of a *rho*⁻ genome to restore a *mit*⁻ tester strain by homologous recombination of the mitochondrial genomes in diploids ***(7)***.

1. The *pet*/*rho*⁻ (*rho*^{0}) mutant is refreshed on a YPD plate and spread for single colonies on a YPD plate.
2. Grid 100–200 *rho*⁻ (*rho*^{0}) colonies on YPD plates and incubate 1 d at 30°C; YPD liquid cultures of *mit*⁻ tester strains (cox2, cox3, cob) of opposite mating type and complementary auxotrophies are run overnight (*see* **Note 10**).
3. The colonies of the YPD grid are cross-replica-plated on YPD lawns spread with 0.1–0.2 mL *mit*⁻ strain cultures.
4. The YPD plates are incubated 1 d at 30°C and replicated on glucose minimum plates to select diploids.
5. After 2 d at 30°C, diploid patches are replica plated on GLY. If one colony gridded as reported in **step 2** contains a *rho*⁻ genome encompassing the *mit*⁻ mutation, then the diploid patches will grow in glycerol after 1–2 d at 30°C.

4. Notes

1. Erythromycin resistant (E^R) mutants are always of mitochondrial inheritance with mutations localizing to the 21S ribosomal ribonucleic acid (RNA) gene, mainly at three positions (GAA, nucleotides 59,965–59,967 on the mtDNA map in the *Saccharomyces* Genome Database at http://www.yeastgenome.org/). A large spectrum of base substitutions is obtained at position A 59,966 (A-to-G or C-to-T transitions, A-to-T or T-to-A and A-to-C or T-to-G transversions) ***(9)***. One base deletion or addition is also observed.
2. E^R mutations are mild and do not affect cellular growth. However, before carrying out experiments on mutator activity, it must be verified that cellular resistance to erythromycin (or any other compound) is similar in parental and mutant strains.
3. A majority of chloramphenicol (C^R) and oligomycin (O^R) mutants are produced by mutations in the nuclear DNA. Mitochondrial C^R mutations localize to the 21S RNA gene, and O^R mutations have been mapped to the *atp6* and *atp9* genes.
4. Mutants in an *arg8* gene targeted to mitochondria by ballistic transformation (*see* Chapter 11) may also be screened. The assay which requires a special strain constructed by T. Fox and col. ***(16)*** measures one base deletion in the mtArg8 gene ***(17)***.

5. To analyze independent mutational events, the mtDNA of only a single *ant*^{R} colony must be sequenced for each rectangle (**Fig. 1**).
6. The frequency of mtDNA mutants is considerably increased in *mip1 exo*$^-$ and *msh1* mutants ***(9–11)***.
7. Sporulation efficiency strongly varies from one strain to another. Although tetrads are easily obtained in W303, BY strains derived from S288c sporulate very poorly. It is necessary to use very fresh plates and adjust sporulation conditions to the strain used.
8. When tested, many *pet/rho*$^-$ mutant strains are completely devoid of mtDNA. This does not mean that these strains are not able to maintain *rho*$^-$ mtDNA under vegetative growth. They may have lost mtDNA during the isolation procedure (mutagenesis, deletion, transformation). It is often possible to introduce in these strains selected *rho*$^-$ genomes by cytoduction using a *kar1* mutant. This method, described previously in **ref. *15***, is based on the capacity of mitochondria to mix and fuse in a heterokaryon and allows determination of the capacity of a strain to maintain a *rho*$^-$ genome stably.
9. It was shown many years ago that the most frequently retained fragments localize to *cox2*, *cox3*, and cytochrome-*b* genes ***(18,19)***.
10. The *mit*$^-$ mutants are rare compared to *rho*$^-$. Manganese is a good mutagen to obtain *mit*$^-$ mutants ***(5)***. Enrichment in *mit*$^-$ mutants can be obtained by introducing a *pet9* (*op1*) mutation ***(20)***. Because *rho*$^-$ mutations are lethal in the *op1* background, only respiratory-deficient mutants with point mtDNA mutations can be obtained. It must be noted, however, that the *op1* mutant itself does not grow on glycerol.

References

1. Ephrussi, B., Hottinguer, H., and Chimenes, A. M. (1949) Action de l'acriflavine sur les levures. I. La mutation "petite colonie." *Ann. Inst. Pasteur* **76,** 351–367.
2. Mounolou, J-C., Jakob, H., and Slonimski, P. P. (1966) Mitochondrial DNA from yeast "petite mutants": specific changes of buoyant density corresponding to different cytoplasmic mutations. *Biochem. Biophys. Res. Commun.* **24,** 218–224.
3. Linnane, A. W., Saunders, G. W., Gingold, E. B., and Lukins, H. B. (1968) The biogenesis of mitochondria. V. Cytoplasmic inheritance of erythromycin resistance in *Saccharomyces cerevisiae. Proc. Natl. Acad. Sci. USA* **59,** 903–910.
4. Thomas, D. Y. and Wilkie, D. (1968) Recombination of mitochondrial drug-resistance factors in *Saccharomyces cerevisiae. Biochem. Biophys. Res. Commun.* **30,** 368–372.
5. Tzagoloff, A., Akai, A., Needleman, R. B., and Zulch, G. (1975) Assembly of the mitochondrial membrane system. Cytoplasmic mutants of *Saccharomyces cerevisiae* with lesions in enzymes of the respiratory chain and in the mitochondrial ATPase. *J. Biol. Chem.* **250,** 8236–8242.
6. Slonimski, P. P. and Tzagoloff, A. (1976) Localization in yeast mitochondrial DNA of mutations expressed in a deficiency of cytochrome oxidase and/or coenzyme QH2-cytochrome *c* reductase. *Eur. J. Biochem.* **61,** 27–41.

7. Dujon, B. (1981) Mitochondrial genetics and function, in *The Molecular Biology of the Yeast* Saccharomyces*: Life Cycle and Inheritance* (Strathern, J. N., Jones, E. W., and Broach, J. R., eds.), Cold Spring Harbor Laboratory Press, Cold Spring Harbor, NY, pp. 505–635.
8. Contamine, V. and Picard, M. (2000) Maintenance and integrity of the mitochondrial genome: a plethora of nuclear genes in the budding yeast. *Microbiol. Rev.* **64,** 281–314.
9. Backer, J., and Foury, F. (1985) Repair properties in yeast mitochondrial DNA mutators. *Curr. Genet.* **10,** 7–13.
10. Vanderstraeten, S., van den Brule, S., Hu, J., and Foury, F. (1998) The role of 3′-5′ exonucleolytic proofreading and mismatch repair in yeast mitochondrial DNA error avoidance. *J. Biol. Chem.* **273,** 23,690–23,697.
11. Reenan, A. R. and Kolodner, R. D. (1993) Characterization of insertion mutations in the *Saccharomyces cerevisiae MSH1* and *MSH2* genes: evidence for separate mitochondrial and nuclear functions. *Genetics* **132,** 975–985.
12. di Rago, J. P., Coppee, J. Y., and Colson, A. M. (1989) Molecular basis for resistance to myxothiazol, mucidin (strobilurin A), and stigmatellin. Cytochrome *b* inhibitors acting at the center *o* of the mitochondrial ubiquinol-cytochrome *c* reductase in *Saccharomyces cerevisiae. J. Biol. Chem.* **264,** 14,543–14,548.
13. Luria, S. E. and Delbrück, M. (1943) Mutations of bacteria from virus sensitivity to virus resistance. *Genetics* **28,** 491–511.
14. Ogur, M., St. John, R., and Nagai, S. (1957) Tetrazolium overlay technique for population studies of respiration deficiency in yeast. *Science* **125,** 928–929.
15. Foury, F. (2002) Yeast nuclear genes for mtDNA maintenance, in *Methods in Molecular Biology Mitochondrial DNA: Methods and Protocols* (Copeland, W. C., ed.), Humana Press, Totowa, NJ, pp. 139–149.
16. He, S. and Fox, T. D. (1999) Mutations affecting a yeast mitochondrial inner membrane protein, pnt1p, block export of a mitochondrially synthesized fusion protein from the matrix. *Mol. Cell. Biol.* **19,** 6598–6607.
17. Strand, M. K. and Copeland, W. C. (2002) Measuring mtDNA mutation rates in *Saccharomyces cerevisiae* using the mtArg8 assay, in *Methods in Molecular Biology Mitochondrial DNA: Methods and Protocols* (Copeland, W. C., ed.), Humana Press, Totowa, NJ, pp. 151–158.
18. Fukuhara, H. and Wesolowski, M. (1977) Preferential loss of a specific region of mitochondrial DNA by *rho*$^-$ mutation, in *Mitochondria 1977: Genetics and Biogenesis of Mitochondria* (Bandlow, W., ed.), De Gruyter, Berlin, pp. 123–131.
19. Mathews, S. Schweyen, R. J., and Kaudewitz, F. (1977) Preferential loss or retention of mitochondrial genes in *rho*$^-$ clones, in *Mitochondria 1977: Genetics and Biogenesis of Mitochondria* (Bandlow, W., ed.), De Gruyter, Berlin, pp. 133–139.
20. Kotylak, Z. and Slonimski, P. P. (1977) Mitochondrial mutants isolated by a new screening method based upon the use of the nuclear mutation op1, in *Mitochondria 1977: Genetics and Biogenesis of Mitochondria* (Bandlow, W., ed.), De Gruyter, Berlin, pp. 83–89.

13

Transcript End Mapping and Analysis of RNA Editing in Plant Mitochondria

Frank Kempken, Nina Bolle, Joachim Forner, and Stefan Binder

Summary

Mitochondria are genetic compartments with their own enzymatic equipment for maintenance and expression of their genetic information. As in all genetic systems, gene expression has to be regulated, and in mitochondria this also has to be coordinated with the expression of nuclear-encoded mitochondrial proteins. Presently, there is virtually no information available about the mechanistic details and the enzymes involved in these processes. There is still much to be learned about how plant mitochondrial gene expression is managed and to what extent the contribution of transcription initiation and posttranscriptional processes, respectively, contribute to this control. As one prerequisite for better understanding of the mechanisms and regulatory controls, more fundamental data on mitochondrial transcription initiation and posttranscriptional RNA processing are necessary. As part of the essential methodology, we present methods for the analysis of the 5′ and 3′ extremities of mitochondrial transcripts and the identification of transcription initiation sites. An *in organello* system is described for the functional investigation of ribonucleic acid editing in plant mitochondria.

Key Words: Electroporation; *in organello* RNA synthesis; in vitro capping; mitochondria isolation; RNA isolation; RT-PCR.

1. Introduction

In plant mitochondria, most genes are represented by a number of different transcripts *(1–3)*. Several factors account for these often-complex ribonucleic acid (RNA) patterns. In addition to the presence of splicing intermediates and semiprocessed RNA accumulating during the disassembly of large multicistronic precursor RNAs, transcripts with different 5′ ends are posttranscriptionally generated even for individually transcribed genes (Forner and S. Binder, 2006, unpublished results). Transcription initiation at multiple sites also can substantially

From: *Methods in Molecular Biology, vol. 372: Mitochondria: Practical Protocols*
Edited by: D. Leister and J. M. Herrmann © Humana Press Inc., Totowa, NJ

contribute to the accumulation of RNAs with various 5′ termini ***(4,5)***. As a consequence, the unambiguous discrimination between 5′ ends generated by posttranscriptional processing from those originating from transcription initiation is extremely difficult but is an important requirement for the analysis of these processes.

To overcome this problem, techniques have been developed that take advantage of a crucial difference between these divergent types of ends. While the 5′ ends directly generated by transcription initiation carry triphosphate groups, only single phosphates are found at those ends created by posttranscriptional endo- or exonucleolytic cleavage. This difference can be exploited because the presence of at least two phosphates is a prerequisite for in vitro capping of RNA by guanylyltransferase. With this so-called capping enzyme, primary transcripts can be specifically labeled with [α-^{32}P]GTP (guanosine 5′-triphosphate). In addition, RNA ligase discriminates between these different 5′ ends. This enzyme accepts only termini with single 5′ phosphate groups ***(6,7)***. This feature allows specific covalent linkage of other single-stranded RNA or deoxyribonucleic acid (DNA) molecules only to such 5′ monophosphate ends and thus specific tagging of RNAs with 5′ ends that have been generated by processing ***(4,8–10)***.

As part of the posttranscriptional maturation processes, primary sequences of plant mitochondrial RNAs are altered by RNA editing. Typically cytidines (C) are converted to uridines (U), but in rare instances (frequently in hornworts) also reverse U-to-C conversions are observed ***(11)***. The dimension of plant mitochondrial RNA editing, which changes the coding information of almost all protein-coding genes as well the primary sequences of distinct transfer RNAs, has been demonstrated by the identification of 441 editing sites in transcripts of the open reading frames in *Arabidopsis thaliana*, which provide a fundamental descriptive data set ***(12)***. Much less, however, is known about the enzymatic mechanism and the specificity features that determine distinct cytidines to be edited. While the former has been suggested to occur by deamination of the cytosine base without breaking the phosphate backbone ***(11)***, the selection of the edited cytidines is suggested to be determined by specific pentatricopeptide, repeat (PPR) proteins in both mitochondria and chloroplasts ***(13,14)***.

The analysis of higher plant mitochondrial RNA editing has so far been hampered by the inability to genetically engineer mitochondria in vivo. Thus, in vitro or *in organello* systems are necessary to dissect the editing mechanism. To this end, we developed a method to introduce foreign DNA into maize and sorghum mitochondria, which allows the uptake and successful integration of DNA fragments of up to 11 kb into the mitochondrial matrix. Subsequent incubation of mitochondria in a specific buffer allows *de novo* transcription, splicing, and editing of the newly synthesized RNAs. This procedure has been successfully applied to homologous genes such as *atp6* and *cox2* as well as for the heterologous *cox2* gene from *A. thaliana* in maize mitochondria ***(15–17)***.

2. Materials

2.1. Transcript End Mapping

1. T4 RNA ligase (Roche Applied Science).
2. Maloney murine leukemia virus (M-MLV) reverse transcriptase (Promega).
3. BD Advantage™ 2 polymerase chain reaction (PCR) enzyme system (BD Bioscience).
4. 10X T4 RNA ligase reaction buffer: 500 m*M* Tris-HCl, pH 7.5, 100 m*M* $MgCl_2$, 100 m*M* dithiothreitol, 10 m*M* adenosine triphosphate, 300 mg/L bovine serum albumin (BSA) (Roche Applied Science).
5. 5X M-MLV reverse transcriptase reaction buffer: 50 m*M* Tris-HCl, pH 8.5, 75 m*M* KCl, 3 m*M* $MgCl_2$, 10 m*M* dithiothreitol (Promega).
6. GFX PCR and gel band elution kit (Amersham Bioscience).
7. pGEM®-T vector kit (Promega).
8. RNasin® ribonuclease inhibitor (Promega).
9. Dimethyl sulfoxide (DMSO).
10. Water, double distilled and autoclaved.
11. Phenol-chloroform-isoamyl alcohol 25:24:1 (v/v/v).
12. Chloroform-isoamyl alcohol 24:1 (v/v).
13. 10 m*M* deoxynucleotide 5′-triphosphates (dNTPs) (Peqlab).
14. 1 *M* NaOH.
15. 1 *M* HCl.
16. 3 *M* sodium acetate, pH 7.0.
17. 100% ethanol (100% EtOH).
18. 70% (v/v) ethanol (70% EtOH).
19. Sterile 1.5-mL reaction tubes.
20. Microcon YM-30 microconcentrator (Amicon).
21. 200-µL thin-wall reaction tubes.

2.2. In Vitro Capping Analysis of Mitochondrial RNA

1. Guanylyltransferase (Ambion) (*see* **Note 1**).
2. 10X in vitro capping buffer as supplied by the manufacturer.
3. Ribonuclease (RNase) inhibitor (Fermentas).
4. [α-^{32}P]GTP (3000 Ci/mmol) (Amersham Bioscience).
5. 500 m*M* ethylenediaminetetraacetic acid (EDTA).
6. 10% (w/v) sodium dodecyl sulfate (SDS).
7. Proteinase K (20 mg/mL).
8. Water, double distilled and autoclaved.
9. Phenol-chloroform-isoamyl alcohol 25:24:1 (v/v/v).
10. Chloroform-isoamyl alcohol 24:1 (v/v).
11. 10 *M* ammonium acetate (NH_4OAc).

12. 100% ethanol (100% EtOH).
13. 70% (v/v) ethanol (70% EtOH).
14. Sterile 1.5-mL reaction tubes.

2.3. *In Organello* Analysis of RNA Editing

2.3.1. Purification of Mitochondria

1. 10X buffer A: 0.05 *M* EDTA, 1% (w/v) BSA, 0.5 *M* Tris-HCl, pH 7.5. Store at 4°C.
2. 1X buffer A with 0.5 *M* sucrose. Store at 4°C.
3. 1X buffer A with 0.5 *M* sucrose but without BSA. Store at 4°C.
4. Solutions for discontinuous sucrose gradients: 1.5 *M* (54% w/v), 1.2 *M* (40% w/v), 1.1 *M* (36% w/v), and 0.6 *M* (20% w/v) sucrose in buffer A. Prepare 100 mL of each solution by adding 10 mL 10X buffer A to the respective amount of sucrose and complete with distilled water.
5. Absorbent gauze: zigzag folded, eight ply, 10 cm wide (Hartmann).
6. Protein assay dye (Bradford) reagent concentrate (Bio-Rad).
7. Meliseptol: 50% (v/v) isopropanol, 0.5% (v/v) glyoxal.
8. 1.5% (w/v) sodium hypochlorite.
9. Waring blender.
10. Potter homogenizer.
11. MitoTracker Orange CM-H_2TMRos (Molecular Probes).
12. 34-mL polyallomer thin-wall centrifuge tubes for swing-out rotor (Kisker).
13. 250-mL polypropylene centrifuge bottles (Beckman Coulter).
14. Glycerol.

2.3.2. Electroporation and *In Organello* Incubation of Mitochondria

1. Bio-Rad Gene Pulser and 0.1-cm electrode-gap cuvettes (Bio-Rad).
2. Expression buffer: 25 m*M* sucrose, 75 m*M* sorbitol, 100 m*M* KCl, 10 m*M* K_2HPO_4, 0.05 m*M* EDTA, 5 m*M* $MgCl_2$, 10 m*M* L-glutamic acid, 2.5 m*M* L-malic acid, 1 mg/mL BSA, 1 m*M* adenosine 5′-diphosphate, 10 m*M* Tris-HCl, pH 7.4. Store at 4°C and add adenosine 5′-diphosphate just before the experiment.
3. 0.6 *M* sucrose solution.

2.3.3. Isolation of Mitochondrial RNA

1. 20% (w/v) SDS.
2. Phenol-chloroform-isoamyl alcohol 25:24:1 (v/v/v).
3. Chloroform-isoamyl alcohol 24:1 (v/v).
4. 8 *M* NH_4OAc.
5. Absolute ethanol.
6. 70% (v/v) ethanol (70% EtOH).
7. 4 *M* LiCl.
8. Diethyl pyrocarbonate (DEPC)-treated water.

2.3.4. PCR and Reverse Transcriptase PCR

1. RNase-free deoxyribonuclease (DNase) I (MBI Fermentas).
2. 10X DNase I reaction buffer (MBI Fermentas).
3. Taq DNA polymerase kit (Eppendorf).
4. One-step reverse transcriptase (RT) PCR kit (Qiagen).
5. 25 m*M* EDTA.

3. Methods

3.1. Transcript End Mapping

RT-PCR analyses of copy DNA (cDNA) derived from 3′ to 5′ ligated RNA allow the simultaneous mapping of both 5′ and 3′ extremities and the detection of nonencoded nucleotides (**Fig. 1**) ***(10)***. The experimental procedure for this analysis is described in the following Subheading. In combination with additional treatments of the RNA, this method allows the identification of primary 5′ ends (*see* **Note 2**).

1. Extract mitochondrial RNA or total cellular RNA from appropriate plant tissues. Any RNA isolation procedure can be applied; however, special care should be taken to avoid degradation of the RNA. Detailed protocols for the isolation of mitochondria and mitochondrial RNA have been described previously ***(18–20)***.
2. For ligation, add the following components for a total volume of 10 µL: 5–15 µg RNA in 7 µL H_2O, 1 µL 10X T4 RNA ligase reaction buffer, 1 µL (= 10 U) T4 RNA ligase, and 1 µL DMSO.
3. Mix gently and incubate over night at 15°C.
4. Add 390 µL H_2O to adjust the total volume to 400 µL.
5. For extraction, gently mix with 400 µL of phenol-chloroform-isoamyl alcohol 25:24:1 (v/v/v) and centrifuge the sample at full speed for 5 min at room temperature in a tabletop centrifuge.
6. Carefully remove the upper aqueous layer and transfer into a new 1.5-mL reaction tube. Extract sample two times with chloroform-isoamyl alcohol 24:1 (v/v).
7. Pipet sample into a Microcon YM 30 microconcentrator (Amicon) and centrifuge in a Beckman F3602 (or F2402) rotor at 12,000 rpm (≤14,000*g*) at 4°C to reduce the volume to about 10 µL (*see* **Note 3**).
8. For cDNA first-strand synthesis, transfer the ligated RNA to a new 1.5-mL reaction tube and adjust with water to a final volume of 14.75 µL.
9. Add 2 µL (= 200 pmol) of primer (P1; *see* **Fig. 1**) and incubate for 5 min at 70°C.
10. After incubation, keep the reaction mixture on ice and add the following components: 1 µL RNasin, 5 µL 5X M-MLV reverse transcriptase reaction buffer, 1.25 µL dNTPs (to 10 m*M*), and 1 µL M-MLV (= 200 U).
11. Reverse transcription is performed for 1 h at 42°C.
12. To stop the reaction and to hydrolyze the RNA, add 5 µL of 1*M* NaOH and incubate sample for 10 min at room temperature.
13. Neutralize the solution with 5 µL 1 *M* HCl and 10 µL H_2O.

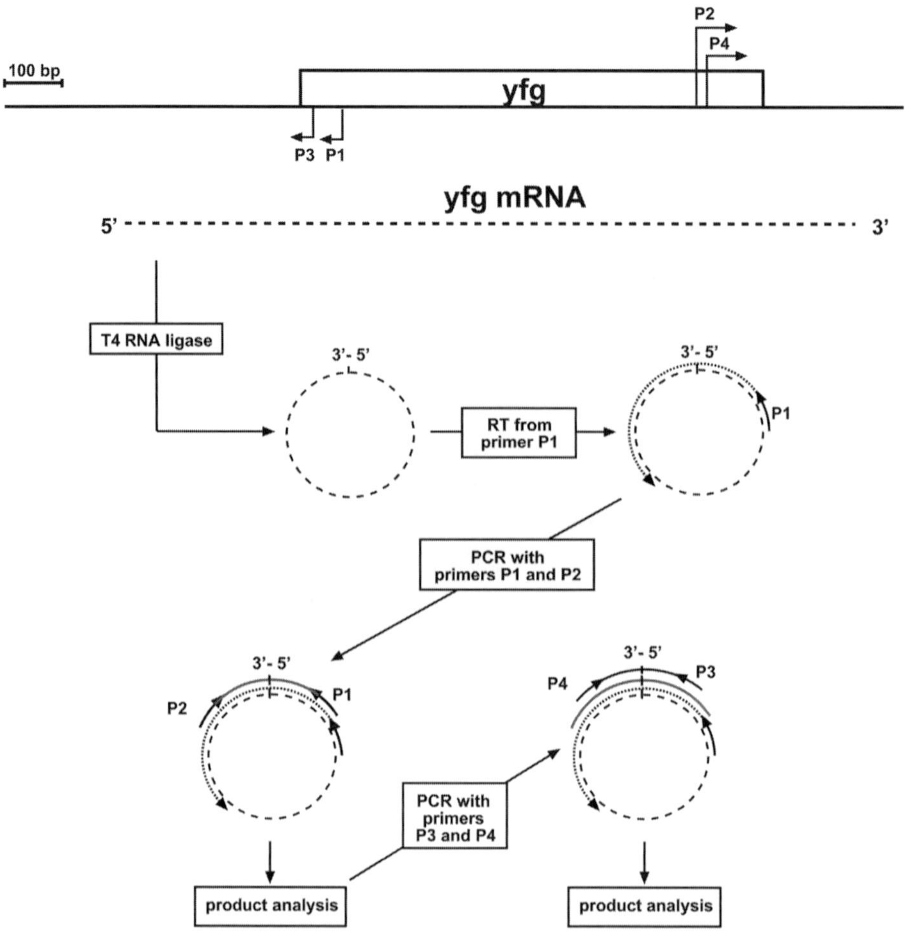

Fig. 1. Flow chart of the CR-RT-PCR analysis. Design oligonucleotides for your favorite gene (yfg). These primers (bent arrows, designated P1 to P4) should be complementary to sequence within the reading frame (box) and oriented to the extremities of the RNA. RNA is then self-ligated, and cDNA synthesis is initiated from a primer (P1) directed to a sequence in the 5′ region of the reading frame. The first PCR is done with primers P1 and P2, which enclose the ligation site. After product analysis by agarose gel electrophoresis, cDNA fragments can be directly cloned and analyzed or used as a template for a second PCR with nested primers P3 and P4. The final products are cloned and sequenced.

14. Precipitate cDNA by the addition of 5 µL (= 1/10 volume) 3 *M* sodium acetate, pH 7.0, and 2 volumes of 100% EtOH (= 100 µL) and by incubation for 30 min at –20°C. Collect cDNA by full-speed centrifugation in a tabletop centrifuge, wash three times with 70% (v/v) EtOH, and resuspend the cDNA in 20 µL H_2O.

15. The cDNA is now ready for PCR amplification.
16. Use a quarter (= 5 µL) of the cDNA as the template for PCR amplification across the ligation site.
17. For the first PCR, add the following components to a 200-µL reaction tube: 5 µL cDNA, 1 µL dNTP (10 m*M* each), 1 µL of 20 µ*M* primers each, 5 µL 10X BD Advantage 2 reaction buffer (supplied by the manufacturer), and 36.5 µL H_2O to a total volume of 45.5 µL. Start reaction by adding 0.5 µL BD Advantage 2 thermostable DNA polymerase.
18. The parameters of the amplification reaction depend on the primers used. BD Advantage 2 polymerase requires 68°C for optimal elongation.
19. Separate PCR products on a standard 1% agarose gel and cut out slices of agarose containing cDNAs. Although clear strong PCR fragments can be directly cloned in the pGEM-T vector, weak fragments should be used for reamplification using nested primers. To this end, cDNA should be recovered from the gel. Cut out several adjacent pieces containing different molecule sizes and use these individually as PCR templates (*see* **Note 4**). Recover cDNA fragments from the gel either by freeze squeeze (*see* **Note 5**) or by using a GFX PCR and gel band elution kit according to the manufacturer's recommendations. Usually, strong fragments are generated in the second PCR with one or more of the recovered cDNA size fractions, which can then be recovered and cloned as described. Finally, PCR fragments are sequenced following standard procedures, and the data are evaluated with standard sequence analysis software packages.

3.2. In Vitro Capping Analysis of Mitochondrial RNA

The in vitro capping analysis takes advantage of the fact that mitochondrial primary 5′ ends are not capped in vivo. Thus, a radioactive cap can selectively be added in vitro by guanylyltransferase with [α-^{32}P]GTP as substrate, and labeled RNA can be further analyzed by different methods (*see* **Note 6**).

1. Extract mitochondrial RNA or total cellular RNA from appropriate plant tissues. Any RNA isolation procedure can be applied; however, special care should be taken to avoid degradation of the RNA. Detailed protocols for the isolation of mitochondria and mitochondrial RNA have been described previously ***(18–20)***.
2. Use about 50–100 µg of mitochondrial or total cellular RNA for a typical in vitro capping reaction in a total volume of 20 µL.
3. Perform the reaction in the presence of 5 U guanylyltransferase (= 1 µL), 650 µ*M* S-adenosyl-methionine (also supplied by Ambion), 12.5 U RNase inhibitor, and 100 µCi [α-^{32}P]GTP.
4. Incubate reaction at 37°C for 75 min.
5. Add another 7.5 U guanylyltransferase and continue incubation for 30 min.
6. Terminate reaction by the addition of 71.5 µL H_2O, 1 µL 500 m*M* EDTA, 5 µL 10% (w/v) SDS, and 1 µL proteinase K (20 mg/mL) and incubation of 15 min at 37°C.
7. Add 100 µL H_2O to the sample and extract with 200 µL phenol-chloroform-isoamyl alcohol (25:24:1).

8. After a short centrifugation, transfer the upper aqueous phase to a new reaction tube and extract twice with 400 µL chloroform-isoamyl alcohol 24:1 (v/v/v).
9. Precipitate RNA by the addition of a quarter volume (= 50 µL) of 10*M* NH_4OAc and 600 µL 100% EtOH and incubate for 30 min at –20°C.
10. Spin down RNA by centrifugation in a tabletop centrifuge at maximum speed and 4°C and wash the RNA pellet three times with 1 mL 70% EtOH.
11. Resuspend capped RNA in water and use as a probe for hybridization assays or dissolve in an appropriate buffer if it is further analyzed by S1 or RNase protection analysis (*see* **Note 7**).

3.3. In Organello *Analysis of RNA Editing*

3.3.1. Purification of Mitochondria

Mitochondria are isolated from 9-d-old etiolated seedlings (*Zea mays* or *Sorghum bicolor*). The seedlings are grown in a dark room at 26°C. The procedure for isolation of mitochondria is modified from that of Newton ***(21)***.

1. Prepare the discontinuous sucrose gradient by layering sucrose solutions in the following order (bottom to top) into 34-mL centrifuge tubes: 1.5 *M* (8 mL), 1.2 *M* (10 mL), 1.1 *M* (10 mL), and 0.6 *M* (4 mL). The gradient should be stored at 4°C for at least 2 h before use.
2. To minimize possible bacterial contamination, spray seedlings with Meliseptol (*see* **Note 8**). After 5-min incubation, harvest seedlings and submerge plants in 1.5% (w/v) sodium hypochlorite for 10 min. Wash seedlings three or four times with sterile distilled water.
3. Perform all further steps at 4°C.
4. Prepare the extraction medium by adding 0.0175% (v/v) β-mercaptoethanol to buffer A (1X, with BSA). A liter of extraction medium is required per 250 g seedlings (fresh weight). Add the extraction medium to the harvested seedlings and grind tissue in a Waring blender by three short pulses of 10–15 s at full speed (*see* **Note 9**).
5. Squeeze extract through four layers of absorbent gauze.
6. Fill the extract into precooled 250-mL centrifuge bottles and centrifuge at 1000*g* for 10 min (e.g., Beckman Coulter J-20 XP, JLA16.250 rotor).
7. Filtrate supernatant through one layer of gauze into clean centrifuge tubes and discard pellet. Sediment mitochondria by differential centrifugation at 15,900*g* (maize) or 12,000*g* (sorghum) for 20 min (Beckman Coulter J-20 XP, JLA16.250 rotor) and resuspend the pellet with a brush (previously sterilized in boiled distilled water) in a small volume of buffer A (usually the remaining buffer A is sufficient).
8. Homogenize suspension in a potter with a single stroke and carefully lay mitochondria on top of the discontinuous sucrose gradient (do not overlay on more than 2 mL of the suspension per centrifuge tube).
9. Centrifuge in a swing-out rotor at 100,000*g* for 1 h (e.g., Kontron TST 28.36 at 24,000 rpm).

10. After centrifugation, the bulk of the mitochondria collect at the interface between the 1.2 *M* and 1.45 *M* sucrose steps (**Fig. 2**). Remove mitochondria from the sucrose gradient by pipeting with a Pasteur pipet and transfer them into a clean tube. Dilute the mitochondrial fractions slowly with buffer A (1X, without BSA) to a final volume of 160 mL. The dilution must proceed rather slowly to avoid osmotic stress for the mitochondria. The dilution process should take about 20–30 min. Finally, sediment diluted mitochondria by centrifugation at 15,900*g* (maize) or 12,000*g* (sorghum) for 20 min at 4°C.
11. Resuspend the organelles in 3–5 mL buffer A (1X, without BSA) and measure concentration of mitochondrial protein with the Bradford reagent. The organelles can be used directly for electroporation or can be stored at –80°C in aliquots after adding glycerol to a final concentration of 10% (v/v). Organelles stored at –80°C can be used for up to 1 month (*see* **Note 10**).

3.3.2. *Electroporation and* In Organello *Incubation of Mitochondria*

All operations are performed at 4°C.

1. Carefully mix mitochondria (equivalent to 300–1200 μg of the mitochondrial protein) in 500 μL 0.6 *M* sucrose solution in a 1.5-mL reaction tube and centrifuge the suspension at 9000*g* for 10 min.
2. Resuspend pellet in 500 μL 0.6 *M* sucrose and centrifuge again at 9000*g* for 10 min.
3. Resuspend the pellet in 25 μL 0.6 *M* sucrose to a final concentration of 10 μg mitochondrial protein/microliter (usually 10% of mitochondria are lost during the washing steps).
4. Add 2.5 μg plasmid DNA to the mitochondrial suspension. Transfer to an electroporation cuvette and incubate for 30–60 s on ice. Electroporation is performed using a Bio-Rad electroporation device with the following conditions: 400 Ω, 25 μF capacity, and an applied field strength of 18–20 kV/cm.
5. Immediately transfer suspension from the cuvette into a 1.5-mL reaction tube and wash cuvette with an additional 100 μL 0.6 *M* sucrose, which is added to the mitochondrial suspension (*see* **Note 11**).
6. Collect mitochondria by centrifugation at 9000*g* for 15 min.
7. Resuspend the pellet in 500 μL expression buffer and centrifuge at 9000*g* for 15 min.
8. Resuspend the pellet once more in 500 μL expression buffer and incubate mitochondria for about 2–3 h at room temperature (*see* **Note 12**).

3.3.3. *Isolation of Mitochondrial RNA*

1. After the *in organello* incubation, recover mitochondria by centrifugation at 9000*g* for 15 min at 4°C. Dissolve the pellet in 180 μL distilled DEPC-treated water. Add 20 μL 20% (w/v) SDS and vortex (*see* **Note 13**).
2. Immediately add 200 μL of phenol-chloroform-isoamyl alcohol 25:24:1 (v/v/v) and mix solution well by vortexing. Centrifuge the sample at 18,000*g* for 5 min at room temperature.

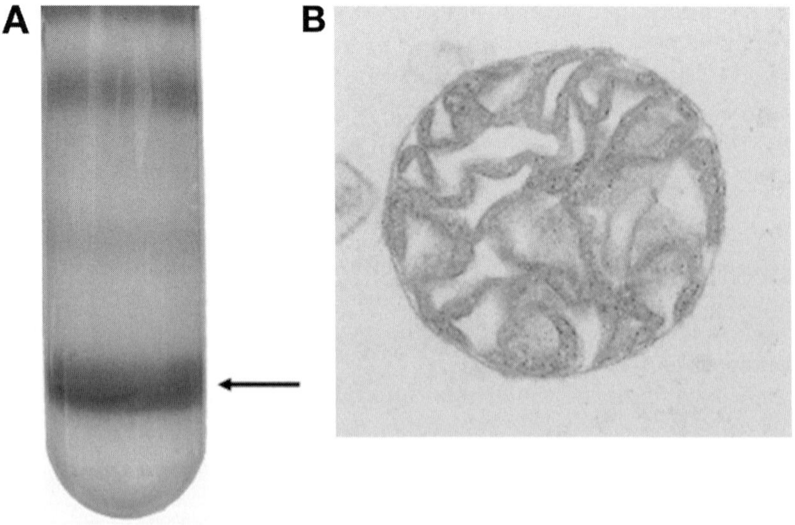

Fig. 2. Preparation of mitochondria. (**A**) The figure shows a sucrose gradient with separated mitochondria. The arrow indicates the position of mitochondria. (**B**) Electron micrograph of a mitochondrion.

3. Carefully remove aqueous layer and pipet into a new tube. Repeat phenol-chloroform extraction once.
4. Add 200 µL chloroform-isoamyl alcohol 24:1 (v/v) to the liquid phase of the second phenol-chloroform extraction and centrifuge at 18,000*g* for 5 min at room temperature.
5. Transfer aqueous layer into a clean tube and precipitate nucleic acids by the addition of 1/10 volume 8 *M* NH_4OAc and 2 volumes absolute ethanol.
6. Mix the sample and incubate for at least 20 min at –80°C.
7. Recover the precipitated RNA by centrifugation at 18,000*g* for 15 min at 4°C.
8. Wash pellet with 400 µL 70% (v/v) EtOH and centrifuge at 18,000*g* for 15 min at 4°C.
9. Remove ethanol, air dry pellet at room temperature, and dissolve pellet in 300 µL DEPC-treated distilled water by vortexing.
10. For selective precipitation of RNA, add 300 µL 4 *M* LiCl.
11. Incubate the sample for at least 4 h at 4°C and centrifuge at 18,000*g* for 15 min at 4°C.
12. Wash the pellet again with 70% (v/v) EtOH.
13. Air dry pellet and dissolve in 20 µL DEPC-treated distilled water. The RNA can be stored at –80°C.

3.3.4. PCR and RT-PCR

1. Before the RT-PCR analysis of the *in organello*-synthesized RNA, treat RNA with RNase-free DNase I (MBI Fermentas). Contaminating DNA could otherwise result in false-positive PCR products in the subsequent RT-PCR.

2. For degradation of DNA, add 2 μL of (RNase-free) DNase I (1 U/μL) and 2 μL 10X reaction buffer to 10 μL RNA.
3. Add DEPC-treated distilled water to a final volume of 20 μL and incubate for 1 h at 37°C.
4. For inactivation of DNase I, add 2 μL of 25 m*M* EDTA and incubate for 10 min at 65°C. The DNA-free RNA can now be used as a template in RT-PCR (Qiagen) amplification (*see* **Fig. 3**). Perform PCR to check for remaining DNA contamination.
5. Perform RT-PCR using primers specific for an endogenous transcript, which is constantly expressed (*see* **Notes 14** and **15**), and with transcript-specific primers for the RNA transcribed *in organello* from the introduced plasmid DNA, respectively. Amplification of the endogenous transcript evaluates the integrity of the mitochondrial RNA.
6. Examine the obtained cDNA pools by agarose gel electrophoresis as shown in **Fig. 3** and then sequence products to analyze RNA editing (*see* **Fig. 4**).

4. Notes

1. Guanylyltransferase previously available from Bethesda Research laboratories was used under conditions detailed in **ref.** ***22***. The enzyme now distributed by Ambion has recently been used for in vitro capping of *A. thaliana* mitochondrial transcripts. The conditions for the capping reaction are given according to this analysis ***(4)***. All following steps are described according to our tested protocols.
2. RT-PCR assay with cDNA derived from 3′ to 5′ ligated RNA allows the simultaneous mapping of both 5′ and 3′ extremities and the detection of nonencoded nucleotides ***(10)***. Because this method provides sequence information about the RNAs analyzed, it is a general step to identify 5′ and 3′ ends unambiguously and exactly. The crucial step of this procedure is the initial self-ligation of mitochondrial RNA or total cellular RNA native 5′ and 3′ termini in an enzymatic step by T4 RNA ligase and the following reverse transcription initiated at an oligonucleotide complementary to a sequence in the 5′ terminal region of a given RNA (**Fig. 1**). Two consecutive PCRs with nested primers, all directed to the 5′ and 3′ extremities of the RNA, respectively, amplify cDNA sequences across the ligation site. Cloning and sequencing of the products will identify both 5′ and 3′ ends. Primary 5′ ends generated by transcription initiation and secondary 5′ termini resulting from endo- or exonucleolytic cleavage can be discriminated by the comparison of two analyses carried out separately with untreated RNA and RNA that has been either dephosphorylated with calf intestine phosphatase and rephosphorylated by polynucleotide kinase (S. Schmidt-Gattung and S. Binder, 2006, unpublished results) or alternatively by tobacco acid pyrophosphatase ***(9)***. These treatments convert nonaccessible primary 5′ ends with di- or triphosphates to monophosphate groups, which are then substrates for ligation by T4 RNA ligase. These ends should only be detectable after such treatments; 5′ ends derived from processing events carrying monophosphate group should be apparent in both assays. A different approach, in which oligonucleotides are ligated to the 5′ ends of

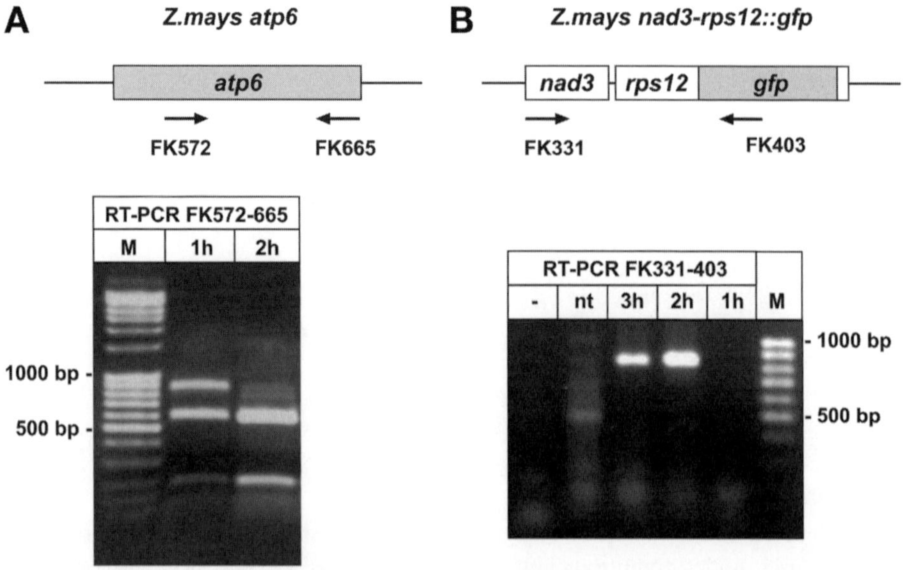

Fig. 3. RT-PCR and enzyme digestion. (**A**) Maize *atp6* gene with location of oligonucleotides (arrows) FK572 (GGGTAGAAGCAAGTCGATT) and FK665 (AGTTCTATGATTGGCTCCTCG). (**B**) RT-PCR of endogenous *atp6* and digestion of amplicons with *Hin*dIII as described in **Note 7**. (**C**) Map of a chimeric *nad3-rps12::gfp* construct introduced in maize mitochondria. (**D**) Detection of chimeric transcripts using oligonucleotides FK331(GCTAGTTTCTTTGATTCCACTCGGTGTTCC) and FK403 (GGGCAGCTTGCCGGTGGTGCA). M, marker; nt, nontransformed control; —, RT-PCR control; 1–3 h, time of *in organello* incubation in hours.

mitochondrial RNAs, has been used to identify and map several transcription initiation sites in *A. thaliana* mitochondria ***(4)***.

3. The *g*-force must not exceed 14,000*g* but should be as high as possible to reduce the time necessary for the reduction of the volume.
4. From the distance of the primers to the 5′ and 3′ ends of the reading frame, one can calculate the minimum size of the expected PCR fragment. To detect far ends, collect cDNA up to sizes of 1–2 kb even if there is no DNA visible in the gel.
5. This is a cheap-and-fast method that allows the efficient recovery of DNA fragments up to 500 bp. For elution, freeze agarose slices for 10 min at –80°C. Place frozen agarose on a clean surface (i.e., aluminum foil) and press with your gloved finger onto the agarose (use fresh gloves for each elution). Now, the agarose starts to thaw and fluid leaks from the gel slice. Collect all seeping fluid with a pipet until the agarose is completely thawed and transfer the solution to a 1.5-mL reaction tube. Spin down agarose debris by centrifugation at maximum speed for 5 min in a tabletop centrifuge. Pipet supernatant to a new tube and precipitate DNA with ethanol.

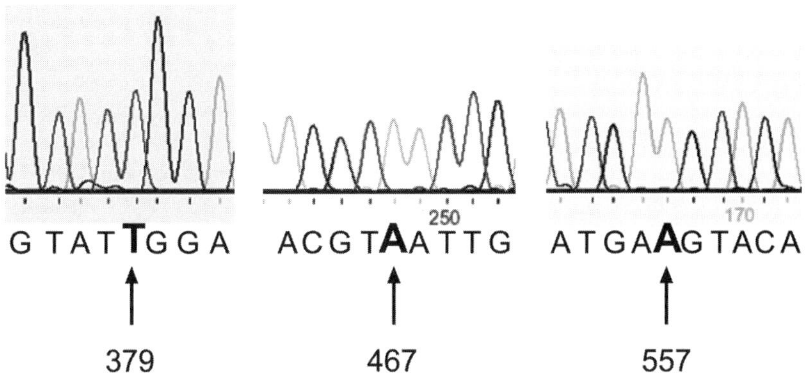

Fig. 4. Sequence analysis of RT-PCR fragments obtained from *in organello* edited transcripts. Usually, sequence analysis is required to check whether RNA editing occurred. In some instances, RNA editing generates or destroys a restriction site, and thus a restriction digest can be employed to analyze such a particular editing event. The figure shows three editing sites in the *A. thaliana cox2* messenger RNA, which was transcribed in maize mitochondria. RNA editing in plant mitochondria typically converts cytidine to uridine. Thus, editing events are indicated by cytidine-to-thymidine or guanine-to-adenine exchanges in the cDNA depending on the strand that is sequenced. Editing positions are highlighted by bold letters and arrows. Positions are given with respect to the ATG (A = +1).

6. In vitro capping has successfully been used for the identification of a number of transcription initiation sites in mitochondria of various plant species ***(1,2,4,5)***. This method takes advantage of the facts that mitochondrial primary 5′ ends are not capped in vivo and that a radioactive cap can thus be added in vitro by guanylyltransferase with [α-^{32}P]GTP as substrate, as indicated in the following reaction: (p)**pp**N(pN)n + pppG g G(5′)**ppp**N(pN)n + ppi + (pi).
   If there is no information about the 5′ ends of the investigated transcripts, then it is recommended to hybridize capped RNA to a Southern blot carrying different DNA fragments of the transcribed region of interest to roughly map the location of the transcription initiation site. The hybridization step can be skipped if 5′ ends have been mapped previously. S1 or RNase protection analyses with capped RNA can then be applied to locate the primary 5′ ends more precisely. Such carefully identified transcription initiation sites can then be used for functional promoter analysis by in vitro transcription systems.
7. Best hybridization results were obtained with Duralon UV membranes (Stratagene). We recommend use of small pieces of membrane and keeping the hybridization volume as small as possible. S1 and RNase protection analysis can be carried out according to standard procedures or with kits available from various companies.
8. It is essential to make sure that most bacteria are destroyed without damaging plant tissue too much. Bacterial contamination can be checked by plating aliquots of mitochondria on Luria-Bertani (LB) plates.

9. Proper disruption of plant tissue is important. Etiolated seedlings give much better results than green plant tissues. We chop seedlings into 2–3 cm sections prior to blending.
10. Activity of mitochondria may be checked by MitoTracker staining and fluorescence microscopy. Active mitochondria exhibit a bright orange fluorescence.
11. After electroporation, the mitochondrial suspension should be clear. When agglutination of mitochondria is observed, the salt concentration of the plasmid DNA has to be reduced by ethanol precipitation or other means.
12. An alternative *in organello* buffer was suggested by Farré and Araya ***(23)***. In addition to RNA editing, either system also allows the analysis of other RNA processing events (e.g., RNA splicing).
13. As an alternative, which has the advantage of largely reducing the risk of RNA degradation, the pellet from the *in organello* incubation can be resuspended in 200 μL boiling RNA lysis buffer (0.2 *M* boric acid or sodium borate, 30 m*M* EDTA, 1% w/v SDS, pH 9.0; autoclave one or two times prior to use).
14. In maize and sorghum mitochondria, the endogenous *atp6* transcript is particularly suitable because the editing status of this transcript can be readily checked. Here, RNA editing creates two *Hind*III sites in the transcript and can thus be checked by restriction digest of the cDNA ***(24,25)***.
15. To analyze if RNA is newly synthesized during *in organello* incubation or pre-existed, biotin-16-uridine-5′-triphospate can be used to label synthesized RNAs *in organello* ***(16)***. The biotin-labeled RNA can be isolated by paramagnetic streptavidin-coated particles.

Acknowledgments

The work described was supported by the Deutsche Forschungsgemeinschaft. We thank Bärbel Weber for her excellent technical advice for RT-PCR analysis of circularized RNA and Matthias Staudinger for his help with the development of the *in organello* RNA editing system.

References

1. Hoffmann, M., Kuhn, J., Däschner, K., and Binder, S. (2001) The RNA world of plant mitochondria. *Prog. Nucleic Acid Res. Mol. Biol.* **70,** 119–154.
2. Binder, S. and Brennicke, A. (2003) Gene expression in plant mitochondria: transcriptional and posttranscriptional control. *Phil. Trans. R. Soc. Lond. B.* **358,** 181–199.
3. Marchfelder, A. and Binder, S. (2004) Plastid and plant mitochondrial RNA processing and RNA stability, in *Molecular Biology and Biotechnology of Plant Organelles* (Daniell, H. and Chase, C., eds.), Springer, Dordrecht, The Netherlands, pp. 261–294.
4. Kühn, K., Weihe, A., and Borner, T. (2005) Multiple promoters are a common feature of mitochondrial genes in *Arabidopsis. Nucleic Acids Res.* **33,** 337–346.
5. Mulligan, R. M., Maloney, A. P., and Walbot, V. (1988) RNA processing and multiple transcription initiation sites result in transcript size heterogeneity in maize mitochondria. *Mol. Gen. Genet.* **211,** 373–380.

6. Sugino, A., Snoper, T. J., and Cozzarelli, N. R. (1977) Bacteriophage T4 RNA ligase. Reaction intermediates and interaction of substrates. *J. Biol. Chem.* **252,** 1732–1738.
7. England, T. E., Gumport, R. I., and Uhlenbeck, O. C. (1977) Dinucleoside pyrophosphate are substrates for T4-induced RNA ligase. *Proc. Natl. Acad. Sci. USA* **74,** 4839–4842.
8. Fromont-Racine, M., Bertrand, E., Pictet, R., and Grange, T. (1993) A highly sensitive method for mapping the 5′ termini of mRNAs. *Nucleic Acids Res.* **21,** 1683–1684.
9. Bensing, B. A., Meyer, B. J., and Dunny, G. M. (1996) Sensitive detection of bacterial transcription initiation sites and differentiation from RNA processing sites in the pheromone-induced plasmid transfer system of *Enterococcus faecalis. Proc. Natl. Acad. Sci. USA* **93,** 7794–7799.
10. Kuhn, J. and Binder, S. (2002) RT-PCR analysis of 5′ to 3′-end-ligated mRNAs identifies the extremities of *cox2* transcripts in pea mitochondria. *Nucleic Acids Res.* **30,** 439–446.
11. Brennicke, A., Marchfelder, A., and Binder, S. (1999) RNA editing. *FEMS Microbiol. Rev.* **23,** 297–316.
12. Giege, P. and Brennicke, A. (1999) RNA editing in *Arabidopsis* mitochondria effects 441 C to U changes in ORFs. *Proc. Natl. Acad. Sci. USA* **96,** 15,324–15,329.
13. Small, I. D. and Peeters, N. (2000) The PPR motif—a TPR-related motif prevalent in plant organellar proteins. *Trends Biochem. Sci.* **25,** 46–47.
14. Kotera, E. Tasaka, M., and Shikanai, T. (2005) A pentatricopeptide repeat protein is essential for RNA editing in chloroplasts. *Nature* **433,** 326–330.
15. Staudinger, M. and Kempken, F. (2004) In organello editing of mitochondrial *atp9, cox2* and *nad9* transcripts. *Endocytobiosis Cell. Res.* **15,** 551–556.
16. Staudinger, M. and Kempken, F. (2003) Electroporation of isolated higher-plant mitochondria: transcripts of an introduced *cox2* gene, but not an *atp6* gene, are edited in organello. *Mol. Genet. Genomics* **269,** 553–561.
17. Staudinger, M., Bolle, N., and Kempken, F. (2005) Mitochondrial electroporation and in organello RNA editing of chimeric *atp6* transcripts. *Mol. Genet. Genomics* **273,** 130–136.
18. Binder, S. (1995) Mitochondrial nucleic acid purification and analysis. *Methods Mol. Biol.* **49,** 383–389.
19. Binder, S. and Grohmann, L. (1995) Isolation of mitochondria. *Methods Mol. Biol.* **49,** 377–381.
20. Klein, M., Binder, S., and Brennicke, A. (1998) Purification of mitochondria from *Arabidopsis. Methods Mol. Biol.* **82,** 49–53.
21. Newton, K. J. (1994) Procedures for isolating mitochondria and mitochondrial DNA and RNA, in *The Maize Handbook* (Freeling, M. and Walbot, V., eds.), Springer, New York, pp. 549–556.
22. Binder, S. and Brennicke, A. (1993) Transcription initiation sites in mitochondria of *Oenothera berteriana. J. Biol. Chem.* **268,** 7849–7855.

23. Farré, J. C. and Araya, A. (2001) Gene expression in isolated plant mitochondria: high fidelity of transcription, splicing and editing of a transgene product in electroporated organelles. *Nucleic Acids Res.* **29,** 2484–2491.
24. Kempken, F., Mullen, J. A., Pring, D. R., and Tang, H. V. (1991) RNA editing of sorghum mitochondrial *atp6* transcripts changes 15 amino acids and generates a carboxy-terminus identical to yeast. *Curr. Genet.* **20,** 417–422.
25. Kumar, R. and Levings, C. S., 3rd (1993) RNA editing of a chimeric maize mitochondrial gene transcript is sequence specific. *Curr. Genet.* **23,** 154–159.

14

In Vitro Analysis of the Yeast Mitochondrial RNA Polymerase

Elizabeth A. Amiott and Judith A. Jaehning

Summary

Understanding the details of how genetic information is expressed from the separate mitochondrial genome requires a detailed description of the properties of the mitochondrial RNA polymerase. This nuclear-encoded enzyme is necessary and sufficient for the transcription of all mitochondrially encoded genes. Mitochondria from yeast to humans use a single-polypeptide catalytic RNA polymerase related to enzymes from bacteriophage. They also require separable transcription factors necessary for initiation at promoter sequences on the mitochondrial DNA template. It has recently become possible to work with highly purified, recombinant forms of the mitochondrial RNA polymerase subunits from yeast. This chapter describes detailed protocols for working in vitro with this purified enzyme in transcription reactions. These assays are critical for elucidating the nature of a mitochondrial promoter and for understanding how the mitochondrial RNA polymerase recognizes these DNA sequences and selectively initiates the transcription cycle, resulting in discrete transcripts.

Key Words: In vitro transcription; Mtf1; mtTFB; RNA polymerase; Rpo41.

1. Introduction

Transcription of mitochondrial DNA (mtDNA), which encodes ribosomal ribonucleic acids (RNAs), mitochondrion-specific transfer RNAs, and a small set of messenger RNAs, is essential for mitochondrial function, particularly oxidative phosphorylation. Mitochondrial functions also depend on proteins encoded in the nuclear compartment; hence, a unique coordination between nuclear and mitochondrial gene expression exists and is crucial for the assembly of energy-generating complexes. Mutations in both nuclear and mtDNA that cause defects in mitochondrial gene expression and oxidative phosphorylation have been implicated

From: *Methods in Molecular Biology, vol. 372: Mitochondria: Practical Protocols*
Edited by: D. Leister and J. M. Herrmann © Humana Press Inc., Totowa, NJ

in aging and disease ***(1–3)***, making the study of mtDNA transcription an important step in understanding mitochondrial pathologies.

The mitochondrial transcription machinery is comprised of a core RNA polymerase (RNAP) and accessory subunits that are encoded by nuclear DNA and imported into the mitochondrion. The core polymerase has significant sequence homology to T7, T3, and SP6 bacteriophage RNA polymerases ***(4)***. However, even though the single-subunit phage RNAPs are self-sufficient, mitochondrial polymerases require at least one accessory factor for efficient initiation both in vivo and in vitro ***(5,6)***.

In the yeast *Saccharomyces cerevisiae*, the accessory factor Mtf1 (sc-mtTFB) pairs with the core polymerase, Rpo41, to create a functional initiation complex ***(7)***. Studies from our lab suggest that Mtf1 facilitates open complex formation on linear DNA templates at an early stage of initiation, and that it may also have inhibitory effects later in the initiation process as the polymerase tries to escape the promoter and transition to an elongating form ***(8)***. Two Mtf1 homologs (mtTFB1 and mtTFB2) have been identified in flies, mice, and humans ***(6,9–12)***, and transcription initiation depends on the presence of either mtTFB1 or mtTFB2, although mtTFB2 has a greater stimulatory effect than mtTFB1 ***(13,14)***. Mammalian mitochondrial transcription requires another factor, mtTFA (TFAM), and disruption of this protein in mice is embryonic lethal, most likely because of loss of mtDNA ***(15)***. The mtTFA homolog in yeast, Abf2, is part of the mtDNA nucleoid and is essential for mitochondrial genome maintenance ***(16,17)***, but it lacks an activation domain present in mtTFA and is not an essential component of the yeast transcription machinery ***(18)***.

Despite homology between polymerase subunits, species-specific differences in mitochondrial genome structure, promoter sequence, transcript organization, and accessory protein requirements exist. Therefore, complete understanding of eukaryotic mitochondrial transcription depends on the establishment of in vitro transcription assays using pure, recombinant polymerase and accessory factors. Such a system has been developed for the yeast mtRNAP ***(19,20)***, and more recently, purification of recombinant mammalian mitochondrial polymerase and other components of the transcription machinery has been achieved ***(6,21)***. This presents an excellent opportunity to translate the knowledge gained from yeast studies into more comprehensive understanding of mitochondrial transcription in complex organisms.

This chapter provides a detailed protocol for in vitro transcription reactions using bacterially expressed recombinant subunits of the yeast mtRNAP, Rpo41 and Mtf1. The availability of recombinant proteins has already changed our understanding of the properties of the yeast mitochondrial polymerase and will allow for more sophisticated kinetic and structural studies. The protocol presented here may be easily modified to allow similar experimentation with homologous systems to gain valuable insight into the detailed mechanisms of mitochondrial transcription from yeast to humans.

2. Materials

2.1. In Vitro Runoff Transcription Reactions

1. Purified mtRNAP subunits Rpo41 and Mtf1 stored in single-use aliquots at –80°C (*see* **Notes 1** and **2**).
2. T50 buffer: 100 m*M* Tris-HCl, pH 7.9, 2 m*M* ethylenediaminetetraacetic acid (EDTA), 5% (w/v) glycerol, 10 m*M* $MgCl_2$, 50 m*M* KCl; store at 4°C. A working solution is prepared by addition of 50 μg/mL bovine serum albumin (BSA) and 1 m*M* dithiothreitol (DTT) and should be stored at –20°C.
3. Stock solutions for reaction buffer (*see* **Note 3**): 1 *M* Tris-HCl, pH 7.9 in diethyl pyrocarbonate (DEPC)-treated water and stored at room temperature; 1 *M* $MgCl_2$ in DEPC-treated water and stored at room temperature; 0.5 *M* DTT stored in aliquots at –20°C; 100 m*M* adenosine triphosphate (ATP), guanosine 5′-triphosphate (GTP), and cytosine triphosphate (CTP) stored in aliquots at –80°C; 10 m*M* uridine triphosphate (UTP) stored in aliquots at –80°C; 10 μCi/μL α-^{32}P-UTP stored at –20°C; DEPC-treated water.
4. Linearized promoter DNA templates.
5. Stop/dye solution: 88% (v/v) formamide, 0.05 *M* EDTA (from 0.5 *M* stock prepared in DEPC water), 0.0025% (w/v) bromophenol blue, 0.0025% (w/v) xylene cyanol. Store at –20°C.

2.2. Polyacrylamide Gel Electrophoresis

1. 40% Acrylamide/bis solution stored at 4°C (*see* **Note 4**).
2. Electrophoresis-grade urea (7 *M* final concentration).
3. *N,N,N,N′*-Tetramethylethylenediamine (TEMED).
4. Ammonium persulfate: prepare a fresh 10% solution in water before each use.
5. 10 X Running buffer: 1.33 *M* Tris-HCl, 0.44 *M* boric acid, 0.025 *M* EDTA(Na_2). Store at room temperature.
6. Silanizing agent (e.g., Sigmacote, Sigma-Aldrich) for treating glass plates (*see* **Note 5**).
7. Optional: Radiolabeled RNA size marker (e.g., Century marker, Ambion) (*see* **Note 6**).

2.3. Detection of Radiolabeled Transcripts

1. Phosphorimager screen or X-ray film and exposure cassette with enhancer screen.
2. Optional: slab gel dryer.

3. Methods

This subheading describes standard conditions for multiple-round in vitro runoff transcription reactions using linear yeast mitochondrial promoter DNA templates and purified recombinant polymerase subunits. The utility of this assay is that almost every parameter can be varied or altered to probe a different question; therefore, the conditions presented here should be considered guidelines rather than absolutes. **Notes 7–11** briefly introduce variations to the standard experimental conditions (based primarily on studies of transcription initiation for *Escherichia coli* and T7 bacteriophage RNAPs) that may be easily applied to study mitochondrial transcription initiation ***(22,23)***.

Yeast mitochondrial promoters have been defined as the nonanucleotide consensus sequence ATATAAGTA, with the final A serving as the +1 initiating nucleotide ***(24,25)***. Of the 20 consensus sequences found in the mitochondrial genome, 12 serve as promoters, driving the synthesis of polycistronic transcription units; the others are believed to be origins for RNA-primed DNA replication ***(26,27)***. Despite near-identical promoter sequences, transcripts are synthesized at different rates in vivo ***(28)***, and consensus promoters exhibit different efficiencies in vitro ***(29)***.

Most in vitro promoter templates are plasmids constructed by cloning polymerase chain reaction (PCR)-amplified promoter-containing segments of the mitochondrial genome into vectors. Plasmid constructs are then linearized by digestion at a unique restriction site downstream of the cloned promoter sequence, creating a linear DNA template. The amount of downstream sequence included in the template should be selected to create a runoff transcript of a unique length (**Fig. 1**). Because the DNA elements that influence promoter recognition and transcription efficiency are not completely understood, we recommend that standard promoter templates be designed to include at least 50–100 bases of endogenous sequence both up- and downstream of the promoter.

3.1. In Vitro Runoff Transcription Reactions

1. Keeping all reagents and samples on ice, prepare the reaction buffer as either a master mix or individual reaction mixes. The total reaction volume is 20 μL, and the Rpo41 polymerase subunit should be added last whenever possible. The final (1X) reaction conditions are as follows: 50 m*M* Tris-HCl, pH 7.9; 20 m*M* $MgCl_2$; 1 m*M* DTT; 250 μ*M* ATP, CTP, GTP (*see* **Note 12**); 50 μ*M* UTP; 1000 cpm α-^{32}P-UTP/pmol UTP (*see* **Note 13**); 20 μg/mL linearized promoter DNA template (*see* **Note 14**); 0.8 pmol Mtf1; 0.8 pmol Rpo41 (*see* **Note 15**); and DEPC-treated water as needed to bring reaction to a final volume of 20 μL.
2. Incubate reactions for approx 7–10 min in a 30°C water bath (*see* **Note 16**).
3. Add 25 μL ice-cold stop/dye solution to each sample to stop the transcription reaction and keep samples on ice.
4. Immediately prior to electrophoresis, denature samples at 75°C for 5 min and quick chill on ice (*see* **Note 17**).

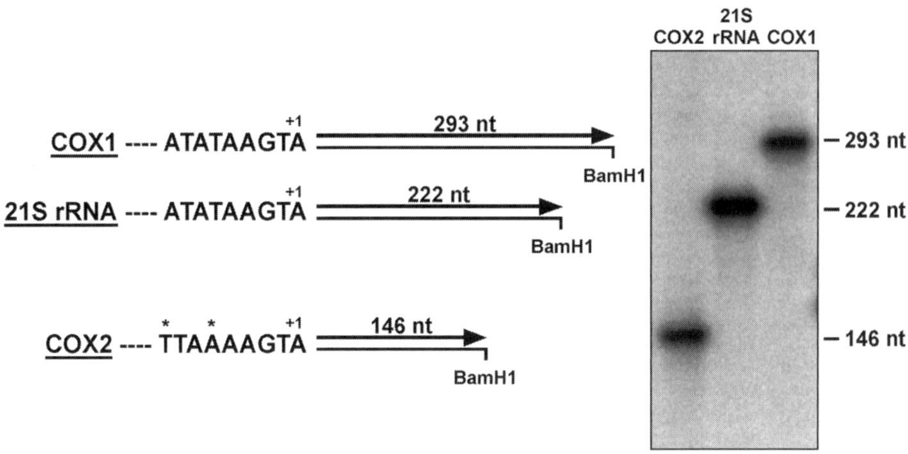

Fig. 1. Linear promoter DNA templates for in vitro transcription reactions. COX1, 21S ribosomal RNA, and COX2 mitochondrial promoter DNA templates for in vitro transcription reactions were generated by PCR amplification of discrete regions of mitochondrial DNA containing consensus or variant (asterisks indicate variation from consensus) promoter and downstream sequences. The PCR fragments were cloned into a pCR2.1 vector using the TOPO TA Cloning Kit (Invitrogen), and the resulting plasmids were digested with BamH1 to linearize the DNA. Standard reaction conditions were used to generate radiolabeled transcripts of 293, 222, and 146 nucleotides. Transcripts were separated by 8% polyacrylamide denaturing gel electrophoresis and visualized with a STORM 860 Phosphorimager (Bio-Rad).

3.2. Polyacrylamide Gel Electrophoresis

1. These instructions are written for use with an Ellard Instruments (Monroe, WA) vertical gel electrophoresis system using 20 × 25 cm glass plates but can be easily modified for other gel formats. Glass plates should be thoroughly cleaned with a rinsable detergent (e.g., Liquinox, Alconox) prior to each use and should occasionally be treated with a silanizing agent (e.g., Sigmacote) to prevent the gel from sticking to the plates.
2. Assemble plates into a gel sandwich. We routinely use 1.5 mm thick Teflon spacers and combs, but thicker or thinner gels can be poured as desired.
3. Prepare 100 mL 8% (w/v) acrylamide/7 *M* urea gel by mixing 48 g urea, 10 mL 10X running buffer, 20 mL 40% (w/v) acrylamide/bis solution, and 33 mL DEPC-treated water (*see* **Note 18**). Working quickly, mix in 80 µL TEMED and 800 µL 10% ammonium persulfate solution, pour the gel, and insert comb into gel assembly. For best results, secure the comb with clamps and allow the gel to polymerize for at least 30 min.
4. Prepare 1X running buffer by diluting 100 mL 10X running buffer with 900 mL water and mix.

5. When ready to run the gel, carefully remove the comb and assemble the gel in the electrophoresis apparatus. Fill the upper and lower chambers of the gel unit with 1X running buffer and use a syringe or Pasteur pipet to remove acrylamide and urea from the wells prior to loading.
6. Load 40 µL of each sample per well. Include one well for an RNA size marker if desired (*see* **Note 19**).
7. Connect the gel unit to a power supply and run at 30–45 W for 2–4 h (*see* **Notes 20** and **21**).
8. Once separation is complete, turn off the power supply and remove the gel sandwich from the gel unit.
9. Take apart the gel sandwich and transfer the gel to a double sheet of sturdy, absorbent paper (e.g., 3M, Whatman) cut slightly larger than the gel (*see* **Note 22**). Wrap the gel twice in plastic wrap (*see* **Note 23**).
10. Check the buffer in the upper and lower chambers for radioactivity and dispose appropriately.

3.3. Detection of Radiolabeled Transcripts

1. Place the wrapped gel face up in a phosphorimaging cassette (or X-ray cassette) (*see* **Note 24**).
2. Expose the gel to the screen or film for the appropriate amount of time for sample detection. We have found that 8–24 h is an adequate exposure time for most transcription reactions on a phosphorimager screen, resulting in good signal without saturation.
3. Phosphorimager samples can be quantified with the appropriate software, and X-ray film can be quantified following densitometer scanning of the developed film (*see* **Note 25**).
4. Wet gels should be air dried and disposed of as solid radioactive waste.

4. Notes

1. Detailed purification protocols have been described for bacterially expressed recombinant His-tagged Rpo41 ***(19)*** and GST-tagged Mtf1 ***(30)***. For space considerations, these procedures have been excluded from this chapter.
2. Purified Rpo41 is unstable, and its activity is reduced by dilution and freeze-thaw cycles. Also, enzyme activity and salt concentration affect in vitro transcription efficiency and will vary from one protein preparation to the next. We recommend that the RNA polymerase enzymes be purified in large quantities so that many experiments can be performed using the same enzyme preparation. Whenever possible, enzymes should be aliquoted and stored at the working concentration in single-use volumes. If additional dilution of the aliquots is necessary, then this should be done in T50 buffer immediately before the enzyme is added to the reaction mix. To minimize loss of enzyme activity, always add diluent (T50 buffer) to the enzyme solution rather than enzyme to the diluent.

3. When working with RNA, it is best to use a separate supply of treated plasticware and glassware to prevent contamination and degradation of the RNA transcripts by ribonucleases. Filtered pipet tips are highly recommended, and whenever possible, solutions should be made with DEPC-treated water. Throughout the text, solutions in which DEPC-treated water is essential are noted.
4. Acrylamide in the nonpolymerized state is a neurotoxin and light sensitive. Acrylamide solutions should be stored in dark containers, and proper safety precautions should be taken when working with this solution.
5. Silanizing agents will create a hydrophobic layer on the glass plate surface to prevent adhesion of the gel matrix to the plate. Working in the hood, apply a thin layer over the entire inner-facing surface of each plate and allow the solution to evaporate at room temperature. Rinse plates with deionized water following treatment and prior to sandwich assembly.
6. A radiolabeled RNA size ladder is useful for determining and comparing transcript sizes. Ambion carries a series of RNA markers ranging from 10 to 1000 nucleotides that can be easily labeled with ^{32}P.
7. Promoter recognition and melting can be studied by making changes to the template sequence or conformation. Mutations in promoter and flanking sequences can be used to define essential promoter elements and the effects of nucleotide substitutions on transcription efficiency and promoter-polymerase complex stability ***(24,29,32,33)***. In addition to template sequence, transcription efficiency is influenced by the conformation of the promoter DNA. For instance, less energy is needed to separate the promoter duplex ***(34)*** on supercoiled plasmid templates than on linear DNA templates. Therefore, defects in transcription that can be rescued by supercoiled templates often indicate that the defect is related to promoter opening ***(8,30)***. Also, fork-junction, bubble, bulge, and partially single-stranded templates can be designed to imitate fully or partially open initiation complexes, facilitating analyses of various stages of transcription initiation, independent of promoter melting ***(8,35–38)***.
8. *De novo* RNA synthesis requires high concentrations of initiating (+1 and +2) nucleotides, but once in the elongating form, RNAPs are active at low nucleotide concentrations ***(39,40)***. The apparent K_M for an NTP on various promoter templates, or in the context of different polymerase mutants, can be determined by varying the concentration of one nucleotide (while the other nucleotides are kept at a constant concentration) in a series of reactions. The amount of transcript synthesized at each concentration is quantified and fit to the Michaelis-Menton equation as a function of nucleotide (substrate) concentration (**Fig. 2**). It is important to keep in mind that this is a measure of steady-state transcription rather than a direct measure of nucleotide affinity (K_d), which is also possible, but requires a more sophisticated experimental setup ***(35,37,38)***.
9. Dinucleotide primers, complementary to the +1 and +2 nucleotides of a transcript, can be added to the reaction mix to eliminate the chemistry of the initial phosphodiester bond formation. Biswas ***(41,42)*** used this strategy to characterize differences in promoter strength conferred by the nature (purine or pyrmidine) of the +2

nucleotide and to confirm start site preferences of the yeast mitochondrial polymerase. Dinucleotide primers may also be used when determining the K_M of one of the initiating nucleotides (+1 and +2) to confirm that high NTP concentration is required only for synthesis of the first bond of the transcript, not for subsequent elongation (**Fig. 3**).

10. Heparin is a compound that binds free RNA polymerase and inhibits its catalytic activity; therefore, heparin is useful for studying the half-life $t_{1/2}$ or stability of polymerase–promoter interactions. Experiments can be designed to study the effects of different components of the transcription reaction on the enzyme–DNA association by comparing transcription efficiency with and without heparin addition. For example, Gaal et al. ***(43)*** used heparin experiments to determine that addition of the initiating nucleotide stabilizes the initiation complex of bacterial RNA polymerase on ribosomal RNA promoters. When heparin was incubated with polymerase and DNA prior to nucleotide addition, RNA synthesis was less efficient than when the +1 nucleotide was added to stabilize the complex before adding heparin to the reaction. This indicates that, in the absence of NTP, a greater proportion of polymerase dissociates from the DNA and is inhibited by heparin ***(43,44)***.
11. Following initial bond formation and synthesis of an RNA strand of about 9–12 nucleotides, RNA polymerase is released from the promoter and transitions to an elongation mode. This process is termed *promoter clearance* and is typically preceded by multiple rounds of abortive synthesis in which the polymerase synthesizes and releases short (approx 2- to 10-nucleotide) transcripts ***(23,45)***. Abortive transcription can be assayed in vitro using a (γ)-^{32}P radionucleotide corresponding to the first NTP of the transcript. The reaction products are separated on a high percentage (20–25%) polyacrylamide gel for single-base resolution of the short end-labeled transcripts.
12. Others have used nucleotide triphosphate (NTP) concentrations as low as 125 µ*M* in their in vitro reactions. We have found that the K_M of the initiating NTP can be greater than 125 µ*M* on some promoter templates; therefore, we feel it is important to use NTPs at a minimal concentration of 250 µ*M*. Experiments using nucleotides at concentrations up to 500 µ*M* give similar results.
13. When working with radioactive material, always wear gloves, eye protection, and lab coat and keep the material appropriately shielded. The work area should be monitored frequently for contamination, and all waste products should be disposed of appropriately.
14. Because subtle differences in pH and salt concentration can significantly affect polymerase activity in vitro, care should be taken to prepare all DNA templates in the same manner. We have found that we obtain the most consistent results if, after digestion and removal of the restriction enzyme, samples are suspended in 10 m*M* Tris-Cl, pH 8.0. As with the protein preparations (*see* **Note 2**), it is best to prepare large quantities of linear templates so that many experiments can be performed using the same material.
15. In our hands, a 1:1 molar ratio of core polymerase (Rpo41) to transcription factor (Mtf1) gives the best in vitro results; however, the exact molar amounts of enzyme may need to be adjusted depending on the activity of the purified enzyme. We

routinely use 0.8 pmol Rpo41 from a preparation with Rpo41 polymerase-specific activity of approx 3500 U/mg ***(19)***.

16. The actual incubation time should be determined by each lab and for each enzyme preparation. Time-course experiments should be set up to determine the linearity of the transcription reaction over time on several promoter templates.
17. If using an RNA ladder, then this should also be mixed with the stop/dye solution and denatured.
18. The gel solution may need to be heated during mixing to allow the urea to go into solution. If so, the acrylamide solution must be cooled to room temperature prior to the addition of TEMED and ammonium persulfate or polymerization will occur too quickly to pour the gel.
19. If empty wells are loaded in between sample wells, then we recommend adding a "mock" or blank sample to the empty wells to create an even migration pattern across the gel. Also, if you experience a "smiling" pattern (samples on the ends of the gel migrate more slowly than samples in the middle), then this can often be corrected by prerunning the gel for 30–60 min prior to loading or by placing an aluminum plate behind the gel sandwich to promote even distribution of heat across the gel.
20. The actual run time will need to be determined based on the polyacrylamide concentration, as well as on the size of the transcripts separated. For reference, an 8% gel is good for separating fragments between 60 and 400 nucleotides long, with bromophenol blue running at approx 19 nucleotides and xylene cyanol at 75 nucleotides ***(31)***.
21. Running the unincorporated radionucleotide off the gel eliminates the need for phenol extraction or spin-column purification of the transcripts. However, care should be taken during disposal of the running buffer from the lower chamber, which will be contaminated with radioactivity. Alternatively, larger transcripts can be run for a shorter period of time, keeping unincorporated nucleotide in the gel matrix but well separated from the transcripts. In this case, the lower portion of the gel, which contains the unincorporated radionucleotide, can be cut off and disposed of as solid radioactive waste before exposing the gel.
22. To transfer the gel from plate to paper, remove the spacers and separate the small plate from the sandwich. Place the paper on top of the gel and invert the entire assembly so that the large plate is now face up. The gel can be removed from the large plate by slowly lifting the plate, leaving the gel attached to the paper. Care should be taken to prevent tearing of the gel and to note the orientation of the gel during the transfer process. If the gel sticks to the plates, then re-treat with silanizing agent.
23. Gels can also be dried on a slab gel dryer to reduce diffusion of the sample bands, particularly during long exposures or when trying to resolve transcripts of similar size. However, we have found that 1.5 mm thick gels do not need to be dried as long as the exposure time is relatively short (about 24 h or less).
24. Phosphorimager screens are sensitive and easily damaged by moisture. Extra precautions should be taken to prevent the screen from getting wet. Undried gels should be carefully wrapped, using at least two layers of plastic wrap.

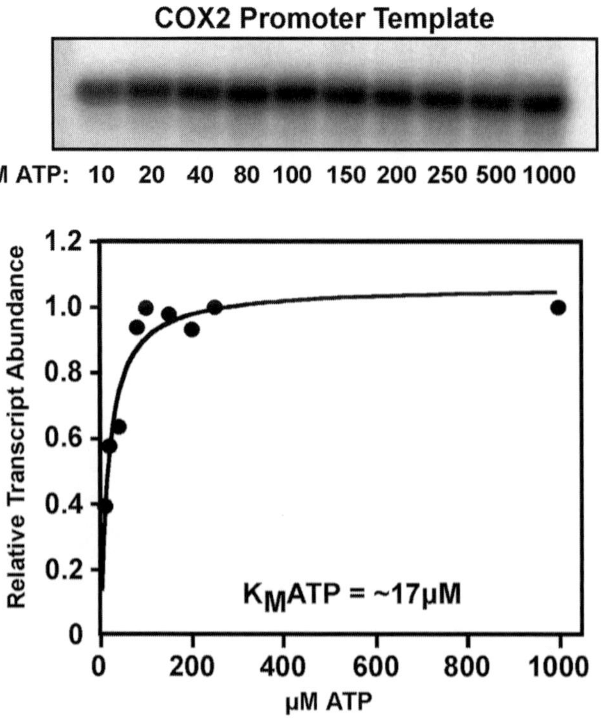

Fig. 2. Changing nucleotide concentration to determine apparent K_MATP. In vitro transcription reactions were performed using a linear COX2 promoter template. The concentration of ATP was varied from 10 to 1000 μ*M*, and the concentration of the other nucleotides was kept constant at 250 μ*M*. The amount of runoff transcript synthesized in each sample was normalized to the maximal transcript signal (relative transcript abundance) and plotted as a function of ATP concentration (μ*M*). Data points were subjected to nonlinear regression analysis using the Michaelis-Menton equation (PRISM, GraphPad Software) to determine the apparent K_MATP from the steady-state reactions.

25. When comparing different length transcripts, each signal must be corrected for the number of UTPs in the transcript. Alternatively, (γ)–^{32}P-ATP can be used to end-label the synthesized transcripts. This eliminates the need to correct for transcript length as each transcript will have just one molecule of ^{32}P at its 5′ end.

Acknowledgments

Special thanks to Michio Matsunaga, Mark Karlok, and Sei-Heon Jang for their efforts in developing and optimizing protein purification strategies and in vitro transcription protocols. Studies in the Jaehning lab have been supported by National Science Foundation grant MCB 0235354.

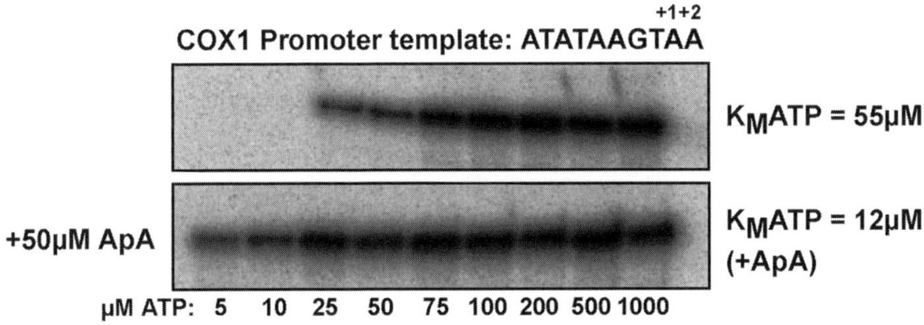

Fig. 3. Adding dinucleotide primer to bypass first bond formation. In vitro transcription reactions were performed as in **Fig. 2** using a linear COX1 promoter template. In the bottom panel, 50 µ*M* of a dinucleotide primer (ApA) corresponding to the +1 and +2 nucleotides was added to each reaction. Samples were analyzed as before to determine the apparent K_MATP with and without dinucleotide. The almost fivefold drop in apparent K_MATP in the presence of dinucleotide indicates that the polymerase requires a high nucleotide concentration for initial phosphodiester bond formation but not for elongation.

References

1. Shadel, G. S. (2004) Coupling the mitochondrial transcription machinery to human disease. *Trends Genet.* **20,** 513–519.
2. Trifunovic, A., Wredenberg, A., Falkenberg, M., et al. (2004) Premature ageing in mice expressing defective mitochondrial DNA polymerase. *Nature* **429,** 417–423.
3. Wallace, D. C. (1999) Mitochondrial diseases in man and mouse. *Science* **283,** 1482–1488.
4. Masters, B. S., Stohl, L. L., and Clayton, D. A. (1987) Yeast mitochondrial RNA polymerase is homologous to those encoded by bacteriophages T3 and T7. *Cell* **51,** 89–99.
5. Schinkel, A. H., Koerkamp, M. J., Touw, E. P., and Tabak, H. F. (1987) Specificity factor of yeast mitochondrial RNA polymerase. Purification and interaction with core RNA polymerase. *J. Biol. Chem.* **262,** 12,785–12,791.
6. Gaspari, M., Larsson, N. G., and Gustafsson, C. M. (2004) The transcription machinery in mammalian mitochondria. *Biochim. Biophys. Acta* **1659,** 148–152.
7. Mangus, D. A., Jang, S. H., and Jaehning, J. A. (1994) Release of the yeast mitochondrial RNA polymerase specificity factor from transcription complexes. *J. Biol. Chem.* **269,** 26,568–26,574.
8. Matsunaga, M. and Jaehning, J. A. (2004) Intrinsic promoter recognition by a "core" RNA polymerase. *J. Biol. Chem.* **279,** 44,239–44,242.
9. Matsushima, Y., Adan, C., Garesse, R., and Kaguni, L. S. (2005) *Drosophila* mitochondrial transcription factor B1 modulates mitochondrial translation but not transcription or DNA copy number in Schneider cells. *J. Biol. Chem.* **280,** 16,815–16,820.

10. Matsushima, Y., Garesse, R., and Kaguni, L. S. (2004) *Drosophila* mitochondrial transcription factor B2 regulates mitochondrial DNA copy number and transcription in Schneider cells. *J. Biol. Chem.* **279,** 26,900–26,905.
11. McCulloch, V., Seidel-Rogol, B. L., and Shadel, G. S. (2002) A human mitochondrial transcription factor is related to RNA adenine methyltransferases and binds S-adenosylmethionine. *Mol. Cell Biol.* **22,** 1116–1125.
12. Rantanen, A., Gaspari, M., Falkenberg, M., Gustafsson, C. M., and Larsson, N. G. (2003) Characterization of the mouse genes for mitochondrial transcription factors B1 and B2. *Mamm. Genome* **14,** 1–6.
13. Falkenberg, M., Gaspari, M., Rantanen, A., Trifunovic, A., Larsson, N. G., and Gustafsson, C. M. (2002) Mitochondrial transcription factors B1 and B2 activate transcription of human mtDNA. *Nat. Genet.* **31,** 289–294.
14. McCulloch, V. and Shadel, G. S. (2003) Human mitochondrial transcription factor B1 interacts with the C-terminal activation region of h-mtTFA and stimulates transcription independently of its RNA methyltransferase activity. *Mol. Cell Biol.* **23,** 5816–5824.
15. Larsson, N. G., Wang, J., Wilhelmsson, H., et al. (1998) Mitochondrial transcription factor A is necessary for mtDNA maintenance and embryogenesis in mice. *Nat. Genet.* **18,** 231–236.
16. Diffley, J. F. and Stillman, B. (1991) A close relative of the nuclear, chromosomal high-mobility group protein HMG1 in yeast mitochondria. *Proc. Natl. Acad. Sci. USA* **88,** 7864–7868.
17. Kaufman, B. A., Newman, S. M., Hallberg, R. L., Slaughter, C. A., Perlman, P. S., and Butow, R. A. (2000) *In organello* formaldehyde crosslinking of proteins to mtDNA: identification of bifunctional proteins. *Proc. Natl. Acad. Sci. USA* **97,** 7772–7777.
18. Dairaghi, D. J., Shadel, G. S., and Clayton, D. A. (1995) Addition of a 29 residue carboxyl-terminal tail converts a simple HMG box-containing protein into a transcriptional activator. *J. Mol. Biol.* **249,** 11–28.
19. Matsunaga, M., Jang, S. H., and Jaehning, J. A. (2004) Expression and purification of wild type and mutant forms of the yeast mitochondrial core RNA polymerase, Rpo41. *Protein Expr. Purif.* **35,** 126–130.
20. Mangus, D. A. and Jaehning, J. A. (1996) Transcription in vitro with *Saccharomyces cerevisiae* mitochondrial RNA-polymerase. *Meth. Enzymol.* **264,** 57–66.
21. Gaspari, M., Falkenberg, M., Larsson, N. G., and Gustafsson, C. M. (2004) The mitochondrial RNA polymerase contributes critically to promoter specificity in mammalian cells. *EMBO J.* **23,** 4606–4614.
22. McClure, W. R., Cech, C. L., and Johnston, D. E. (1978) A steady state assay for the RNA polymerase initiation reaction. *J. Biol. Chem.* **253,** 8941–8948.
23. Hsu, L. M., Vo, N. V., Kane, C. M., and Chamberlin, M. J. (2003) In vitro studies of transcript initiation by *Escherichia coli* RNA polymerase. 1. RNA chain initiation, abortive initiation, and promoter escape at three bacteriophage promoters. *Biochemistry* **42,** 3777–3786.
24. Biswas, T. K., Ticho, B., and Getz, G. S. (1987) In vitro characterization of the yeast mitochondrial promoter using single-base substitution mutants. *J. Biol. Chem.* **262,** 13,690–13,696.

25. Christianson, T. and Rabinowitz, M. (1983) Identification of multiple transcriptional initiation sites on the yeast mitochondrial genome by in vitro capping with guanylyltransferase. *J. Biol. Chem.* **258,** 14,025–14,033.
26. Tracy, R. L., and Stern, D. B. (1995) Mitochondrial transcription initiation: promoter structures and RNA polymerases. *Curr. Genet.* **28,** 205–216.
27. Stohl, L. L. and Clayton, D. A. (1992) *Saccharomyces cerevisiae* contains an RNase MRP that cleaves at a conserved mitochondrial RNA sequence implicated in replication priming. *Mol. Cell Biol.* **12,** 2561–2569.
28. Mueller, D. M., and Getz, G. S. (1986) Transcriptional regulation of the mitochondrial genome of yeast *Saccharomyces cerevisiae. J. Biol. Chem.* **261,** 11,756–11,764.
29. Biswas, T. K. and Getz, G. S. (1986) Nucleotides flanking the promoter sequence influence the transcription of the yeast mitochondrial gene coding for ATPase subunit 9. *Proc. Natl. Acad. Sci. USA* **83,** 270–274.
30. Karlok, M. A., Jang, S. H., and Jaehning, J. A. (2002) Mutations in the yeast mitochondrial RNA polymerase specificity factor, Mtf1, verify an essential role in promoter utilization. *J. Biol. Chem.* **277,** 28,143–28,149.
31. Ausubel, F. M., Brent, R., Kingston, R. E., et al. (1994) *Current Protocols in Molecular Biology*, Vol. 1, Wiley, New York.
32. Biswas, T. K., Edwards, J. C., Rabinowitz, M., and Getz, G. S. (1985) Characterization of a yeast mitochondrial promoter by deletion mutagenesis. *Proc. Natl. Acad. Sci. USA* **82,** 1954–1958.
33. Barker, M. M. and Gourse, R. L. (2001) Regulation of rRNA transcription correlates with nucleoside triphosphate sensing. *J. Bacteriol.* **183,** 6315–6323.
34. deHaseth, P. L. and Helmann, J. D. (1995) Open complex formation by *Escherichia coli* RNA polymerase: the mechanism of polymerase-induced strand separation of double helical DNA. *Mol. Microbiol.* **16,** 817–824.
35. Stano, N. M. and Patel, S. S. (2004) T7 lysozyme represses T7 RNA polymerase transcription by destabilizing the open complex during initiation. *J. Biol. Chem.* **279,** 16,136–16,143.
36. Bandwar, R. P. and Patel, S. S. (2002) The energetics of consensus promoter opening by T7 RNA polymerase. *J. Mol. Biol.* **324,** 63–72.
37. Stano, N. M., Levin, M. K., and Patel, S. S. (2002) The +2 NTP binding drives open complex formation in T7 RNA polymerase. *J. Biol. Chem.* **277,** 37,292–37,300.
38. Bandwar, R. P., Jia, Y., Stano, N. M., and Patel, S. S. (2002) Kinetic and thermodynamic basis of promoter strength: multiple steps of transcription initiation by T7 RNA polymerase are modulated by the promoter sequence. *Biochemistry* **41,** 3586–3595.
39. Lew, C. M. and Gralla, J. D. (2004) Nucleotide-dependent isomerization of *Escherichia coli* RNA polymerase. *Biochemistry* **43,** 12,660–12,666.
40. Nierman, W. C. and Chamberlin, M. J. (1979) Studies of RNA chain initiation by *Escherichia coli* RNA polymerase bound to T7 DNA. Direct analysis of the kinetics and extent of RNA chain initiation at T7 promoter A1. *J. Biol. Chem.* **254,** 7921–7926.
41. Biswas, T. K. and Getz, G. S. (1990) Regulation of transcriptional initiation in yeast mitochondria. *J. Biol. Chem.* **265,** 19,053–19,059.

42. Biswas, T. K. (1990) Control of mitochondrial gene expression in the yeast *Saccharomyces cerevisiae. Proc. Natl. Acad. Sci. USA* **87,** 9338–9342.
43. Gaal, T., Bartlett, M. S., Ross, W., Turnbough, C. L., Jr., and Gourse, R. L. (1997) Transcription regulation by initiating NTP concentration: rRNA synthesis in bacteria. *Science* **278,** 2092–2097.
44. Bartlett, M. S., Gaal, T., Ross, W., and Gourse, R. L. (1998) RNA polymerase mutants that destabilize RNA polymerase-promoter complexes alter NTP-sensing by rrn P1 promoters. *J. Mol. Biol.* **279,** 331–345.
45. Hsu, L. M. (2002) Promoter clearance and escape in prokaryotes. *Biochim. Biophys. Acta* **1577,** 191–207.

15

Functional Analysis by Inducible RNA Interference in *Drosophila melanogaster*

Yuichi Matsushima, Cristina Adán, Rafael Garesse, and Laurie S. Kaguni

Summary

Ribonucleic acid (RNA) interference triggered by double-stranded RNA has become a powerful tool for generating loss-of-function phenotypes. It is used to inactivate genes of interest and represents an elegant approach to genome functional analysis by reverse genetics. In *Drosophila*, RNA interference has been used in both cell culture and animals. We have adopted this approach to reveal the physiological roles of a number of proteins involved in mitochondrial deoxyribonucleic acid metabolism, and present here experimental schemes to induce the stable expression of double-stranded RNA in Schneider cells and in transgenic *Drosophila*.

Key Words: *Drosophila*; RNA interference; RNAi; Schneider cells; transgenesis; UAS-GAL4 system.

1. Introduction

Ribonucleic acid interference (RNAi) triggered by double-stranded RNA (dsRNA) was originally found in plants (posttranscriptional gene silencing; *1*), fungi ("quelling"; *2*) and *Caenorhabditis elegans* *(3)*, but it has been shown to occur in virtually every organism examined, from protozoa to animals *(4)*. The phenomenon of RNAi is triggered by dsRNA molecules that are cleaved into smaller RNA duplexes, 21–27 nucleotides long, by the ribonuclease (RNAse) III-type endonuclease Dicer. In *Drosophila melanogaster*, there are two Dicer isoforms: Dicer-1 (DCR-1) processes microRNAs (miRNAs), and Dicer-2 (DCR-2) is required for long dsRNA cleavage. The small dsRNA molecules, such as short interfering RNAs and miRNAs, are subsequently unwound and rearranged into effector complexes: RNA-induced silencing complex, RNA-induced transcriptional silencing or miRNA ribonucleoprotein particles. The

From: *Methods in Molecular Biology, vol. 372: Mitochondria: Practical Protocols*
Edited by: D. Leister and J. M. Herrmann © Humana Press Inc., Totowa, NJ

RNA-induced silencing complex mediates the posttranscriptional degradation of homologous messenger RNAs, whereas RNA-induced transcriptional silencing promotes the condensation of heterochromatin, and miRNA ribonucleoprotein particles guide translational repression of messenger RNA targets *(5)*. Although the basic mechanisms by which gene expression is suppressed are not completely understood, RNAi has become a powerful tool for generating loss-of-function phenotypes.

In *Drosophila*, RNAi has been used in both cell culture and animals. The transfection of dsRNA into *Drosophila* Schneider cells was found to have a gene-specific silencing function *(6)*, but the transfected dsRNA only works for a short period. However, if the dsRNA is produced from a vector integrated into the genome (i.e., in an established RNAi cell line), then the RNAi effect can be monitored continuously. In living organisms, RNAi can be induced by injecting, feeding, or expressing dsRNA. Injection of dsRNA into *Drosophila* embryos disrupts gene activity efficiently *(7)*, but its effect is transient, not inherited in the next generation, and genes expressed in later stages of development cannot be inactivated. To overcome these limitations, several strategies, mainly the use of the upstream activator sequences (UAS)-GAL4 system to induce controlled expression of the RNAi, have been developed to express dsRNA stably in transgenic *Drosophila*.

We are using a systematic RNAi approach to unravel the function of several essential factors that are constituents of the mitochondrial deoxyribonucleic acid (DNA) replication and transcription machinery in *Drosophila*. In this chapter, we describe protocols to generate RNAi stably in Schneider cell lines using inducible vectors. We also describe the general strategy to induce RNAi in *D. melanogaster* using the UAS-GAL4 system.

2. Materials

2.1. RNAi in Schneider Cells

2.1.1. Construction of the RNAi Vector

1. pMt/Hy DNA *(8)*: this vector can be obtained from the authors, or a variety of metallothionein promoter-based vectors can be obtained from the *Drosophila* Genomics Resource Center (http://dgrc.cgb.indiana.edu/news.html).
2. Restriction enzymes.
3. GeneElute agarose spin column (Sigma).
4. Platinum *Pfx* DNA polymerase (Invitrogen).
5. QIAquick polymerase chain reaction (PCR) purification kit (Qiagen).
6. T4 DNA ligase.
7. *E. coli* SURE cells (Stratagene).
8. Electroporation device such as the *E. coli* pulser (Bio-Rad).

9. Plasmid Midi Kit (Qiagen).
10. 3 *M* sodium acetate, pH 5.2.
11. Phenol/chloroform/isoamyl alcohol (25:24:1, v/v/v).
12. TE buffer: 10 m*M* Tris-HCl, pH 8.0, 1 m*M* ethylenediaminetetraacetic acid (EDTA).

2.1.2. Establishment of the RNAi Cell Line

1. Schneider S2 cells.
2. Schneider's *Drosophila* medium (Gibco) supplemented with 10% (v/v) fetal bovine serum (Gibco).
3. Effectene transfection reagent (Qiagen).
4. Hygromycin B (50 mg/mL) (Invitrogen).
5. 60-mm tissue culture dish (Corning).
6. 25-cm^2 flask (Corning).
7. 75-cm^2 flask (Corning).
8. 125-cm^2 flask (Corning).

2.1.3. Induction of dsRNA Expression

1. 100 m*M* $CuSO_4$.
2. Phosphate-buffered saline: 135 m*M* NaCl, 10 m*M* Na_2HPO_4, 2 m*M* KCl, 2 m*M* KH_2PO_4.
3. Lysis buffer: 10 m*M* Tris-HCl, pH 8.0, 5 m*M* EDTA, 1% (w/v) sodium dodecyl sulfate.
4. BCA protein assay kit (Pierce).

2.2. RNAi in Flies

2.2.1. Construction of the RNAi Vector for Flies

For construction of the RNAi vector for flies, use pUAST DNA ***(9)***. This vector can be obtained from the *Drosophila* Genomics Resource Center (http://dgrc.cgb.indiana.edu/news.html).

2.2.2. Generation of UAS-IR (Inverted Repeat) Lines

There are no specific materials for generation of UAS-IR lines. General *Drosophila* lab supplies can be obtained from LabScientific.

2.2.3. Setting the UAS-IR × GAL4 Cross and RNAi Analysis

1. Plastic vials (75 × 25 mm diameter) (LabScientific).
2. Plastic fly bottles (100 × 25 mm diameter) (LabScientific).
3. Nonabsorbent cotton plugs for fly bottles and vials.
4. Thin brushes and forceps.
5. Anesthetic device (carbon dioxide and related products).
6. Dissecting microscope and halogen lamp.

7. Fly food: 12 g/L agar; 100 g/L yeast (available from bakers' suppliers); 100 g/L sugar; 35 g/L maize meal; and 3 mL/L propionic acid.

2.2.4. Phenotypic Analysis

There are no specific materials for phenotypic analysis. General *Drosophila* lab supplies can be obtained from LabScientific.

3. Methods

3.1. RNAi in Schneider Cells

3.1.1. Construction of the RNAi Vector

We describe the construction of an RNAi vector targeting the *Drosophila mtTFB1* gene as an example. To construct RNAi vectors targeting other genes, use PCR primers appropriate for those genes (*see* **Note 1**). A schematic diagram depicting the structure of a dsRNA expression plasmid targeting the *Drosophila mtTFB1* gene in Schneider cells is shown in **Fig. 1**.

1. Digest 1 μg pMt/Hy plasmid DNA with *Xho*I and *Spe*I in a reaction volume of 20 μL.
2. Electrophorese the reaction mixture in a 0.7% agarose gel.
3. Purify the vector from the agarose gel using a standard method (e.g., using a GeneElute agarose spin column).
4. Amplify the sense fragment using the *mtTFB1* complementary DNA, platinum *Pfx* DNA polymerase and the pair of primers 5′-CGCctcgagactagt ACGGACAA GATAGTCAAGTCG-3′ and 5′-CGCcaattcGGGatcgatTAGCTTCTCAGCAAC CTCCTC-3, and the antisense fragments with 5′-CGCctcgagactagtACGGACA AGATAGTCAAGTCG-3′ and 5′-CGCgaattcAAAaagcttTAGCTTCTCAGCAAC CTCCTC-3′.
5. Purify the PCR products with a PCR purification kit (e.g., QIAquick PCR purification kit).
6. Digest the sense PCR fragment with *Xho*I and *Eco*RI and digest the antisense PCR fragment with *Spe*I and *Eco*RI.
7. Electrophorese the reaction mixtures in a 1% agarose gel.
8. Purify the fragments from the agarose gel using an appropriate method as above.
9. Ligate the vector DNA with the sense and antisense fragments using T4 DNA ligase at 16°C overnight. Use a molecular ratio of vector/sense fragment/antisense fragment of 1:3:3.
10. Transform *E. coli* host cells with 1 μL of ligation mixture using the *E. coli* pulser and plate on an Luria-Bertani (LB) plate containing ampicillin at 100 μg/mL. We use *E. coli* SURE cells as the host.
11. Recover plasmids from the colonies and check their identity and integrity by restriction endonuclease digestion.
12. Select positive colonies and purify the plasmid DNA (e.g., using a Qiagen plasmid Midi Kit).

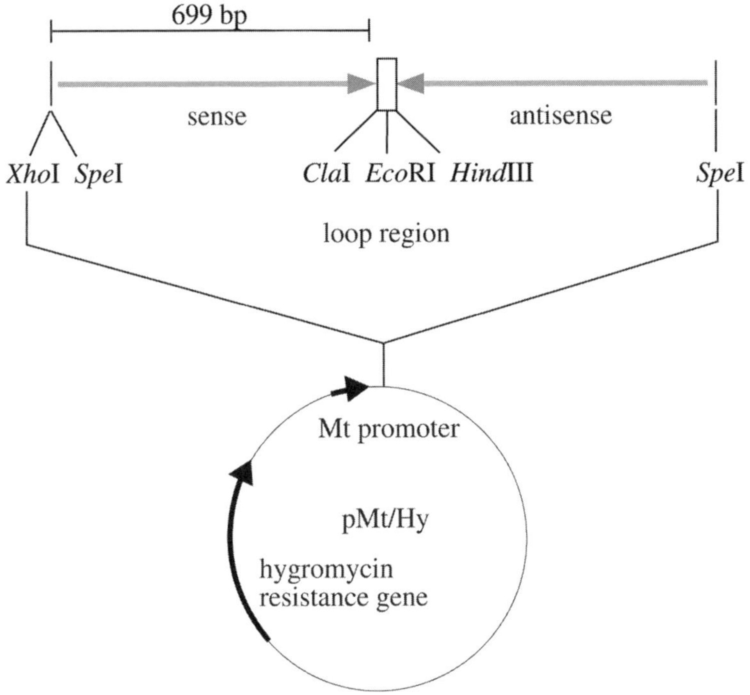

Fig. 1. Schematic representation of a dsRNA expression plasmid targeted to the *Drosophila mtTFB1* gene. Gray arrows show sense and antisense sequences. The open box indicates the loop region. The loop region is 24 nt and contains three restriction enzyme sites, *Cla*I–CCC–*Eco*RI–AAA–*Hin*dIII.

13. Dissolve the DNA in 400 µL TE buffer.
14. Add 400 µL phenol/chloroform/isoamyl alcohol (25:24:1, v/v/v), mix well, and then separate the phases by centrifugation at 12,000*g* for 5 min.
15. Transfer the aqueous phase to a new tube, add 40 µL 3 *M* sodium acetate, pH 5.2, and 800 µL ethanol, mix, and incubate at room temperature for 10 min.
16. Centrifuge at 12,000*g* for 10 min and discard the supernatant. Rinse the pellet with 600 µL 70% ethanol. Centrifuge at 12,000*g* for 5 min and discard the supernatant.
17. Dry the pellet at room temperature for 10 min and dissolve the pellet in 50 µL of TE buffer.

3.1.2. Establishment of the RNAi Cell Line

1. Culture *Drosophila* Schneider S2 cells at 25°C in *Drosophila* Schneider medium supplemented with 10% (v/v) fetal bovine serum. Subculture cells to 3 to 5 × 10^6 cells/mL every third to fifth day.
2. Place 4 mL cells at a density of 3 to 5 × 10^6 cells/mL into a 60-mm dish 24 h prior to transfection.

3. Transfect the cells using Effecten (Qiagen) according to the manufacturer's instructions.
4. Incubate for 24–48 h at 25°C.
5. Transfer the cells from a 60-mm dish into a 25-cm^2 flask containing 4 mL fresh *Drosophila* Schneider medium supplemented with 10% (v/v) fetal bovine serum and 200 μg/mL hygromycin.
6. Incubate at 25°C until cells reach a density of 10 to 15 × 10^6 cells/mL or for 7–10 d.
7. Transfer the cells into a 75-cm^2 flask containing 10 mL fresh *Drosophila* Schneider medium supplemented with 10% (v/v) fetal bovine serum and 200 μg/mL hygromycin.
8. Incubate at 25°C until cells reach a density of 15 to 20 × 10^6 cells/mL or for 5–7 d.
9. Transfer the cells into a 125-cm^2 flask containing 20 mL fresh *Drosophila* Schneider medium supplemented with 10% (v/v) fetal bovine serum and 200 μg/mL hygromycin.
10. Incubate at 25°C until cells reach a density of 15 to 20 × 10^6 cells/mL or for 5–7 d.
11. Transfer 7 mL of the culture into a 125-cm^2 flask containing 20 mL fresh *Drosophila* Schneider medium supplemented with 10% (v/v) fetal bovine serum and 200 μg/mL hygromycin.
12. Incubate at 25°C until cells reach a density of 15 to 20 × 10^6 cells/mL or for 5–7 d.
13. Culture the selected Schneider S2 cells at 25°C in *Drosophila* Schneider medium supplemented with 10% (v/v) fetal bovine serum, subculturing to 3 to 5 × 10^6 cells/mL every third to fifth day.

3.1.3. Induction of dsRNA Expression

1. Dilute the cells to 3 to 5 × 10^6 cells/mL and add $CuSO_4$ to a final concentration of 0.4 m*M*.
2. Incubate at 25°C and subculture to 3 to 5 × 10^6 cells/mL every third day.
3. After 10 d culture, harvest the cells by centrifugation at 2000*g* for 5 min.
4. Wash with phosphate-buffered saline and centrifuge at 2000*g* for 5 min.
5. Add lysis buffer, heat at 100°C for 5 min, and centrifuge at 12,000*g* for 10 min.
6. Transfer the lysate to a fresh tube and assay the protein concentration (e.g., using the BCA protein assay kit).
7. Check the suppression level of the target protein by immunoblot analysis (**Fig. 2**) (*see* **Notes 2** and **3**).

3.2. RNAi in Flies

One of the most valuable tools available for scientists working in *Drosophila* is the availability of the UAS-GAL4 system ***(9)***, which is shown schematically in **Fig. 3**. In this dual system, its two components (UAS and GAL4 transgenic lines) are maintained as independent parental stocks until needed. After crossing, the resulting F1 generation will express the gene (or RNAi) of interest in the pattern driven by GAL4. At present, numerous GAL4 drivers for constitutive or tissue-specific overexpression have been reported in the literature, and a list of

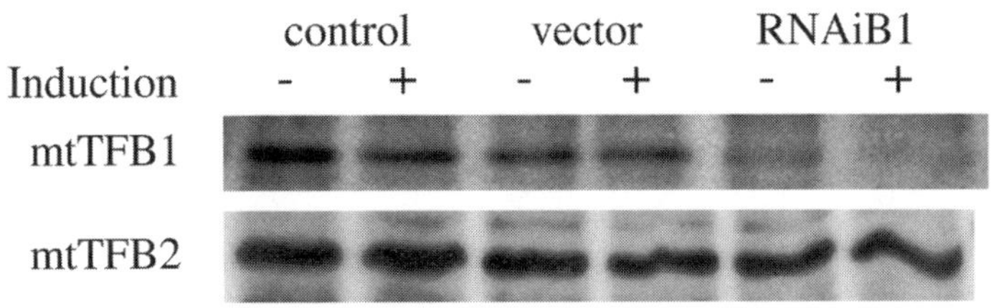

Fig. 2. Expression of *d*-mtTFB1-targeted RNAi in Schneider cells. Immunoblot analysis of mitochondrial extracts probed with affinity-purified rabbit antiserum against *d*-mtTFB1 ***(14)***. Schneider cells with no plasmid (control) or carrying pMt/Hy (vector) or RNAi vector (RNAiB1) were cultured for 10 d in the presence or absence of 0.4 m*M* $CuSO_4$. Protein extracts (20 μg) were fractionated by 10.5% sodium dodecylsulfate-polyacrylamide gel electrophoresis, transferred to nitrocellulose filters, and probed with rabbit antiserum against *d*-mtTFB1 or *d*-mtTFB2 ***(15)*** as indicated (*see* **Notes 2** and **3**).

them can be obtained from public stock centers (http://fly.bio.indiana.edu/gal4.htm). The flexibility of the UAS-GAL4 system has allowed its use for the analysis of many biological processes, including the induction of loss-of-function phenotypes through overexpression of RNAi constructs ***(10)***.

3.2.1. Construction of the RNAi Vector for Flies

Construct the vector as described in **Subheading 3.1.1.**, substituting the pMt/Hy plasmid with the appropriately digested pUAST vector.

3.2.2. Generation of UAS-IR Lines

Transgenesis in *Drosophila* is based on the use of P-element transformation. There are several standard protocols to obtain transgenic animals that are based on the original method described by Spradling and Rubin ***(11)***. Current protocols are similar, with efficiencies ranging from 3 to 12%. We routinely use the protocol of Déjardin and Cavalli, which is available online with a detailed description (http://www.igh.cnrs.fr/equip/cavalli/link.labgoodies.html). It is highly efficient, yielding approx 10 independent transgenic animals from 100 injected embryos. Here, we offer detailed explanatory notes regarding important considerations that we have encountered in our laboratory experience (*see* **Note 4**).

3.2.3. Setting the UAS-IR × GAL4 Cross and RNAi Analysis

1. Collect virgin females from the GAL4 driver stock. Depending on the number of crosses planned, start collecting virgins at the beginning of the week and set the crosses at the end of the week. At the same time, collect newly hatched UAS-IR males and plan to use both parental lines as controls.
2. Set the RNAi crosses using the appropriate number of flies according to the size of the food vials selected. Place no less than 10 males and 20 females in a small

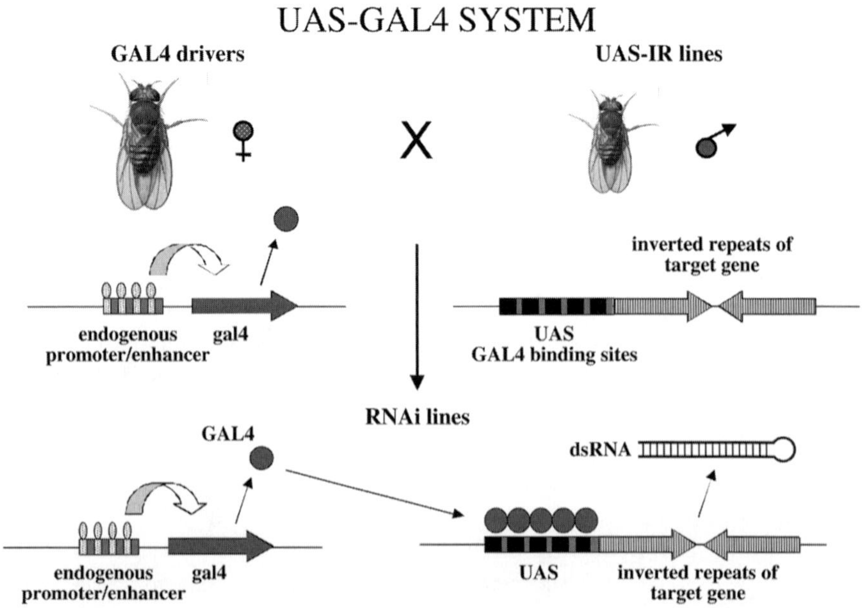

Fig. 3. Schematic representation of the UAS-GAL4 strategy. UAS-IR and GAL4 lines are maintained as independent stocks. GAL4 drivers express the transcriptional regulator GAL4 with a constitutive or tissue-specific pattern depending on the endogenous enhancer that is located in its proximity. There is a large collection of GAL4 lines available from public stock centers. UAS-IR lines are generated by P-element-mediated transformation. A construct containing in tandem several GAL4 binding sites upstream of the inverted repeat of the target gene is inserted randomly in the genome. Leaky expression is generally very low, showing no phenotype. After crossing the GAL4 drivers with the UAS-IR stocks, the F1 generation expresses the dsRNA directed by GAL4 and induces the RNAi. In this way, a knock-down of the gene of interest is obtained in the animal. The scheme is based on that published by Brand and Perrimon ***(9)***.

vial. For a big vial, use 30 males and 50 females and scale it up to 150 flies if needed. When using bottles, use at least 60 males and 100 females.

3. Pass the crossed RNAi lines daily after allowing flies to remain in the same vial for a couple of days. Keep the incubation temperature constant.
4. Isolate total RNA and protein from embryos, larvae, or adults ***(13)***. Place the crosses in suitable cages to collect materials at regular time intervals. If the number of flies is small, then collect several egg lays and store them at 4°C to gather enough material. Collect larvae and proceed with fresh material if possible for best results, although they may be kept frozen at –70°C at the desired stage. Ten larvae from each cross generally provide sufficient amounts of RNA and protein for analysis. Anesthetize and freeze adults at –70°C until use.

5. Check the RNAi-induced suppression level of the target protein by immunoblot analysis or the target RNA by quantitative reverse transcriptase PCR ***(14)***.

3.2.4. Phenotypic Analysis (see **Notes 5–8**)

1. Pass the crossed RNAi lines daily for 1–2 wk.
2. Inspect the vials daily and note relevant events and dates on which they occur, such as the beginning of the first larval instar, third larval instar wandering, prepuparium formation, pupariation, and eclosion.
3. Perform a detailed progeny analysis on selected vials. Count the number of flies (males and females) daily. If the progeny are not expected to have the same genotype, then also count the number of flies of each possible genotype daily.
4. Determine the rate of pupal lethality, life-span, and reproductive capacity in terms of egg laying or perform behavior assays as appropriate.

4. Notes

1. The length of the stem region of a hairpin type of dsRNA is usually longer than 300 bp. Such constructs have been shown to cause more than 90% reduction of expression of the target gene. However, a shorter dsRNA expression vector with less than a 100-bp stem region is also effective ***(12)***.
2. Because of leaky expression from metallothionein promoter, the RNAi construct may repress the target transcript under uninduced conditions (*see* **Fig. 2**).
3. The RNAi cell lines are stable for at least 3 mo. Check the target protein level by immunoblot monthly. If sufficient suppression is not apparent, then it is better to establish a new cell line.
4. Although dsRNA can be obtained by several different strategies, in general two identical fragments of 0.5–1 kb that have sequence identity to the gene to be knocked down are cloned into the pUAST vector in a head-to-head or tail-to-tail orientation. By traditional P-element transformation, a transgenic fly that carries the inverted repeats under UAS control is obtained. Normally, the transgene will remain silent in the absence of GAL4, and only leaky expression of dsRNA will occur.
5. Always work with several UAS-IR independent transgenic lines. Even though dsRNA production will be driven by GAL4, sites of insertion will still have an effect on transgene expression, and some insertions may be lethal because of positional effects.
6. Depending on the driver, RNAi can be set off early in development or not and may be constitutive or tissue specific. Therefore, the use of reporter genes such as green fluorescent protein (GFP) or *lacZ* (using UAS-GFP or UAS-lacZ lines currently available) is recommended to obtain the most accurate spatiotemporal expression pattern given by each GAL4 driver.
7. When mating UAS-IR and GAL4 parental lines, do not underestimate the number of flies needed. Crosses should have at least 10–20 individuals. Setting small crosses can result in lethality not related to the RNAi phenotype. Remember to set the appropriate controls so that, by comparison, all F1 progeny represent the same gene dosage. Mate UAS-IR virgins to GAL4 driver males and the reverse. When

possible, keep parental stocks balanced and maintain each homozygous for IR or GAL4 insertion. When homozygous stocks are used, 100% of the progeny will express dsRNA and show the RNAi phenotype.

8. The function of GAL4 is temperature dependent. Flies can be maintained at a temperature ranging from 16 to 29°C. At low temperature, GAL4 shows little ability to activate transcription; thus, RNAi will have a milder effect. At high temperature, there is increasing activation of transcription by GAL4 that will lead to a more powerful effect of RNAi. Apart from trying different GAL4 drivers, inducing RNAi at different temperatures should also be considered.

Acknowledgments

The work in our laboratories was supported by National Institutes of Health grant GM45295 to L. S. K. and Ministerio de Ciencia y Tecnología, Spain (grant BFU2004–04591) and Instituto de Salud Carlos III, Redes de centros RCMN (C03/08), and Temáticas (G03/011) to R. G.

References

1. Jorgensen, R. A. (2003) Sense cosuppression in plants: past, present and future, in *RNAi : A Guide to Gene Silencing* (Hannon, G. J., ed.), Cold Spring Harbor Laboratory Press, Cold Spring Harbor, NY, pp. 5–22.
2. Cogoni, C., Irelan, J. T., Schumacher, M., Schmidhauser, T. J., Selker, E. U., and Macino, G. (1996) Transgene silencing of the *al-1* gene in vegetative cells of *Neurospora* is mediated by a cytoplasmic effector and does not depend on DNA–DNA interactions or DNA methylation. *EMBO J.* **15,** 3153–3163.
3. Fire, A., Xu, S., Montgomery, M. K., Kostas, S. A., Driver, S. E., and Mello, C. C. (1998) Potent and specific genetic interference by double-stranded RNA in *Caenorhabditis elegans. Nature* **391,** 806–811.
4. Cogoni, C. and Macino, G. (2000) Post-transcriptional gene silencing across kingdoms. *Curr. Opin. Genet. Dev.* **10,** 638–643.
5. Tomari, Y. and Zamore, P. D. (2005) Perspective: machines for RNAi. *Genes Dev.* **19,** 517–529.
6. Hammond, S. M., Bernstein, E., Beach, D., and Hannon, G. J. (2000) An RNA-directed nuclease mediates post-transcriptional gene silencing in *Drosophila* cells. *Nature* **404,** 293–296.
7. Kennerdell, J. R. and Carthew, R. W. (1998) Use of dsRNA-mediated genetic interference to demonstrate that frizzled and frizzled 2 act in the wingless pathway. *Cell* **95,** 1017–1026.
8. Koelle, M. R., Talbot, W. S., Segraves, W. A., Bender, M. T., Cherbas, P., and Hogness, D. S. (1991) The *Drosophila* EcR gene encodes an ecdysone receptor, a new member of the steroid receptor superfamily. *Cell* **67,** 59–77.
9. Brand, A. H. and Perrimon, N. (1993) Targeted gene expression as a means of altering cell fates and generating dominant phenotypes. *Development* **118,** 401–415.
10. Duffy, J. B. (2002) GAL4 system in *Drosophila*: a fly geneticist's Swiss army knife. *Genesis* **34,** 1–15.

11. Spradling, A. C. (1986) P element-mediated transformation, in *Drosophila a Practical Approach* (Roberts, D. B., ed.), IRL Press, Oxford, UK, pp. 175–197.
12. Farr, C. L., Matsushima, Y., Lagina, A. T., 3rd, Luo, N., and Kaguni, L. S. (2004) Physiological and biochemical defects in functional interactions of mitochondrial DNA polymerase and DNA-binding mutants of single-stranded DNA-binding protein. *J. Biol. Chem.* **279,** 17,047–17,053.
13. Lefai, E., Calleja, M., Ruiz de Mena, I., Lagina, A. T., 3rd, Kaguni, L. S., and Garesse, R. (2000) Overexpression of the catalytic subunit of DNA polymerase gamma results in depletion of mitochondrial DNA in *Drosophila melanogaster. Mol. Gen. Genet.* **264,** 37–46.
14. Matsushima, Y., Adan, C., Garesse, R., and Kaguni, L. S. (2005) *Drosophila* mitochondrial transcription factor B1 modulates mitochondrial translation but not transcription or DNA copy number in Schneider cells. *J. Biol. Chem.* **280,** 16,815–16,820.
15. Matsushima, Y., Garesse, R., and Kaguni, L. S. (2004) *Drosophila* mitochondrial transcription factor B2 regulates mitochondrial DNA copy number and transcription in Schneider cells. *J. Biol. Chem.* **279,** 26,900–26,905.

16

Analysis of Replicating Mitochondrial DNA by Two-Dimensional Agarose Gel Electrophoresis

Aurelio Reyes, Takehiro Yasukawa, and Ian J. Holt

Summary

Replication intermediates can be separated on agarose gels in two dimensions to reveal a wealth of data on mechanisms of DNA replication. When applied to mitochondrial DNA of higher vertebrates, this technique unearthed a host of unexpected findings, the full implications of which are still being absorbed. Here, we describe the procedures we use to isolate intact mitochondrial replication intermediates from liver of higher vertebrates and the process of separating DNA fragments on neutral two-dimensional agarose gels.

Key Words: DNA isolation; DNA replication; humans; mammals; mitochondrial DNA; neutral/neutral two-dimensional agarose gel electrophoresis (N2D-AGE); replication intermediates; vertebrates.

1. Introduction

1.1. Neutral Two-Dimensional Agarose Gel Electrophoresis

The study of DNA replication was transformed by the development of a method of separating branched molecules based on their structure. In the first report, Bell and Byers *(1)* demonstrated that a stable recombination intermediate could be separated from linear double-stranded DNA molecules of the same mass. Curiously, its inventors abandoned the technique at birth, and it was left to Bonnie Brewer and Walt Fangman to blaze a trail with this powerful new method *(2–6)*. Such was their mastery of the technique that the gels became widely known as Brewer-Fangman gels, in place of the cumbersome, albeit more informative, neutral/neutral two-dimensional agarose gel electrophoresis (N2D-AGE). Others were quick to appreciate the value of N2D-AGE; a notable milestone was the demonstration that initiation of nuclear DNA replication occurs not from discrete origins, but at multiple sites dispersed over several kilobases of DNA *(7)*.

From: *Methods in Molecular Biology, vol. 372: Mitochondria: Practical Protocols*
Edited by: D. Leister and J. M. Herrmann © Humana Press Inc., Totowa, NJ

N2D-AGE is not only ideal for mapping origins of replication in fragments of DNA ***(2,4,8–10)***, but also it can be used to locate replication pause sites or termini ***(3,5)***, determine the direction of replication fork movement ***(5,11)***, and potentially distinguish multiple mechanisms of replication ***(12)***. The technique is widely applicable, having been used successfully to study replication in eukaryotes, prokaryotes, and viruses.

N2D-AGE is technically straightforward. It entails separating fragments of DNA, first in a low-percentage agarose gel, at low field strength, followed by separation at low temperature and high field strength in a higher percentage agarose gel containing a DNA intercalating dye, such as ethidium bromide (*see* **Fig. 1** and the step-by-step protocol for details). Separation in the first dimension is essentially on the basis of mass, whereas the conditions of second-dimension electrophoresis mean that hydrodynamic volume becomes a key factor. In the presence of ethidium bromide, branched structures such as replication intermediates (RIs) occupy a significantly greater hydrodynamic volume than linear DNA molecules of the same mass; consequently, RIs are more retarded than linear molecules during second-dimension electrophoresis. The mobility of a particular RI depends on the position of the replication fork within the fragment (i.e., the fraction of the fragment that has been replicated). Hence, the passage of a replication fork through a fragment gives rise to a series of intermediates of different mobility, which form a characteristic arc.

The simplest example is a fragment in which a replisome enters at one end and proceeds at a uniform rate to the other end of the fragment, creating a simple fork, or Y, arc (**Fig. 2A**). The arc reaches its apex when the fork is halfway through the fragment; thereafter, the fragment becomes increasingly similar to a linear molecule. Ultimately, just prior to separation of the daughter fragments, the molecule is substantially linear; accordingly, it migrates close to the arc of linear molecules, at a point equivalent to twice the mass of the original fragment (**Fig. 2A**).

If a fragment contains an origin of replication, then a so-called bubble structure will form (**Fig. 2B**); whenever a replication fork exits the fragment, at either end, the restriction enzyme cleaves the bubble. Bubble arcs are readily distinguishable from Y arcs as they are more retarded in the second dimension and end abruptly. The extent of a bubble arc reveals information about the location of the origin. However, it is essential to examine a series of overlapping fragments because a single fragment cannot distinguish differences in the mode of replication (e.g., unidirectional from bidirectional replication); short bubble arcs, with origins located close to the end of a fragment, may be indistinguishable from simple Y arcs ***(13)***. Composite or other more complicated patterns can also be obtained, depending on factors such as the number and distribution of origins (**Fig. 2C**) ***(7,14,15)***, the presence of ribonucleotides in replicating molecules (**Fig. 3**) ***(16)***, or delayed second-strand synthesis ***(17)***.

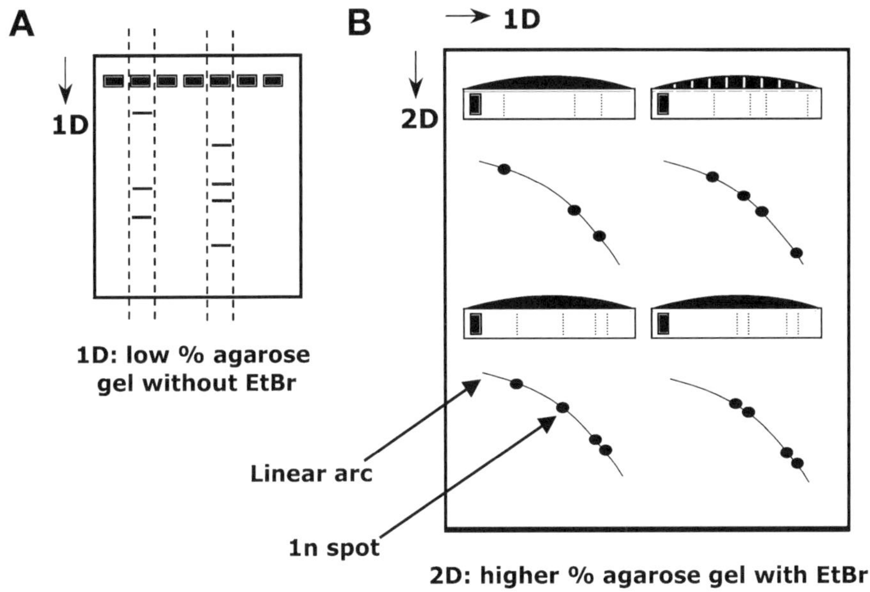

Fig. 1. Schematic diagram of neutral two-dimensional agarose gel electrophoresis. (**A**) A gel after first-dimension electrophoresis; filled black boxes represent wells of the gel, narrow lines indicate linear duplex fragments of DNA, and dotted vertical lines demarcate the lane to be excised. (**B**) Each 1D gel slice is rotated through 90° and a second gel cast around it; four 1D gel slices are shown on a single 2D gel. After second-dimension electrophoresis, the linear duplex fragments, 1n spots (black circles) resolve on a defined arc (narrow unbroken line); the position of each fragment after the first-dimension separation is shown by a broken faint line in the original 1D gel slice.

N2D-AGE is routinely combined with restriction endonuclease digestion, prior to electrophoresis, but a range of pretreatments with DNA-modifying reagents is also possible, which can yield additional information on the nature of the intermediates and thus the mechanism of replication *(18)*. Such treatments have been of immense value in deciphering the process of mitochondrial DNA (mtDNA) replication *(12,16)*. One of the most graphic examples is the application of single-strand nuclease, which greatly simplifies the pattern of mitochondrial RIs in crude preparations of mtDNA (**Fig. 4**). In some contexts, it may also be useful to alkali denature RIs after N2D-AGE and apply a third electrophoresis step *(18)*.

1.2. Considerations Particular to Vertebrate Mitochondrial DNA

An analysis of mtDNA from mammalian cultured cells, conducted over 30 years ago, revealed that it contains sporadic ribonucleotides (up to 10 per molecule) *(19)*. A more recent study *(16)* suggested that the ribonucleotide content of mtDNA of solid tissues was approx fivefold greater than that of cultured cells

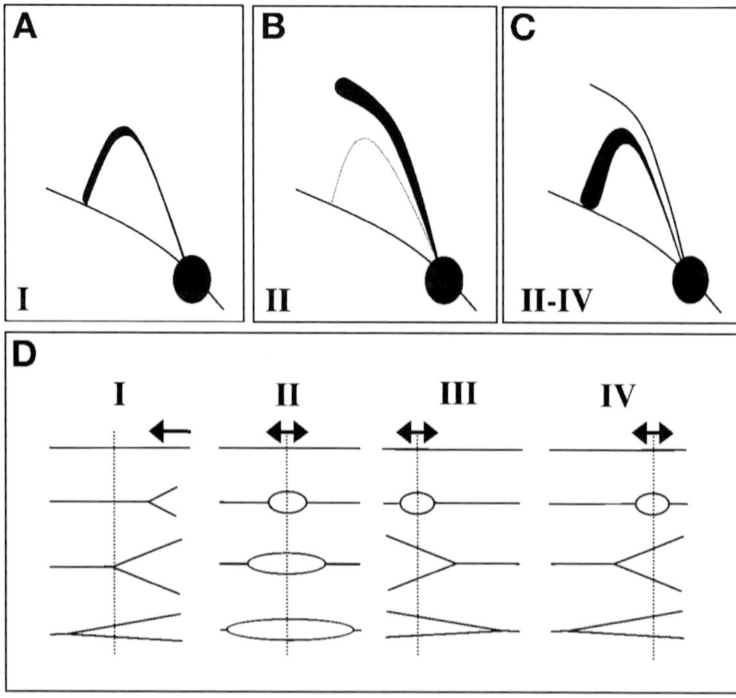

Fig. 2. Different modes of replication produce distinct patterns of replication intermediates on neutral two-dimensional agarose gels. (**A**) to (**C**) Schematic patterns of RIs on N2D-AGE. (**D**) Line drawings of structures of RIs. (**A**) A simple fork or Y arc; replication initiates outside the fragment as interpreted in **D-I**. (**B**) Initiation of bidirectional replication from a discrete origin located at the center of a fragment giving rise to a complete bubble arc and no fork arc (the position of the fork arc is shown as a faint broken line for reference purposes) as the two forks exit the fragment simultaneously, interpreted in **D-II**; note how the bubble arc increases in intensity as it reaches its apex, giving it a "clubheaded" appearance, caused by compression. (**C**) Bidirectional initiation from multiple sites across a zone defined by the fragment; in this case, initiation at the center of the fragment is only one of many possibilities; at other initiation sites, such as those depicted in **D-III** and **D-IV**, one fork exits the fragment well before the other, converting the bubble to a Y structure. Thereafter, the RIs contribute to the Y arc.

(>30 per mtDNA molecule). Moreover, there appear to be extensive ribonucleotide patches in replicating mammalian mtDNA ***(16)***.

The presence of RNA in replicating and nonreplicating mtDNA is problematic. Careful handling can minimize RNA degradation during isolation of mtDNA, but methods such as neutral/alkaline 2D-AGE ***(20)*** and replication initiation point mapping ***(21)*** are inapplicable to mammalian mtDNA as the alkaline and ribonuclease (RNase) treatments, respectively, cause RNA scission.

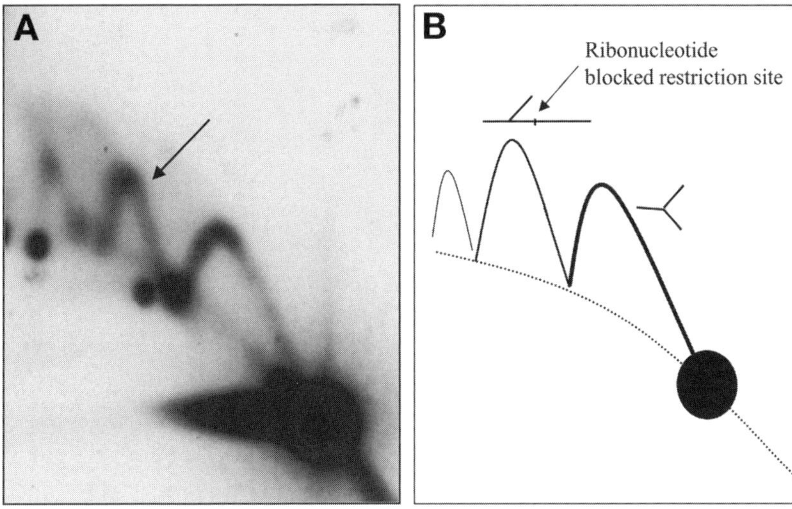

Fig. 3. Novel arcs in mammalian mtDNA. Highly purified vertebrate mtDNA yields slow-moving Y-like arcs (arrow in **A**); these arcs arise because of the presence of ribonucleotides at one (or more) restriction site, thereby linking adjacent restriction fragments (interpreted in **B**). Such arcs are sensitive to RNase H ***(16)***.

Thus, although N2D-AGE is not the only means of dissecting DNA replication, it is undoubtedly the best method for studying DNA with a significant ribonucleotide content.

We provide here detailed protocols of our isolation and separation procedures as a guide to analyzing replicating mtDNA. The protocols include an isolation procedure aimed at extracting intact RIs, a critical point when studying replication. Many standard methods of isolating mtDNA cannot be used for RIs because they are designed for nonreplicating DNA and take no account of its high ribonucleotide content.

2. Materials

2.1. mtDNA Isolation From Solid Tissues

1. Phosphate-buffered saline: 10 m*M* phosphate buffer, pH 7.4, 2.7 m*M* KCl, 137 m*M* NaCl.
2. Homogenization buffer (HB): 225 m*M* mannitol, 75 m*M* sucrose, 10 m*M* HEPES-NaOH, pH 7.8, 10 m*M* ethylenediaminetetraacetic acid (EDTA), 0.1% (w/v) fatty acid-free bovine serum albumin, and 357.5 μ*M* β-mercaptoethanol.
3. Gradient buffer (GB): 10 m*M* HEPES-NaOH, pH 7.8, 10 m*M* EDTA.
4. 1 *M* and 1.5 *M* Sucrose in GB.
5. Lysis buffer (LB): 20 m*M* HEPES-NaOH, pH 7.4, 75 m*M* NaCl, 50 m*M* EDTA.
6. Teflon® homogenizer.

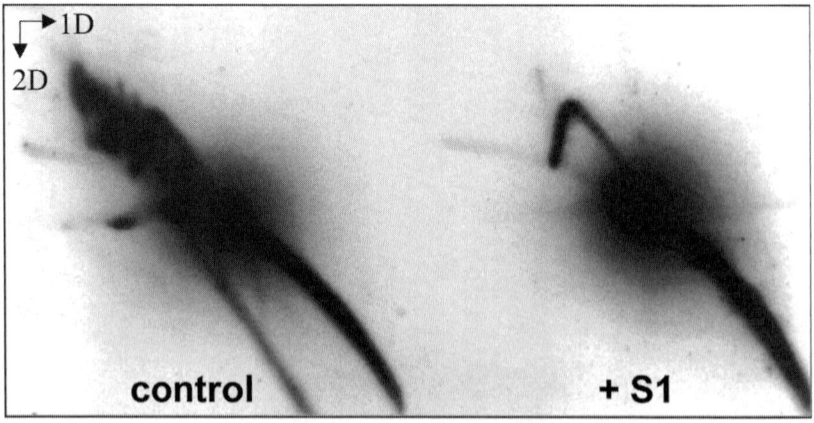

Fig. 4. Effect of single-strand nuclease on crude preparations of mouse mtDNA. A single *Dra*I digest of mouse liver mtDNA was divided in two, treated with and without S1 nuclease, and the products separated by N2D-AGE; after electrophoresis, the gel was blotted and hybridized to a probe detecting the *Dra*I fragment spanning nt 5276–9817.

7. Proteinase K (PK): 20 mg/mL in water.
8. 20% (w/v) sodium lauroyl sarcosinate (sarkosyl).
9. PCIA: equilibrated phenol:chloroform:isoamyl alcohol (25:24:1).
10. TE: 10 m*M* Tris-HCl, pH 8.0, 1 m*M* EDTA.
11. Isopropanol.
12. 70% (v/v) Ethanol.

2.2. mtDNA Treatment With Restriction and Modifying Enzymes

1. Restriction endonucleases.
2. Modifying enzymes: nuclease S1, RNase One, RNase H.
3. 3 *M* Sodium acetate, pH 5.2.
4. 10 m*M* Tris-HCl, pH 8.0
5. TE: 10 m*M* Tris-HCl, pH 8.0, 1 m*M* EDTA.
6. 100% Ethanol.
7. 70% (v/v) Ethanol.
8. 0.5 *M* EDTA, pH 8.0.
9. Dry ice.

2.3. 2D Agarose Gel Electrophoresis

1. TBE: 89 m*M* Tris base, 87 m*M* boric acid, 2 m*M* EDTA, pH 8.0.
2. Low-electroendosmotic agarose.
3. Ethidium bromide: 10 mg/mL.
4. Gel electrophoresis tank with ports for buffer circulation.
5. Power supply.
6. Ultraviolet (UV) light box.

7. Peristaltic pump.
8. DNA molecular weight marker.
9. Loading buffer.

2.4. Southern Blot

1. Depurination buffer (DPB): 0.25 *N* HCl.
2. Denaturing buffer (DNB): 0.5 *M* NaOH, 1.5 *M* NaCl.
3. Neutralizing buffer (NB): 0.5 *M* Tris-HCl, pH 7.4, 1.5 *M* NaCl.
4. Genome-quality nylon membrane.
5. 3MM Whatman filter paper.
6. Paper towels.

2.5. Hybridization

1. Hybridization Buffer: 0.25*M* sodium phosphate, pH 7.2, 7% (w/v) sodium dodecyl sulfate.
2. 20X SSC: 3*M* NaCl, 0.3*M* sodium citrate, pH 7.0.
3. Washing buffer 1 (WB1): 1X SSC.
4. Washing buffer 2 (WB2): 1X SSC/0.1% (w/v) sodium dodecyl sulfate.
5. X-ray film.

3. Methods

3.1. mtDNA Isolation From Liver Tissue

1. Excise fresh liver and place in cold phosphate-buffered saline on ice (*see* **Notes 1** and **2**).
2. Remove any blood clots and contaminating tissue and weigh (*see* **Note 3**).
3. Transfer to an ice-cold beaker containing 5 volumes 1:10 diluted HB/g.
4. Mince the tissues finely with sharp scissors, changing the solution four or five times (*see* **Note 4**).
5. Wash diced tissue with 5 volumes/g wet weight HB and discard as much solution as possible.
6. Add 9 volumes HB/g and homogenize using a motorized tight-fitting Dounce homogenizer until the suspension is smooth (*see* **Notes 5** and **6**).
7. Centrifuge the homogenate at $600g_{max}$ for 10 min at 4°C to pellet nuclei and intact cells.
8. Transfer the supernatant to a clean tube and centrifuge at $5000g_{max}$ for 10 min at 4°C.
9. Discard the supernatant and suspend the mitochondrial pellet in 5 volumes HB/g (*see* **Note 7**).
10. Repeat **steps 7** and **8** once and after the last centrifugation suspend the mitochondrial pellet in 0.5 volumes/g HB (*see* **Note 7**).
11. Prepare single-step sucrose gradients (*see* **Note 8**).
12. Load 2 mL mitochondrial suspension per sucrose gradient and centrifuge in a swing-out rotor at $40,000g_{max}$ for 1 h at 4°C.
13. After centrifugation, mitochondria resolve at the interface of the 1 *M* and 1.5 *M* sucrose solutions.

14. Transfer the mitochondria to a 30-mL tube and add 5 volumes of GB slowly with gentle shaking of the mitochondrial solution (*see* **Note 9**). Centrifuge at 9900g_{max} for 10 min at 4°C.
15. Discard the supernatant and suspend the mitochondrial pellet in 1.6 mL LB/g (*see* **Note 7**). Transfer to a clean 15- or 50-mL disposable tube. Add 0.8 µL PK/g and incubate at 4°C for 45 min (*see* **Note 10**).
16. Add 80 µL 20% sarkosyl/g, mix gently, and incubate at 50°C for 45 min (*see* **Notes 10** and **11**).
17. After incubation, mix thoroughly but gently with 1 volume PCIA and spin at 6000g_{max} for 10 min at room temperature (*see* **Note 12**).
18. Transfer the aqueous (upper) phase to a clean tube, mix carefully with 1 volume Chloroform isoamyl alcohol (CIA), and spin at 6000g_{max} for 10 min at 4°C.
19. Recover the aqueous (upper) phase and make 500-µL aliquots in 1.5-mL Eppendorf tubes. Add 1 volume isopropanol to each tube (*see* **Note 13**), mix gently, and keep at –20°C for at least 30 min (*see* **Note 14**).
20. Centrifuge the sample at 20,000g_{max} for 20 min at 4°C.
21. Discard the supernatant and wash the pellet with 70% ethanol.
22. Air dry the pellet, suspend in TE, and determine the concentration by UV spectrometry (*see* **Note 15**).

3.2. mtDNA Treatment With Restriction and Modifying Enzymes

1. Digest 0.5–3.0 µg mtDNA with the appropriate restriction endonuclease following the conditions recommended by the manufacturer (*see* **Notes 16** and **17**).
2. Add 1/10 volume 3 *M* sodium acetate and 2 volumes 100% ethanol. Mix gently and keep at –20°C for at least 30 min.
3. Centrifuge the sample at 20,000g_{max} for 20 min at 4°C.
4. Discard the supernatant and wash the pellet with 70% ethanol.
5. Air dry the pellet; suspend in TE or in 10 m*M* Tris-HCl at pH 8.0 if it is to be treated with modifying reagents (*see* **Note 18**). If no additional treatment is required, then the sample is ready for loading on the first-dimension agarose gel (*see* **Subheading 3.3.**).
6. Treatments with modifying enzymes require the addition of 1/10 volume of the appropriate 10X buffer (*see* **Note 19**). Standard conditions for the different modifying enzymes are 1 unit of RNase H for 1 h at 37°C; 5 units of RNase One for 10 min at 37°C; and 1 unit S1 nuclease for 1.5 min at 37°C.
7. Stop the reaction by flash-freezing the sample on dry ice/ethanol after adding 1/10 volume of 0.5 *M* EDTA when appropriate (*see* **Note 20**).

3.3. 2D Agarose Gel Electrophoresis

1. Prepare a 0.4% agarose gel in TBE buffer for first-dimension gel electrophoresis (*see* **Notes 21** and **22**).
2. Submerge gel in a tank containing TBE.
3. Load sample and a suitable size marker according to the length of the fragment of interest (*see* **Note 23**).

4. Run the first dimension at 0.7 V/cm for 20 h at room temperature (*see* **Note 24**).
5. After 20 h, cut out the lane containing the sample with the aid of a razor blade and a ruler.
6. Prepare 1% molten agarose in TBE with 500 ng/mL ethidium bromide (*see* **Note 25**).
7. Place gel slice from the first-dimension electrophoresis step in a gel-casting tray. Each slice is rotated 90° counterclockwise with respect to the first-dimension run. A tray 200 × 250 mm can accommodate up to four gel slices (*see* **Note 26** and **Fig. 1**).
8. Remove excess TBE buffer around the gel slice with 3MM Whatman filter paper and adhere the gel slice to the tray with 1% molten agarose (*see* **Note 27**).
9. Pour the remaining 1% molten agarose from the opposite side of the gel slice and wait until it has solidified (*see* **Note 28**).
10. Place the gel in a tank containing cold TBE with 500 ng/mL ethidium bromide (*see* **Note 29**).
11. Circulate the buffer from the positive to the negative electrode by a peristaltic pump (*see* **Note 30**).
12. Run the second dimension at 6 V/cm for 4 h at 4°C (*see* **Note 31**).

3.4. Southern Blot

1. After electrophoresis in the second dimension, remove the gel from the tank and invert into a glass dish (*see* **Note 32**).
2. Add 500 mL DPB for a dish measuring 300 × 400 mm, rock gently for 15–30 min, and discard solution (*see* **Note 33**).
3. Add 500 mL DNB and rock gently for 5 min. Pour off the solution and repeat the treatment for 15 min (*see* **Note 34**).
4. Add 500 mL NB; rock gently for 10 min. Discard the solution and repeat the treatment for 10 min (*see* **Note 35**).
5. Remove excess solution (*see* **Note 36**).
6. Cut a piece of membrane of the same dimensions as the gel and wet in water (*see* **Notes 37** and **38**).
7. Place the wetted membrane face down on the gel, avoiding bubbles (*see* **Note 39**).
8. Soak two sheets of 3MM Whatman filter paper in water and place them on top of the membrane, again avoiding bubbles (*see* **Note 39**).
9. Stack 10–12 cm of paper towel on top of 3MM Whatman filter paper.
10. Place a glass plate or a tray on top and a weight to create some pressure to the stack (*see* **Note 40**). Blot overnight.
11. Remove and discard paper towels and 3MM Whatman filter paper. Place the membrane face up on 3MM Whatman filter paper and let stand for a few minutes.
12. Covalently link the DNA to the membrane in a UV crosslinker at constant energy of 1200 × 100 μJ/cm^2 (*see* **Note 41**).

3.5. Hybridization

1. Place the membrane face up in a hybridization tube and add 15 mL prewarmed hybridization buffer at 65°C (*see* **Note 42**).
2. Incubate in a hybridization oven for at least 30 min at 65°C.

3. Pour off the solution and repeat the incubation with fresh 15 mL prewarmed hybridization buffer at 65°C.
4. Label an appropriate gel-purified mtDNA-specific probe (*see* **Note 43**).
5. Denature the probe at 95°C for 5 min and chill on ice for 2 min (*see* **Note 44**).
6. Add the radiolabeled probe to the hybridization solution and incubate overnight at 65°C (*see* **Note 45**).
7. Discard the hybridization solution containing the probe and wash the membrane four times with 50 mL WB1 at 65°C for 20 min (*see* **Note 46**).
8. Wash the membrane twice with 50 mL WB2 at 65°C for 20 min (*see* **Note 47**).
9. Remove the membrane from the tube and place on top of paper towel to blot excess liquid. Let dry briefly (*see* **Note 48**).
10. Wrap the membrane in cling film. Expose to X-ray film for 0.5–7 d at –80°C.

4. Notes

1. mtDNA isolation has to be done from fresh tissue because the yield of RIs is much lower from frozen material, and the RIs are frequently degraded.
2. All steps are carried out on ice, with ice-cold solutions, in a cold room.
3. It is recommended to start from 10 to 15 g of liver tissue. More starting material yields more mtDNA; however, because the time required for mtDNA isolation also increases, this tends to result in poorer quality DNA.
4. The solution should be changed repeatedly until there is no significant trace of blood or fat because both interfere with the isolation procedure.
5. Homogenization should be thorough (four to five complete strokes) but not excessive because it will damage mitochondria, exposing the mtDNA to contaminating nucleases. We use an IKA Laboratechnik RW 20 motorized homogenizer set at speed 5.
6. Depending on the amount of starting material, 30-mL tubes or 100- to 250-mL bottles can be used.
7. Suspension is achieved by gently swirling the buffer in the tube/bottle on ice or with the aid of a glass homogenizer. It is important to suspend the pellet completely before each centrifugation step.
8. For the sucrose step gradient, take a Beckmann polypropylene tube 25 × 89 mm tube and add 17.5 mL 1.5*M* sucrose. Then overlay, very slowly at first, an equal volume of 1*M* sucrose, taking care not to disturb the interface. To save time, sucrose gradients can be prepared during the earlier centrifugation steps.
9. Carefully remove most of the gradient above the mitochondrial layer. Then, transfer the mitochondria in the minimum volume to a clean tube, using a P1000 tip with the end removed to avoid damaging the mitochondria. This step should be carried out quickly, yet carefully. The final sucrose concentration of the solution is 250 m*M* (isotonic).
10. PK is highly active across the range of 37–50°C; it is nevertheless able to digest protein at 4°C, albeit more slowly.
11. We prefer to incubate at 50°C rather than 37°C after mitochondrial lysis to limit the action of contaminating nucleases before they are digested by PK. This

two-step PK treatment is preferred to the one in which both PK and sarkosyl are added together as it better preserves RIs.

12. After PCIA extraction, a white or cloudy interface indicates an excess of protein. Whenever this is the case, PCIA extraction of the aqueous (upper) phase should be repeated until the interface is clear.
13. Because LB contains 75 m*M* NaCl, there is no need to add extra salt to precipitate DNA.
14. Samples can be stored at this stage in the freezer for a long period of time. Indeed, unless they are to be used immediately, it is better to preserve DNA samples in this way.
15. Many DNA isolation procedures include an RNase treatment; it is imperative to avoid this when extracting mitochondrial RIs as they are highly sensitive to RNase (as shown in **ref. *16***). As RNA is preserved, UV spectrometry will measure total nucleic acid (DNA and RNA) present in the sample.
16. Wherever possible, avoid restriction endonucleases with an optimal temperature greater than 37°C (50–65°C) as this may lead to strand separation of RIs. If it is particularly advantageous to use such an enzyme, then try incubating at 37°C. For example, 50°C is recommended for *Bcl*I; however, it has at least 50% activity at 37°C and is quite capable of giving complete digestion at the lower temperature, although it may be necessary to increase slightly the units of enzyme or the incubation time. Alternatively, try an intermediate temperature of 45 or 48°C.
17. Digestion with restriction endonucleases in small volumes can result in partial digestion. To avoid this problem, perform digestions in a 200- to 400-μL reaction volume.
18. In the case of modifying enzyme treatments after digestion, TE should be avoided because EDTA chelates metal cations required by the enzymes.
19. Most commercial modifying enzymes are supplied with 10X buffer. However, this is not always the case: RNase H buffer must be prepared by the user.
20. A rapid decrease in temperature effectively stops the reaction; hence, reaction times are readily reproduced from experiment to experiment. The addition of EDTA to RNase H and S1 reactions will sequester essential divalent cations, but this will have no effect on RNase One. Notwithstanding, results with RNase One are just as reproducible as those with RNase H and S1 nuclease, indicating the efficacy of rapid cooling.
21. The conditions listed here are designed to give optimal resolution of 3- to 4-kb fragments. In the case of fragments larger than 5 kb, first-dimension electrophoresis is in a 0.35% agarose gel at 1.5 V/cm for 20 h, and second-dimension electrophoresis is at 3 V/cm for 18 h in an 0.875% agarose gel. Fragments shorter than 3 kb are separated in the first dimension at 0.9 V/cm for 20 h on slab gels containing between 0.55 and 1.0% (w/v) agarose at room temperature. Second-dimension electrophoresis is 9 h at 260 mA in gels between 1.5 and 2.0% agarose at 4°C.
22. For the first dimension, we routinely prepare 100 mL agarose to fill a 110 × 140 mm tray fitted with a 14-well comb with 4-mm wide teeth. It is recommended to keep a whole set (tray, comb, and tank) ethidium bromide free for the first dimension.

23. Leave two empty wells between samples to avoid possible cross-contamination and to facilitate gel slice excision after first-dimension electrophoresis.
24. Because the first-dimension gel lacks ethidium bromide, the rate of electrophoresis is estimated from the positions of bromophenol blue and xylene cyanol dyes contained in the loading buffer. Inevitably, the ability to predict the position of fragments accurately comes with experience. More than one lane containing size markers can be loaded; one can be excised and stained in TBE containing 500 ng/mL ethidium bromide and visualized under UV light at intervals to confirm the position of fragments of interest.
25. Molten agarose should be prepared toward the end of the first-dimension electrophoresis step and cooled in a 50°C water bath before adding ethidium bromide and casting.
26. For the second-dimension gel, a small tray (110 × 140 mm) able to fit just one of the gel slices can be used. However, larger trays (200 × 240 mm) can accommodate up to four gel slices.
27. Pour some agarose at the back of the gel slice with the aid of a P1000 tip and allow it to solidify. Avoid agarose in front of the gel slice as it may distort the migration of RIs or even prevent them from entering in the second-dimension gel if it is too dry.
28. Agarose should completely cover the gel slices from the first dimension. Usually, 150 mL agarose is enough for a 110 × 140 mm tray, and 400 mL is sufficient for a 200 × 240 mm tray.
29. The tank with TBE to be used for the second dimension should be precooled to 4°C and ethidium bromide added immediately prior to electrophoresis.
30. Because ethidium bromide migrates to the negative electrode, under the electric field, recirculation of buffer ensures that it will be homogeneously distributed in the gel. Lack of recirculation may result in poor resolution in the second dimension because intercalation of ethidium bromide into DNA is critical for separating DNA molecules on the basis of structure.
31. The second-dimension gel contains ethidium bromide, and hence DNA migration can be visualized under UV light.
32. The 1% agarose gels measuring 200 × 240 mm are inverted by hand routinely, without any aid. When dealing with low-percentage agarose gels (e.g., 0.58%), a piece of X-ray film can be used when inverting the gel to prevent breakage.
33. After the treatment with DPB, bromophenol blue and xylene cyanol change color because of low pH: xylene cyanol turns greenish, and bromophenol blue becomes yellow.
34. After this treatment, both bromophenol blue and xylene cyanol should recover their original colors.
35. It is important not to leave the gel longer than recommended in DPB or DNB. In contrast, the gel can be left in NB solution for up to an hour without any discernible effect. Notwithstanding, slightly longer treatments are appropriate for higher percentage agarose gels as the solution requires longer to permeate the gel.
36. Dry blotting is effective; transfer buffer is not required.

37. Do not touch the membrane without gloves. It is always useful to write down in pencil relevant information on a corner of the membrane and to aid orientation.
38. Use a hydrophilic positively charged nylon membrane with high DNA-binding capacity.
39. Air bubbles between the gel and the membrane are removed by rolling a wet Pasteur pipet or glass rod across the surface. The same applies to bubbles between the membrane and the 3MM Whatman paper.
40. The glass plate or tray should be as large as the gel to transmit pressure evenly across the gel. As weights, empty bottles filled with 0.5 L and 1 L of water can be used for 110 × 140 and 200 × 240 mm gels, respectively.
41. The crosslinked membrane can be kept at room temperature indefinitely.
42. It is convenient to prewarm hybridization tubes and solution in the hybridization oven at 65°C.
43. There are a variety of labeling kits available, and no significant difference has been found between them.
44. Removal of unincorporated label is not essential; probes can be used directly after labeling and denaturation.
45. Add the probe to the hybridization solution, never directly onto the membrane. This avoids strong background in the area where the concentrated probe touched the membrane.
46. Probes can be stored at room temperature and reapplied to a different membrane.
47. Check the washing solution after each wash with a Geiger counter. If it remains hot after the second wash in WB2, then continue washing until there are no appreciable counts in the wash solution. Also, check the membrane with the Geiger counter after the full wash cycle: the unit length fragment, of say, 4 kb, should be readily apparent as the hottest area of the membrane, recording 200–1000 cps, whereas areas without DNA should record few (<5 cps), if any, counts.
48. Do not overdry the membrane. Frequently, it is necessary to strip and reprobe membranes; overdrying impedes stripping of probes.

References

1. Bell, L. and Byers, B. (1983) Separation of branched from linear DNA by two-dimensional gel electrophoresis. *Anal. Biochem.* **130,** 527–535.
2. Brewer, B. J. and Fangman, W. L. (1987) The localization of replication origins on ARS plasmids in *S. cerevisiae*. *Cell* **51,** 463–471.
3. Brewer, B. J. and Fangman, W. L. (1988) A replication fork barrier at the 3′ end of yeast ribosomal RNA genes. *Cell* **55,** 637–643.
4. Brewer, B. J. and Fangman, W. L. (1991) Mapping replication origins in yeast chromosomes. *Bioessays* **13,** 317–322.
5. Brewer, B. J. Lockshon, D., and Fangman, W. L. (1992) The arrest of replication forks in the rDNA of yeast occurs independently of transcription. *Cell* **71,** 267–276.
6. Brewer, B. J. and Fangman, W. L. (1994) Initiation preference at a yeast origin of replication. *Proc. Natl. Acad. Sci. USA* **91,** 3418–3422.

7. Vaughn, J. P., Dijkwel, P. A., and Hamlin, J. L. (1990) Replication initiates in a broad zone in the amplified CHO dihydrofolate reductase domain. *Cell* **61,** 1075–1087.
8. Krysan, P. J. and Calos, M. P. (1991) Replication initiates at multiple locations on an autonomously replicating plasmid in human cells. *Mol. Cell Biol.* **11,** 1464–1472.
9. Brun, C., Dijkwel, P. A., Little, R. D., Hamlin, J. L., Schildkraut, C. L., and Huberman, J. A. (1995) Yeast and mammalian replication intermediates migrate similarly in two-dimensional gels. *Chromosoma* **104,** 92–102.
10. van Brabant, A. J., Hunt, S. Y., Fangman, W. L., and Brewer, B. J. (1998) Identifying sites of replication initiation in yeast chromosomes: looking for origins in all the right places. *Electrophoresis* **19,** 1239–1246.
11. Friedman, K. L. and Brewer, B. J. (1995) Analysis of replication intermediates by two-dimensional agarose gel electrophoresis. *Meth. Enzymol.* **262,** 613–627.
12. Holt, I. J., Lorimer, H. E., and Jacobs, H. T. (2000) Coupled leading- and lagging-strand synthesis of mammalian mitochondrial DNA. *Cell* **100,** 515–524.
13. Linskens, M. H. and Huberman, J. A. (1990) Ambiguities in results obtained with 2D gel replicon mapping techniques. *Nucleic Acids Res.* **18,** 647–652.
14. Little, R. D. and Schildkraut, C. L. (1995) Initiation of latent DNA replication in the Epstein-Barr virus genome can occur at sites other than the genetically defined origin. *Mol. Cell Biol.* **15,** 2893–28903.
15. Bowmaker, M., Yang, M. Y., Yasukawa, T., et al. (2003) Mammalian mitochondrial DNA replicates bidirectionally from an initiation zone. *J. Biol. Chem.* **278,** 50,961–50,969.
16. Yang, M. Y., Bowmaker, M., Reyes, A., et al. (2002) Biased incorporation of ribonucleotides on the mitochondrial L-strand accounts for apparent strand-asymmetric DNA replication. *Cell* **111,** 495–505.
17. Belanger, K. G., Mirzayan, C., Kreuzer, H. E., Alberts, B. M., and Kreuzer, K. N. (1996) Two-dimensional gel analysis of rolling circle replication in the presence and absence of bacteriophage T4 primase. *Nucleic Acids Res.* **24,** 2166–2175.
18. Kalejta, R. F. and Hamlin, J. L. (1996) Composite patterns in neutral/neutral two-dimensional gels demonstrate inefficient replication origin usage. *Mol. Cell Biol.* **16,** 4915–4922.
19. Grossman, L. I., Watson, R., and Vinograd, J. (1973) The presence of ribonucleotides in mature closed-circular mitochondrial DNA. *Proc. Natl. Acad. Sci. USA.* **70,** 3339–3343.
20. Nawotka, K. A. and Huberman, J. A. (1988) Two-dimensional gel electrophoretic method for mapping DNA replicons. *Mol. Cell Biol.* **8,** 1408–1413.
21. Gerbi, S. A., and Bielinsky, A. K. (1997) Replication initiation point mapping. *Methods* **13,** 271–280.

III

Biochemistry

17

The Analysis of tRNA Import Into Mammalian Mitochondria

Anne-Marie Mager-Heckel, Nina Entelis, Irina Brandina, Petr Kamenski, Igor A. Krasheninnikov, Robert P. Martin, and Ivan Tarassov

Summary

Ribonucleic acid (RNA) import into mitochondria occurs in a variety of organisms. In mammalian cells, several small RNAs are imported in a natural manner; transfer RNAs (tRNAs) can be imported in an artificial way, following expression of corresponding genes from another organism (yeast) in the nucleus. We describe how to establish and to analyze such import mechanisms in cultured human cells. In detail, we describe (1) the construction of plasmids expressing importable yeast tRNA derivatives in human cells, (2) the procedure of transfection of either immortalized cybrid cell lines or primary patient's fibroblasts and downregulation of tRNA expression directed by small interfering RNA (siRNA) as a way to demonstrate the effect of import in vivo, (3) the methods of mitochondrial RNA isolation from the transfectants, and (4) approaches for quantification of RNA mitochondrial import.

Key Words: Aminoacylation; mitochondrial import; real-time quantification; siRNA downregulation; tRNA.

1. Introduction

Mitochondrial import of small noncoding ribonucleic acids (RNAs) is now considered a quasi-universal pathway. Found in a variety of species (fungi, protozoans, animals, and plants), it differs dramatically from one system to another in a dramatic manner *(1,2)*. Nuclear-encoded transfer RNAs (tRNAs; ranging from one species to the complete set needed for the organellar translation), but also 5S ribosomal RNA, mitochondrial RNA processing enzyme (MRP), or ribonuclease (RNase) P RNA components have been shown to be encoded in nuclear DNA and targeted into the mitochondrial matrix. In animal cells, no tRNA import was found in vivo, although other small RNAs are imported. Nevertheless, we found that tRNA import may be established in human cells in an artificial way: by

From: *Methods in Molecular Biology, vol. 372: Mitochondria: Practical Protocols*
Edited by: D. Leister and J. M. Herrmann © Humana Press Inc., Totowa, NJ

expressing in cultured human cells "importable" versions of yeast tRNAs *(3,4)*. This artificial import permitted importing functionally active tRNAs that could participate in mitochondrial translation and complement, at least partially, defects caused by mutations of mitochondrial tRNA genes encoded in the mitochondrial genome (mitochondrial deoxyribonucleic acid, mtDNA) *(5)*. This approach may potentially be exploited to develop new gene therapy strategies for mtDNA diseases for which, up to now, no efficient therapy has been possible *(6)*.

To establish and characterize quantitatively the artificial tRNA import into human mitochondria in vivo, we optimized several approaches described here. This chapter also describes how to obtain transgenic lines that import tRNAs into mitochondria, how to downregulate artificial tRNA import in vivo, how to isolate mitochondrial RNA (mtRNA), and how to detect and to quantify the import efficiency.

Various versions of all three yeast $tRNAs^{Lys}$ (cytosolic $tRNA^{Lys}_{CUU}$ or tRK1, cytosolic $tRNA^{Lys}_{UUU}$ or tRK2, and mitochondrially encoded $tRNA^{Lys}_{UUU}$ or tRK3) were shown to be internalized by human mitochondria *(3,4)*. From nearly 40 in vitro importable versions *(7)*, only the versions mentioned in this work were tested in vivo. However, taking into account high flexibility of the import pathway, one can predict that the methods described are exploitable in a wider way (i.e., to import other tRNAs into mammalian mitochondria). Constructs for in vivo expression were based either on the pBK-CMV vector or on pcDNA-3.1, both bearing the gene of resistance to G418 (neomycin derivative) (**Fig. 1**). tRNA genes were polymerase chain reaction (PCR) subcloned using total yeast DNA as the template for amplification and mutagenized by standard procedures (*see* **Note 1**).

The transfection procedure depends on the cell line used. Normal immortalized cell lines (like 143B, HepG2, or HeLa cells) are robust, and transfection can be performed with most of commercial transfection reagents. However, cybrid cells seem to be much more fragile and are sensitive to treatment with lipophilic agents. These reagents proved most toxic for primary patients' cells (fibroblasts or myoblasts). To optimize conditions of transfection, one can assess on dilution of the reagent, on the concentration of DNA, on the confluence of cells, on the concentration of the antibiotic, and finally on the exact composition of the growth medium (*see* **Note 2**).

The assessment of transgene expression in human fibroblasts derived from patients with the myoclonic epilepsy with red ragged fibers (MERRF) syndrome has been complicated by two major problems: (1) a high degree of lethality was caused by the transfection procedure, and (2) the effect of transfection on mitochondrial functions was not observed for several days, being significantly delayed. We tried therefore to optimize the procedure by extending the time for transgene expression and to reduce the cytotoxic effect of the transfection

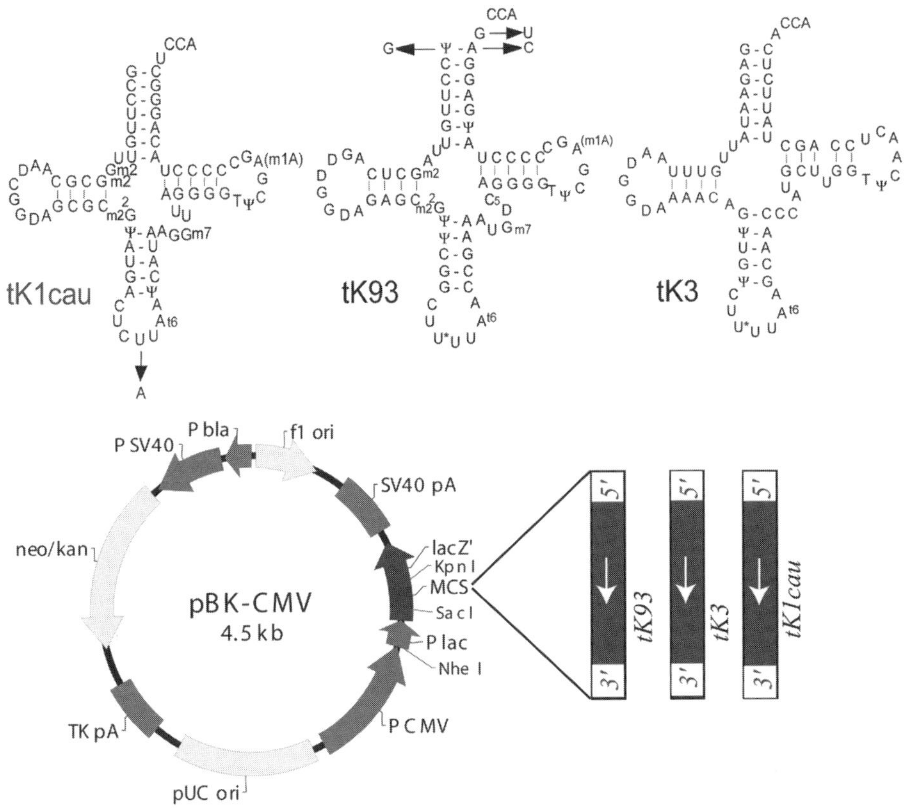

Fig. 1. Schematic drawing of importable tRNAs cloverleaf structures and expressing constructs in pBK-CMV vector. The black arrows indicate base substitutions enabling the import of tK93 version and the mutation in the anticodon of tRK1 (*see* the text for details). (Adapted from **ref. 5**.) The pBK-CMV vector map is adapted from Stratagene. The white arrows indicate the orientation 5′-3′ of the transgenes.

reagents. **Figure 2** shows that the expression of the control luciferase gene was highest 12–24 h after transfection and then decreased because of elimination of the plasmid; the number of cells drops after each transfection procedure because of the cytotoxic effect. These two restrictions necessitated the use of several consecutive transfections before observing the phenotypic effect of transgene expression (*see* **Note 3**).

RNA interference is commonly used to knock down expression of RNA polymerase II-transcribed genes ***(8–10)***. We need to downregulate expression of transgenic tRNAs, which are normally transcribed by RNA polymerase III, to show that the rescued phenotype observed in stable cybrid cell transfectants was caused by the expression of imported tRNAs ***(5)***. We designed several

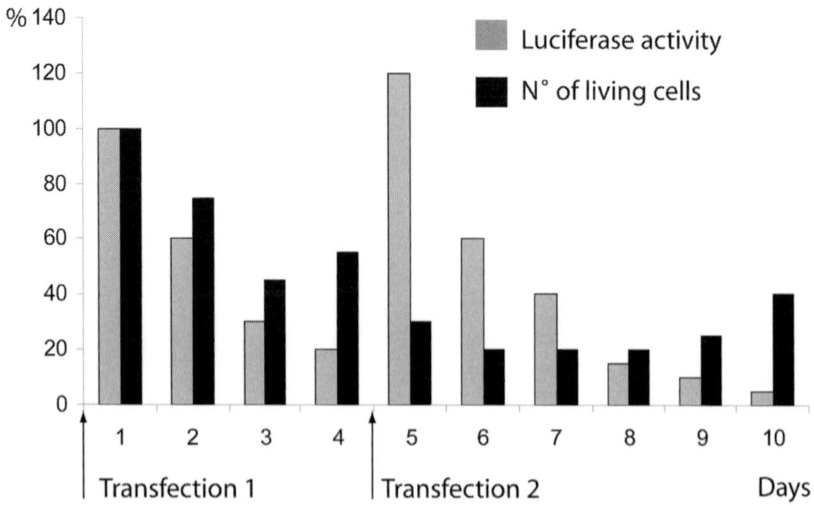

Fig. 2. Assays of transgene expression and cytotoxicity after serial transfection. Here, MERRF cybrid cells (line based on 143B *rho*° line) were transfected twice with the tK3-expressing plasmid, as indicated by the arrows below the graph. The values are presented in percentage of living cells and luciferase (reporter) activity with respect to d 1 after transfection.

RNA duplexes that efficiently inhibited expression of yeast tRNA derivatives in transfected human cells (**Fig. 3**). All these duplexes were 19–21 bases long and, to provide more resistance to nucleases, contained two protruding thymidines at both termini. To follow knockdown experiments, the transfection has to be performed with the reporter gene of luciferase (pGL3 control plasmid) together with the corresponding small interfering RNA (siRNA) duplex.

Mitochondrially imported RNAs are, as a rule, present in low amounts in the cell. MRP RNA and RNase P RNA components seem to be present at 1–2 molecules per mitochondrion ***(11–13)***; imported tRNAs also are underrepresented in the organellar pool of RNAs ***(1,5)***. This means that to detect import and to quantify it one needs to use either extremely sensitive methods or large amounts of mtRNA. We describe two alternative approaches adapted to these possibilities (*see* **Note 4**).

The choice of oligonucleotide probe is crucial for detection of a given RNA. Before beginning large-scale experiments, one needs to optimize oligonucleotide design for each RNA and to test in which conditions the probe reveals a unique band. Then, all mtRNA samples must be checked for cytosolic contamination. It is possible, indeed, to measure this contamination using cytosol-specific probes (*see* **Note 5**) and then to calculate this contamination

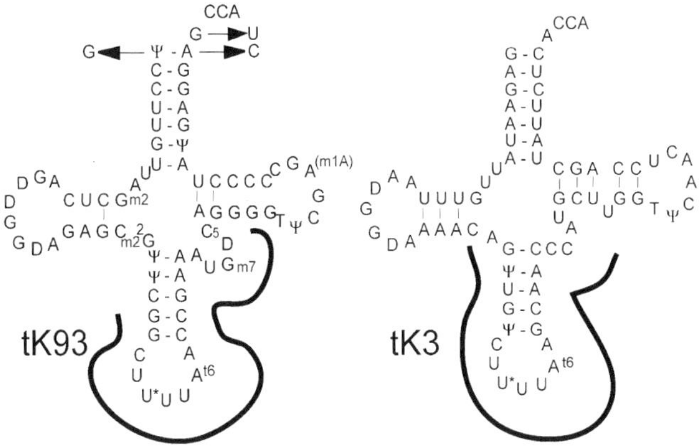

anti-LUC

GGAUUCUAAAACGGAUUACCdTdT
dTdTCCTAAGATTTTGCCTAATGG

anti-tRK93

GGCUUUUAACCGAAAUGUCdTdT
dTdTCCGAAAATTGGCTTTACAG

anti-tRK3

GUUGUCUUUUAAGCAACCCdTdT
dTdTCAACAGAAAATTCGTTGGG

Fig. 3. Design of siRNAs for downregulation assays. Localization of complementary regions on the cloverleaf tRNA structure are indicated in the upper portion (for tK93 and tK3 versions). Sequences of the siRNAs used are indicated at the bottom: anti-LUC is used to control RNA interference by downregulating the reported gene. The presented siRNAs efficiently inhibited (more than 90%) expression of both reporter (luciferase) and tRNA genes in 2–4 d after transfection.

as a percentage of the value obtained for the RNA of interest (**Fig. 4**). However, it is possible that contamination is completely nonspecific, and different RNA species can contaminate mitochondrial preparation in the same way (which is not necessarily true).

The most credible results are obtained when two or more cytosolic RNAs give a negative result; the probe against the imported RNA gives a positive signal (**Fig. 5**). The real-time PCR is an extremely powerful tool to measure absolute amounts of a given nucleic acid in solution. However, the application of this approach to tRNAs raises additional problems that may be resolved using the appropriate controls (*see* **Note 6**).

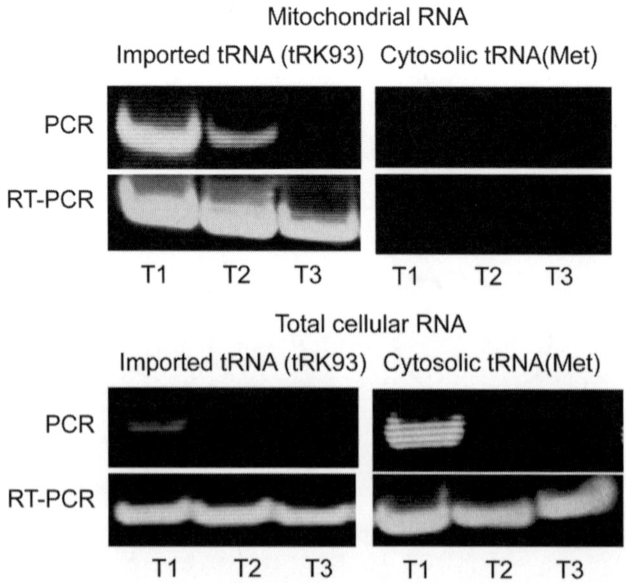

Fig. 4. Serial TRIzol extraction of RNA. Total or mitochondrial RNA isolated from tK93-transfectants was used for PCR and RT-PCR reactions, in parallel. T1, T2, and T3 labels indicate the number of TRIzol extractions. Cytosolic $tRNA^{Met}$-specific primers were used to control the absence of cytosolic contamination in mitochondrial isolate. 10% nondenaturating polyacrylamide gels stained by ethydium bromide are presented.

2. Materials

2.1. Transfection of Human Cultured Cells and siRNA Downregulation Assays

1. pGL3-control vector, luciferase reporter plasmid (Promega), luciferase detection kit (Promega).
2. Transfection reagents SuperFect (Qiagen), LipofectAMINE™ 2000 (Invitrogen) or LyoVec6™ (InvivoGen), OptiMEM Reduced Serum medium for transfection (Invitrogen).
3. Cultured human cells: for these experiences, we used cybrid cells containing the MERRF mutation (A8344G) at greater than 95%; they were based on either HeLa or 143B genetic background.
4. Primary human cells: for the experiments described, we used cultured fibroblasts bearing mtDNA that contained 70% levels of the MERRF mutation.
5. Dulbecco's modified Eagle's medium (DMEM) with 4.5 mg/mL glucose, sodium pyruvate (110 mg/L), and L-glutamine (Sigma).
6. Uridine (Sigma-Aldrich).
7. Eagle's minimum essential medium (EMEM) with nonessential amino acids and 1 m*M* pyruvate (Sigma) and 5 μg/mL uridine (Sigma-Aldrich).

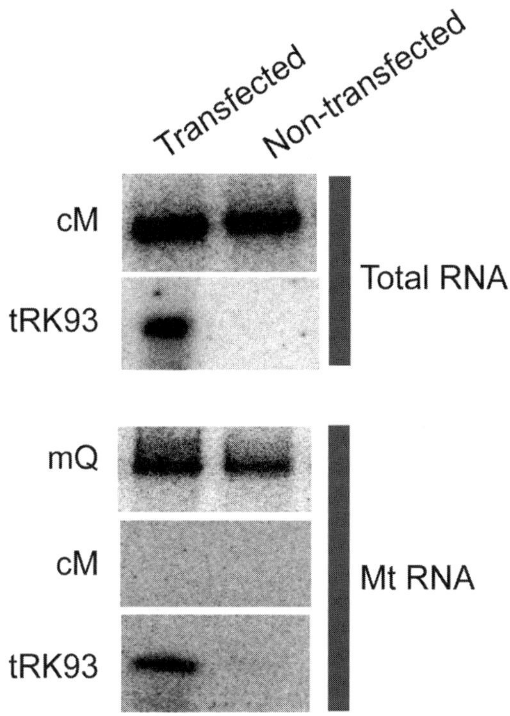

Fig. 5. Typical result of Northern quantitation analysis of import. MERRF cybrid cells and their tK93-stable transfectants were used to isolate mitochondrial RNA (hot phenol protocol). Total and mitochondrial RNA were analyzed for the amount of the transgenic tRNA (tK93) and reference host RNAs: cytosolic $tRNA^{Met}$ (cM) and mitochondrial $tRNA^{Gln}$ (mQ). Equal amounts of cytosolic RNAs (20 µg per slot) and of mitochondrial RNAs (50 µg per slot) were taken for analysis. Autoradiographs of washed membranes are presented.

8. Ham's F14 medium with 6 mg/mL glucose and 1 mg/mL adenosine triphosphate (Vitromex).
9. G418, streptomycin, penicillin, Fungizone (antibiotics) (Sigma or Invitrogen).
10. Synthetic RNA-RNA duplexes 20–21 bases long (siRNAs).

2.2. Isolation of mtRNA for Analysis of Import

1. Mito buffer ± bovine serum albumin (BSA): 0.6 *M* mannitol (or 0.44 *M* sorbitol), 1 m*M* ethylenediaminetetraacetic acid (EDTA), 10 m*M* Na-1, 4-piperazinediethanesulfonic acid (PIPES), pH 6.7, 0.3% (w/v) BSA (to add before use).
2. Bradford protein measurement reactif (Bio-Rad).
3. 2X RNases solution: micrococcale nuclease: 10 U/mL, RNase A: 100 µg/mL, 8 m*M* $MgCl_2$, 2 m*M* $CaCl_2$.

4. RNase stop buffer: Mito buffer without BSA with 2 m*M* EDTA, 2 m*M* ethylene glycol tetraacetic acid (EGTA).
5. Mito-gradient buffer: 0.3 *M* sucrose, 10 m*M* 3-(*N*-morpholino)propanesulfonic acid (MOPS), pH 7.2, 0.1% BSA.
6. TRIzol reagent (Invitrogen).
7. Phenol saturated with water or 0.1 *M* sodium acetate, pH 5.0 (Roth).
8. Diethyl pyrocarbonate (DEPC) (Sigma).

2.3. Quantitative Analysis of Import

1. Standard equipment and reagents for polyacrylamide gel electrophoresis and blotting.
2. 10X Tris borate EDTA (TBE) electrode buffer: 0.89 *M* Tris base, 0.89 *M* boric acid, pH 8.4.
3. Hybond-N membranes (Amersham-Pharmacia).
4. T4-polynucleotide kinase (New England Biolabs).
5. γ-[^{32}P]-Adenosine triphosphate (>4000 Ci/mmol) (Amersham).
6. Equipment for hybridization: rotating oven, ultraviolet (UV) crosslinking chamber.
7. 20X SSC solution: 3 *M* NaCl, 0.3 *M* sodium citrate, 1 m*M* EDTA.
8. 100X Denhardt solution: 2% (w/v) BSA, 2% (w/v) Ficoll, 2% (w/v) polyvinyl-pyrrolidone.
9. Equipment for phosphoimaging and corresponding software (here: Fuji-2000, MacBas).
10. Taq polymerase and its corresponding buffers (reaction buffer, deoxynucleotide 5′-triphosphate solution).
11. One-step reverse transcriptase PCR (RT-PCR) kit (e.g., from Qiagen or Bio-Rad).
12. One-step RT-PCR master mix with SYBR Green (e.g., from Eurogentec or Bio-Rad).
13. Real-time PCR apparatus and corresponding software (here: MyiQ apparatus, Bio-Rad).
14. Synthetic oligonucleotides (hybridization probes corresponding to the studied tRNAs and RT-PCR primers).

3. Methods

The methods described outline procedures of transfection of cultured human cells to establish tRNA import and to downregulate it by siRNAs (**Subheading 3.1.**); methods for isolation of mtRNA from the transfected cells expressing yeast tRNAs (**Subheading 3.2.**); and approaches to quantify RNA mitochondrial import efficiency (**Subheading 3.3.**).

3.1. Transfection of Cultured Human Cells

3.1.1. Establishing Stable Expression and Import of tRNAs in Cybrid Cells

1. Grow the cybrid cells in DMEM or F14 Ham's medium with 20% fetal calf serum (SVF) with penicillin, streptomycin, and Fungizone to the confluence of 30–90%, depending on the line, on Petri dishes for cell culture in a CO_2 incubator (at 37°C, 5% CO_2).
2. At 1 d before transfection, replace the medium with the same medium without antibiotics.

3. Mix 10 μg linearized plasmid DNA expressing the tRNA gene (*Apa*I or any other appropriate restriction enzyme) and 1 μg intact pGL3 in 20 μL of water with 145 μL DMEM without antibiotics and SVF (or, alternatively, OptiMEM medium) and 20 μL SuperFect reagent; incubate 10 min at 20°C.
4. Add 1 mL full DMEM (with antibiotics and serum) and add to cells washed in phosphate-buffered saline (PBS) on Petri dishes; incubate for 12 h in the incubator.
5. Remove the medium with complexes and replace with fresh DMEM medium; incubate for 24 h.
6. Remove the medium and replace with fresh DMEM supplemented by G418 (200–500 μg/mL, depending on your cell line; we routinely used 350 μg/mL for our cybrid MERRF cells); continue growing until clones are visible by eye (0.5–1.0 mm). At this step (36 h after transfection), one can remove a control portion of cells to measure luciferase activity (by standard methods described by the producer of the corresponding kit).
7. Remove the medium and use cloning rings to remove individual clones. To each ring, add 25 μL PBS with 0.1 m*M* EDTA in each tip. Incubate for 5 min at 37°C, save detached cells in a fresh tube, dilute five times with fresh DMEM (with G418), and place in a well of a 24-well plate; continue to incubate.
8. Expand transfected cells for subcellular fractionation. One needs at least one confluent 225-cm^2 flask (10^6 cells) to isolate mitochondria for RNA preparation.

3.1.2. Establishing Transient Expression and Import of tRNA Into Primary Fibroblast Mitochondria

1. Cultivate primary cells in EMEM medium with standard antibiotics until confluence (70–90%) in 225-cm^2 flasks pretreated to increase cell adherence (we suggest using the yellow series of plasticware from Sarstedt, which was optimized for cells with decreased adhesion capacity). One confluent flask may contain 1–3 × 10^6 cells.
2. At 1 d before transfection, replace the medium with the same medium without antibiotics.
3. Mix 50 μg nonlinearized plasmid DNA (expression plasmid) and 5 μg control pGL3 plasmid in 50 μL with 500 μL of OptiMEM medium, incubate for 5 min at 20°C, and combine with 500 μL diluted LipofectAMINE (depending on the cells, the dilution may differ). For MERRF fibroblasts, we used 2 μL LipofectAMINE concentrated solution per 10^6 cells; for control healthy fibroblasts 5 μL, the amount proposed by the manufacturer (Invitrogen), was found too toxic for primary cells. Incubate at 20°C for 10–15 min.
4. Add the DNA-LipofectAMINE mixture to the cells in the flask; gently mix with the medium by rocking flasks back and forth.
5. Incubate at 37°C in a CO_2 incubator for no more than 3 h, remove the medium with DNA-LipofectAMINE complexes, wash with prewarmed PBS, and add a new aliquot of prewarmed fresh medium; continue incubation, changing the medium each 24 h. Each day, take out an aliquot of cells to measure luciferase activity.
6. At d 4, retransfect the cells in the same manner as before, taking into account the decreased number of living cells (with respect to the initially transfected culture; *see* **Fig. 2**).

3.1.3. siRNA Downregulation of Expression and Import of Transgenic tRNAs

1. Cultivate transgenic cybrid cells in F14 Ham's medium containing G418 (or another appropriate antibiotic) to 70–80% of confluence. The day before transfection, replace by the same medium without antibiotic. To be able to perform RNA quantitative analysis and measure respiration or mitochondrial membrane charge, we started with at least four 75-cm^2 flasks for each assay (approx 1–2 × 10^6 cells). When mitochondria are to be isolated, we advise using more cells.
2. For each 75-cm^2 flask, mix 20 µg nonlinearized control pGL3 plasmid, 100 pmol antiluciferase siRNA duplex, and 100 pmol appropriate siRNA duplex (*see* **Fig. 3**) in 50 µL water with 500 µL OptiMEM medium; incubate for 10 min at 20°C and combine with 500 µL diluted LipofectAMINE (for MERRF cybrid transfectants, we used 5 µL LipofectAMINE concentrated solution per 10^6 cells). Incubate at 20°C for 30 min.
3. Add LipofectAMINE-DNA complexes to the cell culture and incubate 24 h in the CO_2 incubator. Remove the medium containing complexes and add a new portion of fresh medium (F14 with G418); continue incubation for another 24 h.
4. At d 4, replace the G418-containing medium with antibiotic-free medium for at least 6 h and repeat the transfection procedure, taking into account the number of living cells, then continue incubation. At regular periods (once every 2 d), take out an aliquot of cells to test luciferase activity.

3.2. Isolation of mtRNA for Analysis of Import

3.2.1. Isolation and Purification of Mitochondria

1. Before starting, refrigerate the Waring blender and the centrifuge.
2. Rinse the cultures in the dishes with 1X PBS.
3. Detach the cells with 1X PBS and 1 m*M* EDTA for 5 min at 37°C.
4. Spin down the cells (10 min at 600*g*).
5. Discard the supernatant and resuspend the pellet in 1X PBS and spin down again.
6. Discard the supernatant, resuspend the cells in 10 mL Mito plus BSA buffer, and keep them on ice.
7. Take 1/10 volume to do a total RNA preparation if needed (1 mL in a 2-mL Eppendorf tube) and transfer the rest to the Waring blender.
8. Break the cells in the blender three times for 10 s at highest speed.
9. Transfer into a 50-mL tube and rinse the blender with 5 mL of Mito plus BSA buffer.
10. Centrifuge for 3 min at 4°C, 1500*g*, and transfer the supernatant into a new tube.
11. Centrifuge the supernatant for 3 min at 4°C and 1500*g*.
12. Centrifuge the supernatant for 20 min or longer at 4°C and 20,000*g*.
13. Discard the supernatant and keep the pellet that contains the mitochondria.
14. Resuspend in 1 mL Mito without BSA buffer, then calculate protein concentration with the Bradford reagent: take 1–5 µL of the mitochondrial suspension, add 50 µL 6% NaOH, add water up to 800 µL total volume, and add 200 µL Bradford reagent (Bio-Rad or another); wait 5 min and measure OD 595 nm (1 OD = 15 µg protein).

15. Add 1 volume 2X RNases and let stand for 5 min at room temperature.
16. Dilute with 2 volumes RNase stop solution to stop the reaction.
17. Centrifuge for 20 min or longer at 4°C and 12,000*g* and resuspend the pellet of mitochondria in 0.2–0.5 mL Mito buffer without BSA.
18. Resuspend pellet of mitochondria in Mito-gradient buffer until a final volume of 0.2–0.5 mL is reached for 5–10 mg mitochondrial protein.
19. Prepare 28% Percoll (sterile) solution in Mito-gradient solution. Put 1–1.2 mL of it in the centrifuge tubes for Beckman 110 TLA ultracentrifuge rotor; keep on ice.
20. Lay solution of mitochondria over the gradient; centrifuge at 30,000*g* for 45 min at 2°C.
21. Take out mitochondria from the gradient: they are visible as the buff-colored band below colorless membranes; avoid taking out the pellet; wash twice with 500 μL Mito buffer.
22. Prepare a concentrated solution of digitonine in Mito buffer without BSA (1–5 mg/mL). Just before use, add 0.2 mg digitonine/mg mitochondrial proteins; let stand for 20 min at room temperature to generate mitoplasts.
23. Dilute with 2–3 volumes of Mito buffer without BSA and centrifuge for 10 min at 4°C and 10,000*g* (in a tabletop microcentrifuge).
24. Wash the pellet of mitoplasts with Mito buffer without BSA and centrifuge for 10 min at 4°C and 10,000*g* (in a tabletop microcentrifuge); rinse and centrifuge again. Freeze 100- to 250-μg aliquots of mitochondrial suspension in Mito buffer in liquid nitrogen, then place at –80°C. Mitoplasts obtained by the procedure described were not contaminated by any visible nuclear small RNAs or cytosolic tRNAs as judged by Northern analysis and RT-PCR (*see* **Note 7**). On average, the treatment described results in 5–10 mg mitochondrial protein from four 225-cm^2 flasks of confluent cells.

3.2.2. Hot Phenol RNA Extraction Protocol

1. Suspend isolated mitoplasts (freshly prepared or frozen in liquid nitrogen and kept at –80°C) in 0.1 *M* sodium acetate, pH 5.0–5.2, and 10 m*M* $MgCl_2$ at 0°C.
2. Add sodium dodecyl sulfate (SDS) to 1%, vigorously mix, and place from ice to 100°C (boiling water or dry incubator) for 2–3 min.
3. Add an equal volume of water-saturated phenol prewarmed at 60°C, mix, and put the mixture at 60°; incubate for 5 min with occasional shaking.
4. Rapidly place the mixtures in ice and incubate for another 5 min at 0°C.
5. Centrifuge for 10 min at 12,500*g* and 0°C and save the upper aqueous phase.
6. Repeat the extraction of the phenol phase with an equal volume of 0.1 *M* sodium acetate, pH 5.0–5.2 and 10 m*M* $MgCl_2$ at room temperature, centrifuge, and combine both aqueous phases.
7. Add 0.1 volume 3 *M* potassium acetate, pH 5.0, and precipitate with 3 volumes of ethanol (2 h at –80°C).
8. To enrich the RNAs with small-size molecules (small ribosomal RNAs, tRNAs), RNA prepared as described is dissolved in DEPC-treated water, 0.1 volume of 3 *M* sodium acetate, pH 5.0, is added, and large RNAs are precipitated by addition

of isopropanol to 20% (10 min at 20°C); after centrifugation, isopropanol is added to the supernatant to reach 60%, and the precipitation step is repeated. This procedure may be repeated twice to eliminate large RNA molecules (**Fig. 6**).

9. RNAs may be stored either under ethanol at –20°C or aliquoted in water at –80°C.

3.2.3. Modified TRIzol Extraction Protocol

1. Detach the cultured cells from plates by treatment with PBS containing 1m*M* EDTA (at 37°C, for 5 min), wash once or twice with PBS, withdraw all liquid, and suspend in TRIzol at a ratio of 1 mL per 10 cm^2 of confluent culture (this ratio must be respected; otherwise, contaminations with DNA can arise).
2. Incubate the homogenate at 30°C for 5 min and add 0.2 mL chloroform per 1 mL TRIzol; mix and incubate at room temperature for 10 min.
3. Centrifuge at 12,000*g* for 10 min at 4°C.
4. Precipitate RNA from the upper phase by adding 0.5 mL isopropanol (*see* **Note 8**) per 1 mL TRIzol used at room temperature (15–20°C) for 10 min, centrifuge, wash the pellet with 80% ethanol (do not use lower concentration of ethanol for washing because it can eliminate small-size RNAs), dry, and dissolve in DEPC-treated water.
5. Add 1 mL of a new portion of TRIzol reagent to 100 μL aqueous solution of RNA; thoroughly mix and repeat all the extraction procedure. This second extraction eliminates traces of DNA in the sample. Sometimes, up to three cycles of extraction are needed to completely remove DNA, which becomes undetectable by PCR (**Fig. 4**; *see* **Note 8**).

3.3. Quantitative Analysis of Import

3.3.1. Quantitative Northern Hybridization

1. Separate RNAs in a 1-mm thick and 20-cm long standard denaturing gel: 13% acrylamide (methylene-bisacrylamide:acrylamide 1:19), 1X TBE buffer, 8 *M* urea at 10 V/cm until the xylene cyanol reaches 3/4 of the gel. One can load up to 50 μg of mtRNA per one 10-mm large well, which may be sufficient to detect under-represented transgenic tRNAs.
2. Soak the gel in the transfer buffer (25 m*M* phosphate, pH 6.5) for 15 min at room temperature.
3. Electrotransfer RNAs onto Hybond-N membrane in a wet transfer camera in the same buffer at 4°C, 200 mA, 10V for 6–12 h.
4. Fix RNAs on the membrane by irradiation in a crosslinking UV chamber for 3 min.
5. Prehybridize the membrane by rotating in a hybridization oven in 6X SSC, 0.1% SDS, 10X Denhardt solution for 1–4 h at 60°C.
6. Discard the prehybridization solution; add the hybridization solution, consisting of 1 volume of prehybridization buffer and 1 volume of 5′-end [^{32}P]-labeled oligonucleotide probe in 1 *M* NaCl (purified before use on a small DEAE-cellulose column). We suggest using at least 10^4 Cpm of labeled probe per 1 cm^2 of membrane.
7. Hybridize overnight at appropriate temperature (depending on the probe, the hybridization temperature normally used is 5°C below the melting point).

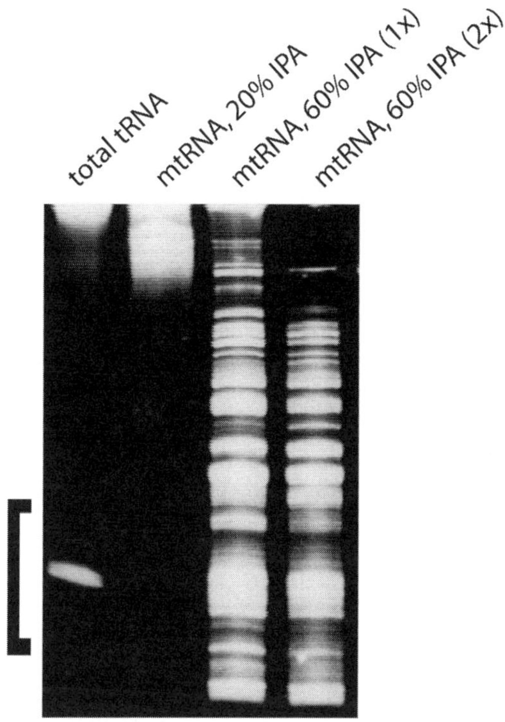

Fig. 6. Isopropanol differential precipitation of small-size mitochondrial RNA. The RNAs were separated on 10-cm 13% denaturating polyacrylamide gel and ethidium bromide stained. Total yeast tRNA (commercial) was used as the reference; the percentage of isopropanol (IPA) used to precipitate RNA is indicated above the gel. The precipitation procedure was done once or twice (as indicated). The bracket indicates location of the transfer RNAs.

8. Remove hybridization solution and wash the membrane three times for 5 min in 2X SSC and 0.1% SDS at the desired temperature (to be optimized for each probe); seal the wet membrane between two thin polyethylene sheets and expose on the phosphoimager.
9. Quantify the signals detecting the individual RNAs and compare ratios between the signal of the RNA of interest and of reference RNAs (cytosolic or mitochondrial) for cellular lines analyzed (**Fig. 5**). Compare also the ratio between mitochondrial and cytosolic reference tRNAs in these lines to be sure that it remains independent of the line. The RNA import efficiency may be calculated as a percentage of the total tRNA species localized in the mitochondria.

3.3.2. Real-Time RT-PCR Analysis

1. Perform preliminary amplification assays by using the Qiagen single-step RT-PCR kit in the following conditions: 50°C, 30 min; 95°C, 15 min; 15 cycles at 95, 55,

72°C; 10 cycles at 95, 58, 72°C; 5 cycles at 95, 62, 72°C (each step for 1 min), final step at 72°C for 10 min. In all cases, include a control PCR reaction performed in similar conditions as RT-PCR by Taq DNA-polymerase (without reverse transcriptase activity) to confirm the absence of DNA contamination in RNA isolates. We advise using, if possible, synthetic T7 transcripts with the same sequence as tRNAs as standards for quantification (*see* **Note 9**).

2. Analyze aliquots on nondenaturing 10% polyacrylamide gel (typical results are presented on **Fig. 4**).
3. Perform quantitative (real-time) RT-PCR using a Bio-Rad i-Cycler with the One-Step RT-qPCR Mastermix for SYBR green following the manufacturer's protocol. For RT-qPCR, conditions used are different from those indicated above: annealing steps are performed at 58°C; the number of cycles is 40–45.
4. For quantitation, serial RNA dilutions are to be done and compared to the calibration curve obtained in parallel reactions with a series of diluted gel-purified T7 transcripts (tK1, tK3, or tK93), ranging from 1 pg to 10 ng per reaction. All qPCR samples have to be done in triplicate. In each series, the corresponding reference (T7 transcript) has to be included, and the same holds for serial dilutions. Blank controls without RNA or oligonucleotides are also to be included in each series.

4. Notes

1. To choose primers, we checked the sequences of the tRNAs available in the Munich Information Center for Protein Sequences database (http://mips.gsf.de/), and the amplicons included the complete tRNA sequence (76 bases). To optimize expression, all three versions of genes cloned (tRK1cau, tRK3, and tRK93) were flanked by short sequences homologous to the flanks of one of the expressed tRK1 copies in yeast *(14)*, although expression of the tRNA genes is normally driven by the internal promoter for RNA polymerase III. The following sequences were used: 5′ACATATTAAACCTGAGAGGTCAGATTTCCAATAACAGAATA (-1) … and … TTCTTTTTTTTTTTAAAACACGATGACATAAATTTCC-3′. The presence of these flanking sequences enhances the expression. tRK1cau version corresponds to the tRK1 with one-base substitution in the anticodon (U35A), which does not inhibit its import but prevents recognition of the lysine codons. tRK93 corresponds to the mutant version of tRK2. tRK2 is normally not imported into yeast or human mitochondria. However, introducing Ψ1G, A72C, and G73U mutations makes this version importable (tRK93). In addition, all seven corresponding gene copies contain a 28-base intron, which was removed by standard PCR-cloning methods *(14)* Yeast tRK3 is normally encoded by mtDNA and resides exclusively in the mitochondria; however, human mitochondria were found to import it as well. This tRNA, because it is mitochondrially encoded, has no need of internal RNA polymerase III promoter, but it possesses the required sequence motifs *(15)* and therefore may be expressed without additional mutations. PCR and cloning procedures were performed in a standard way. To obtain plasmid constructions based on pBK-CMV vector in *Escherichia coli*, we used selection on media containing kanamycin. Using pBK-CMV and pcDNA3.1/Neo(±) vectors gives

similar expression/transfection results. When, for one or another reason, the G418 marker cannot be used (e.g., the line to be transfected is already G418 resistant), one can exploit the vector pcDNA3.1/Zeo(±) (Invitrogen). This vector contains the Zeocine resistance gene, for selection in human cells (to be used at 5–50 µg/mL, depending on the cell line), and the ampicillin resistance gene, to select clones in *E. coli*. When selecting in *E. coli* for kanamycin resistance, it is important to grow cells after the transfection procedure for at least 1.5 h in a rich medium, preferably with low salt (2X YT or SOB).

2. For cybrid lines based on 143B or HeLa rho° cells, we found optimal efficiency with the rich Ham's F14 medium and LipofectAMINE 2000 or LyoVec transfection reagents. However, for those cybrids that are more robust, SuperFect reagent and DMEM medium may be used as well. For primary fibroblasts, most reproducible results were obtained with EMEM medium and LipofectAMINE 2000. In all cases, transfection was effective, but the cytotoxic effect was important. To check for the efficiency of transfection, an internal control is to be included. We used the commercially available pGL3 plasmid bearing the genes coding for luciferase. For the control experiment, an aliquot of transfected cells was always plated in one 10-cm^2 well; 24–48 h after the transfection procedure, the luciferase activity was measured using the protocol for the commercialized luciferase detection kit. With the large number of transfection procedures now available, one is confronted with the choice of deciding the most appropriate method. The most easy to use are lipophilic agents, proposed by most leading manufacturers. However, when abnormal cells are to be transformed, it becomes a more complicated task because of different cytotoxic effects of the same product on two different cell lines. For example, for a similar pair of cybrid lines both bearing the MERRF mutation at 95–100% of heteroplasmy but one based on 143B cells and the second on HeLa cells, we had to use different transfection protocols: for the first one, the best results were obtained with 50% confluent cells and the use of OptiMEM I Reduced Serum Medium (Invitrogen) with DNA:LipofectAMINE ratio proposed by Invitrogen; for the second, with 90% confluent cells, the method was optimal with a LipofectAMINE/DNA ratio three times reduced and normal DMEM medium but supplemented with serum. The best expression results with the third independent cybrid cell line were with LyoVec reagent (used as follows: 8 µg linearized expressing plasmid, 2 µg pGL3 with 600 µL LyoVec reagent for one 6-well plate of 70% confluent cells), and LipofectAMINE did not give any detectable expression. Finally, for several lines, one can obtain very different results in stable and transient expression experiments: for MERRF cybrids, we always obtained best stable transfections with the SuperFect reagent and the best transient expression with LipofectAMINE. Unfortunately, there is no way to predict *a priori* the best way to transfect your cell line, and the optimization procedure is a necessary step before any set of stable or transient expression experiments.
3. The main problem of consecutive transfections when working with primary cells is the amount of material needed to purify mitochondria at the final step. For example, to show that imported tRK3/93 versions lead to an increase

of mitochondrial respiration and membrane charge along with detection of import into mitochondria, we initially used four 225-cm^2 flasks of confluent culture for transfection. With this initial amount and after three rounds of transfection, we obtained the equivalent of the amount in only one flask (10^6 cells), which was roughly the minimum needed to isolate mitochondria and to perform respiration, membrane charge, and RNA import analysis.

4. Both Northern and RT-qPCR methods have advantages and disadvantages. For Northern analysis, one needs a significant amount of pure mtRNA, which is complicated when transient expression experiments are performed. We suggest using it for analysis of stable transfectants, which can be expanded to obtain more cells. In addition, Northern analysis is extremely important when one needs to quantify the aminoacylation level of a given tRNA. In this case, the RNA isolation and separation system is different from that described: all manipulations are performed in acid conditions to avoid deacylation ***(16)***. On the other hand, when quantifying Northern experiments, we mainly obtain relative values, that is, ratios between the tRNA of interest (e.g., the imported one) and another, present either in cytosol or in the mitochondrion. It should be assumed, therefore, that the concentrations of these reference RNAs are similar in different cell lines analyzed, which is not always the case. To make data more representative, one needs to quantify several different reference tRNAs and to compare ratios, which may vary if the balance between different tRNAs in the total pool varies from line to line. Real-time PCR seems to offer a good alternative to Northern analysis because it gives absolute values of RNA concentration, which can be normalized to the number of cells, mitochondrial protein, and total or mtRNA. Second, the amount of RNA needed to perform quantification is at least one order less than that needed for Northern analysis. On the other hand, RT-PCR is extremely sensitive, which is not only an advantage, but also any trace of DNA in the reaction completely negates the result, and isolation of mtRNA needs supplementary efforts to eliminate all DNA contamination. In our experiments, we preferred using hot phenol-extracted mtRNA for Northern experiments and the modified TRIzol extraction protocol to isolate templates for RT-qPCR.
5. In our experiments with transgenic human cells, we used the following: to detect tK3, the oligonucleotide probe antiK3(1–39): CTTAAAAGACAACT-GTTTTACCATTAAACAATATTCTC; for tK93, the probe antiK2(2–32): GCCGAACGCTCTACCAACTCAGCTAACAAGG; for tK1cau, the probe antitK1(met): CTTATGATTATGAGTCAT; for human cytosolic $tRNA^{Met}_i$, the probe anti-cM: TGGTA GCAGAGGATGGTTTCG, for human mitochondrial $tRNA^{Gln}$, anti-mQ: CTAGGACTATGAGAATCG. For RT-PCR detection of tRNAs, we used the following pairs of oligonucleotides: for tK93, CTTGTT-AGCTCAGTTGGT and TGGAGCCTCATAGGGGGC; for tK3, GAGAATATT-GTTTAATGGTAAAAC and GGTGAGAATAGCTGGAGTTG; for tK1cau, GCCTTGTTGGCGCAATCGG and GGAGCCCTGTAGGGGGCTCG; for mitochondrial $tRNA^{Gln}$, TAGGACTATGAGAATCG and AGGATGGGGTGT-GATAG; for cytoplasmic $tRNA^{Met}_i$, GGTAGCAGAGGATGGTTTCG and CAGAGTGGCGCAGCGGAAG.

6. One of the problems of using RT-qPCR for tRNA amplification is the limited amplifiable sequence (the tRNA is 76 nucleotides long, and the oligonucleotide probes are at least 15 nucleotides; the polymerized region is less than 50 nucleotides). The second problem is that, contrary to the case of messenger RNAs, the choice of the region of the tRNA to hybridize with the oligonucleotide probes is restricted and may differ only by several bases; otherwise, there would not be any sequence to amplify. These restrictions suggest thorough preliminary optimization of primers in ordinary RT-PCR assays (like experiments presented on **Fig. 4**). The fact that different pairs of probes may require different conditions of PCR results in the need to perform sequential real-time quantifications for different tRNAs present in the same isolate. The current market proposes a number of reagents for real-time PCR and RT-PCR. We found that, when the preliminary optimization work was performed, the more economical and reproducible results were obtained with the single-step RT-PCR procedure, nonlabeled oligonucleotide probes, and CYBR green labeling.
7. The purification procedure depends on the way that the RNA isolated from the mitochondria will be analyzed. When Northern hybridization was used, the main objective was to eliminate contaminations with cytosolic tRNAs; for real-time RT-PCR, contamination with DNA has to be avoided. In the case of Northern analysis, the presence of small amounts of DNA is not harmful.
8. The TRIzol reagent is sold by Invitrogen and is based on extraction with a monophasic solution of phenol and guanidinisothiocyanate according to the protocol described elsewhere ***(17)***. This method gives excellent results on whole cells, but not as good results (with respect to yield and purity) on mitochondria, probably because of lipid contaminations. Another problem when using the manufacturer's protocol is that it usually gives less RNA than the hot phenol extraction, and often the RNA contains trace amounts of DNA, which becomes an important problem when RT-PCR is used to detect (or to quantify) a given RNA species. As a rule, deoxyribonuclease treatment does not permit full elimination of these contaminants. Proposed modifications aim to avoid this problem. The TRIzol extraction is also compatible with the differential precipitation of RNA by isopropanol. Precipitation of high molecular weight RNAs at 20% isopropanol may be performed at the first extraction step of TRIzol treatment. Small-size RNAs are then precipitated by 60% isopropanol as described in **Subheading 3.2.2.**
9. To synthesize a tRNA transcript in vitro, we suggest amplifying the target gene by PCR, including in the primers the promoter for T7 RNA polymerase upstream and a site for *Bst*NI downstream, which gives, on cleavage, the 3′-terminal CCA sequence. The tRNA transcript is then obtained by T7 transcription in vitro followed by gel purification of the RNA. This approach, used in a number of tRNA studies, is complicated when the first 5′-nucleotide of the tRNA is a U or C, which makes the T7 RNA polymerase nonefficient. For the studies we describe here, this is not a problem because the first nucleotide may be replaced in the primer by the optimal G; however, when the tRNA transcript is studied *per se* and not as a reference for quantification, one can use the alternative approach ***(18)***: to synthesize a longer T7 transcript including a ribozyme, which can, on self-cleavage in the presence of Mg^{2+}, release the desired transcript.

Acknowledgments

We thank R. N. Lightowlers, Z. Chrzanowska-Lightowlers, D. Turnbull, J. Hayashi, and M. Zeviani for providing cell lines and for advice concerning their cultivation and transfection. We also thank O. Kolesnikova for her help with manipulating cybrid cells. This work was supported by the CNRS, ULP, AFM, and EU (MitEURO concerted action), RFFI, and GIS Institut de Maladies Rares.

References

1. Entelis, N. S., Kolesnikova, O. A., Martin, R. P., and Tarassov, I. A. (2001) RNA delivery into mitochondria. *Adv. Drug Deliv. Rev.* **49,** 199–215.
2. Schneider, A., and Marechal-Drouard, L. (2000) Mitochondrial tRNA import: are there distinct mechanisms? *Trends Cell. Biol.* **10,** 509–513.
3. Kolesnikova, O. A., Entelis, N. S., Mireau, H., Fox, T. D., Martin, R. P., and Tarassov, I. A. (2000) Suppression of mutations in mitochondrial DNA by tRNAs imported from the cytoplasm. *Science* **289,** 1931–1933.
4. Entelis, N. S., Kolesnikova, O. A., Dogan, S., Martin, R. P., and Tarassov, I. A. (2001) 5S rRNA and tRNA import into human mitochondria. Comparison of in vitro requirements. *J. Biol. Chem.* **276,** 45,642–45,653.
5. Kolesnikova, O. A., Entelis, N. S., Jacquin-Becker, C., et al. (2004) Nuclear DNA-encoded tRNAs targeted into mitochondria can rescue a mitochondrial DNA mutation associated with the MERRF syndrome in cultured human cells. *Hum. Mol. Genet.* **13,** 2519–2534.
6. Smith, P. M., Ross, G. F., Taylor, R. W., Turnbull, D. M., and Lightowlers, R. N. (2004) Strategies for treating disorders of the mitochondrial genome. *Biochim. Biophys. Acta* **1659,** 232–239.
7. Kolesnikova, O., Entelis, N., Kazakova, H., Brandina, I., Martin, R. P., and Tarassov, I. (2002) Targeting of tRNA into yeast and human mitochondria : the role of anticodon nucleotides. *Mitochondrion* **2,** 95–107.
8. Arenz, C., and Schepers, U. (2003) RNA interference: from an ancient mechanism to a state of the art therapeutic application? *Naturwissenschaften* **90,** 345–359.
9. Lavery, K. S., and King, T. H. (2003) Antisense and RNAi: powerful tools in drug target discovery and validation. *Curr. Opin. Drug Discov. Dev.* **6,** 561–569.
10. Matzke, M., and Matzke, A.J. (2003) RNAi extends its reach. *Science* **301,** 1060–1061.
11. Kiss, T., and Filipowicz, W. (1992) Evidence against a mitochondrial location of the 7-2/MRP RNA in mammalian cells. *Cell* **70,** 11–16.
12. Topper, J. N., Bennett, J. L., and Clayton, D. A. (1992) A role for RNase MRP in mitochondrial RNA processing. *Cell* **70,** 16–20.
13. Puranam, R. S., and Attardi, G. (2001) The RNase P associated with HeLa cell mitochondria contains an essential RNA component identical in sequence to that of the nuclear RNase P. *Mol. Cell Biol.* **21,** 548–561.
14. Entelis, N. S., Kieffer, S., Kolesnikova, O. A., Martin, R. P., and Tarassov, I. A. (1998) Structural requirements of tRNALys for its import into yeast mitochondria. *Proc. Natl. Acad. Sci. USA* **95,** 2838–2843.

15. Huang, Y., and Maraia, R. J. (2001) Comparison of the RNA polymerase III transcription machinery in *Schizosaccharomyces pombe*, *Saccharomyces cerevisiae* and human. *Nucleic. Acids Res.* **29,** 2675–2690.
16. Varshney, U., Lee, C. P., and RajBhandary, U. L. (1991) Direct analysis of aminoacylation levels of tRNAs in vivo. Application to studying recognition of *Escherichia coli* initiator tRNA mutants by glutaminyl-tRNA synthetase. *J. Biol. Chem.* **266,** 24,712–24,718.
17. Chomczynski, P. (1993) A reagent for the single-step simultaneous isolation of RNA, DNA and proteins from cell and tissue samples. *Biotechniques* **15,** 532–534, 536–537.
18. Fechter, P., Rudinger, J., Giege, R., and Theobald-Dietrich, A. (1998) Ribozyme processed tRNA transcripts with unfriendly internal promoter for T7 RNA polymerase: production and activity. *FEBS Lett.* **436,** 99–103.

18

Analysis of Mitochondrial Protein Synthesis in Yeast

Soledad Funes and Johannes M. Herrmann

Summary

Mitochondrial biogenesis is an intricate process that requires the coordinated function of two separate genetic systems: one in the organelle and one in the nucleus. The study of mitochondria requires the analysis of both genetic systems and their protein products. We describe the general procedures used to label mitochondrially encoded proteins in the baker's yeast *Saccharomyces cerevisiae,* a starting point for the investigation of various aspects of organelle biogenesis, such as folding and assembly, sorting, and degradation of proteins.

Key Words: [^{35}S]-Methionine; mitochondrial DNA; OXPHOS; ribosomes; SDS-PAGE; translation.

1. Introduction

Mitochondria evolved from bacteria that established an intracellular symbiosis with ancestral cells. Over evolutionary time, most of the genes contributed by the endosymbiont were lost, presumably to eliminate redundant processes. Most of the residual genes were transferred from the organellar genome to the nucleus. As a consequence, present-day mitochondria are composed essentially of components encoded in the nucleus. However, mitochondria still retain small genomes encoding mainly or exclusively components of the respiratory chain as well as the ribosomal ribonucleic acids (rRNAs) and transfer ribonucleic acids (tRNAs) required for their translation *(1,2)*. The biogenesis of mitochondria requires the coordinated action of both nuclear and mitochondrial genomes.

The size, structure, and gene content of mitochondrial genomes can differ considerably between different organisms. The majority of the mitochondrial genomes characterized so far are circular molecules; however, there are also many linear genomes. The size of mitochondrial genomes can vary from around 6 kb in parasites like *Plasmodium falciparum* to 2400 kb in some

From: *Methods in Molecular Biology, vol. 372: Mitochondria: Practical Protocols*
Edited by: D. Leister and J. M. Herrmann © Humana Press Inc., Totowa, NJ

species of cucurbit plants. These enormous differences in size are mainly because of large variability in the extent of noncoding sequences. Despite this large size diversity, the gene content is essentially conserved: in most cases, the mitochondrial genomes code for some core components of complexes I, III, IV, and V of the respiratory chain and for some components of the mitochondrial translation machinery like rRNAs and tRNAs.

The mitochondrial genome of *Saccharomyces cerevisiae* has a size of 85,779 bp ***(3,4)***. It contains the genes for 8 proteins, 2 rRNAs, 24 tRNAs, and, depending on the genetic background, around 10 endonucleases, reverse transcriptases, and messenger RNA maturases. The encoded proteins are cytochrome-*b* of the bc_1 complex; the subunits Cox1, Cox2, and Cox3 of the cytochrome oxidase; the subunits Atp6, Atp8, and Atp9 of the F_0 sector of the adenosine triphosphate (ATP) synthase; and Var1, a protein of the small ribosomal subunit.

Radioactive labeling of the mitochondrial translation products is a powerful tool for analyzing mitochondrial protein synthesis *per se* or studying the membrane insertion, folding, assembly, or degradation of mitochondrially encoded proteins. Depending on the question to be addressed, translation reactions in complete cells (in vivo) or in isolated mitochondria (*in organello*) can be used ***(5–8)***. In general, in vivo reactions are used when the presence of an active cellular cytosol and nuclear-encoded proteins is needed. Moreover, in vivo labeling can be easily performed with a large set of strains as it does not require the time-consuming isolation of mitochondria. On the other hand, the *in organello* labeling allows the subfractionation of mitochondria after the labeling to monitor the intramitochondrial location of the synthesized translation products. In this chapter, we describe the basic protocols for both techniques. The synthesized proteins are typically visualized by autoradiography after sodium dodecyl sulfate-polyacrylamide gel electrophoresis (SDS-PAGE), for which a standard protocol is also provided here.

2. Materials

2.1. In Vivo Labeling

1. Yeast synthetic media: 6.7 g/L yeast nitrogen base (with ammonium sulfate, without amino acids), 20 mg/L adenine, 20 mg/L uracil, 20 mg/L tryptophan, 20 mg/L histidine, 100 mg/L leucine, 30 mg/L lysine. As carbon source, add glucose, galactose, or lactic acid to a final concentration of 2% (w/v) (*see* **Note 1**).
2. 7.5 mg/mL Cycloheximide stock solution: dissolve 7.5 mg cycloheximide in 1 mL water. Prepare this stock fresh right before use.
3. 200 m*M* Cold methionine stock solution: dissolve 0.298 g of methionine in 10 mL water. Divide the solution in 500-µL aliquots and store them at –20°C.
4. [^{35}S]-Methionine, 10 mCi/mL, approx 1000 Ci/mmol.
5. 1 mg/mL Puromycin stock solution: dissolve 1 mg puromycin in 1 mL water. Store the solution at –20°C.

6. Lysis buffer: 1.8 *M* NaOH, 1.0 *M* β-mercaptoethanol, 1 m*M* phenylmethylsulfonyl fluoride.
7. Trichloroacetic acid (TCA) stock solution: dissolve 72 g TCA in 100 mL water.
8. Cold acetone: before starting the experiment, prechill acetone at –20°C.
9. Sample buffer: 2% (w/v) SDS, 5% (v/v) β-mercaptoethanol, 10% (v/v) glycerol, 0.02% (w/v) bromophenol blue, 60 m*M* Tris-HCl, pH 6.8.

2.2. In Organello Labeling

1. SH buffer: 0.6 *M* sorbitol, 20 m*M* HEPES-KOH, pH 7.4.
2. SEH buffer: 0.6 *M* sorbitol, 20 m*M* HEPES-KOH, pH 7.4, 1 m*M* ethylenediaminetetraacetic acid.
3. 1.5X *in organello* translation buffer: 375 μL 2.4*M* sorbitol, 225 μL 1*M* KCl, 22.5 μL 1 *M* potassium phosphate buffer, pH 7.2, 19 μL 1*M* magnesium sulfate, 45 μL 100 mg/mL bovine serum albumin (essentially fatty acid free; *see* **Note 2**), 30 μL 200 m*M* ATP, pH 7.0, 15 μL 50 m*M* guanosine triphosphate, 9.1 μL amino acid stock solution (2 mg/mL of all proteinogenic amino acids except tyrosine, cysteine, and methionine), 10 μL 10 m*M* cysteine, 18.2 μL 1 mg/mL tyrosine, 1.7 mg α-ketoglutarate, and 3.5 mg phosphoenol pyruvate. Add distilled water to 1 mL.
4. 0.5 mg/mL Pyruvate kinase.

2.3. Sodium Dodecyl Sulfate-Polyacrylamide Gel Electrophoresis

2.3.1. Gel Casting

1. Acrylamide stock solution: 30% (w/v) acrylamide, 0.2% (w/v) bisacrylamide.
2. Tris buffer, pH 8.8: 1.875 *M* Tris-HCl, pH 8.8.
3. Tris buffer, pH 6.8: 0.6 *M* Tris-HCl, pH 6.8.
4. 10% (w/v) SDS stock solution.
5. 10% (w/v) Ammonium persulfate (APS) solution. Make fresh before use.
6. *N,N,N′,N′*-Tetramethylethylenediamine (TEMED).

2.3.2. Gel Running

For gel running, the running buffer is 10 g/L SDS, 30 g/L Tris base, 144 g/L glycine. Do not adjust pH.

2.3.3. Coomassie Blue Staining

1. Coomassie staining solution: 0.1% (w/v) Coomassie brilliant blue, 50% (v/v) methanol, 10% (v/v) acetic acid.
2. Destaining solution: 50% (v/v) methanol, 10% (v/v) acetic acid.

2.3.4. Semidry Protein Transfer

1. Blotting buffer: 0.02% (w/v) SDS, 20 m*M* Tris, 150 m*M* glycine, 20% (v/v) methanol. Do not adjust pH.
2. Ponceau S red staining solution: 0.2% (w/v) Ponceau S, 3% (w/v) TCA.

3. Methods

3.1. *In Vivo Labeling*

1. Make an overnight culture of the desired strains in synthetic media supplemented with the appropriate carbon sources and markers.
2. Take cells equivalent to 1 mL of an OD_{600} of 1.
3. Collect cells by centrifugation for 5 min at 20,000*g*.
4. Resuspend the cell pellet in 1 mL of the same synthetic media (to obtain a cell suspension of 1 OD/mL).
5. Incubate the cell suspension for 10 min at 30°C.
6. Add 50 μL cycloheximide solution.
7. Incubate the cells for 5 min at 30°C.
8. Add 5 μL [^{35}S]-methionine to each sample.
9. Incubate 30 min under constant agitation at around 800 rpm.
10. Terminate the labeling reaction by the addition of 10 μL cold methionine stock solution and 30 μL puromycin stock solution.
11. Centrifuge the samples 10 min at 20,000*g*. Discard the supernatant (*see* **Note 3**).
12. Resuspend the pellet in 500 μL water.
13. Centrifuge 10 min at 20,000*g*. Discard the supernatant.
14. Repeat this wash step once and resuspend the pellet in 500 μL water.
15. Add 75 μL lysis buffer.
16. Incubate the samples 10 min on ice.
17. Add 100 μL TCA stock solution.
18. Incubate the samples for 30 min at –20°C.
19. Centrifuge the samples for 30 min at 33,000*g* at 4°C. Discard the supernatant.
20. Wash the pellet with 1 mL cold acetone.
21. Centrifuge the samples for 30 min at 26,000*g* at 4°C. Discard the supernatant.
22. Let the pellet dry at room temperature.
23. Resuspend the pellet in 20 μL sample buffer (*see* **Note 4**) and subject solution to SDS-PAGE (*see* **Subheading 3.3.**).

3.2. In Organello *Labeling*

1. Mix 20 μL freshly prepared 1.5X *in organello* translation buffer with 0.5 μL pyruvate kinase and 6.5 μL water.
2. Add 20 μg isolated mitochondria resuspended in 2 μL of an iso-osmotic buffer such as SH or SEH (*see* **Note 5**). For isolation of mitochondria from yeast cells, *see* Chapter 6.
3. Incubate the reaction for 5 min at 30°C.
4. Add 1 μL [^{35}S]-methionine and incubate for 20 min at 30°C (*see* **Note 6**).
5. Stop the labeling by addition of 10 μL cold methionine or 30 μL of the puromycin stock solution (*see* **Note 7**).
6. Incubate the reaction for 5 min at 30°C.
7. Collect the mitochondria by centrifugation for 10 min at 20,000*g* at 4°C. Discard the supernatant. Wash the pellet in 500 μL SH or SEH buffer.

Table 1
Composition of SDS-PAGE Gels

	Base gel	Separation gel				Stacking gel
	20%	14%	16%	17.5%	19%	5%
30% acrylamide stock solution	6.7 mL	7.9 mL	9.0 mL	9.5 mL	10.5 mL	830 μL
1.875*M* Tris-HCl, pH 8.8	2.0 mL	3.5 mL	3.5 mL	3.5 mL	3.5 mL	—
0.6*M* Tris-HCl, pH 6.8	—	—	—	—	—	500 μL
10% (w/v) SDS	100 μL	167 μL	167 μL	167 μL	167 μL	50 μL
Water	1.1 mL	5.3 mL	4.2 mL	3.5 mL	2.3 mL	3.6 mL
10% (w/v) APS	50 μL	100 μL	100 μL	100 μL	100 μL	25 μL
TEMED	25 μL	10 μL	10 μL	10 μL	10 μL	5 μL

8. Collect the mitochondria by centrifugation for 10 min at 20,000*g* at 4°C. Discard the supernatant. Resuspend the pellet in 20 μL sample buffer (*see* **Note 4**) and subject it to SDS-PAGE (*see* **Subheading 3.3.**).

3.3. Sodium Dodecyl Sulfate-Polyacrylamide Gel Electrophoresis

3.3.1. Gel Casting

Gels are cast between a pair of glass plates separated by two plastic spacers. We normally use gels consisting of a separation gel of 90 × 150 × 1 mm and a stacking gel of 10 × 150 × 1 mm. In principle, different sizes and gels systems can be used, but the running behavior and the separation of the different translation products might strongly vary from gel system to gel system.

1. Wash the two glass plates, the spacers, and the comb carefully with water and ethanol.
2. Place the spacers between both glass plates. Fix them tightly with metal or plastic clamps and place them into a gel-pouring tray.
3. Prepare the three different gel mixtures. The acrylamide/bisacrylamide concentration of the separation gel might be varied to influence the separation of the individual migration speeds of the various translation products on the gel (*see* **Table 1**). APS and TEMED are added to each solution right before it is poured between the glass plates.
4. Pour the base gel about 1.5–2 cm high into the gel tray so that it completely covers the bottom of the glass plates.
5. After the base gel has polymerized, pour the separation gel mixture between the two glass plates until approx 0.5 cm below the lower end of the comb. Add 1 mL isopropanol on top of the acrylamide mixture.

6. When the separation gel has polymerized, wash off the isopropanol with water and dry the gel carefully. Pour the stacking gel and insert the comb carefully. Wait until the polymerization is complete. Gels can be used immediately or stored for several days at 4°C inside a plastic bag covered with paper towels soaked in water.

3.3.2. Gel Running

1. Load the samples on the gel.
2. Electrophoresis is performed in a vertical chamber at constant 25 mA until the blue bromophenol blue front reaches the bottom gel. After electrophoresis, either the gel can be stained or the proteins can be transferred to a nitrocellulose membrane.

3.3.3. Coomassie Blue Staining

1. Disassemble the glass plates carefully and transfer the gel to a tray filled with Coomassie staining solution. Incubate it for 30–60 min at room temperature under gentle agitation.
2. Pour off the Coomassie staining solution, wash the gel with water, and incubate it with destaining solution with gentle agitation until the background is completely destained.
3. Rinse two cellophane papers with water. Place gel on one cellophane sheet and sandwich it with the second piece. Remove all air bubbles between the cellophane sheets and seal the edges of the sheets with plastic frames or clamps. Air dry the gel overnight or use a gel drying unit.

3.3.4. Semidry Protein Transfer

1. Prepare a semidry blot chamber unit. Wet the lower plate with blotting buffer.
2. Prepare four Whatman papers and one nitrocellulose membrane of sizes similar to that of the gel and soak them with blotting buffer. Put two Whatman papers onto the lower plate of the blot chamber. Remove carefully any air bubbles between the plate and the Whatman papers.
3. Place the nitrocellulose membrane onto the Whatman papers and remove air bubbles.
4. Disassemble the glass plates from the gel and transfer the gel to a tray filled with blotting buffer for at least 2 min. Then, place the gel on the nitrocellulose membrane. Cover the gel with two additional pieces of Whatman paper. Remove air bubbles.
5. Wet the upper plate of the blotting chamber. Close the blotting chamber. Place a weight of about 500 g on top of the blotting chamber and apply a constant current of approx 200 mA for 90 min.
6. Disassemble the blotting chamber. Remove the nitrocellulose membrane and stain it with Ponceau S red staining solution for 10 min. Wash the membrane with water to remove background staining. Dry the membrane completely. Drying can be accelerated by putting the membrane under a red lightbulb or by using a hair drier.

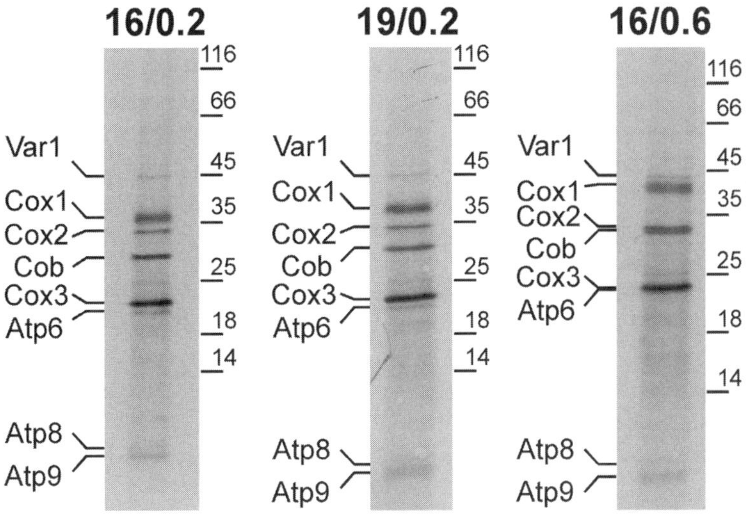

Fig. 1. Mitochondrial translation products after synthesis *in organello*. Translation products were radiolabeled in isolated mitochondria of a yeast wild-type strain (W303) and subjected to SDS-PAGE using different gel composition. The left and middle lanes were prepared with an acrylamide solution containing 30% (w/v) acrylamide and 0.2% (w/v) bisacrylamide as described in **Subheadings 2.** and **3**. The right lane was prepared with a stock solution containing 30% (w/v) acrylamide and 0.6% (w/v) bisacrylamide. The final acrylamide concentrations in the gels were 16, 19, and 16%, respectively. Radiolabeled translation products were detected by autoradiography. The names of the synthesized products and positions of the molecular weight standards are indicated.

3.3.5. Analysis of Mitochondrial Translation Products

By exposure of films or phosphoimager screens onto the dried gels or the nitrocellulose membranes, the radiolabeled mitochondrial translation products can be detected. The running behavior of these hydrophobic proteins can vary considerably depending on the gel systems used. The migration of highly hydrophobic proteins (like those encoded on mitochondrial genomes) can be completely different than predicted. In addition, the pattern observed in the SDS-PAGE can vary considerably when the acrylamide composition of a gel is changed. **Figure 1** shows some examples of how the migration of the translation products of yeast mitochondria varies between different gel systems.

The eight translation products of yeast mitochondria can typically be identified by their specific running behavior and appearance on gels:

Var1 is the only hydrophilic protein encoded by the mitochondrial genome of yeast. It runs as a sharp band of about 45 kDa. In contrast to most of the other translation products, the migration of Var1 is not significantly influenced by the gel composition. However, the size of the Var1 protein varies significantly between different yeast strains ***(9)***. It therefore received its name (protein of variable size 1).

Cox1 is the largest subunit of the cytochrome oxidase. It can be easily identified because it runs as a fuzzy, broad band. The synthesis of this protein depends considerably on the strain, and many petite mutants fail to produce this protein.

Cox2 is produced as a precursor protein with a small amino-terminal extension that is cleaved after the insertion of the protein into the inner membrane. The precursor and the mature forms of the protein run as sharp bands below the Cox1 signal. The maturation normally occurs cotranslationally; therefore, in wild-type cells only the mature form of Cox2 can be seen.

Cytochrome-*b* is the only subunit of the bc_1 complex that is mitochondrially encoded. It is one of the strongest signals observed after the labeling reaction.

Cox3 and Atp6: the third of the core subunits of the cytochrome oxidase and subunit 6 of ATP synthase both migrate very close to each other, and on many gel systems, both proteins cannot be distinguished.

Atp8 and Atp9: these two very small proteins migrate quickly. On our gel system, they run close to the front of the gel.

4. Notes

1. The appropriate carbon source depends on the strain and its ability to grow on nonfermentable conditions. When working with strains that are unable to respire, the use of galactose is preferred in comparison to glucose as glucose repression of many mitochondrial genes leads to rather poor labeling. Marker amino acids or nucleotides might be omitted to select for plasmids.
2. Bovine serum albumin normally helps to stabilize the mitochondria during the labeling. Its presence is, however, not absolutely required. The conditions for the *in organello* labeling procedure were initially established by Poyton and coworkers ***(8)***. In the original publication, the influence of the buffer composition on the labeling reaction was well documented.
3. Be careful when removing the supernatant. The pellets are often not tightly attached to the walls of the tube.
4. Boiling of the samples must be avoided. Incubation of the highly hydrophobic proteins leads to the formation of protein aggregates, which do not enter the SDS gels. After boiling for several minutes, especially the signals of Cox1, Cox3, and cytochrome-*b* disappear. Very short boiling steps (between 30 and 45 s) can, however, be used to inactivate proteases when mitochondria were proteolytically treated after the translation reaction.
5. The concentration of the mitochondria is not very critical and may be adjusted to the specific conditions of the experiment.
6. Under these conditions, the mitochondria translate proteins efficiently for about 30–60 min. At longer incubation times, the translation levels off. The labeling reaction can be performed at different temperatures from about 10–40°C, with an optimum at 25–35°C.

7. If only cold methionine is added, then the translation reaction proceeds without labeling of the products. This reduces the amount of radiolabeled incompleted nascent chains and clears the background between the bands of the full-length translation products.

References

1. Burger, G., Gray, M. W., and Lang, B. F. (2003) Mitochondrial genomes: anything goes. *Trends Genet.* **19,** 709–716.
2. Kurland, C. G. and Andersson, S. G. E. (2000) Origin and evolution of the mitochondrial proteome. *Microbiol. Mol. Biol. Rev.* **64,** 786–820.
3. Foury, F., Roganti, T., Lecrenier, N., and Purnelle, B. (1998) The complete sequence of the mitochondrial genome of *Saccharomyces cerevisiae. FEBS Lett.* **440,** 325–331.
4. Borst, P. and Grivell, L. A. (1978) The mitochondrial genome of yeast. *Cell* **15,** 705–723.
5. Tzagoloff, A., Akai, A., and Needleman, R. B. (1975) Assembly of the mitochondrial membrane system. Characterization of nuclear mutants of *Saccharomyces cerevisiae* with defects in mitochondrial ATPase and respiratory enzymes. *J. Biol. Chem.* **250,** 8228–8235.
6. Groot, G. S., Rouslin, W., and Schatz, G. (1972) Promitochondria of anaerobically grown yeast. VI. Effect of oxygen on promitochondrial protein synthesis. *J. Biol. Chem.* **247,** 1735–1742.
7. Westermann, B., Herrmann, J. M., and Neupert, W. (2001) Analysis of mitochondrial translation products in vivo and *in organello* in yeast. *Methods Cell Biol.* **65,** 429–438.
8. McKee, E. E. and Poyton, R. O. (1984) Mitochondrial gene expression in *Saccharomyces cerevisiae*. I. Optimal conditions for protein synthesis in isolated mitochondria. *J. Biol. Chem.* **259,** 9320–9331.
9. Strausberg, R. L. and Butow, R. A. (1981) Gene conversion at the *VAR*1 locus on yeast mitochondrial DNA. *Proc. Natl. Acad. Sci. USA* **78,** 494–498.

19

Large-Scale Isolation of Mitochondrial Ribosomes From Mammalian Tissues

Linda L. Spremulli

Summary

The preparation of mammalian mitochondrial ribosomes in sufficient quantities for biochemical studies is best done beginning with whole tissue rather than from cells in culture. This issue arises because of the low abundance of these ribosomes in cells, making their isolation a challenge. Crude mitochondrial preparations are made by differential centrifugation and are treated with digitonin to remove the outer membrane. This treatment also effectively removes most contamination by cytoplasmic ribosomes. Purification of mammalian mitochondrial ribosomes requires treatment with detergents to release the ribosomes from their association with the membrane. Sucrose density gradient centrifugation is used to separate the ribosomes from other large oligomeric complexes from this organelle.

Key Words: Elongation; initiation; mammal; protein synthesis; ribosome.

1. Introduction

Mitochondria contain a translational system that is distinct from that of the cell cytoplasm. Mitochondrial ribosomes from different organisms are structurally quite diverse and range in size from 55S to 80S *(1,2)*. Mammalian mitochondrial ribosomes sediment at about 55S in sucrose gradients and consist of 28S and 39S subunits *(3)*. They have molecular masses of about 2.6×10^6 Da, just slightly larger than the size of the *Escherichia coli* ribosome. Animal mitochondrial ribosomes have only two ribosomal ribonucleic acid (RNA) species, 12S in the small subunit and 16S in the large subunit. About 70% of the mass of the mammalian mitochondrial ribosome consists of proteins *(4)*. Proteomic studies have led to the identification of 29 proteins in the small subunit of the bovine mitochondrial ribosome and 48 proteins in the large subunit *(5–11)*. About half

From: *Methods in Molecular Biology, vol. 372: Mitochondria: Practical Protocols*
Edited by: D. Leister and J. M. Herrmann © Humana Press Inc., Totowa, NJ

of these proteins are homologs of bacterial ribosomal proteins; the remainder appear to be unique to mitochondrial ribosomes.

Pioneering work on the large-scale preparation of mammalian mitochondrial ribosomes was initiated in the laboratory of Tom O'Brien ***(12)***. The preparation of these ribosomes is challenging because they are present only in small amounts in most cells. Thus, for biochemical studies, they are generally prepared from tissues that can be obtained in significant amounts. Most studies have been carried out with bovine liver mitochondrial ribosomes because the liver is a reasonably rich source of mitochondria and does not have too much connective tissue. Mitochondrial ribosomes cannot be isolated successfully from whole cell extracts as can bacterial or cytoplasmic ribosomes. This problem leads to the necessity of preparing large amounts of reasonably clean mitochondria as the starting material. To improve the purity of the mitochondria, crude mitochondrial preparations are treated with digitonin, which strips the outer membrane, producing mitoplasts (mitochondria from which the outer membrane has been removed), thus reducing the amount of contamination by cytoplasmic ribosomes.

A significant fraction of mitochondrial ribosomes is associated with the inner membrane ***(13)***. This association is in agreement with the observation that all of the products of mammalian mitochondrial protein biosynthesis are hydrophobic proteins, which are components of the large oligomeric protein complexes involved in oxidative phosphorylation. Treatment of crude mitoplast preparations with a detergent such as Triton is used to release the ribosomes, which are subsequently collected by high-speed centrifugation. Preparation of 55S ribosomes or separated 28S and 39S subunits is completed by sucrose density gradient centrifugation.

2. Materials

2.1. Preparation of Liver Mitochondria (From 4 kg Liver)

1. 0.25 *M* sucrose (18 L): prepare by dissolving 1539 g sucrose in enough deionized water to give a final volume of 18 L. Store at 4°C. This solution should be prepared the day before the liver preparation. Do not store for long periods of time because bacterial contamination can develop.
2. Isolation medium (30 L): This solution is prepared by combining 15 L 0.14 *M* sucrose and 15 L HEM. Prepare the 0.14 *M* sucrose solution by dissolving 719 g commercial grade sucrose in enough water to give a final volume of 15 L. Prepare HEM (4 m*M* HEPES-KOH, pH 7.6, 0.44 *M* mannitol, 2 m*M* ethylenediaminetetraacetic acid, EDTA) by mixing 60 mL 1.0 *M* HEPES-KOH, pH 7.6, 1202 g mannitol, and 60 mL 0.5 *M* EDTA, pH 8.0, and diluting it to a final volume of 15 L. Prepare the isolation medium by thoroughly mixing 15 L 0.14 *M* sucrose with 15 L HEM. Store at 4°C. Do not store for long periods of time because bacterial contamination can develop.

3. Preparation of digitonin: heat isolation medium (615 mL) to just before boiling on a hot plate set up on aluminum foil in the hood with continuous stirring. When the isolation medium is hot but not boiling, add 800 mg of digitonin. Heat the solution with continuous stirring until the solution is clear. Cool the solution slowly in the hood, then transfer it to 4°C, and finally pack it in ice before use (*see* **Note 1**).
4. Large carboys for solution preparation and storage.
5. Knives, sharpeners, and cutting boards.
6. Two chests filled with ice.
7. Strainer.
8. Rubber spatulas.
9. Six 1-L centrifuge bottles.
10. Food processor.
11. Aspirator set up with an 8-L bottle.
12. Waring blending with speed control (Variac controller).
13. Four 4-L beakers.
14. Cheesecloth.
15. Two 2-L graduated cylinders.
16. Tissue homogenizer (described in **Subheading 3.**).
17. Rubber policemen.
18. 20 centrifuge tubes (40 cc).
19. Kilogram weight scale.

2.2. Preparation of Crude Mitochondrial Ribosomes From 20–25 g Mitochondria

1. Preparation buffer 1: prepare as follows for 500 mL: 0.26 *M* sucrose (44.6 g), 40 m*M* KCl (20 mL 1*M*), 15 m*M* $MgCl_2$ (7.5 mL 1 *M*), 15 m*M* Tris-HCl at pH 7.6 (7.5 mL 1*M*), 7 m*M* β-mercaptoethanol (0.25 mL 14 *M*), 0.8 m*M* EDTA, (0.8 mL of 0.5*M* of a pH 8.0 stock), 0.05 m*M* spermine (0.25 mL 0.1 *M*), 0.05 m*M* spermidine (0.25 mL 0.1 *M*), 1.6% Triton X-100. Mix all the reagents except the Triton and adjust the volume to 400 mL. Add 80 mL of a 10% Triton X-100 solution in water and adjust the final volume to 500 mL. The β-mercaptoethanol should be added just before the solution is to be used because this reagent will oxidize readily because of dissolved O_2. This procedure should be used for all buffers containing β-mercaptoethanol.
2. Preparation buffer 2: The quantities are given for the preparation of 150 mL of this buffer, which is the amount required for a routine preparation beginning with 20–25 g mitoplasts: 100 m*M* KCl (15 mL 1*M*), 20 m*M* $MgCl_2$ (3 mL 1 *M*), 20 m*M* Tris-HCl at pH 7.6 (3 mL 1*M*), 7 m*M* β-mercaptoethanol (0.075 mL 14 *M*), 1% Triton (15 mL 10% solution). Adjust the volume to 150 mL and then add 77.3 g sucrose, giving a 34% solution.
3. Preparation buffer 3: the amounts given are for 10 mL of this buffer. Mix 100 m*M* KCl (1 mL 1 *M*), 20 m*M* $MgCl_2$ (0.2 mL 1 *M*), 20 m*M* Tris-HCl at pH 7.6 (0.2 mL 1 *M*), 7 m*M* β-mercaptoethanol (0.005 mL 14 *M*); adjust the volume to 10 mL.

4. 0.1 *M* Phenylmethylsulfonyl fluoride (PMSF) (dissolved in acetone) (0.174 g/10 mL) (*see* **Note 2**).
5. 10 m*M* guanosine diphosphate (GDP) in water: 44.3 mg/10 mL; store frozen at –20°C.

2.3. Preparation of Ribosomes and Ribosomal Subunits

1. Preparation buffer 4: the amounts given are for 10 mL of this buffer; 100 m*M* KCl (1 mL 1 *M*), 20 m*M* $MgCl_2$ (0.2 mL 1 *M*), 20 m*M* Tris-HCl at pH 7.6 (0.2 mL 1 *M*), 7 m*M* β-mercaptoethanol (0.005 mL 14 *M*), 10% glycerol (1 mL). Adjust the volume to 10 mL.
2. Preparation buffer 5: the amounts given are for 10 mL of this buffer; 20 m*M* Tris-HCl at pH 7.6 (0.2 mL 1 *M*), 10 m*M* $MgCl_2$ (0.1 mL 1 *M*), 40 m*M* KCl (0.4 mL 1 *M*), 2 m*M* dithiothreitol (0.01 mL 2 *M*), 10% glycerol (1 mL).
3. 10 m*M* Puromycin: 27 mg in enough water to give a final volume of 5 mL.
4. 55S Gradient buffer: 50 m*M* Tris-HCl at pH 7.6, 100 m*M* KCl, 20 m*M* $MgCl_2$, 2 m*M* dithiothreitol.
5. Subunit gradient buffer: 50 m*M* Tris-HCl at pH 7.6, 100 m*M* KCl, 2 m*M* $MgCl_2$, 2 m*M* dithiothreitol.
6. 10% Sucrose solution: 10 g sucrose plus 90 mL appropriate gradient buffer.
7. 30% Sucrose solution: 30 g sucrose plus 70 mL appropriate gradient buffer.

3. Methods

The preparation of bovine liver mitochondria must be carried out on fresh liver. The quality of the livers, and therefore of the preparations, depends to a significant extent on the individual animal from which the liver is removed. It is important to keep all materials as cold as possible throughout preparation. Once prepared, the mitoplasts can be stored at –70°C prior to the extraction of the ribosomes.

3.1. Preparation of Bovine Liver Mitoplasts

1. Obtain bovine liver from a slaughterhouse located as close as possible to the laboratory. For the best reproducibility, livers are preferably obtained from male animals about 2 years of age. They should be obtained within 30 min of the slaughter of the animal. Place the liver on ice immediately after receipt. Cut the liver into slices (about 5 × 5 × 2 cm) and put into ice-cold 0.25 *M* sucrose on ice. Use about 1 L of 0.25 *M* sucrose for each kilogram of liver. The standard preparation uses 4 kg of liver, requiring 1 or 2 livers. Maintain the liver in the ice-cold sucrose for transport back to the laboratory (*see* **Note 3**).
2. On returning to the laboratory, cut the liver into smaller pieces (about 2 × 2 × 1 cm) using standard kitchen knives, making sure to avoid as much of the connective tissue and blood vessels as possible. As each 1-kg batch of liver pieces is ready, wash it twice with about 1.5 L 0.25 *M* sucrose each time. To change the wash solution, pour the slices and bloody solution through a strainer into a 4-L beaker.

Process each 1-kg batch of liver separately using a team of about four people to keep the process moving efficiently with subsequent batches.

3. After each 1-kg batch is washed, transfer it at 4°C to a standard kitchen food processor fitted with a steel blade, such as a Hamilton Beach food processor. Grind the liver pieces using short bursts of about 5 s at a low-to-intermediate setting on the processor. Continue this process for each 1-kg batch until the material has a consistency somewhere between hamburger and paté. With the aid of a rubber spatula, scrape the liver out of the processor into a 2-L beaker. Add a total of about 1500 mL isolation medium to each 1-kg sample and keep the slurry stirred and well chilled.
4. Homogenize the finely ground liver in isolation medium in a large-scale, continuous flow tissue homogenizer ***(14)*** (*see* **Note 4**). Adjust the flow rate to about 60 mL/min. Filter the homogenate through several layers of cheesecloth as it emerges from the homogenizer and collect it into a 6-L Erlenmeyer flask packed in ice. Rinse the container for each 1-kg sample with 500 mL isolation medium and use it to dilute the sample as it is applied to the homogenizer.
5. Transfer the filtered slurry from each 1-kg batch of liver to three bottles (1 L each) and add isolation medium to dilute the samples, filling the bottles. When 2 kg of liver have been processed to this stage, there should be 6 L solution. While the remaining 2 kg liver is homogenized, centrifuge this material at $600g_{max}$ for 15 min at 4°C with a Sorvall H6000A rotor using the RC-3 centrifuge and 1-L swinging buckets. This step pellets unbroken cells, many membrane fragments, and nuclei. A fat layer floating on the surface is often present at this stage. Remove this fat layer using a rubber spatula and wipe the walls of the centrifuge bottles with a tissue to remove any adhering fat. Remove the supernatant from this step by aspiration into a 6-L flask packed in ice. Avoid the gelatinous red-brown layer at the bottom of the centrifuge bottles. Leave some of the liquid on this pellet rather than disturb it. Subject the supernatant containing the mitochondria to centrifugation at $5900g_{max}$ for 24 min at 4°C with a Sorvall H6000A rotor using the RC-3 centrifuge and 1-L swinging buckets. This step pellets the mitochondria along with additional cellular debris.
6. Remove the supernatant from this step by aspiration and discard it. The pellets are generally brown with red blood cells visible on the bottom. Resuspend the pellets in about 800 mL isolation medium for this 2-kg batch. Resuspend the sample using a rubber policeman. Take care not to resuspend red blood cells that are found on the bottom. Filter the resuspended mitochondria through two layers of cheesecloth to remove clumps of red blood cells inadvertently taken. Further resuspend the sample using a 1-L Waring blender at a very low speed, controlling the speed with a Variac (variable autotransformer). Carry out this step for about 30 s to resuspend any clumps in the mitochondrial pellet. Take care not to use a high speed, and little or no vortex should be observed at the top of the sample. Rinse the blender with about 400 mL isolation medium. Divide this material into three 1-L bottles and fill each bottle gently, mixing the crude mitochondrial sample with the isolation medium.

7. When both 2-kg preparations have reached this stage, the material should be in six 1-L bottles in a total of about 6 L isolation medium. From this point, the material from the full 4 kg is processed as a unit. Centrifuge this material at $600g_{max}$ for 15 min at 4°C with a Sorvall H6000A rotor using the RC-3 centrifuge and 1-L swinging buckets. Remove the supernatant from this step by aspiration into a 6-L flask packed in ice. The red pellet should be carefully avoided (*see* **Note 5**).
8. Divide the supernatant into six of the 1-L bottles and centrifuge at $5900g_{max}$ for 24 min at 4°C with a Sorvall H6000A rotor using the RC-3 centrifuge. Remove the supernatant from this centrifugation by aspiration and discard it. Resuspend each mitochondrial pellet as described in **step 6** using about 400 mL isolation medium per pellet.
9. Transfer the resuspended mitochondria into a 4-L beaker on ice and stir gently. The volume at this point should be about 2.4 L (*see* **Note 6**). Add the digitonin solution to the beaker and then fill it to the 4-L mark with isolation medium. Stir this sample very gently for 15 min at 4°C (*see* **Note 7**).
10. Following digitonin treatment, adjust the mitoplast suspension to 6 L with isolation medium and collect the mitoplasts by centrifugation at $5900g_{max}$ for 24 min at 4°C. Remove the supernatant by aspiration and resuspend the mitoplast pellet very gently in 2 L isolation medium as described in **step 6**. Collect the mitoplasts by centrifugation at $5900g_{max}$ for 24 min.
11. Resuspend the mitoplast pellets in a total of 700 mL isolation medium and transfer the sample to 24 centrifuge tubes (40 cc each). Fill the tubes using isolation medium. Collect the mitoplasts by centrifugation at $27{,}000g_{max}$ for 15 min at 4°C. Discard the supernatant and remove any residual liquid as fully as possible using a Pasteur pipet.
12. Fast freeze the pellets in a dry ice 2-propanol bath. Warm the sides of the centrifuge tubes slightly in a room temperature water bath and pop the frozen pellets out of the tubes using a curved metal spatula. Wrap the pellets in aluminum foil and store at –70°C. Around 80–120 g of mitoplasts are generally obtained.

3.2. Preparation of Crude Mitochondrial Ribosomes

1. Remove 20–25 g of frozen mitoplast pellets from the –70°C freezer and hit them with a hammer several times while still frozen in the aluminum foil to break them into smaller pieces. Transfer the pieces to a handheld 50-cc glass homogenizer tube and thaw the sample in about 30 mL preparation buffer 1. Homogenize the sample by hand with about 5 strokes of a Teflon™ pestle. Alternatively, attach the homogenizing pestle to a drill press and carry out the homogenization using the drill at very low speed while running the pestle up and down in the homogenizing tube about five times. Adjust the volume of the sample to 200 mL using preparation buffer 1 and add 0.25 mL 0.1 *M* PMSF and 0.25 mL 10 m*M* GDP. Stir the sample for 15 min at 4°C to complete the lysis of the mitoplasts (*see* **Note 8**).
2. Centrifuge the lysed material for 45 min at $27{,}000g_{max}$ at 4°C. Decant the supernatant carefully from the pellet and discard the pellet. Measure the volume of the supernatant. Adjust the KCl concentration to about 0.30 *M* and the final volume

to 250 mL. To do this step, dissolve solid KCl in the amount of preparation buffer 1 required for the volume adjustment. Remember that preparation buffer 1 already contains 40 m*M* KCl. Add this solution to the supernatant with mixing so that no part of the sample is exposed to a high concentration of KCl other than momentarily. Note that this volume adjustment brings the PMSF to a final concentration of 0.1 m*M* and the GDP concentration to 10 μ*M*.

3. Layer aliquots (26 mL) on 11 mL cushions of preparation buffer 2 in centrifuge tubes. Adjust the volumes based on the available rotors. Centrifuge the sample at 148,000g_{max} at 4°C overnight (generally 14–15 h).
4. Discard the supernatant. The pellets are generally relatively small and yellow-brown. Rinse the pellets carefully in preparation buffer 3 to remove residual Triton X-100. Use a Pasteur pipet to help remove remaining buffer and briefly invert the tubes to drain any residual liquid.
5. Resuspend each pellet in 50 μL preparation buffer 3, giving a total volume of about 400 μL. Use an additional 100 μL preparation buffer 3 to rinse out the centrifuge tubes and combine this wash with the original material.
6. Transfer the sample to an Eppendorf tube and centrifuge for 5 min at 14,000*g*. The consistency of the pellet is variable at this point; if the pellet is loose, then repeat the centrifugation step. Remove the final supernatant carefully using a Pasteur pipet and transfer it to a fresh tube (*see* **Note 9**).
7. Fast freeze the crude ribosome preparation in a dry ice/2-propanol bath and store the sample at –70°C.

3.3. Sucrose Density Gradient Purification of 55S Ribosomes

1. Adjust the volume of the entire crude ribosome sample from 20–25 g mitoplasts to approx 1.5 mL with preparation buffer 3. Add puromycin (10 m*M* stock) to give a final concentration of 0.5 m*M*. Incubate the sample for 15 min at 27°C.
2. Load the entire sample (about 1.6 mL) onto a single linear 36-mL gradient of 10–30% sucrose in 55S gradient buffer (*see* **Note 10**).
3. Centrifuge the gradient in a swinging bucket rotor for 16 h at 87,000g_{max} at 4°C. Fractionate the gradient at a flow rate of 2 mL/min and collect fractions of 1.2 mL (*see* **Note 11**).
4. Fractions containing 55S monosomes are generally located about two-thirds into the gradient. A typical pattern is shown in **Fig. 1**. Pool the fractions containing 55S particles and collect the monosomes by centrifugation at 159,000g_{max} for 6 h at 4°C. Discard the supernatant (*see* **Note 12**). Remove any excess liquid from the pellet using a Pasteur pipet. Resuspend the pellets in a final volume of 200 μL preparation buffer 4. Measure the concentration by making a 1/160 dilution in water and reading the A_{260} (*see* **Note 13**).

3.4. Sucrose Density Gradient Purification of 28S and 39S Subunits

1. The method used for the preparation of 28S and 39S subunits is similar to that used for 55S ribosomes with a few minor changes.

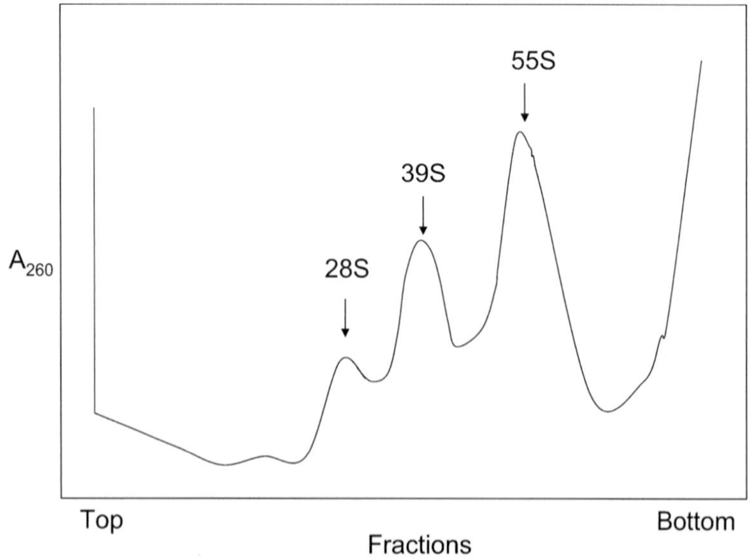

Fig. 1. Sucrose density gradient pattern of bovine mitochondrial ribosomes. Crude ribosomes were centrifuged on a 10–30% sucrose gradient and the A_{260} monitored as 1.2-mL fractions were collected. The gradient was displaced from the bottom using a 50% sucrose solution containing blue dextran to allow easy visualization. The A_{260} was monitored during fractionation using a 0.5-A_{260} scale and a 0.5-cm flow cell. Fractions containing 55S monosomes were pooled, and concentration by ultracentrifugation was performed as described in the text.

   a. Following puromycin treatment, dialyze the sample for 2–3 h in subunit gradient buffer. Apply the dialyzed sample to a 10–30% sucrose gradient prepared in subunit preparation buffer.
   b. Resuspend the subunit pellets in preparation buffer 5.
2. Determine the concentration by making a 1/160 dilution in water and reading the A_{260} (*see* **Note 14**).
3. Subunits prepared by this method have considerable contamination by large complexes, such as the pyruvate dehydrogenase complex. To obtain purer subunit preparations, dissociate 55S ribosomes by dialysis into subunit gradient buffer and then treat them as described in this subheading.

4. Notes

1. Digitonin is toxic. It should be handled with gloves and used in the hood. Do not inhale the vapors or expose the eyes to the vapors.
2. PMSF is toxic and should be handled with care. When preparing, the powder should not be inhaled. Once prepared, it is stored at –20°C.

3. Some animals brought to the slaughter house are healthier than others. It is important to try to get a liver that shows no signs of disease in the animal. Livers to avoid include those with excess fatty tissue and livers with a greenish or light gray hue.
4. The homogenizer should be firmly mounted on a bench model drill press table. The liver homogenate is disrupted by the shearing force exerted on it by a rotating inner cone and the wall of the homogenizer. The homogenization step breaks the cells but leaves subcellular organelles intact as long as a sufficiently high osmotic strength is provided by the buffer solution. The tissue homogenizer was designed many years ago for the large-scale preparation of mitochondria. It was built in-house by the machine shop. Details are described in **ref. *14***. Plans are available on request. It is also possible to disrupt the liver tissue using a commercially available instrument such as the Tekmar SD-45K unit (Tekmar Co., Cincinnati, OH) as described in **ref. *12***.
5. The preparation of the digitonin solution should be started during this centrifugation step.
6. The outer membrane of the mitochondria is removed at this stage by treatment with digitonin. This detergent solubilizes membranes based on their lipid composition. The outer membrane of the mitochondria and contaminating membrane vesicles from the endoplasmic reticulum are solubilized at lower digitonin concentrations than is the inner membrane.
7. This procedure treats the mitochondrial preparation with 800 mg digitonin for 4 kg of liver. This formula works well for general preparations. However, it should be noted that too much digitonin can reduce the yield, and too little digitonin can result in the incomplete removal of the outer membrane ***(15)***. If greater accuracy is required, then the mitochondria from 4 kg liver are resuspended in approx 2 L isolation medium. An aliquot of the sample is removed and mixed with room temperature isolation medium to dilute it as necessary. The absorbance of sample at 550 nm is read within 1 min. The concentration of the mitochondria is calculated using the relationship that 1 A_{550} is equal to 250 μg/mL. The mitochondrial suspension is then adjusted to 25 mg/mL using isolation medium, and digitonin is added to a final concentration of 100 μg/mL.
8. The addition of GDP does not appear to be important for the isolation of active ribosomes. It has been included as a precaution because there is a guanine nucleotide-binding site on the 28S subunit ***(16)***.
9. This sample generally has a yellow tinge and consists of ribosomes contaminated with various soluble and membrane complexes. Particularly abundant are the pyruvate dehydrogenase and glutamate dehydrogenase complexes, which are large enough to sediment during centrifugation.
10. The gradients are prepared using room temperature solutions and then cooled for about 2 h prior to use. This process reduces problems arising from the solubility of air dissolved in the buffers.
11. It is useful to use a flow system that allows the A_{260} of the displaced sample to be monitored throughout the fractionation process. A good system is made by Teledyne Isco. This system uses a 50% sucrose solution applied to the bottom of

the gradient to displace the gradient through an optical flow cell and then directly into a fraction collector. This 50% sucrose solution contains blue dextran to allow its position to be observed during the fractionation process.

12. The pellets are generally small and quite transparent. It is useful to note the orientation of the tube in the rotor to facilitate locating the pellet.
13. The concentration of the 55S ribosomes can be calculated by using the conversion that 12 A_{260}/mL is equal to 1 mg/mL, or that 1 A_{260} is equal to 32 pmol of 55S ribosomes ***(17)***. The yield is somewhat variable and is related to the quality of the original mitoplast preparation. Generally, about 1 mg of 55S ribosomes is obtained from 20–25 g mitoplasts.
14. The concentration of the subunits can be calculated based on the observation that 1 A_{260} is equal to 77 pmol of 28S subunits or 55 pmol of 39S subunits. Yields are generally about 400 pmol of subunits for 20 g of mitoplast starting material.

Acknowledgment

This work was supported in part by National Institutes of Health grant GM32734.

References

1. Pel, H. and Grivell, L. (1994) Protein synthesis in mitochondria. *Mol. Biol. Rep.* **19,** 183–194.
2. Kitakawa, M. and Isono, K. (1991) The mitochondrial ribosome. *Biochimie* **73,** 813–825.
3. O'Brien, T. W., Denslow, N. D., Faunce, W., Anders, J., Liu, J., and O'Brien, B. (1993) Structure and function of mammalian mitochondrial ribosomes, in *The Translational Apparatus: Structure, Function Regulation and Evolution* (Nierhaus, K., Franceschi, F., Subramanian, A., Erdmann, V., and Wittmann-Liebold, B., eds.), Plenum Press, New York, pp. 575–586.
4. Pietromonaco, S., Denslow, N., and O'Brien, T. W. (1991) Proteins of mammalian mitochondrial ribosomes. *Biochimie* **73,** 827–836.
5. Goldschmidt-Reisin, S., Kitakawa, M., Herfurth, E., Wittmann-Liebold, B., Grohmann, L., and Graack, H.-R. (1998) Mammalian mitochondrial ribosomal proteins: N-terminal amino acid sequencing, characterization, and identification of corresponding gene sequences. *J. Biol. Chem.* **273,** 34,828–34,836.
6. Graack, H.-R., Bryant, M., and O'Brien, T. W. (1999) Identification of mammalian mitochondrial ribosomal proteins (MRPs) by N-terminal sequencing of purified bovine MRPs and comparison to data bank sequences: the large subribosomal particle. *Biochem.* **38,** 16,569–16,577.
7. Koc, E. C., Blackburn, K., Burkhart, W., and Spremulli, L. L. (1999) Identification of a mammalian mitochondrial homolog of ribosomal protein S7. *Biochem. Biophys. Res. Commun.* **266,** 141–146.
8. Koc, E. C., Burkhart, W., Blackburn, K., Moseley, A., Koc, H., and Spremulli, L. L. (2000) A proteomics approach to the identification of mammalian mitochondrial small subunit ribosomal proteins. *J. Biol. Chem.* **275,** 32,585–32,591.

9. Koc, E. C., Burkhart, W., Blackburn, K., Moseley, A., and Spremulli, L. L. (2001) The small subunit of the mammalian mitochondrial ribosome: identification of the full complement of ribosomal proteins present. *J. Biol. Chem.* **276,** 19,363–19,374.
10. Koc, E. C., Burkhart, W., Blackburn, K., Koc, H., Moseley, A., and Spremulli, L. L. (2001) Identification of four proteins from the small subunit of the mammalian mitochondrial ribosome using a proteomics approach. *Protein Sci.* **10,** 471–481.
11. Koc, E. C., Burkhart, W., Blackburn, K., Schlatzer D.M., Moseley, A., and Spremulli, L. L. (2001) The large subunit of the mammalian mitochondrial ribosome: analysis of the complement of ribosomal protein present. *J. Biol. Chem.* **276,** 43,958–43,969.
12. O'Brien, T. W. and Denslow, N. D. (1996) Bovine mitochondrial ribosomes, in *Methods in Enzymology: Mitochondrial Biogenesis and Genetics, Part B* (Attardi, G., and Chomyn, A., eds.), Academic Press, San Diego, CA, pp. 237–248.
13. Liu, M. and Spremulli, L. L. (2000) Interaction of mammalian mitochondrial ribosomes with the inner membrane. *J. Biol. Chem.* **275,** 29,400–29,406.
14. Ziegler, D. M. and Pettit, F. H. (1966) Microsomal oxidases. I. The isolation and dialkylarylamine oxygenase activity of pork liver microsomes. *Biochem.* **5,** 2932–2938.
15. Greenawalt, J. W. (1974) The isolation of outer and inner mitochondrial membranes. *Meth. Enzymol.* **31,** 310–323.
16. Denslow, N., Anders, J., and O'Brien, T. W. (1991) Bovine mitochondrial ribosomes possess a high affinity binding site for guanine nucleotides. *J. Biol. Chem.* **266,** 9586–9590.
17. Matthews, D. E., Hessler, R. A., Denslow, N. D., Edwards, J. S., and O'Brien, T. W. (1982) Protein composition of the bovine mitochondrial ribosome. *J. Biol. Chem.* **257,** 8788–8794.

20

Protein Import Into Isolated Mitochondria

Dejana Mokranjac and Walter Neupert

Summary

Import of proteins into mitochondria is a pivotal process in the biogenesis of mitochondria. Only about 1% of the 1000–2000 different proteins constituting the mitochondrion are encoded in the mitochondrial deoxyribonucleic acid (DNA). All others are specified by nuclear genes. They are translated in the cytosol and released from ribosomes as precursor proteins, which are then translocated into the various mitochondrial subcompartments. In the past, a variety of methods has been developed to study the process of import. An important tool is the use of in vitro import systems using isolated mitochondria and precursor proteins synthesized in cell-free systems. Together with the use of genetic and biochemical methods, this led to the identification of several translocation machineries consisting of a large number of components.

Key Words: Cell-free protein synthesis; in vitro import; mitochondria; protein translocation; *Saccharomyces cerevisiae.*

1. Introduction

The majority of the mitochondrial proteins are encoded by nuclear deoxyribonucleic acid (DNA) and must be imported into the organelle following their synthesis in the cytosol as preproteins (for reviews of protein import into mitochondria, *see* **refs. *1–4***). In the past, assays for protein import into the isolated mitochondria were established. In combination with the analysis of mutants of *Saccharomyces cerevisiae* and *Neurospora crassa* affected in import, these in vitro studies of import made possible the dissection of the complex reactions and the identification of components involved.

Preproteins are targeted to mitochondria by virtue of their targeting signals. Mitochondrial targeting signals are diverse but in most cases represent positively charged sequences that have the potential to form amphipathic helices and are

From: *Methods in Molecular Biology, vol. 372: Mitochondria: Practical Protocols*
Edited by: D. Leister and J. M. Herrmann © Humana Press Inc., Totowa, NJ

present at the N terminus of preproteins. Targeting signals are recognized by the components of the translocases present in the outer and inner mitochondrial membranes, Toms and Tims, respectively.

Whichever type of targeting signal a preprotein possesses, it is recognized at the cytosolic surface of the mitochondrial outer membrane by the receptor proteins of the TOM complex. The receptor proteins deliver the precursor to the translocation channel in the outer membrane through which the preprotein crosses the outer membrane. The TOM complex itself is sufficient for import of some outer membrane proteins and for import of some proteins of the intermembrane space. For import of most mitochondrial proteins, the TOM complex cooperates with other mitochondrial translocases. Precursors of β-barrel outer membrane proteins require for their import, integration, and assembly, in addition to the TOM complex, the topogenesis of outer membrane β-barrel proteins (TOB)/sorting and assembly machinery (SAM) complex ***(5,6)***. A recently discovered disulfide relay system in the intermembrane space, in cooperation with the TOM complex, mediates the import of small, metal-coordinating proteins of the intermembrane space ***(6a)***.

Precursors of hydrophobic inner membrane proteins, such as the members of the carrier family, use the TOM complex to cross the outer membrane, soluble small Tim proteins to traverse the intermembrane space, and the TIM22 complex for integration in the inner membrane. The TIM22 complex uses the membrane potential to drive integration of precursor proteins into the membrane. The TIM23 complex is the major translocase of the inner membrane. Using adenosine triphosphate (ATP) and the membrane potential as energy sources, it mediates the translocation across and integration into the inner membrane of preproteins, which contain N-terminally located, positively charged presequences. This type of signal is present in the precursors of the majority of matrix and inner membrane proteins.

Once in the matrix, presequences are usually proteolytically removed by the mitochondrial processing peptidase (MPP). Some inner membrane proteins are conservatively sorted. Their import pathway comprises complete translocation into the matrix via TOM and TIM23 complexes, followed by their export from the matrix into the inner membrane, a process in which the oxidase assembly (OXA) complex is involved.

Elucidation and characterization of these various mitochondrial import and sorting pathways was mainly achieved by in vitro import studies. In such experiments, isolated mitochondria are incubated with in vitro synthesized, ^{35}S-labeled precursor proteins under various conditions, and their import characteristics and requirements are analyzed. Typical in vitro import experiments using isolated yeast mitochondria are described.

2. Materials

2.1. Synthesis of Preproteins

2.1.1. In Vitro Transcription

1. Template DNA: purified plasmid (1 μg/μL) containing the cloned gene for the mitochondrial precursor protein downstream of the Sp6 ribonucleic acid (RNA) polymerase transcription initiation site (*see* **Note 1**).
2. 5X transcription buffer: 200 m*M* Tris-HCl, pH 7.5, 30 m*M* $MgCl_2$, 50 m*M* NaCl, 10 m*M* spermidine. Store in aliquots at –20°C.
3. Dithiothreitol: make a 0.1 *M* solution in water and store in aliquots at –20°C.
4. RNAsin ribonuclease inhibitor (40 U/μL). Store at –20°C.
5. NTPs (2.5 m*M* each): mix 1 μL of each ATP, UTP, CTP, and guanosine triphosphate solution (0.1 *M*, pH 7.5) and add 36 μL water. Store in single-use aliquots at –20°C.
6. $m^7G(5')ppp(5')G$: to make a 2.5 m*M* solution, add 483 μL sterile water to 25 A_{250} units of $m^7G(5')ppp(5')G$ (*see* **Note 2**). Store in aliquots at –20°C.
7. Sp6 RNA polymerase (25 U/μL). Store at –20°C.
8. LiCl: make a 10 *M* solution in water and store in aliquots at –20°C.
9. 70% (v/v) Ethanol, store at –20°C.

2.1.2. In Vitro Translation

1. Commercially available rabbit reticulocyte lysate. Store at –80°C (*see* **Note 3**).
2. Amino acid mixture, minus methionine, 1 m*M*. Store at –20°C.
3. RNAsin ribonuclease inhibitor.
4. Mg-acetate: make a 15 m*M* solution in water and store at –20°C.
5. RNA produced in the in vitro transcription reaction.
6. ^{35}S-Methionine, specific activity 1175 Ci/mmol, concentration 10 mCi/mL. Store in aliquots at –80°C (*see* **Note 4**).
7. Cold methionine: make a 58 m*M* solution in water and store at –20°C.
8. Sucrose: 1.5 *M* solution in water. Store at –20°C.

2.2. In Vitro Import Into Isolated Mitochondria

1. 2X Import buffer: 1 mg/mL fatty acid-free bovine serum albumin (BSA), 1.2 *M* sorbitol, 160 m*M* KCl, 20 m*M* Mg-acetate, 4 m*M* KH_2PO_4, 5 m*M* ethylenediaminetetraacetic acid, 5 m*M* $MnCl_2$, 100 m*M* HEPES-KOH, pH 7.2 (*see* **Note 5**). Store at –20°C.
2. ATP: make a 0.2 *M* solution in water and adjust to pH 7.0 with KOH. This is an 80X stock and is kept in aliquots at –20°C.
3. NADH: 0.2 *M* solution in water. This is a 40X stock and is kept in single-use aliquots at –20°C.
4. CP: 1*M* solution of creatine-phosphate in water. This 100X stock is kept in aliquots at –20°C.
5. CK: 10 mg/mL solution of creatine kinase in water. This is 100X stock and is kept in single-use aliquots at –20°C.

6. Valinomycin: make a 0.1 m*M* solution in ethanol and store at –20°C. This is a 100X stock.
7. Oligomycin: make a 2 m*M* solution in ethanol and store at –20°C. This is a 100X stock.
8. PK: make a 10 mg/mL solution of proteinase K in water and store in single-use aliquots at –20°C.
9. PMSF: make a 0.2*M* solution of phenylmethylsulfonyl fluoride in ethanol and store at 4°C.
10. SH buffer: 0.6*M* sorbitol, 20 m*M* HEPES-KOH, pH 7.2. Store in 50-mL aliquots at –20°C.
11. ^{35}S-Labeled mitochondrial precursor protein prepared in in vitro translation reaction and kept in single-use aliquots at –80°C.
12. Isolated yeast mitochondria at 10 mg/mL kept in single-use aliquots at –80°C (*see* **Note 6**).

3. Methods

Most in vitro import reactions into mitochondria are performed with in vitro synthesized, radiolabeled precursor proteins. They can be obtained relatively easily and detected by autoradiography. It is, however, also possible to use mitochondrial preproteins as recombinantly expressed and purified proteins (*see* **Note 7**). As most mitochondrial precursors are translocated across the outer membrane, their import into mitochondria is easily assessed by their becoming inaccessible to an externally added protease. Furthermore, most preproteins are also translocated across or inserted into the inner membrane and therefore require a membrane potential for their import. N-Terminal targeting signal is usually removed in the matrix to give rise to the mature form of the protein, which can be easily separated from the precursor form by sodium dodecyl sulfate-polyacrylamide gel electrophoresis (SDS-PAGE) (*see* **Fig. 1**).

3.1. Synthesis of Preproteins

3.1.1. In Vitro Transcription

1. To make a 100 µL transcription reaction, mix 20 µL 5X transcription buffer, 10 µL dithiothreitol, 4 µL RNAsin, 20 µL NTPs, 5.2 µL m^{7}G, 3 µL Sp6 RNA polymerase, 27.8 µL water, and 10 µL DNA template (*see* **Note 8**).
2. Incubate for 1 h at 37°C.
3. Add 10 µL LiCl and 300 µL ethanol to pellet RNA.
4. Incubate at least 30 min at –20°C.
5. Centrifuge in tabletop centrifuge at 37,000*g* at 4°C for 20 min.
6. Wash RNA pellet with 200 µL ice-cold 70% ethanol.
7. Centrifuge in tabletop centrifuge at 37,000*g* at 4°C for 10 min.

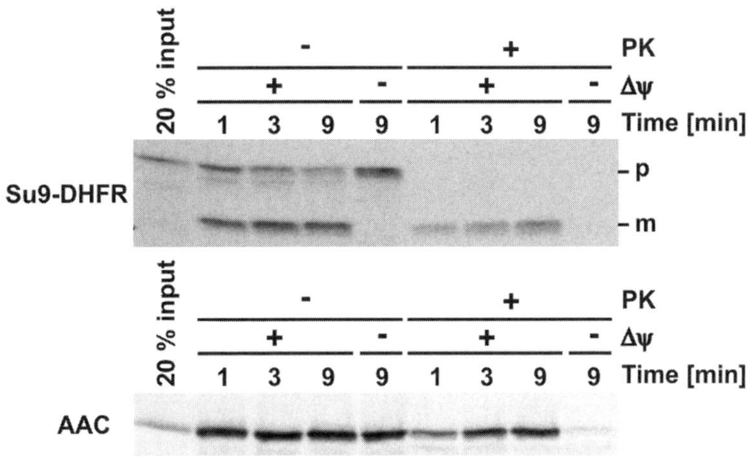

Fig. 1. Import of Su9-DHFR and AAC into isolated yeast mitochondria. Fusion protein between the presequence of subunit 9 of *Neurospora crassa* ATPase and mouse dihydrofolate dehydrogenase (Su9-DHFR) and ATP/ADP carrier (AAC) were synthesized in vitro in the presence of ^{35}S-methionine. They were then incubated with isolated yeast mitochondria in the presence or in the absence of membrane potential ($\Delta\Psi$). Aliquots were taken at the depicted time-points, and half of the samples were treated with proteinase K (PK). After reisolation of mitochondria, samples were analyzed by SDS-PAGE and autoradiography. Su9-DHFR is translocated into the matrix, while AAC is integrated into the inner membrane. Therefore, both preproteins require membrane potential for their import. In addition, the presequence of Su9-DHFR is proteolytically processed. It is detectable as the shorter, mature form of the protein. p, precursor; m, mature form of Su9-DHFR.

8. Remove ethanol completely and resuspend RNA pellet in 100 μL water containing 40 U RNAsin.
9. Make single-use aliquots and store at –80°C (*see* **Note 9**).

3.1.2. In Vitro Translation

1. To make an in vitro translation reaction, mix 100 μL rabbit reticulocyte lysate, 3.5 μL amino acid mixture minus methionine, 1 μL RNAsin, 7 μL Mg-acetate, 12 μL ^{35}S-Met, and 25 μL RNA.
2. Incubate for 1 h at 37°C.
3. Stop the reaction by adding 12 μL cold methionine.
4. Make the reaction mixture isotonic with mitochondria by adding 24 μL sucrose.
5. Remove ribosomes and aggregated proteins by centrifugation at 100,000*g* at 4°C for 1 h.
6. Make single-use aliquots of the lysate and store them at –80°C.
7. To assess the quality of the reaction, analyze 1 μL lysate by SDS-PAGE and autoradiography (*see* **Note 10**).

3.2. In Vitro Import Into Isolated Mitochondria

This is a basic protocol for following the kinetics of import of precursors into mitochondria (*see* **Notes 11–13**). A typical control is to incubate precursor with mitochondria with membrane potential that was dissipated, leading to complete block of import across the inner membrane. The experiment is designed so that there is enough material to take out samples at three time-points and split them so that one-half can be treated with protease. The sample incubated without membrane potential is usually performed only for the longest time-point and split in two so that one-half can be treated with protease.

1. Prepare on ice 8 tubes with 810 µL SH buffer (*see* **Note 14**).
2. In tube 1, mix in the following order: 300 µL 2X import buffer, 235.5 µL water, and 30 µL mitochondria (*see* **Note 15**). This will be the tube in which the precursor is going to be imported in the presence of membrane potential.
3. In tube 2, mix in the following order: 100 µL 2X import buffer, 2 µL oligomycin, 2 µL valinomycin, 86 µL water, and 10 µL mitochondria. This will be the tube in which the precursor is going to be imported in the absence of membrane potential.
4. Incubate both tubes for 5 min at 25°C (*see* **Note 16**).
5. Into tube 1, add 7.5 µL ATP, 15 µL NADH, and 6 µL each of CP and CK.
6. Incubate both tubes for an additional 3 min at 25°C (*see* **Note 17**).
7. Add 6 µL preprotein-containing lysate to tube 1 and 2 µL to tube 2 (*see* **Note 18**).
8. Incubate at 25°C.
9. After 1, 3, and 9 min, twice take out 90 µL from tube 1 and dilute them into ice-cold SH buffer from **step 1** (*see* **Note 19**).
10. After 9 min, twice take out 90 µL from tube 2 and dilute them into ice-cold SH buffer from **step 1**.
11. Add 5 µL PK to one sample from each time-point (*see* **Note 20**).
12. Incubate for 15 min on ice.
13. To stop proteolysis, add 5 µL PMSF to all samples.
14. Reisolate mitochondria by centrifugation at 18,000*g* for 10 min at 4°C.
15. Remove the supernatants completely and resuspend the mitochondrial pellets in 20 µL 1X Laemmli buffer.
16. Heat the samples *immediately* to 95°C for 5 min (*see* **Note 21**).
17. Analyze the samples by SDS-PAGE, blotting, and autoradiography (*see* **Note 22**). Two examples of such an experiment are given in **Fig. 1**.

4. Notes

1. We routinely clone open reading frames into pGEM vectors (Promega) downstream of the Sp6 RNA polymerase transcription initiation site. We observed that Sp6 RNA polymerase generally produces transcripts more efficiently than T3 or T7 RNA polymerases. However, the latter enzymes were successfully used as well. It is useful to check the coding sequence of the gene prior to cloning for the presence of methionines in the corresponding protein. If the start methionine is

the only one present in the sequence, then it is helpful to introduce several ATG codons (we generally introduce up to three) at the 3′ end of the gene by the cloning procedure. This will increase labeling efficiency in the translation reaction. Introduction of methionines at the C terminus of the protein is less likely to interfere with the import reaction and prevents their possible removal by MPP on import into mitochondria.

2. m7G(5′)ppp(5′)G is a messenger RNA cap analog. It is used during in vitro transcription reactions to yield RNA with a cap at the 5′ end. The cap is required for efficient translation in in vitro translation systems.
3. Avoid repeated freezing and thawing of the lysate as it significantly reduces the efficiency of translation. We never freeze the same aliquot more than twice. Lysate should be rapidly thawed by hand warming and immediately placed on ice.
4. Refer to the general safety precautions when dealing with radioactivity.
5. We normally make the 2X import buffer by weighing in all the ingredients, except BSA, as powders, dissolve them in water, and then adjust the pH with KOH. BSA is added after pH adjustment. The purpose of BSA in the buffer is to stabilize the precursor protein. Although other protocols suggest the use of 3% (w/v) BSA, we found that this is generally not necessary. In fact, the majority of precursors we tested do not need addition of BSA to be imported. Using less BSA has the advantage that the quality of subsequent SDS polyacrylamide gels is higher because of the absence of huge blobs of BSA.
6. We keep mitochondria frozen at –80°C in single-use aliquots. It is advisable to thaw the needed aliquots by hand warming and place them immediately on ice. Use them as soon as possible after thawing. Keeping them for long, even on ice, strongly reduces the quality of the import reaction.
7. The concentration of a preprotein synthesized in the reticulocyte lysate is estimated to be approx 2 pmol/mL *(7)*. These amounts are too small to saturate mitochondrial import sites. When the saturation of mitochondrial import sites is necessary in the experiment, precursor protein has to be present in excess compared to the available translocases. For this purpose, mitochondrial precursors are expressed in bacteria and purified as recombinant proteins. It should be noted, however, that this was successfully done for only a few precursors. In general, fusion proteins are used consisting of N-terminal mitochondrial-targeting signals (e.g., subunit 9 of the *Neurospora crassa* ATPase or yeast cytochrome-b_2) and mouse dihydrofolate dehydrogenase ***(8–10)***.
8. Sometimes, it is possible to improve the yield of RNA by linearization of the DNA template prior to transcription. This is particularly advisable when T7 or T3 polymerases are used. Also, when working with RNA, the water used should be sterile and highest quality.
9. RNA produced and stored in this way can be usually kept for longer than 1 yr.
10. Lower molecular weight translation products are often observed in addition to the desired full-length product. This is because of the initiation of translation at the internal methionine residues. These internal initiation reactions can be suppressed by increasing the concentration of Mg^{2+} and K^+ ions in the translation

mix. On the other hand, in most cases internal initiation is not a problem as N-terminally truncated products usually lack functional mitochondrial targeting information and will not even bind to mitochondria. Sometimes it is difficult to produce full-length proteins. This is frequently observed when very large or very hydrophobic proteins are to be synthesized. It is then advisable to optimize RNA and salt concentrations in the translation reaction. Furthermore, it is often helpful to make fresh plasmid preparations and carefully control their quality. However, sometimes none of these manipulations work. In these cases, one can use coupled transcription/translation systems commercially available from many suppliers. They usually yield higher amounts of proteins but are also significantly more expensive.

11. This protocol is established for purified yeast mitochondria. It can also be employed without any changes for purified *N. crassa* mitochondria. It should be noted that *Neurospora* mitochondria are usually freshly prepared for import experiments. Freezing them can lead to the rupture of the outer membrane. It is then difficult to analyze import into any mitochondrial subcompartment except matrix. The same is true for mitochondria isolated from mammalian cells. In the latter case, Na-succinate has to be added to the import mix to maintain membrane potential ***(11,12)***.
12. Wild-type yeast mitochondria can be used to determine the import kinetics of a particular precursor and its general import requirements. The power of the method, however, is in the use of mutants, which enable the dissection of the import pathway of a precursor. On the other hand, one can also use the precursors for which the import pathways are known and establish the potential involvement of new import components in the particular pathway (*see*, e.g., ***13–15***).
13. This a basic protocol for protein import into isolated mitochondria. For modifications of the method, such as depletion of mitochondrial ATP, translocation arrest at the outer membrane followed by the chase into the matrix, generation of two-membrane-spanning intermediates, and assessing protein assembly after import, we refer to the methods described in the original literature (*see*, e.g., ***5,13,16***).
14. Samples incubated for various time-periods are diluted into ice-cold SH buffer and kept on ice until the last sample is collected. The logic is that the import process will be drastically slowed by dilution and cooling. It is also possible to add valinomycin to dissipate membrane potential and thereby stop import completely (at least in the case of precursors that need a membrane potential), but this is usually not necessary.
15. Make sure that the content of the tube is mixed thoroughly after addition of each solution at each step. However, do not vortex vigorously as this will lead to breakage of mitochondria.
16. We usually perform import reactions at 25°C. Some precursors are imported rapidly at this temperature, so to be in the linear time range of import we decrease the temperature to 12°C. One can also follow the temperature dependence of the import reaction by performing the experiment at several temperatures.

17. During this incubation, mitochondria in tube 1 are energized, and the membrane potential is dissipated in tube 2.
18. The amount of lysate to be added depends on its quality as well as on the ability of the precursor to be imported. We normally use 1% (v/v) of lysate in the import reaction, but up to 20% can be used without major deleterious effects.
19. Make sure to pipet the content of the tube at least once up and down before taking the samples out. Mitochondria have a tendency to sediment in suspension.
20. In this step, translocation across the outer membrane is assessed. All the precursors bound at the mitochondrial surface will be accessible to the added protease and therefore degraded. Only precursors translocated at least across the outer membrane will be protected. PK can also be substituted by trypsin, if desired. We generally use PK as it has a broader specificity and therefore is capable of degrading more proteins.
21. It is essential to heat the samples to 95°C immediately. We observed that PK is never stopped completely by addition of PMSF, and any remaining active PK will start degrading proteins rapidly on addition of SDS-containing buffer. Incomplete inactivation usually leads to degradation of proteins larger than approx 60 kDa, which can be observed on blots after SDS-PAGE.
22. We find it useful to load a sample on the SDS polyacrylamide gel that contains 20% of the lysate input. First, it serves as an additional lysate quality control; second, it allows quantification of the import efficiency. As the samples are to be analyzed by autoradiography, it is not necessary to transfer the proteins from the gel onto the nitrocellulose membrane, one can easily stain the gel with Coomassie brilliant blue, dry it, and immediately expose to phosphoimaging plates or X-ray films. However, we usually do the blotting step in between as it allows for subsequent analysis of mitochondrial proteins by immunodecoration. In this way, it is possible to check the integrity of mitochondria, for example, if the outer membrane was broken.

Acknowledgments

This work was supported by grants from the Deutsche Forschungsgemeinschaft, SFB 594 (B3, B13) and the Fonds der Chemischen Industrie.

References

1. Neupert, W. (1997) Protein import into mitochondria. *Annu. Rev. Biochem.* **66,** 863–917.
2. Rehling, P., Brandner, K., and Pfanner, N. (2004) Mitochondrial import and the twin-pore translocase. *Nat. Rev. Mol. Cell. Biol.* **5,** 519–530.
3. Endo, T., Yamamoto, H., and Esaki, M. (2003) Functional cooperation and separation of translocators in protein import into mitochondria, the double-membrane bounded organelles. *J. Cell Sci.* **116,** 3259–3267.
4. Koehler, C. M. (2004) New developments in mitochondrial assembly. *Annu. Rev. Cell Dev. Biol.* **20,** 309–335.

5. Paschen, S. A., Waizenegger, T., Stan, T., et al. (2003) Evolutionary conservation of biogenesis of β-barrel membrane proteins. *Nature* **426,** 862–866.
6. Wiedemann, N., Kozjak, V., Chacinska, A., et al. (2003) Machinery for protein sorting and assembly in the mitochondrial outer membrane. *Nature* **424,** 565–571.
6a. Mesecke, N., Terziyska, N., Kozany, C., et al. (2005) A disulfide relay system in the intermembrane space of mitochondria that mediates protein import. *Cell.* **121,** 1059–1069.
7. Rassow, J., Guiard, B., Wienhues, U., Herzog, V., Hartl, F. U., and Neupert, W. (1989) Translocation arrest by reversible folding of a precursor protein imported into mitochondria. A means to quantitate translocation contact sites. *J. Cell Biol.* **109,** 1421–1428.
8. Stan, T., Ahting, U., Dembowski, M., et al. (2000) Recognition of preproteins by the isolated TOM complex of mitochondria. *EMBO J.* **19,** 4895–4902.
9. Dekker, P. J., Martin, F., Maarse, A. C., et al. (1997) The Tim core complex defines the number of mitochondrial translocation contact sites and can hold arrested preproteins in the absence of matrix Hsp70-Tim44. *EMBO J.* **16,** 5408–5419.
10. Yamamoto, H., Esaki, M., Kanamori, T., Tamura, Y., Nishikawa, S., and Endo, T. (2002) Tim50 is a subunit of the TIM23 complex that links protein translocation across the outer and inner mitochondrial membranes. *Cell* **111,** 519–528.
11. Ishihara, N., and Mihara, K. (1998) Identification of the protein import components of the rat mitochondrial inner membrane, rTIM17, rTIM23, and rTIM44. *J. Biochem. (Tokyo)* **123,** 722–732.
12. Rothbauer, U., Hofmann, S., Muhlenbein, N., et al. (2001) Role of the deafness dystonia peptide 1 (DDP1) in import of human Tim23 into the inner membrane of mitochondria. *J. Biol. Chem.* **276,** 37,327–37,334.
13. Mokranjac, D., Sichting, M., Neupert, W., and Hell, K. (2003) Tim14, a novel key component of the import motor of the TIM23 protein translocase of mitochondria. *EMBO J.* **22,** 4945–4956.
14. Frazier, A. E., Dudek, J., Guiard, B., et al. (2004) Pam16 has an essential role in the mitochondrial protein import motor. *Nat. Struct. Mol. Biol.* **11,** 226–233.
15. Ishikawa, D., Yamamoto, H., Tamura, Y., Moritoh, K., and Endo, T. (2004) Two novel proteins in the mitochondrial outer membrane mediate β-barrel protein assembly. *J. Cell Biol.* **166,** 621–627.
16. Mokranjac, D., Paschen, S. A., Kozany, C., et al. (2003) Tim50, a novel component of the TIM23 preprotein translocase of mitochondria. *EMBO J.* **22,** 816–825.

21

Analyzing Import Intermediates of Mitochondrial Proteins by Blue Native Gel Electrophoresis

Thomas Waizenegger and Doron Rapaport

Summary

Blue native gel electrophoresis (BNGE) is a powerful tool for analyzing native protein complexes from biological membranes as well as water-soluble proteins. It can be used for determining relative molecular masses of protein complexes and their subunit composition and for the detection of subcomplexes. We describe the analysis by BNGE of in vitro import reactions composed of radiolabeled precursor proteins and isolated mitochondria. Such an analysis is a powerful tool to follow import intermediates and to study assembly of protein complexes. Analysis of import reactions by BNGE provides information on the molecular mass of the complex with which the imported precursor is associated. In addition, components of such a complex can be identified by incubating the mitochondrial lysate with either soluble antibodies or antibodies coupled to protein A matrix. The binding of soluble antibodies to specific complexes results in an observed shift in their apparent molecular mass (antibody shift). Alternatively, addition of matrix-bound antibodies followed by removal of the matrix from the mixture will result in depletion of the specific complex from the mitochondrial lysate (antibody depletion). The experimental details of these techniques are described.

Key Words: Antibody depletion; antibody shift; blue native gel electrophoresis; in vitro import; membrane proteins; protein complexes; TOB complex; Tom40.

1. Introduction

Relative molecular masses of membrane protein complexes can be determined by different methods. Sucrose density gradient centrifugation, size exclusion chromatography, and blue native gel electrophoresis (BNGE) are commonly used. The advantage of BNGE in comparison to other methods is the ability to separate complexes with higher resolution while using smaller protein amounts. Proteins or oligomeric complexes from 10 kDa to more than 1 MDa in size can

From: *Methods in Molecular Biology, vol. 372: Mitochondria: Practical Protocols*
Edited by: D. Leister and J. M. Herrmann © Humana Press Inc., Totowa, NJ

be analyzed. BNGE was established to study mitochondrial membrane protein complexes ***(1,2)***. Membrane proteins are solubilized by nonionic detergents, and native protein complexes are resolved in a nondenaturing polyacrylamide gel electrophoresis in the presence of Coomassie brilliant blue G250. The electrophoretic mobility of the proteins is determined by the size of the complex and by the negative charges of bound Coomassie dye.

The original protocol was extended to analyze the assembly of mitochondrial outer membrane protein complexes. Radiolabeled mitochondrial precursor proteins are synthesized in a cell-free transcription/translation system. The precursor proteins are incubated with isolated mitochondria in an import reaction in which the temperature, length of incubation, and reaction buffer can be varied. After this in vitro import reaction, mitochondria are solubilized with detergent, and the proteins are analyzed by BNGE followed by Western blotting or autoradiography. In this way, assembly intermediates of precursor proteins can be identified.

Furthermore, addition of specific antibodies to the mitochondrial lysate before the analysis by BNGE allows the identification of constituents of the various complexes to which the imported radiolabeled precursors are associated. In the case of antibody shift, an antibody against one component of a complex binds to its antigen, thus causing a shift to a higher apparent molecular mass of the complex. An alternative method is the antibody depletion procedure. In the latter method, solubilized mitochondria are incubated with antibodies coupled to protein A-Sepharose. If a component of the complex is recognized by the matrix-bound antibody, then the whole complex can be pulled down with the antibody. Thus, complexes that contain this specific protein are depleted from the supernatant analyzed by BNGE.

2. Materials

2.1. Pouring the Gradient Gel

1. AA-mix 1: Acrylamide 30% (w/v)/bisacrylamide 0.2% (w/v); store at 4°C.
2. AA-mix 2: Acrylamide 48% (w/v)/bisacrylamide 1.5% (w/v); store at room temperature.
3. Gel buffer: 1.5 *M* ε-amino-caproic acid, 0.15 *M* BisTris-HCl, pH 7.0; store at 4°C.
4. Glycerol.
5. Ammonium peroxodisulfate (APS), 10% (w/v) in double-distilled H_2O.
6. *N,N,N′,N′*-tetramethylethylenediamine (TEMED).

2.2. Solubilizing the Mitochondria

1. Buffer N: 20 m*M* Tris-HCl, pH 7.4, 0.1 m*M* ethylenediaminetetraacetic acid, 50 m*M* NaCl, 10% (v/v) glycerol; store at –20°C.
2. Detergents: 10% (w/v) digitonin (*see* **Note 1**) in buffer N or 10% (w/v) n-dodecyl-β-D-maltosid in buffer N or 20% (v/v) Triton X-100 in buffer N.
3. 10X sample buffer: 5% (w/v) Coomassie brilliant blue G250 (Serva), 100 m*M* BisTris-HCl, pH 7.0, 0.5 *M* ε-amino-caproic acid; aliquots are stored at –20°C.
4. 200 m*M* Phenylmethylsulfonyl fluoride in ethanol.

2.3. Running the Gel

1. Cathode buffer A: 50 m*M* Tricine, 15 m*M* BisTris-HCl, pH 7.0, 0.02% (w/v) Serva Blue G; store at 4°C.
2. Cathode buffer B: 50 m*M* Tricine, 15 m*M* BisTris-HCl, pH 7.0; store at 4°C.
3. Anode buffer: 50 m*M* BisTris-HCl, pH 7.0; store at 4°C.

2.4. Transfer of Proteins to Polyvinylidene Fluoride Membrane

1. Blotting buffer: 20 m*M* Tris-HCl (pH not adjusted), 150 m*M* glycine, 20% (v/v) methanol, 0.02% (w/v) sodium dodecyl sulfate.
2. Staining solution: 0.1% (w/v) Coomassie brilliant blue R250 (Serva), 40% (v/v) methanol, 10% (v/v) acetic acid.
3. Destaining solution: 30% (v/v) methanol, 10% (v/v) acetic acid.
4. TBS: 10 m*M* Tris-HCl, pH 7.5, 154 m*M* NaCl.

2.5. Analyzing Assembly of Radiolabeled Precursor Proteins

1. Import buffer: 3% (w/v) bovine serum albumin, 250 m*M* sucrose, 80 m*M* KCl, 5 m*M* $MgCl_2$, 2 m*M* adenosine triphosphate, 2 m*M* nicotinamide adenine dinucleotide (NADH), 100 µg/mL creatine kinase, 5 m*M* creatine phosphate, 5 m*M* methionine, 10 m*M* 3-(*N*-morpholino)propanesulfonic acid (MOPS)-KOH, pH 7.2.
2. SEM: 250 m*M* sucrose, 1 m*M* ethylenediaminetetraacetic acid, 10 m*M* MOPS-KOH, pH 7.2.

3. Methods

3.1. Pouring the Gradient Gel

The given recipe is calculated for a gel with the following dimensions: bottom gel $1 \times 15 \times 0.1$ cm, separating gel $9 \times 15 \times 0.1$ cm, and stacking gel $2 \times 15 \times 0.1$ cm.

1. Bottom gel: mix 6.7 mL AA-mix 1, 3.3 mL gel buffer, 25 µL TEMED, and 50 µL APS and pour the bottom gel.
2. Separating gel: the separating gel is poured by the help of a gradient mixer. For this purpose, pour the high-percentage gel mixture into the chamber closer to the exit of the gradient mixer and the low-percentage gel mixture into the second chamber (for specific acrylamide concentrations, *see* **Table 1**). Stir the solutions by adding a magnetic stirrer into each chamber of the gradient mixer, add APS and TEMED to both chambers, and pour the gel mixture between the glass plates (*see* **Note 2**). Overlay the mixture with isopropanol; let the gel polymerize.
3. Stacking gel: mix 0.6 mL AA-mix 2, 2.5 mL gel buffer, 4.33 mL double-distilled H_2O, 6 µL TEMED, and 60 µL APS and pour the stacking gel.

3.2. Solubilizing the Mitochondria

1. Mitochondria (50–150 µg) are pelleted either directly or after in vitro import experiments and resuspended in 30–45 µL of buffer N supplemented with 1 m*M* phenylmethylsulfonyl fluoride and 1% (w/v) digitonin, 0.5% (v/v) Triton X-100, or 0.2% (w/v) n-dodecyl-β-D-maltosid (*see* **Note 3**).

Table 1
Acrylamide Concentrations Used for Separating Gels

	16.5%	13%	12%	11%	10%	6%	5%
AA-mix 2	2.38 mL	1.76 mL	1.63 mL	1.5 mL	1.36 mL	0.938 mL	0.782 mL
Gel buffer	2.31 mL	2.25 mL	2.08 mL	1.9 mL	1.73 mL	2.625 mL	2.19 mL
Glycerol	1.38 mL	1.35 mL	1.25 mL	1.14 mL	1.038 mL	—	—
Water	0.85 mL	1 mL	1.44 mL	1.84 mL	2.24 mL	4.275 mL	4.86 mL
TEMED	2.3 µL	2.3 µL	2.3 µL	2.3 µL	2.3 µL	3.4 µL	3.4 µL
APS	23 µL	23 µL	23 µL	23 µL	23 µL	34 µL	34 µL
Total volume	6.945 mL	6.385 mL	6.425 mL	6.405 mL	6.393 mL	7.875 mL	7.869 mL

2. The mixture is incubated on ice for 15 min.
3. Unsolubilized material is pelleted by centrifugation (36,700*g*, 20 min, 4°C).
4. The supernatant of this centrifugation step is mixed with the appropriate volume of 10X sample buffer.

3.3. Running the Gel

1. Use cold buffers and load the gel in a cold room.
2. Put the gel into the gel chamber and use cathode buffer A and anode buffer to fill the buffer tanks.
3. Load the samples with a Hamilton syringe. For estimating the size of the protein complexes, different markers can be loaded on the gel. Mix 2–4 µL of a 25 µg/µL solution of bovine serum albumin (66-kDa monomer; 132-kDa dimer), or apoferritin (440 kDa), or thyreoglobulin (669 kDa) with 30 µL buffer N and 4 µL 10X sample buffer and load the mixture on the gel (*see* **Note 4**).
4. Let the gel run for 1 h at 100 V, 15 mA. This step allows the proteins to enter the separating gel.
5. After 1 h, increase voltage to 500 V (limited to 15 mA).
6. When the blue dye has reached the middle of the gel, soak out cathode buffer A and replace it by cathode buffer B.
7. Run the gel until the blue dye front enters the bottom gel.
8. Transfer proteins to polyvinylidene fluoride (PVDF) membrane (*see* **Note 5**).

3.4. Transfer of Proteins to PVDF Membrane

1. Wet PVDF membrane for 15 s in methanol, wash it for 2 min in water, and preincubate the membrane for few minutes in blotting buffer.
2. Transfer the proteins to the PVDF membrane at 4°C for 1 h at 220 mA.
3. Stain the membrane for 15 min in staining solution and transfer it into destaining solution. As soon as you see the markers, label them and destain the membrane completely in methanol. If you want to detect radioactive signals, then wash membrane once in TBS after blotting without staining it, dry the membrane for 30 min under red light or several hours at room temperature, and expose to X-ray film.

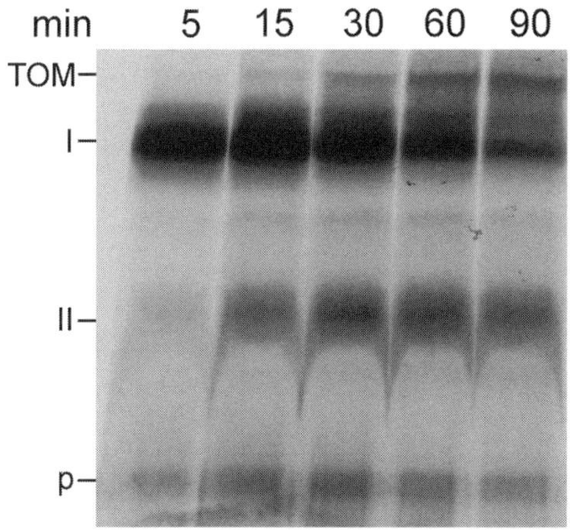

Fig. 1. Assembly of Tom40 precursor into the TOM complex. Isolated yeast mitochondria were incubated at 25°C with radiolabeled Tom40 precursor for the indicated time-periods. Mitochondria were lysed with 1% (w/v) digitonin and analyzed on a 6–13% blue native gel followed by autoradiography. P, precursor protein; I and II, assembly intermediates I and II, respectively; TOM, assembled TOM complex.

4. If you stain or destain the membrane, then wash it extensively several times in TBS before immunodecoration.

3.5. Analyzing Assembly of Radiolabeled Precursor Proteins

In vitro import of radiolabeled precursor proteins in combination with BNGE is used to analyze the assembly of the protein of interest into its oligomeric complex *(3–6)*. For this purpose, mitochondria (50 µg) in a total volume of 200 µL import buffer are incubated with 5–20 µL reticulocyte lysate containing the radiolabeled precursor protein for various time periods at the desired temperature. The import reaction is stopped by diluting the reaction mixture with 800 µL cold SEM buffer; mitochondria are reisolated by centrifugation and solubilized as described in **Subheading 3.2**.

This method was used to study the import and assembly of different precursor proteins such as porin *(4)*, small intermembrane space proteins *(5)*, the adenosine 5′-diphosphate/adenosine triphosphate carrier *(7)*, and Oxa1 *(8)*. BNGE was most extensively used to investigate the assembly of Tom40, the major component of the translocase of the mitochondrial outer membrane (TOM complex) *(3,9–12)*. During import of Tom40 precursor, two intermediates of 250 kDa (assembly intermediate I) and 100 kDa (assembly intermediate II) precede the assembly of the precursor into preexisting TOM complex (**Fig. 1**).

To analyze the individual components of these import intermediates, BNGE combined with either antibody shift or antibody depletion can be utilized. Furthermore, complexes can also be depleted by additional means (*see* **Note 6**).

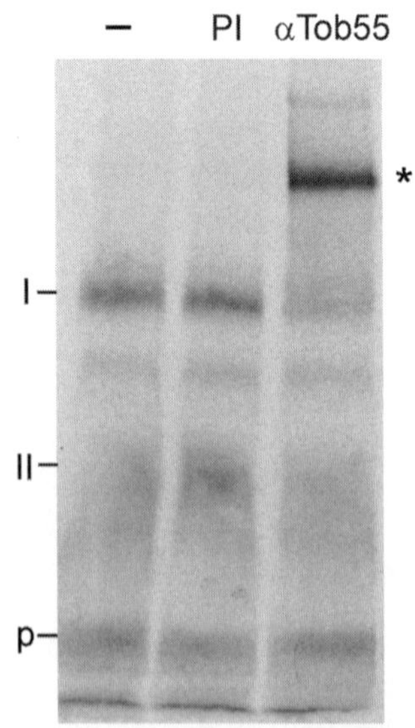

Fig. 2. Antibody shift BNGE. Isolated mitochondria were incubated at 25°C with radiolabeled Tom40 precursor for 20 min. Mitochondria were lysed with 1% (w/v) digitonin and incubated further without (–) or with antibodies either against Tob55 (αTob55) or from preimmune serum (PI). The lysate was analyzed on a 5–12% blue native gel followed by autoradiography. P, I, and II, assembly intermediates as in legend to Fig. 1. The supercomplex of intermediate I with antibodies against Tob55 is indicated with an asterisk.

3.5.1. Antibody Shift BNGE

Antibodies used in combination with BNGE can be monoclonal, affinity-purified polyclonal, or immunoglobulin Gs (IgGs) gained from serum. Typically, 1 μg of monoclonal antibodies or 10 μg of IgGs or affinity-purified antibodies are added to the reaction mixture after solubilizing the mitochondria, and the mixture is incubated for an additional 60 min under mild agitation at 4°C. Afterward, unsolubilized material is pelleted as described in **Subheading 3.2.3**. If the added antibodies recognize and bind to a protein that is part of the investigated import intermediate, then the latter will migrate slower in BNGE and be shifted to a higher apparent molecular mass. It is recommended to include IgG from preimmune serum or against unrelated protein as a control.

This method has been applied for characterizing the import intermediates of precursors of Tim9 ***(5)*** and Tom40 ***(3,6,11)***. For example, addition of antibodies against the mitochondrial outer membrane protein Tob55, which is the major component of the TOB complex (topogenesis of outer membrane β-barrel proteins) resulted in a shift of the band corresponding to intermediate I to a higher apparent molecular mass. Hence, it was concluded that Tob55 is a constituent of this intermediate ***(3,13)*** (**Fig. 2**).

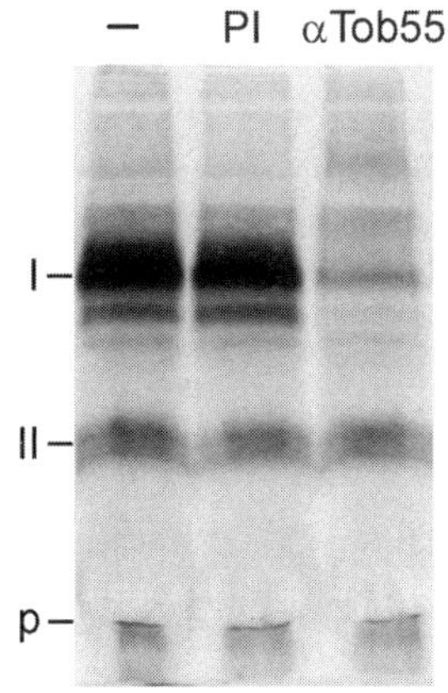

Fig. 3. Antibody depletion BNGE. Isolated mitochondria were incubated at 25°C with radiolabeled Tom40 precursor for 15 min. Mitochondria were lysed with 1% (w/v) digitonin and incubated with protein A-Sepharose beads to which antibodies against Tob55 (αTob55) or from preimmune serum (PI) were previously coupled. After pelleting the beads, the supernatant was analyzed on a 6–13% blue native gel followed by autoradiography. Assembly intermediates are indicated as described in the legend to **Fig. 1**.

3.5.2. Antibody Depletion BNGE

Import intermediates can also be depleted by using antibodies against components of the intermediate. For this purpose, antibodies are coupled to protein A-Sepharose beads, and the lysed mitochondria after the clarifying spin (**Subheading 3.2.3.**) are incubated with the beads for 1 h at 4°C in an overhead shaker. Next, the beads are pelleted; the supernatant is mixed with 10X sample buffer and analyzed by BNGE. This method was used to study import intermediates of Tom40 *(3)*. For example, in this way it could be confirmed that Tob55 is present in import intermediate I of Tom40 (**Fig. 3**).

4. Notes

1. Digitonin is a natural product with quality that varies from batch to batch. It is recommended to recrystallize the commercially available digitonin. Digitonin recrystallization:
   a. Add 20 mL ethanol (p.a. grade) per gram digitonin; heat the mixture (in a water bath) until all the ethanol is cooking, and the digitonin is completely dissolved.
   b. Filtrate the hot solution very fast (be aware that digitonin starts to precipitate immediately when ethanol cools) into a 50-mL centrifuge tube; let the solution cool to room temperature and afterward leave the tubes overnight at –20°C.
   c. Reisolate the digitonin by centrifugation (5 min, 4000*g*). Dry the digitonin under vacuum.
   d. The resulting powder can be stored at 4°C.

2. The gradient is chosen according to the size of the complexes analyzed. For the TOM and TOB complexes with apparent molecular mass of 410 and 210 kDa, respectively, a 6–13% gradient gel is most commonly used. It shows good separation in the size range of 40–700 kDa.
3. Membrane protein complexes differ in their stability. Some are only stable in the mild detergent digitonin, whereas others are also stable in Triton X-100. Therefore, different detergents should be tested for each membrane protein complex. The detergent-to-protein ratio can also influence the stability of complexes or alter their running behavior on blue native gels. For solubilizing with digitonin, we recommend a detergent:protein ratio of 3:1 to 10:1, whereas for Triton X-100 a ratio of 2:1 is sufficient. Note that the concentration of each detergent should be above its critical micelle concentration.
4. Do not add any detergent to the markers.
5. PVDF membranes have to be used for blotting because nitrocellulose membranes are destroyed in methanol.
6. Employment of affinity chromatography is an additional method to deplete complexes before analyzing mitochondria by BNGE. For example, to show the presence of a certain protein in an import intermediate, mitochondria containing a His-tagged version of this protein can be incubated with Ni-NTA-agarose after solubilization. If the protein with the His-Tag is present in the import intermediate of the radiolabeled precursor, then this intermediate will be retained on the Ni beads and will not be detected on the BNGE. In this way, Tob38 was shown to be a constituent of intermediate I in the assembly pathway of Tom40 ***(14)***.

Acknowledgments

This work was supported by the Deutsche Forschungsgemeinschaft-Sonderforschungsbereich 594 (D. R.) and a predoctoral fellowship from the Boehringer Ingelheim Fonds (T. W.).

References

1. Schägger, H., Cramer, W. A., and von Jagow, G. (1994) Analysis of molecular masses and oligomeric states of protein complexes by blue native electrophoresis and isolation of membrane protein complexes by two-dimensional native electrophoresis. *Anal. Biochem.* **217,** 220–230.
2. Schägger, H., Bentlage, H., Ruitenbeek, W., et al. (1996) Electrophoretic separation of multiprotein complexes from blood platelets and cell lines: technique for the analysis of diseases with defects in oxidative phosphorylation. *Electrophoresis* **17,** 709–714.
3. Paschen, S. A., Waizenegger, T., Stan, T., et al. (2003) Evolutionary conservation of biogenesis of β-barrel membrane proteins. *Nature* **426,** 862–866.
4. Krimmer, T., Rapaport, D., Ryan, M. T., et al. (2001) Biogenesis of the major mitochondrial outer membrane protein porin involves a complex import pathway via receptors and the general import pore. *J. Cell Biol.* **152,** 289–300.

5. Chacinska, A., Pfannschmidt, S., Wiedemann, N., et al. (2004) Essential role of Mia40 in import and assembly of mitochondrial intermembrane space proteins. *EMBO J.* **23,** 3735–3746.
6. Ishikawa, D., Yamamoto, H., Tamura, Y., Moritoh, K., and Endo, T. (2004) Two novel proteins in the mitochondrial outer membrane mediate beta-barrel protein assembly. *J. Cell Biol.* **166,** 621–627.
7. Ryan, M. T., Muller, H., and Pfanner, N. (1999) Functional staging of ADP/ATP carrier translocation across the outer mitochondrial membrane. *J. Biol. Chem.* **274,** 20,619–20,627.
8. Frazier, A. E., Chacinska, A., Truscott, K. N., Guiard, B., Pfanner, N., and Rehling, P. (2003) Mitochondria use different mechanisms for transport of multispanning membrane proteins through the intermembrane space. *Mol. Cell. Biol.* **23,** 7818–7828.
9. Model, K., Meisinger, C., Prinz, T., et al. (2001) Multistep assembly of the protein import channel of the mitochondrial outer membrane. *Nat. Struct. Biol.* **8,** 361–370.
10. Rapaport, D. and Neupert, W. (1999) Biogenesis of Tom40, core component of the TOM complex of mitochondria. *J. Cell Biol.* **146,** 321–331.
11. Wiedemann, N., Kozjak, V., Chacinska, A., et al. (2003) Machinery for protein sorting and assembly in the mitochondrial outer membrane. *Nature* **424,** 565–571.
12. Hoppins, S. C. and Nargang, F. E. (2004) The Tim8-Tim13 complex of *Neurospora crassa* functions in the assembly of proteins into both mitochondrial membranes. *J. Biol. Chem.* **279,** 12,396–12,405.
13. Kozjak, V., Wiedemann, N., Milenkovic, D., et al. (2003) An essential role of Sam50 in the protein sorting and assembly machinery of the mitochondrial outer membrane. *J. Biol. Chem.* **278,** 48,520–48,523.
14. Waizenegger, T., Habib, S. J., Lech, M., et al. (2004) Tob38, a novel essential component in the biogenesis of β-barrel proteins of mitochondria. *EMBO Rep.* **5,** 704–709.

22

In Vitro and In Vivo Protein Import Into Plant Mitochondria

Pavel F. Pavlov, Charlotta Rudhe, Shashi Bhushan, and Elzbieta Glaser

Summary

In plants, the majority of mitochondrial and chloroplast proteins are nuclear encoded, synthesized on cytosolic polyribosomes, and then imported into the organelle. Most of the nuclear encoded precursor proteins contain an N-terminal extension called signal or targeting peptide that directs the protein to the correct organelle. Here, we describe in vitro and in vivo methods to study mitochondrial protein import. In a common single-organelle in vitro import procedure, transcribed/translated precursor proteins are imported into isolated mitochondria. A novel semi-in vivo system for simultaneous import of precursor proteins into isolated mitochondria and chloroplasts, called a dual-import system, is superior to the single-import system as it abolishes mistargeting of chloroplast precursors into mitochondria as observed in a single-organelle import system. Precursor proteins can also be imported into the organelles in vivo using an intact cellular system. In vivo approaches include import of transiently expressed fusion constructs containing a targeting peptide or a precursor protein fused to a reporter gene, most commonly the green fluorescence protein in protoplasts or in an *Agrobacterium*-mediated system in intact tobacco leaves.

Key Words: *Agrobacterium*; GFP fusion; precursor; presequence; protein import; protoplasts; signal; targeting peptide; transient expression.

1. Introduction

Most mitochondrial and chloroplast proteins are nuclear encoded and synthesized on cytosolic polyribosomes as precursor proteins with an N-terminal extension called signal or targeting peptide that directs precursor proteins to the correct organelle. Cleavable targeting peptides are commonly called a presequence for mitochondrial proteins and a transit peptide for chloroplast proteins. A series of molecular chaperones and other cytosolic factors interact with newly

From: *Methods in Molecular Biology, vol. 372: Mitochondria: Practical Protocols*
Edited by: D. Leister and J. M. Herrmann © Humana Press Inc., Totowa, NJ

synthesized precursor proteins to maintain an import-competent conformation of the precursor and facilitate organellar protein import *(1,2)*.

Targeting peptides are recognized by import receptors on the organellar outer membrane, and precursors are imported into the organelle through translocase complexes located on the outer and inner membranes of the organelles, called TOM and TIM (translocase of the outer or inner mitochondrial membrane, respectively) in mitochondria and TOC and TIC (translocase of the outer or inner chloroplast membrane, respectively) in chloroplasts. After translocation into the mitochondrial matrix or chloroplast stroma, presequences are cleaved off by the mitochondrial processing peptidase, which in plants is an integral part of the cytochrome-bc_1 complex of the respiratory chain *(3)*, and transit peptides by the stromal-processing peptidase (SPP) *(4)*. The mature proteins are transported within the organelle to the final destination and assembled with their partner proteins either spontaneously or on action of molecular chaperones. The cleaved targeting peptides, potentially harmful to biological membranes, are degraded inside the organelles by a newly identified presequence peptide degrading zinc metalloprotease, PreP, that is dually targeted and executes its function both in mitochondria and in plastids *(5–9)*.

Mitochondrial and chloroplast protein import is conventionally viewed as a posttranslational process; however, cotranslational import has also been suggested for a subset of proteins *(10)*. The targeting of proteins to mitochondria and chloroplasts can be studied using both in vitro and in vivo approaches. Early in vitro import studies of several mitochondrial proteins *(11,12)*, as well as a number of in vivo studies using transgenic approaches *(13–16)*, showed that import is organelle specific. However, several more recent in vitro import studies reported mistargeting of chloroplast proteins into mitochondria *(17–21)*, whereas no missorting of mitochondrial proteins into chloroplasts has been reported so far. Furthermore, there is an increasing number of proteins encoded by a single gene expressed as a single precursor protein that is dually imported into both mitochondria and chloroplasts *(7,22–24)*. This demands high-stringency study of the specificity of the organellar protein import.

Both in vitro and in vivo approaches have several limitations. In vitro import approaches have disadvantages because of a lack of an intact cellular system. It is possible to obtain incorrect targeting not seen in vivo, and competition between organelles is absent. In vivo approaches use an intact cellular system and obviously reflect the in vivo targeting capacity of a signal. However, they also have some disadvantages. Precursor proteins are usually expressed as fusion proteins containing the native targeting peptide and a reporter protein, ignoring the role of the mature protein. The investigated proteins are usually expressed at very high levels, which might affect interpretation of results. Additional weaknesses are that no kinetics or efficiency of targeting can be assessed, and dissection of the mechanisms involved in protein recognition and import is not possible.

To overcome the limitations of the in vitro import system, to elucidate the mechanisms involved in dual targeting, and to ensure the correct specificity of targeting, we have developed a novel in vitro dual-import system for simultaneous targeting of precursor proteins into mitochondria and chloroplasts ***(21)***. We have used purified organelles that are mixed, incubated with precursors, and repurified after import. This allows the determination of the targeting specificity into either organelle. This also allows the use of authentic precursors, so that the role of the mature protein in import can be assessed. Furthermore, mistargeting of chloroplast precursors into mitochondria is abolished.

Here, we describe three methodologies to study protein import into plant mitochondria, a single in vitro mitochondrial import, dual-import system to study mitochondrial import in the presence of chloroplasts, and in vivo import studies of chimeric precursors containing green fluorescent protein (GFP) in protoplasts and tobacco leaves.

2. Materials

2.1. Single In Vitro Mitochondrial Import

2.1.1. Preparation of Potato Tuber Mitochondria

1. Potatoes (*Solanum tuberosum* cv. King Edward) bought in a local shop.
2. Grinding medium: 0.6 *M* mannitol, 40 m*M* 3-(*N*-morpholino)propanesulfonic acid (MOPS)-KOH, pH 7.5, 10 m*M* ethylenediaminetetraacetic acid (EDTA), 10 m*M* cysteine, 0.4% (w/v) bovine serum albumin (BSA).
3. 2X wash medium: 0.8 *M* mannitol, 10 m*M* MOPS-KOH, pH 7.2, 0.2% (w/v) BSA.
4. Percoll solution: 20 mL 2X wash medium mixed with 12 mL Percoll and 8 mL distilled water (dH_2O) to obtain 30% (v/v) Percoll solution.
5. Juice centrifuge (Moulinex).
6. Nylon net (60-µm mesh).
7. Type Ti-21 rotor tubes (Beckman Instruments Inc., Palo Alto, CA, USA).

2.1.2. Preparation of Spinach Leaf Mitochondria

1. Spinach (*Spinacia oleracea* L.cv. Medania) grown hydroponically for 6 wk under artificial light at 25°C with a light period of 10 h.
2. Grinding medium: 0.3 *M* sucrose, 50 m*M* MOPS-KOH, pH 7.8, 5 m*M* $MgCl_2$, 2 m*M* EDTA, 4 m*M* cysteine, 0.6% (w/v) polyvinylpyrrolidone (MW 40,000), 0.2% (w/v) BSA.
3. 2.5X Wash medium: 0.75 M, 25 m*M* MOPS-KOH, pH 7.2, 0.25% (w/v) BSA.
4. Percoll solution: 4 mL 2.5X wash medium mixed with 6 mL Percoll to obtain 60% (v/v) Percoll solution; 4 mL 2.5X wash medium mixed with 4 mL Percoll and 3 mL dH_2O to obtain 30% (v/v) Percoll solution; 4 mL 2.5X wash medium mixed with 2 mL Percoll and 4 mL dH_2O to obtain 20% (v/v) Percoll solution; apply 60, 30, and 20% Percoll solutions stepwise into a centrifuge tube at the ratio 1:3:1.
5. Waring blender (Braun model KSM-2B, Woburn, MA).

2.1.3. In Vitro Synthesis of Radiolabeled Precursor Proteins

1. Coupled in vitro transcription/translation rabbit reticulocyte lysate system (Promega).
2. [^{35}S]-Methionine (specific activity > 1000 Ci/mmol).
3. Ribonuclease RNase inhibitor (Amersham).
4. Plasmids encoding proteins of interest (*see* **Note 1**).

2.1.4. Protein Import Into Potato Tuber and Spinach Leaf Mitochondria

1. Import medium: 0.25 *M* mannitol, 10 m*M* MOPS-KOH, pH 7.2, 50 m*M* KCl, 1 m*M* succinate, 1 m*M* methionine, 2 m*M* $MgCl_2$, 2 m*M* KH_2PO_4, 1 m*M* adenosine triphosphate, 10 μm adenosine 5′-diphosphate, 0.1% (w/v) BSA (make fresh prior to use).
2. 10 mg/mL Proteinase K.
3. 0.1 *M* Phenylmethylsulfonyl fluoride in ethanol.
4. 10 μm Valinomycin in ethanol.

2.1.5. Sodium Dodecyl Sulfate Electrophoresis and Image Processing

1. 4X Separating buffer: 1.5*M* Tris-HCl, pH 8.8, 0.4% (w/v) sodium dodecyl sulfate (SDS). Store at room temperature.
2. 4X Stacking buffer: 0.5*M* Tris-HCl, pH 6.8, 0.4% (w/v) SDS. Store at room temperature.
3. 2X SDS loading buffer: 1*M* Tris-HCl, pH 6.8, 0.8% (w/v) SDS, 20% (v/v) glycerol, 400 m*M* β-mercaptoethanol, 0.04% (w/v) bromophenol blue.
4. 30% (w/v) Acrylamide (AA) solution (27.3% w/v AA, 2.7% w/v bisacrylamide) (Hintze AB) (*see* **Note 2**).
5. *N,N,N,N′*-Tetramethylethylenediamine (TEMED).
6. Ammonium peroxodisulfate (APS): 10% (w/v) solution in water; freeze immediately in small aliquots and store at –20°C.
7. Butanol.
8. 5X SDS running buffer: 125 m*M* Tris-HCl, 960 m*M* glycine, 0.5% (w/v) SDS.
9. Fix solution: 50% (v/v) methanol, 5% (v/v) acetic acid.
10. Power supply for electrophoresis model 1000/500 (Bio-Rad, Hercules, CA, USA).
11. Vacuum gel drier (Savant).
12. BAS-MP phosphoimaging plates (FujiFilm, Japan).
13. FLA-3000 phosphoimaging scanner (FujiFilm, Japan).

2.2. Dual In Vitro Import to Mitochondria and Chloroplasts

2.2.1. Preparation of Organelles

2.2.1.1. Preparation of Pea Mitochondria

1. Pea seeds (*Pisum sativum* L. Greenfeast)
2. Grinding medium, pH 7.5: 0.3*M* sucrose, 25 m*M* tetrasodium pyrophosphate, 2 m*M* EDTA, 10 m*M* potassium dihydrophosphate, 1% (w/v) polyvinylpyrrolidone 40,000 (PVP-40), 1% (w/v) BSA, 20 m*M* ascorbic acid (add the ascorbic acid just prior to use).

3. 2X Wash medium, pH 7.5: 0.6*M* sucrose, 20 m*M* TES (*N*-[Tris(hydroxymethyl) methyl]-2-amino-ethanesulfonic acid), 0.2% (w/v) BSA.
4. Percoll gradient: 8.75 mL 2X wash medium, 4.9 mL Percoll, 3.85 mL 20% (w/v) PVP-40 (heavy gradient solution); 8.75 mL 2X wash medium, 4.9 mL Percoll, 3.85 mL dH_2O (light gradient solution).

2.2.1.2. Preparation of Pea Chloroplasts

1. 2X Grinding medium, pH 7.3: 100 m*M* HEPES, 660 m*M* sorbitol, 0.2% (w/v) BSA, 2 m*M* $MgCl_2$, 2 m*M* $MnCl_2$, 4 m*M* EDTA, 60 mg/L ascorbic acid (add the ascorbic acid just prior to use).
2. Import buffer, pH 8.0: 50 m*M* HEPES, 330 m*M* sorbitol.
3. Percoll gradient: 25 mL Percoll, 25 mL 2X grinding medium.

2.2.2. In Vitro Transcription/Translation

For in vitro transcription/translation, proceed as in single in vitro import, **Subheading 2.1.3**.

2.2.3. Dual Import Into Pea Mitochondria and Pea Chloroplasts

1. 2X Import medium, pH 7.4: 0.6 *M* sucrose, 30 m*M* HEPES, 10 m*M* KH_2PO_4, 1% (w/v) BSA.
2. Import master mix: 2X import medium, 1 *M* $MgCl_2$, 100 m*M* methionine, 100 m*M* adenosine triphosphate, 100 m*M* guanosine 5′-triphosphate, 100 m*M* adenosine 5′-diphosphate, 0.5*M* succinate, 0.5*M* dithiothreitol, 1*M* potassium acetate, 1*M* $NaHCO_3$ (make up fresh prior to use).
3. 5 m*M* $CaCl_2$, 2 mg/mL thermolysin, 0.5*M* EDTA.
4. Percoll gradient: Percoll, 2X import medium.
5. Centrifuge tubes (400 μL) (Beckman).

2.2.4. SDS Electrophoreses

Perform SDS electrophoreses as for single in vitro import, **Subheading 2.1.5**.

2.3. In Vivo Import Using Fusions With GFP

2.3.1. Transient Expression of GFP Fusion Constructs in Tobacco Protoplasts

2.3.1.1. Cloning

Suitable binary vector containing GFP construct in an eukaryotic expression cassette under a strong promoter such as 35S promoter ***(26)***.

2.3.1.2. Preparation of Protoplasts

1. 4- to 6-week-old plants of *Nicotiana tabacum* cv. SRI grown at 23°C under normal conditions.
2. Digestion medium (must be freshly prepared): 0.5*M* sucrose, 0.125% (w/v) macerozyme R-10 (Onozuka, Yakult Pharmaceutical, Tokyo, Japan), 0.2% (w/v)

cellulase R-10 (Onozuka), 5 m*M* $CaCl_2$, 0.1% (w/v) BSA; adjust to pH 5.5 with HCl and sterilize by filtration.

3. Flotation medium: 0.6 *M* sucrose, 0.15 *M* 2[N-morpholino]ethanesulfonic acid (MES)-NaOH, pH 6.0, 15 m*M* $CaCl_2$, autoclave.
4. Washing medium: 154 m*M* NaCl, 125 m*M* $CaCl_2$, 5 m*M* KCl, 5 m*M* glucose; sterilize by filtration.
5. Electroporation medium: 4 m*M* $CaCl_2$, 80 m*M* KCl, 8% (w/v) mannitol, 2 m*M* NaH_2PO_4, pH 7.0, 0.1% (w/v) BSA; sterilize by filtration.
6. K3M medium: 10X diluted macroelements K3, 1000X diluted microelements K3, 500X diluted Morel vitamins, 100 μ*M* Fe-EDTA, 0.5 *M* glucose, 100 mg/L thiamine; adjust to pH 5.7 with KOH.
7. Macroelements K3 (200 mL 10X): 1.2 g NH_4NO_3, 3.8 g KNO_3, 1.2 g $CaCl_2 \cdot 2H_2O$, 0.6 g $MgSO_4 \cdot 7H_2O$, 0.34 g KH_2PO_4, 0.6 g KCl; autoclave.
8. Microelements K3 (100 mL 1000X): 0.62 g H_3BO_3, 2.23 g $MnSO_4 \cdot H_2O$, 0.86 g $ZnSO_4 \cdot 7H_2O$, 0.025 g $Na_2MoO_4 \cdot 2H_2O$, 2.5 mg $CuSO_4 \cdot 5H_2O$, 2.5 mg $CoCl_2 \cdot 6H_2O$; autoclave.
9. Morel vitamins (100 mL 500X): 16.85 mg thiamine-HCl, 0.9 g myoinositol, 10.25 mg pyridoxine-HCl, 23.8 mg Ca^{2+}-pantothenate, 0.122 mg biotin, 6.15 mg nicotinic acid; sterilize by filtration.
10. Fluorescein diacetate (FDA; 20X): 5 mg/mL in acetone (keep at –20°C).
11. Nylon net (60-μm mesh).

2.3.1.3. Electroporation of Protoplasts

For electroporation of protoplasts, use gene pulser transfection apparatus and 0.4-cm gap cuvettes (Bio-Rad).

2.3.1.4. Confocal Microscopy

1. Confocal microscopy was performed using a Bio-Rad MRC-1024 laser scanning confocal imaging system.
2. MitoTracker Red CM-H2Xros (Molecular Probes, Eugene, OR, USA): 10 μ*M* in dimethyl sulfoxide.

2.3.2. Transient Expression Into Tobacco Leaves

2.3.2.1. Preparation of Electrocompetent Cells of *Agrobacterium*

1. 2YT plates: Add 2% (w/v) agar in 2YT medium and autoclave.
2. 2YT medium: 16 g/L Bacto™ tryptone, 10 g/L Bacto yeast extract, 5 g/L NaCl. Adjust the to pH 7.0 with 5*N* NaOH and make the volume to 1 L with deionized water. Sterilize by autoclaving.
3. Antibiotics: kanamycin (Kan), gentamicin (Gent), rifampicin (Rif).
4. 50 m*M* HEPES buffer: 16.4 g/L NaCl, 11.9 g/L HEPES, 2.1 g/L Na_2HPO_4. Adjust to pH 7.05 with NaOH and make the volume to 1 L with deionized water. Sterilize by autoclaving.

2.3.2.2. Transformation of Electrocompetent *Agrobacterium* Cells With Plasmid DNA by Electroporation (*see* **Note 1**)

1. *Agrobacterium tumefaciens*-competent cells.
2. YEB medium: 0.5% (w/v) beef extract, 0.1% (w/v) yeast extract, 0.5% (w/v) peptone, 0.5% (w/v) sucrose, 2 m*M* $MgSO_4$.
3. Acetosyringone (Sigma): make 200 m*M* stock in dimethyl sulfoxide, aliquot, and store at –20°C.
4. Bio-Rad gene pulser.
5. 0.2-cm Bio-Rad electroporation cuvettes.

2.3.2.3. Infiltration of the *Agrobacterium* Suspension Into Tobacco Leaves

1. 4- to 6-week-old plants of *N. tabacum* grown at 23°C under normal conditions.
2. Infiltration medium: 50 m*M* MES, pH 5.6, 0.5% (w/v) glucose, 2 m*M* Na_3PO_4, 100 μ*M* acetosyringone (add from stock prior to use).

2.3.2.4. Microscopy

For microscopy, *see* **Subheading 2.3.1.4**.

3. Methods

3.1. Single In Vitro Mitochondrial Import

3.1.1. Preparation of Potato Tuber Mitochondria

1. Grind potato tubers (1 kg) in a juice centrifuge in 400 mL grinding medium (*see* **Notes 3** and **4**).
2. Adjust to pH 7.0–7.5.
3. Filter the slurry through four layers of nylon net (60-μm mesh).
4. Sediment starch and cell wall fragments at 4000*g* for 5 min.
5. Recentrifuge the supernatant at 9000*g* for 10 min. Resuspend the mitochondrial pellet in 2.5 mL 1X wash medium.
6. Apply crude mitochondrial suspension onto 40 mL of Percoll solution containing 30% (v/v) Percoll solution and centrifuge at 32,000*g* for 45 min.
7. After centrifugation, collect mitochondrial band positioned close to the bottom of the tube; dilute 10 times in wash medium.
8. Centrifuge at 9000*g* for 15 min. Dilute mitochondrial pellet in fresh medium; wash to a final protein concentration of 30 mg/mL. Keep mitochondria on ice until use.

3.1.2. Preparation of Spinach Leaf Mitochondria

1. Homogenize depetiolated spinach leaves (300 g) in 300 mL grinding medium using Waring blender two times for 3 s at high speed.
2. Filter the slurry through four layers of nylon net (60-μm mesh).
3. Sediment chloroplasts, starch, and cell wall fragments at 5000*g* for 10 min.
4. Recentrifuge the supernatant at 9000*g* for 15 min.

5. Resuspend the mitochondrial precipitate in small volume of 1X wash medium and load on top of the same medium containing step gradient of Percoll: 2 mL 60% (v/v), 4 mL 28% (v/v), and 2 mL 21% (v/v) for type Ti-21 rotor tubes.
6. Centrifuge at 30,000*g* for 50 min. Collect the yellowish band containing mitochondria; dilute 10 times with 1X wash medium.
7. Centrifuge at 12,000*g* for 10 min. Dilute the mitochondrial pellet in fresh wash medium to a final protein concentration of 30 mg/mL. Keep mitochondria on ice until use.

3.1.3. In Vitro Synthesis of Radiolabeled Precursor Proteins

All precursor proteins are synthesized with the TNT-coupled reticulocyte lysate system (Promega). For the coupled transcription/translation reaction, mix 1 µg plasmid with components of the TNT-coupled reticulocyte lysate system; 1 µL methionine-free amino acid mixture, 2 µL TNT reaction buffer, 1 µL T7 polymerase, 1 µL ribonuclease inhibitor, 4 µL [^{35}S]-methionine; adjust volume to 50 µL with distilled water and incubate for 1.5 h at 30°C.

*3.1.4. Protein Import Into Potato Tuber and Spinach Leaf Mitochondria (see **Fig. 1**)*

1. Import experiments are carried out using isolated potato tuber or spinach leaf mitochondria (200 µg protein; *see* **Subheading 3.1.1.** or **3.1.2.**) resuspended in 200 µL import medium. To dissipate mitochondrial membrane potential, add 0.1 µm valinomycin to import mixture.
2. Import reaction is started by addition of 2 µL radiolabeled precursor protein, and the mixture is incubated for 30 min at 15°C using potato mitochondria or 20 min at 25°C using spinach mitochondria (*see* **Note 5**).
3. Stop reaction by placing tubes on ice. Split samples in two; leave one sample on ice and treat the second sample with 10 µg/mL proteinase K for 20 min at 4°C. Stop the proteolytic reaction by addition of phenylmethylsulfonyl fluoride to 1 m*M*. Reisolate mitochondria by centrifugation in Eppendorf microcentrifuge for 5 min at 15,000*g* at 4°C and solubilize mitochondrial pellet with SDS loading buffer.

3.1.5. SDS Electrophoresis and Image Processing

1. Boil samples for 5 min in 40 µL SDS loading buffer and analyze them by SDS-PAGE (polyacrylamide gel electrophoresis) in the presence of 4*M* urea.
2. To prepare 12% (w/v) polyacrylamide gel, mix 6.5 mL 4X separation buffer, 10.5 mL 30% (w/v) acrylamide/bis solution, 9 mL H_2O, 6.3 g urea, and 20 µL TEMED. Custom-made glass plates (150 mm × 150 mm) with 1.5-mm spacers are used for gel formation. AA polymerization begins after addition of 75 µL APS, and solution is immediately applied into gel chamber and overlaid with 1 mL butanol to achieve even gel formation.
3. After polymerization of separating gel (polymerization takes ~30 min; can be visualized as solution/gel partitioning), top solution containing butanol is removed,

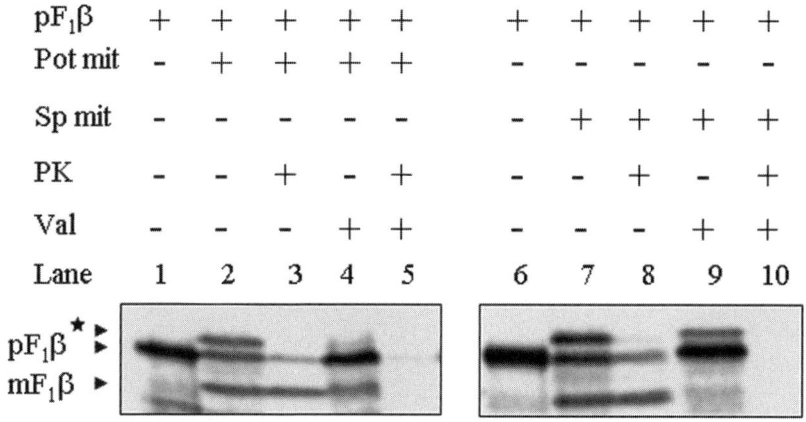

Fig. 1. Phosphoimage of the SDS-PAGE of the $pF_1\beta$ after import into potato tuber (Pot mit) and spinach leaf mitochondria (Sp mit). Lane 1, precursor protein $pF_1\beta$ alone. Lane 2, precursor protein incubated with potato tuber mitochondria. Lane 3, same as lane 2 but with the addition of proteinase K (PK). Lanes 4 and 5, same as lanes 2 and 3 but with the addition of valinomycin (Val). Lanes 6–10, same as lanes 1–5, but spinach leaf mitochondria were used. Asterisk indicates the presence of a modified form of $pF_1\beta$ ***(25)***.

and stacking solution is applied. Stacking gel (4% w/v AA) is prepared by mixing 6.5 mL 4X stacking buffer, 5 mL 30% (w/v) acrylamide/bis solution, 13.5 mL H_2O, 6 g urea, 20 µL TEMED, and 100 µL APS. The comb is immediately inserted, and stacking gel is polymerized for 30 min at room temperature. After removal of comb, pockets of 80-µL loading volume are formed.

4. Glass plates containing polyacrylamide gels are fixed in custom-made electrophoresis apparatus, and upper and lower chambers are filled with 500 mL SDS running buffer.
5. Apply samples (40 µL) into each pocket and perform gel electrophoresis overnight at 10 mA constant current at 4°C.
6. After completion of electrophoresis, when the bromophenol blue dye front is seen at the bottom of the gel, take the gels out of the glass plates, soak in fix solution for 30 min, and dry using vacuum gel dryer at 70°C for 1 h.
7. Subsequently, dried gels are exposed overnight to BAS-MP phosphoimaging plates and scanned using a Fujix BAS 1000 MacBAS bioimaging analyzer system.

3.2. Dual In Vitro Import to Mitochondria and Chloroplasts

3.2.1. Preparation of Organelles (see ***Note 6***)

3.2.1.1. Preparation of Pea Mitochondria

1. Approximately 80 g pea seeds are sown in plastic trays with drainage holes over a layer of grade 3 vermiculite and covered with a thin layer of vermiculite. Plants are grown at 28°C with a 16-h light and 8-h dark cycle. The seeds should be watered daily.

2. Pour two Percoll gradients using a gradient former (Bio-Rad, model 385) and keep on ice until used.
3. Harvest 100–200 g pea leaves. Homogenize the tissue in 300 mL grinding medium using a mortar and pestle.
4. Filter the homogenate through four layers of wet and chilled nylon net, transfer to 8 × 50 mL centrifugation tubes, and centrifuge for 5 min at 2400*g* in a refrigerated centrifuge.
5. Transfer the supernatant to two clean centrifugation tubes and discard the pellet. Centrifuge for 20 min at 17,400*g*.
6. Discard the supernatant and disperse pellet in residual supernatant using a chilled fine-hair brush. Combine into two clean centrifugation tubes and fill with 1X wash medium. Centrifuge for 5 min at 2400*g*.
7. Discard the supernatant, disperse the pellet in residual supernatant, and carefully overlay the two gradients. Centrifuge the gradients for 40 min at 35,000*g* with the brakes turned off.
8. Aspire green top layer (broken chloroplasts and thylakoids) and collect lower white-yellowish band (mitochondria). Transfer to clean centrifugation tubes and fill with 1X wash medium.
9. Carefully aspire the supernatant (the pellet is very loose at this stage). Fill with 1X wash medium and centrifuge for 15 min at 27,200*g*. Repeat this step once.
10. Carefully resuspend the mitochondria in a small volume of 1X wash medium. Keep mitochondria on ice until use.

3.2.1.2. Preparation of Pea Chloroplasts

1. Before starting the isolation, two 50% (v/v) Percoll gradients are prepared of equal parts of Percoll and 2X grinding medium (25 mL Percoll and 25 mL medium) and centrifuged for 30 min at 39,000*g* with the brakes turned off. Keep the gradients on ice until use.
2. Harvest approx 100 g pea leaves from the dark period. Homogenize the tissue in 300 mL 1X grinding medium using a blender with five short bursts. Filter the homogenate through one layer of chilled and wet nylon net and transfer into 8 × 50 mL centrifugation tubes and centrifuge for 5 min at 2000*g*.
3. Discard the supernatant and resuspend the pellet with fine-hair brush (*see* **Note 7**). Gently overlay the crude chloroplasts on the gradients. Centrifuge for 10 min at 12,100*g* with the brakes turned off.
4. Aspirate upper green band (broken chloroplasts and thylakoids); collect lower band (lighter green) using a Pasteur pipet and transfer to a clean centrifugation tube. Add 3 volumes of import medium and centrifuge for 4 min at 2000*g*.
5. Discard the supernatant and resuspend the pellet in a small volume of wash medium. Keep chloroplasts on ice and in dark until use.

3.2.2. In Vitro Transcription/Translation

In vitro transcription/translation is the same as for single in vitro import, **Subheading 3.1.3**.

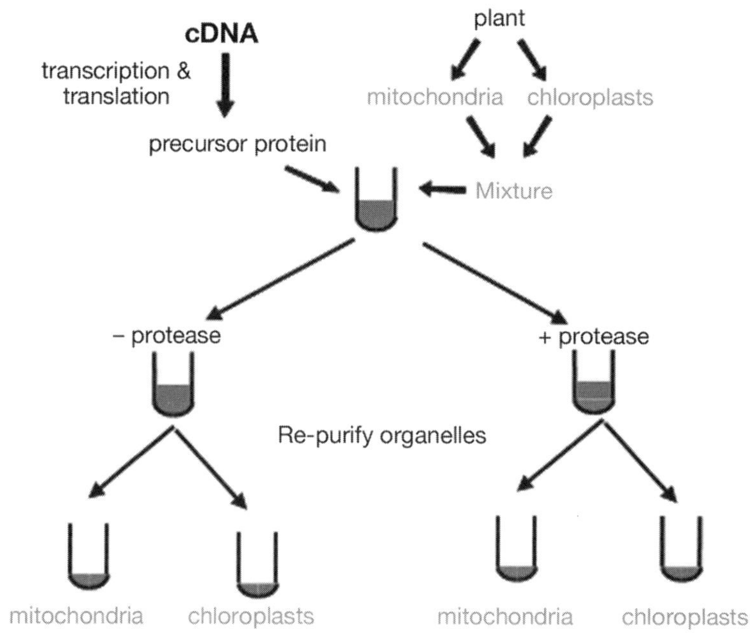

Fig. 2. Overview of the procedure of the dual-import assay. Chloroplasts and mitochondria are purified separately, mixed, and incubated under conditions that support import into both organelles. Immediately after import, the reaction is divided into two aliquots, and one is treated with protease. After protease treatment, the chloroplasts and mitochondria are repurified using a 4% Percoll gradient.

*3.2.3. Dual Import Into Pea Mitochondria and Pea Chloroplasts (**Figs. 2 and 3**)*

1. Measure the concentration of isolated organelles and adjust mitochondria to 10 mg/mL and chloroplasts to 1 mg chlorophyll/mL.
2. Add 65 µL master mix to two (for each precursor protein) prechilled round-bottom Falcon tubes (Eppendorf tubes can be used). Add 10 µLL mitochondria, 25 µL chloroplasts, and precursor and incubate for 20 min with gentle agitation and with a light source.
3. Place the reaction on ice and divide into two aliquots. Supplement one aliquot with 0.1 m*M* $CaCl_2$ (1 µL of a 5 m*M* stock) and 120 µg/mL thermolysin (3 µL of a 2 mg/mL stock) and incubate for 30 min on ice. Inhibit the thermolysin activity with 10 m*M* EDTA (1.25 µL of a 0.5*M* stock).
4. Carefully load each reaction on a 4% (v/v) Percoll gradient prepoured into a 400-µL elongated microfuge tube (*see* **Note 8**) (Eppendorf tubes can be used) and centrifuge for 30 s at 4000*g*. Collect the fractions separately, chloroplasts at the bottom of the gradient as a pellet and mitochondria at the top. Wash the fractions in 1 mL import medium. Recover the organelles by centrifugation for 2 min at 830*g* for chloroplasts and 20,800*g* for mitochondria (*see* **Note 9**).

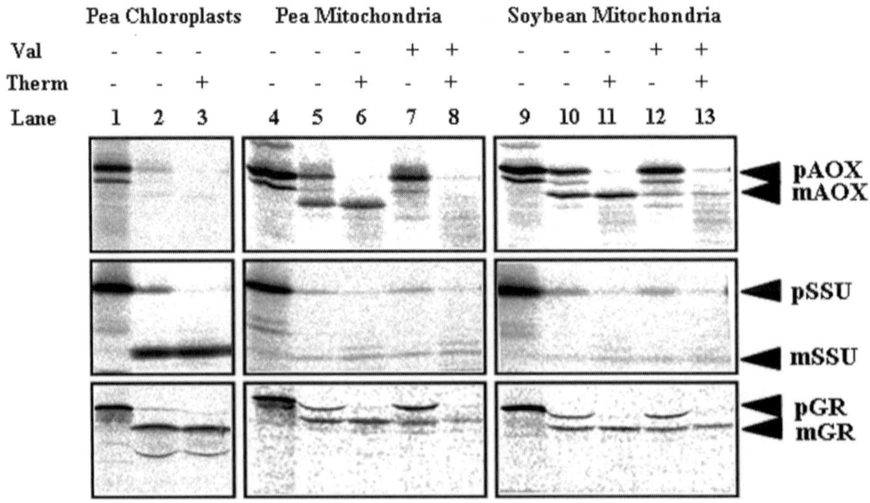

Fig. 3. Phosphoimage of the SDS-PAGE of the import of precursor proteins into chloroplasts and mitochondria. Alternative oxidase (AOX), small subunit of Rubisco (SSU), and glutathione reductase (GR) were imported into pea leaf chloroplasts, pea leaf mitochondria, and soybean cotyledon mitochondria using the dual-organelle import system. Lane 1, precursor protein alone. Lane 2, precursor protein import into pea leaf chloroplasts. Lane 3, same as lane 2 but with the addition of protease. Lane 4, precursor protein alone. Lane 5, precursor protein import into pea leaf mitochondria. Lane 6, same as lane 5 but with the addition of protease. Lane 7, same as lane 5 but with the addition of valinomycin (Val). Lane 8, same as lane 7 but with the addition of protease. Lanes 9–13, same as lanes 4–8 but using soybean cotyledon mitochondria.

3.2.4. SDS Electrophoresis

SDS electrophoresis is performed as for single in vitro import, **Subheading 3.1.5**.

3.3. In Vivo Import Using Fusions With GFP

3.3.1. Transient Expression of GFP Fusion Constructs in Tobacco Protoplasts

3.3.1.1. Cloning

Use standard cloning techniques to clone the full-length or targeting sequence of the protein of interest fused to GFP in an appropriate plasmid containing an eukaryotic expression cassette ***(26)***.

3.3.1.2. Preparation of Protoplasts (*see* **Note 10**)

1. Abrade the lower face of 30 tobacco leaves with no. 1200 A sandpaper and place them in a sterile Petri dish containing 10 mL digestion medium. The lower (abraded) face is in contact with liquid. Seal the dishes with parafilm.

2. Leave to digest for 15 h at 25°C in the dark without shaking. To free the protoplasts into the liquid medium, carefully shake the dishes for 30 min (*see* **Note 11**).
3. Remove the undigested pieces of leaf and add 4 mL flotation medium.
4. Filter the protoplast suspension in a 50-mL Falcon tube through a 60-μm nylon net.
5. Centrifuge at 110*g* for 7 min in the swinging rotor. Transfer the protoplasts (floating band) into another tube (take ~2.5 mL).
6. Dilute the protoplasts three times with dilution medium. Centrifuge at 110*g* for 7 min in the swinging rotor; remove supernatant with vacuum.
7. Wash the protoplasts with 40 mL electroporation medium and collect by centrifugation at 110*g* for 7 min. Resuspend the protoplasts in 20 mL electroporation medium.
8. Centrifuge at 110*g* for 7 min in the swinging rotor, remove supernatant with vacuum, and resuspend the protoplasts in electroporation medium to obtain suspension of $1–2 \times 10^6$/mL.
9. To determine protoplast viability with FDA, dilute 5 μL 20X FDA stock solution in 95 μL electroporation medium. Pour 10 μL diluted FDA onto a microscopic slide. Transfer 10 μL protoplast suspension into the diluted FDA solution. Observe the protoplasts under ultraviolet light (375–425 nm) without anything covering it. Count the green (viable) and red (dead) protoplasts and estimate the percentage of dead protoplasts. A normal preparation should contain between 10 and 15% dead protoplasts.

3.3.1.3. Electroporation of Protoplasts (*see* Notes 12 and 13)

1. Set 30 μg plasmid DNA in a sterile Eppendorf tube.
2. Add 800 μL well-homogenized protoplasts ($1–2 \times 10^6$/mL in electroporation medium).
3. Mix by slowly inverting the tube and incubate for 15 min at room temperature.
4. Transfer 800 μL of mixture into an electroporation cuvette (0.4 cm). The protoplasts are then electroporated at 250 μF and 0.32 kV (time const = 5).
5. Leave the protoplasts for 20 min.
6. Transfer into 15-mL Falcon tubes. Add 6 mL K3M medium.
7. Incubate the tubes in the dark at 25°C for 24 h.

3.3.1.4. Confocal Microscopy Analysis of the GFP Fusion Expression (Fig. 4A)

Perform confocal microscopy using a Bio-Rad MRC-1024 laser scanning confocal imaging system. For GFP detection, excitation was at 488 nm and detection between 506 and 538 nm. Chloroplast autofluorescence was detected between 664 and 696 nm, with an excitation at 488 nm. For staining of mitochondria, protoplast suspension was incubated with 0.1 μ*M* MitoTracker Red CM-H2Xros for 15 min. MitoTracker Red CM-H2Xros fluorescence was detected between 589 and 621 nm with an excitation at 568 nm.

3.3.2. Transient Expression Into Tobacco Leaves

3.3.2.1. Preparation of Electrocompetent Cells of *Agrobacterium*

1. Streak out *Agrobacterium* on 2YT plates containing Rif (50 μg/mL) and Gent (20 μg/mL) and grow for 2–3 d at 28°C.

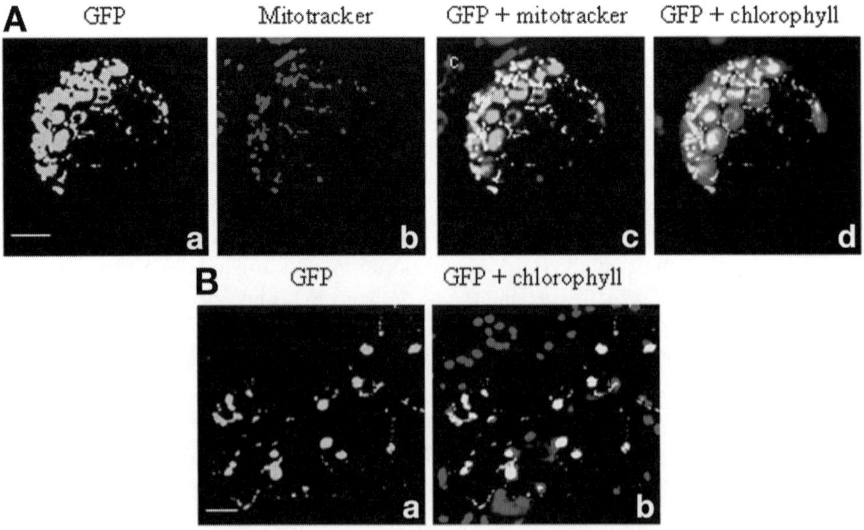

Fig. 4. In vivo targeting analysis of *Arabidopsis thaliana* presequence protease, *At*PreP2-GFP construct. (**A**) Transient expression of the *At*PreP2-GFP fusion constructs into *Nicotiana tabacum* protoplasts. GFP column shows the signal detected in the green channel (a); MitoTracker column shows the signal detected in the red channel (b); GFP + MitoTracker corresponds to the merging of the two previous columns, in which yellow represents the superposition of green and red (c); GFP + chlorophyll corresponds to the merging of the green channel and the chlorophyll signal detected in the far red channel (d). (**B**) *Agrobacterium*-mediated transient expression into *N. tabacum* leaves. GFP column shows the signal detected in the green channel (a); GFP + chlorophyll corresponds to the merging of the green channel and the chlorophyll signal detected in the far-red channel (b). Scale bars: 10 mm.

2. Inoculate a single colony into 10 mL 2YT medium containing the Rif (50 μg/mL) and Gent (20 μg/mL) antibiotics and grow at 28°C with shaking for 2 d.
3. Inoculate 100 mL fresh 2YT medium containing the Rif and Gent antibiotics with 1 mL preculture and grow (8–10 h) at 28°C until the OD_{600} reaches 0.4–0.6.
4. Place the culture on ice for 15–30 min.
5. Centrifuge the culture at 8000*g* for 10 min at 4°C.
6. Pour off supernatant and dissolve the pellet in 20 mL ice-cold 1 m*M* HEPES, pH 7.0.
7. Centrifuge again and dissolve the pellet in 20 mL ice-cold 1 m*M* HEPES, pH 7.0, 10% (v/v) glycerol.
8. Centrifuge again and dissolve the pellet in 10 mL ice-cold 1 m*M* HEPES, pH 7.0, 10% (v/v) glycerol.
9. Aliquot the cells in 50-μL batches, freeze immediately in liquid nitrogen, and store at –80°C.

3.3.2.2. Transformation of Electrocompetent *Agrobacterium* Cells With Plasmid DNA by Electroporation

1. Use standard cloning techniques to clone the full-length or targeting sequence of the protein of interest fused to GFP in an appropriate eukaryotic expression cassette into the *Agrobacterium* pBi plasmid.
2. Thaw electrocompetent cells on ice.
3. Incubate 50 μL cells with 1 μL plasmid DNA for 2 min on ice. The plasmid DNA may be in either distilled water or a low-salt buffer such as Tris-EDTA (TE) buffer. High-salt buffers should be avoided.
4. Transfer the cell-DNA mixture to a chilled 0.2-cm Bio-Rad electroporation cuvette. Tap the mixture to the bottom of the cuvette.
5. Set the Bio-Rad gene pulser apparatus to the 25 μF capacitor and the pulser controller unit to 200 Ω.
6. Transfer the cuvette to a chilled Bio-Rad gene pulser slide.
7. Set voltage to 2.5 KV.
8. Apply a single pulse, immediately add 1 mL 2YT broth (room temperature) to the cuvette, and gently resuspend the cells.
9. Transfer the cell suspension to a tube and incubate the culture at 28°C with shaking for 2 h.
10. Plate the suspension on Rif, Gen, and Kan plates and incubate the plates at 28°C. Transformed colonies will be visible after 2–3 d of incubation at 28°C.

3.3.2.3. Infiltration of the Agrobacterium Suspension Into Tobacco Leaves (see Notes 14–17)

1. Culture the 2 mL *Agrobacterium* (transformed with your construct) in YEB medium containing 100 μL/mL Kan, 10 μL/mL Gent, and 50 μL/mL of Rif at 28°C with shaking to stationary phase (24–48 h).
2. Take 1 mL of the culture, pellet (6000*g* for 2 min), and wash twice in infiltration medium; then, resuspend at OD_{600} of 0.1–0.3.
3. Inject bacterial suspension into the abaxial epidermis of healthy plant leaves from a 1-mL plastic syringe by pressing the nozzle against the leaf surface (*see* **Note 18**). Spread of liquid entering the leaf via stomata can be visualized by darkening of the leaf. Mark boundaries of the infiltrated area with an indelible pen.
4. Incubate plants for 2–3 d at 20–25°C under normal growing conditions.

3.3.2.4. Microscopy (Fig. 4B)

Check for expression by analyzing a piece of the infiltrated area under fluorescence microscope capable of detecting GFP fluorescence. The abaxial surface of the leaf should be facing the incoming light.

4. Notes

1. Plasmids encoding proteins of interest were purified from *Escherichia coli* night cultures and isolated using QIAprep® spin miniprep kit (Quiagen) according to manufacturer's instructions.

2. Unpolymerized acrylamide is neurotoxic; care should be taken to avoid exposure.
3. As a rule, try to keep operating temperature during isolation of mitochondria close to 0°C and minimize time between isolation of mitochondria and import of precursor proteins.
4. The yield and import competence of potato tuber mitochondria can depend on quality of potatoes.
5. Thaw precursor protein on ice just prior to use.
6. Make sure to keep mitochondria and chloroplast cold during all steps of the isolation procedure. Precool buffers, centrifuge, centrifuge tubes, pipets, and beakers before use.
7. Be gentle when resuspending chloroplasts to avoid breakage.
8. Prepare the 4% (v/v) Percoll gradients before starting the import procedure; keep on ice until use.
9. If mitochondria do not pellet when recovered, then extend the centrifugation time to 10 min. The mitochondrial pellet can be quite loose, so remove supernatant as soon as the centrifugation is finished.
10. All the manipulations should be performed under sterile laminar flow.
11. Protoplasts are very fragile and should be handled with care; for example, use blunt pipet tips to avoid damage.
12. Transient expression of a GFP fusion construct can also be performed using polyethylene glycol-mediated transformation of a plasmid into protoplasts ***(27)***.
13. Transient expression of GFP fusion constructs has also been achieved in soybean suspension cells using biolistic transformation ***(28)***.
14. Transient expression of GFP constructs in tobacco leaves can also be performed by biolistic transformation ***(28,29)***.
15. Expression of GFP varies with the infiltrated concentration of *Agrobacterium,* which offers the opportunity to control GFP expression levels.
16. Plant species so far tested are tobacco, tomato, potato, and petunia, with expression obtained in all cases.
17. Expression of GFP varies with the infiltrated concentration of *Agrobacterium,* which offers the opportunity to control GFP expression levels.
18. Avoid the bottom 2 leaves and the top 2 unfolded leaves for infiltration because of the variability in infection rate.

Acknowledgments

We would like to thank Prof. J. Whelan and Dr. O. Chew for a fruitful collaboration on the development of the dual import system and Prof. M. Boutry and Dr. B. Lefebvre for instructions and cooperation using in vivo import procedures. This work was supported by a grant from the Swedish Research Council to E. G.

References

1. Zhang, X. P. and Glaser, E. (2002) Interaction of plant mitochondrial and chloroplast signal peptides with Hsp70 molecular chaperone. *Trends Plant Sci.* **7,** 14–21.

2. Glaser, E. and Soll, J. (2005) Targeting signals and import machinery of plastids and plant mitochondria, in *Molecular Biology and Biotechnology of Plant Organelles: Chloroplasts and Mitochondria* (Daniell, H., and Chase, C., eds.), Springer, Dordrecht, The Netherlands, pp. 385–418.
3. Glaser, E. and Dessi, P. (1999) Integration of the mitochondrial processing peptidase into the bc1 complex of the respiratory chain in plants. *J. Bioenerg. Biomembr.* **31,** 259–274.
4. Richter, S. and Lamppa, G. K. (1998) A chloroplast processing enzyme functions as the general stromal processing peptidase. *Proc. Natl. Acad. Sci. USA* **95,** 7463–7468.
5. Ståhl, A., Moberg, P., Ytterberg, J., Panfilov, O., Brockenhuus von Löwenhielm, H., Nilsson, F., and Glaser, E. (2002) Isolation and identification of a novel mitochondrial metalloprotease (PreP) that degrades targeting presequences. *J. Biol. Chem.* **277,** 41,931–41,939.
6. Moberg, P., Ståhl, A., Bhushan, S., et al. (2003) Characterization of a novel zinc metalloprotease involved in degrading signal peptides in mitochondria and chloroplasts. *Plant J.* **36,** 616–628.
7. Bushan, S., Lefebvre, B., Ståhl, A., Boutry, M., and Glaser, E. (2003) Dual targeting and function of a protease in mitochondria and chloroplasts. *EMBO Rep.* **4,** 1073–1078.
8. Ståhl, A., Nilsson, S., Lundberg, P., et al. (2005) Two novel targeting peptide degrading proteases, PrePs, in mitochondria and chloroplasts, so similar and still different. *J. Mol. Biol.* **349,** 849–860.
9. Bhushan, S., Stahl, A., Nilsson, S., et al. (2005) Catalysis, subcellular localization, expression and evolution of the targeting peptides degrading protease, AtPreP2. *Plant Cell Physiol.***46,** 985–996.
10. Marc, P., Margeot, A., Devaux, F., Blugeon, C., Corral-Debrinski, M., and Jacq, C. (2002). Genome-wide analysis of mRNAs targeted to yeast mitochondria. *EMBO Rep.* **3,** 159–164.
11. Glaser, E., Sjoling, S., Tanudji, M., and Whelan, J. (1998) Mitochondrial protein import in plants. *Plant Mol. Biol.* **38,** 311–338.
12. Soll, J. and Tien, R. (1998) Protein translocation into and across the chloroplastic envelope membranes. *Plant Mol. Biol.* **38,** 191–207.
13. Boutry, M., Nagy, F., Poulsen, C., Aoyagi, K., and Chua, N.H. (1987) Targeting of bacterial chloramphenicol acetyltransferase to mitochondria in transgenic plants. *Nature* **328,** 340–342.
14. Whelan, J., Knorpp, C., and Glaser, E. (1990) Sorting of precursor proteins between isolated spinach leaf mitochondria and chloroplasts. *Plant Mol. Biol.* **14,** 977–982.
15. Schmitz, U. K. and Lonsdale, D. M. (1989) A yeast mitochondrial presequence functions as a signal for targeting to plant mitochondria in-vivo. *Plant Cell* **1,** 783–791.
16. Silva Filho, M. d. C., Wieers, M.-C., Flugge, U.-I., Chaumont, F., and Boutry, M. (1997) Different in vitro and in vivo targeting properties of the transit peptide of a chloroplast envelope inner membrane protein. *J. Biol. Chem.* **272,** 15,264–15,269.

17. Hugosson, M., Nurani, G., Glaser, E., and Franzen, L.G. (1995) Peculiar properties of the PsaF photosystem I protein from the green alga *Chlamydomonas reinhardtii*: presequence independent import of the PsaF protein into both chloroplasts and mitochondria. *Plant Mol. Biol.* **28,** 525–535.
18. von Stedingk, E. (1999) Sorting and import of plant mitochondrial precursors. PhD thesis. Department of Biochemistry, Stockholm University, Stockholm.
19. Lister, R., Chew, O., Lee, M., and Whelan, J. (2001) *Arabidopsis thaliana* ferrochelatase-I and -II are not imported into *Arabidopsis* mitochondria. *FEBS Lett.* **506,** 291–295.
20. Cleary, S. P., Tan, F.-C., Nakrieko, K.-A., et al. (2002) Isolated plant mitochondria import chloroplast precursor proteins in vitro with the same efficiency as chloroplasts. *J. Biol. Chem.* **277,** 5562–5569.
21. Rudhe, C., Chew, O., Whelan, J., and Glaser, E. (2002) A novel in vitro system for simultaneous import of precursor proteins into chloroplast and mitochondria. *Plant J.* **30,** 213–220.
22. Creissen, G., Reynolds, H., Xue, Y., and Mullineaux, P. (1995) Simultaneous targeting of pea glutathione reductase and of a bacterial fusion protein to chloroplasts and mitochondria in transgenic tobacco. *Plant J.* **8**, 167–175.
23. Small, I., Wintz, H., Akashi, K., and Mireau, H. (1998) Two birds with one stone: genes that encode products targeted to two or more compartments. *Plant Mol. Biol.* **38,** 265–277.
24. Hedtke, B., Börner, T., and Weihe, A. (2000) One RNA polymerase serving two genomes. *EMBO Rep.* **1,** 435–440.
25. Von Stedingk, E., Pavlov, P. F., Grinkevich, V. A., and Glaser, E. (1999) The precursor of $F_1\beta$ subunit of the ATP synthase is covalently modified upon binding to plant mitochondria. *Plant Mol. Biol.* **41,** 505–515.
26. Duby, G., Oufattole, M., and Boutry, M. (2001) Hydrophobic residues within the predicted N-terminal amphiphilic α-helix of a plant mitochondrial targeting presequence play a major role in in vivo import. *Plant J.* **27,** 539–549.
27. Datta, K. and Datta, S. K. (1999) Transformation of rice via PEG-mediated DNA uptake into protoplasts. *Methods Mol. Biol.* **111,** 335–347.
28. Chew, O., Rudhe, C., Glaser, E., and Whelan, J. (2003) Characterization of the targeting signal of dual-targeted pea glutathione reductase. *Plant Mol. Biol.* **53,** 341–356.
29. Lukaszewicz, M., Jerouville, B., and Boutry, M. (1998) Signs of translational regulation within the transcript leader of a plant plasma membrane H^+-ATPase gene. *Plant J.* **14,** 413–423.

23

Two-Dimensional Blue Native/Blue Native Polyacrylamide Gel Electrophoresis for the Characterization of Mitochondrial Protein Complexes and Supercomplexes

Stephanie Sunderhaus, Holger Eubel, and Hans-Peter Braun

Summary

Blue native polyacrylamide gel electrophoresis (BN-PAGE) employs the dye Coomassie for the labeling of proteins and protein complexes under native conditions. Electrophoresis under native conditions subsequently allows resolution of proteins and protein complexes according to their molecular mass. BN-PAGE can be combined with second gel dimensions. Best known is the two-dimensional (2D)-BN/sodium dodecyl sulfate (SDS)-PAGE system, which allows resolution of subunits of protein complexes. A 2D-BN/BN-PAGE system was developed that proved useful for investigating the substructure of protein complexes and protein supercomplexes. The basis of this 2D system is a variation of the conditions used for the two BN gel dimensions. Here, we present a basic protocol for the analysis of mitochondrial fractions by 2D-BN/BN-PAGE. Because both gel dimensions are carried out under native conditions, the 2D-BN/BN system is compatible with in-gel enzyme activity staining.

Key Words: Blue native polyacrylamide gel electrophoresis (BN-PAGE); digitonin; in-gel enzyme assay; membrane proteins; protein complexes; protein supercomplexes; respiratory chain.

1. Introduction

Blue native polyacrylamide gel electrophoresis (BN-PAGE) is based on incubation of proteins with Coomassie blue. Coomassie belongs to the trimethylmethane dye family and has been used as a wool dye since the late 19th century. Because of its efficient and specific binding to proteins, it was introduced in biochemistry to visualize proteins on gels after electrophoresis and later to determine protein concentrations photometrically ***(1,2)***. Coomassie

From: *Methods in Molecular Biology, vol. 372: Mitochondria: Practical Protocols*
Edited by: D. Leister and J. M. Herrmann © Humana Press Inc., Totowa, NJ

is a negatively charged compound that carefully introduces charges into proteins and protein complexes. Because of this property, it was suggested to be used for protein labeling prior to electrophoretic separations *(3)*. Coomassie does not denature proteins. In combination with nonionic detergents, BN-PAGE proved to be an ideal tool for the analysis of membrane-bound protein complexes, especially for those of mitochondria *(3)*. Even supramolecular assemblies of membrane-bound mitochondrial protein complexes were shown to be stable during BN-PAGE if the membrane solubilization step is carried out under very mild conditions *(4)*. Classically, a first blue native gel dimension is combined with sodium dodecyl sulfate (SDS)-PAGE as a second gel dimension, which allows separation of subunits of protein complexes. On the resulting gels, subunits of protein complexes form vertical rows. Detailed protocols for two-dimensional (2D)-BN/SDS-PAGE have been published *(5–7)*.

A novel 2D gel electrophoresis system was developed that is based on BN-PAGE for both gel dimensions *(4)*. If the two gel dimensions are carried out under identical conditions, then protein complexes form a diagonal line on the resulting 2D gels. However, if conditions of the second-dimension BN-PAGE are slightly less gentle than conditions of the first-dimension BN-PAGE, then protein complexes are dissected into subcomplexes, which are visible underneath this diagonal line. 2D-BN/BN-PAGE proved to be a very powerful tool for the investigation of the protein complex composition of supercomplexes or the subcomplex composition of protein complexes. It was successfully used to characterize the supramolecular structure of the protein complexes of the oxidative phosphorylation system in mitochondria *(4,6,8–14)* and of the photosynthetic electron transport system in chloroplasts *(15)*. Conditions to be varied between the two BN gel dimensions can refer to detergent type (e.g., digitonin and dodecylmaltoside), detergent concentration (e.g., 0.5 and 2.0% dodecylmaltoside), temperature (e.g., 4 and 20°C), or presence of chaotropic compounds (absence and presence of urea). 2D-BN/BN-PAGE is compatible with in-gel enzyme staining, allowing determination of the activity of supramolecular assemblies of protein complexes as well as the activities of subcomplexes of protein complexes (**Fig. 1**). We present a basic protocol for the analysis of mitochondrial protein complexes by BN/BN-PAGE; the protocol is based on detergent variation between the two BN gel dimensions. The presented protocol also is suitable for the analysis of other organelle fractions or of bacteria.

2. Materials

2.1. Preparation of BN Gels for First and Second Gel Dimensions

1. Acrylamide solution: 49.5% (w/v), acryl/bisacryl 32:1 (AppliChem, Darmstadt, Germany).
2. 6X gel buffer BN: 1.5*M* amino caproic acid, 150 m*M* BisTris, pH 7.0 (adjust at 4°C).

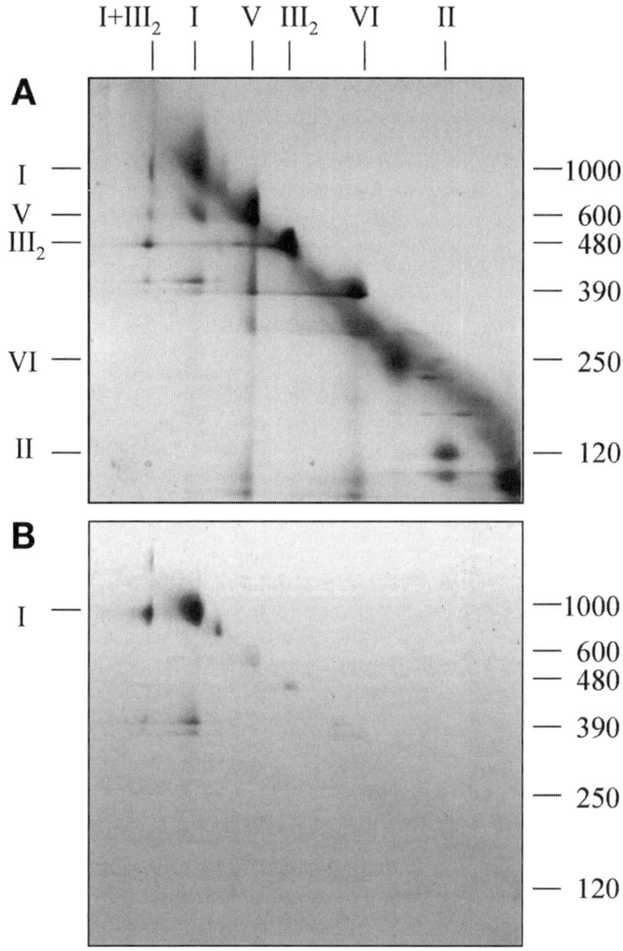

Fig. 1. Resolution of mitochondrial protein complexes from *Arabidopsis* by 2D-BN/BN-PAGE. Proteins were solubilized by digitonin. (**A**) 2D-BN/BN gel after Coomassie staining; (**B**) 2D-BN/BN gel after activity staining for NADH dehydrogenase (complex I). The numbers to the right refer to the molecular masses of standard proteins (in KDa). I+III_2, supercomplex composed of complexes I and III_2; I, NADH dehydrogenase (complex I); V, adenosine triphosphate synthase (complex V); III_2, dimeric cytochrome-*c* reductase (complex III); IV, cytochrome-*c* oxidase (complex IV); II, succinate dehydrogenase (complex II). Note that activity staining allows visualization of the singular and the supercomplex-bound forms of complex I. Furthermore, complex I is partially dissected into fragments of 600 kDa (hydrophobic arm) and 400 kDa (matrix-exposed arm). The 400-kDa arm includes the NADH-oxidizing domain and therefore is also labeled by activity staining, whereas the 600-kDa arm is not.

2.2. Sample Preparation

1. Digitonin solubilization solution: 5.0% (w/v) digitonin (e.g., Fluka, Buchs, Switzerland), 30 m*M* HEPES, 150 m*M* potassium acetate, 10% (v/v) glycerol, pH 7.4 (adjust at 4°C). This buffer should be freshly prepared and shortly heated to 98°C to dissolve the detergent. Add phenylmethylsulfonyl fluoride directly before use (2 m*M* final concentration; stock solution 200 m*M* phenylmethylsulfonyl fluoride in ethanol).
2. Coomassie blue solution: 5% (w/v) Coomassie G 250 (e.g., Merck, Darmstadt, Germany), 750 m*M* amino caproic acid.

2.3. First-Dimension BN PAGE

1. 5X Cathode buffer BN: 250 m*M* Tricine, 75 m*M* BisTris, 0.1% (w/v) Coomassie G 250 (e.g., Merck), pH 7.0 (adjust at 4°C)
2. 6X Anode buffer BN: 300 m*M* BisTris, pH 7.0 (adjust at 4°C)

2.4. Transfer of Gel Stripes of First Gel Dimensions Onto Second Gel Dimensions

1X Cathode buffer BN plus dodecylmaltoside: 50 m*M* Tricine, 15 m*M* BisTris, 0.03% (w/v) dodecylmaltoside, 0.02 (w/v) Coomassie G 250 (e.g., Merck), pH 7.0 (adjust at 4°C).

2.5. Second-Dimension BN PAGE

1. Agarose solution: 1.5% (w/v) agarose.
2. Cathode buffer BN plus dodecylmaltoside: (*see* **Subheading 2.4.**).
3. 6X Anode buffer BN: (*see* **Subheading 2.3.**).

2.6. Enzyme Activity Staining Procedures of 2D-BN/BN Gels

1. Phosphate buffer stock solution: 0.5 *M* KH_2PO_4, 0.5 *M* K_2HPO_4, pH 7.4.
2. DAB stock solution: 0.1 *M* 3,3′-diaminobenzidine tetrahydrochloride dehydrate.
3. KCN stock solution: 1 *M* KCN.
4. Ethylenediaminetetraacetic acid (EDTA) stock solution: 0.1 *M* EDTA, pH 8.0.
5. Tris-HCl stock solution: 2 *M* Tris-HCl, pH 7.4.
6. Cytochrome-*c* oxidase staining solution: 10 m*M* phosphate buffer, pH 7.4, 0.1% (w/v) DAB, 7.5% (w/v) sucrose, 19 U/mL catalase, and 16 m*M* cytochrome-*c* (Sigma-Aldrich, St. Louis, MO, USA).
7. Succinate dehydrogenase staining solution: 50 m*M* phosphate buffer, pH 7.4, 84 m*M* succinic acid, 0.2 m*M* *N*-methylphenazonium methyl sulfate (Fluka), 0.2% (w/v) nitroblue tetrazolium (Fluka), 4.5 m*M* EDTA, 10 m*M* KCN.
8. Nicotinamide adenine dinucleotide (NADH) dehydrogenase staining solution: 0.1 *M* Tris-HCl, pH 7.4, 0.14 m*M* NADH, 0.1% (w/v) nitroblue tetrazolium (Fluka).
9. Fixing solution: 40% (v/v) methanol, 10% (w/v) acetic acid.

3. Methods

Digitonin currently is considered the mildest detergent for mitochondrial research, stabilizing better most supramolecular structures than Triton X-100, dodecylmaltoside, or other nonionic detergents. In our protocol, first-dimension BN-PAGE is therefore carried out after membrane solubilization using digitonin and second dimension BN-PAGE after incubation of the gel stripe of the first gel dimension with dodecylmaltoside. However, numerous variations of this experimental design are possible (*see* **Note 1**).

3.1. Preparation of BN Gels for First and Second Gel Dimensions

Best resolution capacity of BN gels is achieved if the electrophoretic separation distance is more than 12 cm. The following instructions refer to the Protean II electrophoresis unit (Bio-Rad, Richmond, CA, USA; dimensions 0.1 × 16 × 20 cm for the gel of the first dimension, 0.15 × 16 × 20 cm for the gel of the second dimension). However, units from other manufacturers are of comparable suitability for BN-PAGE (e.g., the Hoefer SE-400 or SE-600 gel systems, Amersham Biosciences, Uppsala, Sweden). Usage of gradient gels is recommended because molecular masses of protein complexes can vary between 50 kDa and several thousand kilodaltons (*see* **Note 2**).

For the first-dimension gel (thickness 0.1 cm):

1. Prepare a 4.5% separation gel solution by mixing 1.2 mL acrylamide solution with 2.2 mL gel buffer BN and 10.0 mL double-distilled water (ddH_2O).
2. Prepare a 16% separation gel solution by mixing 4.5 mL acrylamide solution with 2.2 mL gel buffer BN, 4.0 mL ddH_2O, and 2.7 mL glycerol.
3. Transfer the two gel solutions into the two chambers of a gradient former and connect the gradient former via a hose and a needle with the space between two glass plates, which are preassembled in a gel-casting stand. Gradient gels can be poured either from the top (16% gel solution has to enter the gel sandwich first) or from the bottom (4.5% gel solution has to enter first). Poring gradient gels at 4°C is recommended to avoid premature polymerization.
4. Add ammonium persulfate (APS) and *N,N,N′*-tetramethylethylenediamine (TEMED) to the two gel solutions (60 μL 10% APS/6 μL TEMED to the 4.5% gel solution, 40 μL 10% APS/4 μL TEMED to the 16% gel solution).
5. Pour the gradient gel, leaving space for the stacking gel, and overlay with double-distilled water. The gel should polymerize in about 60 min.
6. Pour off the double-distilled water.
7. Prepare the stacking gel solution by mixing 1.2 mL acrylamide solution, 2.5 mL gel buffer BN, and 11.3 mL ddH_2O.
8. Add 65 μL 10% APS and 6.5 μL TEMED and pour the stacking gel around an inserted comb. The stacking gel should polymerize within 30 min.

For the 2D gel (thickness 0.15 cm):

1. Prepare a 5.0% separation gel solution by mixing 2.0 mL acrylamide solution with 3.3 mL gel buffer BN and 14.7 mL ddH_2O.
2. Prepare a 20% separation gel solution by mixing 8.0 mL acrylamide solution with 3.3 mL gel buffer BN, 4.7 mL ddH_2O, and 4.0 mL glycerol.
3. Transfer the two gel solutions into the two chambers of a gradient former as described for the first-dimension gel.
4. Add APS and TEMED to the two gel solutions (90 µL 10% APS/9 µL TEMED to the 5% gel solution, 50 µL 10% APS/5 µL TEMED to the 20% gel solution).
5. Pour the gradient gel, leaving space for the stacking gel, and overlay with double-distilled water. The gel should polymerize in about 60 min.
6. Pour off the double-distilled water.
7. Prepare the stacking gel solution by mixing 1.2 mL acrylamide solution, 2.5 mL gel buffer BN, 1.7 mL glycerol, and 9.6 mL ddH_2O.
8. Add 65 µL 10% APS and 6.5 µL TEMED and cast the stacking gel until 1 cm below the upper edge of the inner glass plate (do not insert a comb). Overlay with double-distilled water. The stacking gel should polymerize within 30 min. Finally, remove the overlaying water.

Gels for both gel dimensions should be prepared 1 d before usage and should be stored at 4°C.

3.2. Sample Preparation

All steps of the sample preparation should be carried out at 4°C.

1. Prepare mitochondria of interest (*see* **Note 3**).
2. Determine the protein concentration (e.g., according to Lowry; ***16***).
3. Adjust the protein concentration to 10 µg/µL.
4. Centrifuge 50-µL fractions (about 0.5 mg protein) for 10 min at 15,000*g* to sediment organelles.
5. Resuspended pellets in 50 µL digitonin solubilization solution.
6. Incubate the fractions for 20 min on ice.
7. Centrifuge the fractions for 20 min at 20,000*g* to remove insoluble material.
8. Supplement the supernatants with 2.5 µL Coomassie blue solution.
9. Load 30–50 µL of the fractions (corresponding to about 0.3–0.5 mg mitochondrial protein) directly into the wells of a BN gel (protein amounts are adjusted to allow staining of gels by Coomassie; if silver staining will be applied, then protein amounts can be reduced by a factor of 10).

3.3. First-Dimension BN-PAGE

1. Prepare 1X anode and 1X cathode buffers BN by diluting the corresponding stock solutions.
2. Carefully remove the comb of the first-dimension BN gel.
3. Assemble the gel electrophoresis unit; add 1X cathode and 1X anode buffers BN to the upper and lower chambers of the gel unit. Cool the unit to 4°C.

4. Load Coomassie blue pretreated protein samples (*see* **Subheading 3.2.**) into the gel wells.
5. Connect the gel unit to a power supply. Start electrophoresis at constant voltage (100 V for 45 min) and continue at constant current (15 mA for about 5 h; *see* **Note 4**). Electrophoresis should be carried out at 4°C. Blue gel bands should already be visible during the electrophoresis run.

3.4. Transfer of Gel Stripes of First Gel Dimensions Onto Second Gel Dimensions

1. Cut out a lane of the first-dimension BN gel.
2. Incubate the gel stripe for 10 min in 1X cathode buffer BN plus dodecylmaltoside at 4°C.

3.5. Second-Dimension BN-PAGE

1. Assemble the gel electrophoresis unit and transfer the gel stripe of the first gel dimension onto a 2D gel. Make sure it is placed centrally and has close contact to the 2D gel. Fix the gel stripe with 1.5% agarose solution (boil the solution before use and allow it to cool to approx 45°C) (*see* **Note 5**).
2. Prepare 1X anode buffer BN by diluting the corresponding stock solution.
3. Add 1X cathode buffer BN plus dodecylmaltoside and 1X anode buffer BN to the upper and lower chambers of the gel unit.
4. Connect the gel unit to a power supply. Start electrophoresis at constant voltage (100 V for 45 min) and continue at constant current (15 mA for 6–12 h; *see* **Note 6**). Electrophoresis should be carried out at 4°C. Blue gel spots should already be visible during the electrophoresis run.

3.6. Enzyme Activity Staining Procedures for 2D-BN/BN Gels

After completion of the electrophoresis run, 2D-BN/BN gels can be stained using Coomassie colloidal ***(17,18)*** or silver ***(19)*** (*see* **Notes 7** and **8**). However, because both gel dimensions are carried out under native conditions, 2D-BN/BN-PAGE also is compatible with in-gel enzyme activity staining ***(20,21)***. Three classical in-gel staining procedures for enzymes of the respiratory chain are given (activity staining procedures for several other enzymes can be found in the literature).

1. Incubate the gel with double-distilled water for 10 min twice.
2. Incubate the gel with 100 mL freshly prepared staining solution (cytochrome-*c* oxidase staining solution, succinate dehydrogenase staining solution, or NADH dehydrogenase staining solution). Staining takes minutes to hours, depending on the abundance of the stained enzyme.
3. Stop the reaction by transferring the gel into fixing solution (*see* **Notes 9** and **10**).

4. Notes

1. Conditions to be varied in the two gel dimensions of BN/BN-PAGE can refer to detergent type, detergent concentration, temperature, presence of chaotropic

compounds, and others. Conditions for dissection of supercomplexes into protein complexes or dissection of protein complexes into subcomplexes should be optimized for the mitochondria of interest.

2. If very large protein complexes (>3 MDa) have to be resolved, then the acrylamide gradient gel of the BN gel dimension can be substituted by a 2.5% agarose gel prepared in gel buffer BN *(22)*.
3. Alternatively, mitochondrial subfractions can be used as starting material for BN/BN-PAGE (e.g., an inner membrane or a matrix fraction).
4. Electrophoresis should not be carried out for more than 5 h because protein complexes might get stuck in the pores of the gradient polyacrylamide gel, which might prevent migration of the complexes into the second gel dimension. In general, two-thirds completion of the electrophoresis run is sufficient for protein complex resolution on the first gel dimension.
5. Transfer of a lane of a BN gel onto a second gel dimension is proposed to be carried out by fixing a lane of a first-dimension BN gel onto a prepoured BN gel for second gel dimension. By using this procedure, time between first and second gel dimension is minimized, which is advantageous for activity staining of 2D-BN/BN gels. However, physical contact between the lane of the first-dimension gel and the second-dimension gel might be better if the lane of the second gel dimension is embedded into the stacking gel of the second gel dimension (*see* **refs. *5–7*** for corresponding protocols). On the other hand, TEMED and APS of the gel solution for the second gel dimension can diffuse into the gel stripe of the first gel dimension, which usually greatly reduces enzymatic activities. Therefore, this procedure should only be applied for 2D-BN/BN gels in combination with Coomassie and silver staining.
6. Electrophoresis of the second gel dimension should be carried out for 6–12 h or longer. Often, sharpness of protein spots is best after long electrophoresis because protein complexes get stuck into the pores of the gradient gel of the second gel dimension at defined polyacrylamide concentrations.
7. Protein complexes resolved on 2D-BN/BN gels also can be blotted onto membranes. Short preblots should be carried out to destain gels from excess of Coomassie blue electrophoretically *(23)*. Alternatively, the cathode buffer BN can be replaced by a cathode buffer BN without Coomassie blue after 50% completion of the electrophoresis run of the second BN gel dimension. Furthermore, protein complexes excised from 2D-BN/BN gels also can be separated on a third gel dimension, which is carried out in the presence of SDS *(4)*.
8. Protein complexes resolved by 2D-BN/BN-PAGE also can be cut out and prepared for analysis by mass spectrometry. Usually, the subunits of protein complexes are first fragmented by trypsin. Afterward, peptides are best analyzed by coupled liquid choromatography and electrospray tandem mass spectrometry.
9. Alternatively, in-gel activity staining also can be stopped by adding inhibitors of the monitored enzymes.
10. If 2D-BN/BN gels will be Coomassie or silver stained after activity staining, then wash the gels several times with double-distilled water before fixation to remove proteins of the activity staining solutions. This will reduce background staining of the gels.

Acknowledgments

This work was supported by the Deutsche Forschungsgemeinschaft (grant BR 1829-7/1).

References

1. Groth, S. F., Webster, R. G., and Datyner, A. (1963) Two new staining procedures for quantitative estimation of proteins on electrophoretic stripes. *Biochim. Biophys. Acta* **71,** 377–391.
2. Bradford, M. (1976) A rapid and sensitive method for the quantitation of microgram quantities of protein utilizing the principle of protein-dye binding. *Anal. Biochem.* **72,** 248–254.
3. Schägger, H. and von Jagow, G. (1991) Blue native electrophoresis for isolation of membrane protein complexes in enzymatically active form. *Anal. Biochem.* **199,** 223–231.
4. Schägger, H. and Pfeiffer, K. (2000) Supercomplexes in the respiratory chain of yeasts and mammalian mitochondria. *EMBO J.* **19,** 1777–1783.
5. Schägger, H. (2001) Blue-native gels to isolate protein complexes from mitochondria. *Methods Cell Biol.* **166,** 231–244.
6. Schägger, H. (2003) Blue native electrophoresis, in *Membrane Protein Purification and Crystallization: A Practical Guide* (Hunte, C., von Jagow, G., and Schägger, H., eds.), Academic Press, London, pp. 105–130.
7. Heinemeyer, J., Lewejohann, D., and Braun, H. P. (2005) Blue-native gel electrophoresis for the characterization of protein complexes in plants, in *Methods in Molecular Biology: Plant Proteomics* (Thiellement, H., Méchin, V., Damerval, C., and Zivy, M., eds.), Humana Press, Totowa, NJ, pp. 343–352.
8. Schägger, H., and Pfeiffer, K. (2001) The ratio of oxidative phosphorylation complexes I–V in bovine heart mitochondria and the composition of respiratory chain supercomplexes. *J. Biol. Chem.* **276,** 37,861–37,867.
9. Schägger, H. (2001) Respiratory chain super complexes. *IUBMB Life* **52,** 119–128.
10. Schägger, H. (2002) Respiratory supercomplexes of mitochondria and bacteria. *Biochim. Biophys. Acta* **1555,** 154–159.
11. Eubel, H., Jänsch, L., and Braun, H. P. (2003) New insights into the respiratory chain of plant mitochondria: supercomplexes and a unique composition of complex II. *Plant Physiol.* **133,** 274–286.
12. Eubel, H., Heinemeyer, J., and Braun, H. P. (2004) Identification and characterization of respirasomes in potato mitochondria. *Plant Physiol.* **134,** 1450–1459.
13. Millar, A. H., Eubel, H., Jänsch, L., Kruft, V., Heazlewood, L., and Braun, H. P. (2004) Mitochondrial cytochrome-*c* oxidase and succinate dehydrogenase contain plant-specific subunits. *Plant Mol. Biol.* **56,** 77–89.
14. Krause, F., Reifschneider, N. H., Vocke, D., Seelert, H., Rexroth, S., and Dencher, N. A. (2004) "Respirasome"-like supercomplexes in green leaf mitochondria of spinach. *J. Biol. Chem.* **279,** 48,369–48,375.

15. Heinemeyer, J., Eubel, H., Wehmhöner, D., Jänsch, L., and Braun, H. P. (2004) Proteomic approach to characterize the supramolecular organization of photosystems in higher plants. *Phytochemistry* **65,** 1683–1692.
16. Lowry, O. H., Rosebrough, N. J., Farr, A. L., and Randall, R. J. (1951) Protein measurement with the Folin phenol reagent. *J. Biol. Chem.* **193,** 265–275.
17. Neuhoff, V., Stamm, R., and Eibl, H. (1985) Clear background and highly sensitive protein staining with Coomassie blue dyes in polyacrylamide gels: a systematic analysis. *Electrophoresis* **6,** 427–448.
18. Neuhoff, V., Stamm, R., Pardowitz, I., Arold, N., Ehrhardt, W., and Taube, D. (1990) Essential problems in quantification of proteins following colloidal staining with Coomassie brilliant blue dyes in polyacrylamide gels, and their solution. *Electrophoresis* **11,** 101–117.
19. Heukeshoven, J. and Dernick, R. (1986) Silver staining of proteins, in *Elektrophoresis Forum '86* (Radola, B. J., ed.), Technische Universität München, Munich, pp. 22–27.
20. Grandier-Vazeille, X., and Guerin, M. (1996) Separation by blue native and colorless native polyacrylamide gel electrophoresis of the oxidative phosphorylation complexes of yeast mitochondria solubilized by different detergents: specific staining of the different complexes. *Anal. Biochem.* **242,** 248–254.
21. Zerbetto, E., Vergani, L., and Dabbeni-Sala, F. (1997) Quantitation of muscle mitochondrial oxidative phosphorylation enzymes via histochemical staining of blue native polyacrylamide gels. *Electrophoresis* **18,** 2059–2064.
22. Henderson, N. S., Nijtmans, L. G., Lindsay, J. G., Lamantea, E., Zeviani, M., and Holt, I. J. (2000) Separation of intact pyruvate dehydrogenase complex using blue native agarose gel electrophoresis. *Electrophoresis* **21,** 2925–2931.
23. Jänsch, L., Kruft, V., Schmitz, U. K., and Braun, H. P. (1996) New insights into the composition, molecular mass and stoichiometry of the protein complexes of plant mitochondria. *Plant J.* **9,** 357–368.

24

Investigation of Iron-Sulfur Protein Maturation in Eukaryotes

Oliver Stehling, Paul M. Smith, Annette Biederbick, Janneke Balk, Roland Lill, and Ulrich Mühlenhoff

Summary

Iron-sulfur (Fe-S) clusters are cofactors of many proteins that are involved in central biochemical pathways, such as oxidative phosphorylation, photosynthesis, and amino acid biosynthesis. The assembly of these cofactors and the maturation of Fe-S proteins require complex cellular machineries in all kingdoms of life. In eukaryotes, Fe-S protein biogenesis is an essential process, and mitochondria perform a primary role in synthesis. Defects in Fe-S protein maturation in yeast result in respiratory deficiency and auxotrophies for certain amino acids and vitamins that require Fe-S proteins for their biosynthesis. Frequently, heme biosynthesis is also affected. The present compendium describes assays for the analysis of *de novo* Fe-S cluster and heme formation, cellular iron homeostasis, and the activity of Fe-S cluster- and heme-containing enzymes. These approaches are crucial to elucidate the mechanisms underlying the maturation of Fe-S proteins and may aid in the identification of new members of this evolutionary ancient process.

Key Words: Biosynthesis; iron-sulfur proteins; heme; iron homeostasis; mammalian cell culture; *Saccharomyces cerevisiae*.

1. Introduction

Iron-sulfur (Fe-S) clusters are versatile, ancient cofactors of proteins that are involved in electron transport, enzyme catalysis, and regulation of gene expression *(1)*. Recent years have shown that the synthesis of these cofactors and their insertion into apo-proteins involves the function of complex cellular machineries in all kingdoms of life (reviewed in **refs.** ***2–4***). The budding yeast *Saccharomyces cerevisiae* has served as the key model organism for the study of this novel biochemical pathway in eukaryotes. Mitochondria play a central role in this pathway as they are essential for the maturation of all cellular

From: *Methods in Molecular Biology, vol. 372: Mitochondria: Practical Protocols*
Edited by: D. Leister and J. M. Herrmann © Humana Press Inc., Totowa, NJ

Fe-S proteins. The so-called mitochondrial iron-sulfur cluster (ISC) assembly machinery is responsible for the *de novo* synthesis of Fe-S clusters and the insertion of these cofactors into mitochondrial Fe-S apo-proteins. This system is also involved in the maturation of Fe-S proteins that are located outside the mitochondria in the cytosol or nucleus. A mitochondrial export system and a recently discovered cytosolic Fe-S protein assembly (CIA) system specifically participate in the maturation of cytosolic and nuclear Fe-S proteins. Of the approx 20 assembly components known to date, many are encoded by essential genes, including several components of mitochondria. This indicates that the process is indispensable for life. In fact, the maturation of cellular Fe-S proteins is so far the only mitochondrial function that is essential for eukaryotes ***(3)***.

Defects in Fe-S protein maturation result in respiratory deficiency caused by a collapse of the respiratory chain or the citric acid cycle and auxotrophies for certain amino acids, lipids, and vitamins, which require Fe-S proteins for their biosynthesis. In *S. cerevisiae*, these include leucine, methionine, lysine, and glutamate ***(5,6)***. Severe defects result in cell death. The first known reason for the indispensable character of Fe-S protein biogenesis in yeast is the essential Fe-S protein Rli1p, which plays a key role in ribosomal biogenesis ***(7,8)***. In yeast, defects in the mitochondrial Fe-S protein maturation also affect the biosynthesis of heme, the second major iron-consuming process of the cell ***(9)***. The mechanistic linkage between these two mitochondrial processes is currently unclear. Moreover, mitochondria play a key role in the regulation of cellular iron homeostasis. Defects in the mitochondrial ISC assembly and export apparatus elicit the induction of iron-uptake genes (iron regulon) and result in the accumulation of iron within mitochondria ***(10)***. In *S. cerevisiae*, the iron regulon is under the control of the transcription factors Aft1p and Aft2p. Analyses have shown that these proteins are regulated by a component that is generated by the ISC assembly machinery and is exported from mitochondria rather than directly by cytosolic iron levels ***(11)***. In mammalian cells, iron homeostasis is at least in part under the control of the cytosolic iron regulatory proteins 1 and 2 (IRP1 and IRP2, respectively) ***(12)***. IRP1 makes use of an Fe-S cluster for cellular iron regulation, suggesting that mitochondrial or cytosolic Fe-S protein assembly machineries might be directly linked to mammalian iron metabolism.

These phenotypes show that cellular Fe-S protein maturation is tightly connected to mitochondrial respiration, heme synthesis, and cellular iron homeostasis. A full understanding of Fe-S cluster maturation thus requires an integrative approach that takes all of these processes into account. This chapter provides a comprehensive compilation of the most important routine methods used for the analysis of these processes. Most of these assays were initially established for the yeast *S. cerevisiae*. The procedures can be adapted for other fungi, necessitating only minor adjustments. In case of mammalian tissue or cell cultures or of pathogenic protists, experiments may be complicated because of low amounts of available sample. Nevertheless, most techniques used in yeast can be applied in principle. Wherever possible, we provide assays that have been established in our laboratory for the analysis of Fe-S protein biogenesis in cell culture models, and we provide hints for the adaptation of our standard yeast assays to the analysis of mammalian cells.

Subheading 3.1. describes our standard methods for analyzing the *de novo* synthesis of Fe-S cofactors on cellular Fe-S apo-proteins by radiolabeling of cells with ^{55}Fe in vivo. The labeled ^{55}Fe-S proteins of interest are subsequently isolated by immunoprecipitation, and the amount of copurified radioactive iron is determined by liquid scintillation counting. We include two assays for the analysis of Fe-S cluster synthesis activities in mitochondrial extracts in vitro. The first is based on radiolabeling of yeast mitochondria containing an over-produced endogenous Fe-S protein with radioactive iron ***(13)***. This assay is similar to the in vivo assay and specific for *S. cerevisiae*. In the second assay, a soluble apo-ferredoxin is added to a mitochondrial extract, and the fully reconstituted holo-ferredoxin is subsequently purified either by anion exchange or by native gel electrophoresis ***(14)***. This method can be employed in the form of a radioassay or in nonradioactive form and is suitable for mitochondria or cell extracts from a variety of organisms with only minor adjustments.

Subheading 3.2. includes our routine methods for the determination of cellular heme synthesis by radiolabeling with ^{55}Fe in vivo or with isolated mitochondria in vitro. Because of its high solubility in hydrophobic solvents, heme is quantitatively extracted from a cell extract into the organic phase. The formation of ^{55}Fe-labeled heme is determined by liquid scintillation counting of the organic phase. In **Subheading 3.3.**, we describe our routine methods for the determination of mitochondrial and cellular iron content that are based on the formation of colored iron complexes with the chelators bathophenantroline or nitro-PAPS [2-(5-nitro-2-pyridyl-azo)-5-(*N*-propyl-*N*-sulfopropylamino)phenol]. Furthermore, an assay for the determination of iron uptake into the cell by radiolabeling experiments with ^{55}Fe in vivo is included. This assay is useful for cultivated mammalian cells but is cell density dependent and might require data correction. Similar iron uptake experiments can be performed in yeast, yet they are not very reliable.

Subheading 3.4. represents a compilation of routine enzyme procedures that were optimized for the analysis of Fe-S cluster- and heme-containing enzymes in mammalian cell cultures. Among others, this subheading includes a method for determination of the activity of the mammalian IRP1. This cytosolic Fe-S protein, when lacking its Fe-S cluster, binds to specific messenger ribonucleic acid (RNA) stem-loop structures called iron-responsive elements (IREs). RNA binding can be analyzed by an RNA electrophoretic mobility shift assay (REMSA; ***15***).

2. Materials

If not otherwise stated, all reagents are dissolved in water.

2.1. *Analysis of* De Novo *Fe-S Cluster Formation*

1. "Iron-free" minimal medium for growth of *S. cerevisiae*. This medium corresponds to regular synthetic complete (SC) medium but lacks added iron chloride ***(16)***. A ready-made powder is commercially available (Formedium, UK).
2. Mitochondria from yeast cells grown in iron-free medium. We routinely use mitochondria containing an overproduced version of biotin synthase (Bio2p) for this type of experiment. Mitochondria are isolated from Zymolyase-treated *S. cerevisiae* as described in **ref.** ***17*** or in Chapter 6 of this book.
3. Citrate buffer: 50 m*M* sodium citrate, 1 m*M* ethylenediaminetetraacetic acid (EDTA), pH 7.0.
4. TNETG buffer: 20 m*M* Tris-HCl, pH 7.4, 2.5 m*M* EDTA, 150 m*M* NaCl, 0.5% (w/v) Triton X-100, 10% (w/v) glycerol.
5. TNETG-200 buffer: 20 m*M* Tris-HCl, pH 7.4, 2.5 m*M* EDTA, 200–250 m*M* NaCl, 0.1% (w/v) Triton X-100, 10% (w/v) glycerol.
6. In vitro experiments for the *de novo* formation of Fe-S proteins necessitate anoxygenic conditions. We work in an anaerobic chamber (Coy Laboratories) filled with 95% (v/v) nitrogen and 5% (v/v) hydrogen gas. Solutions in **items 7–11** have to be oxygen free.
7. Mitobuffer (2X, anaerobic): 40 m*M* HEPES-KOH, pH 7.4, 100 m*M* KCl, 2 m*M* $MgSO_4$, 1.2 *M* sorbitol.
8. Anaerobic solutions of the following chemicals: 10% (w/v) Triton X-100; 2-mercaptoethanol; 25% (v/v) HCl; 1 *M* Tris-HCl, pH 8.3; 0.5 *M* EDTA. These solutions are introduced to anaerobic conditions at least 1 wk in advance. They can be stored under anaerobic conditions indefinitely.
9. The following stock solutions are prepared freshly with oxygen-free water: 0.1 *M* sodium ascorbate; 0.1 *M* dithiothreitol (DTT); 10 m*M* cysteine; 1 m*M* pyridoxal phosphate; 0.1 *M* reduced nicotinamide adenine dinucleotide (NADH); 0.3 m*M* ferric ammonium citrate; $^{55}FeCl_3$ (NEN/Perkin-Elmer).
10. An acidic, low molecular weight [2Fe-2S] ferredoxin. These can be obtained in recombinant form from *Escherichia coli* or are commercially available (Sigma).
11. Additional reagents not requiring deoxygenation: 20 m*M* HEPES-KOH, pH 7.4; saturated phenylmethylsulfonyl fluoride (PMSF) in ethanol; 25% trichloroacetic acid (optional); appropriate antibodies coupled to protein-A Sepharose; Bromphenol blue 0.2% (w/v).

2.2. *Analysis of Heme Formation*

1. Cells or mitochondria from any source.
2. Mitobuffer: 20 m*M* HEPES-KOH, pH 7.4, 50 m*M* KCl, 1 m*M* $MgSO_4$, 0.6 *M* sorbitol.
3. Stop solution: 100 m*M* $FeCl_3$ in 5 *M* HCl.
4. Phosphate-buffered saline (PBS; pH 7.4) supplemented with 2.7 m*M* EDTA.
5. Additional reagents required: 10 m*M* deuteroporphyrine; 3% (w/v) β-dodecylmaltoside; 0.1 *M* NADH; 0.1 *M* sodium ascorbate; carbonyl cyanide m-chlorophenyl

hydrazone (CCCP; 10 m*M* in ethanol); 2.5 m*M* nitrilotriacetic acid; butyl acetate; glass beads (0.5 mm diameter).

2.3. Analysis of Cellular Iron Levels

1. 100 m*M* Bathophenanthroline disulfonic acid disodium salt (freshly prepared).
2. 1 *M* Sodium dithionite (freshly prepared).
3. 2.5% (w/v) and 10% (w/v) sodium dodecyl sulfate (SDS).
4. 1 *M* Tris-HCl, pH 7.4.
5. 1% (w/v) Hydrochloric acid.
6. 7.5% (w/v) Ammonium acetate solution.
7. 4% (w/v) Ascorbic acid (freshly prepared).
8. 1.3 m*M* nitro-PAPS.
9. Iron standard: 0.2 m*M* $(NH_4)_2Fe(SO_4)_2 \cdot 6H_2O$, formula weight (FW) = 392.14 (Mohr's salt, 0.008% (w/v), ~40 mg/500 mL, freshly prepared). Alternatively, use a commercial iron standard.

2.4. Enzyme Assays

2.4.1. Spectrophotometric Enzyme Assays

1. Cell suspension buffer: 5 m*M* Tris-HCl, pH 7.4, 250 m*M* sucrose, 1 m*M* EDTA, 1 m*M* ethylene glycol bis 2-aminoethyl tetraacetic acid (EGTA), 1.5 m*M* $MgCl_2$, 1 m*M* PMSF.
2. Cytochrome cyclooxidase (COX) buffer: 15 m*M* KH_2PO_4, pH 7.2, 0.1% (w/v) BSA.
3. Aconitase buffer: 100 m*M* triethanolamine, pH 8.0, 1.5 m*M* $MgCl_2$, 0.1% Triton X-100.
4. Succinate dehydrogenase (SDH) buffer: 50 m*M* Tris-H_2SO_4, pH 7.4, 0.1 m*M* EDTA, 1 m*M* KCN, 0.1% (w/v) Triton X-100.
5. Citrate synthase (CS) buffer: 50 m*M* Tris-HCl, pH 8.0, 50 m*M* NaCl, 0.1% (w/v) Triton X-100.
6. Malate dehydrogenase (MDH) buffer: 50 m*M* Tris-H_2SO_4, pH 7.4, 100 m*M* NaCl, 0.1% (w/v) Triton X-100.
7. 20 m*M* *cis*-Aconitate.
8. 100 m*M* NAD phosphate oxidized ($NADP^+$).
9. Isocitrate dehydrogenase (IDH; 1 U/25 µL; *see* **Note 1**).
10. 20% (w/v) Sodium succinate.
11. 20% (w/v) Sodium malonate.
12. 5 m*M* Decylubiquinone (in ethanol).
13. 20 mg/mL Oxidized or reduced cytochrome-*c* (*see* **Note 2**).
14. 0.1 *M* KCN.
15. 50 m*M* Dithio-*bis*-nitrobenzoic acid (DTNB; in dimethyl sulfoxide).
16. 10 mg/mL Acetyl-coenzyme A.
17. 15 mg/mL Oxalacetate.
18. 10 mg/mL NADH.

2.4.2. RNA Electrophoretic Mobility Shift Assay

All reagents should be essentially ribonuclease (RNase) free.

1. Plasmid pSPT-Fer: This vector contains the 5′-untranslated region of the human apo-ferritin heavy chain (bases 31–58) downstream of the promoter for T7 RNA polymerase ***(15)***.
2. 5X transcription buffer (Promega): 200 m*M* Tris-HCl, pH 7.9, 30 m*M* $MgCl_2$, 50 m*M* NaCl, 10 m*M* spermidine.
3. 2X Precipitating (PPT) buffer: 40 µL of 0.5*M* EDTA, pH 8.0, 660 µL of 7.5 *M* ammonium acetate, 300 µL water.
4. 5X TBE buffer: 450 m*M* Tris-HCl, 450 m*M* borate, 10 m*M* EDTA.
5. Munro buffer: 10 m*M* HEPES-KOH, pH 7.6, 3 m*M* $MgCl_2$, 40 m*M* KCl, 5% (w/v) glycerol, 1 m*M* DTT, 0.2% (w/v) Nonident P-40 (NP40).
6. Sample buffer: 30 m*M* Tris-HCl, pH 7.5, 40% (w/v) sucrose, 0.2% (w/v) bromophenol blue, 12.5 µg/µL heparin.
7. Additional reagents required: 2-mercaptoethanol (20% (w/v) in Munro buffer); [α-^{32}P] cytosine 5′triphosphate (CTP) (10 µCi/µL); 100 m*M* DTT; 10 m*M* acetylated bovine serum albumin (BSA; 50X); 10 m*M* RNAsin; 10 m*M* stock solutions of adenosine 5′triphosphate (ATP), guanosine 5′-triphosphate, and uridine triphosphate; T7 RNA polymerase (New England Biolabs); RNase T1; 10% (w/v) ammonium peroxodisulfate; 40 µg/µL glycogen; 96 and 70% (v/v) ethanol.

3. Methods

3.1. Analysis of De Novo *Fe-S Protein Formation*

3.1.1. Determination of De Novo *Fe-S Protein Formation by Radioactive Labeling of Yeast Cells In Vivo*

1. Harvest yeast cells from a 50 mL overnight culture by centrifugation (1500*g* for 5 min), wash once in 10 mL sterile water, and dilute cells in 100 mL iron-free SC medium supplemented with the appropriate carbon source at $OD_{600} = 0.2$. Incubate overnight at 30°C (*see* **Note 3**).
2. Collect cells by centrifugation, wash once in 10 mL sterile water, and determine the wet weight. Resuspend 0.5 g cells in 10 mL iron-free medium in a 50-mL culture flask and incubate for 10 min at 30°C.
3. Dilute 10 µCi $^{55}FeCl_3$ in 100 µL of 0.1 *M* sodium ascorbate and add mix to the preincubated cells. The final concentration of ascorbate in the medium is 1 m*M*. Incubate for 1–2 h at 30°C (*see* **Note 4**).
4. Transfer the radiolabeling mixture to a 15 mL Falcon tube and harvest the cells by centrifugation (1500*g* for 5 min). Wash the cells once in 10 mL citrate buffer and once in 1 mL 20 m*M* HEPES-KOH, pH 7.4. Collect the cells by centrifugation.
5. Resuspend the cells in 0.5 mL TNETG buffer. Add 10 µL saturated PMSF and 0.5 volume glass beads. Lyse the cells by three 1 min bursts on a vortex at maximum speed with intermediate cooling for 1 min on ice. Tubes are best vortexed upside down. All subsequent steps are carried out at 4°C.
6. Remove the coarse cell debris by centrifugation for 5 min at 1500*g*. Transfer the supernatant to a 1.5-mL reaction tube and centrifuge at 17,000*g* for 10 min. Transfer the supernatant to a fresh tube. At this point, avoid carrying over any membrane debris. Remove 5 µL of the extract for scintillation counting. This sample gives

a crude measure of the cellular iron uptake. Another 25 µL of the extract are precipitated with TCA for immunoblotting (optional). The remaining 250 µL are used for immunoprecipitation of the proteins of interest. Usually, two immunoprecipitations can be carried out with one labeling reaction.

7. Add 20–40 µL of immunoglobulin G-coupled immunobeads or 10 µL of commercially available, coupled anti-hemagglutinin A (HA) or anti-Myc beads (Santa Cruz) to 250 µL cell extract. Cut off the end of the pipet tip when handling the immunobeads. Do not vortex the beads. Incubate the reaction tubes in a rotating shaker at 4°C for 1 h (*see* **Notes 5** and **6**).
8. Collect the beads by centrifugation at 3000*g* for 5 min. Remove all of the supernatant with a syringe. Wash the beads three times in 500 µL TNETG buffer and collect the beads by centrifugation. It is essential that virtually no supernatant remains in the reaction tubes after each washing step.
9. Add 50 µL water and 1 mL scintillation cocktail to the beads, vortex briefly, and determine the radioactivity associated with the beads in a scintillation counter (*see* **Note 4**; for an example, *see* ***18***).

3.1.2. Determination of De Novo *Fe-S Cluster Formation in Isolated Yeast Mitochondria Containing an Overproduced Fe-S Protein by Radioactive Labeling With* ^{55}Fe

Steps 1–4 are carried out under anaerobic conditions (*see* **Note 7**).

1. Store isolated mitochondria (*see* **Note 8**) and all other solutions in an anaerobic chamber with reaction cups open for 1 h on ice to remove oxygen.
2. In a standard reaction tube, add 125 µL 2X mitobuffer, 2.5 µL 100 m*M* sodium ascorbate, 2.5 µL 0.1 *M* DTT, 2.5 µL 0.1 *M* NADH (1 m*M* final), 2.5 µL 1 m*M* pyridoxal phosphate (10 µ*M* final), 5 µL 10 m*M* cysteine (0.2 m*M* final), 2.5 µL 10% (w/v) Triton X-100; fill to 250 µL with anaerobic water.
3. Add mitochondria isolated from iron-starved cells (corresponding to 100 µg protein) and 5 µCi $^{55}FeCl_3$ (reduced in 10 m*M* ascorbate). Incubate with gentle shaking for 2.5 h at 25°C under anaerobic conditions (*see* **Note 4**).
4. Terminate the labeling reaction by addition of 2.5 µL 0.5 *M* EDTA on ice. All further steps are carried out aerobically at 4°C.
5. Remove membrane debris by centrifugation at 17,000*g* for 10 min. Transfer the supernatant to a fresh tube. Using a pipetman with the ends of the pipet tips cut off, add 20–40 µL immunobeads or 10 µL commercially coupled anti-HA or anti-Myc beads. Incubate the tubes in a rotating shaker for 1 h (*see* **Note 6**).
6. Collect the beads by centrifugation at 3000*g* for 5 min. Carefully remove all of the supernatant with a syringe. Wash the beads three times in 500 µL TNETG buffer. In between the washing steps, collect the beads by centrifugation. It is essential that virtually no supernatant remains in the reaction tubes after each wash.
7. Add 50 µL water and 1 mL scintillation cocktail; vortex and count the radioactivity associated with the beads in a scintillation counter using the counter settings for ^{3}H (*see* **Note 4**).

3.1.3. Analysis of Fe-S Cluster Formation in Mitochondrial Extracts Using Recombinant [2Fe-2S] Ferredoxins

Subheadings 3.1.3.1. and **3.1.3.2.** necessitate anaerobic conditions and oxygen-free solutions.

3.1.3.1. Preparation of Apo-Ferredoxin

1. Incubate a recombinant ferredoxin (4–10 mg/mL concentration) under anaerobic conditions for at least 1 h on ice to remove oxygen (*see* **Notes 9** and **10**).
2. In a standard reaction tube, add 250 µL ferredoxin and 10 µL 2-mercaptoethanol; fill with anaerobic water to a final volume of 1 mL. Cool the sample on ice, add concentrated HCl to a final concentration of 0.5*M*, mix gently, and incubate on ice for 5 min. A white precipitate forms immediately.
3. Collect the precipitated protein by centrifugation at 12,000*g* for 10 min at 4°C. Remove the supernatant completely and rinse the pellet briefly with 500 µL cold water containing 0.1% 2-mercaptoethanol. Remove water completely.
4. Add 250 µL 50 m*M* Tris-HCl, pH 8.3. Resuspend the pellet carefully with a pipet and store the sample on ice. Do not vortex. The protein should dissolve within 10–15 min. If necessary, add 1 µL droplets of unbuffered 1*M* Tris-HCl until the solution clears completely.
5. Repeat **steps 1–4**. For details, *see* **Note 10** and **ref. *19***.

3.1.3.2. Reconstitution Assay

1. Under anaerobic conditions, combine 125 µL 2X mitobuffer, 2.5 µL 100 m*M* sodium ascorbate, 2.5 µL 0.1 *M* DTT, 2.5 µL 0.1 *M* NADH (1 m*M* final), 2.5 µL 1 m*M* pyridoxal phosphate (10 µ*M* final), 5 µL 10 m*M* cysteine (0.2 m*M* final), and 2.5 µL 10% (w/v) Triton X-100; fill to 250 µL with anaerobic water (*see* **Note 11**). When holo-ferredoxin formation is detected by native polyacrylamide gel electrophoresis, smaller reaction volumes (50–100 µL) should be used.
2. Add isolated mitochondria or cell extracts corresponding to at least 100 µg protein and 20 µg apo-ferredoxin. In case of radioassays, add 5 µCi $^{55}FeCl_3$ (reduced in 10 m*M* ascorbate) or 5 µCi ^{35}S-cysteine. Control reactions lacking mitochondria and added apo-ferredoxin should be analyzed in parallel. For nonradioactive assays, add ferric ammonium citrate to a final concentration of 0.3 m*M* and cysteine to 4 m*M*. Higher amounts of ferredoxin (50 µg) may be used. Incubate with gentle shaking for 2 h at 25°C under anaerobic conditions.
3. Terminate the reconstitution reaction by addition of 2.5 µL 0.5 *M* EDTA on ice. All further steps are carried out aerobically at 4°C.
4. Remove membrane debris by centrifugation at 17,000*g* for 10 min. Transfer the supernatant to a fresh cap. Continue with the protocol in **Subheading 3.1.3.3.** or **3.1.3.4.** (*see* **Note 12**).

3.1.3.3. Isolation of Radiolabeled Holo-Ferredoxin by Binding to Ion Exchange Resin

1. Add 25 µL of a 1:1 slurry of anion exchanger beads (Q-Sepharose, Amersham) in TNETG-200 buffer and incubate in a rotating shaker for 10 min at 4°C.

2. Collect the beads by centrifugation at 3000*g* for 5 min. Carefully remove all of the supernatant with a syringe. Wash the beads three times in 500 μL TNETG-200 buffer containing 1 m*M* ascorbate. In between the washing steps, collect the beads by centrifugation. Remove the supernatant as completely as possible.
3. Add 50 μL water and 1 mL scintillation cocktail, vortex, and count the radioactivity associated with the beads in a scintillation counter.

3.1.3.4. Separation of Holo-Ferredoxin by Native Polyacrylamide Gel Electrophoresis

1. Prepare a 17.5% polyacrylamide gel with a 6% stacking gel and cold standard electrophoresis buffer according to the standard Laemmli procedure, except omit SDS.
2. Add 1 μL 0.2% bromophenol blue to each 100 μL reaction mix; load 50 μL of the samples onto the chilled native polyacrylamide gel. The electrophoresis is carried out at 4°C at 30 mA and 200 V until the dye front reaches the bottom of the gel.
3. For detection of radiolabeled ferredoxin, the gel is fixed with cold 20% ethanol at 4°C for 30 min, dried under vacuum, and analyzed by autoradiography. Because Fe-S cofactors are acid labile, the gels should not be stained. In case of non-radioactive assays, the red color of the reconstituted holo-ferredoxin may be visible by eye. For higher sensitivity, the gel may be stained with Coomassie brilliant blue. Holo-ferredoxin forms a sharp band slightly above the dye front and is well separated from the majority of proteins.

3.2. Analysis of Heme Formation

3.2.1. Determination of Heme Formation by Radioactive Labeling of Yeast Cells In Vivo

1. Perform **steps 1–4** of an in vivo ^{55}Fe radiolabeling reaction as described in **Subheading 3.1.1.** (*see* **Note 13**).
2. Resuspend the washed, radiolabeled cells in 0.5 mL water and divide the suspension in two.
3. To each half, add 25 μL stop solution, 800 μL butyl acetate, and 0.5 volume glass beads at 0°C. Lyse the cells by three 1-min bursts on a vortex at high speed with intermediate cooling for 1 min on ice. The tubes should be vortexed upside down.
4. Separate the organic phase by centrifugation at 10,500*g* for 10 min at 20°C. Transfer two 250 μL aliquots (for duplicate values) of the upper organic phase into two reaction tubes. Carefully avoid touching the membrane debris or parts of the lower aqueous phase.
5. Add 1 mL scintillation cocktail to the organic phase, mix, and determine the ^{55}Fe radioactivity in the organic phase by scintillation counting using the standard settings for ^{3}H (*see* **Notes 4**, **14**, and **15**).

3.2.2. Determination of Heme Formation by Radioactive Labeling of Mammalian Cell Culture Cells In Vivo

1. Cultivate mammalian cells in standard medium to the desired density (not exceeding half confluency); the minimal protein amount required is 150–200 μg (*see* **Note 16**).

2. Mix $^{55}FeCl_3$ (1 µCi/µL in 0.1 *M* HCl) with half of the volume 2.5 m*M* nitrilotriacetic acid. Add the mixture (final concentration 1 µCi/mL) to standard medium supplemented with only half the amount of fetal calf serum (FCS) and 150 µ*M* ascorbate.
3. Replace the standard medium with the ^{55}Fe-containing incubation medium and cultivate cells for another 18 h.
4. Harvest the cells by trypsinization, collect them in a 15-mL tube, and wash them twice with 14 mL PBS/2.7 m*M* EDTA.
5. Resuspend the cells in approx 1 mL PBS/EDTA and determine the final volume exactly.
6. Withdraw an aliquot (10–20 µL) for protein determination and use half of the remaining cells for determination of total ^{55}Fe uptake (*see* **Subheading 3.3.4.**).
7. To the remaining sample, add 50 µL stop solution and 800 µL butyl acetate and lyse the cells by three 1-min bursts on a vortex at high speed with intermediate cooling for 1 min on ice.
8. Separate the organic phase by centrifugation at 10,500*g* for 10 min at 16°C. Transfer two 250-µL aliquots of the upper organic phase into two reaction tubes. Carefully avoid touching the membrane debris or parts of the lower aqueous phase.
9. Add 1 mL scintillation cocktail to the aliquots, mix, and determine the radioactivity in the organic phase by scintillation counting using the standard settings for ^{3}H (*see* **Note 4**).

3.2.3. Determination of Heme Formation in Isolated Mitochondria

3.2.3.1. Assay 1: Intact Mitochondria

1. To a 1.5 mL reaction tube, add 500 µL mitobuffer, 10 µL 0.1 *M* NADH (2 m*M* final), 5 µL 0.1 *M* ascorbate (1m*M* final), and 12.5 µL 0.1 m*M* deuteroporphyrine (2.5 µ*M* final). Add 10 µL isolated mitochondria corresponding to 10 µg protein and incubate for 3 min at 25°C with gentle shaking.
2. Add 1 µCi of ^{55}Fe-chloride (in 0.1 *M* ascorbic acid) and incubate for 10 min at 25°C with gentle shaking.
3. Store the samples on ice; add 25 µL stop solution and 500 µL butyl acetate. Vortex twice for 30 s and centrifuge for 5 min at 17,000*g*. Transfer 250 µL of the organic phase (upper layer) to a fresh tube for scintillation counting. For proper determination, a control reaction containing 50 µ*M* CCCP and lacking NADH is required.

3.2.3.2. Assay 2: Mitochondrial Extracts

1. Dilute mitochondria corresponding to 10–25 µg protein in 50 µL mitobuffer on ice. Add 2.5 µL 3% (w/v) dodecylmaltoside (0.15% final) and incubate for 1 min on ice.
2. Add 450 µL mitobuffer, 5 µL 0.1*M* ascorbate (1 m*M* final), 12.5 µL 0.1 m*M* deuteroporphyrine (2.5 µ*M* final), and 1 µCi ^{55}Fe-Cloride (in 0.1*M* ascorbic acid) and incubate for 10 min at 25°C with gentle shaking (*see* **Note 4**).
3. Put the samples on ice; add 25 µL stop solution and 500 µL butyl acetate. Vortex twice for 30 s and centrifuge for 5 min at 18,000*g*. Transfer 250 µL of the organic phase (upper layer) to a fresh tube for scintillation counting. For further details, *see* **ref. *20*** and **Note 14.**

3.3. Analysis of Cellular Iron Homeostasis

3.3.1. Determination of Cellular or Mitochondrial Iron Content

We routinely use two different iron chelators to spectrophotometrically determine the iron content of isolated mitochondria or cell lysates. The bathophenantroline assay is a rapid assay useful for the determination of artificially increased mitochondrial iron levels. The nitro-PAPS assay is more sensitive but more time consuming and may be used to determine the physiological iron levels of cell lysates.

3.3.2. Bathophenantroline Assay

1. Mix 100 µL 1 *M* Tris-HCl, pH 7.4, with 60 µL 10% SDS (0.6% final); 20 µL 1 *M* dithionite solution; 100 µL bathophenantroline (10 m*M* final); and isolated mitochondria (at least 200 µg protein). Dilute with distilled water to a final volume of 1 mL. Incubate at room temperature for 5 min.
2. Remove membrane debris by centrifugation for 5 min at 10,000*g*. Record the absorption spectrum between 700 and 500 nm of the sample using a reference sample that contains all chemicals but no mitochondria. Determine the difference OD_{540} minus OD_{700} (to account for light scattering). The absorption coefficient $\varepsilon_{540\ nm}$ is approx 23,500 $M^{-1}\ cm^{-1}$ ***(21)***.

3.3.3. Nitro-PAPS Assay

The nitro-PAPS test is based on the release of iron by treatment of samples with hydrochloric acid. Excess acid is neutralized with ammonium acetate, and Fe^{3+} is converted to Fe^{2+} by reduction with ascorbic acid. Finally, the iron chelator nitro-PAPS is added to form a blue complex.

1. Dilute samples (including blank and iron standard) to 100 µL with water in standard reaction tubes.
2. Add 100 µL 1% hydrochloric acid, mix by gentle shaking, and incubate at 80°C for 10 min. Allow the tubes to cool (keep closed) and centrifuge for 1 min.
3. Add (in the following order) 500 µL 7.5% ammonium acetate, 100 µL 4% ascorbic acid, and 100 µL 2.5% SDS. Vortex after each addition.
4. Centrifuge for 5 min at 9000*g*, transfer supernatant (855 µL) into a fresh tube, and add 95 µL of the iron chelator nitro-PAPS (130 µ*M* final concentration). Measure the absorbance at 585 nm against the blank sample in quartz cuvettes. The absorption coefficient $\varepsilon_{585\ nm}$ is approx 94,000 $M^{-1}\ cm^{-1}$ ***(22)*** (*see* **Note 17**).

3.3.4. Determination of Cellular Iron Uptake in Mammalian Cell Culture Cells by Radioactive Labeling In Vivo

1. Perform **steps 1–6** of the protocol in **Subheading 3.2.2.** (*see* **Notes 16** and **18**).
2. Add 1 volume 2X TNETG to the corresponding aliquot and vortex.
3. Centrifuge at 16,000*g* for 5 min to remove membrane debris.

4. Use 10–75 µL of the supernatant for scintillation counting (*see* **Note 2**).

3.4. Assays for the Analysis of Fe-S and Heme-Containing Enzymes in Mammalian Cells

The spectroscopic assays in **Subheadings 3.4.1.** to **3.4.7.** are adaptations of routine assays that can be found in various original publications. Here, they are compiled in an easily accessible collection. These assays have been optimized for the analysis of cultured mammalian cells, which are frequently more difficult to assay because low amounts of sample often preclude the isolation of mitochondria. For spectroscopic assays, cells are washed twice with PBS prior to analysis, and the dry cell pellets may be shock-frozen in liquid nitrogen. They are stored at –80°C and resuspended in ice-cold cell suspension buffer for analysis of enzyme activities.

3.4.1. Aconitase

This coupled assay is based on the formation of isocitrate by the Fe-S protein aconitase. Isocitrate is then used by IDH for the reduction of NADP.

1. Prepare a sample cuvette containing 950 µL aconitase buffer, 200 µ*M* *cis*-aconitate, 1.3 m*M* $NADP^+$, 400 µU IDH (*see* **Note 1**), and a reference cuvette that lacks *cis*-aconitate and IDH.
2. To each cuvette, add the cell suspension (> 25 µg protein) and determine the absorption increase at 340 nm in a double-beam spectrophotometer. The absorption coefficient is $\varepsilon_{340\text{ nm}} = 6220\ M^{-1}\ cm^{-1}$ ***(23–25)***.

3.4.2. Succinate Dehydrogenase

1. Prepare two cuvettes containing 950 µL SDH buffer, 0.25% succinate, 70 µ*M* dichlorophenol-indophenol, and 60 µ*M* decylubiquinone.
2. Add 0.25% malonate to the reference cuvette and start the assay by adding the cell suspension (>25 µg protein). Determine the absorption decrease at 600 nm in a double-beam spectrometer ($\varepsilon_{600\text{ nm}} = 21{,}000\ M^{-1}\ cm^{-1}$) ***(26)***.

3.4.3. Coupled SDH-Cytochrome-c Reductase

This assay is suitable for isolated intact mitochondria only.

1. Prepare two cuvettes containing approx 950 µL enzyme buffer, 0.25% succinate, and 50 µL cytochrome-*c*. Add 0.25% malonate to the reference cuvette.
2. Add mitochondria (>25 µg protein) and determine the absorption decrease at 550 nm for 2 min in a double-beam spectrometer ($\varepsilon_{550\text{ nm}} = 20{,}000\ M^{-1}\ cm^{-1}$).

3.4.4. Cytochrome Oxidase

1. Prepare two cuvettes containing 950 µL COX buffer and approx 50 µL 20 mg/mL reduced cytochrome-*c* ($OD_{550\text{ nm}}$ ~1) (*see* **Note 2**).

2. Add 10 μL 0.1 *M* KCN to the reference cuvette. Add more than 10 μg protein to both cuvettes and determine the absorption decrease at 550 nm in a double-beam spectrophotometer ($\varepsilon_{550\ nm}$ = 20,000 M^{-1} cm^{-1}) ***(27)***. When cell suspensions are used instead of isolated mitochondria, include 0.025% (w/v) dodecylmaltoside, and mix vigorously.

3.4.5. Citrate Synthase

1. Prepare a cuvette containing 950 μL CS buffer and 0.5 m*M* dithio-*bis*-nitrobenzoic acid.
2. Add 10 μL 10 mg/mL acetyl-coenzyme A and cell lysate (>10 μg protein). After 1 min, add 15 μL oxaloacetate (15 mg/mL) and determine the absorption increase at 412 nm ($\varepsilon_{412\ nm}$ = 13,300 M^{-1} cm^{-1}) ***(28)***. This is a reliable mitochondrial marker enzyme.

3.4.6. Malate Dehydrogenase

1. Prepare a sample cuvette containing 950 μL MDH buffer, 15 μL NADH (10 mg/mL), and cell suspension (>25 μg protein) and equilibrate for approx 1 min.
2. Add 15 μL oxaloacetate (15 mg/mL) and determine the absorption decrease at 340 nm ($\varepsilon_{340\ nm}$ = 6,220 M^{-1} cm^{-1}) ***(29)*** (*see* **Note 19**).

3.4.7. Gel Shift Mobility Assay for the Determination of the IRE-Binding Activity of IRP1

IRP1 is a cytosolic Fe-S protein that binds to specific messenger RNA stem-loop structures called IREs when it lacks its Fe-S cluster ***(15,30)***. Binding can be analyzed by REMSA. Because RNA integrity is crucial for IRP binding, any contamination by RNase has to be avoided.

1. *Preparation of the IRE probe:* Transcribe an [α-^{32}P]CTP-labeled IRE probe from 1 μg of a *Hin*dIII-linearized pSPT-Fer plasmid by mixing 5 μL 5X transcription buffer (Promega); 2.5 μL DTT; 0.5 μL acetylated BSA; 40 U RNAsin; 1.3 μL each ATP, guanosine 5′-triphosphate (GTP), and uridine triphosphate (UTP); 20 U T7 polymerase; and 10 μL [α-^{32}P]CTP. Incubate at 37°C for 2.5 h. Precipitate the RNA by adding 25 μL water, 50 μL 2X PPT buffer, 2 μL glycogen, and 250 μL ethanol (96%) and incubate for 5 min at room temperature. Spin down at 17,000*g* for 15 min, rinse the pellet with 300 μL 70% ethanol, spin down again, and resuspend the ethanol-free pellet in 200 μL water.
2. *Preparation of native polyacrylamide gels:* Gel-shift experiments are usually performed in 1.5 mm thick 6% polyacrylamide gels. Mix 7 mL acrylamide-bisacrylamide solution with 2.1 mL 5X TBE, 25.9 mL water, 210 μL ammonium peroxodisulfate, and 10 μL TEMED and pour the solution between sealed 14 × 16 cm glass plates. Store the apparatus horizontally with the comb inserted until the gel is polymerized. The gel may be stored for up to 3 d at 4°C in the assembled electrophoresis apparatus with 0.3X TBE as running buffer.
3. *Labeling of IRP1:* Lyse snap-frozen cell pellets in Munro buffer and pellet nuclei by spinning for 6 min at 3300*g*. Determine the protein concentration of the

clarified extracts and dilute them to a concentration of 100 µg/mL protein. To 18 µL of cell lysate, add either 2 µL 2-mercaptoethanol (20% in Munro buffer) to achieve maximal IRE binding or 2 µL Munro buffer to determine the IRE-binding capacity. Incubate for 30 min at room temperature. Add 2 µL of the IRE probe (~350,000 cpm/µL by Cherenkov counting) and incubate for another 30 min. Unbound free IRE is digested by 1 U RNase T1 (1 U/µL) for 10 min, and the samples are equilibrated in 10 µL sample buffer (*see* **Note 20**).

4. *REMSA and probe detection:* Prerun the gels in 0.3X TBE for 30 min at 14 V/cm (~200 V total), then load the samples onto the gel. The electrophoresis is carried out at the stated voltage for 2 h at 4°C. The gels are dried (e.g., in a vacuum gel dryer) and subjected to autoradiography or phosphoimaging (*see* **Note 21**).

4. Notes

1. IDH is reconstituted in 100 m*M* triethanolamine/10% glycerol at a concentration of 40 mU/µL and can be stored at –80°C after shock-freezing in liquid nitrogen.
2. For the preparation of reduced cytochrome-*c*, add 25 µL 1*M* fresh sodium dithionite (10 m*M* final) to 2.5 mL of 25 mg/mL cytochrome-*c* solution in COX buffer. Incubate for 2–5 min on ice and desalt on a small gel filtration column (PD10, Amersham) equilibrated with Cox buffer. The solution can be shock-frozen in aliquots and stored indefinitely at –80°C.
3. In all labeling experiments involving ^{55}Fe, it is essential that all solutions and glassware be iron free. Standard dishwasher detergent and laboratory glassware frequently contain iron. Glassware should be acid washed in 1*M* HCl. Double-distilled water of highest quality should be used throughout. The contaminated glass flasks used for in vivo labeling of yeast are incubated with citrate buffer and washed in distilled water to remove remnant radioactivity. The flasks are rinsed with 70% ethanol for sterilization.
4. Reduction of $^{55}FeCl_3$ is essential as oxidized Fe^{3+} is virtually insoluble at neutral pH. Therefore, labeling reactions with $^{55}FeCl_3$ in vivo or in vitro are always carried out in the presence of 1 m*M* fresh ascorbate to avoid precipitation of ferric iron. The radiation safety conditions for ^{55}Fe (an electron capture radiation) are similar to those for radioactive ^{3}H. For the quantification of ^{55}Fe, the counter setting for ^{3}H is usually appropriate.
5. In vitro experiments for the *de novo* formation of Fe-S proteins necessitate anoxygenic conditions. We use an anaerobic chamber (Coy Laboratories) filled with 95% nitrogen and 5% hydrogen.
6. For yeast, best results are obtained with cells overproducing the Fe-S protein of interest from a high-copy plasmid under the control of a strong promoter. If antibodies are not available, then HA-tagged versions of the Fe-S protein can frequently be used. In *S. cerevisiae*, the endogenous levels of aconitase, Yah1p (ferredoxin), and Leu1p are sufficient for analysis without overexpression. For other organisms, a suitable reporter protein has to be determined empirically.
7. Antibodies are coupled to protein A-Sepharose for immunoprecipitation as follows:
   a. Resuspend 50 mg dried protein A-Sepharose in cold 500 µL TNETG. The beads are swollen by incubation for at least 30 min in the cold room; mix occasionally.

b. Collect the beads by centrifugation at 850*g* for 5 min. Add 500 µL antibody serum and incubate in the cold room using a rotating shaker for at least 1 h.
c. Collect the beads by centrifugation. Wash five times in 500 µL TNETG buffer; spin down in between washes. The beads are resuspended in 500 µL TNETG and stored at 4°C.

8. Mitochondria are isolated from iron-starved yeast cells grown in iron-free medium ***(17)***. We prefer mitochondria containing an overproduced version of the biotin synthase, Bio2p, for this type of experiment ***(13)***.
9. The preparation of apo-ferredoxin and the Fe-S cluster reconstitution necessitate anaerobic, reducing conditions and oxygen-free solutions.
10. These protocols take advantage of the fact that most low molecular mass [2Fe-2S] ferredoxins are soluble in their apo-form, which can be generated by acid precipitation. Care should be taken that the Fe-S cluster is removed completely, and that no oxidation occurs. Optimally, the ultraviolet/visible spectrum of the apo-ferredoxin should lack any absorption above 300 nm. If this is not the case, then the procedure should be repeated.
11. Biochemical Fe-S protein reconstitution requires ATP. For mitochondria from *S. cerevisiae*, the endogenous levels are sufficient, and the addition of ATP is not recommended because ATP is an effective chelator of iron. For mitochondria from other sources, however, the addition of low amounts of ATP (0.2–0.5 m*M*) may improve the reconstitution. For accurate results, control reactions lacking added apo-ferredoxins or mitochondria or the like should be analyzed in parallel.
12. The quantitative estimation of holo-ferredoxin formation in **Subheadings 3.1.3.3.** and **3.1.3.4.** takes advantage of the acidic character of this type of ferredoxin, which allows its binding to either an anion exchange resin at relatively high ionic strength or fast movement through an electric field. It is therefore essential that a model protein with a low p*I* be used. We use either yeast ferredoxin Yah1p or plant-type ferredoxins. For further details, *see* the original references to **Subheadings 3.1.3.3.** and **3.1.3.4.** ***(13,14,31,32)***.
13. Assays in **Subheadings 3.1.1.** and **3.2.1.** can be performed in parallel on the same sample.
14. The heme biosynthesis assay takes advantage of the high solubility of protonated heme in organic solvents at low pH. Butyl acetate is the preferred solvent as it does not interfere with scintillation counting.
15. Because of the short incubation times (usually 1 h), the in vivo heme formation assay gives a measure mainly for the activity of ferrochelatase. The steady-state heme content of a yeast strain in vivo can be determined using cells cultivated in medium supplemented with $^{55}FeCl_2$ for longer time periods. To this end, cells from a preculture are diluted in 50 mL iron-free SC medium supplemented with the appropriate carbon source, 1 m*M* ascorbate, and 10 µCi ^{55}Fe at $OD_{600} = 0.1$. The cultures are incubated overnight at 30°C and analyzed as described in **Subheading 3.2.1.** For details, *see* **ref. *9***.
16. We noted that cellular uptake of ^{55}Fe and its incorporation into heme depends on the cell density. Thus, it is essential to analyze cell populations at comparable densities or to correct the data appropriately for cell density effects ***(32)***.

17. The same protocol may also be used for the dye ferene (3-(2-pyridyl)-5,6,bis (2-[5furyl sulfonic acid])-1,2,4-triazine). The absorption coefficient $\varepsilon_{593\ nm}$ is 35,000 $M^{-1}\ cm^{-1}$.
18. Assays in **Subheadings 3.2.2.** and **3.3.2.** can be performed in parallel on the same sample.
19. The specific activity of MDH depends on protein concentration in the sample cuvette. Thus, always use more than 25 μg cell lysate to obtain reproducible results.
20. The analysis of the IRE-binding capacity of human IRP1 is complicated by the fact that the non-Fe-S protein IRP2 is also binding the IRE probe and is running at the same position as IRP1 during gel electrophoresis. The incubation with an anti-IRP2 antibody added to the sample buffer 30 min before loading the gel induces an IRP2 supershift and allows the discrimination between human IRP1 and IRP2.
21. Take care that no air bubbles are trapped during the gel-drying process if gel-drying films are used. This would affect probe detection. For autoradiography, use an X-ray screen and expose for at least 1 h at –80°C.

Acknowledgments

We thank E. W. Müllner for pSPT-fer plasmid containing the ferritin IRE. Our work was supported by grants from the Sonderforschungsbereich 593, Deutsche Forschungsgemeinschaft (Gottfried Wilhelm Leibniz program), European Union, and Fonds der chemischen Industrie.

References

1. Beinert, H., Holm, R. H., and Munck, E. (1997) Iron-sulfur clusters: nature's modular, multipurpose structures. *Science* **277,** 653–659.
2. Johnson, D. C., Dean, D. R., Smith, A. D., and Johnson, M. K. (2004) Structure, function, and formation of biological iron-sulfur clusters. *Annu. Rev. Biochem.* **74,** 247–281.
3. Balk, J. and Lill, R. (2004) The cell's cookbook for iron-sulfur clusters: recipes for fool's gold? *Chembiochem.* **5,** 1044–1049.
4. Lill, R. and Muhlenhoff, U. (2005) Iron-sulfur-protein biogenesis in eukaryotes. *Trends Biochem. Sci.* **30,** 133–141.
5. Kispal, G., Csere, P., Prohl, C., and Lill, R. (1999) The mitochondrial proteins Atm1p and Nfs1p are essential for biogenesis of cytosolic Fe/S proteins. *EMBO J.* **18,** 3981–3989.
6. Jensen, L. T. and Culotta, V. C. (2000) Role of *Saccharomyces cerevisiae ISA1* and *ISA2* in iron homeostasis. *Mol. Cell. Biol.* **20,** 3918–3927.
7. Kispal, G., Sipos, K., Lange, H., et al. (2005) Biogenesis of cytosolic ribosomes requires the essential iron-sulphur protein Rli1p and mitochondria. *EMBO J.* **24,** 589–598.
8. Yarunin, A., Panse, V. G., Petfalski, E., Dez, C., Tollervey, D., and Hurt, E. C. (2005) Functional link between ribosome formation and biogenesis of iron-sulfur proteins. *EMBO J.* **24,** 580–588.
9. Lange, H., Muhlenhoff, U., Denzel, M., Kispal, G., and Lill, R. (2004) The heme synthesis defect of mutants impaired in mitochondrial iron-sulfur protein

biogenesis is caused by reversible inhibition of ferrochelatase. *J. Biol. Chem.* **279,** 29,101–29,108.
10. Kispal, G., Csere, P., Guiard, B., and Lill, R. (1997) The ABC transporter Atm1p is required for mitochondrial iron homeostasis. *FEBS Lett.* **418,** 346–350.
11. Rutherford, J. C., Ojeda, L., Balk, J., Muhlenhoff, U., Lill, R., and Winge, D. R. (2005) Activation of the iron regulon by the yeast Aft1/Aft2 transcription factors depends on mitochondrial but not cytosolic iron-sulfur protein biogenesis. *J. Biol. Chem.* **280,** 10,135–10,140.
12. Eisenstein, R. S. (2000) Iron regulatory proteins and the molecular control of mammalian iron metabolism. *Annu. Rev. Nutr.* **20,** 627–662.
13. Muhlenhoff, U., Richhardt, N., Gerber, J., and Lill, R. (2002) Characterization of iron-sulfur protein assembly in isolated mitochondria. A requirement for ATP, NADH, and reduced iron. *J. Biol. Chem.* **277,** 29,810–29,816.
14. Lutz, T., Westermann, B., Neupert, W., and Herrmann, J. M. (2001) The mitochondrial proteins Ssq1 and Jac1 are required for the assembly of iron sulfur clusters in mitochondria. *J. Mol. Biol.* **307,** 815–825.
15. Mullner, E. W., Neupert, B., and Kuhn, L. C. (1989) A specific mRNA binding factor regulates the iron-dependent stability of cytoplasmic transferrin receptor mRNA. *Cell* **58,** 373–382.
16. Guthrie, C. and Fink, G. R. (1991) Guide to yeast genetics and molecular biology. *Meth. Enzymol.* **194,** 1–863.
17. Diekert, K., de Kroon, A. I., Kispal, G., and Lill, R. (2001) Isolation and subfractionation of mitochondria from the yeast *Saccharomyces cerevisiae. Methods Cell Biol.* **65,** 37–51.
18. Muhlenhoff, U., Richhardt, N., Ristow, M., Kispal, G., and Lill, R. (2002) The yeast frataxin homolog Yfh1p plays a specific role in the maturation of cellular Fe/S proteins. *Hum. Mol. Genet.* **11,** 2025–2036.
19. Meyer, J., Moulis, J. M., and Lutz, M. (1986) High-yield chemical assembly of [2Fe-2X] (X = S, Se) clusters into spinach apoferredoxin: product characterisation by Raman spectroscopy. *Biochim. Biophys. Acta* **871,** 243–249.
20. Lange, H., Kispal, G., and Lill, R. (1999) Mechanism of iron transport to the site of heme synthesis inside yeast mitochondria. *J. Biol. Chem.* **274,** 18,989–18,996.
21. Li, J., Kogan, M., Knight, S. A., Pain, D., and Dancis, A. (1999) Yeast mitochondrial protein, Nfs1p, coordinately regulates iron-sulfur cluster proteins, cellular iron uptake, and iron distribution. *J. Biol. Chem.* **274,** 33,025–33,034.
22. Makino, T., Kiyonaga, M., and Kina, K. (1988) A sensitive, direct colorimetric assay of serum iron using the chromogen, nitro-PAPS. *Clin. Chim. Acta* **171,** 19–27.
23. O'Connell, I. A. R. a. E. L. (1967) Mechanism of aconitase action. I. The hydrogen transfer reaction. *J. Biol. Chem.* **242,** 1870–1879.
24. Drapier, J. C. and Hibbs, J. B., Jr. (1996) Aconitases: a class of metalloproteins highly sensitive to nitric oxide synthesis. *Meth. Enzymol.* **269,** 26–36.
25. Hausladen, A. and Fridovich, I. (1996) Measuring nitric oxide and superoxide: rate constants for aconitase reactivity. *Meth. Enzymol.* **269,** 37–41.
26. Hatefi, Y. and Stiggall, D. L. (1978) Preparation and properties of succinate: ubiquinone oxidoreductase (complex II). *Meth. Enzymol.* **53,** 21–27.

27. Birch-Machin, M. A. and Turnbull, D. M. (2001) Assaying mitochondrial respiratory complex activity in mitochondria isolated from human cells and tissues. *Methods Cell Biol.* **65,** 97–117.
28. Ellman, G. L. (1959) Tissue sulfhydryl groups. *Arch. Biochem. Biophys.* **82,** 70–77.
29. Siegel, L. and Englard, S. (1962) Beef-heart malic dehydrogenases. III. Comparative studies of some properties of M-malic dehydrogenase and S-malic dehydrogenase. *Biochim. Biophys. Acta* **64,** 101–110.
30. Leibold, E. A., and Munro, H. N. (1988) Cytoplasmic protein binds in vitro to a highly conserved sequence in the 5′ -untranslated region of ferritin heavy- and light-subunit mRNAs. *Proc. Natl. Acad. Sci. USA.* **85,** 2171–2175.
31. Takahashi, Y., Mitsui, A., and Matsubara, H. (1991) Formation of the Fe-S cluster of ferredoxin in lysed spinach chloroplasts. *Plant Physiol.* **95,** 97–103.
32. Suzuki, S., Izumihara, K., and Hase, T. (1991) Plastid import and iron-sulphur cluster assembly of photosynthetic and nonphotosynthetic ferredoxin isoproteins in maize. *Plant Physiol.* **97,** 375–80.
33. Stehling, O., Elsasser, H. P., Bruckel, B., Muhlenhoff, U., and Lill, R. (2004) Iron-sulfur protein maturation in human cells: evidence for a function of frataxin. *Hum. Mol. Genet.* **13,** 3007–15. Epub October 27, 2004.

25

Studying Proteolysis Within Mitochondria

Takashi Tatsuta and Thomas Langer

Summary

Mitochondria are dynamic organelles with activities that adjust to altering physiological conditions and variable metabolic demands. A conserved proteolytic system present within the organelle exerts essential functions during the biogenesis of mitochondria and ensures the maintenance of organellar activities under varying conditions. Proteases dependent on adenosine triphosphate, in concert with oligopeptidases, degrade nonassembled or damaged proteins in various subcompartments of mitochondria, preventing their accumulation and possibly deleterious effects on mitochondrial functions. Although an increasing number of mitochondrial peptidases are characterized and functionally linked to diverse cellular processes, only limited information is available on the stability of the mitochondrial proteome and the turnover rates of individual proteins. We describe experimental approaches in the yeast *Saccharomyces cerevisiae* and in mice, allowing analysis of the proteolytic breakdown of mitochondrial proteins individually or on a proteomewide scale.

Key Words: ATP-dependent protease; oligopeptidase; proteolysis; quality control.

1. Introduction

Mitochondria contain a number of conserved and often ubiquitously distributed peptidases essential for the maintenance of their activity and homeostasis. One essential proteolytic function is the processing and maturation of nuclear-encoded mitochondrial precursor proteins on import into the organelle *(1)*. Processing peptidases include the well-characterized mitochondrial processing peptidase, the mitochondrial intermediate peptidase, and the innermembrane protease, but also the rhomboid-like protease Pcp1 and the membrane-bound adenosine triphosphate (ATP)-dependent *m*-AAA protease *(2–4)*.

Mitochondrial peptidases also ensure the quality control of mitochondrial proteins and the removal of nonassembled or misfolded polypeptides, like oxidatively damaged proteins in aged cells or excess subunits of mitochondrial

From: *Methods in Molecular Biology, vol. 372: Mitochondria: Practical Protocols*
Edited by: D. Leister and J. M. Herrmann © Humana Press Inc., Totowa, NJ

multiprotein complexes *(5,6)*. Main players in this process are ATP-dependent proteases, including Lon- and Clp-like proteases in the matrix or AAA proteases in the inner membrane, which expose their catalytic domains to the matrix (*m*-AAA protease) or intermembrane space (*i*-AAA protease). They are thought to degrade polypeptides to peptides, which are released from the organelle *(7,8)* or further degraded to amino acid residues by oligopeptidases within mitochondria *(9)*. It remains to be determined whether the pleiotropic phenotypes observed on inactivation of ATP-dependent proteases in yeast and mammals *(5,6)* reflect the deleterious effect of nonnative proteins accumulating within mitochondria or the impaired proteolysis of mitochondrial proteins with regulatory functions.

Although mitochondria are dynamic organelles with protein composition that varies in different tissues or under different physiological conditions, next to nothing is known about the stability of the mitochondrial proteome. Only 5–10% of the mitochondrial proteins were found to be degraded per hour in logarithmically growing *Saccharomyces cerevisiae* cells indicating a high stability of the mitochondrial proteome as well as a high efficiency of mitochondrial biogenesis *(8)*.

We introduce current working protocols for the analysis of proteolytic processes in yeast and murine mitochondria. The overall stability of the mitochondrial proteome in yeast can be assessed by the quantification of degradation products released from radiolabeled, isolated mitochondria (*see* **Subheading 3.2.**). The turnover of individual proteins on a proteomic scale can be analyzed by two-dimensional gel electrophoresis combined with mass spectrometric methods.

If individual proteins are analyzed, then various experimental approaches are feasible that, however, should be interpreted very carefully. In case specific antibodies are available, the turnover rate of individual proteins can be analyzed in pulse-chase experiments (*see* **Subheading 3.3.**). This approach is laborious but allows determining the turnover rate of a protein under in vivo conditions. On the other hand, the stability of radiolabeled mitochondrial proteins can be assessed after their posttranslational import into isolated mitochondria (*see* **Subheading 3.4.**).

Though easy to perform, this approach bears the disadvantage that the assembly of newly imported proteins is often impaired causing their immediate degradation within mitochondria. Therefore, peptidases involved in the proteolytic breakdown of a given protein can be identified, but the turnover rate of the corresponding, endogenous mitochondrial protein cannot be deduced from these studies. This has also to be taken into account when the degradation of mitochondrial-encoded proteins is analyzed after their synthesis in isolated mitochondria (*see* **Subheadings 3.5.** and **3.7.**). They often do not assemble with nuclear-encoded proteins into functional respiratory complexes and are therefore efficiently degraded. This approach, however, allows analyzing the degradation of hydrophobic nonassembled subunits of respiratory complexes in the inner membrane.

2. Materials

2.1. Isolation of Mitochondria From Yeast Saccharomyces cerevisiae

1. YP medium supplemented with galactose and lactate (YP-gal-lac): 2% (w/v) Bacto™ peptone, 1% (w/v) yeast extract, 2% (w/v) galactose, 0.5% (w/v) lactate; sterile autoclave. Carbon sources should be added to medium after autoclaving from the stock solutions (30% w/v galactose and 40% v/v lactate, respectively). The pH of the medium and the stock solution of lactate are adjusted to 5.5 with HCl or NaOH, respectively, before autoclaving (*see* **Note 1**).
2. Lactate medium: 0.3% (w/v) yeast extract, 0.5 g/L NaCl, 0.6 g/L $CaCl_2 \cdot 2H_2O$, 0.5 g/L $MgSO_4 \cdot 7H_2O$, 1 g/L KH_2PO_4, 1 g/L NH_4Cl, 3 mg/L $FeCl_3$, 2% (w/v) lactate, 0.1% (w/v) glucose, pH 5.5; sterile autoclave. All ingredients are dissolved in water, and the pH of the medium is adjusted to 5.5 by NaOH before autoclaving (*see* **Note 1**).
3. 1 *M* Tris base: pH is not adjusted; sterile autoclave.
4. 1 *M* Tris-HCl, pH 7.4; sterile autoclave.
5. 1 *M* Potassium phosphate buffer, pH 7.4: 800 m*M* K_2HPO_4, 200 m*M* KH_2PO_4; check pH; sterile autoclave.
6. 2.4 *M* Sorbitol; sterile autoclave.
7. 0.5 *M* Ethylenediaminetetraacetic acid (EDTA), pH 8.0; sterile autoclave (*see* **Note 2**).
8. 5X SEM buffer: 1.25 *M* sucrose, 5 m*M* EDTA, 50 m*M* 3-(*N*-morpholino) propanesulfonic acid (MOPS)-KOH, pH 7.2; sterile autoclave (*see* **Note 2**).
9. Tris dithiothreitol (DTT) buffer: 100 m*M* Tris base, 10 m*M* dithiothreitol. Prepare freshly.
10. 1.2 *M* Sorbitol. Diluted freshly from the 2.4 *M* stock solution.
11. Sorbitol phosphate buffer: 20 m*M* potassium phosphate buffer, pH 7.4, 1.2 *M* sorbitol. Prepare freshly from stock solutions.
12. Homogenization buffer: 10 m*M* Tris-HCl, pH 7.4, 1 m*M* EDTA, 0.2% (w/v) bovine serum albumin (BSA; fatty acid free), 1 m*M* phenylmethylsulfonyl fluoride (PMSF), 0.6 *M* sorbitol. Prepare freshly and handle with care (*see* **Note 3**).
13. Lyticase (Sigma).
14. Glass homogenizer (B. Braun).
15. Bradford protein assay reagent (Bio-Rad).

2.2. Overall Stability of Mitochondrial Proteome In Vivo

1. Synthetic complete (SC) medium supplemented with glucose or galactose: 0.67% (w/v) of yeast nitrogen base with ammonium sulfate, 92 mg/L arginine, aspartic acid, asparagine, cysteine, glutamic acid, glutamine, glycine, isoleucine, methionine,

phenylalanine, proline, serine, threonine, tyrosine, valine, and myoinositol; 23 mg/L alanine; 180 mg/L leucine; 30 mg/L lysine; 40 mg/L tryptophan; 20 mg/L histidine; 40 mg/L adenine; 20 mg/L uracil; 9.2 mg/L 4-aminobenzoic acid; 2% (w/v) glucose or galactose; sterile autoclave. Carbon sources, methionine, tryptophan, and histidine should be added to medium after autoclaving from stock solutions. Stock solutions are 30% (w/v) glucose, 30% (w/v) galactose, 10 mg/mL tryptophan, 10 mg/mL histidine, and 10 mg/mL methionine, respectively. Tryptophan and histidine stock solutions should be sterile filtrated. The pH of the medium is adjusted to 5.5 by NaOH before autoclaving. Omit methionine from the media for radiolabeling of the protein by [^{35}S]-methionine.

2. [^{35}S]-Methionine, 10 µCi/µL.
3. 100 mg/mL Cycloheximide in ethanol. Prepare freshly and handle with care.
4. 0.2 *M* Methionine. Store in 200-µL aliquots at –20°C.
5. Solutions and materials for isolation of mitochondria from yeast cell (*see* **Subheading 2.1.**).
6. SHKCl buffer: 0.6 *M* sorbitol, 50 m*M* HEPES-KOH, pH 7.4, 80 m*M* KCl.
7. 1 *M* $MgSO_4$, sterile filtrated.
8. 1 *M* KCl, sterile autoclaved.
9. 10% (w/v) BSA (fatty acid free). Store in 200-µL aliquots at –20°C (*see* **Note 3**).
10. 0.2 *M* ATP, pH 7.0: dissolve 60.5 mg of ATP in 470 µL sterile water. Add 30 µL 5 *M* KOH drop by drop. Check pH. Freeze in 20-µL aliquots in liquid nitrogen and store at –20°C.
11. 50 m*M* Guanosine 5′-triphosphate (GTP): dissolve 2.8 mg GTP in 100 µL water. Store in 20-µL aliquots at –20°C.
12. Amino acid stock mix: each of 0.89 g/L alanine, 1.74 g/L arginine, 1.33 g/L aspartic acid, 1.50 g/L asparagine-monohydrate, 1.47g/L glutamic acid, 1.46 g/L glutamine, 0.75 g/L glycine, 1.55 g/L histidine, 1.31 g/L isoleucine, 1.31 g/L leucine, 1.46 g/L lysine, 1.65 g/L phenylalanine, 1.15 g/L proline, 1.05 g/L serine, 1.19 g/L threonine, 2.04 g/L tryptophan, and 1.17 g/L valine are dissolved at 10 m*M* in water. Do not filtrate. Store in 100-µL aliquots at –20°C.
13. 10 m*M* Cysteine. Do not filtrate. Store in 20-µL aliquots µL at –20°C.
14. 5 m*M* Tyrosine. Dissolve 0.9 mg tyrosine in 900 µL water. Adjust to pH 7.0 by adding 1 *M* KOH drop by drop; add water to a total volume of 1 mL. Do not filtrate. Store in 20-µL aliquots at –20°C.
15. Buffer A: 0.6 *M* sorbitol, 150 m*M* KCl, 15 m*M* potassium phosphate, pH 7.4, 20 m*M* Tris-HCl, pH 7.4, 13 m*M* $MgSO_4$, 0.3% (w/v) BSA, 4 m*M* ATP, 0.5 m*M* GTP, 6 m*M* α-ketoglutarate, 5 m*M* phosphoenolpyruvate, 0.1 m*M* each amino acid except methionine. Omit BSA if the samples are analyzed by mass spectrometry. Prepare freshly.
16. Sodium dodecyl sulfate-polyacrylamide gel electrophoresis (SDS-PAGE) sample buffer: 50 m*M* Tris-HCl, pH 6.8, 2% (w/v) SDS, 10% (v/v) glycerol, 1% (v/v) β-mercaptoethanol, 0.01% (w/v) bromophenol blue. β-Mercaptoethanol should be added freshly.
17. Ultima Gold, scintillation cocktail (PerkinElmer).

2.3. Stability of Mitochondrial Proteins In Vivo (Pulse-Chase)

1. SC medium supplemented with glucose or galactose (*see* **Subheading 2.2., item 1**).
2. [^{35}S]-Methionine, 10 µCi/µL.
3. 100 mg/mL Cycloheximide in ethanol: *see* **Subheading 2.2., item 3**.
4. 150 mg/mL Chloramphenicol in ethanol. Prepare freshly and handle with care.
5. 0.2 *M* Methionine: *see* **Subheading 2.2., item 4**.
6. 100 m*M* PMSF in ethanol. Store in 500-µL aliquots at –20°C. Handle with care.
7. Alkaline extraction mix: 1.85 *M* NaOH, 10 m*M* PMSF, 7.4% (v/v) β-mercaptoethanol. Prepare freshly.
8. TE buffer: 10 m*M* Tris-HCl, pH 8.0, 1 m*M* EDTA.
9. 72% (v/v) Trichloroacetic acid (TCA). Store at 4°C. Avoid exposure to light.
10. SHKCl buffer: *see* **Subheading 2.2., item 6**.
11. SDS-PAGE sample buffer: *see* **Subheading 2.2., item 16**.
12. Ice-cold acetone. Store at –20°C.
13. Glass beads, 0.4–0.6 mm (B. Braun).
14. Immunoprecipitation (IP) buffer: 0.3 *M* NaCl, 10 m*M* Tris-HCl, pH 7.5, 1% (v/v) Triton X-100, 0.5 m*M* PMSF.
15. Ultima Gold, scintillation cocktail (PerkinElmer).
16. Antibody specific for the protein of your interest.
17. Protein A-Sepharose beads.
18. SDS-solubilization buffer: 50 m*M* Tris-HCl, pH 7.5, 2% (w/v) SDS.

2.4. Degradation of Newly Imported, Radiolabeled Polypeptides in Isolated Yeast Mitochondria (Import-Chase Assay)

1. 200 m*M* Spermidine: 290 mg spermidine is dissolved in 10 mL water. Store in 1-mL aliquot at –20°C. Air sensitive.
2. 10X Buffer for premix: 400 m*M* HEPES-KOH, pH 7.4, 60 m*M* $Mg(CH_3COO)_2$, 20 m*M* spermidine; sterile filtrate. Store in 1-mL aliquot at –20°C. Air sensitive.
3. Premix for in vitro transcription: 40 m*M* HEPES-KOH, pH 7.4, 6 m*M* $Mg(CH_3COO)_2$, 2 m*M* spermidine, 0.1 mg/mL BSA, 10 m*M* DTT, 0.5 m*M* ATP, 0.5 m*M* cytidine 5′-triphosphate (CTP), 0.1 m*M* GTP, 0.5 m*M* uridine 5′-triphosphate (UTP); sterile filtrate. Store in 200-µL aliquot at –20°C. Air sensitive.
4. SP6 or T7 ribonucleic acid (RNA) polymerase (Promega).
5. RNasin, ribonuclease (RNase) inhibitor (Promega).
6. 2.5 m*M* m7G(5′)ppp(5′)G, sodium salt.
7. Rabbit reticulocyte lysate system, nuclease treated (Promega).
8. [^{35}S]-Methionine, 10 µCi/µL.
9. 0.2 *M* Methionine: *see* **Subheading 2.2., item 4**.
10. 1.5 *M* Sucrose. Store in 200-µL aliquots at –20°C.
11. 2X Import buffer: 100 m*M* HEPES-KOH, pH 7.2, 6% (w/v) BSA (fatty acid free), 1 *M* sorbitol, 160 m*M* KCl, 20 m*M* $Mg(CH_3COO)_2$, 2 m*M* $MnCl_2$; sterile filtrate. Check pH. Freeze in 500-µL aliquots in liquid nitrogen and store at –20°C (*see* **Note 3**).

12. 0.2 *M* β-nicotinamide adenine dinucleotide reduced (NADH). Dissolve 7.6 mg NADH in 50 μL sterile water. Make freshly.
13. 0.2 *M* ATP, pH 7.0: *see* **Subheading 2.2., item 10**.
14. 10 mg/mL Creatine kinase (CK) in 50% (v/v) glycerol. Dissolve 50 mg CK in 500 μL sterile 50% (v/v) glycerol. Store in 20-μL aliquots at –20°C.
15. 1 *M* Creatine phosphate (CP). Dissolve 127 mg phosphocreatine in 500 μL sterile water. Freeze in 20-μL aliquots in liquid nitrogen and store at –20°C.
16. 10 mg/mL Trypsin: dissolve 10 mg trypsin in 1 mL 20 m*M* HEPES-KOH, pH 7.4. Freeze in 25-μL aliquots by liquid nitrogen and store at –20°C.
17. 20 mg/mL Soybean trypsin inhibitor (STI). Dissolve 100 mg STI in 2.5 mL 20 m*M* HEPES-KOH, pH 7.4. Freeze in 100-μL aliquots in liquid nitrogen and store at –20°C.
18. SHKCl buffer: *see* **Subheading 2.2., item 6**.
19. SDS-PAGE sample buffer: *see* **Subheading 2.2., item 16**.

2.5. Degradation of Nonassembled Mitochondrial-Encoded Proteins in Isolated Yeast Mitochondria

For preparation of the following solutions, *see* **Subheading 2.1., items 4–6**; **Subheading 2.2., items 4–16**; and **Subheading 2.3., items 9** and **12**.

1. 2.4 *M* Sorbitol.
2. 1 *M* KCl.
3. 1 *M* Potassium phosphate buffer, pH 7.4.
4. 1 *M* Tris-HCl, pH 7.4.
5. 1 *M* $MgSO_4$.
6. 10% (w/v) BSA (fatty acid free).
7. 0.2 *M* ATP, pH 7.0.
8. 50 m*M* GTP.
9. Amino acid stock mix.
10. 10 m*M* Cysteine.
11. 5 m*M* Tyrosine.
12. 1.5X Translation buffer: 0.9 *M* sorbitol, 225 m*M* KCl, 22.5 m*M* potassium phosphate, pH 7.4, 30 m*M* Tris-HCl, pH 7.4, 19 m*M* $MgSO_4$, 0.45% (w/v) BSA, 6 m*M* ATP, 0.75 m*M* GTP, 9 m*M* α-ketoglutarate, 7.5 m*M* phosphoenolpyruvate, 0.15 m*M* each amino acid except methionine. Prepare freshly.
13. 10 mg/mL Pyruvate kinase.
14. [^{35}S]-Methionine, 10 μCi/μL.
15. 0.2 *M* Methionine.
16. SHKCl buffer.
17. SDS-PAGE sample buffer.
18. 72% (v/v) TCA.
19. Ice-cold acetone.
20. Ultima Gold, scintillation cocktail (PerkinElmer).

2.6. Isolation of Mitochondria From Murine Liver

1. 0.5 *M* HEPES-KOH, pH 7.4; sterile autoclave.

2. Isolation buffer: 220 m*M* manitol, 70 m*M* sucrose, 2 m*M* EGTA, 0.1% (w/v) BSA, 20 m*M* HEPES-KOH, pH 7.4. Prepare freshly (*see* **Notes 2** and **3**).
3. Freezing buffer: 500 m*M* sucrose, 10 m*M* HEPES-KOH, pH 7.4; sterile filtrate. Store at 4°C.
4. Teflon homogenizer (B. Braun).
5. Gauze bandage.
6. Bradford protein assay reagent (Bio-Rad).

2.7. Degradation of Mitochondrial-Encoded Proteins in Murine Liver Mitochondria

For preparation of the following solutions, *see* **Subheadings 2.1.–2.6.**

1. 1 *M* Mannitol, sterile autoclaved.
2. 2 *M* Sucrose, sterile autoclaved.
3. 1 *M* KCl.
4. 1 *M* Potassium phosphate buffer, pH 7.4.
5. 1 *M* $MgCl_2$, sterile filtrated.
6. 0.5 *M* HEPES-KOH, pH 7.4.
7. 0.2 *M* ATP, pH 7.0.
8. 50 m*M* GTP.
9. 10 mg/mL CK.
10. 1 *M* CP.
11. Amino acid stock mix.
12. 10 m*M* Cysteine.
13. 5 m*M* Tyrosine.
14. 1 *M* Sodium succinate. Store in 20-μL aliquots at –20°C.
15. Translation buffer M: 100 m*M* mannitol, 80 m*M* sucrose, 10 m*M* sodium succinate, 80 m*M* KCl, 5 m*M* $MgCl_2$, 1 m*M* potassium phosphate, pH 7.4, 25 m*M* HEPES-KOH, pH 7.4, 0.45 m*M* each amino acid except methionine, tyrosine, and cysteine, 0.3 m*M* tyrosine and cysteine, 5 m*M* ATP, 30 μ*M* GTP, 6 m*M* CP, 60 μg/mL CK. Prepare freshly from stock solutions.
16. [^{35}S]-Methionine, 10 μCi/μL.
17. 0.2 *M* Methionine.
18. Washing buffer: 200 m*M* mannitol, 70 m*M* sucrose, 10 m*M* HEPES-KOH, pH 7.4.
19. SDS-PAGE sample buffer.
20. 72% (v/v) TCA.
21. Ice-cold acetone.
22. Ultima Gold, scintillation cocktail (PerkinElmer).

3. Methods

3.1. Isolation of Mitochondria From Yeast Saccharomyces cerevisiae

Related protocols are described in **refs. *10–12***.

1. Growth conditions can affect the quality of isolated mitochondria. Use logarithmically growing cells in standard experiments. Grow the cells in logarithmic phase for at least 2 d before preparation of mitochondria. *See* **Note 1** for media of choice.

2. Dilute overnight culture to 100 mL of the appropriate medium. Grow cells to OD_{600} ~ 2 and then dilute them in 500 mL medium. After further incubation (OD_{600} ~ 2), the culture is diluted in the prewarmed medium (up to 20 L). Grow cells to OD_{600} ~ 1.5.
3. Isolate cells by centrifugation (3000*g*, 5 min) and resuspend cell pellets in water (500 mL). Reisolate cells again (3000*g*, 5 min) and measure the weight of the cell pellet.
4. Resuspend the cell pellet in Tris DTT buffer (2 mL per gram pellet) and incubate the suspension at 30°C for 10 min with shaking. Start preparation of homogenization buffer and cool it on ice.
5. Isolate cells by centrifugation (2000*g*, 5 min). Resuspend the cell pellet in 1.2 *M* sorbitol (2 mL per gram pellet) and reisolate cells by centrifugation (2000*g*, 5 min). Cool the centrifuge to 4°C after this centrifugation step.
6. Spheroplast formation: resuspend the cell pellet in sorbitol phosphate buffer (6.7 mL per gram pellet) and add lyticase powder (2 mg per gram pellet). Incubate at 30°C for 30 min with shaking. To check for the formation of spheroplast, measure the OD_{600} of (A) 50 μL suspension plus 2 mL water and (B) 50 μL suspension plus 2 mL 1.2 *M* sorbitol. The value of (A) should be 10–20% of (B) (*see* **Note 4**).
7. Keep solutions on ice in all subsequent steps. Cool centrifuge tubes and the glass homogenizer on ice.
8. Isolate cells by centrifugation (1200*g*, 5 min, 4°C). The pellet will be very soft and sticky. Resuspend the soft cell pellet carefully in ice-cold homogenization buffer (13.3 mL per gram pellet).
9. Homogenize the cell suspension by 12 strokes in the glass homogenizer. Rinse the homogenizer with homogenization buffer to recover the remaining material.
10. Precipitate unbroken cells by centrifugation (1500*g*, 5 min, 4°C). *Caution: mitochondria are in the supernatant*. Recover supernatant to new centrifuge tube and discard pellet. After an additional centrifugation (2000*g*, 5 min, 4°C), transfer the supernatant to a new centrifuge tube.
11. Isolate mitochondrial fraction by centrifugation (17,500*g*, 12 min, 4°C).
12. Remove the supernatant (*see* **Note 5**) and resuspend the pellet very carefully in ice-cold SEM buffer (10 mL) (*see* **Note 6**). After another centrifugation step (3000*g*, 5 min, 4°C), transfer the supernatant to a new centrifuge tube.
13. Mitochondria are reisolated by centrifugation (17,500*g*, 12 min, 4°C) and resuspended carefully in ice-cold SEM buffer (0.5 mL) (*see* **Note 6**).
14. Determine the protein concentration of the suspension using the Bradford assay according to the manufacturer's instructions.
15. Dilute the mitochondrial suspension to 10 mg protein/mL with SEM buffer. Freeze in small aliquots (30–50 μL/tube) in liquid nitrogen and keep aliquots at –80°C.
16. This step is optional (*see* **Note 7**). Place mitochondrial suspension in SEM buffer on the top of sequential sucrose gradient (1.5 mL 60% w/v; 4 mL 32% w/v; 1.5 mL 23% w/v; 1.5 mL 15% w/v sucrose in 10 m*M* MOPS-KOH, pH 7.4) in an ultracentrifuge tube (13 mL). After an centrifugation step with a swing rotor (138,000*g*, 1 h, 4°C), recover a band containing mitochondria between the 32 and 60% sucrose segment carefully and mix it with 2 mL SEM. Reisolate

mitochondria by centrifugation (17,500*g*, 12 min, 4°C), resuspend them in 0.5 mL SEM, determine the protein concentration, and store in aliquot at −80°C as described in **steps 14** and **15**.

3.2. Overall Stability of the Mitochondrial Proteome In Vivo

1. Dilute overnight culture of yeast cells in 20 mL SC medium supplemented with 2% (w/v) glucose at OD_{600} = 0.2 and incubate the culture at 30°C to midlogarithmic phase (OD_{600} ~ 1.0). Dilute the culture in SC medium (40 mL) supplemented with 2% (w/v) galactose at OD_{600} = 0.1 and incubate until the cell density reaches OD_{600} ~ 0.8.
2. Isolate the cell from 10 mL of the culture by centrifugation (1200*g*, 5 min). Dissolve the cell pellet in 1.5 mL SC medium without methionine supplemented with galactose and incubate cells for 15 min at 30°C (*see* **Note 8 and 9**).
3. Add 30 µL [^{35}S]-methionine to cultures and further incubate them for 10 min (*see* **Note 8 and 9**).
4. After stop labeling by adding 75 µL 0.2 *M* methionine and 15 µL cycloheximide (100 mg/mL), recover sample on ice. Isolate cells by centrifugation (3000*g*, 5 min).
5. Isolate mitochondria from the radiolabeled cell. The procedures are essentially the same as described in **Subheading 3.1.** with following modifications:
   a. Treat the cell with Tris DTT buffer and lyticase at 24°C and do not elongate the lyticase treatment to avoid degradation of protein.
   b. Purify mitochondria with sucrose gradient sedimentation (*see* **Subheading 3.1., step 16**).
6. Isolate mitochondria (800 µg) by centrifugation (7000*g*, 7 min, 4°C), discard the supernatant, and add 500 µL SHKCl buffer. After resuspending the pellet carefully, mitochondria are isolated by centrifugation (7000*g*, 7 min, 4°C). Repeat the washing step three times (*see* **Note 6**).
7. Resuspend the mitochondrial pellet in 1.2 mL buffer A carefully and divide it into four aliquots (300 µL). Incubate three aliquots for 10, 20, and 30 min at 37°C, respectively, and isolate the mitochondria immediately by centrifugation (16,100*g*, 4 min).
8. Transfer supernatant to new tubes (*see* **Note 10**). Dissolve the pellet in 20 µL SDS-PAGE sample buffer and incubate the sample at 95°C for 10 min.
9. Add 1 mL scintillation cocktail to both supernatant and pellet and determine the radioactivity present in both fractions by scintillation counting. Total radioactivity of supernatant and pellet at each point is set to 100%.

3.3. StabiliFty of Mitochondrial Proteins In Vivo

For stability of mitochondrial proteins in vivo, *see also* **ref. *13***.

1. Cultivate yeast cell as described in **Subheading 3.2., item 1**.
2. Isolate the cell from 4 mL of the culture by centrifugation (1200*g*, 5 min). Dissolve the cell pellet in 750 µL SC medium without methionine supplemented with galactose and incubate cells for 15 min at 30°C.
3. Add 15 µL [^{35}S]-methionine to cultures and further incubate them for 10 min (*see* **Note 9**).

4. Stop labeling by adding 37.5 µL 0.2 *M* methionine, 7.5 µL cycloheximide (100 mg/mL), and 30 µL chloramphenicol (150 mg/mL). Recover sample on ice and divide it into 250-µL aliquots. Incubate two aliquots for 30 min or 60 min at desired temperature to allow proteolysis to occur. Recover samples on ice. Isolate cells by centrifugation (3000*g*, 5 min) (*see* **Note 11**).
5. Alkaline extraction of proteins from yeast cell. Resuspend the cell pellet in 0.5 mL TE buffer. Reisolate the cell by centrifugation (12,000*g*, 1 min), discard the supernatant, and resuspend the cell in 0.5 mL water. Add 75 µL alkaline extraction mix, vortex vigorously, and incubate the samples for 10 min on ice.
6. TCA precipitation of the protein: add 300 µL ice-cold 72% (v/v) TCA to the samples and incubate at least 15 min on ice. Isolate acid-insoluble material by centrifugation (15,000*g*, 15 min, 4°C). Discard the supernatant, add 500 µL ice-cold acetone without resuspending the pellet, and centrifuge again (15,000*g*, 5 min, 4°C). Repeat the washing step twice using ice-cold acetone.
7. Dry the precipitate (30°C, 10 min) and resuspend it in 30 µL SDS-solubilization buffer. Incubate the samples for 1 h at 25°C with mixing and for 10 min at 95°C. Precipitate insoluble material by centrifugation (20,000*g*, 10 min) and transfer the supernatant to a new tube.
8. To check the incorporation of radioactivity, mix 1 µL supernatant and 1 mL scintillation cocktail (Ultima Gold) and measure the radioactivity (*see* **Note 12**).
9. Immunoprecipitation: mix 5 µL supernatant with 1 mL IP buffer, add antibody-conjugated protein A-Sepharose beads to the sample, and continue immunoprecipitation according to the standard protocol. Analyze the samples on SDS-PAGE and then blot the protein on nitrocellulose membrane. Detect the radioactive bands by autoradiography using a phosphorimaging system and quantify the stability of the protein.

3.4. Degradation of Newly Imported, Radiolabeled Polypeptides in Isolated Yeast Mitochondria (Import-Chase Assay)

1. The gene encoding the protein of interest has to be cloned into a plasmid under the control of SP6- or T7-polymerase-driven promoters (i.e., pGEM4). Kozak consensus sequence [(GCC)(A/G)CCATGG] preceding the start codon (underlined) should be added for efficient translation in the reticulocyte lysate ***(14)***. The last G is variable. *See* **Note 13** for the substrates used in the literature.
2. In vitro transcription: mix 60 µL water, 120 µL premix, 10 µL m7G(5′)ppp(5′)G, 10 µL plasmid DNA-encoding protein (1 µg/µL, highly pure and RNase free), and 4.7 µL RNasin. Add 1.3 µL polymerase and incubate the mixture for 60 min at 37°C. After incubation, add 20 µL 10 *M* LiCl and 600 µL ethanol and incubate for 30 min at –80°C. Precipitate the RNA by centrifugation (20,000*g*, 30 min, 4°C) and remove the supernatant completely using a micropipet. Dry the RNA pellet (2 min at 30°C) and dissolve the pellet in 50 µL water and 1 µL RNasin. Store in 16-µL aliquots at –80°C.
3. Cell-free synthesis of radiolabeled precursor proteins: mix 70 µL rabbit reticulocyte lysate, 2 µL amino acid mix minus methionine (supplemented with lysate), 2 µL RNasin, 8 µL [^{35}S]-methionine, and 16 µL RNA. After the incubation for

60 min at 30°C, add 2.5 µL 0.2 *M* methionine and 20 µL 1.5 *M* sucrose. Ribosome and aggregated proteins can be removed by ultracentrifugation (100,000*g*, 30 min, 4°C) (optional). Freeze in 10-µL aliquots in liquid nitrogen and store at –80°C.

4. In vitro import of precursor proteins into mitochondria. A detailed description of various in vitro import assays can be found in other chapters of this book. Here, we describe a protocol for the analysis of the degradation of the imported protein. *See* **Note 14** for possible modification.
5. Mix 200 µL 2X import buffer, 4 µL 0.2 *M* NADH, 4 µL 0.2 *M* ATP, 4 µL 10 mg/mL CP, 4 µL 1 *M* CK, and 164 µL water. Add 20 µL mitochondria (10 mg/mL in SEM; final concentration of protein is 0.5 mg/mL) and mix gently.
6. Import reaction: incubate the mix for 3 min at 25°C. Add 6.7 µL lysate of precursor protein. Mix gently and incubate for 20 min at 25°C. Recover the tube on an ice-cold metal block and incubate for 3 min (*see* **Note 14**).
7. Trypsin treatment to remove the nonimported proteins: add 2 µL 10 mg/mL trypsin to the sample (50 µg/mL final concentration) (*see* **Note 15**) and incubate it for 20 min on ice. Stop trypsin treatment by adding 20 µL 20 mg/mL STI (1 mg /mL final concentration). Incubate for 5 min on ice.
8. Incubate six aliquots (60 µL) for 0, 2.5, 5, 10, 20, and 30 min at 37°C, respectively, to allow proteolysis to occur. Transfer tubes on an ice-cold metal block.
9. Isolate mitochondria by centrifugation (12,000*g*, 10 min, 4°C). Remove supernatant and add ice-cold SHKCl buffer without resuspending the pellet. Repeat this washing step twice.
10. Resuspend mitochondria in 20 µL SDS-PAGE sample buffer. Incubate samples for 3 min at 95°C. Analyze samples by SDS-PAGE and transfer proteins onto a nitrocellulose membrane. Detect the radioactive bands either by autoradiography or using a phosphorimaging system and quantify radiolabeled proteins within mitochondria at various time-points (*see* **Fig. 1**).

3.5. Degradation of Nonassembled Mitochondrial-Encoded Proteins in Isolated Yeast Mitochondria

For this process, *see also* **ref. *15***.

1. *In organello* translation of the mitochondrial-encoded polypeptides: mix 30.5 µL water, 80 µL 1.5X translation buffer, and 0.5 µL 10 mg/mL pyruvate kinase in a 1.5-mL Eppendorf tube.
2. Add 8 µL mitochondria (10 mg/mL in SEM buffer; final concentration of protein is 0.67 mg/mL) to the translation mix and mix gently.
3. Incubate the sample at 30°C for 3 min to adjust the temperature. Add 1 µL [^{35}S]-methionine (10 µCi/µL), mix gently, and incubate further at 30°C for 5 min.
4. Add 30 µL 0.2 *M* methionine and mix gently to stop the incorporation of radioactivity (*see* **Note 16**). Place the sample on ice and incubate for 3 min. When you want to examine the degradation of specific translation products, go to **step 5**. If you focus on the overall turnover rate of the labeled polypeptides, then go to **step 9**.

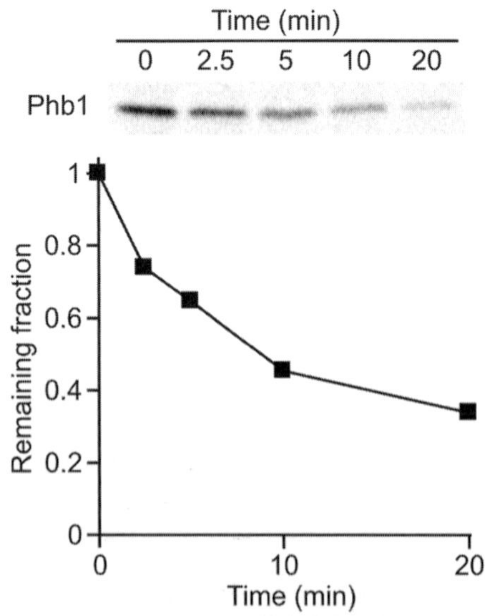

Fig. 1. Degradation of Phb1 on import into mitochondria isolated from *S. cerevisiae*. Phb1 was synthesized in the presence of [^{35}S]-methionine in reticulocyte lysate and imported into mitochondria for 10 min at 25°C. After proteolytic removal of non-imported precursor proteins by trypsin, mitochondria were incubated at 37°C for indicated times to allow the degradation of newly imported proteins. Samples were subjected to SDS-PAGE, and the radioactivity remaining at each time-point was quantified by phosphorimaging. Value at time 0 was set to 1.

5. Divide the sample into three aliquots (37.5 µL) and incubate two aliquots at 37°C, one for 30 and one for 60 min. Transfer samples on ice.
6. Isolate mitochondria by centrifugation (16,000*g*, 5 min, 4°C) and add 180 µL SHKCl buffer without resuspending the pellet. Repeat the washing step once.
7. Resuspend mitochondria in 20 µL SDS-PAGE sample buffer. Incubate samples for 10 min at 30°C. Analyze samples by SDS-PAGE and transfer proteins onto a nitrocellulose membrane. Detect the radioactive bands by autoradiography using a phosphorimaging system (*see* **Fig. 2A**).
8. To determine the overall stability of newly synthesized translation products, [^{35}S]-methionine must be removed completely. Isolate mitochondria by centrifugation (7000*g*, 7 min, 4°C), discard the supernatant, and add 500 µL SHKCl buffer. After resuspending the pellet carefully, mitochondria are isolated by centrifugation (7000*g*, 7 min, 4°C). Repeat the washing step three times. During the washing steps, prepare 1X translation mix by mixing 39.5 µL water, 80 µL 1.5X translation buffer, and 0.5 µL pyruvate kinase (10 mg/mL) (*see* **Note 6**).
9. Resuspend the pellet very carefully in 120 µL 1X translation mix. Divide the sample into three aliquots (37.5 µL) and incubate two aliquots at 37°C, one for 30 and one for 60 min. Transfer them on ice.

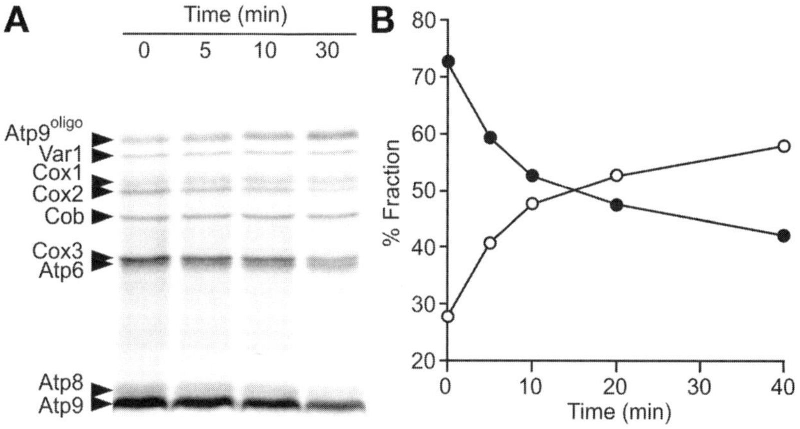

Fig. 2. Degradation of mitochondrial translation products in isolated yeast mitochondria. Mitochondrial translation products were synthesized and labeled for 5 min at 30°C in isolated mitochondria, which were then incubated for indicated times. (**A**) Degradation of mitochondrial encoded proteins. All eight mitochondrial encoded proteins (Var1, Cox1, Cox2, Cob, Cox3, Atp6, Atp8, and Atp9) and an SDS-resistant oligomeric form of Atp9 are indicated. (**B**) Overall turnover of newly synthesized translation products. Total radioactivity present in supernatant and pellet at each time-point is set to 100%. Open circle, radioactivity in the supernatant fraction; closed circle, radioactivity in the pellet fraction. Cob, cytochrome-*b*.

10. TCA precipitation: add 7.5 μL 72% (v/v) TCA to each sample (12% final concentration of TCA). Incubate samples on ice for at least 15 min and follow by a centrifugation step (25,000*g*, 30 min, 4°C).
11. Transfer the supernatant to new tubes. Dissolve the pellet in 20 μL SDS-PAGE sample buffer and incubate the sample at 95°C for 10 min.
12. Add 1 mL scintillation cocktail to both supernatant and pellet and determine the radioactivity present in both fractions by scintillation counting. Total radioactivity of supernatant and pellet at each point is set to 100% (*see* **Fig. 2B**).

3.6. Isolation of Mitochondria From Murine Liver

For isolation of mitochondria from murine liver, *see also* **ref. *16***.

1. Adjust buffer, centrifuge, and the homogenizer to 4°C.
2. Isolate liver from a young mouse (*see* **Note 17**). Remove gallbladder. Wash liver with 10 mL isolation buffer and determine the weight.
3. Cut the liver in pieces in 5 mL isolation buffer. Homogenize the tissue by 10 strokes with Teflon homogenizer.
4. Centrifuge the homogenate (1000*g*, 10 min, 4°C). Take the supernatant and then pass them through gauze bandage to remove the lipid layer.

5. Isolate mitochondrial fraction by centrifugation (8000*g*, 10 min, 4°C). Remove the fat layer by a paper before discarding the supernatant and then resuspend the pellet in 300 μL isolation buffer carefully.
6. Determine the protein concentration using the Bradford assay according to the instructions of the manufacturer.
7. Dilute the mitochondrial suspension to 10 mg protein/mL with isolation buffer.
8. Mitochondria can be frozen and stored at –80°C; however, freezing will retard the activity of mitochondria severely. To freeze mitochondria, isolate mitochondria by centrifugation (8000*g*, 10 min, 4°C) and resuspend the pellet carefully in freezing buffer at 10 mg/mL. Freeze in small aliquots (30 μL/tube) in liquid nitrogen.

3.7. Degradation of Mitochondrial Encoded Proteins in Murine Liver Mitochondria

1. Translation of polypeptides in isolated mitochondria: isolate mitochondria from 40 μL of a mitochondrial suspension (10 mg/mL) by centrifugation (8000*g*, 10 min, 4°C) and resuspend mitochondria in 400 μL translation buffer M gently (*see* **Note 6**).
2. Incubate the sample at 30°C for 3 min and add 8 μL [^{35}S]-methionine (10 μCi/μL). After gentle mixing, mitochondria are further incubated at 30°C for 20 min.
3. Add 100 μL 0.2 *M* methionine and mix gently. Transfer the tube on ice and incubate it for 5 min (*see* **Note 16**).
4. To remove nonincorporated [^{35}S]-methionine, isolate mitochondria by centrifugation (10,000*g*, 5 min, 4°C), discard the supernatant, and add 500 μL washing buffer. Resuspend the mitochondrial pellet carefully and reisolate mitochondria by centrifugation (10,000*g*, 5 min, 4°C). Repeat the washing step three times.
5. Resuspend the mitochondrial pellet very carefully in 400 μL translation buffer M. Divide the sample into four aliquots (90 μL) and incubate three aliquots at 37°C for 10, 30, and 60 min to allow proteolysis to occur. Transfer the samples on ice.
6. Precipitate proteins by TCA (*see* **Subheading 3.5., step 10**).
7. Transfer the supernatant to new tubes. Dissolve the pellet in 30 μL SDS-PAGE sample buffer and incubate the samples for 3 min at 95°C. Analyze 15 μL of sample on SDS-PAGE and transfer proteins onto nitrocellulose membrane. Detect and quantify the radioactive bands (*see* **Subheading 3.5., step 7**).
8. Add 1 mL scintillation cocktail to both supernatant and pellet fractions and determine the radioactivity present in both fractions by scintillation counting. Radioactivity present in the pellet fraction at time-point 0 is set to 100% (*see* **Note 18**).

4. Notes

1. We usually use lactate medium to cultivate the *S. cerevisiae* cells for mitochondrial preparation to guarantee a high yield. Mutant strains with a *petite* phenotype that cannot grow on nonfermentable carbon sources like lactate are grown on YP-gal-lac medium. Alternatively, SC media can be used. In general, medium containing glucose as a carbon source should be avoided because glucose represses many genes encoding mitochondrial proteins *(17)*. It should be noted that a large amount of NaOH is

required to adjust the pH of lactate media or of stock solutions. Please control the temperature of the medium during the pH adjustment to avoid boiling and damage of the pH meter.

2. EDTA and EGTA are not soluble at acidic pH.
3. You should use fatty acid-free BSA to avoid lysis of mitochondrial membranes.
4. If spheroplast formation is not efficient after incubation for 30 min, then you may add additional 2 mg per gram cells of lyticase and elongate the incubation time up to 1 h. However, do not extend the incubation time beyond this point and continue with preparation (perhaps resulting in reduced yields).
5. The supernatant is almost free of mitochondrial proteins. You may analyze a small amount of this fraction by SDS-PAGE to exclude a cytosolic localization of your target protein (e.g., in the case of exogenously expressed, tagged protein).
6. Mitochondria are labile. Harsh treatment will damage the outer membrane. Always use a cut tip and avoid air bubbles when you resuspend the mitochondrial pellet using a micropipet.
7. Mitochondrial preparations are considerably contaminated by membranes derived from other cellular compartments like vacuole, plasma membrane, or endoplasmic reticulum. We further purify mitochondria by sucrose density gradient sedimentation (**Subheading 3.1.**, **step 16**) to remove these contaminants if high purity is required.
8. The complete mitochondrial proteome can be labeled by growing cells overnight in the presence of [^{35}S]-methionine. Briefly, isolate the cell from 2 mL culture after first cultivation on SC with galactose (1200*g*, 5 min), dilute the cell in 40 mL SC medium without methionine and supplemented with galactose containing 2 mCi of [^{35}S]-methionine, and incubate the culture for 14 h. Isolate mitochondria as described in **Subheading 2.1.** and analyze the degradation of the protein of interest *in organello* as in **Subheading 3.2.** *(8)*. The turnover of a protein can be analyzed in vivo using labeled cells as described in **Subheading 3.3**. Alternatively, proteolysis can be monitored without radiolabeling after cycloheximide treatment and immunoblot analysis of cell extracts.
9. If cells are treated for 15 min with cycloheximide (1 mg/mL final concentration) before labeling, then [^{35}S]-methionine is only incorporated into mitochondrial encoded proteins, and their stability can be analyzed by SDS-PAGE as described for *in organello* translated polypeptides (*see* **Subheading 3.5.** and **Fig. 2A**).
10. The mitochondrial supernatant contains small peptides and amino acids generated and exported from the organelle on proteolysis by ATP-dependent proteases and oligopeptidases in mitochondria *(8,9)*. Thus, the release of radioactivity from the mitochondria reflects the degradation of radiolabeled proteins within mitochondria. Characterization of the exported peptides in the supernatant using sizing column chromatography and mass spectrometry allows identification of the origin of the peptides *(8)*.
11. As an alternative to alkaline extraction of whole cell proteins, mitochondria can be isolated at a small scale after proteolysis has occurred in vivo essentially as described in **Subheading 2.1**. To ensure efficient homogenization of a small volume

of cell suspension, place the suspension in homogenization buffer in an Eppendorf tube, put a micropipet tip very close to the wall of the tube, and pass the suspension 10 times through the narrow opening between the tube wall and the micropipet tip. Mitochondria are analyzed by two-dimensional electrophoresis on immobilized pH gradient gels, and alterations in the intensity of individual protein spots are monitored ***(8)***.

12. In routine experiments, up to 20% of the radioactivity is incorporated. Five million cpm per sample is usually enough to detect the target protein on immunoprecipitation.
13. Substrates listed in this note have been used for ATP-dependent proteases. Yeast protein substrates are temperature-sensitive variant of Oxa1 (*Oxa1^{ts}*) ***(18)***; Yme2 ***(19)***; nonassembled Phb1 and Phb2 ***(9)*** and Nde1 ***(8)***. Chimeric fusion proteins are also commonly used, such as unfolded variant of mouse dihydrofolate reductase (DHFR) fused to mitochondrial targeting sequence of cytochrome-b_2 (b_2-DHFR^{ds}) ***(20)*** or yta10 [Yta10(161)-DHFR^{mut}] ***(21)*** or with domains of Yme2 and mitochondrial targeting sequence of subunit 9 of ATP synthase ***(19)***. The common features of these substrates are efficient translation in reticulocyte lysate; efficient targeting to mitochondria in vitro; presence of (an) unfolded domain. Recombinant precursor proteins purified from *Escherichia coli* can also be used as a substrate in import-chase experiments ***(20)***. Up to 50 pmol precursor protein per milligram mitochondria can be added to the import mix, leading to a saturation of the proteolytic machinery within mitochondria. In this case, the proteolysis of newly imported proteins in mitochondria can be monitored by immunoblotting.
14. The kinetics of in vitro import varies between different precursor proteins. You should manipulate the temperature, time, and amount of precursor protein (up to 15% lysate in import buffer) to ensure efficient import. To minimize proteolysis during the import reaction, the temperature should be as low as possible. The time-course for proteolysis should be adjusted according to the stability of the respective protein.
15. We prefer to use trypsin to remove nonimported precursor protein because it can be irreversibly inactivated by STI. However, if the precursor protein of interest is resistant to trypsin, then increase the concentration of protease or use another protease (i.e., proteinase K). When you use proteinase K, mitochondria must be reisolated and washed with SHKCl buffer containing PMSF (2 m*M* final concentration).
16. You can incubate the samples for 5 min at 30°C after addition of methionine to allow the completion of the synthesis of labeled nascent polypeptides. Transiently synthesized polypeptides appear as smear bands on the gel. Puromycin can be added to a final concentration at 50 μg/mL to release nascent polypeptides from the ribosome after labeling. Addition of a limited concentration of puromycin (~25 μg/mL) during the labeling increases the fraction of incompletely translated polypeptides. Chloramphenicol (300 μg/mL final concentration) is used to block the translation in mitochondria.

17. Use relatively young (less than 12-month-old) mice for isolation of mitochondria to avoid age-dependent effects.
18. Using murine liver mitochondria, we always detect high background radioactivity in the TCA-soluble fraction even in the sample at time-point 0.

Acknowledgments

We thank S. Augustin, S. Ehses, M. Graef, C. Günther, M. Metodiev, and M. Nolden for providing working protocols and figures and for critical reading of the manuscript.

References

1. Gakh, O., Cavadini, P., and Isaya, G. (2002) Mitochondrial processing peptidases. *Biochim. Biophys. Acta* **1592,** 63–77.
2. Esser, K., Tursun, B., Ingenhoven, M., Michaelis, G., and Pratje, E. (2002) A novel two-step mechanism for removal of a mitochondrial signal sequence involves the *m*-AAA complex and the putative rhomboid protease Pcp1. *J. Mol. Biol.* **323,** 835–843.
3. Herlan, M., Vogel, F., Bornhövd, C., Neupert, W., and Reichert, A.S. (2003) Processing of Mgm1 by the rhomboid-type protease Pcp1 is required for maintenance of mitochondrial morphology and of mitochondrial DNA. *J. Biol. Chem.* **278,** 27,781–27,788.
4. McQuibban, G. A., Saurya, S., and Freeman, M. (2003) Mitochondrial membrane remodelling regulated by a conserved rhomboid protease. *Nature* **423,** 537–541.
5. Van Dyck, L. and Langer, T. (1999) ATP-dependent proteases controlling mitochondrial function in the yeast *Saccharomyces cerevisiae*. *Cell Mol. Life Sci.* **55,** 825–842.
6. Bota, D. A. and Davies, K. J. A. (2001) Protein degradation in mitochondria: implications for oxidative stress, aging and disease: a novel etiological classification of mitochondrial proteolytic disorders. *Mitochondrion* **1,** 33–49.
7. Young, L., Leonhard, K., Tatsuta, T., Trowsdale, J., and Langer, T. (2001) Role of the ABC transporter Mdl1 in peptide export from mitochondria. *Science* **291,** 2135–2138.
8. Augustin, S., Nolden, M., Müller, S., Hardt, O., Arnold, I., and Langer, T. (2005) Characterization of peptides released from mitochondria: evidence for constant proteolysis and peptide efflux. *J. Biol. Chem.* **280,** 2691–2699.
9. Kambacheld, M., Augustin, S., Tatsuta, T., Müller, S., and Langer, T. (2005) Role of the novel metallopeptidase MOP112 and saccharolysin for the complete degradation of proteins residing in different subcompartments of mitochondria. *J. Biol. Chem.* **280,** 20,132–20,139.
10. Daum, G., Gasser, S. M., and Schatz, G. (1982) Import of proteins into mitochondria. Energy-dependent, two-step processing of the intermembrane space enzyme cytochrome b2 by isolated yeast mitochondria. *J. Biol. Chem.* **257,** 13,075–13,080.

11. Herrmann, J. M., Fölsch, H., Neupert, W., and Stuart, R. A. (1994) Isolation of yeast mitochondria and study of mitochondrial protein translation, in *Cell Biology: A Laboratory Handbook*, vol. 1 (Celis, D. E., ed.), Academic Press, San Diego, CA, pp. 538–544.
12. Meisinger, C., Sommer, T., and Pfanner, N. (2000) Purification of *Saccharomcyes cerevisiae* mitochondria devoid of microsomal and cytosolic contaminations. *Anal. Biochem.* **287,** 339–342.
13. Brandt, A. (1991) Pulse labeling of yeast cells as a tool to study mitochondrial protein import. *Methods Cell Biol.* **34,** 369–376.
14. Kozak, M. (1987) An analysis of 5′-noncoding sequences from 699 vertebrate messenger RNAs. *Nucleic Acids Res.* **15,** 8125–8148.
15. Black-Schaefer, C. L., McCourt, J. D., Poyton, R. O., and McKee, E. E. (1991) Mitochondrial gene expression in *Saccharomyces cerevisiae*. Proteolysis of nascent chains in isolated yeast mitochondria optimized for protein synthesis. *Biochem. J.* **274,** 199–205.
16. Mattiazzi, M., D'Aurelio, M., Gajewski, C. D., et al. (2002) Mutated human *SOD1* causes dysfunction of oxidative phosphorylation in mitochondria of transgenic mice. *J. Biol. Chem.* **277,** 29,626–29,633.
17. Gancedo, J. M. (1998) Yeast carbon catabolite repression. *Microbiol. Mol. Biol. Rev.* **62,** 334–361.
18. Käser, M., Kambacheld, M., Kisters-Woike, B., and Langer, T. (2003) Oma1, a novel membrane-bound metallopeptidase in mitochondria with activities overlapping with the *m*-AAA protease. *J. Biol. Chem.* **278,** 46,414–46,423.
19. Leonhard, K., Guiard, B., Pellechia, G., Tzagoloff, A., Neupert, W., and Langer, T. (2000) Membrane protein degradation by AAA proteases in mitochondria: extraction of substrates from either membrane surface. *Mol. Cell* **5,** 629–638.
20. Rottgers, K., Zufall, N., Guiard, B., and Voos, W. (2002) The ClpB homolog Hsp78 is required for the efficient degradation of proteins in the mitochondrial matrix. *J. Biol. Chem.* **277,** 45,829–45,837.
21. Leonhard, K., Stiegler, A., Neupert, W., and Langer, T. (1999) Chaperone-like activity of the AAA domain of the yeast Yme1 AAA protease. *Nature* **398,** 348–351.

IV

Physiology

26

Methods to Determine the Status of Mitochondrial ATP Synthase Assembly

Sharon H. Ackerman and Alexander Tzagoloff

Summary

The adenosine triphosphate (ATP) synthase (F_1-F_0 complex) of the mitochondrial inner membrane is responsible for making nearly all of the ATP utilized by eukaryotic organisms. The enzyme is an oligomer of more than 20 different subunits, 14 of which are essential for its catalytic activity. The other subunits function in the regulation and structure of the complex. Subunits essential for catalytic activity make up the proton pore, the bulk of the F_1 headpiece, and the two stalks that physically and functionally couple the catalytic and proton-translocating activities of the ATP synthase. *Saccharomyces cerevisiae* provides an excellent model system for studying mutations that affect assembly of the complex because of the ability of this organism to survive on the ATP produced from fermentation in the absence of mitochondrial respiration or oxidative phosphorylation. Studies of such mutants have been instrumental in identifying novel molecular chaperones that act at discrete steps of F_1-F_0 assembly. Here, we describe some experimental approaches useful in assessing the status of F_1-F_0 assembly.

Key Words: ATPase assays; ATPase mutants; F_1-F_0 assembly; isopycnic gradients; mitochondrial ATP synthase; reconstitution of F_1-F_0; sedimentation of F_1 and F_1-F_0.

1. Introduction

The adenosine triphosphate (ATP) synthase (F_1-F_0 complex) of mitochondria is a hetero-oligomeric constituent of the inner membrane that catalyzes the synthesis of ATP from adenosine 5′-diphosphate (ADP) and inorganic phosphate using electrochemical energy derived from the oxidation, by the electron transfer chain, of nicotinamide adenine dinucleotide (NADH) and other substrates such as succinate *(1)*. This important component of the mitochondrial oxidative phosphorylation system consists of several morphologically distinguishable parts (**Fig. 1**) that are combined in discrete steps during assembly of the complex *(2)*.

From: *Methods in Molecular Biology, vol. 372: Mitochondria: Practical Protocols*
Edited by: D. Leister and J. M. Herrmann © Humana Press Inc., Totowa, NJ

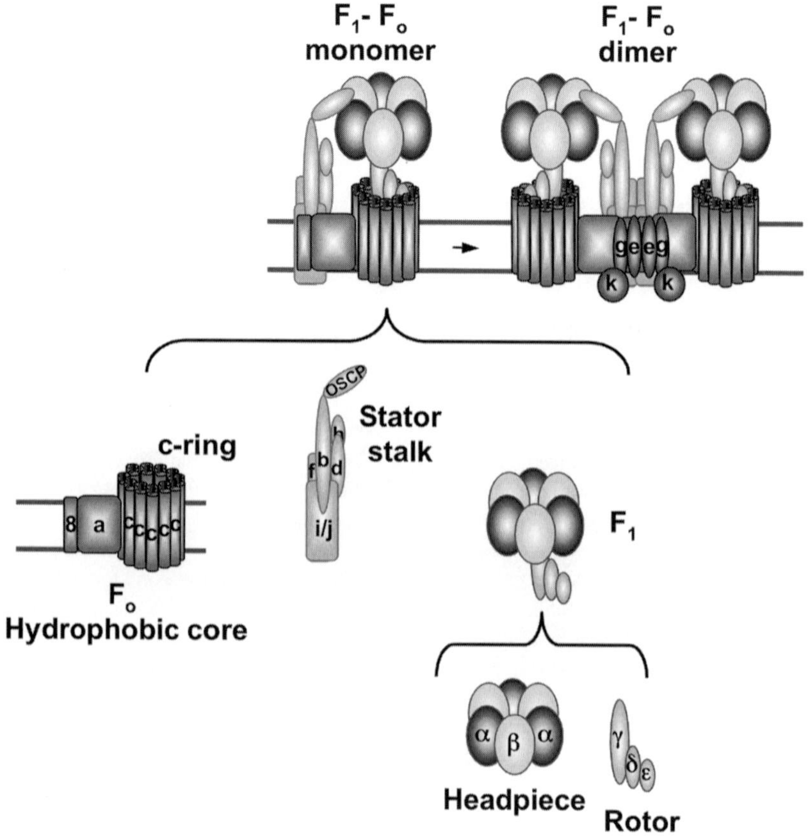

Fig. 1. Structure of mitochondrial F_1-F_0. Cartoon representations of the yeast mitochondrial F_1-F_0 monomer and dimer are given at the top of the figure. Below the bracket are groupings of F_0 and F_1 subunits according to four different functional units recognized in the holoenzyme (*see* text for details). Such groupings are not meant to indicate actual structural intermediates in the pathway for assembly or degradation of the enzyme.

The F_1 unit consists of five different subunits, designated α, β, γ, δ, and ε, in a 3:3:1:1:1, respectively, stoichiometry. The $(\alpha\beta)_3$ headpiece of F_1 contains the catalytic sites and surrounds the γδε central stalk (rotor element). The hydrophobic core of the membrane F_0 unit is composed of three proteins, subunits **a** and **c** and subunit 8. In *Saccharomyces cerevisiae*, these subunits are encoded by mitochondrial DNA. Subunit **c** is present in 10 copies, which combine to form a ring-like structure that is embedded in the phospholipid bilayer. F_1 is attached to F_0 by means of the central stalk of F_1 and a second peripheral stalk (stator) comprised of six polypeptides, referred to as **b**, **d**, **h**, **f**, **i/j**, and **OSCP**, each of which is present in single copy. The topology of the F_0 and stator subunits and their locations vis-à-vis each other and other subunits are still not clear at present. Recent evidence indicates that the ATP synthase exists as a dimer in the membrane *(3)*. Dimerization of the complex has been shown to be mediated by the F_0-associated subunits **e**, **g**, and **k** *(3)*.

The channel through which protons are translocated is thought to be formed sequentially by an interface between subunit **a** and each subunit **c** as the subunit $\mathbf{c}_{10}$-ring rotates during ATP synthesis or hydrolysis *(4)*. The energy of the proton gradient is coupled to ATP synthesis via the central stalk of F_1, which rotates in unison with the **c**-ring. Although in vivo the F_1-F_0 complex functions principally to synthesize ATP, because of the reversible nature of the reaction catalyzed by F_1 it has the additional capacity to convert energy released from the hydrolysis of ATP into an electrochemical gradient. Both ATP synthesis and hydrolysis by the F_1-F_0 complex are inhibited by oligomycin, which blocks the coupled reaction by binding to F_0 and preventing proton translocation. When detached from the membrane, F_1 functions as an oligomycin-insensitive ATP hydrolase that exhibits properties typical of water-soluble globular proteins.

A substantial number of nuclear mutations in *S. cerevisiae* elicit a defect in the mitochondrial ATP synthase *(2)*. Although some mutations are in the structural genes themselves, others are in genes coding for accessory factors that promote different events during assembly of the complex. The methods described in this chapter measure the enzymatic and physical properties of the ATP synthase, thereby providing insight into the status of F_1-F_0 assembly.

2. Materials

2.1. Yeast Mitochondria

1. Galactose medium (YPGal): 2% (w/v) galactose, 2% (w/v) Bacto™ peptone, and 1% (w/v) yeast extract.
2. Sorbitol wash solution: 1.2 *M* sorbitol. Store at 4°C (*see* **Note 1**).
3. Digestion buffer: 1.2 *M* sorbitol (dilute from a 2 *M* stock), 40 m*M* potassium phosphate buffer, pH 7.5, 1 m*M* ethylenediaminetetraacetic acid (EDTA), 0.14 *M* β-mercaptoethanol, and 1 mg/mL zymolyase 20T (MP Biochemicals, Aurora, OH, USA). Prepare fresh as needed.
4. Lysis buffer: 0.6 *M* sorbitol, 20 m*M* Tris-HCl, pH 7.5, 1 m*M* EDTA. Store at 4°C.
5. Mitochondrial wash buffer: 0.6 *M* sorbitol, 10 m*M* Tris-HCl, pH 7.5. Store at 4°C.
6. 0.2 *M* Phenylmethylsulfonyl fluoride (PMSF) in ethanol. Store at –20°C.

2.2. Adenosine Triphosphatase Assay by Chemical Determination of Inorganic Phosphate Release From Adenosine Triphosphate

1. Adenosine triphosphatase (ATPase) assay buffer (2X): 0.1 *M* Tris-sulfate, pH 8.5, 8 m*M* $MgSO_4$.
2. 0.1 *M* Adenosine triphosphate (ATP), adjusted to pH 7.2 with NaOH. Store at –20°C in 1- to 2-mL aliquots.

3. 5% (w/v) Ammonium molybdate.
4. ANS reagent: 15% (w/v) $NaHSO_3$, 6% (w/v) Na_2SO_3, and 0.25% (w/v) amino naphthol sulfonic acid. This reagent should be made fresh every 2–4 wk and stored at 4°C in an amber bottle (*see* **Note 2**).
5. Stop reagent: 50% (w/v) trichloroacetic acid (TCA).
6. Phosphate assay solution: 0.5% (w/v) TCA.
7. 2 mg/mL Oligomycin in 100% ethanol. Store at –20°C.

2.3. ATPase Assay Monitored Enzymatically by NADH Oxidation Coupled to ATP Hydrolysis

1. 5X Assay buffer: 250 m*M* Tris-acetate, pH 8.0, 25 m*M* MgOAc, 25 m*M* KOAc. Dissolve the Tris-acetate with the salts and adjust the pH of the solution with acetic acid. Store in small aliquots at –20°C.
2. 20 m*M* Phosphoenolpyruvate dissolved in 50 m*M* Tris-acetate, pH 8.0. Store at –20°C.
3. 0.1 *M* ATP, adjusted to pH 7.2 with NaOH and stored in 1- to 2-mL aliquots at –20°C.
4. 7.5 m*M* NADH. Store at –20°C.
5. Pyruvate kinase (PK; 10 mg/mL ammonium sulfate suspension; Boehringer-Mannheim).
6. Lactate dehydrogenase (LDH; 5 mg/mL ammonium sulfate suspension; Boehringer-Mannheim).

2.4. Sodium Dodecyl Sulfate-Polyacrylamide Gels for Resolution of F_1 Subunits

1. 10% (w/v) Sodium dodecyl sulfate (SDS) (*see* **Note 3**).
2. Acrylamide:bisacrylamide (29:1) solution. Store at 4°C (*see* **Note 4**).
3. Separation gel solution: 1.5 *M* Tris-HCl, 0.4% SDS, pH 8.0.
4. Stacking gel solution: 0.5 *M* Tris-HCl, 0.4% SDS, pH 6.8.
5. 10% (w/v) ammonium persulfate. Store at 4°C.
6. *N,N,N,N′*-tetramethyl ethylenediamine (TEMED).
7. 5X Running buffer: Dissolve 60 g Tris base, 288 g glycine, and 10 g SDS in a final volume of 2 L water.
8. 4X Sample buffer: mix 4 g glycerol and 0.8 g SDS, 2.5 mL 1 *M* Tris-HCl, pH 6.8, 2 mL β-mercaptoethanol, and 1.5 mL water. Add 20 mg bromophenol blue and mix.

2.5. SDS-Polyacrylamide Gels for Separation of F_0 Subunits

1. 3 *M* Tris-HCl, pH 8.8.
2. 1 *M* Tris-HCl, pH 6.8.
3. SDS (10%), acrylamide:bisacrylamide, ammonium persulfate (10%), and TEMED as described in **Subheading 2.4.** except that the acrylamide stock is 30:0.8 acrylamide:bisacrylamide (*see* **Notes 3** and **4**).
4. 10X Running buffer: dissolve 30.3 g Tris-HCl, 144 g glycine, and 10 g SDS in a final volume of 1 L distilled water.
5. 4X Sample buffer: 2 mL 1 *M* Tris-HCl, pH 6.8, 4 mL 10% SDS, 4 g glycerol, 0.4 mL β-mercaptoethanol, 20 mg bromophenol blue. Adjust to 10 mL with water.

2.6. Preparation of Submitochondrial Particles

Sonication buffer: 0.25 *M* sucrose, 20 m*M* Tris-HCl, pH 7.5, 0.5 m*M* EDTA.

2.7. Solubilization of F_1

F_1 release buffer: 20 m*M* Tris-HCl, pH 8.5, 2 m*M* ATP, 0.1 m*M* EDTA.

2.8. Solubilization of F_1-F_0

1. TEA buffer: 20 m*M* Tris-HCl, pH 8.0, 0.1 m*M* EDTA, 2 m*M* ATP (*see* **Note 5**).
2. 10% (w/v) Triton X-100.
3. 10% (w/v) Potassium deoxycholate solution (*see* **Note 6**).
4. 10% (w/v) Lauryl/dodecyl maltoside.

2.9. Sedimentation of F_1 and F_1-F_0 in Linear Sucrose Gradients

1. Gradient buffer for F_1 sedimentation: 20 m*M* Tris-HCl, pH 7.5, 2 m*M* ATP, 0.1 m*M* EDTA, 0.05% Triton X-100 (*see* **Note 7**).
2. Gradient buffer for F_1-F_0 sedimentation: Same as in **item 1** except that the concentration of Triton X-100 is increased to 0.1%.
3. Sucrose solutions: 7% (w/v) and 20% (w/v) sucrose in gradient buffer.
4. β-Galactosidase: (Aldrich, grade 8 purified from *Escherichia coli*).
5. β-Galactosidase reaction buffer: 20 m*M* sodium phosphate buffer, pH 7.3, 1 m*M* $MgCl_2$, 0.1 m*M* β-mercaptoethanol.
6. *o*-Nitrophenol β-D-galactopyranoside: 34 m*M* stock solution, made fresh.
7. Equipment for sucrose gradients centrifugation and collection (*see* **Note 8**).

2.10. Preparation of F_1-Depleted SMPs and Reconstitution With Exogenously F_1

1. Extraction buffer: 0.25 *M* sucrose, 10 m*M* Tris-acetate, pH 7.5. Store at 4°C.
2. 6 *M* NaBr.
3. 2 *M* Tris base.
4. 0.1 *M* EDTA adjusted to pH 7.5 with NaOH.
5. F_1: purified F_1 or crude extract of F_1 prepared by sonically irradiating SMPs as described in **Subheading 3.7**.

2.11. Sedimentation of Sonically Disrupted Mitochondria in Discontinuous Sucrose Gradients

1. Sucrose solutions: 20, 30, 50, 60, 80% (w/v) sucrose in 10 m*M* Tris-HCl, pH 7.0. Store at 4°C.
2. Equipment necessary for running sucrose gradients (*see* **Note 8**).
3. Antibodies against α- and β-subunits of F_1 and against cytochrome-c_1 or any other integral protein of the mitochondrial inner membrane.

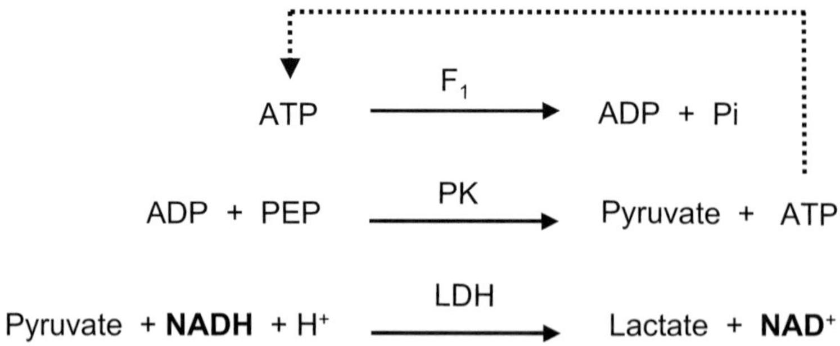

Fig. 2. Coupled enzyme assay. PK, pyruvate kinase; LDH, lactate dehydrogenase.

3. Methods

Most of procedures described for the analysis of the yeast mitochondrial ATP synthase encompass slight modifications of previously published methods for the isolation of mitochondria ***(5)***, preparation of SMPs and solubilization of F_1 and F_1-F_0 ***(6)***. In general, nuclear or mitochondrial mutants of *S. cerevisiae* impaired in the assembly of the ATP synthase present one of two distinct enzymatic phenotypes. Mutations interfering with assembly of F_1 display the absence of measurable ATPase activity in isolated mitochondria. Mutations that abort assembly of either the intrinsic (hydrophobic core) or extrinsic (stator stalk) components of F_0 block assembly of F_1-F_0 but not of F_1. Mitochondria obtained from such mutants have oligomycin-insensitive ATPase activity.

Two different methods are described for assaying ATPase activity, either of which is suitable for preliminary screening to determine if F_1 is physically and functionally coupled to F_0. One measures the amount of inorganic phosphate ***(7)*** released during ATP hydrolysis. The other is the spectrophotometric coupled enzyme assay of Pullman et al. ***(8)*** in which F_1-catalyzed ATP hydrolysis is enzymatically coupled to NADH oxidation through the actions of PK and LDH. In this assay, the concentration of ATP is maintained at a constant value during the catalysis (**Fig. 2**).

To visualize the individual subunits of F_1 and F_0 in various mitochondrial fractions, two SDS-polyacrylamide gel electrophoresis (PAGE) systems are described. One is a modification of the standard Laemmli method ***(9)***, which has been optimized to increase the resolution of F_1 α- and β-subunit proteins ***(10)***. The second incorporates urea in the separation gel to increase the resolution of the F_0 subunits ***(11)***.

Also described are methods to examine the sedimentation properties of ATP synthase subunits in sucrose gradients. Linear sucrose gradients are used to resolve F_1 (360-kDa) and F_1-F_0 (550-kDa) complexes from smaller precursor complexes or monomers (**Fig. 3**). The isopycnic discontinuous sucrose gradient permits

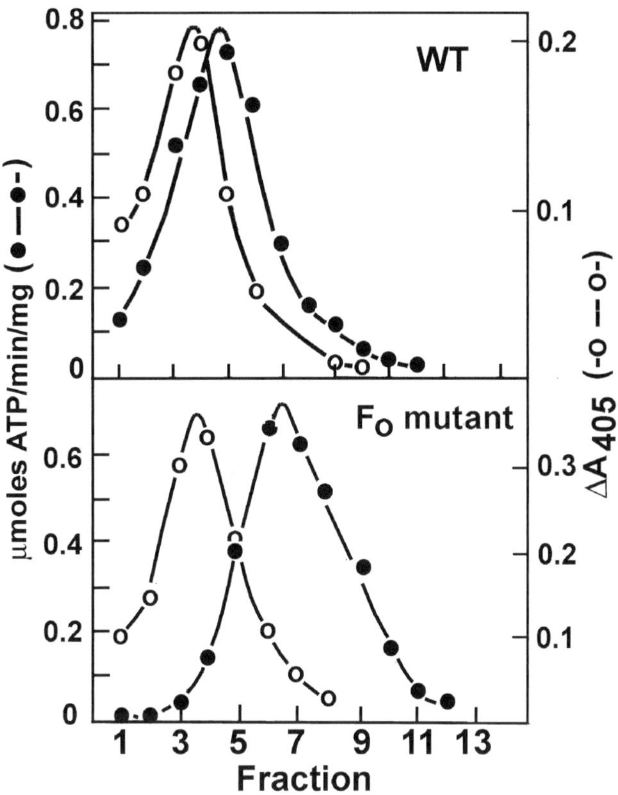

Fig. 3. Sedimentation of F_1 and F_1-F_0 in linear sucrose gradients. Wild-type and F_0 mutant mitochondria were suspended in 20 m*M* Tris-HCl, pH 7.5, 2 m*M* ATP, and 0.1 m*M* EDTA and extracted with 0.25% Triton X-100. Following clarification, the extracts were first supplemented with β-galactosidase and applied to 7–20% linear sucrose gradient prepared in the presence of 20 m*M* Tris-HCl, pH 7.5, 2 m*M* ATP, and 0.1 m*M* EDTA and were centrifuged at 4°C (wild type) or 23°C (mutant) for 3.5 h in a Beckman SW65 rotor. Fifteen fractions were collected and assayed for ATPase and β-glactosidase. The 550-kDa F_1-F_0 complex that is observed in the wild-type sample (upper panel) is resolved from the 360-kDa F_1 unit, which is present in the mutant sample that lacks the F_0 component (lower panel).

multiple forms (if present) of F_1 subunit proteins to be detected (**Fig. 4**) ***(12)***. F_1 bound to F_0 is detected at the margin of 30–50% sucrose and comigrates with a marker for the mitochondrial inner membrane (e.g., cytochrome-c_1) (**Fig. 4**, top panel). The assembled F_1 oligomer is detected in the region of low sucrose density (20–30%) because the *g*-force employed is not sufficient to move the 360-kDa F_1 to the region corresponding to the density of pure protein (ρ ~ 1.2–1.3) (**Fig. 4**, top and middle panels). In contrast, macromolecular aggregates of F_1 protein accumulate

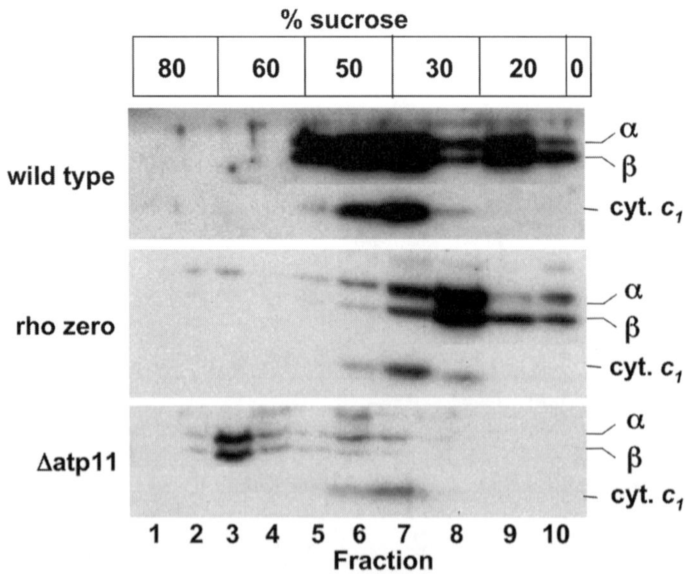

Fig. 4. Western blots showing the sedimentation behavior of F_1 α- and β-subunits from various yeast strains in discontinuous isopycnic sucrose gradients. Experimental details for isopycnic centrifugation of sonically irradiated mitochondria are given in **Subheading 3.11**. Mitochondrial samples were from D273-10B/A1 (wild type), D273-10B/A1ρ^0 (rho zero), and **a**W303Δatp11 (Δatp11). Proteins in the sucrose gradient fractions were separated by SDS-PAGE on 10% polyacrylamide gels (*see* **Subheading 3.4.**), electrophoretically transferred to nitrocellulose, and probed with a mixture of antibodies against F_1 α-subunit, F_1 β-subunit, and cytochrome-c_1 (cyt c_1). The concentration of

at the 60–80% sucrose interface (**Fig. 4**, bottom panel). Finally, a mix-and-match protocol is described for determining if an assembly defect resides in the F_1 or F_0 unit ***(10)***. In brief, F_1 is stripped from SMPs, and the depleted membranes are mixed with a preparation of F_1. Detection of oligomycin-sensitive ATPase activity in the reconstituted SMP fraction is indicative of the presence of functional F_0.

3.1. Yeast Mitochondria

1. Pregrowth: inoculate 50 mL YPGal medium in a 250-mL flask with a fresh culture of yeast and grow 20 h at 30°C.
2. Inoculate 5–15 mL of the pregrown culture into 800 mL YPGal medium in a 2-L flask and grow for 20 h in a rotary shaker (250 rpm) at 30°C (*see* **Note 9**).
3. Centrifugation at 1000–2000g_{av} for 10 min is sufficient for collecting cells (*see* **Note 10**). Discard the supernatant and suspend cells in 150 mL 1.2 *M* sorbitol. Transfer the suspension to preweighed 250-mL bottles and centrifuge at 2600g_{av} for 10 min. Discard supernatant and record wet weight of the cell pellet.

4. Suspend washed cells in the digestion buffer at a concentration of 10 g cells per 30 mL buffer. Incubate at 30°C with moderate shaking until most cells have been converted to spheroplasts. This usually requires 30–90 min and can be checked by adding a few microliters of digestion mixture to a drop of water on a glass slide and viewing under the microscope. The presence of mostly debris indicates hypotonic destruction of spheroplasts. Overdigestion of the cells should be avoided as it will affect the quality of the mitochondria.
5. Add 100 mL cold 1.2 *M* sorbitol to the digested cells and centrifuge at $5000g_{av}$ for 10 min at 4°C. Discard supernatant and wash the spheroplasts one additional time with 200 mL 1.2 *M* sorbitol.
6. Resuspend spheroplasts in lysis buffer using 3 mL buffer per gram of cells. Add 1/200 volume of the PMSF solution. First suspend spheroplasts with a loose-fitted glass-Teflon homogenizer, then transfer to a small Waring blender cup, and blend for 20 s at top speed. Transfer blended material to a 250-mL centrifuge bottle.
7. Centrifuge lysed spheroplasts at $900g_{av}$ for 5 min to remove nuclei and cell debris. Carefully transfer the supernatant to a fresh bottle and centrifuge a second time at $900g_{av}$ for 5 min to remove any remaining cell debris. Transfer the supernatant to 50-mL centrifuge tubes.
8. Centrifuge supernatant from **step 7** at $15,000g_{av}$ for 15 min at 4°C. Pour off supernatant and suspend the mitochondrial pellet in 20–30 mL of mitochondrial wash buffer. Centrifuge at $15,000g_{av}$ for 10 min. Wash the mitochondrial pellet two additional times.
9. Suspend mitochondria in the mitochondrial wash buffer at a protein concentration of 10–20 mg/mL. Add PMSF to a final concentration of 1 m*M*. Store at –80°C.

3.2. ATPase Assay by Chemical Determination of Inorganic Phosphate Release From ATP

1. To a 1 × 13 cm glass tube add 0.5 mL 2X ATPase assay buffer, 0.39 mL water, and 10 µL mitochondria at a protein concentration of 10 mg/mL. To test for oligomycin sensitivity, set up a duplicate tube and add 5 µL of the oligomycin solution. A third tube containing everything except mitochondria serves as a reagent blank.
2. Incubate at 37°C for 1 min before starting the reaction with 0.1 mL of the 0.1 *M* ATP solution. Incubate for an additional 6 or 12 min. Terminate the reaction by adding 0.2 mL stop reagent. If necessary, clarify the sample by centrifuging at $140g_{av}$ in a clinical centrifuge for 5 min.
3. To a 1.5 × 15 cm glass tube containing 5 mL 0.5% TCA, add 0.2 mL of the deproteinized assay solution followed by 0.5 mL 5% ammonium molybdate and 0.15 mL ANS reagent. Mix well after addition of molybdate and again after ANS and incubate at room temperature for 10 min.
4. Record the absorbance at 660 nm against the reagent blank (*see* **Note 11**).
5. For a 12-min assay, the absorbance reading is equivalent to the micromoles of P_i liberated per minute. The specific activity in micromoles per minute per milligram protein is calculated by dividing the P_i liberated per minute by the milligrams of mitochondrial protein added to the assay.

Table 1
Recipes for SDS-Polyacrylamide Minigels for Separating F_1 Subunits

Solution	Stacking gel (5%) (mL)	Separating gel (10%) (mL)
	V_f = 10 mL each	
1.5*M* Tris-HCl, 0.4% SDS, pH 8.0	—	2.5
0.5 *M* Tris-HCl, 0.4% SDS, pH 6.8	2.5	—
29:1 Acrylamide:bisacrylamide	1.7	3.3
Water	5.8	4.1
10% Ammonium persulfate	0.03	0.05
TEMED	0.0013	0.0042

3.3. ATPase Assay by Spectrophotometric Determination of NADH Oxidation Enzymatically Coupled to ATP Hydrolysis

1. Prepare a sufficient amount of fresh reaction mix (*see* **Note 12**).
2. Equilibrate a thermostated cuvette holder of the spectrophotometer at 37°C using a circulating water bath. Set the wavelength at 340 nm (*see* **Note 13**).
3. Add 1 mL reaction mix to a 1-mL cuvette and incubate 1 min before starting the assay. To assay for oligomycin sensitivity, include 5 μL of the oligomycin solution in the assay. A 1 mL solution of 50 m*M* Tris-acetate, pH 8.0, serves as the reaction blank.
4. Start the reaction by adding 5 μL mitochondria at 10 mg protein/mL to the cuvette. The mitochondria are applied to a plastic or glass rod bent at the tip to form a small platform that can be inserted and mixed in the cuvette seated in the spectrophotometer (Sarstedt Ruhr-Spatel, item 81.970, is well suited for this purpose). Record the absorbance at 340 nm for 3 min.
5. With a 1-cm path length, the micromoles NADH oxidized in 1 min are calculated from the slope of the trace by dividing the change in absorbance over 1 min (ΔA) by 6.23, the extinction coefficient ε for NADH expressed as $mM^{-1}\ cm^{-1}$. This value is equal to the units (μmole/min) of ATP hydrolyzed. The specific activity is calculated by dividing the number of units by the milligrams mitochondrial protein added to the assay.

3.4. Preparation of SDS-Polyacrylamide Gels for Separation of F_1 Subunits

1. Follow the guide in **Table 1** for the preparation of 10% polyacrylamide minigels.
2. Prepare samples such that the final concentration of protein is 2 mg/mL and of the sample buffer is 1X. Heat for 5 min at 90°C. For mitochondria or SMPs, apply 10–20 μg protein per lane.
3. Use a 1:5 dilution of the 5X running buffer. Run the gel at constant 150 V.

3.5. SDS-Polyacrylamide Gels for Separation of F_0 Subunits

1. Follow the guide in **Table 2** for the preparation of 12% polyacrylamide gels to separate F_0 subunits. Use 14 × 16 cm glass plates to pour the gel.

Table 2
Recipes for Resolution of F_0 Subunits on a Maxigel

	Stacking gel (6%) (mL)	Separating gel (11.4%) (mL)
Solution	V_f = 10 mL	V_f = 30 mL
1.5 *M* Tris-HCl, pH 8.6	—	7.6
1 *M* Tris-HCl, pH 6.8	1.25	—
10% SDS	0.1	—
30:0.8 Acrylamide:bisacrylamide	2	12.5
Urea		6.5 (g)
Glycerol		8
Water	6.6	—
10% Ammonium persulfate	0.05	0.1
TEMED	0.015	0.02

2. Prepare samples as described in **Subheading 3.4.** but omit heating at 90°C.
3. Use a 1:10 dilution of the 10X running buffer. Run the gel at constant 80 V.

3.6. Preparation of SMPs

1. Suspend mitochondria at a protein concentration of 10 mg/mL in sonication buffer. Sonically irradiate the sample on ice using an appropriately sized probe (*see* **Note 14**).
2. Centrifuge at 110,000g_{av} for 30 min at 4°C.
3. Suspend SMPs in appropriate buffer.

3.7. Solubilization of F_1 From SMPs

1. Adjust the protein concentration of SMPs to 10 mg/mL in 20 m*M* Tris-HCl, pH 8.5, 2 m*M* ATP, and 0.1 m*M* EDTA.
2. Starting with the SMP suspension at 4°C, sonically irradiate the sample until the temperature reaches 25°C for small samples (5 mL or less) or 35°C for large samples (100 mL or more) to achieve maximal release of F_1 from the membrane. All subsequent steps are carried out at room temperature. The progress of F_1 release can be monitored using the chemical or coupled enzyme assay described in **Subheading 3.3.** to follow the conversion of F_1-catalyzed ATP hydrolysis from an oligomycin-sensitive F_1-F_0 to oligomycin-insensitive F_1 activity.
3. Centrifuge the sonically irradiated suspension at 110,000g_{av} for 1 h at 25°C. The supernatant contains soluble F_1, which can be further purified on a sizing column (*see* **Note 15**).

3.8. Solubilization of F_1-F_0

1. For extraction of the ATP synthase complex with Triton X-100, the SMPs are diluted to a protein concentration of 6.7 mg/mL with 20 m*M* Tris-HCl, pH 8.0, 0.1 m*M* EDTA. If deoxycholate or lauryl maltoside are used, then the SMPs are diluted with the same buffer to a concentration of 10 mg/mL, and solid KCl is added to a concentration of 1 *M*.

2. Add either 10% Triton X-100 to a final concentration of 0.25% or 10% potassium deoxycholate (or 10% lauryl maltoside) to a final concentration of 0.5%.
3. Centrifuge at 110,000g_{av} for 60 min if the extraction is with Triton X-100 and 20 min if deoxycholate or lauryl maltoside are used.
4. Collect the clear supernatant fraction.

3.9. Sedimentation of F_1 and F_1-F_0 in Linear Sucrose Gradients

1. Using the 7 and 20% sucrose solutions, prepare a 4.6-mL gradient in the Beckman ultraclear centrifuge tube (*see* **Note 8**).
2. To 200 µL solubilized F_1 or F_1-F_0, add 5 units β-galactosidase and carefully overlay the sample on the gradient.
3. Load the swinging bucket rotor and centrifuge the sample at 300,000g_{av} for 3.5 h at 25°C for an F_1 sample or at 300,000g_{av} for 3 h at 4°C for an F_1-F_0 sample.
4. Collect 0.3-mL fractions from the bottom of the tube.
5. Assay the gradient fractions for F_1 or F_1-F_0 by Western assay or using one of the ATPase assays described in **Subheadings 3.2.** and **3.3.** (**Figs. 3** and **4**).
6. Determine the peak position of the β-galactosidase marker by assaying for its activity spectrophotometrically (**Fig. 4**). Set the wavelength of the spectrophotometer to 405 nm. Transfer 0.93 mL β-galactosidase reaction mix to a 1-mL cuvette. Add 50-µL gradient sample to the cuvette and mix. Start reaction with 33 µL *o*-nitrophenol β-D-galactopyranoside and record the increase in absorbance for 2 min. A duplicate cuvette containing everything except the gradient sample serves as reagent blank. Calculate relative activities of the fractions from the slope of the absorbance changes.

3.10. Preparation of F_1-Depleted SMPs and Reconstitution With Exogenously Added F_1

1. Combine and mix the SMPs (20–25 mg/mL in extraction buffer) with an equal volume of 6 *M* NaBr. Adjust to pH 7.5 with 2 *M* Tris-HCl.
2. Centrifuge the mixture at 110,000g_{av} for 30 min at 4°C. Three different fractions will be observed in the tube: a film of particles floating on the surface, a clear solution (infranatant), and a firmly packed pellet at the bottom of the tube.
3. Using a Pasteur pipet, move the surface film to the side, aspirate the infranatant, and discard.
4. Add a minimal volume of fresh extraction buffer and carefully transfer the filmy material on the side of the tube to a fresh centrifuge tube.
5. Suspend the film in one-half the original volume of buffer and add an equal volume of 6 *M* NaBr.
6. Centrifuge as described in **step 2**. The film of particles should be thicker and easier to manipulate. Remove the infranatant and transfer the film to a clean tube.
7. Suspend the film in 5–10 mL extraction buffer and centrifuge. All the material should now pellet at the bottom of the tube. Suspend the F_1-depleted membrane pellet in extraction buffer. The membranes can be used for the reconstitution at this stage or stored at –80°C.

8. For the reconstitution, combine depleted membranes at a protein concentration of 1.5 mg/mL with different amounts of F_1 (0.1–0.5 mg/mL range for pure enzyme or 0.3–1.5 mg/mL of crude extract) in 50 m*M* Tris-acetate, pH 7.5, and 1 m*M* EDTA and incubate the mixture for 10 min at room temperature.
9. Add 2 volumes of extraction buffer and centrifuge at 110,000g_{av} for 10 min at 4°C.
10. Resuspend the pellet in one-half the starting volume of extraction buffer.
11. Assay the membranes for oligomycin-sensitive ATPase activity as described in **Subheading 3.2.** or **3.3**.

3.11. Sedimentation of Sonically Disrupted Mitochondria in Discontinuous Sucrose Gradients

1. Transfer 600 µL diluted mitochondria (3–4 mg/mL in 10 m*M* Tris-HCl, pH 7.5, 2 m*M* ATP, 1 m*M* EDTA) to a 1.5-mL Eppendorf tube. Using the microtip probe, sonically irradiate the sample at low power for 2–3 s (*see* **Notes 14–16**). Keep the sample at 25°C to avoid cold denaturation of any F_1 oligomer that may be present.
2. Prepare the gradient by first adding 1.2 mL buffered 80% sucrose to the bottom of the tube. Carefully overlay 0.9 mL each of the remaining buffered sucrose solutions in the order of decreasing concentration. The total volume in the tube should be 4.8 mL.
3. Carefully overlay 200 µL of the sonically disrupted mitochondria on top of the gradient.
4. Centrifuge the tube at 180,000g_{av} for 3 h at 25°C.
5. Collect 0.5-mL fractions from the bottom of the tube.
6. Run 15 µL each fraction on a minigel using the modified Laemmli SDS gel system described in **Subheading 2.4**. Electroblot to nitrocellulose and probe with antibodies against F_1 α or F_1 β and an inner membrane protein (e.g., cytochrome-c_1). It is not necessary to remove the sucrose prior to gel electrophoresis. The migration of samples in the 80% sucrose fractions will be somewhat retarded.

4. Notes

1. Unless indicated otherwise, all solutions are made in distilled or deionized water and are stored at room temperature.
2. To make the ANS reagent, first add anhydrous sodium sulfite slowly to the water. Then, add the $NaHSO_3$. Bring the solution to volume, add the ANS, and stir until completely dissolved.
3. Wear a mask when weighing SDS powder.
4. Wear gloves when handling acrylamide.
5. This solution should be made before use because some hydrolysis of ATP will occur during storage.
6. Adjust pH of the deoxycholic acid solution with a solution of KOH to pH 8.
7. The Triton X-100 is included to reduce surface tension of the solution and permit a continuous flow of the sucrose solution down the side of the tube during preparation of the gradient.
8. These protocols assume the necessary equipment is on hand for the preparation, centrifugation, and fractionation of the gradients. This includes a 5-mL gradient

Table 3
Sonication Conditions to Minimize F_1 Loss From Mitochondrial Inner Membrane

Vessel	Volume	mg protein/mL	Time	Watts
1.5-mL Eppendorf tube	1 mL	3–4	5 s	20
100-mL Beaker	50 mL	10	20 s	50

maker, 13 × 51 mm ultraclear centrifuge tubes (Beckman part 344057 or equivalent), rotor for 5-mL gradients (e.g., Beckman SW55Ti or SW65Ti), and a gradient collector with capabilities for collecting fractions from the bottom of the tube.

9. The volume of the inoculum depends on the density of cells. Typically, 5–8 mL of a wild-type or 10–15 mL of a mutant pregrowth at stationary phase will be in the range to reach late log phase or early stationary phase after overnight growth.
10. The use of a high capacity centrifuge (IEC or Sorvall) capable of accommodating 1-L bottles is recommended for large-scale or multiple cultures.
11. The blank should be pale blue. A dark blank indicates a potential problem with the ATP solution or contaminating inorganic phosphate in a reagent or glassware.
12. For 10 mL reaction mix, combine 2 mL buffer/salt mix, 1 mL phosphoenolpyruvate solution, 0.4 mL ATP solution, and 0.4 mL NADH solution and bring the volume to 10 mL with water. Transfer 320 μg PK and 130 μg LDH as ammonium sulfate suspensions to an Eppendorf tube, centrifuge in a microfuge for 5 min, and remove the supernatant. The pellet is dissolved with 200–300 μL of the buffer solution and added to the remaining solution to complete the reaction mixture.
13. The starting absorbance at 340 nm should be approx 1.8.
14. Conditions for sonic irradiation are empirical and depend on the volume of the sample, the geometry of the vessel, and the concentration of protein in the sample. The information in **Table 3** is a guide for converting mitochondria to SMPs with minimal loss of F_1 from the membrane using a Branson Sonifier equipped with a microprobe.
15. Soluble F_1 is cold labile and should be maintained at 25°C.
16. The idea is to disrupt the mitochondrial inner membrane gently and release matrix components that can be separated from the membranes in the isopycnic gradient.

References

1. Boyer, P. D. (1997) The ATP synthase—a splendid molecular machine. *Annu. Rev. Biochem.* **66,** 717–749.
2. Ackerman, S. H. and Tzagoloff, A. (2005) Function, structure, and biogenesis of mitochondrial ATP synthase. *Prog. Nucl. Acids Res. Mol. Biol.* **80,** 95–133.
3. Arnold, I., Pfeiffer, K., Neupert, W., Stuart, R. A., and Schagger, H. (1998) Yeast mitochondrial F_1F_0-ATP synthase exists as a dimer: identification of three dimer-specific subunits. *EMBO J.* **17,** 7170–7178.
4. Nakamoto, R. K., Ketchum, C. J., and al-Shawi, M. K. (1999) Rotational coupling in the F_0F_1 ATP synthase. *Annu. Rev. Biophys. Biomol. Struct.* **28,** 205–234.

5. Faye, G., Kujawa, C., and Fukuhara, H. (1974) Physical and genetic organization of petite and grande yeast mitochondrial DNA. IV. In vivo transcription products of mitochondrial DNA and localization of 23 S ribosomal RNA in petite mutants of *Saccharomyces cerevisiae*. *J. Mol. Biol.* **88,** 185–203.
6. Tzagoloff, A. (1969) Assembly of the mitochondrial membrane system. II. Synthesis of the mitochondrial adenosine triphosphatase, F_1. *J. Biol. Chem.* **244,** 5027–5033.
7. King, E. J. (1932) The colorimetic determination of phosphorus, *Biochem. J.* **26,** 292–297.
8. Pullman, M. E., Penefsky, H. S., Datta, A., and Racker, E. (1960) Partial resolution of the enzymes catalyzing oxidative phosphorylation. I. Purification and properties of soluble dinitrophenol-stimulated adenosine triphosphatase. *J. Biol. Chem.* **235,** 3322–3329.
9. Laemmli, U. K. (1970) Cleavage of structural proteins during the assembly of the head of bacteriophage T4. *Nature* **227,** 680–685.
10. Ackerman, S. H. and Tzagoloff, A. (1990) *ATP10*, a yeast nuclear gene required for the assembly of the mitochondrial F_1-F_0 complex. *J. Biol. Chem.* **265,** 9952–9959.
11. Velours, J., Arselin de Chateaubodeau, G., Galante, M., and Guerin, B. (1987) Subunit 4 of ATP synthase (F_0F_1) from yeast mitochondria. Purification, amino-acid composition and partial N-terminal sequence. *Eur. J. Biochem.* **164,** 579–584.
12. Ackerman, S. H. and Tzagoloff, A. (1990) Identification of two nuclear genes (*ATP11*, *ATP12*) required for assembly of the yeast F_1-ATPase. *Proc. Natl. Acad. Sci. U. S. A.* **87,** 4986–4990.

27

ATP Production in Isolated Mitochondria of Procyclic *Trypanosoma brucei*

André Schneider, Nabile Bouzaidi-Tiali, Anne-Laure Chanez, and Laurence Bulliard

Summary

This chapter describes a luciferase-based protocol to measure adenosine triphosphate (ATP) production in isolated mitochondria of *Trypanosoma brucei*. The assay represents an excellent method to characterize the functionality of isolated mitochondria. Comparing the ATP production induced by substrates for oxidative phosphorylation to the one induced by substrates for substrate-level phosphorylation allows conclusions regarding the integrity of the outer and inner mitochondrial membranes. Furthermore, the assay is a valuable tool for characterization of RNA interference cell lines suspected to affect mitochondrial functions.

Key Words: ATP; digitonin extraction; luciferase; oxidative phosphorylation; substrate-level phosphorylation; *Trypanosoma brucei*; trypanosome.

1. Introduction

The single mitochondrion of insect stage *Trypanosoma brucei* has three partly overlapping adenosine triphosphate (ATP) production pathways *(1,2)* (**Fig. 1**). First, as in mitochondria from other organisms, ATP is produced by oxidative phosphorylation (OXPHOS) in a cyanide-sensitive electron transport chain. Second, as expected, one step of substrate-level phosphorylation (SUBPHOS) catalyzed by succinyl-coenzyme A synthetase (SCoAS) occurs in the citric acid cycle. In higher eukaryotes, it is guanosine 5′-triphosphate (GTP), which is synthesized at this step, whereas the *T. brucei* enzyme directly produces ATP. Finally, mitochondrial ATP can be produced anaerobically by SUBPHOS coupled to acetate formation using the acetate:succinate coenzyme A (CoA) transferase/SCoAS cycle (ASCT cycle) *(3)*. This pathway consists of two enzymes, the acetate:succinate CoA transferase *(4)* and the same SCoAS,

From: *Methods in Molecular Biology, vol. 372: Mitochondria: Practical Protocols*
Edited by: D. Leister and J. M. Herrmann © Humana Press Inc., Totowa, NJ

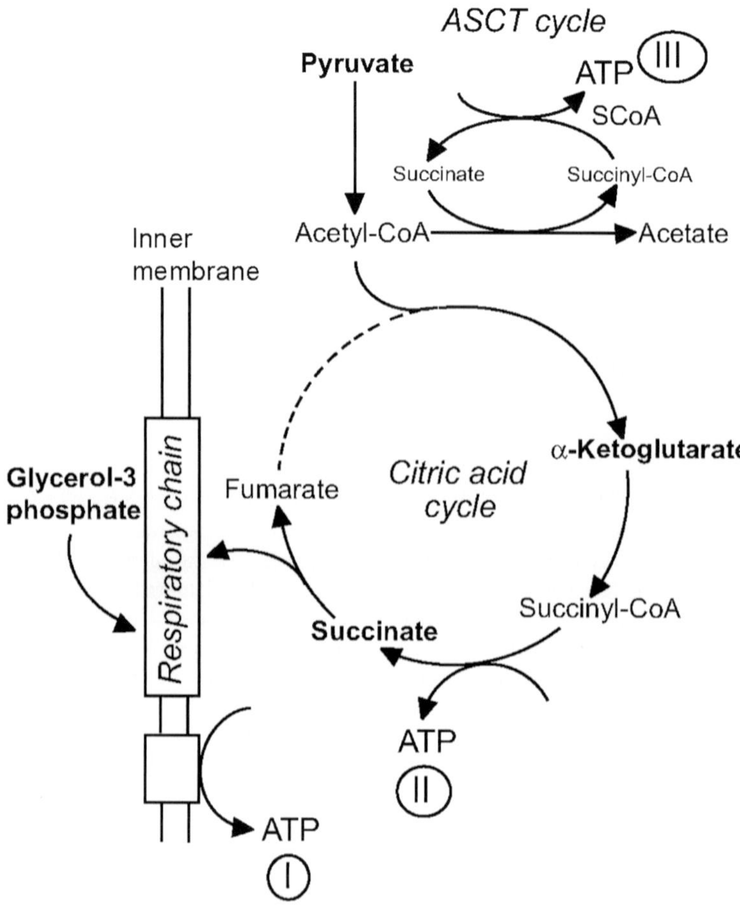

Fig. 1. Overview of mitochondrial ATP production in procyclic *T. brucei*. The three ATP production pathways correspond to the respiratory chain, the citric acid cycle, and the ASCT cycle. The three sites of ATP production are indicated by roman numerals: I corresponds to OXPHOS; II and III correspond to SUBPHOS. The substrates used in the *in organello* ATP production assay are indicated in bold.

which is found in the citric acid cycle *(5)*. Occurrence of the ASCT cycle in mitochondria is restricted; it has only been found in trypanosomatid and some parasitic helminths. Interestingly, however, the ASCT cycle is found in the hydrogenosome of trichomonads and some fungi.

Although mitochondrial ATP production is of interest on its own, it also provides an excellent tool to assay the integrity and functionality of isolated mitochondria of *T. brucei* and thus has many potential applications. Assaying OXPHOS indirectly monitors the presence of the membrane potential and, depending on the substrate used, allows testing for the presence of an intact outer membrane ***(6)***. The two modes of SUBPHOS, on the other hand, do not depend on the membrane potential and require an intact inner membrane only. Ribonucleic acid interference (RNAi) is an efficient and rapid method for the generation of conditional mutants in *T. brucei* ***(7,8)***. Using digitonin extractions in combination with *in organello* ATP production assays, it is possible to rapidly identify RNAi strains that interfere with mitochondrial functions ***(5,9)***. Furthermore, OXPHOS, in contrast to the two modes of SUBPHOS, requires some mitochondrially encoded proteins and thus indirectly depends on organellar protein synthesis. Thus, RNAi cell lines impaired in OXPHOS but at the same time show normal levels of mitochondrial SUBPHOS are of special interest because they include the ones that are ablated for proteins required for mitochondrial translation, a process that in trypanosomes is notoriously difficult to study ***(10)***.

It is the aim of this review to provide a simple protocol to measure mitochondrial ATP production in isolated mitochondria or in digitonin-extracted *T. brucei* cells and discuss some of its applications.

2. Materials

2.1. Digitonin Extraction

1. Procyclic *T. brucei* cells (*see* **Note 1**).
2. SDM-79 medium supplemented with 5% (v/v) heat-inactivated fetal bovine serum ***(11)***.
3. Wash buffer: 20 m*M* phosphate buffer, pH 7.9, 20 m*M* glucose, 0.15 *M* NaCl. Prepare as 4X stock and sterilize by filtration.
4. SoTE buffer: 20 m*M* Tris-HCl, pH 7.5, 0.6 *M* sorbitol, 2 m*M* ethylenediaminetetraacetic acid, pH 7.5. Prepare as 2X stock and sterilize by filtration.
5. Digitonin stock (cat. no. 37006, Fluka): 0.8% (w/v), dissolve in SoTE buffer. Digitonin does not dissolve well; thus, heat the suspension to 95°C until it clears, then cool to room temperature. The stock solution will now stay dissolved even at ambient temperature. Keep the stock at –20°C; if a precipitate forms after defrosting, then repeat the procedure.
6. ATP assay buffer: 20 m*M* Tris-HCl pH 7.4, 15 m*M* KH_2PO_4, 0.6 *M* sorbitol, 10 m*M* $MgSO_4$, 2.5 mg/mL fatty acid-free bovine serum albumin (cat. no. A-6003, Sigma); sterilize by filtration.

2.2. In Organello *ATP Production Assay*

1. Atractyloside stock (cat. no. A6882, Sigma): 10 m*M*, dissolve in dimethylsulfoxide. Concentration used in assay: 10 μ*M*.
2. Antimycin stock (cat. no. A8674, Sigma): 0.2 m*M*; dissolve in ethanol. Concentration used in assay: 2 μ*M*.
3. Malonate stock: 0.5 *M*; dissolve in water. Concentration used in assay: 7 m*M*.

4. Carbonyl cyanide(*p*-trifluoro-methoxy)-phenylhydrazone (FCCP) (cat. no. C2920, Sigma) stock: 20 m*M*; dissolve in ethanol. Concentration used in assay: 5 μ*M*.
5. Substrate stocks: 0.2 *M* each of succinate, glycerol-3 phosphate, α-ketoglutarate, and pyruvate dissolved in water. Concentration used in assay: 5 m*M*.
6. Adenosine 5′-diphosphate (ADP) stock: 4.5 m*M*; dissolve in water. Concentration used in assay: 60 μ*M*.

2.3. Processing and Luciferase Assay

1. 60% Perchloric acid.
2. 1*N* KOH.
3. 0.5 *M* Tris-acetate, pH 7.75.
4. ATP bioluminescence assay kit CLS II (cat. no. 1699695, Roche). Prepare and store the luciferase reagent as described by the manufacturer.
5. Luminometer.

3. Methods

Principle: Mitochondrial fractions are incubated with ADP and the corresponding OXPHOS or SUBPHOS substrates. After incubation, the reaction is stopped by perchloric acid, and the produced ATP is quantified using a luciferase-based ATP bioluminescence kit ***(12)***.

Organellar fractions: Mitochondrial ATP production assays in *T. brucei* can be performed using mitochondria isolated by the hypotonic or the isotonic isolation protocol (*see* Chapter 5). However, the assay is especially useful (e.g., for the phenotypic analysis of RNAi cell lines) when it is combined with digitonin extraction of whole cells (*see* **Subheading 3.1.**) because this allows rapid analysis of multiple samples using low cell numbers only ***(5,9)***. Low concentrations of the detergent digitonin selectively permeabilize the cell membrane but leave (at least the inner) mitochondrial membrane intact. Thus, a single centrifugation step of digitonin-extracted *T. brucei* cells will yield a pellet enriched for mitochondria.

Substrates: We routinely use four substrates: succinate and glycerol-3 phosphate, which induce OXPHOS, and α-ketoglutarate and pyruvate, which induce mitochondrial SUBPHOS (**Fig. 1**). Each of the substrates should be measured in the absence and the presence of the inhibitors indicated in **Table 1** to make sure that the correct mode of ATP production is measured. The different possible outcomes of the ATP production assays and the conclusions regarding the state of the mitochondria that can be drawn from the results are illustrated in **Table 2**.

3.1. Digitonin Extraction

1. Use 10^8 cells for each planned digitonin extraction.
2. Spin cells in 15-mL Falcon tubes at 23°C for 7 min at approx 700*g*.

Table 1
Substrates Used to Induce Mitochondrial ATP Production

Substrate	Efficiency of ATP production[a]	Mode of ATP production	Sensitivity toward inhibitors			
			Atractyloside[b]	Antimycin[c]	Malonate[d]	FCCP[e]
Succinate	100%	OXPHOS	+	+	+	+
Glycerol-3 phosphate	~300%	OXPHOS	+	+	–	+
α-Ketoglutarate[f]	100–200%	~90% SUBPHOS (citric acid cycle)[f]	+	–	–	–
Pyruvate[g] (cosubstrate succinate)	100–200%	~80% SUBPHOS (ASCT cycle)[g]	+	–	–	–

[a]Comparison of ATP production induced by the four substrates, as measured in isolated mitochondria having intact outer and inner membranes. ATP production induced by succinate was set to 100%. In intact mitochondria, glycerol-3 phosphate is for unknown reasons by far the most efficient substrate ***(6)***.

[b]Atractyloside is a specific inhibitor of the adenine nucleotide translocater of the mitochondrial inner membrane. It prevents access of ADP to the matrix and thus serves to show that a measured ATP production is indeed mitochondrial ***(5,6)***.

[c]Antimycin is an inhibitor of respiratory complex III and thus inhibits OXPHOS.

[d]Malonate is a competitive inhibitor of succinate dehydrogenase and therefore specifically inhibits OXPHOS induced by succinate.

[e]FCCP is a potassium ionophore; thus, by disrupting the membrane potential, it affects OXPHOS.

[f]The antimycine-resistant part of the α-ketoglutarate-induced ATP production (~90%) can be attributed to SUBPHOS. Analysis of RNAi strains has shown that it is the SUBPHOS in the citric acid cycle that is detected ***(5)***. α-Ketoglutarate is converted into succinate in the citric acid cycle, a small fraction of which is apparently used for OXPHOS, explaining the approx 10% of antimycine-sensitive ATP production that is observed ***(5)***.

[g]Pyruvate on its own is not able to induce ATP production. However, by adding succinate as a cosubstrate, efficient ATP production is observed, to which approx 80% is due to SUBPHOS and approx 20% to OXPHOS. RNAi analysis has shown that the antimycine-resistant part of the pyruvate-induced ATP production can be attributed to the SUBPHOS in the ASCT cycle ***(5)***. Thus, in *T. brucei*, unlike in other organisms, pyruvate does not enter the citric acid cycle.

Table 2
Condition of Isolated Mitochondria and Expected Outcomes of ATP Production Assays in Wild-Type *T. brucei*

	Mitochondrial ATP production induced by			
Condition of isolated mitochondria	Succinate[a]	Glycerol-3 phosphate[b]	α-Ketoglutarate	Pyruvate
Intact membrane potential: OM intact, IM intact	+	+	+	+
Intact membrane potential: OM disrupted, IM intact	+	–	+	+
No membrane potential: OM disrupted, IM intact	–	–	+	+
No intact mitochondria present in the tested fraction	–	–	–	–

[a]Cytochrome-*c* is a peripheral membrane protein associated with the outer face of the inner mitochondrial membrane (IM). Experiments have shown that even in the absence of the outer membrane (OM) enough of cytochrome-*c* remains associated with the inner membrane to support OXPHOS ***(6)***.

[b]Glycerol-3 phosphate dehydrogenase is a soluble protein of the intermembrane space. In the absence of an intact outer mitochondrial membrane, it is rapidly lost ***(6)***.

3. Wash in an equal volume of wash buffer.
4. Prepare the required volume of 0.03% digitonin-containing SoTE buffer (*see* **Note 2**) by dilution of the 0.8% digitonin stock. Warm to ambient temperature.
5. Resuspend the pellet in 0.5 mL SoTE buffer (prewarmed to ambient temperature) and transfer to 1.5-mL Eppendorf tube.
6. Add 0.5 mL 0.03% digitonin-containing SoTE buffer (prewarmed to ambient temperature) (*see* **Note 3**).
7. Invert once and incubate 5 min on ice (*see* **Note 4**).
8. Spin in Eppendorf centrifuge at 4°C for 3 min at 5000*g*.
9. Remove supernatant.
10. Resuspend pellet in 80–120 μL ATP assay buffer.

3.2. In Organello *ATP Production Assay*

1. Decide how many reactions you want to perform (*see* **Note 5**).
2. For each reaction, resuspend 25–75 μg of isolated mitochondria or 10 μL of the resuspended digitonin pellet (*see* **Subheading 3.1., step 10**) in a total volume of 75 μL ATP assay buffer.
3. Set up the required number of 75-μL reactions. Add inhibitors to control reactions (*see* **Note 6**); incubate on ice for 5 min.
4. Add 2 μL of the 0.2 *M* stocks of substrate (succinate, glycerol-3 phosphate, α-ketoglutarate, or pyruvate) (*see* **Note 7**).
5. Start reaction by adding 1 μL 4.5 m*M* ADP.
6. Incubate for 30 min at 27°C.

3.3. Processing and Luciferase Assay

1. After incubation, add 1.75 μL 60% perchloric acid and mix immediately on vortex.
2. Incubate on ice for at least 10 min. A white precipitate will form.
3. Spin in Eppendorf centrifuge for 5 min at full speed.
4. Transfer 60 μL of the supernatant to a new tube.
5. Add 11.5 μL 1*N* KOH; the pH of the resulting mixture should be between 7.0 and 8.0.
6. Mix on vortex and incubate on ice for 3 min.
7. Spin in Eppendorf centrifuge for 5 min at full speed; keep supernatant and discard pellet (*see* **Note 8**).
8. Set up luciferase reaction: use 10 μL of supernatant (*see* **step 7**), 40 μL 0.5 *M* Tris-acetate, pH 7.75, and 50 μL luciferase reagent (*see* **Note 9**).
9. Measure chemiluminescence in a luminometer.
10. Analyze results by comparing ATP production in the presence and the absence of inhibitors for the different substrates. The conclusions that can be drawn from the different possible outcomes are listed in **Table 2** (*see* **Note 10**).

4. Notes

1. The procedure appears to work for any *T. brucei* cell line. We have used it for the *T. brucei* 427 ***(6)*** and 29-13 strains, as well as for many transgenic cell lines, including induced RNAi strains ***(5,9)***.

2. The digitonin concentration is the most important parameter of this experiment. Both the concentration of digitonin as well as the concentration of cells are important. The indicated final concentration of 0.015% digitonin has been optimized for a cell density of 10^8 cells/mL. At this concentration, we are able to detect ATP production in response to glycerol-3 phosphate, indicating that both the outer and the inner mitochondrial membranes remain intact (**Table 2**). However, if higher concentrations are used, then the outer membrane will be disrupted, and no glycerol-3 phosphate-induced ATP production can be detected. Moreover, further increasing the digitonin concentration, prior to affecting the inner membrane barrier, will remove the cytochrome-*c*, which normally, even after the disruption of the outer membrane, remains associated with the inner membrane ***(6)***. The concentration of digitonin to be used also depends on the cell line; thus, some transgenic cell lines may behave differently.
3. The correct temperature is important because it influences solubilizing properties of digitonin.
4. Invert only once; more mixing will result in lower ATP production activities.
5. This depends mainly on how many substrates will be tested and how many control reactions with inhibitors are performed.
6. It is mandatory to test, as a control, at least one inhibitor that is expected to interfere with the corresponding mode of ATP production. We routinely test OXPHOS substrates in the presence and absence of antimycin and the SUBPHOS substrates in the presence and absence of atractyloside ***(5)***. For a more complete list of inhibitors, *see* **Table 1**. A detected ATP production is considered to be significant if it is at least 10-fold reduced in the presence of the corresponding inhibitors.
7. To measure pyruvate-induced ATP production, 5 m*M* succinate has to be added as a cosubstrate ***(5)***.
8. The samples are stable for at least 24 h when kept at 4°C.
9. It is important that the chemiluminescence be measured in the linear range of the assay. Thus, using 5 μL supernatant (*see* **Subheading 3.3., step 7**) in the luciferase reaction should give half the signal obtained for 10 μL. If this is not the case, then dilute the sample accordingly.
10. When analyzing digitonin-extracted pellets, we have sometimes encountered sample-to-sample variations in the absolute levels of ATP production induced by a given substrate. However, it was generally possible to reproduce the relative values. Thus, to control for this potential variation, we routinely replicate each experiment at least three times and compare the relative efficiencies of ATP production (e.g., by setting the ATP production in noninduced RNAi cell lines to 100%; **5**).

Acknowledgments

This study was supported by grant 31-067906.02 from the Swiss National Foundation and by a grant from the Novartis Foundation.

References

1. Tielens, A. G., Rotte, C., van Hellemond, J. J., and Martin, W. (2002) Mitochondria as we don't know them. *Trends Biochem. Sci.* **27,** 564–572.

2. Besteiro, S., Barrett, M. P., Riviere, L., and Bringaud, F. (2005) Energy generation in insect stages of *Trypanosoma brucei*: metabolism in flux. *Trends Parasitol.* **21,** 185–191.
3. van Hellemond, J. J., Opperdoes, F. R., and Tielens, A. G. M. (1998) Trypanosomatides produce acetate via a mitochondrial acetate:succinate CoA transferase. *Proc. Natl. Acad. Sci. U. S. A.* **95,** 3036–3041.
4. Riviere, L., van Weelden, S. W., Glass, P., et al. (2004) Acetyl:succinate CoA-transferase in procyclic *Trypanosoma brucei*. Gene identification and role in carbohydrate metabolism. *J. Biol. Chem.* **279,** 45,337–45,346.
5. Bochud-Allemann, N., and Schneider, A. (2002) Mitochondrial substrate level phosphorylation is essential for growth of procyclic *Trypanosoma brucei*. *J. Biol. Chem.* **277,** 32,849–32,854.
6. Allemann, N., and Schneider, A. (2000) ATP production in isolated mitochondria of procyclic *Trypanosoma brucei*. *Mol. Biochem. Parasitol.* **111,** 87–94.
7. Wang, Z., Morris, J. C., Drew, M. E., and Englund, P. T. (2000) Inhibition of *Trypanosoma brucei* gene expression by RNA interference. *J. Biol. Chem.* **275,** 40,174–40,179.
8. Shi, H., Djikeng, A., Mark, T., Wirtz, E., Tschudi, C., and Ullu, E. (2000) Genetic interference in *Trypanosoma brucei* by heritable and inducible double-stranded RNA. *RNA* **6,** 1069–1076.
9. Charrière, F., Tan, T. H. P., and Schneider, A. (2005) Mitochondrial initiation factor 2 of *Trypanosoma brucei* binds imported formylated elongator-type methionyl-tRNA. *J. Biol. Chem.* **280,** 15,659–15,665.
10. Horvath, A., Berry, E. A., and Maslov, D. A. (2000) Translation of the edited mRNA for cytochrome *b* in trypanosome mitochondria. *Science* **287,** 1639–1640.
11. Brun, R., and Schönenberger, M. (1979) Cultivation an in vitro cloning of procyclic culture forms of *Trypanosoma brucei* in a semi-defined medium. *Acta Tropica* **36,** 289–292.
12. Glick, B. S., Wachter, C., Reid, G. A., and Schatz, G. (1993) Import of cytochrome b_2 to the mitochondrial intermembrane space: the tightly folded heme-binding domain makes import dependent upon matrix ATP. *Prot. Science* **2,** 1901–1917.

28

Oxidative Stress and Plant Mitochondria

Nicolas L. Taylor and A. Harvey Millar

Summary

Mitochondria not only are a source of reactive oxygen species (ROS) but also are sites of oxidative damage. In plants, mitochondria must normally operate when there are high levels of ROS produced during photosynthesis and photorespiration. These levels are further enhanced during biotic and abiotic stress of plants. Excessive stress can lead to mitochondrial damage, which may then lead to induction of programmed cell death in plants. We outline methods for imposing oxidative stress in plants, provide methods for measurements of its severity, and then explain assays for assessing plant mitochondrial oxidative damage and measuring the capacity of key stress defense and response pathways.

Key Words: Aconitase; alternative oxidase; glycine decarboxylase; lipid peroxidation; lipoic acid; reactive oxygen species; superoxide dismutase.

1. Introduction

Mitochondria form a focus for much oxidative stress research as not only they are the sites of oxygen consumption and a significant source of cellular reactive oxygen species (ROS), but also oxidative damage of the organelle perturbs the cell's energy supply required for repair mechanisms. Consequently, the nature of oxidative damage to mitochondria is under investigation in a variety of organisms. These studies are providing information on the general susceptibilities of these organelles to damage, as well as uncovering a range of defense mechanisms specific to experimental conditions and the mitochondrial protein profile found in different organisms. Concomitant with imposed oxidative damage, specific proteins are either synthesized or lost from mitochondria. This includes loss of, or replacement of, tricarboxylic acid (TCA) cycle enzymes and selected subunits of the respiratory chain and induction of peroxiredoxins and defense machinery.

Significant manipulation of mitochondrial functions can also influence oxidative damage elsewhere in the cell, and this can have wide-reaching

From: *Methods in Molecular Biology, vol. 372: Mitochondria: Practical Protocols*
Edited by: D. Leister and J. M. Herrmann © Humana Press Inc., Totowa, NJ

consequences for whole cell/tissue oxidative stress tolerance. For example, overexpression of the mitochondrial superoxide dismutase (SOD) can increase plant stress tolerance ***(1)***; its knockout in yeast resulted in the specific oxidation of an array of mitochondrial proteins in the absence of this enzyme ***(2)***. Mammalian and yeast mitochondria contain thioredoxin and peroxiredoxin systems ***(3,4)***; plants contain both of these systems ***(5)*** as well as an ascorbate/glutathione cycle able to respond to chloroplast-dependent ROS production ***(6)***. Mammalian mitochondria contain uncoupling proteins to lower membrane potential and alleviate high levels of ubiquinone (UQ) reduction, and plants contain both uncoupling proteins and nonphosphorylating respiratory bypass proteins such as the alternative oxidase (AOX), allowing nonclassical entry and exit of electrons from the respiratory chain ***(7)***.

Increasingly, it will be important to gage the significance of these oxidative loss-of-function and gain-of-function processes by quantifying their response while accurately mimicking real-life stresses that an organism may experience. We present protocols on how to impose oxidative stress on plants and plant cell cultures, how to measure the severity of this stress, and how to assay the impact on oxidative stress-sensitive components and antioxidant defense of plant mitochondria.

2. Materials

2.1. Imposing Oxidative Stress on Plant Cells and Plant Organs

2.1.1. Mitochondrial Inhibitors

1. 1 *M* Salicylhydroxamic acid made up in dimethyl sulfoxide.
2. 25 m*M* Antimycin A (AA) made up in 100% EtOH.
3. 0.1 *M* Cyanide (KCN) made up in 100 m*M* *N*-Tris (hydroxymethyl) methyl-2-aminoethane sulfonic acid (TES)-KOH, pH 7.5.

2.1.2. General Oxidants Known to Affect Mitochondria

1. 30% (w/v) H_2O_2.
2. 80 m*M* Menadione (vitamin K3, 2-methyl-1,4-naphthoquinone).
3. Paraquat (often sold as Tryquat, a commercially available herbicide combination of paraquat [437.5 mg/L] and diquat [225 mg/L]).

2.1.3. Environmental Stresses Leading to Oxidative Stress

1. Plant or cell culture growth faculties with temperature adjustment.
2. 1 *M* NaCl solution.

2.2. Assessment of Severity of Cell Oxidative Stress

2.2.1. Fluorescent Stains for Measuring Superoxide Formation in Plants

1. Spectrofluorometer.
2. Fluorescence microscope.

3. DCF-DA (e.g., Calbiochem reagent 2′,7′-dichlorofluorescein diacetate, cat. no. 287810).

2.2.2. Colorimetric Assay for Measuring H_2O_2 in Plants

1. Mortar and pestle.
2. 0.45-µm Nylon filter.
3. Spectrophotometer to measure absorbance at 508 nm.
4. Liquid nitrogen.
5. 5% (v/v) Trichloroacetic acid.
6. Activated charcoal.
7. 17 *M* (w/w) Ammonia solution.
8. 0.6 m*M* 4-(2-Pyrdylazo)resorcinol (use the sodium salt from Sigma).
9. 20% (w/v) $TiCl_2$ (made up in concentrated HCl).
10. Assay reagent: 1:1 (v/v) 0.6 m*M* 4-(2-pyrdylazo)resorcinol and 2% $TiCl_2$.

2.2.3. Thiobarbituric Acid Reactive Substances Assay

1. Mortar and pestle.
2. 2-mL plastic cryogenic vial with external threads (such as Iwaki, cat. no. 2712-002).
3. Dry block heater.
4. Spectrophotometer to measure absorbances at 440, 532, and 600 nm.
5. −TBA solution: 20% (v/v) trichloroacetic acid (make up fresh before use).
6. +TBA solution: 20% (v/v) trichloroacetic acid, 0.65% (w/v) thiobarbituric acid. Requires mild heating to dissolve thiobarbituric acid; thiobarbituric acid is particularly sensitive to oxidation, so store at room temperature under inert gas.
7. 10% (w/v) Butylated hydroxytoluene: make up in methanol; light sensitive. Store at room temperature for less than 1 mo.
8. 80% (v/v) EtOH.

2.2.4. Spectrophotometric Assay for Free Malondialdehyde

1. Mortar and pestle.
2. Dry block heater.
3. Spectrophotometer to measure absorbance at 586 nm.
4. 13.33 m*M* 1-Methyl-2-phenylindole in 3:1 acetonitrile:methanol.
5. 37% (v/v) Hydrochloric acid.
6. 20 m*M* Tris-HCl, pH 7.4.
7. Methanol.
8. 500 m*M* Butylated hydroxytoluene in acetonitrile.
9. 10 m*M* 1,1,3,3-Tetramethoxypropane (TMOP) in 10 m*M* Tris-HCl, pH 7.0.

2.3. Mitochondrial Assays for Damaged Proteins

2.3.1. Antilipoic Acid Antibodies Assay

1. Western blotting apparatus.
2. Nitrocellulose.
3. Chemiluminescence detection kit.

4. Antilipoic acid antibodies available from Calbiochem (rabbit anti-lipoic acid polyclonal antibody, unconjugated, cat. no. 437695).

2.3.2. Aconitase Activity Assay

1. Visible wavelength spectrophotometer.
2. 0.1 *M* HEPES-NaOH, pH 7.5.
3. 10% (w/v) Triton X-100.
4. 20 m*M* Nicotinamide adenine dinucleotide phosphate (NADP).
5. 0.1 *M* $MnCl_2$.
6. 2000 U/mL NADP-isocitrate dehydrogenase (ICDH) (porcine heart).
7. Reaction master mix: 80 m*M* HEPES-NaOH, pH 7.5, 0.05% (w/v) Triton X-100, 0.5 m*M* NADP, 0.5 m*M* $MnCl_2$, 2 U NADP-ICDH (porcine heart). One needs 900 μL master mix per assay (i.e., to fill a 1-mL spectrophotometric cuvette after subsequent additions of sample and substrate), but normally one should make up a master mix of about 10 mL and use for a series of 8–10 assays immediately.
8. 0.2 *M* Aconitate.

2.3.3. Pyruvate Dehydrogenase Complex Activity Assay

1. Visible-wavelength spectrophotometer.
2. 0.1 *M* TES-NaOH, pH 7.5.
3. 10% (w/v) Triton X-100.
4. 12 m*M* Coenzyme A (CoA).
5. 1 *M* $MgCl_2$.
6. 50 m*M* Thiamine pyrophosphate (TPP).
7. 0.1 *M* Nicotinamide adenine dinucleotide (NAD^+).
8. Reaction master mix: 70 m*M* TES-NaOH (pH 7.5), 0.05% (w/v) Triton X-100, 0.12 m*M* CoA, 2 m*M* $MgCl_2$, 0.2 m*M* TPP, and 2 m*M* NAD^+ at the same volume as discussed in **Subheading 2.3.2**.
9. 0.1 *M* Pyruvate.

2.3.4. Glycine Decarboxylase Activity Assay

1. Isolated mitochondria.
2. Oxygen electrode.
3. Reaction medium: 0.3 *M* sucrose, 5 m*M* KH_2PO_4, 10 m*M* TES-KOH, pH 7.2, 10 m*M* NaCl, 2 m*M* $MgSO_4$, 0.1% (w/v) bovine serum albumin.
4. Sodium hydrosulfite.
5. 0.5 *M* glycine.
6. 100 m*M* NAD^+.
7. 100 m*M* Adenosine triphosphate in 100 m*M* TES, pH 7.
8. 50 m*M* Adenosine 5′-diphosphate (ADP): make up in water for the salt and in 100 m*M* TES, pH 7, for the acid.
9. 100 m*M* TPP.
10. 12 m*M* CoA.

2.4. Measuring Mitochondrial Antioxidant Defenses

2.4.1. Superoxide Dismutase Activity

1. Isolated mitochondria (freeze-thaw mitochondria prior to use).
2. Spectrophotometer to measure the absorbance at 560 nm.
3. 50 m*M* KH_2PO_4/K_2HPO_4, pH 7.5.
4. 40 m*M* Xanthine in 50 m*M* KH_2PO_4/K_2HPO_4, pH 7.5.
5. 6 m*M* Nitroblue tetrazolium (NBT) in 70% (v/v) dimethylformamide.
6. Xanthine oxidase.

2.4.2. Alternative Oxidase Activity

1. Isolated mitochondria.
2. Oxygen electrode.
3. Reaction medium: 0.3 *M* sucrose cubster, 5 m*M* KH_2PO_4, 10 m*M* TES-KOH, pH 7.2, 10 m*M* NaCl, 2 m*M* $MgSO_4$, 0.1% (w/v) bovine serum albumin.
4. Sodium hydrosulfite.
5. 100 m*M* NADH.
6. 50 m*M* ADP: make up in water for the salt and in 100 m*M* TES-KOH, pH 7, for the acid.
7. 500 μ*M* Myxothiazol in 100% EtOH.
8. 500 m*M* Dithiothreitol (DTT; made up fresh each day).
9. 50 m*M* *n*-Propylgallate (nPG) in 100% EtOH.

3. Methods

3.1. Imposing Oxidative Stress on Plant Cells and Plant Organs (see Note 1)

3.1.1. Mitochondrial Inhibitors (see **Note 2**)

	Concentrations applied	
Inhibitor	Cells/protoplasts	Intact plant tissues
Salicylhydroxamic acid (SHAM)	1–5 m*M*	1–25 m*M*
Antimycin A (AA)	5–25 μ*M*	10–100 μ*M*
Cyanide (KCN)	0.1–1 m*M*	1–5 m*M*

3.1.2. General Oxidants Known to Affect Mitochondria (see **Note 3**)

General oxidant	Concentration applied	Application method
H_2O_2	5–100 m*M*	Added to cell culture
Menadione	400 μ*M*	Added to cell culture
Paraquat	650 mg/L	Plants sprayed

3.1.3. Environmental Stresses Leading to Oxidative Stress (see **Note 4**)

	Method	Timing
Drought	No water	4–7 d
Chilling	4–10°C	12 h to 2 d
Heating	45°C	Several minutes
Salinity	250 m*M* NaCl	4–7 d

3.2. Assessment of Severity of Cell Oxidative Stress (see Note 5)

3.2.1. Fluorescent Stains for Measuring Superoxide Formation in Plants (see **Note 6**)

1. For spectrofluorometry: DCF-DA (0.2 μ*M*) is added to cells diluted in 3 mL fresh cell culture medium in a stirred 3-mL cuvette. Fluorescence is initiated by 488-nm excitation, and emission is measured at 525 nm. Intensity recorded at time intervals over 30 min to get rates of DCF production (emission [Em] 525 units/min).
2. For fluorescence microscopy: DCF-DA can be used at a higher concentration (10 μ*M*) to ensure visibility for photography. It is important that the microscope slide not be continuously exposed to blue light. This causes photo-oxidation of DCF and very high background fluorescence. Cells should only be exposed when you need to take the pictures.

3.2.2. Colorimetric Assay for Measuring H_2O_2 in Plants (see **Note 7**)

1. Dilute 20% (w/v) $TiCl_2$ to 2% and adjust to pH 8.4.
2. Mix 1:1 (v/v) 0.6 m*M* 4-(2-pyrdylazo)resorcinol and 2% (w/v) $TiCl_2$ to make assay reagent. Keep on ice until use.
3. For the standard curve, pipet the indicated volumes of 100 μ*M* H_2O_2 and water to the reaction tube to give a total of 500 μL. The final concentration value is the H_2O_2 in the entire reaction mixture. The following is the recommended addition table for the H_2O_2 standard curve:

Volume of 100 μ*M* H_2O_2 (μL)	0	100	200	300	400	500
Volume of water (μL)	500	400	300	200	100	0
Final concentration (μ*M*)	0	10	20	30	40	50

4. Grind 150–300 mg tissue in liquid N2 to a fine powder using a mortar and pestle.
5. Add 1.5 mL 5% (v/v) trichloroacetic acid and 45 mg activated charcoal and mix by vortexing.
6. Centrifuge the homogenate at 18,000*g* for 10 min at 4°C.
7. Collect the supernatant and pass through a 0.45-μm nylon filter and adjust the filtrate to pH 8.4 with 17 *M* (w/w) ammonia solution.
8. Repeat filtration through a fresh 0.45-μm filter, take a 500-μL aliquot, and add 500 μL assay reagent.
9. Incubate at 45°C for 60 min.

10. Measure absorbance at 508 nm.
11. Using the standard data, perform a linear regression of A_{508} on $[H_2O_2]$:

$$A_{508} = a[H_2O_2] + b$$

12. Calculate the concentration of analyte in a sample:

$$[H_2O_2] = [(A_{508} - b)/a] \times df$$

where $[H_2O_2]$ is the concentration of H_2O_2 in the sample; A_{508} is the absorbance at 508 nm of sample; a is the regression coefficient (slope); b is the intercept; and df is the sample dilution factor.

3.2.3. Thiobarbituric Acid Reactive Substances Assay (see **Note 8**)

1. Homogenize leaves with 80% (v/v) EtOH at a ratio of 1:25 (grams fresh weight:mL), adding inert acid-washed sand to aid grinding using a chilled 4°C mortar and pestle on ice. It is best to grind at a ratio of 1:20 and then use remaining 80% (v/v) EtOH to wash mortar and pestle and to aid in transfer to tube.
2. Transfer homogenate (including sand) to an appropriate size tube and centrifuge at 3000*g* for 10 min at 4°C.
3. Collect supernatant and add 1 mL to each of two 2-mL plastic cryogenic vials with external threads.
4. Add 975 µL of –TBA solution to 1 cryovial and 975 µL +TBA solution to the other cryovial. Add 25 µL 10% (w/v) butylated hydroxytoluene to each cryovial. An emulsion will form on the surface after this addition.
5. In a fume cabinet, vortex samples and heat to 95°C in a dry block heater for 25 min.
6. Carefully remove heated samples and cool on ice for 5 min.
7. Transfer samples to 2-mL microfuge tubes and centrifuge at 3000*g* for 10 min at room temperature.
8. Transfer supernatant (be very careful not to disturb or collect pellet) to spectrophotometer cuvette and record absorbance readings at 440, 532, and 600 nm. Plant tissues may contain a number of other substances that can inflate thiobarbituric acid reactive substances (TBARS) estimation, leading to an overestimation of lipid peroxidation levels. This includes sugars, which is corrected in this assay by subtracting the sugar absorbance maximum at 440 nm from that at 532 nm. In addition, phenylpropanoid-type pigments, such as flavonoids (like anthocyanins), may also contribute to overestimations of lipid peroxidation products. Again, this assay corrects for this by subtracting $A_{532\ -TBA}$ from $A_{532\ +TBA}$ values.
9. Calculate MDA equivalents (nmol mL^{-1}) from the following equation:

$$A_{max} = 532\text{nm}\ \varepsilon = 157000$$

$$\text{MDA equivalents (nmol mL}^{-1}) = (A - B/157000)\ 10^6$$

$$A = [(A_{532\ +TBA}) - (A_{600\ +TBA}) - (A_{532\ -TBA} - A_{600\ -TBA})]$$

$$B = [(A_{440\ +TBA} - A_{600\ +TBA})\ 0.0571]$$

$$(A_{532\ -TBA} - A_{600\ -TBA}) = \text{Correction for anthocyanin content}$$

$$(A_{600\ +\text{TBA}}) = \text{Correction for nonspecific turbidity}$$

$$[(A_{440\ +\text{TBA}} - A_{600\ +\text{TBA}})\ 0.0571] = \text{Correction for sucrose}$$

3.2.4. Spectrophotometric Assay for Free MDA

1. Homogenize leaves in liquid N_2 and transfer to a weighed 15-mL tube (keep frozen). Weigh and add 2.970 mL 20 m*M* Tris-HCl, pH 7.4, and 30 µL 500 m*M* butylated hydroxytoluene/g fresh weight. Mix by vortexing.
2. Centrifuge at 3000*g* for 10 min.
3. Dilute 13.33 m*M* 1-methyl-2-phenylindole 3:1 with methanol for use in the assay. Prepare this solution immediately before use. Do not leave the methyl-2-phenylindole bottle uncapped (open to the atmosphere); it will turn yellow.
4. The MDA standard is TMOP because MDA is not stable. The TMOP is hydrolyzed during the acid incubation step at 45°C, which will generate MDA. Just prior to use, dilute the stock 1/500 (v/v) in water to give a 20 µ*M* stock solution. Place at 0–4°C until use.
5. For the standard curve, pipet the indicated volumes of standard and water to the reaction tube to give a total of 200 µL. The final concentration value is the [MDA] in the MDA entire reaction mixture. The following is the recommended addition table for the MDA standard curve:

Volume of 20 µ*M* standard (µL)	0	25	50	100	150	200
Volume of water (µL)	200	175	150	100	50	0
Final concentration (µ*M*)	0	0.5	1.0	2.0	3.0	4.0

6. Add 650 µL diluted 1-methyl-2-phenylindole reagent to each tube.
7. Mix by briefly by vortexing each tube.
8. Add 150 µL hydrochloric acid.
9. Stopper the tubes and mix well by vortexing each sample.
10. Incubate at 45°C for 60 min.
11. To remove any turbidity, centrifuge at 10,000*g* for 10 min at 4°C.
12. Transfer the clear supernatant to a cuvette.
13. Measure absorbance at 586 nm vs a sample blank (replacing 1-methyl-2-phenylindole reagent with 3:1 acetonitrile:methanol). The color is stable for at least 2 h at room temperature or 4°C.
14. Calculations: using the standard data, perform a linear regression of A_{586} on [MDA]:

$$A_{586} = a[\text{MDA}] + b$$

Calculate the concentration of analyte in a sample:

$$[\text{MDA}] = [(A_{586} - b)/a] \times df$$

where [MDA] is the concentration of MDA in the sample; A_{586} is the absorbance at 586 nm of sample; a is the regression coefficient (slope); b is the intercept; and df is the sample dilution factor.

3.3. Mitochondrial Assays for Damaged Proteins

3.3.1. Antilipoic Acid Antibodies Assay (see **Note 9**)

1. Separate 20–50 μg of mitochondrial protein samples by sodium dodecyl sulfatepolyacrylamide gel electrophoresis and transfer to nitrocellulose for Western blotting.
2. Follow commercial chemiluminescence detection kit protocols; use antilipoic acid antibodies at 1/15,000 dilution.

3.3.2. Aconitase Activity Assay (see **Note 10**)

1. Add mitochondrial protein sample (10–100 μg protein) to 900 μL reaction master mix.
2. Add aconitate to start reaction to a final concentration of 8 m*M*.
3. Progression of the reaction is measured as NADP reduction to NADPH at 340 nm ($\varepsilon = 6.22\ mM^{-1}$).

3.3.3. Pyruvate Dehydrogenase Complex Activity Assay (see **Note 11**)

1. Add mitochondrial sample (10–50 μg protein) to 900 μL reaction master mix.
2. Add pyruvate to start the reaction to a final concentration of 1 m*M*.
3. Progression of the reaction is measured as NAD^+ reduction to NADH at 340 nm ($\varepsilon = 6.22\ mM^{-1}$).

3.3.4. Glycine Decarboxylase Activity Assay (see **Note 12**)

1. Set up the O_2 electrode with a circulating water bath set to 25°C. To calibrate the O_2 electrode, establish a constant O_2 concentration with 1 mL reaction media in the O_2 electrode chamber. Add a few small grains of sodium hydrosulfite to remove all oxygen; this will establish your zero O_2 concentration. Wash the electrode with several water washes and add back 1 mL reaction buffer; when the rate of O_2 consumption or evolution returns to zero, this establishes a maximum O_2 concentration of 253 μ*M* at 25°C.
2. Rinse chamber and add 1 mL reaction buffer and approx 10 μg mitochondria; place lid on to seal chamber. Once the electrode settles and no oxygen consumption or evolution is occurring, add 20 μL 0.5 *M* glycine, 5 μL 100 m*M* NAD^+, 5 μL 0.1 *M* adenosine triphosphate, 5 μL 50 m*M* ADP, 5 μL 100 m*M* TPP, and 5 μL 12 m*M* CoA. This will establish a derestricted rate of oxygen consumption (state 4) in the presence of ADP.
3. Once this ADP is depleted, a constant slower rate of O_2 consumption (state 4) will be established. Allow this to proceed for approx 2 min, then add 3 μL 50 m*M* ADP; this de-restricts electron transfer and allows for faster oxygen consumption (state 3). Within approx 2 min, the ADP will be depleted, and the O_2 consumption rate will slow again (state 4). Once this slower rate has occurred for approx 2 min, then a second addition of ADP may allow the return to state 3 (*see* **Fig. 1**).

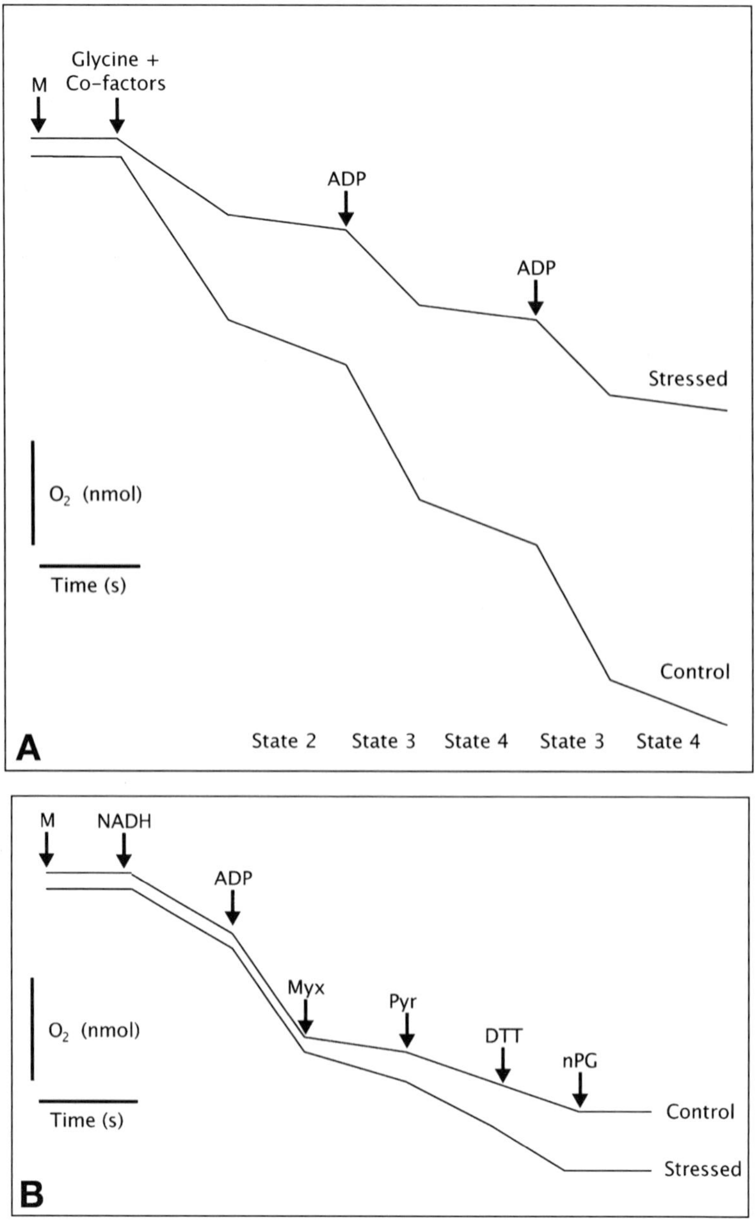

Fig. 1. Representative oxygen electrode traces showing the effect of oxidative stress on glycine-stimulated oxygen consumption (**A**) and oxygen consumption by the alternative oxidase (**B**) M, mitochondria; Myx, myxothiazol; Pyr, pyruvate; DTT, dithiothreitol; nPG, n-propylgallate.

4. The rate of oxygen consumption should be calculated from the state 3 rates by the following equation:

   O_2 consumption rate (nmol O_2/min/mg protein) = nmol O_2/time (min)/mg protein

3.4. Measuring Mitochondrial Antioxidant Defenses

3.4.1. Superoxide Dismutase Activity (see **Note 13**)

1. Set up the following reaction mix in a 1-mL cuvette: 10 µL 6 m*M* NBT, 100 µL 40 m*M* xanthine, 890 µL 50 m*M* KH_2PO_4/K_2HPO_4, pH 7.5. Add 0.025 U xanthine oxidase to start the reaction and follow the reduction of NBT at 560 nm. This is the control rate.
2. Set up the following reaction mix in a 1-mL cuvette: 10 µL 6 m*M* NBT, 100 µL 40 m*M* xanthine, 890 µL 50 m*M* KH_2PO_4/K_2HPO_4, pH 7.5 plus 50 µg mitochondrial protein. Add 0.025 U xanthine oxidase to start the reaction and follow the reduction of NBT at 560 nm. This is the inhibited rate.
3. Calculations: 1 unit of activity is defined as the amount of enzyme required to inhibit NBT reduction by 50%:

   Unit of SOD = (Control rate)/(Inhibited rate * 0.5)

3.4.2. Alternative Oxidase Activity (see **Note 14**)

1. Setup the O_2 electrode with a circulating water bath set to 25°C. To calibrate the O_2 electrode, establish a constant O_2 concentration with 1 mL reaction media in the O_2 electrode chamber. Add a few small grains of sodium hydrosulfite to remove all oxygen; this will establish your zero O_2 concentration. Wash the electrode with several water washes and add back 1 mL reaction buffer; when the rate of O_2 consumption or evolution returns to zero, this establishes a maximum O_2 concentration of 253 µ*M* at 25°C.
2. Rinse chamber and add 1 mL reaction buffer and approx 10 µg mitochondria; place lid on to seal chamber. Once the electrode settles and no oxygen consumption or evolution is occurring, add 20 µL 100 m*M* NADH. This will establish a rate of oxygen consumption (state 2).
3. Allow this to proceed for approx 2 min, then add 10 µL 50 m*M* ADP; this de-restricts electron transfer and allows for faster oxygen consumption (state 3), which will continue for the entire trace. Allow this to proceed for approx 2 min, then add 5 µL 500 µ*M* myxothiazol; this will inhibit complex III and the cytochrome pathway and leave only the O_2 consumption occurring at AOX. To maximize AOX activity, the allosteric activator pyruvate is added, followed by the reductant DTT. The rate after these additions is the maximal AOX rate of O_2 consumption. Finally, the AOX inhibitor nPG is added to abolish AOX activity (*see* **Fig. 1**).
4. Calculations. The rate of oxygen consumption should be calculated from rates after the addition of pyruvate and DTT by the following equation:

   O_2 consumption rate (nmol O_2/min/mg protein) = nmol O_2/time (min)/mg protein

4. Notes

1. Oxidative stress occurs when the rate of ROS production exceeds the rate of breakdown, leading to increasing ROS concentrations and increased levels of unrepaired oxidative damage to cellular components. Such stress can be imposed on intact plants or isolated plant cells by chemical treatments that generate ROS or by physical stresses that lead to enhanced endogenous ROS production.
2. Inhibitors of mitochondrial function can initiate oxidative stress in plant cells by generating ROS from the electron transport chain through overreduction of the ubiquionone pool. The complex III inhibitor AA can do this very specifically in plants cells; complex IV inhibitors like KCN will also work but will inhibit a range of other peroxidases and oxidases in cells, potentially complicating this effect. Inhibition of the alternative oxidase of the plant electron transport chain is also known to increase ROS production from mitochondria, but the salicylhydroxamic acids that penetrate whole cells to inhibit AOX are also metal chelators (and as weak acids also act as mild uncouplers of membrane potentials), so may have other effects in plant cells.
3. Direct addition of hydrogen peroxide (H_2O_2) is widely used in plant research of stress response. Because of the high rates of catalase operation in plant cells, large amounts of H_2O_2 are generally required to induce stress (millimolar range). Menadione is also used as this quinone derivative induces lipid peroxidation in plants through the generation of ROS in membranes. Paraquat is a photosystem II inhibitor and leads to excessive superoxide generation under lighted conditions.
4. The length of time plants are exposed to simulations of environmental stress will need to be varied depending on the hardiness of the plant species studied; however, we have given a guide to approximate values for the length of time required for each stress scenario.
5. Before the effect of such oxidative stress on mitochondria can be considered, the extent of the oxidative stress imposed on cells needs to be measured. This can be done by measuring the elevation of the rate of ROS formation or by measuring the accumulation of peroxidation end products in cells.
6. Dichlorofluorescein (DCF) or its diacetate form, 2′,7′-dichlorofluorescein-diacetate (DCF-DA) can be used to measure the rate of superoxide formation in living cells. Nonfluorescent 2′,7′-dichlorofluorescein is converted to the fluorescent 2′,7′-dichlorofluorescein in an oxidation reaction catalyzed by superoxide. DCF-DA accumulates in cells and cannot be oxidized until intracellular esterases cleave the diacetate group. The use of DCF-DA ensures low fluorescence in the cell culture medium and maximizes the measurement of intracellular superoxide formation. The fluorescence can be accurately measured by spectrofluorometry or can be observed in cells by fluorescence microscopy.
7. Most ROS are short-lived and hard to measure directly in plant tissues. The exception is H_2O_2, which can be measured to give a broad assessment of changes in ROS abundance.

8. The TBARS assay is accepted as an index of oxidative stress; however, it should be noted that this method quantitates malondialdehyde-like material (including other aldehydes produced from the peroxidation of membrane lipids such as 4-hydroxy-2-nonenal and 4-hydroxy-2-hexenal) and does not specifically measure malondialdehyde, although results are often presented as MDA equivalents.
9. These antibodies react with the lipoic acid groups attached to the E2 subunits of pyruvate dehydrogenase complex (PDC), 2-oxoglutarate dehydrogenase complex, and the branched chain 2-oxoacid dehydrogenase complex as well as the H protein subunit of the glycine dehydrogenase complex ***(8)***. As oxidative stress increases, the lipoic acid groups present on these enzyme complexes become modified and no longer bind to the antibody. This decrease in immunoreactivity is a sign of oxidative damage to the protein-bound cofactor that is essential for metabolic function of each enzyme.
10. Aconitase activity can be determined through the measurement of isocitrate production from citrate. Isocitrate production rate is measured by activity of an isocitrate and NADP-dependent enzyme. Aconitase contains an Fe-S center that is readily damaged by H_2O_2-inhibiting activity of the protein, and the protein itself has been shown to be degraded during prolonged oxidative stress ***(5,7)***.
11. PDC activity can be measured as pyruvate-dependent NAD^+ reduction in mitochondrial samples. PDC activity is inhibited by oxidative stress through the modification of lipoic acid cofactors on its E2 subunits by lipid peroxidation products ***(9)***. The rate becomes nonlinear over time because of NADH inhibition, so the initial rate should be used for calculations. The CoA stock should be made up in 0.1 *M* cysteine and can be frozen in aliquots for several weeks. Pyruvate stock solution can be supplemented with 1 m*M* HCl to help prevent polymerization of the pyruvate, which will reduce effective pyruvate concentrations and can act as inhibitors of PDC activity. Pyruvate can be stored as frozen aliquots and thawed once. The reaction solution can also be supplemented with 1 m*M* $NaSO_3$ to inhibit any contaminating lactate dehydrogenase that will otherwise cause an underestimation of the PDC rate of flux.
12. Glycine decarboxylase (GDC) presents a large proportion of the protein in mitochondria isolated from photosynthetic plant tissues, where it has a major role in photorespiration. Measuring GDC is best done using an O_2 electrode assay, in which GDC function is monitored as glycine-dependent respiration by intact mitochondria. Rupture of mitochondria and attempts to measure directly the operation of the GDC enzyme require complex cofactors and anoxic assays and are difficult to perform. GDC activity is very sensitive to oxidative stress, again not only through modification of lipoic acid on its H protein component ***(8)***, but also by selective degradation of GDC proteins during prolonged environmental stress (N. L. Taylor, unpublished data). To measure GDC activity, mitochondria must be isolated in the presence of 1 m*M* glycine to maintain activity; this needs to be added to both the isolation and wash media throughout isolation.

13. Mitochondrial SOD is the primary defense against ROS, helping to prevent lipid peroxidation and forming H_2O_2, which is then degraded by a series of pathways, including those containing catalase, peroxiredoxins, and the ascorbate/glutathione cycle.
14. The cyanide-insensitive alternative oxidase is induced in mitochondria from many plants following chemical inhibition of the cytochrome respiratory chain, oxidative treatment, or environmental stress of plants ***(10,11)***. By lowering the redox poise of the ubiquionone pool, alternative oxidase is thought to contribute to the prevention of mitochondrial superoxide formation from the electron transport chain.

Acknowledgments

N. L. T. is funded by a long-term EMBO fellowship. A. H. M. is funded by grants from the Australian Research Council (ARC) and is an ARC Queen Elizabeth II Fellow.

References

1. Bowler, C., Slooten, L., Vandenbranden, S., et al. (1991) Manganese superoxide dismutase can reduce cellular damage mediated by oxygen radicals in transgenic plants. *EMBO J.* **10,** 1723–1732.
2. O'Brien, K. M., Dirmeier, R., Engle, M., and Poyton, R. O. (2004) Mitochondrial protein oxidation in yeast mutants lacking Mn or CuZn superoxide dismutase: evidence that MnSOD and CuZnSOD have both unique and overlapping functions in protecting mitochondrial proteins from oxidative damage. *J. Biol. Chem.* **279,** 51,817–51,827.
3. Pedrajas, J. R., Miranda-Vizuete, A., Javanmardy, N., Gustafsson, J. A., and Spyrou, G. (2000) Mitochondria of Saccharomyces cerevisiae contain one-conserved cysteine type peroxiredoxin with thioredoxin peroxidase activity. *J. Biol. Chem.* **275,** 16,296–16,301.
4. Rabilloud, T., Heller, M., Rigobello, M. P., Bindoli, A., Aebersold, R., and Lunardi, J. (2001) The mitochondrial antioxidant defence system and its response to oxidative stress. *Proteomics* **1,** 1105–1110.
5. Sweetlove, L. J., Heazlewood, J. L., Herald, V., et al. (2002) The impact of oxidative stress on *Arabidopsis* mitochondria. *Plant J.* **32,** 891–904.
6. Chew, O., Whelan, J., and Millar, A. H. (2003) Molecular definition of the ascorbate-glutathione cycle in *Arabidopsis* mitochondria reveals dual targeting of antioxidant defenses in plants. *J. Biol. Chem.* **278,** 46,869–46,877.
7. Møller, I. M. (2001) Plant mitochondria and oxidative stress: electron transport, NADPH turnover, and metabolism of reactive oxygen species. *Annu. Rev. Plant Physiol. Plant Mol. Biol.* **52,** 561–591.
8. Taylor, N. L., Day, D. A., and Millar, A. H. (2002) Environmental stress causes oxidative damage to plant mitochondria leading to inhibition of glycine decarboxylase. *J. Biol. Chem.* **277,** 42,663–42,668.

9. Millar, A. H., and Leaver, C. J. (2000) The cytotoxic lipid peroxidation product, 4-hydroxy-2-nonenal, specifically inhibits decarboxylating dehydrogenases in the matrix of plant mitochondria. *FEBS Lett.* **481,** 117–121.
10. Vanlerberghe, G. C. and McIntosh, L. (1997) Alternative oxidase: from gene to function. *Annu. Rev. Plant Physiol. Plant Mol. Biol.* **48,** 703–734.
11. Considine, M. J., Holtzapffel, R. C., Day, D. A., Whelan, J., and Millar, A. H. (2002) Molecular distinction between alternative oxidase from monocots and dicots. *Plant Physiol.* **129,** 949–953.

29

Measuring Mitochondrial Shape Changes and Their Consequences on Mitochondrial Involvement During Apoptosis

Christian Frezza, Sara Cipolat, and Luca Scorrano

Summary

Mitochondria are key players in cell death following intrinsic and, in some cell types, extrinsic stimuli. The recruitment of the mitochondrial pathway results in mitochondrial dysfunction and release of intermembrane space proteins like cytochrome-c that are required in the cytosol for complete activation of effector caspases. Apoptotic shape changes of this organelle and the role of "mitochondria-shaping" proteins in cell death has attracted considerable attention. We present protocols to investigate how morphological changes of the mitochondrial reticulum regulate release of cytochrome-c, as evaluated quantitatively by an in situ approach, and changes in mitochondrial membrane potential measured in real time.

Key Words: Apoptosis; cytochrome-*c* release; fission; fusion; imaging; membrane potential; OPA1.

1. Introduction

Besides providing most cellular ATP, mitochondria participate in the early stages of programmed cell death or apoptosis. Apoptosis is essential for successful development and tissue homeostasis of all multicellular organisms, and it is accomplished by evolutionarily conserved pathways that result in an orderly process of cell demise with distinct morphological and biochemical parameters *(1)*. Dysregulation of apoptosis contributes to a variety of human diseases, including cancer *(2)*. In mammalian cells, there are two main pathways downstream of death signals that appear to be linked in certain cell types: the death receptor pathway and the mitochondrial pathway *(3)*. Both culminate in the activation of caspases, cysteine proteases that cleave a number of substrates

From: *Methods in Molecular Biology, vol. 372: Mitochondria: Practical Protocols*
Edited by: D. Leister and J. M. Herrmann © Humana Press Inc., Totowa, NJ

involved in maintenance of cytoskeletal and nuclear integrity, cell cycle progression, and deoxyribonucleic acid (DNA) repair, resulting in the orderly demise of the cell.

Mitochondria participate in the competent activation of caspases by releasing cytochrome-*c* and additional apoptogenic factors from the intermembrane space into the cytosol ***(3)***. Cytochrome-*c* in complex with Apaf-1 activates caspase-9 and other downstream caspases ***(4)***. The release of cytochrome-*c* is preceded by changes in the structure of the mitochondrial network and of mitochondrial cristae ***(5,6)***.

Besides mitochondrial shape changes during cell death, a vast variety of physiological and pathological conditions, ranging from elevated intracellular Ca^{2+} levels ***(7,8)*** to mitochondrial uncoupling ***(9)*** and inhibition of autophagocytosis ***(10)***, have been reported to affect morphology of the organelle. Mitochondrial shape is regulated by the balance between fusi on and fission processes ***(11)***. Several mitochondria-shaping proteins have been identified through genetic screens in yeast; their mammalian counterparts are less characterized ***(11)***.

Mitochondrial fission in mammalian cells is regulated by dynamin-related protein (DRP-1), a cytosolic dynamin that translocates to fission sites, where it interacts with its molecular adapter homolog fission (hFis1) ***(12)***, an integral protein of the outer mitochondrial membrane ***(13)***. Fusion is regulated by optic atrophy 1 (OPA1) and mitofusin (MFN) 1 and 2. MFNs are outer membrane proteins required for mitochondrial fusion ***(9,14–16)***. Interestingly, MFN1 seems to cooperate with the inner mitochondrial membrane protein OPA1 to fuse mitochondria ***(17)***.

DRP-1 has been shown to mediate mitochondrial fragmentation during developmental cell death of *Caenorhabditis elegans* ***(18)***. Moreover, an interesting crosstalk between "BH3-only" members (BCL-2 homology domain 3) of the B-cell lymphoma 2 (BCL-2) family, DRP-1, and remodeling of the cristae has been described ***(19)***.

Thus, considerable interest has developed in the relationship between mitochondrial shape and mitochondrial and cellular function, in particular, but not only in the course of apoptosis. Researchers exploiting these avenues face the major challenge of having to combine quantitative analysis of mitochondrial morphology and pathophysiology during apoptosis. How to reliably follow components of the latter process, such as depolarization and cytochrome-*c* release, is still a matter of debate (for a review, *see* **ref. *20***). A safe way that is less artifact prone is to use in situ methods accompanied by quantitative analyses. Here, we present protocols to address morphological changes of mitochondria and to verify if these changes play any role in controlling release of cytochrome-*c* and in mitochondrial depolarization in response to intrinsic apoptotic stimuli.

2. Materials

2.1. Mitochondrial Morphology

2.1.1. Seeding of Cells for Morphological Analysis

1. Sterile 75-cm^2 tissue culture flasks and six-well sterile tissue culture plates.
2. Sterile Dulbecco's modified Eagle's medium (DMEM) supplemented under sterile conditions with sterile 10% (v/v) fetal bovine serum (FBS), 50 U/mL penicillin, 50 µg/mL streptomycin, 100 µ*M* minimum essential medium (MEM) nonessential amino acids, and 2 m*M* glutamine. Filter sterilize through a 22-µm filter and store at 4°C.
3. Sterile phosphate-buffered saline (PBS): 2.7 m*M* KCl, 1.5 m*M* KH_2PO_4, 140 m*M* NaCl, 8 m*M* Na_2HPO_4. Alternatively, prepare working solution by dilution of one part of sterile 10X PBS (Gibco) with nine parts of sterile deionized water. Filter sterilize through a 22-µm filter and store at 4°C.
4. Sterile trypsin/ethylenediaminetetraacetic acid (EDTA) solution: sterile 0.25% (w/v) trypsin, 1 m*M* EDTA, pH 7.4. Divide under sterile conditions into 2-mL aliquots and store at 4°C.
5. Sterile Hanks' balanced salt solution (HBSS): prepare working solution by diluting one part sterile 10X HBSS (Gibco) with nine parts sterile deionized water; add 0.1 part sterile 100X HEPES (Gibco) and adjust to pH 7.4 with NaOH if necessary. Filter sterilize through a 0.22-µm filter and store at 4°C.
6. 24-mm Round glass grade 0 or 1 coverslips: coverslips must be ultraviolet (UV) sterilized by placing them under sterile conditions vertically inside the wells of a six-well plate (without the cover). Plates must be exposed to the UV source of a laminar flux hood for 45 min, with coverslips oriented toward the lamp (*see* **Note 1**).

2.1.2. Transfection of Cells for Morphological Analysis

1. Cationic lipid and colipid vehicle TransFectin lipid reagent (Bio-Rad) (*see* **Note 2**).
2. Plasmids for the expression of mitochondrially targeted DsRED. Always cotransfect it with the negative control plasmid (e.g., empty plasmid of the one containing the complementary deoxyribonucleic acid [cDNA] of your protein of interest) or with the one containing the cDNA of your protein of interest. In our experiments, we use pMSCV (BD-Clontech) and pMSCV containing murine *OPA1* cDNA (corresponding to human transcript variant 1) ***(17)*** (*see* **Note 3**).
3. Sterile DMEM (*see* **Subheading 2.1.1.**).

2.1.3. Confocal Imaging of Mitochondrial Morphology

1. HBSS: prepare as described in **Subheading 2.1.1., item 5** (*see* **Note 4**).
2. Coverslip holder: 25-mm round Attofluor stainless steel coverslip holders (Molecular Probes).

3. An inverted confocal microscope with HeNe laser light line and appropriate emission filters and photomultipliers, with a motorized *z*-axis connected to a computer for image storage and analysis.
4. Image analysis software: the freeware ImageJ (National Institutes of Health [NIH]) is suitable for all postacquisition image editing and analysis and three-dimensional (3D) reconstruction.

2.2. Cytochrome-c Release Immunofluorescence Assay

2.2.1. Seeding of Cells for Cytochrome-c Release Immunofluorescence Assay

1. Sterile 75-cm^2 tissue culture flasks and 6- and 24-well sterile tissue culture plates.
2. Sterile DMEM supplemented under sterile conditions with sterile 10% (v/v) FBS, 50 U/mL penicillin, 50 µg/mL streptomycin, 100 µ*M* MEM nonessential amino acids and 2 m*M* glutamine. Filter sterilize through a 22-µm filter and store at 4°C.
3. Sterile trypsin/EDTA solution: sterile 0.25% (w/v) trypsin, 1 m*M* EDTA, pH 7.4. Divide under sterile conditions into 2-mL aliquots and store at 4°C.
4. Sterile HBSS: prepare working solution by diluting one part sterile 10X HBSS (Gibco) with nine parts of sterile deionized water; add 0.1 part sterile 100X HEPES (Gibco) and adjust to pH 7.4 with NaOH if necessary. Filter sterilize through a 0.22-µm filter and store at 4°C.
5. 13-mm Round glass grade 0 or 1 coverslips: coverslips must be UV sterilized by placing them under sterile conditions vertically inside the wells of a 24-well plate (without the cover). Plates must be exposed to the UV source of a laminar flux hood for 45 min, with coverslips oriented toward the lamp (*see* **Note 1**).

2.2.2. Transfection of Cells for Cytochrome-c Release Immunofluorescence Assay

1. Cationic lipid and colipid vehicle TransFectin lipid reagent (Bio-Rad) (*see* **Note 2**).
2. Plasmids for the expression of mitochondrially targeted DsRED. Always cotransfect it with the negative control plasmid (e.g., empty plasmid of the one containing the cDNA of your protein of interest) or with the one containing the cDNA of your protein of interest. In our experiments, we use pMSCV (BD-Clontech) and pMSCV containing murine *OPA1* cDNA (corresponding to human transcript variant 1) *(17)* (*see* **Note 3**).
3. Sterile DMEM (*see* **Subheading 2.1.1.**).

2.2.3. Treatment of Cells With an Apoptosis Inducer and Immunostaining and Confocal Immunofluorescence of Cytochrome-c

1. Freshly prepared hydrogen peroxide (Sigma) dissolved in sterile HBSS at a final concentration of 1 m*M* (*see* **Note 5**).
2. PBS: 2.7 m*M* KCl, 1.5 m*M* KH_2PO_4, 140 m*M* NaCl, 8 m*M* Na_2HPO_4. Alternatively, prepare working solution by dilution of one part 10X PBS (Gibco) with nine parts of deionized water.

3. Fixing solution: prepare working solution by diluting one part 37% (v/v) formaldehyde solution (Sigma) in nine parts PBS; adjust to pH 7.4 using NaOH. Store at 4°C and prepare fresh every 4 wk (*see* **Note 6**).
4. Permeabilization solution: 0.01% (v/v) Nonidet P-40 (Sigma) in PBS; adjust to pH 7.4 using HCl or NaOH as required.
5. Blocking solution: 0.5% (w/v) bovine serum albumin (BSA) in PBS; divide into 10-mL aliquots and store at –20°C.
6. Primary antibody: purified anti-cytochrome-*c* mouse monoclonal antibody (BD-Pharmingen clone 6H2.B4), 1:200 in PBS.
7. Secondary antibody: antimouse immunoglobulin G, fluorescein isothiocyanate conjugated (Calbiochem), 1:200 in PBS.
8. Mounting medium: Prolong Antifade Gold (Molecular Probes).
9. 76 × 26 mm rectangular microscope slides.
10. An upright confocal microscope with HeNe and Xe laser light lines and appropriate emission filters and photomultipliers and connected to a computer for image storage and analysis.
11. Image analysis software: the freeware ImageJ (NIH) is suitable for all postacquisition image processing and analysis.

2.3. Imaging of Mitochondrial Membrane Potential

2.3.1. Seeding of Cells for Analysis of Mitochondrial Membrane Potential

1. Sterile 75-cm^2 tissue culture flasks and 6- and 24-well sterile tissue culture plates.
2. Sterile DMEM supplemented under sterile conditions with sterile 10% (v/v) FBS, 50 U/mL penicillin, 50 μg/mL streptomycin, 100 μ*M* MEM nonessential amino acids, and 2 m*M* glutamine. Filter sterilize through a 22-μm filter and store at 4°C.
3. Sterile trypsin/EDTA solution: sterile 0.25% (w/v) trypsin, 1 m*M* EDTA, pH 7.4. Divide under sterile conditions into 2-mL aliquots and store at 4°C.
4. Sterile HBSS: prepare working solution by diluting one part of sterile 10X HBSS (Gibco) with nine parts of sterile deionized water; add 0.1 part of sterile 100X HEPES (Gibco) and adjust to pH 7.4 with NaOH if necessary. Filter sterilize through a 0.22-μm filter and store at 4°C.
5. 24-mm round glass grade 0 or 1 coverslips: coverslips must be UV sterilized by placing them under sterile conditions vertically inside the wells of a 6- or 24-well plate (without the cover), respectively. Plates must be exposed to the UV source of a laminar flux hood for 45 min, with coverslips oriented toward the lamp (*see* **Note 1**).

2.3.2. Transfection of Cells for Analysis of Mitochondrial Membrane Potential

1. Cationic lipid and colipid vehicle TransFectin lipid reagent (Bio-Rad) (*see* **Note 2**).
2. Plasmids for the expression of cytosolic green fluorescent protein (GFP) (like pEGFP, BD-Clontech). Always cotransfect it with the negative control plasmid (e.g., empty plasmid of the one containing the cDNA of your protein of interest) or with the one containing the cDNA of your protein of interest. In our experiments,

we use pMSCV (BD-Clontech) and pMSCV containing murine *OPA1* cDNA (corresponding to human transcript variant 1) ***(17)*** (*see* **Note 3**).

2.3.3. Imaging of Mitochondrial Membrane Potential

1. HBSS: 1.3 m*M* $CaCl_2$, 0.5 m*M* $MgCl_2$, 0.4 m*M* $MgSO_4$, 5.3 m*M* KCl, 4.4 m*M* KH_2PO_4, 138 m*M* NaCl, 0.3 m*M* Na_2HPO_4, 1000 mg/L D-glucose; add 0.1 part 100X HEPES (Gibco) and adjust to pH 7.4 with NaOH if necessary. Alternatively, prepare working solution by dilution of one part 10X HBSS (Gibco) with nine parts deionized water; add 0.1 part 100X HEPES (Gibco) and adjust to pH 7.4 with NaOH if necessary.
2. 0.1 m*M* Tetramethylrodhamine methyl ester (TMRM) (Molecular Probes) in dimethyl sulfoxide. Store at –20°C in the dark.
3. 10 mg/mL Cyclosporine H (CsH) (Sigma) in dimethyl sulfoxide. Store at –20°C (*see* **Note 7**).
4. 2 m*M* carbonyl cyanide(*p*-trifluoromethoxy)-phenylhydrazone (FCCP) (Sigma) in absolute ethanol. Store at –20°C (*see* **Note 8**).
5. 1 m*M* H_2O_2 prepared freshly as described in **Subheading 2.2**.
6. An imaging workstation including an inverted microscope equipped with a fluorescent light source, proper excitation and emission filters, a shutter to avoid photobleaching of the samples, and a 12-bit charge coupled device camera for image acquisition. All must be connected to a computer with imaging software (usually provided with the imaging workstation) to set up the acquisition routine and to store the imaging sequence.
7. Image analysis software to analyze gray levels in the selected regions of interest (ROIs). The freeware ImageJ (NIH) with the MultiMeasure plugin is suitable.

3. Methods

3.1. Mitochondrial Morphology

Our method of choice to analyze the effect of a putative mitochondria-shaping protein on mitochondrial morphology is to cotransfect it with a mitochondrially targeted fluorescent protein and to compare the shape of the mitochondrial reticulum with that of cells transfected with the mitochondrially targeted fluorescent protein alone (*see* **Note 9**). We prefer to image the mitochondria in living cells confocally to avoid possible fixation artifacts. It must be kept in mind that when performing confocal imaging, tubular structures that move in and out of the focal plane can be easily mistaken for individual rod or spherical organelles. We therefore strongly advise acquiring stacks of mitochondrial images along the *z*-axis of the entire cell, followed by 3D image reconstruction to confirm the single-plane confocal images.

3.1.1. Seeding of Cells for Morphological Analysis

1. Check a 75-cm^2 flask containing mouse embryonic fibroblasts (MEFs) by using a standard inverted, transmitted light microscope. If cells are approaching confluence,

then open the flask in a laminar flux hood, sterile aspirate the medium, and wash the cells three times with sterile PBS.

2. Detach cells from flasks using a 0.25% (v/v) sterile trypsin/EDTA solution. For a 75-cm^2 flask, evenly distribute 1 mL sterile solution on top of the cells, gently swirl the flasks, and incubate for 3 min at 37°C.
3. Check cells for detachment using a standard inverted microscope. Gently tap the bottom of the flask if cells are still attached. After complete detachment, inactivate trypsin by adding 10 mL complete DMEM.
4. Count the cells using a hemocytometer (Burker chamber).
5. Seed 10^5 cells in each well of a six-well plate containing the sterile 22-mm round coverslips (*see* **Note 10**).
6. Place the plate in the tissue culture incubator and leave for 24 h.
7. Check confluence after 24 h. A 50–60% confluence will yield optimal transfection efficiency. Proceed with transfection if confluence is optimal.

3.1.2. Transfection of Cells for Morphological Analysis

1. For each well, 3 μg plasmid DNA in 250 μL serum-free medium are required: 1.5 μg of the fluorescent protein plasmid DNA and 1.5 μg of plasmid DNA of the protein of interest or empty vector for the control transfection.
2. For each well, add 3 μL TransFectin transfection reagent to 250 μL serum-free medium.
3. Mix the DNA and TransFectin solutions together. Gently mix by tapping or pipeting.
4. Incubate for 20 min at room temperature.
5. Take the plate containing cells grown on coverslips from the incubator.
6. Add 500 μL DNA–TransFectin complexes directly to cells in serum-containing medium. Swirl gently.
7. Place the plate in the tissue culture incubator and leave for 4 h.
8. Change the medium with complete DMEM 4 h after the addition of the DNA–TransFectin complexes.
9. Place the plate in the tissue culture incubator and leave for 20 h.

3.1.3. Confocal Imaging for Morphological Analysis

1. At 24 h after transfection, place coverslips with transfected cells in the coverslip holder.
2. Wash the cells free of medium, add HBSS, and place cells on the stage of a confocal microscope (*see* **Note 11**).
3. Choose the appropriate objective. Good images with a great degree of definition can be acquired using a 60×, 1.4-numerical aperture (NA) Plan Apo objective (*see* **Note 12**).
4. Using the binocular and epifluorescence illumination, rapidly find a field with transfected cells.
5. Regulate the power of the laser beam to obtain contrasted images and at the same time to minimize photobleaching and phototoxicity. It is advisable not to exceed 10% of the maximum power of the laser.

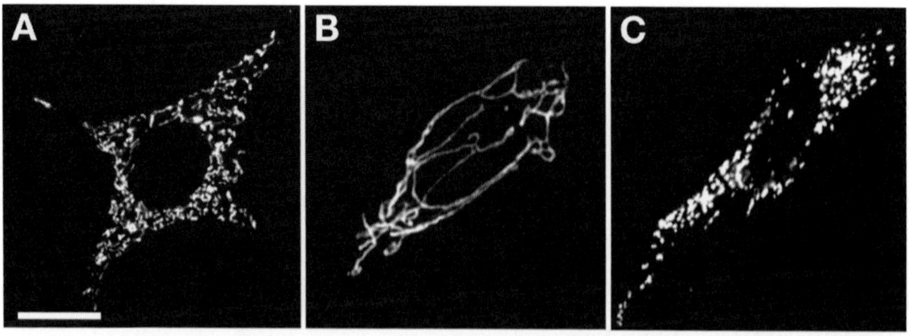

Fig. 1. Overexpression of OPA1 promotes mitochondrial elongation. Mouse embryonic fibroblasts (MEFs) grown on coverslips were cotransfected with mtRFP and empty vector (**A**), WT OPA1 (**B**), K301A OPA1 (**C**). After 24 h, confocal images of mtRFP fluorescence from randomly selected cells were acquired and stored. Bar: 10 μm.

6. Acquire and store images of transfected cells. If cells are expressing mitochondrially targeted red fluorescent protein (mtRFP), then excite using the 543-nm line of the HeNe laser and acquire emitted light through a 600-nm long-pass filter. Examples of images of MEFs expressing mtRFP and wild type (WT) or K301A *OPA1* are shown in **Fig. 1**.
7. Acquire and save stacks of images separated by 0.5 μm along the *z*-axis by using the appropriate function of your confocal microscope.
8. Open the acquired stacks with ImageJ and use the 3D reconstruction function of the program to reconstruct them.

3.2. Immunofluorescence Analysis of Cytochrome-c Release

Several methods are available to estimate the release of cytochrome-*c* from mitochondria during apoptosis. Most rely on the preparation by differential centrifugation of cytosolic and mitochondrial fractions, followed by semiquantitative determination of cytochrome-*c* content in each fraction, performed by enzyme-linked immunosorbent assay or immunoblotting. Separation of subcellular fractions by differential centrifugation requires the mechanical rupture of the plasma membrane, which can cause unspecific mitochondrial disruption with cytochrome-*c* release *(21)*. Moreover, it is always difficult to assess the effect of a transiently transfected protein at a bulk population level. On the other hand, these approaches are far more quantitative than the analysis of cytochrome-*c* subcellular localization by immunofluorescence. We therefore modified a double-immunofluorescence protocol coupled to a quantitative analysis of cytochrome-*c* distribution devised by Petronilli et al. *(22)* to evaluate quantitatively the effects of a transfected mitochondria-shaping protein on the release of cytochrome-*c*.

3.2.1. Seeding of Cells for Analysis of Cytochrome-c Release

1. Check a 75-cm^2 flask containing MEFs using a standard inverted, transmitted light microscope. If cells are approaching confluence, then open the flask in the hood, aspirate the medium, and wash the cells three times with sterile PBS.
2. Detach cells from flasks using the sterile trypsin/EDTA solution. For a 75-cm^2 flask, evenly distribute 1 mL sterile solution on top of the cells, gently swirl the flasks, and incubate for 3 min at 37°C.
3. Check cells for detachment. Gently tap the bottom of the flask if cells are still attached. After complete detachment, inactivate trypsin by adding 10 mL complete DMEM.
4. Count the cells using a hemocytometer (Burker chamber).
5. Seed 10^4 cells in each well of a 24-well plate containing the sterile 13-mm round coverslips (*see* **Note 10**).
6. Place the plate in the tissue culture incubator and leave for 24 h.
7. Check confluence after 24 h. A 50–60% confluence will yield optimal transfection efficiency. Proceed with transfection if confluence is optimal.

3.2.2. Transfection of Cells for Analysis of Cytochrome-c Release

1. For each well, 0.5 μg plasmid DNA in 50 μL serum-free medium is required: 0.25 μg of the fluorescent protein plasmid DNA and 0.25 μg of plasmid DNA of the protein of interest or empty vector for the control transfection.
2. For each well, add 1 μL TransFectin transfection reagent to 50 μL serum-free medium.
3. Mix the DNA and TransFectin solutions together. Gently mix by tapping or pipeting.
4. Incubate 20 min at room temperature.
5. Take the plate containing cells grown on coverslips from the incubator.
6. Add 100 μL DNA–TransFectin complexes directly to cells in serum-containing medium. Swirl gently.
7. Change the medium with complete DMEM 4 h after the addition of the DNA–TransFectin complexes.
8. Place the plate in the tissue culture incubator and leave for 20 h.

3.2.3. Treatment of Cells With an Apoptosis Inducer and Immunostaining and Confocal Immunofluorescence of Cytochrome-c

1. Seeded, transfected cells are now ready to be treated with the apoptotic stimulus of choice. We use H_2O_2, which at 1 m*M* is an intrinsic, mitochondria-utilizing apoptotic stimulus ***(23)***.
2. Aspirate medium and wash twice with PBS.
3. Add the solution of 1 m*M* H_2O_2 (freshly prepared; *see* **Note 5**) in HBSS.
4. Treat cells for 30, 60, and 90 min by placing the plate back in the tissue culture incubator (*see* **Note 13**).
5. Discard medium.
6. Add 0.3 mL 3.7% (v/v) ice-cold formaldehyde to each well.
7. Fix cells by leaving for 30 min at room temperature (*see* **Note 14**).
8. Discard formaldehyde by following your local hazardous waste regulations.

9. Wash samples twice with PBS.
10. Permeabilize cells by incubating with 0.3 mL 0.01% (v/v) ice-cold Nonidet NP40 for 20 min at room temperature.
11. Wash samples twice with PBS.
12. Block by adding 0.3 mL 0.5% (w/v) BSA for 15 min at room temperature.
13. Discard the blocking solution.
14. Add anti-cytochrome-*c* antibody (1:200) in PBS at room temperature for 30 min or at 4°C overnight.
15. Recover the primary antibody.
16. Wash samples twice with PBS.
17. Block by adding 0.3 mL 0.5% (w/v) BSA for 15 min at room temperature.
18. Add secondary antibody (1:200) in PBS at room temperature for 30 min at room temperature (*see* **Note 15**).
19. Wash samples twice with PBS and then with deionized water.
20. Add a drop (~5 µL) of mounting medium Prolong Antifade Gold to the microscopy slides.
21. Mount the coverslip on the slide with cells facing the mounting medium.
22. Remove any remaining water by blotting the coverslip against clean kimwipes.
23. When samples are completely dry, seal the coverslips with nail polish.
24. The sample can be viewed immediately after the nail polish dries or be stored in the dark at 4°C for up to a month.
25. Place slides on the stage of a confocal microscope.
26. For detection of mtRFP and of cytochrome-*c* immunodecorated with fluorescein isothiocyanate-conjugated antibodies, red and green channel images can be acquired simultaneously using two separate color channels on the detector assemblies of most confocal microscopes. Check that your microscope is equipped with 605-nm long-pass and 522- (± 25) nm band-pass filters, respectively.
27. Acquire and store RGB (red-green-blue) images of transfected, treated, and untreated cells for subsequent analysis.
28. Open the images using ImageJ.
29. Draw a line across the cell (**Fig. 2** illustrates such lines).
30. Using the Analyze >Plot Profile function of ImageJ, measure the fluorescence intensity of each pixel along the line in both the green and the red channels (**Fig. 2A′, B′** illustrates fluorescence intensity profiles along the lines drawn in **Fig. 2A,B**).
31. Export data to a spreadsheet program such as Excel™.
32. Calculate the localization index, defined as the ratio between the normalized standard deviations (SDs) of the fluorescence intensities of each channel: $(SD_{cyt\text{-}c}/\Sigma_{cyt\text{-}c})/(SD_{mtRFP}/\Sigma_{mtRFP})$. A punctuate distribution results in a higher SD; normalization allows correction for different fluorescence intensities in the two channels. A localization index of 1 indicates that cytochrome-*c* follows a mitochondrial distribution; an index lower than 1 means that cytochrome-*c* is randomly distributed (i.e., released cytochrome-*c*). In the example of **Fig. 2**, the localization index is 1 for the cell in panel A and 0.6 for the cell in panel B.
33. A macro can be conveniently recorded to repeat this calculation on several lines from different cells in separate experiments

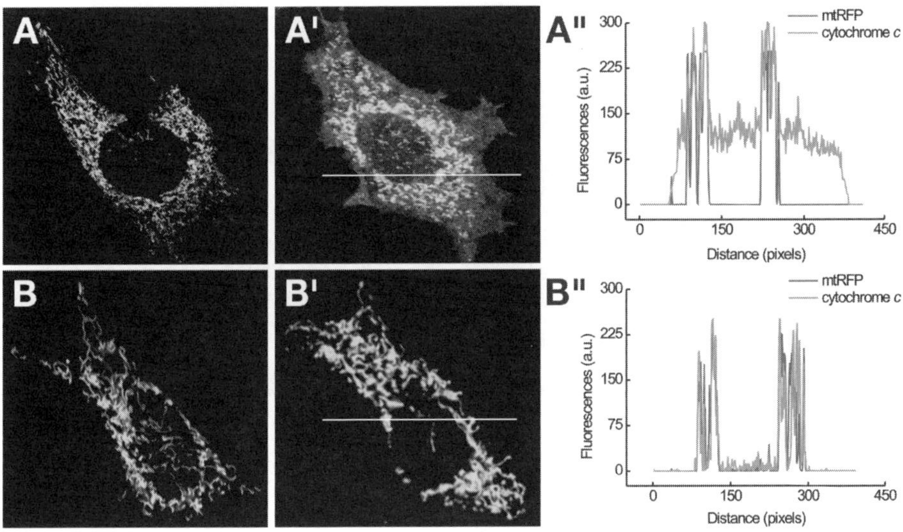

Fig. 2. Effect of the mitochondria-shaping protein OPA1 on cytochrome-*c* release evaluated by a quantitative *in situ* approach. MEFs were cotransfected with mtRFP and empty vector (**A**) or OPA1 (**B**). After 24 h, the cells were treated for 60 min with H_2O_2 (1 m*M*). Cells were then fixed, immunostained with an anti-cytochrome-*c* antibody (green), and imaged using a confocal microscope. Images of randomly selected cells before (**A**), (**B**) and after (**A′**), (**B′**) H_2O_2 treatment are shown. Sample lines are shown for the calculation of the localization index. Their fluorescence intensity profiles in the red and green channels of the lines drawn in panels **A′** and **B′** are reported in **A′** and **B′**, respectively. Bar: 10 μm.

3.3. Real-Time Imaging of Mitochondrial Membrane Potential During Apoptosis

Mitochondrial dysfunction accompanies cytochrome-*c* release during apoptosis. One of its aspects is the decrease in the mitochondrial membrane potential, which can be imaged using cationic lipophilic fluorescent dyes. To assess if overexpression of a protein of interest interferes with the apoptotic loss of mitochondrial membrane potential, transient cotransfection with a fluorescent protein such as GFP is needed to identify cells expressing the protein of interest.

3.3.1. Seeding and Transfection of Cells for Analysis of Mitochondrial Membrane Potential

Proceed exactly as indicated in the **Subheadings 3.1.1.** and **3.1.2.**

3.3.2. Imaging of Mitochondrial Membrane Potential

1. At 24 h after transfection, place coverslips with transfected cells in the coverslip holder.

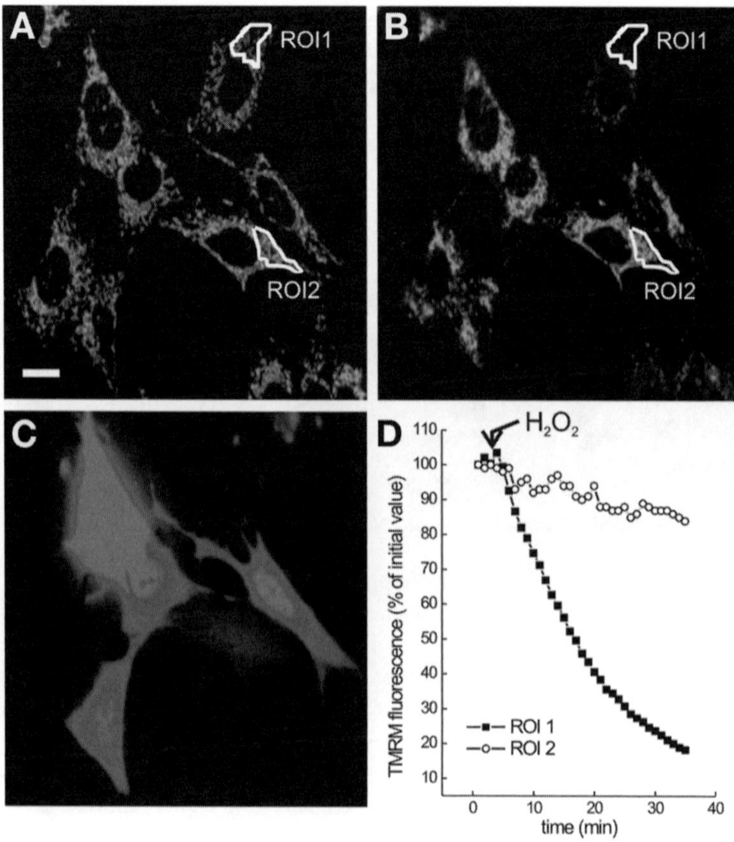

Fig. 3. Effect of the mitochondria-shaping protein OPA1 on apoptotic mitochondrial depolarization evaluated by a real-time approach. MEFs were cotransfected with GFP and OPA1. After 24 h, cells were loaded with TMRM and placed on the stage of an Olympus CellR Imaging system, and images of GFP fluorescence (**C**) were acquired and stored to identify cotransfected cells. Images of TMRM fluorescence were then acquired every 60 s for 40 min; after 3 min, cells were treated with 1 m*M* H_2O_2. Representative ROIs are drawn in images taken before (**A**) and 35 min after (**B**) addition of H_2O_2 in untransfected (ROI1) and transfected (ROI2) cells. The fluorescence intensity in the depicted ROIs was calculated, background subtracted, and normalized and is reported in (**D**). Where indicated, 1 m*M* H_2O_2 was added. Bar: 15 μm.

2. Add 1 mL 20 n*M* TMRM in HBSS supplemented with 2 μg/mL CsH.
3. Incubate for 30 min at 37°C in the dark.
4. Place coverslips on the stage of an inverted microscope (*see* **Note 11**).
5. Using the binocular and epifluorescence illumination, rapidly find a field with multiple GFP-positive cells. Check that TMRM fluorescence is stable before starting the experiment.

6. Regulate exposure times to obtain contrasted images and at the same time minimize photobleaching and phototoxicity. It is advisable not to exceed 50-ms exposure times.
7. Acquire and store the image of GFP fluorescence. This will be needed to identify the transfected cells. **Figure 3** shows GFP fluorescence in a field of transfected, TMRM-loaded MEFs.
8. Set up your imaging workstation to acquire sequential frames of TMRM fluorescence, one each 30 s to 1 min, for a total of 1–2 h.
9. After 5 min, add the apoptotic inducer.
10. Save the time series stack of images. **Figure 3** shows TMRM fluorescence before (panel B) and 35 min after (panel C) the addition of 1 m*M* H_2O_2.
11. Import the time series stack in ImageJ and proceed to analyze quantitatively the changes in mitochondrial TMRM fluorescence.
12. Open the MultiMeasure Plugin and freehand draw regions of interest on cytosolic areas comprising 10–20 mitochondria in both transfected and untransfected cells. **Figure 3** shows such ROIs. Draw a ROI on an area without cells, which will be identified as the background fluorescence.
13. Measure average fluorescence intensity values of the selected ROIs in the whole time series stack using the MultiMeasure function of ImageJ.
14. Copy the results in a spreadsheet, subtract the background, and normalize the values for the initial fluorescence. **Figure 3D** shows the quantitative analysis of changes of TMRM fluorescence in the depicted ROIs in response to 1 m*M* H_2O_2.

4. Notes

1. UV sterilization is essential for larger, 22-mm round coverslips. 13-mm round coverslips can also be sterilized by submerging them in a 1:3 isopropanol:ethanol mixture, followed by fast passage on a Bunsen flame. This protocol, however, is less safe, and free flames will disrupt the laminar flux of your sterile hood, increasing the risk of bacterial contamination.
2. Our personal experience is that this is the optimal transfection reagent for MEFs Other reagents, as well as alternative methods, such as Ca^{2+}-phosphate-mediated transfection, adenoviral infection, or electroporation, can be used successfully with this and other cell types.
3. Fluorescent proteins are essential for the morphological and functional analysis of the transfected cells. One should obtain mitochondrially targeted DsRED (mtRFP, BD-Clontech) and pEGFP (BD-Clontech) for the identification of mitochondria and the analysis of the mitochondrial morphology and for the visualization of the cotransfected cells, respectively; these fluorescent markers should be cotransfected with plasmids encoding the protein of interest. In our experiments, we use empty pMSCV and pMSCV containing murine *OPA1* cDNA. These fluorescent proteins are selected to minimize spectral interaction with other fluorescent molecules and probes exploited in the protocols presented here. Users can choose other spectral variants of GFP (like cyan and yellow fluorescent protein, for example), but it should be always kept in mind that their spectra should not overlap with those of

the other fluorescent probes used, and that the imaging workstation available to the user should have appropriate filters.

4. HBSS is used in imaging experiments to avoid spectral interference of emitting components of tissue culture media, such as phenol red. It can be replaced with phenol red-free complete media. FBS can also interfere with probes and fluorescent proteins with emission maxima around 560 nm, like TMRM and mtRFP.
5. H_2O_2 must be prepared fresh the day of the experiment as it tends to dismute spontaneously. Failure to do this will alter the formal concentration of the solution, having an impact on the reproducibility of the experiment.
6. 37% (w/v) Formaldehyde is highly toxic and a potential carcinogen, so always handle it very carefully and in a chemical hood. The use of free amines (like Tris-HCl) in the buffer will decrease the efficiency of formaldehyde, which reacts with amino groups. Efficiency will fade with time, dictating preparation of fresh solutions every month. The pH of the 3.7% (v/v) formaldehyde solution is crucial for the success of the cytochrome-*c* immunolocalization. The pH should be checked the day of the experiment.
7. CsH is an inhibitor of the P-glycoprotein multidrug resistance pump, of which all rhodamine derivatives are substrates ***(20)***. Failure to inhibit these pumps will introduce additional variables in the equilibrium distribution of TMRM, complicating the interpretation of any recorded changes. Alternatively, other multidrug resistance inhibitors, like verapamil, can be used ***(24)***.
8. FCCP is dissolved in absolute ethanol: always keep the 2 m*M* stock solution at 4°C (in an ice bath) during the whole experiment to avoid ethanol evaporation and consequent concentration of FCCP.
9. All the protocols presented in this chapter have been thoroughly tested with adherent mammalian cell lines, such as MEFs, HeLa, PC3, DU145 and several other cell lines. With minimal adjustments, they can be adapted to cells grown in suspension, such as Jurkat cells, which can adhere to coverslips in the absence of serum or once plated on polylysine-coated coverslips. We are unaware of the suitability of these protocols in cell lines derived from different organisms (i.e., insects or plants).
10. When passaging cell cultures, accurately resuspend MEFs by pipeting the suspension a few times. This will ensure an even distribution of the plated cells on the coverslips.
11. This is a delicate procedure because the coverslip is fragile. Always check that the cleft where the coverslip is placed is free of debris and use extreme caution when sealing the Attofluor chamber. Once HBSS is added, check for sealing by wiping the bottom of the coverslip with a dry kimwipe.
12. A detailed description of the optical limitations of confocal microscopy is beyond our scope, but the user should always remember that the resolution (i.e., the ability to image two adjacent fluorescent emitters as separate objects) of a confocal microscope depends on several factors, including the excitation-emission wavelength, the numerical aperture of the objective, the refraction index of the medium used by the objective (air, oil, water). When using a 60×, 1.4-NA oil immersion objective and probes emitting in the red zone of the light spectrum, the resolution can be around 200–300 nm.

13. Treatment with hydrogen peroxide may be influenced by intrinsic susceptibility of the cell type used; to assess the proper concentration and timing for treatment, a titration curve is needed; moreover, treatment with hydrogen peroxide may be influenced by cell density.
14. Immunofluorescence can be paused at this step if needed; after fixation, wash coverslips with PBS and keep at 4°C for no longer than 24 h.
15. During incubation with secondary antibody, wrap the plate with aluminum foil to protect conjugated fluorophores from light.

References

1. Hengartner, M. O. (2000) The biochemistry of apoptosis. *Nature* **407,** 770–776.
2. Thompson, C. B. (1995) Apoptosis in the pathogenesis and treatment of disease. *Science* **267,** 1456–1462.
3. Gross, A., McDonnell, J. M., and Korsmeyer, S. J. (1999) BCL-2 family members and the mitochondria in apoptosis. *Genes Dev.* **13,** 1899–1911.
4. Zou, H., Henzel, W. J., Liu, X., Lutschg, A., and Wang, X. (1997) Apaf-1, a human protein homologous to *C. elegans* CED-4, participates in cytochrome c-dependent activation of caspase-3. *Cell* **90,** 405–413.
5. Frank, S., Gaume, B., Bergmann-Leitner, E. S., et al. (2001) The role of dynamin-related protein 1, a mediator of mitochondrial fission, in apoptosis. *Dev. Cell* **1,** 515–525.
6. Scorrano, L., Ashiya, M., Buttle, K., et al. (2002) A distinct pathway remodels mitochondrial cristae and mobilizes cytochrome *c* during apoptosis. *Dev. Cell* **2,** 55–67.
7. Breckenridge, D. G., Stojanovic, M., Marcellus, R. C., and Shore, G. C. (2003) Caspase cleavage product of BAP31 induces mitochondrial fission through endoplasmic reticulum calcium signals, enhancing cytochrome c release to the cytosol. *J. Cell Biol.* **160,** 1115–1127.
8. Scorrano, L. (2003) Divide et impera: Ca^{2+} signals, mitochondrial fission and sensitization to apoptosis. *Cell Death. Differ.* **10,** 1287–1289.
9. Legros, F., Lombes, A., Frachon, P., and Rojo, M. (2002) Mitochondrial fusion in human cells is efficient, requires the inner membrane potential, and is mediated by mitofusins. *Mol. Biol. Cell* **13,** 4343–4354.
10. Terman, A., Dalen, H., Eaton, J. W., Neuzil, J., and Brunk, U. T. (2003) Mitochondrial recycling and aging of cardiac myocytes: the role of autophagocytosis. *Exp. Gerontol.* **38,** 863–876.
11. Yaffe, M. P. (1999) The machinery of mitochondrial inheritance and behavior. *Science* **283,** 1493–1497.
12. Yoon, Y., Krueger, E. W., Oswald, B. J., and McNiven, M. A. (2003) The mitochondrial protein hFis1 regulates mitochondrial fission in mammalian cells through an interaction with the dynamin-like protein DLP1. *Mol. Cell Biol.* **23,** 5409–5420.
13. James, D. I., Parone, P. A., Mattenberger, Y., and Martinou, J. C. (2003) hFis1, a novel component of the mammalian mitochondrial fission machinery. *J. Biol. Chem.* **278,** 36,373–36,379.

14. Chen, H., Detmer, S. A., Ewald, A. J., Griffin, E. E., Fraser, S. E., and Chan, D. C. (2003) Mitofusins Mfn1 and Mfn2 coordinately regulate mitochondrial fusion and are essential for embryonic development. *J. Cell Biol.* **160,** 189–200.
15. Koshiba, T., Detmer, S. A., Kaiser, J. T., Chen, H., McCaffery, J. M., and Chan, D. C. (2004) Structural basis of mitochondrial tethering by mitofusin complexes. *Science* **305,** 858–862.
16. Ishihara, N., Eura, Y., and Mihara, K. (2004) Mitofusin 1 and 2 play distinct roles in mitochondrial fusion reactions via GTPase activity. *J. Cell Sci.* **117,** 6535–6546.
17. Cipolat, S., de Brito, O. M., Dal Zilio, B., and Scorrano, L. (2004) OPA1 requires mitofusin 1 to promote mitochondrial fusion. *Proc. Natl. Acad. Sci. U. S. A.* **101,** 15,927–15,932.
18. Jagasia, R., Grote, P., Westermann, B., and Conradt, B. (2005) DRP-1-mediated mitochondrial fragmentation during EGL-1-induced cell death in *C. elegans. Nature* **433,** 754–760.
19. Germain, M., Mathai, J. P., McBride, H. M., and Shore, G. C. (2005) Endoplasmic reticulum BIK initiates DRP1-regulated remodelling of mitochondrial cristae during apoptosis. *EMBO J.* **24,** 1546–1556.
20. Bernardi, P., Scorrano, L., Colonna, R., Petronilli, V., and Di Lisa F. (1999) Mitochondria and cell death. Mechanistic aspects and methodological issues. *Eur. J. Biochem.* **264,** 687–701.
21. Adachi, S., Gottlieb, R. A., and Babior, B. M. (1998) Lack of release of cytochrome *c* from mitochondria into cytosol early in the course of Fas-mediated apoptosis of Jurkat cells. *J. Biol. Chem.* **273,** 19,892–19,894.
22. Petronilli, V., Penzo, D., Scorrano, L., Bernardi, P., and Di Lisa, F. (2001) The mitochondrial permeability transition, release of cytochrome c and cell death. Correlation with the duration of pore openings in situ. *J. Biol. Chem.* **276,** 12,030–12,034.
23. Hockenbery, D. M., Oltvai, Z. N., Yin, X. M., Milliman, C. L., and Korsmeyer, S. J. (1993) Bcl-2 functions in an antioxidant pathway to prevent apoptosis. *Cell* **75,** 241–251.
24. Cornwell, M. M., Pastan, I., and Gottesman, M. M. (1987) Certain calcium channel blockers bind specifically to multidrug-resistant human KB carcinoma membrane vesicles and inhibit drug binding to P-glycoprotein. *J. Biol. Chem.* **262,** 2166–2170.

30

Assessing Mitochondrial Potential, Calcium, and Redox State in Isolated Mammalian Cells Using Confocal Microscopy

Sean M. Davidson, Derek Yellon, and Michael R. Duchen

Summary

Mitochondria play a vital role in the regulation of intracellular calcium dynamics. Fluorescent dyes can be used to provide a direct measurement of the redox state, mitochondrial membrane potential, and mitochondrial calcium content. The simplicity of this approach lends itself to high-throughput assays and time-resolved analyses; however, care must be taken to avoid artifactual results. We outline general methods using confocal microscopy for analysis of the redox state, mitochondrial membrane potential, and mitochondrial calcium content in adult cardiomyocytes. We demonstrate how these parameters can be analyzed in parallel using the emission spectra "fingerprinting" method even when emission spectra overlap.

Key Words: Calcium; cardiomyocytes; confocal microscopy; membrane potential; NADH; redox state.

1. Introduction

Since the early 1990s we have witnessed a renaissance in the study of mitochondrial physiology as awareness has increased of its relevance to many pathological situations. It is now generally appreciated that mitochondria play a vital role in the regulation of intracellular calcium dynamics, and that $[Ca^{2+}]_{mito}$ can affect mitochondrial metabolism. Hence, $[Ca^{2+}]_{mito}$, mitochondrial membrane potential (also called $\Delta\psi_m$), and redox state are important, interrelated indicators of cellular physiology *(1)*.

With the parallel development of fluorescent dyes that can report these parameters and of confocal technology, the resolution of the signals that can be detected has improved in terms of both intracellular spatial resolution and temporal resolution. However, there are many potential pitfalls with these

From: *Methods in Molecular Biology, vol. 372: Mitochondria: Practical Protocols*
Edited by: D. Leister and J. M. Herrmann © Humana Press Inc., Totowa, NJ

methods that, without careful precautions and controls, can easily cause artifactual results. For example, particularly when using live cells, the use of lasers at high power is precluded by the oxidative damage that may occur. Inevitably, there must be a trade-off between temporal and spatial resolution.

The following subheadings describe methods to measure mitochondrial membrane potential, calcium content, and redox state accurately.

2. Materials

2.1. General Method: Measurement of Mitochondrial Membrane Potential

2.1.1. Low-Concentration Mode

1. Plating medium: laminin solution of 1 μg/mL laminin in distilled water (dH_2O). Store in aliquots at –80°C and thaw just before use. Do not refreeze.
2. Glass coverslips (round, 22-mm diameter).
3. Fine forceps.
4. A suitable ring to hold the coverslip in the microscope stage (*see* **Note 1**).
5. A fresh preparation of adult rat cardiomyocytes (*see* **Note 2**).
6. Imaging buffer: 116 m*M* NaCl, 5.4 m*M* KCl, 0.4 m*M* $MgSO_4$, 20 m*M* HEPES, 0.9 m*M* Na_2HPO_4, 1.2 m*M* $CaCl_2$, 10 m*M* glucose, 20 m*M* taurine, 5 m*M* pyruvate. Adjust to pH 7.4 with NaOH.
7. 20 μ*M* TMRM prepared in distilled water.
8. Inhibitor stock solutions: 1 *M* KCN; 1 m*M* carbonyl cyanide(*p*-trifluoro-methoxy)-phenylhydrazone (FCCP). Prepared in distilled water. As per most drugs that inhibit mitochondria, these drugs are extremely toxic and should be handled wearing gloves.

2.1.2. Dequench Mode

Materials for the dequench mode are the same as per **Subheading 2.1.1.**, except that the imaging buffer should contain 3 μ*M* tetramethylrodhamine methyl ester (TMRM) (diluted from a 3 m*M* stock of TMRM).

2.2. Intrinsic Fluorescence as a Measure of Redox State

Materials for this measurement are the same as **Subheading 2.1.1.**, with the exception of TMRM.

2.3. Measurement of Mitochondrial Calcium Using rhod-2

Materials are the same as per **Subheading 2.1.1.** and the following: Rhod-2 AM, X-rhod-1 AM (both from Molecular Probes/Invitrogen). Add 50 μL anhydrous dimethyl sulfoxide to one 50-mg aliquot of the dye just before use for a 1 m*M* stock solution (*see* **Note 3**). This should be stored at –20°C and used within 1 wk.

2.4. Simultaneous Measurement of $\Delta\psi_m$ and Mitochondrial Calcium

Materials are the same as in **Subheading 2.3**.

3. Methods

3.1. General method: Measurement of Mitochondrial Membrane Potential

3.1.1. Low Concentration Mode

The measurement of mitochondrial potential is based on the simple principle of the Nernstian distribution of lipophilic cations such as TMRM (*see* **Note 4** and **ref. 2**) into the negatively charged mitochondria. To minimize phototoxicity and ensure a linear response, it is desirable to use concentrations as low as 15–30 n*M*. TMRM will concentrate approx tenfold across the plasma membrane and approx 400- to 600-fold across the mitochondrial membrane assuming a mitochondrial membrane potential $\Delta\psi_m$ of about –150 mV. The time necessary to reach this equilibrium state varies between cell types, and although 30 min is sufficient for some cell types, cardiomyocytes require over 60 min (**Figs. 1, 2A**).

On a decrease in $\Delta\psi_m$, TMRM will rapidly redistribute into the cytosol. With prolonged depolarization, this will result in gradual re-equilibration of the increased cytosolic TMRM into the extracellular buffer. Importantly, a change in the plasma membrane potential will also cause a redistribution of TMRM from cytosol to buffer and may cause a change in the mitochondrial signal, even though $\Delta\psi_m$ remains unchanged. These issues have been discussed at length *(3–5)*.

1. Pipet 1 drop of laminin solution on the center of the coverslip and leave it in a laminar flow hood until it dries. Use on the day of preparation.
2. Prepare a suspension of adult cardiomyocytes at a concentration of approx 20,000 cells per 100 µL plating medium.
3. Pipet 100-µL drops onto laminin-treated glass coverslips and allow the cells to attach for 1 h (*see* **Note 5**).
4. Carefully remove the coverslip from the plate with fine forceps and place it in a suitable ring that can be held in the microscope stage. All of the following steps are performed with room lights off.
5. Add imaging buffer containing 15–30 n*M* TMRM (*see* **Note 6**).
6. Incubate for 60 min at room temperature (*see* **Note 7**).
7. Place the ring and coverslip on the microscope stage. Focus on the cells using phase contrast (set at a low power to avoid bleaching) and locate an appropriate field.
8. Excitation at 543 nm (helium neon) laser results in emission from TMRM with a peak at 577 nm; therefore, a band-pass filter about this wavelength or long-pass filter from about 560 nm is effective.
9. Set the HeNe (543-nm) laser to the lowest setting. Adjust the gain (and offset, if necessary) so that the signal does not saturate. In addition, if it is anticipated that

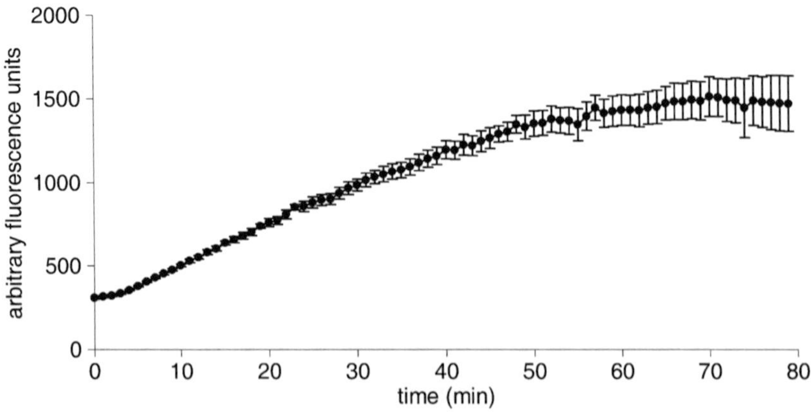

Fig. 1. Adult cardiomyocytes incubated in 30 n*M* TMRM require approx 70 min for the TMRM distribution to reach equilibrium at 37°C. Every minute, a field containing three cells was scanned with the 543-nm laser at 0.05% intensity. The average emission intensity (using a 560-nm long-pass filter) was plotted, with standard deviation.

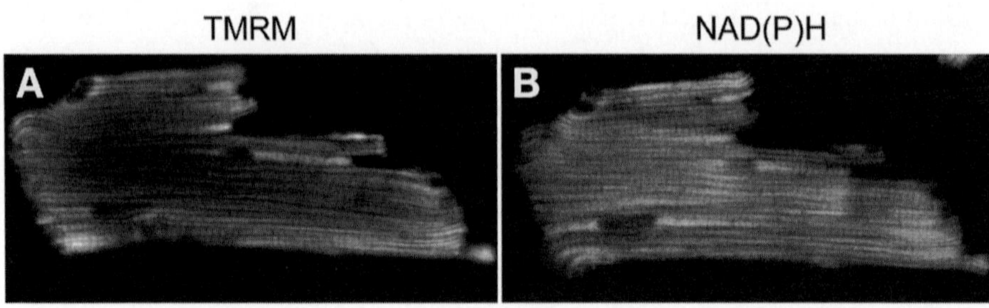

Fig. 2. The same cardiomyocyte loaded with TMRM, then scanned for TMRM fluorescence (**A**) and NAD(P)H autofluorescence (**B**). The pattern of NAD(P)H autofluorescence overlaps with the pattern of mitochondria detected using TMRM.

the signal will increase, then additional "overhead" must be left so that the signal will not saturate later.

10. Finally, take a single image. Taking into account the desired spatial and temporal resolution and the constraints of possible phototoxicity, adjust the scan speed.
11. A line average setting of 2 is useful to reduce noise without excessive laser stimulation.
12. At this point, it is often useful to move to a "fresh" field that has not been exposed to the potentially damaging effects of the laser.
13. Capture the definitive image or series of images.
14. At the end of the experiment, addition of FCCP (a mitochondrial uncoupler) to a final concentration of 10 μ*M* will completely depolarize mitochondria within a few

minutes. This can be useful for establishing background levels of fluorescence (*see* **Note 8**).

3.1.2. Dequench Mode

The dequench variation of the standard method uses a high concentration of TMRM, which accumulates within the mitochondria to such an extent that one molecule of TMRM will quench fluorescence emissions from neighboring TMRM molecules. The advantage of this approach is that even small changes in $\Delta\psi_m$ will result in large apparent changes in fluorescence as TMRM redistributes and the signal is "dequenched" (*see* **Note 9**), but the disadvantage is that the changes are not quantitative. Thus, this protocol lends itself particularly to high throughput screening of chemicals that alter $\Delta\psi_m$ because it has been found to be two orders of magnitude more sensitive than the standard method *(6)*.

In the dequench method, the only alteration to the general method (**Subheading 3.1.1.**) is that in **step 5** the imaging buffer contains 3 µ*M* TMRM (diluted from a 3 m*M* stock of TMRM).

3.2. Intrinsic Fluorescence as a Measure of Redox State

The pyridine nucleotide NADH (and NADPH [nicotinamide adenine dinucleotide phosphate]) is excited by ultraviolet (UV) light at a peak of approx 350 nm and emits in blue with a peak at approx 450 nm *(7)*; the oxidized form of NADH, NAD^+ (and $NADP^+$) is nonfluorescent. Thus, the fluorescent signal indicates the $NAD(P)H/NAD(P)^+$ ratio. The source of this fluorescent signal is primarily mitochondrial NAD(P)H (**Fig. 2B**). Flavoprotein fluorescence is excited at approx 450 nm, and peak emission is green at approx 550 nm. In contrast to NADH, it is the oxidized FAD that is fluorescent, so the total fluorescence is inversely proportional to the ratio of reduced to oxidized flavoprotein.

Follow **steps 1–4** of the general method (**Subheading 3.1.1.**). Then, with the cells in imaging buffer:

5. Use an excitation wavelength of 364 from a UV laser (*see* **Note 10**).
6. Select an appropriate emission filter, for example, 425–485 nm (peak fluorescence 450 nm).
7. Set the pinhole to maximum as the signal is fairly weak.
8. To maximize the signal intensity, it is often useful to vary the pinhole position and setting empirically. With Zeiss LSM 510 software, this is done through the Maintain menu option.
9. Capture the definitive image or series of images.
10. To establish the maximum range of NADH fluorescence, add KCN to a final concentration of 5 m*M* [inhibiting the electron transport chain will result in accumulation of NAD(P)H]. After several minutes, add an uncoupler such as FCCP to a concentration of 10 µ*M*; this will maximize O_2 consumption and therefore result in oxidation of all NAD(P)H to $NAD(P)^+$ within a few minutes.

3.3. Measurement of Mitochondrial Calcium Using rhod-2

Rhod-2 was developed as a long-wavelength calcium indicator *(8)*. Esterification with AM (acetoxymethyl) results in a cationic molecule that will accumulate in mitochondria and therefore can be used as a reporter of $[Ca^{2+}]_{mito}$. Cleavage by intracellular esterases will cause it to be trapped in the cell. However, care must be taken because some dye will remain in the cytosol, and signals from $[Ca^{2+}]_{cyt}$ may even obscure the mitochondrial signal.

Follow **steps 1–4** of the general method (**Subheading 3.1.1.**). Then, with the cells in imaging buffer:

5. Add imaging buffer containing 5 μ*M* rhod-2 AM to the coverslip.
6. Incubate the coverslip for 30 min at room temperature.
7. Replace the buffer with imaging buffer alone and return the cover-slip to the incubator for 1 h. This allows intramitochondrial esterases to cleave the AM moiety and allows time for some cytosolic dye to be extruded from the cell (*see* **Note 11**).
8. Place the ring and coverslip on the microscope stage.
9. Focus on the cells using white light and locate an appropriate field.
10. Excitation at 543-nm (HeNe) laser results in emission from rhod-2 with a peak at 581 nm; therefore, a band-pass filter about this wavelength or long-pass filter from about 560 nm is effective.
11. Set the laser to a low setting, about 1%. Adjust the gain (and offset, if necessary) so that the signal does not saturate (*see* **Note 12**). In addition, if it is anticipated that the signal will increase, then additional overhead must be left so that the signal will not saturate later.
12. Finally, take a single image. Taking into account the desired spatial and temporal resolution and the constraints of possible phototoxicity, adjust the scan speed.
13. Set line average to 2.
14. At this point, it is often useful to move to a "fresh" field that has not been exposed to the potentially damaging effects of the laser.
15. Capture the definitive image or series of images (*see* **Note 13**).

3.4. Simultaneous Measurement of $\Delta\psi_m$ and Mitochondrial Calcium

Because TMRM and rhod-2 have overlapping emission spectra, it is not possible to distinguish their fluorescence emissions using standard optical systems. However, recently developed systems perform digital analyses of the emission "fingerprints" over the entire spectrum and can transform the data into separate TMRM and rhod-2 signals. This procedure is further simplified with the use of X-rhod-1, a variant with excitation and emission maxima that are approx 25 nm further separated from the maxima of TMRM (**Fig. 3A**). To establish the fingerprints of each fluorochrome, cells must first be stained separately with TMRM and with X-rhod-1 (**Fig. 3B,C**). A third coverslip of unstained cells is used to determine the background fluorescence. Finally, after configuration, the test samples can be analyzed (**Fig. 3D,E**). The following method describes such

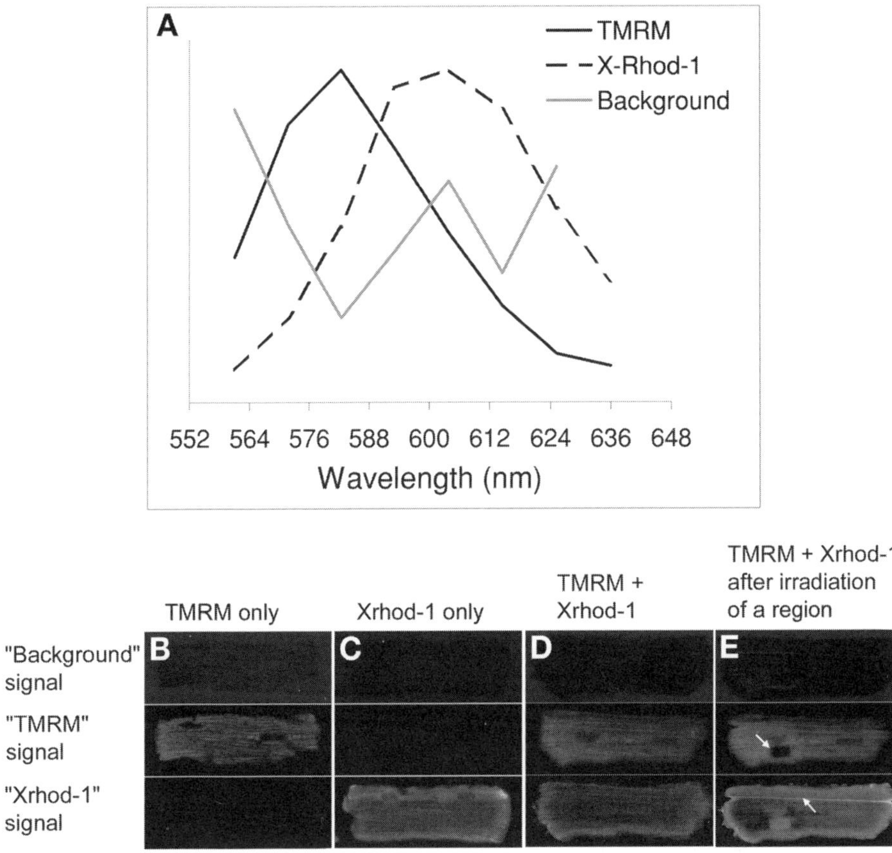

Fig. 3. The use of online fingerprinting can distinguish the fluorescent signal from TMRM and X-rhod-1 in the same cell. (**A**) The total emitted signal is separated into three components (background, TMRM, and X-rhod-1) using individual spectra. (**B**) Cardiomyocytes loaded with 30 n*M* TMRM. (**C**) Cardiomyocytes loaded with 5 μ*M* X-rhod-1 shortly after the addition of KCl to cause an increase in intracellular calcium. (**D**) Cardiomyocytes loaded with 30 n*M* TMRM and 5 μ*M* X-rhod-1. (**E**) The same cell from (**C**) after irradiating a region with repeated laser scanning (downward arrow). The TMRM signal has been lost from the region; the X-rhod-1 signal has increased in the region. Note also the contraction wave of calcium (upward arrow), here captured in the X-rhod-1 panel as a bright horizontal line because of the slow scan rate.

an analysis using the Zeiss LSM 510 META confocal microscope, but the same principle is applicable to other confocal microscopes with this capability.

Follow **steps 1–4** of the general method (**Subheading 3.1.1.**).

5. Add imaging buffer to four cover slips containing as follows:
   a. 30 n*M* TMRM.

   b. 5 μ*M* X-rhod-1 AM.
   c. Nothing.
   d. 30 n*M* TMRM and 5 μ*M* X-rhod-1 AM.
6. Incubate the cover slips for 30 min at room temperature.
7. Replace the buffer with imaging buffer containing 30 n*M* TMRM (**steps 5a** and **5d**) or imaging buffer alone (**steps 5b** and **5c**) and return the coverslip to the incubator for 1–3 h.
8. Place coverslip of **step 5d** on the microscope stage. Focus on the cells using white light and locate an appropriate field.
9. In λ-mode, configure the confocal to use the 543-nm laser. Create a λ stack from 557.2 to 621.4 nm at 10-nm intervals. Adjust the gain so that the signal does not saturate at any wavelength (it is important to leave some headroom in the signal to accommodate any anticipated increase in fluorescence).
10. Change to the coverslip of **step 5a**. Create a λ stack from 557.2 to 621.4 nm at 10-nm intervals. Save this as the TMRM spectrum (*see* **Note 14**).
11. Change to the cover slip of **step 5b**. Create a λ stack from 557.2 to 621.4 nm at 10-nm intervals. Save this as the X-rhod-1 spectrum (*see* **Note 15**).
12. Change to the coverslip of **step 5c**. Create a λ stack from 557.2 to 621.4 nm at 10-nm intervals. Save this as the background spectrum.
13. Using the Online Fingerprinting Mode of the Zeiss LSM software, assign TMRM, X-rhod-1, and background spectra to their own virtual channel with a unique color. Now, when scanning, the TMRM and X-rhod-1 signals should be digitally separated and displayed.
14. Verify that each of the four test coverslips gives the correct signal with little spillover into the other dye, and that the signal does not saturate (*see* **Note 16**). If it is reasonably certain that the signal will not saturate, then the display for the background spectrum can be disabled because it imparts no useful information.
15. The system is now configured for analysis of the experimental samples (*see* **Note 17**).

4. Notes

1. Any device may be used that is able to hold the coverslip in place and prevent leakage of buffer (e.g., a small Petri dish with the center cut out, with silicon grease to prevent leakage).
2. The method presented here assumes the cells to be analyzed are live, primary adult cardiomyocytes because their abundant and highly ordered mitochondria can simplify analysis of mitochondrial signals. If other cell types are to be used, then it is important to establish the optimal dye-loading concentration and the time required to reach equilibrium of the TMRM signal. In some cases, the use of another cell type that happens to be particularly flat may mean that it is possible to obtain good resolution of fluorescent signals from single mitochondria using a fluorescent (as opposed to confocal) microscope.
3. If bright, fluorescent, artifactual spots are visible when imaging, then the addition to the dye/dimethyl sulfoxide stock solution of 50 μL 20% Pluronic (Molecular Probes), a nonionic detergent, can assist in dispersal of the AM ester.

4. TMRM is regarded as one of the best potentiometric dyes for various reasons, including its (relatively) low toxicity *(2)*. Related dyes such as trimethyl rhodamine ethyl ester (TMRE) and rhodamine 123 may be used instead; however, optimal loading times and concentrations may be different. In addition, the rhodamine 123 signal may result in artifactual results in cardiomyocytes because we have observed that mitochondrial uncouplers do not alter its fluorescence. Some dyes are less suitable for measuring mitochondrial potential, particularly the MitoTracker® series, which is designed to react with mitochondrial thiol groups and causes their retention after fixation, meaning the entry of MitoTracker into mitochondria is essentially one way.
5. Imaging of cells directly in tissue culture plates results in very poor images because of the diffractive properties of the plastic.
6. It may be necessary to reduce the concentration of TMRM or the laser intensity if there is any sign of mitochondrial toxicity (e.g., alterations in morphology, rapid disappearance of signal in individual mitochondria in untreated cells). It is particularly important to be aware of this when using high concentrations of TMRM.
7. An initial experiment can be performed taking images every 5 min to observe when the mitochondrial signal reaches a steady state.
8. This may not be possible if using an alternative dye such as rhodamine 123 (*see* **Note 4**). Also, other control treatments can be useful at this point, such as the addition of potassium to 50 m*M*, which will depolarize the plasma membrane to approx 0 mV and should not affect the mitochondrial signal.
9. It is important to remember that when using the dequench mode, mitochondrial depolarization will be evidenced by an *increase* in fluorescence intensity as TMRM is dequenched. Furthermore, addition of FCCP will cause an increase in fluorescence, preceding a decrease as the last of the TMRM is extruded from the mitochondria. Note that TMRM will remain within the cytosol at this stage and will only slowly equilibrate with the buffer.
10. For detection of NADH fluorescence, the excitation wavelength should be below 390 nm to prevent contamination with flavoprotein autofluorescence and detected with a band-pass filter with a peak about 450 nm and a bandwidth of $\pm$ 20 to $\pm$ 40 nm. For detection of flavin autofluorescence, the excitation wavelength should be 450 $\pm$ 20 nm or similar, and emission collected using a band pass at 550 $\pm$ 40 nm or long pass above 510 nm.
11. Some groups incubate the cells at 37°C for several hours or overnight to eliminate cytosolic loading. However, in certain situations we have observed that this reduces the Ca^{2+} responsiveness of the dye *(5)*.
12. The K_d of rhod-2 for Ca^{2+} is 570 n*M*, and this may saturate in some models. An alternative dye is X-rhod-1, which has slightly longer excitation/emission maxima of approx 580/602 nm. X-Rhod-1 has a K_d for Ca^{2+} of 700 n*M*.
13. To demonstrate that rhod-2 (or X-rhod-1) fluorescence originates primarily from the mitochondria, mitochondrial Ca^{2+} uptake can be inhibited, and this should alter the fluorescent signal. Two approaches are to inhibit mitochondrial Ca^{2+} uptake and efflux using ruthenium red and clonazepam, respectively, or to collapse $\Delta\psi_m$ (e.g., using FCCP), thus preventing mitochondrial Ca^{2+} uptake. It is useful

to include 2.5 μg/mL oligomycin in this approach to prevent adenosine triphosphate depletion by mitochondrial ATPase. Alternatively, it is possible to verify that the majority of the signal is mitochondrial by permeabilizing the plasma membrane with digitonin or by adding 1 m*M* $CoCl_2$, which will quench only the cytosolic signal.

14. If using Zeiss LSM 510 META, then click on Mean of ROI to determine the reference spectrum, then Save to Dye Database to record the spectrum.
15. Healthy cells typically have a low $[Ca^{2+}]_{mito}$, making it difficult to record an accurate X-rhod-1 spectrum. One solution is to add 10 μ*M* FCCP to the well. Although a large proportion of the signal may be from X-rhod-1 binding calcium in the cytosol, it will be suitable for establishing the X-rhod-1 spectrum. Alternatively, one can locate and scan an "unhealthy" or contracting cell and determine the X-rhod-1 spectrum from the region with high $[Ca^{2+}]_{mito}$.
16. If the signal does saturate, then it can cause artifactual results as the signal will not be correctly separated into its separate components. This can often be detected as a strong signal in the background panel. For this reason, it can be useful to set the color of the background signal to white, so that the presence of saturation artifacts will be obvious in the overlay panel.
17. If the configuration of the confocal microscope (including the objective used), is altered, then it is necessary to regenerate the spectral fingerprints from scratch; otherwise, the digital separation will generate artifactual results. On the other hand, the spectra can be recalled for use in future sessions as long as the exact same configuration is used.

References

1. Duchen, M. R. (2000) Mitochondria and calcium: from cell signalling to cell death. *J. Physiol.* **529**(Pt. 1), 57–68.
2. Bernardi, P., Scorrano, L., Colonna, R., Petronilli, V., and Di Lisa, F. (1999) Mitochondria and cell death. Mechanistic aspects and methodological issues. *Eur. J. Biochem.* **264,** 687–701.
3. Nicholls, D. G. and Ward, M. W. (2000) Mitochondrial membrane potential and neuronal glutamate excitotoxicity: mortality and millivolts. *Trends Neurosci.* **23,** 166–174.
4. Rottenberg, H. and Wu, S. (1998) Quantitative assay by flow cytometry of the mitochondrial membrane potential in intact cells. *Biochim. Biophys. Acta* **1404**, 393–404.
5. Duchen, M. R., Surin, A., and Jacobson, J. (2003) Imaging mitochondrial function in intact cells. *Methods Enzymol.* **361,** 353–389.
6. Voronina, S. G., Barrow, S. L., Gerasimenko, O. V., Petersen, O. H., and Tepikin, A. V. (2004) Effects of secretagogues and bile acids on mitochondrial membrane potential of pancreatic acinar cells: comparison of different modes of evaluating DeltaPsim. *J. Biol. Chem.* **279,** 27,327–27,338.
7. Chance, B., Schoener, B., Oshino, R., Itshak, F., and Nakase, Y. (1979) Oxidation-reduction ratio studies of mitochondria in freeze-trapped samples. NADH and flavoprotein fluorescence signals. *J. Biol. Chem.* **254,** 4764–4771.
8. Minta, A., Kao, J. P., and Tsien, R. Y. (1989) Fluorescent indicators for cytosolic calcium based on rhodamine and fluorescein chromophores. *J. Biol. Chem.* **264,** 8171–8178.

V

Dynamics and Morphology

31

Fluorescence Imaging of Mitochondria in Yeast

Theresa C. Swayne, Anna C. Gay, and Liza A. Pon

Summary

The budding yeast *Saccharomyces cerevisiae* has many advantages as a model system, but until recently high-resolution microscopy was not often attempted in this organism. Its small size, rounded shape, and rigid cell wall were obstacles to exploring the cell biology of this model eukaryote. However, it is now feasible for laboratories to acquire and analyze high-resolution, multidimensional images of yeast cell biology, including the mitochondria. As a result, imaging of yeast has emerged as an important tool in eukaryotic cell biology. This chapter describes labeling methods and optical approaches for visualizing yeast mitochondria using fluorescence microscopy.

Key Words: Deconvolution; fluorescent proteins; immunofluorescence; live-cell imaging; microscopy; vital staining; yeast.

1. Introduction

Although unstained mitochondria in some cells can be visualized with transmitted-light microscopy (phase contrast or differential interference contrast), this is not possible in budding yeast. However, mitochondria can be easily visualized in living cells with vital fluorescent dyes, immunofluorescence, or targeted fluorescent proteins (FPs). Applications and outcomes for each of these approaches are summarized in **Table 1** and **Fig. 1**.

1.1. Vital Staining of Yeast Mitochondria

The main advantages of vital staining are speed and functional readout (**Table 2**). Because vital dyes are commercially available and stain mitochondria with short incubation times, they are useful for rapidly assessing mitochondrial distribution, morphology, and dynamics in live cells. In addition, because vital dyes work by sensing membrane potential or binding to DNA, they also provide

From: *Methods in Molecular Biology, vol. 372: Mitochondria: Practical Protocols*
Edited by: D. Leister and J. M. Herrmann © Humana Press Inc., Totowa, NJ

Table 1
Recommended Applications for Methods for Fluorescence Imaging of Mitochondria in Budding Yeast

	Imaging method		
Sample	Vital dye	IF	Targeted GFP
Fixed cells			
Wild-type cells	*	✔	*
Cells with respiratory defects	*	*	*
Cells with import defects	*	*	*
Live cells			
Wild-type cells	✔	x	✔
Cells with respiratory defects	*	x	✔
Cells with import defects	✔	x	*

Choices for visualization:✔, recommended; *, use with caveats; x, not recommended/not possible; IF, immuno fluorescene

information regarding mitochondrial integrity and function and stain independent of the cell's ability to express or import foreign proteins into mitochondria.

1.1.1. Deoxyribonucleic Acid—Binding Dyes

In yeast and other eukaryotes, mitochondrial DNA (mtDNA) assembles into punctate structures, mtDNA nucleoids, that are associated with the inner leaflet of the inner mitochondrial membrane. mtDNA nucleoids contain multiple copies of mtDNA and proteins that contribute to the organization, replication, and expression of mtDNA. Because mitochondria are the only extranuclear organelles in animal cells and fungi that contain DNA, cytoplasmic DNA staining can be diagnostic for mitochondria. In yeast, a species in which mtDNA is dispensable, DNA-binding dyes are also used to determine if a strain is *rho*^{0} (lacks mtDNA).

DAPI (4′,6′-diamidino-2-phenylindole) is the most common DNA-binding dye used in yeast. On binding to nucleic acids, DAPI fluorescence increases greatly, and the increase is more pronounced with DAPI binding to DNA than with ribonucleic acid (RNA). These characteristics make DAPI a strong nuclear and mtDNA marker with little cytoplasmic background staining. Another advantage is that it stains mtDNA independent of the metabolic state of the mitochondria. Consequently, it can be used in cells with mitochondrial function that may be impaired. Finally, DAPI stains DNA in living and fixed cells and produces a robust, persistent fluorescent signal.

There are two issues to bear in mind when working with DAPI. First, because DAPI also stains nuclear DNA, mtDNA nucleoids in close proximity to the nucleus are not well resolved. Second, because DAPI is visualized with ultraviolet illumination, sustained imaging of DAPI in live cells results in phototoxicity. Indeed, mitochondrial fragmentation or rupture can occur in

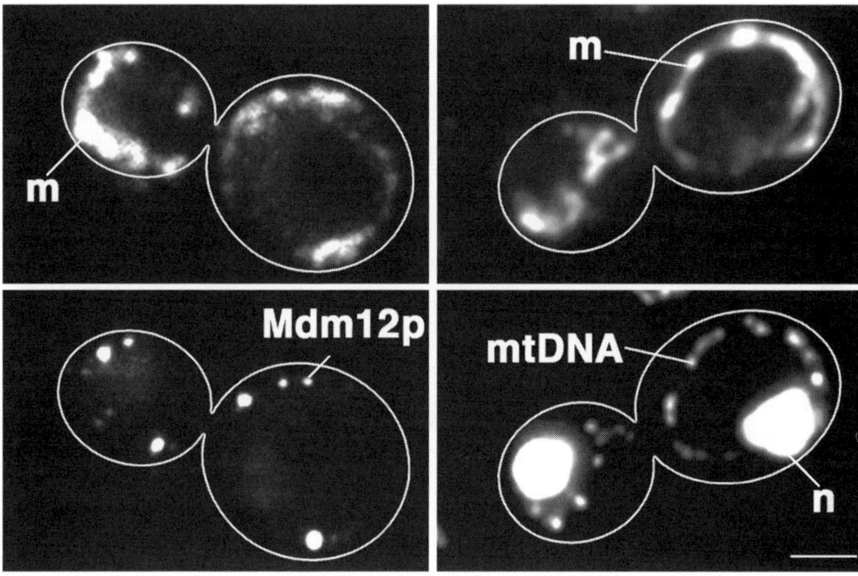

Fig. 1. Mitochondria visualized using different imaging approaches in fixed and living yeast cells. Left panels: localization of mitochondria (m; top) and myc-tagged Mdm12p, an integral mitochondrial outer membrane protein (bottom), by indirect immunofluorescence. Right panels: visualization of mitochondria (m) using ectopically expressed GFP fused to the *CIT1* signal sequence (top) and DNA in the nucleus (n) and mitochondria (mtDNA) using the DNA binding dye DAPI (bottom). Bar: 1 μm.

DAPI-stained cells after 1–2 min of continuous illumination with conventional fluorescent light sources.

1.1.2. Lipophilic Membrane Potential-Sensing Dyes

Membrane potential-sensing dyes are lipophilic, positively charged fluorophores that accumulate in cellular compartments that have a membrane potential. Because functioning mitochondria have the strongest membrane potential in the cell, these dyes accumulate in the mitochondria more readily than in any other compartment. Moreover, in contrast to DNA-binding dyes, which stain punctate intramitochondrial structures, membrane potential-sensing dyes stain the entire mitochondrial membrane. As a result, they are excellent tools for investigating mitochondrial distribution and morphology. Finally, these dyes provide information regarding mitochondrial function and integrity.

The membrane potential-sensing dyes that work well in yeast are the carbocyanine $DiOC_6(3)$, the styryl dye DASPMI [4-(4-(dimethylamino)styryl)-*N*-methylpyridinium iodide (4-Di-1-ASP)], the cationic rhodamine derivative rhodamine 123, and the fixable stains of the MitoTracker family (**Table 2**). Of the dyes described here, rhodamine 123 is least lipophilic and consequently the

Table 2
Vital Dyes for Yeast Mitochondria

Dye	Full name	λ_{ex}	λ_{em}
$DiOC_6(3)$	3,3′-Dihexyloxacarbocyanine iodide	484	501
DASPMI	4-(4-(Dimethylamino)styryl)-*N*-methylpyridinium iodide (4-Di-1-ASP)	475	605[a]
Rhodamine 123	2-(6-Amino-3-imino-3H-xanthen-9-yl) benzoic acid methyl ester	505	534[a]
MitoTracker[b]	Various	Various	Various
DAPI	4′,6′-Diamidino-2-phenylindole	358	461

[a]Broad emission range; not recommended for green/red double-label studies.
[b]Proprietary dyes from Invitrogen, Inc.

most sensitive to membrane potential. $DiOC_6(3)$ and various MitoTrackers are useful for double-label experiments because they have narrow excitation and emission spectra. Finally, the orange and red MitoTracker dyes persist in mitochondria after aldehyde fixation and permeabilization by acetone or Triton X-100 and thus are the only membrane potential-sensing dyes that can be used together with immunofluorescence staining.

1.2. Targeted FPs

Mitochondria contain two membranes (the outer and the inner) and two soluble compartments (the intermembrane space and the matrix). They also contain submitochondrial structures, including contact sites (where outer and inner membranes are closely apposed) and mtDNA nucleoids. FPs can be targeted to each of these compartments within mitochondria. For studies of morphology and dynamics of mitochondrial membranes, our laboratory uses FPs targeted to the matrix or the inner surface of the inner mitochondrial membrane. Appropriately targeted FPs can also be used to determine whether outer membrane, inner membrane, or both have fused.

FPs can be targeted to mitochondria by two methods, both of which employ fusion proteins. One approach relies on ectopic expression of fusion proteins consisting of mitochondrial signal sequences fused to FPs. Methods have been developed to insert an FP gene into the chromosomal locus of any nuclear gene. Targeting of FPs to mitochondria requires more of a time investment compared to staining with vital dyes. However, plasmids for expression of mitochondria-targeted FPs and cassettes for insertion of FP genes into the yeast genome are readily available. Moreover, targeted FPs produce a signal that is stronger, more persistent, and more specific than that produced by vital dyes. Finally, the fluorescence of targeted FPs can persist after fixation, and targeted FPs can be detected using immunofluorescence with commercially available anti-FP antibodies.

Table 3
Mitochondria-Targeted Signal Sequence-FP fusion proteins[a]

Site	Targeting	Promoter	Vector	FP	Reference
Matrix	*OLI1*[b] signal sequence	*ADH1*	2-μ pRS426 derivative	HcRed	***24***
	CIT1[c] signal sequence	*CIT1*	CEN-*URA3*	bGFP (F99S, M153T, V163A)	***25***
	CIT1 signal sequence	*GAL1*	CEN-*URA3*	bGFP	***25***
	OLI1	*GAL1*	CEN-*URA3*	DsRed	***26***
OM	*TOM6* signal sequence	*GAL1*	CEN-*URA3*	bGFP	***25***
IM	*YTA10*	*GAL1*	CEN-*URA3*	bGFP	***25***
mtDNA	*ABF2*	*GAL1*	CEN-*URA3*	bGFP	***25***

OM, mitochondrial outer membrane; IM, mitochondrial inner membrane.
[a]For other mitochondria-targeted fusion proteins, *see* **refs.** ***27*** and ***28***.
[b]F_0ATP synthase subunit 9.
[c]Citrate synthase 1.

1.2.1. Ectopic Expression of Mitochondria-Targeted FP Fusion Proteins

Over 95% of the proteins that are present in mitochondria are encoded in the nucleus, synthesized in the cytoplasm, and imported into the organelle. The targeting information for import of proteins into mitochondria resides in signal sequences, which may be in the N-terminus or C-terminus or within nuclear-encoded mitochondrial proteins ***(1)***. Mitochondrial signal sequences or full-length proteins containing mitochondrial signal sequences have been used to target FPs to mitochondria and to specific compartments within mitochondria. We have used several plasmid-borne targeted FPs to label yeast mitochondria (**Table 3**). All of the targeted FPs used produce a robust fluorescent signal that is specific for mitochondria, and they have no deleterious effect on cell growth or on mitochondrial morphology, motility, or respiratory activity.

1.2.2. Tagging Endogenous Proteins

Immunofluorescence is a powerful tool. However, localization by immunofluorescence may be compromised by fixation or staining artifacts and relies on the availability of antibodies that are specific and can bind to antigens in fixed cells. Thus, the development of methods to (1) insert DNA-encoding FPs or epitope tags into yeast genes at their chromosomal locus, (2) express tagged genes at wild-type levels from their endogenous promoters, (3) determine whether the tags are deleterious, and (4) visualize tagged proteins allows for more reliable determination of protein localization. Chromosomal tagging technology has also enabled imaging of the localization, movements, and expression of FP-tagged proteins in living cells. Finally, because FP and epitope tags can be used in antibody-based techniques including affinity purification, immunoprecipitation, Western blot analysis, and immunofluorescence, chromosomal tagging technology has facilitated biochemical characterization of proteins.

For chromosomal tagging, an insertion cassette, double-stranded linear DNA that encodes the tag of interest plus a selectable marker, is inserted into a target site in the genome by homologous recombination. Tagging vectors have been developed for insertion of a variety of FPs (e.g., green fluorescent protein [GFP], GFP color variants, and GFPs that have been optimized for expression in yeast); epitopes (e.g., hemagglutinin [HA] or myc); affinity tags (e.g., glutathione-*S*-transferase [GST], tandem affinity purification [TAP], 6x histidine [6x His]); and various combinations of FPs, epitopes, and affinity tags. Some readily available tagging vectors are shown in **Table 4**.

Insertion cassettes are produced by polymerase chain reaction (PCR) using tagging vectors as templates and primers that hybridize both to the insertion cassette within the tagging vector and the target site within the yeast chromosome. The amplified DNA contains the desired tag and a selectable marker, flanked by DNA that is homologous to the desired insertion site (**Fig. 2**). The amplified DNA is transformed directly into yeast using a standard protocol *(2)*. Recombinants that carry the inserted tag are identified using the selectable marker in the insertion cassette and characterized.

Tagging vectors are versatile as different tags can be inserted into a target gene using a single set of primers and different cassettes from the same family. In addition, vectors are available for expression of tagged genes from their endogenous promoter or from the *GAL1* regulatable promoter. Finally, variations in the tagging cassettes have been developed in which the selectable markers can be excised from the tagged gene (**Fig. 2B**). As a result, tags can be inserted anywhere within the coding region of the gene of interest, and the tagged gene can be expressed at wild-type levels under control of the endogenous promoter. Moreover, the same selectable marker can be used for multiple rounds of insertion.

1.3. Visualizing Yeast Mitochondria by Immunostaining

Several proteins can serve as markers for immunofluorescence visualization of mitochondria in *Saccharomyces cerevisiae*. These include proteins targeted to each of the submitochondrial compartments (**Table 5**). In addition to these antibodies to specific proteins, a polyclonal antibody raised against outer mitochondrial membranes has been used successfully for immunofluorescence *(3,4)*. Finally, with the advent of chromosomal tagging, proteins can be epitope tagged and visualized by immunofluorescence using commercially available, well-characterized antibodies.

Table 4
Yeast Tagging Cassette Vectors

Plasmid family	Tag position	Promoter	Tags	Markers
pFA6a ***(29)***	C-terminal	Endogenous	GFP(S65T) 3xHA 13xMyc GST	*TRP1* *kanMX6* *HIS3MX6*
pFA6a-PGAL1 ***(29)***	N-terminal or internal	*GAL1*	GFP(S65T) 3xHA GST	*TRP1* *kanMX6* *HIS3MX6*
pUR ***(30)***	C-terminal	Endogenous	DsRed	*HIS3* *URA3(K.l.)*
pYM ***(31)***[a]	C-terminal	Endogenous	yEGFP EGFP EBFP ECFP EYFP DsRed, DsRedI RedStar, RedStar2 eqFP611 FlAsH 1xHA, 3xHA, 6xHA 3xMyc, 9xMyc 1xMyc+7xHis TAP Protein A	*kanMX4* *hphNT1* *natNT2* *HIS3MX6* *klTRP1*
pKT ***(32)***[b]	C-terminal	Endogenous	yEGFP yECFP yEVenus yECitrine yESapphire yEmCFP[c] yEmCitrine tdimer2[d] yECitrine+3xHA yECitrine+13xMyc yECFP+3xHA yECFP+13xMyc	*KanMX* *SpHIS5* *CaURA3*
pOM (***33***)[a]	N-terminal or internal	Endogenous[e]	yEGFP 6xHA 9xMyc Protein A TEV-ProteinA	*kanMX6* *URA3(K.l.)* *LEU2(K.l.)*

(*Continued*)

Table 4 (*Continued*)

Plasmid family	Tag position	Promoter	Tags	Markers
			TEV-GST-6xHis	
			TEV-ProteinA-7xHis	

[a]Available through EUROSCARF.
[b]Available through EUROSCARF or Harvard University.
[c]Monomeric version.
[d]Tandem dimer of DsRed.
[e]After Cre-mediated removal of auxotrophic marker.

Table 5
Useful Marker Antigens for Yeast Mitochondria

Protein	Location	References
Porin	OM	***34,35***
Cytochrome oxidase subunit III	IM	***36,37***
Citrate synthase I	MAT	***29***
OM14	OM	***3,38***
Abf2p	mtDNA	***39***

OM, mitochondrial outer membrane; IM, mitochondrial inner membrane; MAT, mitochondrial matrix; mtDNA, mitochondrial DNA.

For indirect immunofluorescence staining, yeast cells are typically grown to mid-log phase and fixed with paraformaldehyde. Because antibodies will not penetrate the yeast cell wall, the cell wall is removed from fixed cells enzymatically (e.g., with zymolyase or lyticase). Spheroplasts are then permeabilized using a nonionic detergent and immobilized on a microscope cover slip using a polycation, polylysine. The sample-coated cover slip is then incubated with the primary antibody, which binds to the antigen of interest, and the secondary antibody, which binds to the invariant (Fc) region of the primary antibody and is tagged with a fluorophore. Finally, the stained cover slip is applied to a microscope slide using mounting solution.

1.4. Imaging Strategies for Visualization of Yeast Mitochondria

The revolution in biological imaging that followed the development of GFP and the popularization of confocal microscopy has produced a powerful collection of fluorescence techniques: imaging modalities such as deconvolution, confocal, and two-photon excitation microscopy; manipulations of fluorescence such as photoactivation and photobleaching; and the use of fluorescence as a tool beyond simple imaging (e.g., as a molecular ruler; Förster resonance energy transfer). Further developments, including 4π microscopy and structured illumination, have only recently been commercialized but have been useful in some situations. Having more options is good, but the full arsenal of technology is not needed for every

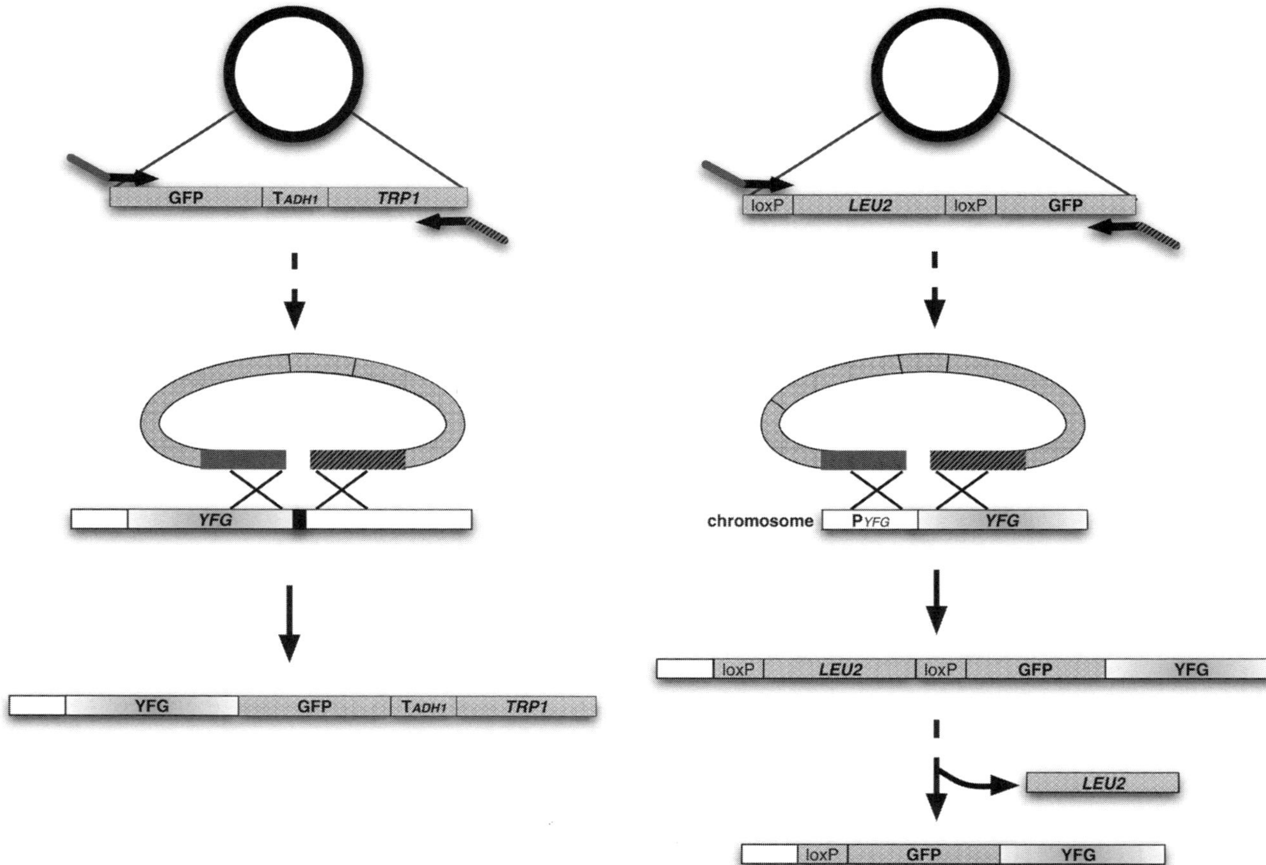

Fig. 2. Approaches for insertion of FPs or epitope tags into target sites in the yeast genome. *See* main text for description.

experiment. The investigator must determine which technology can best provide the information needed in a given study.

1.4.1. Defining Imaging Needs

Imaging is essentially a two-dimensional (2D) process; biological processes are intrinsically three-dimensional. In addition, biological structures are dynamic and change position and morphology with time. As a result, a complete picture of a biological process requires at least four dimensions of information. However, illumination diminishes the fluorescence output of FPs (photobleaching) and can damage the sample (phototoxicity). Therefore, carefully defining the question to be answered will allow the investigator to choose the most efficacious route to getting the needed information.

2D imaging is widely used for rapid assessment of mitochondrial morphology and membrane potential. The round shape of the yeast cell can be a concern when taking single 2D images. However, mitochondria in budding yeast tend to lie near the cell cortex. As a result, a focal plane that captures that cortical region of the cell can provide interpretable images of mitochondria.

3D imaging is the method of choice for high-resolution imaging of mitochondrial morphology and distribution. As described, the appearance of mitochondria in a normal yeast cell varies with focal plane. Because mitochondria are cortically distributed, a focal plane at the center of the cell will show cortical dots, representing cross-sections through mitochondrial tubules. Moreover, in yeast with abnormally aggregated mitochondria, the organelle may not be detectable in some optical planes. To characterize mitochondrial morphology and distribution fully, we collect images at a series of focal planes (a z-series) through the entire cell. This can be performed on living or fixed cells. The 3D image data can be reconstructed for viewing at any angle and used for quantitative analysis of mitochondrial volume, morphology, and distribution ***(5–7)***.

Three-dimensional imaging is also essential for colocalization studies (e.g., ***5,8,9***). With 2D imaging, two particles that are in different planes within a 3D cell may appear to colocalize. Thus, viewing structures of interest at all angles in 3D reconstructions is the only way to assess colocalization accurately.

This approach is used to study mitochondrial motility and plasticity. Like many other organelles, mitochondria are highly dynamic. They are motile ***(10)*** and undergo fission and fusion ***(11)***, and their overall morphology changes during processes such as sporulation, transition to stationary phase, and changes in carbon source. The inheritance of mitochondria during budding and their mixing and resorting during mating ***(12)*** are dynamic events in the life of mitochondria that have been followed with time-lapse fluorescence microscopy.

Four-dimensional imaging (3D imaging over time) tracks mitochondria through multiple focal planes over time and reveals the complete dynamic shape of the organelle as it evolves. However, optical sectioning increases sample illumination times and the resulting photodamage sustained by the cells. Thus, spatial resolution, temporal resolution, and label persistence/cell viability are in essential conflict.

1.4.2. Imaging Technologies

There are three major strategies for collecting serial optical sections of fluorescent yeast mitochondria: wide-field microscopy with deconvolution, spinning disk confocal microscopy, and scanning confocal microscopy.

1.4.2.1. Wide-Field Microscopy With Deconvolution

Deconvolution is a general term for computational techniques that increase the contrast and resolution of digital images. There are four primary sources for image degradation: noise, scatter, glare, and blur ***(13)***. Noise is semirandom image degradation produced by the signal or the digital imaging system. Scatter and glare are random disturbances of light produced by passage through areas with different refractive indices in the sample and through the lenses or filters of the imaging system. Finally, blur is the nonrandom spreading of light after passage through the lens. With any lens with a finite depth of field viewing a 3D sample, some features in the image are in focus, and others are out of focus because they are at a different focal depth. Light from out-of-focus focal planes is the most significant cause of image blurring in fluorescence microscopy. The other source of degradation is diffraction. Although fluorescence emanates from point sources (individual fluorescent molecules), no optical system can perfectly resolve them because the diffraction of light waves blurs the image.

Because the spreading of blurred light is nonrandom, methods were developed to determine the point spread function (PSF), the pattern of light spreading from a point source. A generic (theoretical) PSF may be calculated from data such as the objective lens magnification and numerical aperture, dye emission wavelength, and camera pixel size. Alternatively, an empirical PSF can be determined from a z-series of images of subresolution (<0.2-μm diameter) fluorescent beads. We prefer the empirical PSF, but the theoretical PSF can also yield acceptable results. In a wide-field microscope, the PSF looks like a set of concentric rings in the x-y plane that extend from the original point source and like an hourglass in the x-z or y-z planes with the point source at the crossover point in the hourglass.

Deconvolution algorithms apply the PSF to every point of light in a digital image to distinguish blurred light from the actual signal and to eliminate or reduce blurred light from the acquired image. Nearly any image acquired by a digital fluorescence microscope can be deconvolved. However, most users apply deconvolution to enhance images obtained using a wide-field fluorescence microscope (i.e., a standard epifluorescence microscope equipped with a CCD [charge-coupled device] camera for digital image acquisition]. For 3D and 4D studies, a z-series of wide-field images is collected. There are several algorithms for deconvolution of fluorescence images ***(13)***. The algorithm of choice depends on the sample, computation time, and software implementation. Our system (Volocity, Improvision, Inc., Coventry, UK) uses a maximum entropy iterative restoration algorithm.

1.4.2.2. Spinning Disk Confocal Microscopy

Confocal microscopy and wide-field deconvolution remove blur by opposite means. In confocal microscopy, pinholes in the light path prevent out-of-focus light from reaching the detector. The feature that distinguishes spinning disk confocal microscopy from other forms of confocal microscopy is simultaneous illumination of multiple points in the sample using a spinning disk that is perforated with strategically placed pinholes. Light delivery is enhanced using lasers, which provide high initial light flux, and using microlenses on each pinhole ***(14)***. Light capture is enhanced using increasingly sensitive cooled CCD cameras. As a result, large fields can be imaged rapidly, in some cases up to 30 fps (standard video rate). In practice, the actual temporal resolution depends on the sensitivity of the camera and the efficiency of light delivery through the pinholes. Nonetheless, because spinning disk confocal imaging allows for rapid acquisition of high-resolution images, many users are applying this approach to live cell imaging.

1.4.2.3. Scanning Confocal Microscopy

Scanning confocal imaging also uses pinholes in the excitation and emitted light paths to block detection of blurred light. However, in contrast to spinning disk methods, a single laser beam illuminates the sample and scans across the field, pixel by pixel. As a result of the long image acquisition times, this method is not widely used for imaging very rapid processes such as mitochondrial motility. Moreover, the photomultiplier tubes used to detect emitted light in current laser scanning systems are significantly less sensitive compared to CCD detectors. As a result, although scanning confocal microscopy can be used to image yeast mitochondria (e.g., *7*), it requires a strong fluorescent signal. The lasers in scanning confocal systems can also be used for techniques that require high-intensity illumination of select regions, including fluorescence recovery after photobleaching and localized photoactivation. Thus, laser scanning systems are widely used to study protein and organelle dynamics ***(15)***.

2. Materials

2.1. Staining Yeast Mitochondria With Vital Dyes

1. Lactate medium: 3.0 g yeast extract, 0.5 g glucose, 0.5 g $CaCl_2$, 0.5 g NaCl, 0.6 g $MgCl_2$, 1 g KH_2PO_4, 1 g NH_4Cl, 22 mL 90% lactic acid, 7.5 g NaOH. Dissolve ingredients in 800 mL distilled water (dH_2O). Adjust to pH 5.5 with NaOH. Bring

volume to 1 L with distilled water. Optional: for growth of *ade2*$^-$ cells, add 0.1 mg/mL adenine.

2. Raffinose medium: 10 g yeast extract, 20 g Bacto™ peptone, 20 g raffinose. Dissolve ingredients in 800 mL dH_2O. Bring volume to 1 L with distilled water. Optional: for growth of *ade2*$^-$ cells, add 0.1 mg/mL adenine.
3. SE (synthetic glutamate) medium: 1.0 g glutamic acid, monosodium salt, 1.7 g yeast nitrogen base *without* amino acids or ammonium sulfate, 20.0 g dextrose, amino acid supplements as needed. Dissolve ingredients in 800 mL dH_2O. Adjust to pH 5.5 with $NaHCO_3$. Bring volume to 1 L with distilled water. Optional: supplement after autoclaving with 200 μg/mL G418.

2.2. Transformation of Yeast With Plasmid-Borne Targeted FPs

1. Polyethylene glycol 3350: 50% w/v in water. Autoclave.
2. Lithium acetate: 1.0 *M* in water. Sterile filter.
3. Single-stranded calf thymus DNA: 2.0 mg/mL in water. Store frozen. After every four freeze-thaw cycles, boil aliquot 5 min.

2.3. Methods for Modification of Nuclear-Encoded Mitochondrial Genes in Yeast

1. PCR buffer: 3.3X XL buffer II stock, containing no Mg^{2+}.
2. $Mg(OAc)_2$: 25 m*M* stock.
3. Deoxynucleotide 5′-triphosphate (dNTP) mix: 2.5 m*M* each deoxyadenosine 5′-triphosphate, deoxycytidine 5′-triphosphate, deoxyguanidine 5′-triphosphate, deoxythymidine 5′-triphosphate.
4. Primers: 2×10^{-5} *M* stock.
5. Vector template: approx 50 ng/μL stock.
6. DNA polymerase: r*Tth* (Applied Biosystems).

2.4. Immunofluorescence Methods

1. Mounting solution: 10 mL 1X phosphate-buffered saline (PBS), 100 mg *p*-phenylenediamine. Stir vigorously until dissolved. Adjust to pH 9.0 with NaOH. Add 90 mL glycerol and mix thoroughly. Optional: for 4,6-diamidino-2-phenylindole (DAPI) (*see* **Note 1**) counterstaining of nuclear and mitochondrial DNA, add 100 μL of a stock solution of 1 mg/mL in distilled water. Store at –20°C in aliquots.
2. NS: 10 mL 1 *M* Tris-HCl, pH 7.5, 21.4 g sucrose, 1 mL 0.5 *M* ethylenediaminetetraacetic acid, 0.5 mL 1 *M* $MgCl_2$, 0.05 mL 1 *M* $ZnCl_2$, 0.05 mL 0.5 *M* $CaCl_2$. Adjust volume to 500 mL with distilled water. Filter sterilize and store in 40-mL aliquots at –20°C.
3. NS^+: On day of use, supplement 10 mL NS with 40 μL 200 m*M* phenylmethylsulfonylfluoride in ethanol, 5 μL β-mercaptoethanol, and 20 μL 10% NaN_3.
4. Paraformaldehyde (*see* **Note 2**), electron microscopy grade (Electron Microscopy Sciences, Hatfield, PA), 16% stock.
5. PBS^+: 1X PBS supplemented on day of use with 1% bovine serum albumin and 0.1% sodium azide. Sterilize by passing through a 0.2-μm filter.

6. PBT: 1X PBS^+ supplemented on day of use with 0.1% Triton X-100. Optional: to reduce nonspecific antibody binding, add 0.913 g lysine hydrochloride (final concentration 0.1 *M*). Sterilize by passing through a 0.2-µm filter.
7. Polylysine: dissolve 0.5 mg/mL polylysine in distilled water. Sterilize by passing through a 0.2-µm filter. Store in aliquots at –20°C.
8. Wash solution: 12.5 mL 1 *M* potassium phosphate, pH 7.5, 400 mL 1 *M* KCl. Bring volume to 500 mL with distilled water. Autoclave.
9. Tris/dithiothreitol: make on day of use. 10 mL 1 *M* Tris-SO_4, pH 9.4, 1 mL 1 *M* dithiothreitol. Bring volume to 100 mL with distilled water.
10. Zymolyase: make on day of use. 0.125 mg/mL Zymolyase 20T (Seikagaku, Inc., Tokyo, Japan) in wash solution.

2.5. Visualization of Yeast Mitochondria

1. Agarose bed growth chamber: 5.0 mL growth medium (*see* **Note 3**), 0.1 g 2% agarose (low melting). Combine ingredients in 50-mL Falcon tube and boil approx 5 min to ensure contents are fully dissolved. Dispense 200-µL aliquots into Eppendorf tubes (*see* **Note 4**). Store at room temperature in dark.
2. VALAP: 1:1:1 petrolatum (Vaseline), lanolin, and paraffin (hard). Combine ingredients. Melt by submerging in a 70°C water bath. Aliquot melted VALAP into 60 × 15 mm Petri dishes. Store at room temperature.

3. Methods

3.1. Staining Yeast Mitochondria With Vital Dyes (Table 6)

1. Grow yeast cells to mid-log phase in liquid media.
2. Concentrate cells by centrifugation at approx 13,000*g* for 5 s (*see* **Note 5**).
3. Wash cell pellet in 1X PBS or growth media. Although PBS has lower background fluorescence, it should be used only if cells will be observed for less than 30 min.
4. Resuspend cells to approx 2×10^8 cells/mL in 1X PBS or growth medium (*see* **Note 6**).
5. Add appropriate volume of dye stock solution and mix thoroughly (*see* **Note 7**).
6. Incubate in the dark.
7. Pipet gently to resuspend cells (*see* **Note 8**).
8. Mount cells for short-term or long-term observation as described in **Subheading 3.5.2**.

3.2. Transformation of Yeast With Plasmid-Borne Targeted FPs

The lithium acetate method *(2)* is commonly used for yeast transformation. The following protocol is for one transformation reaction. A mock transformation containing no DNA should always be carried out in parallel.

1. Grow yeast to mid-log phase.
2. Remove 10^8 cells from the culture (*see* **Note 6**).
3. Wash cells with 100 m*M* lithium acetate (*see* **Note 9**) and resuspend in 50 µL 100 m*M* lithium acetate.

Table 6
Vital Dyes for Yeast Mitochondria

		Staining conditions	
Dye	Stock[a]	Dye	Incubation
$DiOC_6(3)$ ***(10)***	10 mg/mL in ethanol	10–100 ng/mL	15–30 min RT
DASPMI ***(38)***	1 mg/mL in ethanol	10–100 µg/mL	30 min RT
Rhodamine 123 ***(40)***[b]	10 mg/mL in DMSO[c]	5–10 µg/mL	15–30 min RT
MitoTracker ***(11)***	1 m*M* in DMSO[c]	25–500 n*M*	15–30 min RT
DAPI	1 mg/mL in H_2O	0.1 µg/mL	15 min RT

DMSO, dimethyl sulfoxide; RT, room temperature.
[a]All stock solutions should be stored in the dark.
[b]Broad emission range; not recommended for green/red double-label studies.
[c]DMSO can be toxic and inhibits partitioning of the dye into the aqueous environment of the cells. If DMSO is the solvent, then use a stock concentration that is at least 100X.

4. Add the following reagents *in order, without mixing*:
   240 µL Polyethylene glycol 3350 (50% w/v stock)
   36 µL 1.0 *M* Lithium acetate
   25 µL Single-stranded calf thymus DNA, 2.0 mg/mL
   50 µL dH_2O and plasmid DNA (1.0 µg minimum)
5. Vortex vigorously and incubate in a water bath at 30°C for 30 min.
6. Heat shock at 42°C for 15–25 min.
7. Concentrate cells by mild centrifugation (30 s, 7000*g*).
8. Resuspend cell pellet gently in 200 µL sterile water.
9. On nonselective medium (yeast extract/peptone/dextrose, YPD), plate 1 µL of the reaction (*see* **Note 10**). On selective medium, plate 10 µL, 50 µL, and the remainder of the reaction.
10. Check targeted FPs for toxicity, localization, and effect on mitochondrial morphology and function (*see* **Note 11**).

3.3. Methods for Modification of Nuclear-Encoded Mitochondrial Genes in Yeast

3.3.1. Primer Design for PCR Amplification of Insertion Cassette

Primers should be 60 or more bases long. The 5′ end of each primer should contain 40–45 bases of perfect homology to the target site. At the 3′ ends of the primers, 18–25 bases are complementary to and enable amplification of sequences encoding the epitope tag and the selectable marker from the plasmid (*see also* **Fig. 2**).

We find that GFP fluorescence can often be optimized by varying the length of the linker region between the target gene and the GFP molecule. For C-terminal tagging, this is accomplished by introducing DNA sequence encoding extra amino acids between the 40 bases of DNA homologous to the target gene and the 20 bases corresponding to the plasmid template in the forward primer. We suggest starting with five alanines.

3.3.2. Amplification of Insertion Cassette From Tagging Vector

1. Amplify the cassette using standard PCR conditions. Eight to ten 100-µL PCR reactions should be pooled to obtain DNA sufficient for a single yeast transformation (1 µg DNA per 1×10^8 cells). Thermocycler conditions: 1 cycle of 94°C (5 min); 30 cycles of 94°C (1 min), 55°C (1 min), 72°C (2–4 min, about 1 min per kilobase product length); 1 cycle of 72°C (10 min).

Reagent	Volume (µL)
PCR buffer (3.3X stock, contains no Mg^{2+})	30
$Mg(OAc)_2$ (25 m*M* stock) (*see* **Note 12**)	8
dNTP mix (2.5 m*M* each dNTP)	8
Forward primer (2×10^{-5} *M* stock) (*see* **Note 13**)	1
Reverse primer (2×10^{-5} *M* stock) (*see* **Note 13**)	1
Vector template (~50 ng/µL stock)	1
r*Tth* DNA polymerase	1
Distilled water	50

2. Purify the PCR product by ethanol precipitation and 1% agarose gel electrophoresis to remove contaminants and undesired products.
3. Cut out the band containing the desired PCR product and isolate and concentrate the product using the Qiaex II gel extraction kit (Qiagen).

3.3.3. Transformation of Yeast With Amplified Insertion Cassette

1. Transform the purified PCR product into the target strain as described in **Subheading 3.2**. For transformations using auxotrophic markers, a single reaction typically yields 10–20 transformed clones for further characterization. For transformations using drug resistance markers (e.g., kanMX6), plate all cells from a single transformation on a nonselective (YPD) plate.
2. Incubate for 2–3 d and replica plate onto a plate containing the drug (e.g., 200 mg/L G418 for kanMX6; *see* **Note 14**). Incubate 2–3 d and replica plate onto a fresh plate containing the drug. Colonies can be picked from this plate for further analysis.

3.3.4. PCR Screening, Protein Expression, and Functional Analysis

1. To validate insertion of the tag into the target locus, isolate genomic DNA from transformants and perform PCR using a pair of primers (20–25 bases) that hybridize on both sides of the insertion. Optional PCR can be carried out using a third primer that anneals to a region within the integration cassette.
2. Cells with correct tag integration at the target locus should be validated using Western blots to analyze protein expression.
3. Finally, tagged constructs should be characterized for functionality given knowledge about the gene of interest. For example, if an inserted tag compromises function of

a nonessential gene, then the tagged strain may show a phenotype resembling that of cells mutated in that gene.

3.3.5. Marker Excision by Cre Recombinase

The tagging vectors were first developed for C-terminal tag insertion. Similar vectors for one-step, N-terminal tagging insert the selection marker and tag into the 5′ end of genes of interest. However, because the selection marker in these cassettes separates the tagged gene from its promoter, the tagged gene is expressed from a nonendogenous promoter included in the tagging cassette. These tagging vectors are used for regulated expression or overexpression studies. Until recently, tagging vectors were not available that could both (1) insert tags in the N-terminus or internal regions of proteins and (2) express tagged proteins from their endogenous promoter (i.e., at wild-type levels).

The pOM family of tagging vectors ***(33)***, which were designed for removal of the selection marker after tag insertion, solved this problem (*see* **Fig. 2B**). In this set of vectors, the selection marker is flanked by loxP sites, and can be removed using bacteriophage Cre recombinase ***(16)***. Insertion and validation of the tagging cassette is carried out as described. Yeast bearing integrated tags are then transformed with a plasmid that encodes Cre recombinase under control of the galactose-inducible promoter.

1. To induce Cre expression, cells are grown in liquid medium selecting for the plasmid, incubated for 4 h in medium containing 1% raffinose plus 1% galactose, and plated on nonselective (YPD) media.
2. Colonies are then screened for loss of the tagging cassette marker by replica plating or patching on selective plates.
3. Further characterization is performed by PCR, Western blot analysis, and visualization, as described in this chapter. We use pSH47 ***(16)***, which expresses Cre from a *GAL* promoter and contains the *URA3* selectable marker. We prefer *URA3* as a marker because it allows curing cells of the plasmid by counterselection on 5-fluoro-orotic acid plates after marker excision.

3.4. Immunofluorescence Methods

3.4.1. Pretreatment of Antibodies With Yeast Cell Walls

Rabbit antisera, even after affinity purification, often contain antibodies that recognize the yeast cell wall, bind to residual cell wall on spheroplasts, and generate background staining, which may be punctate or uniformly distributed over the surface of the spheroplast. We describe a simple method for removing these contaminating antibodies by preadsorbing them to intact yeast cells. A batch of antibody may be pretreated in this way and stored for later use.

1. Grow yeast to late log phase (24–30 h).
2. Dilute antiserum to 1/25 in PBS^+.

3. Remove an aliquot of cells from the late log phase culture: 250 µL of cell culture/milliliter diluted antibody.
4. Concentrate cells by centrifugation (5 s at ~12,500*g*). Resuspend cell pellet in 1 mL PBS$^+$. Repeat this wash procedure three times.
5. Resuspend washed cell pellet in diluted antibody and incubate with gentle mixing at 4°C for 2 h.
6. Concentrate cells by centrifugation and transfer pretreated antibody to a fresh tube.
7. Carry out a second round of pretreatment. Repeat **steps 3** and **4** and resuspend cell pellet in pretreated antibody recovered from **step 6**. Incubate with gentle mixing at 4°C for 2 h.
8. Concentrate cells by centrifugation and transfer pretreated antibody to a fresh tube. Store preadsorbed antibody in aliquots at –20°C.

3.4.2. Fixation and Spheroplast Formation

1. Grow a 5-mL culture (*see* **Note 15**) to mid-log phase (10^6–10^7 cells/mL) (*see* **Note 6**).
2. Add paraformaldehyde (*see* **Note 16**) to the culture medium to a final concentration of 3.7%.
3. Incubate cells with fixative under normal culture conditions for 1 h (*see* **Note 17**).
4. Concentrate cells by centrifugation (5 s at ~12,500*g*) and resuspend cell pellet in 1 mL Tris-HCl/dithiothreitol and incubate for 20 min at 30°C.
5. Concentrate cells by centrifugation. Resuspend cell pellet in 1 mL zymolyase solution and incubate for 1.5 h at 30°C (*see* **Note 18**).
6. Concentrate cells by centrifugation and resuspend cell pellet in 1 mL NS$^+$. Repeat this wash twice and resuspend the final cell pellet in 2 volumes NS$^+$.
7. Store fixed spheroplasts at 4°C for up to 1 wk.

3.4.3. Preparation of a Staining Chamber for Immunofluorescence

For all incubations, the sample-coated cover slip is placed on 20–40 µL of incubation solution on a platform of parafilm sheets in a dark, humid chamber (**Fig. 3**). The staining chamber protects fluorescent dyes from unnecessary exposure to light and the sample from desiccation. To make the staining chamber, a platform consisting of ten 10 × 15 cm sheets of parafilm is placed on a stack (~2 cm) of damp paper towels, and both are covered by an inverted opaque tray or pan. The parafilm platform can be reused if it is washed after use.

3.4.4. Indirect Immunofluorescent Staining: The Coverslip Method

1. Place 20–40 µL of polylysine on the parafilm platform.
2. Using forceps, lay a 22-mm^2 cover slip on the polylysine drop, taking care to avoid creating bubbles at the interface of the cover slip and slide (*see* **Note 19**).
3. After at least 10 s, remove the cover slip from the polylysine. To do so, apply 200–300 µL distilled water under one edge of the cover slip. This will float the cover slip off the parafilm so it can be lifted off the drop with forceps.
4. Rinse by dipping the cover slip in a beaker containing distilled water 5–10 times, draining excess liquid with filter paper. Allow cover slip to air dry.

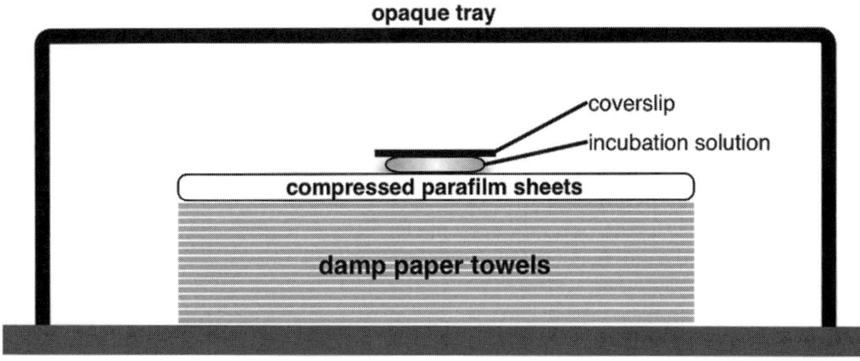

Fig. 3. Chamber for immunofluorescence staining. *See* main text for description.

5. Mix 10 µL of the fixed spheroplast suspension with 100 µL 1X PBS.
6. Place coated cover slip on parafilm platform, coated side facing up. Apply the spheroplast mixture to the coated side of the cover slip and incubate for 30 min at room temperature in the staining chamber.
7. Remove unbound spheroplasts by dipping cover slip in beaker containing PBT 5–10 times. Drain excess liquid with filter paper.
8. Place the 20–40 µL diluted primary antibody in a drop on the parafilm platform. Lay the spheroplast-coated side of the cover slip on the drop of antibody and incubate for 2 h at room temperature in staining chamber.
9. Lift the cover slip by introducing 200–300 µL PBT under the edge of the cover slip with a micropipet. Rinse cover slip in PBT as for **step 7**. Drain excess liquid with filter paper.
10. Place the cover slip on a 20- to 40-µL drop of secondary antibody and incubate 1 h at room temperature in the staining chamber (*see* **Note 20**).
11. Remove cover slip from secondary antibody, rinse, and drain as for **step 9**.
12. Place 1–2 µL mounting solution (*see* **Note 21**) on a microscope slide. Lower the cover slip, sample side down, onto the mounting solution. Dry any residual liquid from the edges of the cover slip. Seal the edges with clear nail polish and let dry.
13. Rinse the cover slip surface with distilled water to remove residual salt deposits from staining and dry gently with a kimwipe or Q-tip. View samples as soon as possible, not more than a week after preparation.

3.5. Visualization of Yeast Mitochondria

3.5.1. Equipment for Wide-Field Deconvolution

Selection of an appropriate microscope and camera to detect fluorescence is essential for imaging mitochondria in living or fixed yeast cells. The small size of yeast cells requires a microscope equipped with a high-magnification objective lens and a high-resolution camera. Phototoxicity and photobleaching can be greatly reduced by image acquisition with a sensitive digital camera that can detect fluorescence with relatively weak, short illumination. Data-processing speed and memory capacity of the computer linked to the digital camera can affect imaging speed and therefore should be tested thoroughly before committing to a particular imaging system.

An imaging system that works well for visualizing mitochondria in yeast is based on an upright epifluorescence microscope (e.g., Zeiss Axioskop 2 or Nikon 80i) equipped with a Plan-Apochromat 100×, 1.4-NA (numerical aperture) objective lens and a cooled CCD camera (e.g., Orca ER, Hamamatsu, or HQ Monochrome, Roper). A shutter driver (Uniblitz D122, Vincent Associates) is synchronized with the camera shutter to control excitation light from the 100-W mercury arc lamp. A software package (e.g., OpenLab, Improvision, Inc., or Metamorph, Universal Imaging) is used to control the camera and shutter, capture images at defined time intervals, and export them to TIFF format for further analysis. Filter wheels (e.g., Sutter or Ludl) or beam splitters (DualView, Optical Insights) are used for automated two-color imaging. We control sample temperature using a heating collar on the objective lens (Bioptechs). Finally, for *z*-axis control, we use stepper motors coupled to the microscope fine focus drive (Ludl) or piezoelectric focus motors on the objective lens (Physik Instrumente).

3.5.2. Imaging Mitochondria in Living Yeast Cells

Time-lapse recordings of mitochondria yield useful data on the velocity and direction of mitochondrial movement if the time interval between images is compatible with the rate of mitochondrial movement. For example, short-term recordings of mitochondrial movements in cells treated with Latrunculin-A and in cells bearing temperature-sensitive mutations in actin or actin-binding proteins uncovered the requirement for an intact actin cytoskeleton for the control of mitochondrial organization and movement ***(4,10,17)***.

3.5.2.1. Short-Term Imaging

Short-term mitochondrial visualization is carried out by adding cells directly to a glass microscope slide and imaging immediately for a period of no longer than 10 min. After 10 min, cells experience a significant decrease in viability, and the use of a growth chamber becomes necessary. A protocol for the preparation of yeast and staining of mitochondria for short-term visualization is described next, followed by a protocol useful for those visualizations requiring longer imaging periods.

1. Grow yeast cells to mid-log phase in liquid medium (*see* **Note 3**). Concentrate to density appropriate for visualization (*see* **Note 22**).
2. Pipet 2.7–3.0 μL of sample onto a microscope slide and cover with a 22-mm^2 cover slip. Avoid creating bubbles between slide and cover slip.
3. View immediately without sealing.

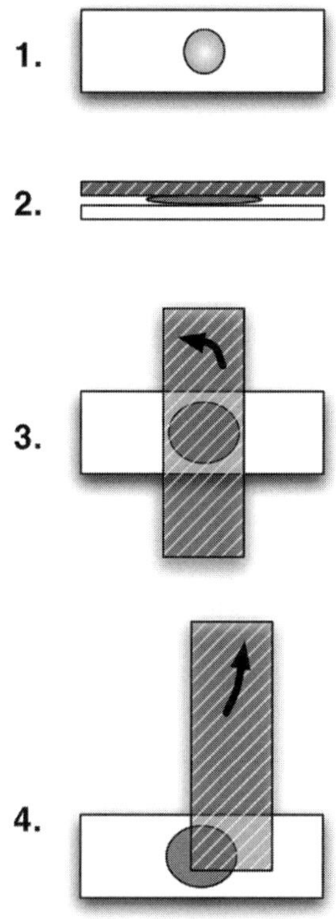

Fig. 4. Preparation of agarose bed for long-term live cell imaging: (**1**) 35 μL of agarose bed material on a microscope slide. (**2**) Side view of agarose sandwiched between two microscope slides. (**3**), (**4**) Rotation and removal of top slide.

3.5.2.2. Long-Term Imaging

The growth chamber described next supports cell growth at wild-type levels for up to 5 h. The growth chamber also immobilizes cells in low-melting agarose, reducing cellular movement that can decrease image resolution. Finally, the chamber has minimal autofluorescence and remains transparent and thin enough for observation with oil immersion lenses. For time-sensitive imaging trials, such as those involving the addition and removal of dyes or drugs, note that it takes roughly 5 min to make an agarose bed (**Fig. 4**).

1. Grow yeast cells to mid-log phase in liquid medium. Concentrate cells to appropriate density for imaging, as needed.
2. Melt an aliquot of agarose bed material in a boiling water bath (~2 min).
3. Pipet 35 μL agarose bed material onto a glass microscope slide. Cool slightly (5 s).
4. Place a second microscope slide on top of agarose, applying light pressure to distribute the agarose bed until it is the size of a standard cover slip.
5. Let bed harden between microscope slides (~2 min).

6. Gently remove top slide by rotating the slide 90° degrees and then sliding it past the bottom slide.
7. Pipet 1.5 µL imaging sample onto surface of the agarose bed (*see* **Note 23**).
8. Cover with a 22-mm^2 cover slip and seal with VALAP. (To seal with VALAP, heat metal spatula in flame, dip into solid VALAP, and drag spatula along edges of cover slip.)

3.5.2.3. Optimizing Imaging Conditions and Preventing Toxicity

During live-cell imaging, photons react with cellular molecules to produce free radicals and reactive oxygen species. Mitochondrial health and cell viability can suffer as a result. We use the following specific criteria to confirm that our imaging conditions are not harmful to mitochondria:

1. Intact mitochondrial structure. Intense or long excitation can cause cleavage or fragmentation.
2. Mitochondrial movements: behavior of photo-damaged mitochondria has not been fully investigated; however, photodamage should be suspected if mitochondria show extremely low velocity of movement in wild-type cells.
3. If potential-sensitive dyes are used, maintain membrane potential (dye retention) throughout the imaging period.

If phototoxicity or photobleaching occurs, then the following techniques (in order of preference) should be tried to reduce harmful light exposure:

1. Reduce excitation intensity (e.g., with a neutral density filter). Generally, a longer exposure time at lower intensity gives a comparable quality image with less photo-damage.
2. Reduce exposure time.
3. Increase time interval or z sectioning interval.

To increase signal-to-noise ratio without increasing light exposure:

1. Apply binning in the camera (if necessary, preserve spatial resolution by adding a projection tube before the camera).
2. Increase gain in the camera.
3. If these techniques fail, then consider trying alternative filter sets with higher throughput: either a broader spectral window or more efficient coating processes.

4. Notes

1. DAPI is a possible carcinogen. It is harmful if inhaled, swallowed, or absorbed through the skin. Wear gloves, face mask, and safety glasses when working with the dry compound.
2. Paraformaldehyde is toxic and carcinogenic. It is absorbed through the skin and is irritating to the eyes, skin, and the respiratory tract. Always wear gloves and safety glasses when working with this fixative. When possible, work in a chemical hood.
3. Some media, such as YPD, autofluoresce. Therefore, we recommend either synthetic complete dropout or lactate medium.

4. Aliquots of 200 μL or less are recommended to avoid multiple rounds of boiling.
5. Extended centrifugation can cause breakage of tubular mitochondria.
6. A culture with OD_{600} = 1.0 contains approx 10^7 cells/mL.
7. The dye concentrations suggested in **Table 6** should be considered only as a starting point. Specificity is highly dependent on concentration and mitochondrial membrane potential. Titration of dye concentration and incubation times should be performed when working with a new strain or mitochondrial mutant. In general, dye concentration should be kept as low as possible because excessive accumulation of dye in mitochondria results in organelle swelling and respiratory defects.
8. Washing cells is generally not necessary as the fluorescence of most vital dyes is low until the dye binds to cell lipids. If high background is observed, then cells can be washed once in medium.
9. The lithium acetate wash should be as brief as possible because this reagent is toxic to cells.
10. When plating less than 100 μL, add 50–100 μL sterile water to facilitate spreading.
11. Mislocalization of FP fusion markers may occur because targeting information is masked or insertion of the fusion protein into membranes is inhibited. For example, fusion proteins consisting of GFP fused to the C terminus of *QRC1* (subunit 6 of the cytochrome-bc_1 complex) or to porin localize to the cytosol ***(18)***. Localization of FP fusion proteins may be assessed by visual inspection of the FP in cells in which mitochondria are counterstained using vital dyes or by biochemical methods for subcellular and submitochondrial fractionation. Possible effects of fusion proteins on mitochondrial morphology or respiratory activity can be assessed by visual inspection of mitochondria in fixed or living cells and by analysis of cell growth and growth rates on nonfermentable carbon sources.
12. For different plasmid templates, we suggest first titrating the $Mg(OAc)_2$ concentration; for example, try four reactions with final $Mg(OAc)_2$ concentrations of 0.5 m*M*, 1.0 m*M*, 2.0 m*M*, and 3.0 m*M*.
13. To assure PCR specificity, for new primers we suggest three control reactions: no primers in the reaction, forward primer only, and reverse primer only. These can be omitted in future PCR runs if nonspecific reactions do not occur.
14. Synthetic medium containing ammonium sulfate interferes with G418 activity. To make dropout plates containing G418, use SE (synthetic media with glutamate) rather than synthetic complete medium.
15. Although it is sometimes possible to perform immunofluorescent staining on cells picked from colonies on solid media, internal structures, including mitochondria, of mid-log phase cells from liquid culture are better preserved and more reproducible. The choice of carbon source for growth affects the abundance and morphology of mitochondria ***(19–21)***. Traditional glucose-based medium is not typically used because glucose represses mitochondrial biogenesis and consequently makes the organelles more difficult to visualize. Lactate medium selects for cells with mtDNA and mitochondrial metabolic potential; it is also inexpensive and easy to prepare. For strains that cannot grow on nonfermentable carbon sources, we use a raffinose-based medium. Another consideration in choice of medium is autofluorescence. Lactate

and raffinose media as described produce little general autofluorescence. However, in any carbon source, the red material that accumulates in vacuoles of *ade2*⁻ cells is fluorescent and can interfere with microscopy in the red or green emission channels. Addition of extra adenine (two to five times the amount in normal medium) can prevent this problem.

16. There are two main classes of fixatives: crosslinkers and precipitants. The protocol described here uses paraformaldehyde, a crosslinking fixative that forms hydroxymethylene bridges between spatially adjacent amino acid residues. Methanol and acetone, a commonly used precipitating fixative mixture, is suitable for some immunofluorescence staining, including staining of the actin cytoskeleton. However, we find it to be a poor choice for mitochondria. It solubilizes many membranes and can cause extraction of antigens from mitochondria ***(22)***.
17. The fixation conditions can affect the quality of the immunofluorescence results. Common manipulations, including centrifugation and increasing or decreasing temperature, can alter internal structures in *S. cerevisiae* ***(23)***. Therefore, we add the fixative directly to a liquid culture under growth conditions to minimize such disruptions. The other important variables in paraformaldehyde fixation are the concentration of fixative, duration of fixation, and pH of the fixative solution. High paraformaldehyde concentrations, low pH, and a long fixation period increase the number of crosslinks formed and thereby improve structural preservation. Excessive crosslinking, however, can decrease antibody binding to antigens of interest and result in crosslinking of fluorescent substances in the medium to cells, which increases background fluorescence surface. The optimal protocol strikes a balance between structural preservation and antigen accessibility.
18. The optimum time for cell wall removal varies among strains and can be determined by visual inspection using phase-contrast or bright-field microscopy. Intact cells have prominently refractile edges, and spheroplasts, which do not have cell walls, are less refractile. Excessive zymolyase treatment can damage cellular structures.
19. If desired, the cover slip may be labeled with a fine-point marker in one corner of the uncoated side.
20. Detection of two antigens simultaneously can be achieved by combining the two primary antibodies in the first step and the two secondary antibodies in the second step. For such an experiment, the primary antibodies must be raised in different species, and the fluorophores used for detection must have sufficiently separated excitation and emission spectra.
21. The mounting solution used here consists of glycerol, which reduces spheroplast movement during imaging and prevents freezing of the samples during storage at −20°C. It also contains *p*-phenylenediamine, an antiphotobleaching agent, which helps to prevent destruction of fluorophores by oxygen radicals generated during illumination.
22. The volume used is important because excess volume can cause cells to float, and too little volume can affect cell structure.
23. If cells are not immobilized, then decrease the volume of sample added to the agarose bed. In addition, the water content of the agarose bed can be adjusted by letting the solidified bed stand, uncovered, for a few seconds before adding sample.

References

1. Koehler, C. M. (2004) New developments in mitochondrial assembly. *Annu. Rev. Cell. Dev. Biol.* **20,** 309–335.
2. Gietz, R. D., Schiestl, R. H., Willems, A. R., and Woods, R. A. (1995) Studies on the transformation of intact yeast cells by the LiAc/SS-DNA/PEG procedure. *Yeast* **11,** 355–360.
3. Riezman, H., Hase, T., van Loon, A. P., Grivell, L. A., Suda, K., and Schatz, G. (1983) Import of proteins into mitochondria: a 70-kDa outer membrane protein with a large carboxy-terminal deletion is still transported to the outer membrane. *EMBO J.* **2,** 2161–2168.
4. Smith, M. G., Simon, V. R., O'Sullivan, H., and Pon, L. A. (1995) Organelle-cytoskeletal interactions: actin mutations inhibit meiosis-dependent mitochondrial rearrangement in the budding yeast *Saccharomyces cerevisiae. Mol. Biol. Cell.* **6,** 1381–1396.
5. Fehrenbacher, K. F., Boldogh, I. R., and Pon, L. A. (2005) A role for Jsn1p in recruiting the Arp2/3 complex to mitochondria in budding yeast. *Mol. Biol. Cell.* **16,** 5094–5102.
6. Boldogh, I. R., Ramcharan., S. L., Yang, H.-C., and Pon, L. A. (2004) A type V myosin (Myo2p) and a Rab-like protein (Ypt11p) are required for retention of newly inherited mitochondria in yeast cells during cell division. *Mol. Biol. Cell.* **15,** 3994–4002.
7. Yang, H.-C., Palazzo, A., Swayne, T. C., and Pon, L. A. (1999) A retention mechanism for distribution of mitochondria during cell division in budding yeast. *Curr. Biol.* **9,** 1111–1114.
8. Boldogh, I. R., Nowakowski, D. W., Yang, H-C., et al. (2003) A protein complex containing Mdm10p, Mdm12p, and Mmm1p links mitochondrial membranes and DNA to the cytoskeleton-based segregation machinery. *Mol. Biol. Cell.* **14,** 4618–4627.
9. Hobbs, A. E., Srinivasan, M., McCaffery, J. M., and Jensen, R. E. (2001) Mmm1p, a mitochondrial outer membrane protein, is connected to mitochondrial DNA (mtDNA) nucleoids and required for mtDNA stability. *J. Cell. Biol.* **152,** 401–410.
10. Simon, V. R., Swayne, T. C., and Pon, L. A. (1995) Actin-dependent mitochondrial motility in mitotic yeast and cell-free systems: identification of a motor activity on the mitochondrial surface. *J. Cell Biol.* **130,** 345–354.
11. Nunnari, J., Marshall, W. F., Straight, A., Murray, A., Sedat, J. W., and Walter, P. (1997) Mitochondrial transmission during mating in *Saccharomyces cerevisiae* is determined by mitochondrial fusion and fission and the intramitochondrial segregation of mitochondrial DNA. *Mol. Biol. Cell.* **8,** 1233–1242.
12. Azpiroz, R. and Butow, R. A. (1993) Patterns of mitochondrial sorting in yeast zygotes. *Mol. Biol. Cell.* **4,** 21–36.
13. Wallace, W., Schaefer, L. H., and Swedlow, J. R. (2001) A workingperson's guide to deconvolution in light microscopy. *Biotechnology* **31,** 1076–1078.
14. Inoué, S. and Inoué, T. (2002) Direct-view high-speed confocal scanner: the CSU-10. *Meth. Cell Biol.* **70,** 87–127.
15. Lippincott-Schwartz, J., Altan-Bonnet, N., and Patterson, G. H. (2003) Photobleaching and photoactivation: following protein dynamics in living cells. *Nat. Cell Biol.* **5,** S7–S14.

16. Gueldener, U., Heinisch, J., Koehler, G. J., Voss, D., and Hegemann, J. H. (2002) A second set of loxP marker cassettes for Cre-mediated multiple gene knockouts in budding yeast. *Nucleic Acids Res.* **30,** e23.
17. Boldogh, I., Vojtov, N., Karmon, S., and Pon, L. A. (1998) Interaction between mitochondria and the actin cytoskeleton in budding yeast requires two integral mitochondrial outer membrane proteins, Mmm1p and Mdm10p. *J. Cell Biol.* **141,** 1371–1381.
18. Okamoto, K., Perlman, P. S., and Butow, R. A. (2001) Targeting of green fluorescent protein to mitochondria. *Meth. Cell Biol.* **65,** 277–283.
19. Visser, W., van Spronsen, E. A., Nanninga, N., Pronk, J. T., Kuenen, J. G., and van Dijken, J. P. (1995) Effects of growth conditions on mitochondrial morphology in *Saccharomyces cerevisiae. Antonie van Leeuwenhoek* **67,** 243–253.
20. Stevens, B. (1977) Variation in number and volume of the mitochondria in yeast according to growth conditions. A study based on serial sectioning and computer graphics reconstitution. *Biol. Cell.* **28,** 37–56.
21. Damsky, C. H. (1976) Environmentally induced changes in mitochondria and endoplasmic reticulum of *Saccharomyces carlsbergensis* yeast. *J. Cell Biol.* **71,** 123–135.
22. Boldogh, I. R., Yang, H-C., Nowakowski, W. D., et al. (2001) Arp 2/3 complex and actin dynamics are required for actin-based mitochondrial motility in yeast. *Proc. Natl. Acad. Sci. U. S. A.* **98,** 3162–3167.
23. Lillie, S. H., and Brown, S. S. (1994) Immunofluorescence localization of the unconventional myosin, Myo2p, and the putative kinesin-related protein, Smy1p, to the same regions of polarized growth in *Saccharomyces cerevisiae. J. Cell Biol.* **125,** 825–842.
24. Fehrenbacher, K. L., Yang, H.-C., Gay, A. C., Huckaba, T. M., and Pon, L. A. (2004) Live cell imaging of mitochondrial movement along actin cables in budding yeast. *Curr. Biol.* **14,** 1996–2004.
25. Okamoto, K., Perlman, P. S., and Butow, R. A. (1998) The sorting of mitochondrial DNA and mitochondrial proteins in zygotes: preferential transmission of mitochondrial DNA to the medial bud. *J. Cell Biol.* **142,** 613–623.
26. Mozdy, A. D., McCaffery, J. M., and Shaw, J. M. (2001) Dnm1p GTPase-mediated mitochondrial fission is a multi-step process requiring the novel integral membrane component Fis1p. *J. Cell Biol.* **151,** 367–380.
27. Nunnari, J., Wong, E. D., Meeusen, S., and Wagner, J. A. (2002) Studying the behavior of mitochondria. *Methods Enzymol.* **351,** 381–393.
28. Westermann, B. and Neupert, W. (2000) Mitochondria-targeted green fluorescent proteins: convenient tools for the study of organelle biogenesis in *Saccharomyces cerevisiae. Yeast* **16,** 1421–1427.
29. Longtine, M. S., McKenzie, A., III, Demarini, D. J., et al. (1998) Additional modules for versatile and economical PCR-based gene deletion and modification in *Saccharomyces cerevisiae. Yeast* **14,** 953–961.
30. Rodrigues, F., van Hemert, M., Steensma, H. Y., Corte-Real, M., and Leao, C. (2001) Red fluorescent protein (DsRed) as a reporter in *Saccharomyces cerevisiae. J. Bacteriol.* **183,** 3791–3794.

31. Janke, C., Magiera, M. M., Rathfelder, N., et al. (2004) A versatile toolbox for PCR-based tagging of yeast genes: new fluorescent proteins, more markers and promoter substitution cassettes. *Yeast* **21,** 947–962.
32. Sheff, M. A. and Thorn, K. S. (2004) Optimized cassettes for fluorescent protein tagging in *Saccharomyces cerevisiae*. *Yeast* **21,** 661–670.
33. Gauss, R., Trautwein, M., Sommer, T., and Spang, A. (2005) New modules for the repeated internal and N-terminal epitope tagging of genes in *Saccharomyces cerevisiae*. *Yeast* **22,** 1–12.
34. Roeder, A. D., Hermann., G. J., Keegan, B. R., Thatcher, S. A., and Shaw, J. M. (1998) Mitochondrial inheritance is delayed in *Saccharomyces cerevisiae* cells lacking the serine/threonine phosphatase PTC1. *Mol. Biol. Cell.* **9,** 917–930.
35. Mihara, K. and Sato, R. (1985) Molecular cloning and sequencing of cDNA for yeast porin, an outer mitochondrial membrane protein: a search for targeting signal in the primary structure. *EMBO J.* **4,** 769–774.
36. Taanman, J. W. and Capaldi, R. A. (1993) Subunit VIa of yeast cytochrome *c* oxidase is not necessary for assembly of the enzyme complex but modulates the enzyme activity. Isolation and characterization of the nuclear-coded gene. *J. Biol. Chem.* **268,** 18,754–18,761.
37. Poot, M., Zhang, Y. Z., Kramer, J. A., et al. (1996) Analysis of mitochondrial morphology and function with novel fixable fluorescent stains. *J. Histochem. Cytochem.* **44,** 1363–1372.
38. McConnell, S. J., Stewart, L. C., Talin, A., and Yaffe, M. P. (1990) Temperature-sensitive yeast mutants defective in mitochondrial inheritance. *J. Cell Biol.* **3,** 967–976.
39. Diffley, J. F. and Stillman, B. (1991) A close relative of the nuclear, chromosomal high-mobility group protein HMG1 in yeast mitochondria. *Proc. Natl. Acad. Sci. U. S. A.* **88,** 7864–7868.
40. Skowronek, P., Krummeck, G., Haferkamp, O., and Rodel, G. (1990) Flow cytometry as a tool to discriminate respiratory-competent and respiratory-deficient yeast cells. *Curr. Genet.* **18,** 265–267.

32

Mitochondrial Fusion In Vitro

Shelly L. Meeusen and Jodi Nunnari

Summary

The field of mitochondrial dynamics has received a great deal of attention as a result of a number of studies linking mitochondrial fission and fusion machinery to apoptosis. Specifically, elevated levels of mitochondrial fission or compromised mitochondrial fusion can sensitize cells to apoptotic stimuli. Conversely, stimulation of mitochondrial fusion can render cells resistant to apoptotic stimuli. In addition, the machinery involved in fission and fusion has been spatially linked to Bax, a pro-apoptotic protein. However, the mechanistic implications of interactions between the machinery of mitochondrial fission and fusion and apoptotic effectors are largely unknown. Our understanding of the pathways of mitochondrial fission and fusion have come from genetic studies coupled with direct observation of both fission and fusion components and mitochondrial organelle morphology and behavior in vivo in *Saccharomyces cerevisiae*. These approaches have identified the key players in both mitochondrial fission and fusion and have generated good models for their roles in mitochondrial dynamics. However, the lack of in vitro systems for studying these processes has impeded a deeper investigation of the mechanism. We have recapitulated the process of mitochondrial fusion in vitro ***(5)***. Using this in vitro fusion assay, we have separated outer mitochondrial membrane fusion from inner and identified the mechanistic requirements for each step.

Key Words: Assay; dynamics; fusion; in vitro.

1. Introduction

Defects in mitochondrial fusion have recently been linked to the onset of neurodegenerative diseases. Mutations in Mfn2 (mammalian ortholog of yeast Fzo1) and Opa1 (mammalian ortholog of yeast Mgm1) are associated with the neurodegenerative disorders Charcot-Marie-Tooth neuropathy and dominant optic atrophy, respectively ***(1–3)***. Understanding the roles of these proteins in mitochondrial fusion will enhance our understanding of the fusion mechanism as well as aid in the development of potential therapeutic agents.

From: *Methods in Molecular Biology, vol. 372: Mitochondria: Practical Protocols*
Edited by: D. Leister and J. M. Herrmann © Humana Press Inc., Totowa, NJ

The mechanism of mitochondrial fusion is unique. No paradigm exists for fusion of a double-membrane system. In addition, heterotypic fusion between mitochondria and other membranes has not been observed, and proteins required for secretory vesicle fusion are not required. Components known to be required for mitochondrial fusion are also unique, making it likely that the mechanism underlying fusion is distinct and evolved independently. The recent development of a fluorescent-based in vitro fusion assay has allowed us to dissect the unique process of mitochondrial fusion. The recapitulation of mitochondrial fusion in vitro requires that mitochondria be brought into contact for a critical duration of time (by centrifugation and incubation) and resuspended in a reaction buffer containing exogenous guanosine 5′-triphosphate (GTP) and an energy regeneration system ***(4)***. Cytosol is not required. By varying energetic conditions, we were able to resolve outer mitochondrial membrane fusion from inner.

Fusion of outer and inner mitochondrial membranes can be directly observed by using confocal microscopy or wide-field deconvolution microscopy with fluorescent proteins targeted to specific mitochondrial compartments or membranes. Specifically, outer membrane fusion is analyzed by mixing mitochondria isolated from cells expressing an outer membrane-targeted green fluorescent protein (om-GFP) and a matrix-targeted DsRed (m-DsRed) with mitochondria isolated from cells expressing a matrix-targeted blue fluorescent protein (m-BFP). Fusion of outer and inner membranes results in a single continuous green outer membrane encircling a single colocalized blue and red matrix compartment. In the event that outer membrane fusion occurs but inner membrane fusion is blocked, two tightly juxtaposed distinct red and blue matrix compartments encircled by a single continuous green outer membrane is observed. Because of the low quantum fluorescence yield and sensitivity to photobleaching of m-BFP, detection can be difficult with CCD (charge-coupled device) cameras on most conventional microscopes, especially after it is diluted by matrix content mixing. To address this experimental problem, inner membrane fusion can be assessed by using mitochondria from cells expressing m-DsRed with mitochondria from cells expressing matrix-targeted GFP (m-GFP). On inner membrane fusion, matrix contents will mix, and one single matrix compartment will contain colocalized red and green fluorophores. The ability to dissect and directly observe the process of mitochondrial fusion in vitro will continue to contribute to our mechanistic understanding of this complex and unique process.

2. Materials

1. The m-GFP used was pVT100UGFP, a gift from Dr. B. Westermann, University of Bayreuth, Germany.
2. The m-DsRed was created by replacing GFP in pVT100U-mtGFP with DsRed2, a gift from Dr. B. Glick, University of Chicago, Chicago, Illinois.

3. The om-GFP used was pOM45-GFP, a gift from Dr. R. Jensen, Johns Hopkins University, Baltimore, Maryland.
4. The m-BFP used was pYES-mtBFP, a gift from Dr. B. Westermann, University of Bayreuth, Germany.

2.1. Cell Culture and Isolation of Fusogenic Mitochondria

1. Starter cultures: complete minimal dextrose or complete minimal galactose media lacking uracil, leucine, or uracil and leucine is used to maintain episomal plasmids.
2. Tris-β-me solution: 0.1 *M* Tris-HCl, pH 9.4, 50 m*M* β-mercaptoethanol.
3. Spheroplasting solution: dissolve 3 mg/mL (ICN Biomedical) *Arthrobacter luteus* yeast lytic enzyme into 1.2 *M* sorbitol immediately prior to use.
4. NMIB: 0.6 *M* sorbitol, 5 m*M* $?gCl_2$, 50 m*M* KCl, 100 m*M* KOAc, 20 m*M* HEPES-KOH, pH 7.4 (no serine protease inhibitors).
5. Dounces (Wheaton) must be labeled "tight" for efficient homogenization.

2.2. Fusion In Vitro

1. Fusion reaction buffer: 0.6 *M* sorbitol, 20 m*M* PIPES-KOH, pH 6.8, 150 m*M* KOAc, 5 m*M* $Mg(OAc)_2$.
2. Creatine phosphokinase (Sigma): *This enzyme is unstable*, and its activity is critical to support mitochondrial fusion. On opening a new bottle of lyophilized enzyme, make dry aliquots containing 2–4 mg per tube and desiccate. Resuspend in reaction buffer immediately prior to use. Discard unused resuspended enzyme.
3. Creatine phosphate (Sigma): 1 *M* creatine phosphate can be resuspended in reaction buffer, aliquoted, and stored at –80°C.
4. Nucleotide stocks: 0.5 *M* GTP and 1 m*M* adenosine triphosphate (Sigma) can be resuspended in reaction buffer, aliquoted, and stored at –80°C. The pH of the GTP stock may need to be adjusted to 7.0.

2.3. Analysis of Fusion In Vitro

1. Fixative: 8% (w/v) paraformaldehyde in phosphate-buffered saline, pH 7.4.
2. Mitochondria are viewed with an Olympus IX70 Deltavision Microscope using a 60× objective (Olympus) and a 100-W mercury lamp (Applied Precision, Inc.). Two- and three-dimensional light microscopy data are collected using an integrated, cooled CCD-based Princeton Micromax camera equipped with a Sony Interline chip. Other microscopes, such as laser scanning or spinning disk confocals, may also be used to assess fusion.

3. Methods

The reconstitution of efficient mitochondrial fusion in vitro is sensitive to a number of different parameters. This subheading explains in detail the protocols used for (1) isolating fusogenic mitochondria, (2) conducting fusion in vitro, and (3) analyzing completed fusion reactions. For electron microscopic analysis of mitochondrial fusion reactions, please see Chapter 33.

3.1. Cell Culture and Isolation of Fusogenic Mitochondria

1. At 2 d prior to the in vitro assay, 2- to 100-mL starter cultures of oppositely labeled *S. cerevisiae* W303 cells are inoculated into synthetic minimal complete media (*see* **Note 1**). A day later, two 2-L Erlenmeyer flasks containing 1 L rich yeast extract/peptone/glycerol medium is inoculated at OD_{600} 0.1 and placed on a rotary shaker at 250 rpm to reach log phase (*see* **Note 2**).
2. Cells between OD_{600} 0.6 and OD_{600} 1.0 are harvested by centrifugation at 1500*g* for 5 min and washed once in distilled H_2O (*see* **Note 3**).
3. Cells are then resuspended at 20 OD_{600}/mL in Tris-β-me buffer and incubated at 30°C for 20 min to compromise cell walls (*see* **Note 4**).
4. Cells are collected by centrifugation at 1500*g* for 5 min, washed in 100 mL 1.2 *M* sorbitol, and collected as in **step 2**.
5. Cells are then resuspended at 50 OD_{600}/mL in 1.2 *M* sorbitol containing 3 mg/mL yeast lytic enzyme and incubated at less than 100 rpm and 30°C for 30 min to generate spheroplasts.
6. Spheroplasts are then collected by centrifugation at 1500*g* for 5 min, washed in 100 mL 1.2 *M* sorbitol, and collected as in **step 2**.
7. All residual 1.2 *M* sorbitol is removed from cell pellets using a P1000 micropipetman, and the spheroplast pellet is then resuspended in a minimal volume of ice-cold NMIB (*see* **Note 5**). (For example, for 1000 OD_{600}, 4 mL NMIB is sufficient.)
8. Spheroplast resuspensions are immediately transferred to chilled tight dounces that are prewashed with cold NMIB. Spheroplast resuspensions are then dounced a minimum of 100 times until cell lysis approximates 60% by phase microscopy.
9. Extracts are then diluted threefold in cold NMIB and centrifuged at 3000*g* for 5 min at 4°C to remove unlysed cells and large cellular debris.
10. Supernatant fraction is then centrifuged at 10,170*g* for 10 min at 4°C to yield a mitochondrial enriched pellet.
11. Mitochondrial pellets are then immediately resuspended in a minimal volume of cold NMIB to a final protein concentration of 5–10 mg/mL. (For example, mitochondrial pellets from 1000 OD_{600} of cells should be resuspended in 0.25–0.3 mL and quantified using a Bradford assay.)

3.2. Fusion In Vitro

1. Equal amounts of freshly prepared (*see* **Note 6**), oppositely labeled mitochondria are thoroughly mixed by pipeting (total mitochondrial protein should equal 0.5 mg/reaction).
2. Mixed mitochondria are then centrifuged at 10,170*g* or 10 min at 4°C to concentrate mitochondria.
3. Centrifuged reactions are then incubated at 25°C on ice for 10 min to promote associations between outer mitochondrial membranes.
4. Supernates from centrifuged reactions are removed by aspiration.
5. Mitochondrial pellets are immediately resuspended in 40 μL of either stage 1 reaction mix [20 m*M* PIPES-KOH, pH 6.8, 150 m*M* KOAc, 5 m*M* $Mg(OAC)_2$, 0.6 *M* sorbitol] or stage 2 reaction mix [20 m*M* PIPES-KOH, pH 6.8, 150 m*M*

KOAc, 5 m*M* $Mg(OAC)_2$, 0.6 *M* sorbitol, 0.2 mg/mL creatine phosphokinase, 40 m*M* creatine phosphate, 1 m*M* adenosine triphosphate, 0.5 m*M* GTP] (*see* **Note 7**).

6. Standard fusion reactions are carried out at 22°C for up to 1 h.

3.3. Analysis of Fusion In Vitro

1. Aliquots of stage 1 and 2 reactions are fixed by resuspension in 2 volumes of 8% paraformaldehyde in phosphate-buffered saline, pH 7.4, for a minimum of 20 min.
2. Aliquots of fixed reactions are immobilized on microscope slides by mixing with an equal volume of 4% low melting point agarose in NMIB.
3. Three-dimensional data should be collected to distinguish between outer and inner mitochondrial membrane fusion events. Given that the average diameter of isolated mitochondria is 0.5 μm, a total distance of 3 μm in the z direction is sufficient to analyze fusion intermediates and fused structures (*see* **Note 8**).
4. Fusion efficiency is quantified by dividing the number of fused mitochondria (mitochondria with colocalized fluorophores) by the total number of mitochondria in a given field.

4. Notes

1. Cells expressing m-BFP are maintained on minimal complete galactose plates lacking uracil. Starter culture for this strain is grown in synthetic minimal galactose medium lacking uracil to induce expression of m-BFP.
2. It is critical that the cultures are shaken vigorously in at least a 2-L flask to allow sufficient aeration; respiratory competence of cells affects the ability of their mitochondria to fuse. In addition, we have observed that less aeration impedes growth rates contributing to thicker cell walls and ultimately less cell lysis diminishing yield from the mitochondrial preparation.
3. From this step until isolated mitochondria are placed in reaction buffer, all steps must be conducted as quickly as possible; ability to fuse is compromised by excessive preparation time (the entire preparation should not exceed 4 h).
4. If isolating mitochondria from temperature-sensitive mutants, then it may be important to conduct all steps described at 30°C at lower temperatures instead to prevent manifestations of mutant phenotypes prior to the in vitro assay. When lowering the temperature of a given step, it is important to extend the incubation time.
5. NMIB should not contain any serine protease inhibitors; these and other protease inhibitors negatively affect fusion. Protease inhibitors should be omitted, but *all* steps between cell lysis and fusion must be conducted on ice to minimize protease activity.
6. Mitochondria must be freshly prepared immediately prior to resuspension in fusion reactions.
7. Stage 1 reaction mix supports only outer membrane fusion. Stage 2 reaction mix supports fusion of both membranes.
8. When collecting images, it is best to focus on areas of the slide where mitochondria are a fairly dispersed monolayer. Imaging highly concentrated mitochondria can interfere with quantification of outer membrane tethering and other early steps of fusion.

References

1. Alexander, C., Votruba, M., Pesch, U. E., et al. (2000) OPA1, encoding a dynamin-related GTPase, is mutated in autosomal dominant optic atrophy linked to chromosome 3q28. *Nat. Genet.* **26,** 211–215.
2. Delettre, C. L., Lenaers, G., Griffoin, J. M., et al. (2000) Nuclear gene OPA1, encoding a mitochondrial dynamin-related protein, is mutated in dominant optic atrophy. *Nat. Genet.* **26,** 207–210.
3. Zuchner, S., Mersiyanova, I. V., Muglia, M., et al. (2004) Mutations in the mitochondrial GTPase mitofusin 2 cause Charcot-Marie-Tooth neuropathy type 2A. *Nat. Genet.* **36,** 449–451.
4. Meeusen, S., McCaffery, M. J., and Nunnari, J. (2004) Mitochondrial fusion intermediates revealed in vitro. *Science* **305,** 1747–1752.

33

Conventional and Immunoelectron Microscopy of Mitochondria

Edward M. Perkins and J. Michael McCaffery

Summary

Electron microscopy (EM) has been a central tool in delineating the subcellular organization and function of the eukaryotic cell. It has provided valuable information on the organization of the Golgi complex; the polarized distribution of proteins on the plasma membrane; and fundamental insights into the essential structure and function of mitochondria beginning with the first EM observations of Claude and Fullam on isolated mitochondria in 1944. Most significant for this volume is the contribution immunoelectron microscopy (IEM) has made in the study of mitochondrial dynamics and in demonstrating the localizations of key mitochondrial proteins in yeast, including, though not limited to, Dnm1p, Fiz1p, and Mgm1p. This chapter is not intended to provide a comprehensive review of all EM and IEM methods as there are a number of excellent books and reviews already available on these topics. Rather, this chapter provides detailed protocols of conventional EM and IEM methods successfully utilized in our center for the examination and analysis of mitochondria in yeast and mammalian cells.

Key Words: Electron microscopy; immunoelectron microscopy; immunocytochemistry; immunogold labeling.

1. Introduction

Many cell biologists have come to rely solely on fluorescence microscopy for the subcellular localization/analysis of proteins *(1,2)*. This has been largely because of the ever-increasing use of confocal microscopy coupled with advances in vital dye design. In addition, the discovery of green fluorescent protein (GFP) has led to the development of an abundance of genetically encoded fluorescent proteins with which to probe the "real-time" dynamics of living cells/organisms at the light microscopy level *(3,4)*. Although remarkable, these advances in fluorescence microscopy have furthered a disturbing trend in

From: *Methods in Molecular Biology, vol. 372: Mitochondria: Practical Protocols*
Edited by: D. Leister and J. M. Herrmann © Humana Press Inc., Totowa, NJ

the declining use of electron microscopy (EM) and a lack of appreciation for its unique capability to complement, extend, and validate the results obtained by traditional immunofluorescence microscopy, biochemical (cell fractionation), immunochemical (immunoprecipitation, immunoblotting), and molecular biological (Northern blotting) methods. For many researchers, unfortunately, it has become commonplace to utilize confocal microscopy exclusively as a substitute for immunoelectron microscopy (IEM) in cell biology research in spite of the fact that the confocal microscope, with a spatial resolution greater than 500 nm and an axial resolution approx three to five times greater (under even the most optimal conditions for biological imaging), is incapable of reliably discriminating most subcellular structures, including mitochondria.

For decades, EM has been a central tool used in establishing our current concepts of cell structure and function. It first established unambiguously the role of endosomes in receptor and ligand traffic ***(5–8)***; it has provided valuable information on the organization of the Golgi complex ***(9–12)***; and it conclusively demonstrated the polarized distribution of proteins on the plasma membrane ***(13,14)***. Most important for this volume, EM has revealed essential insights into the fundamental structure/organization of the mitochondrion, proceeding from the first EM observations of isolated mitochondria by Albert Claude and Ernest Fullam in 1944, to the first ultrastructural analysis of the mitochondrion by George Palade in 1953, through the study by Koshiba et al. in 2004, which characterized the structural basis of tethering in mitochondrial fusion ***(15–17)***.

Moreover, the results obtained by IEM have provided key insights into the localization and function of many proteins implicated in mitochondrial structure/dynamics. For example, IEM first revealed the definitive association of Dnm1p with mitochondrial constriction sites and its role in the fission process ***(18)***. Still later, IEM demonstrated the broadly dispersed distribution of Fiz1p around the mitochondrial outer membrane, which is necessary for proper Dnm1p assembly, membrane distribution, and function during fission ***(19)***. IEM was used by Meeusen et al. in 2004 to delineate the multistep process of mitochondrial fusion by assaying for the distribution of DsRed and GFP at the EM level in an isolated mitochondrial-matrix mixing assay ***(20)*** (*see also* Chapter 32). This work monitored, in a time-dependent manner utilizing double indirect immunogold labeling of ultrathin cryosections, the step-by-step redistribution of these fluorescent proteins from initially segregated populations of mitochondria at $T_{0'}$ to a progressively mixed/homogeneously labeled population at $T_{+30'}$.

Consequently, although fluorescence microscopy has proven invaluable for live cell experiments, immunoreagent screening, and the crucial sampling of broad cell populations required to identify and establish recognizable phenotypes, any comprehensive study of mitochondria should necessarily include complementary morphologic data obtained by conventional EM or immunocytochemical localizations whenever possible. This chapter provides detailed conventional EM and IEM methods successfully utilized in our center for the in vivo and in vitro examination/analysis of mitochondria in yeast and mammalian cells.

2. Materials

2.1. Conventional EM of Mammalian Cells in Culture

1. Glutaraldehyde stock, 10% ampules, EM grade (Electron Microscopy Sciences [EMS], Ft. Washington, PA, USA, cat. no. 16120). Sealed ampules prepared in double-distilled water (ddH_2O); store under inert N_2 at 4°C.
2. 4% Osmium tetroxide (OsO_4) stock: under a fume hood in an acid-cleaned/light-tight reagent bottle, place 2 g crystalline OsO_4 (EMS, cat. no. 19110) in 50 mL ddH_2O; sonicate in a water bath 10–15 min or until dissolved. Do not allow solution to warm during sonication; store at 4°C.
3. Na cacodylate stock, 0.2 *M*: sodium cacodylate trihydrate (EMS, cat. no. 12300). Dissolve 21.4 g in 400 mL ddH_2O; bring to pH 7.4 with 1.0 *N* HCl; bring volume to 500 mL with double-distilled water; store at 4°C.
4. Thiocarbohydrazide powder (MW 106.15; EMS, cat. no. 21900). Utilized in osmium-thiocarbohydrazide-osmium (OTO) (*see* **Subheading 3.2.2.**).
5. Veronal-acetate stock (store at 4°C): add 1.15 g anhydrous sodium acetate (Sigma, cat. no. S-2289) and 2.943 g sodium barbiturate (Sigma, cat. no. B-0500) to 100 mL ddH_2O and dissolve. Store at 4°C.
6. Kellenberger's uranyl acetate (UA; store in the dark): to 20 mL veronal-acetate stock, add 51 mL ddH_2O, 28 mL 0.1 *N* HCl, and 0.5g UA. Sonicate to dissolve and store at room temperature in brown bottle.
7. 2% UA: mix 1 g UA (EMS, cat. no. 22400) in 50 mL ddH_2O; sonicate to dissolve; store in light-tight 60-mL syringe, and filter with 0.22-μm filter prior to use.
8. Lead citrate: dissolve 100 mg lead citrate (EMS, cat. no. 17800) in 5 mL 1 *N* NaOH; bring volume to 50 mL in freshly degassed double-distilled water; store in light-tight/air-tight 60-mL syringe and filter with 0.22-μm filter prior to use.
9. Palade's 1% OsO_4: To prepare 5 mL, mix 1 mL acetate-veronal stock, 1.25 mL 4% OsO_4 stock, 1 mL 0.1 *N* HCl, and 1.75 mL ddH_2O.
10. EPON (EMbed 812 resin) (EMS, cat. no. 14120).

2.2. Conventional EM of Yeast

1. Reduced OsO_4: dissolve 0.1 g KFeCN (potassium ferrocyanide; Sigma, cat. no. P-9387) in 5 mL 200 m*M* cacodylate, pH 7.4; add 2.5 mL ddH_2O to bring the volume to 7.5 mL; add 2.5 mL 4% OsO_4 stock to bring final volume to 10 mL. This fixative should be prepared just before use. Used in OTO fixation of yeast.
2. 4% $KMnO_4$ (potassium permaganate; EMS, cat. no. 20200): dissolve 4 g $KMnO_4$ in 100 mL ddH_2O. Do not adjust pH.
3. 2% UA: mix 1 g UA (EMS; cat. no. 22400) in 50 mL ddH_2O; sonicate to dissolve; store in light-tight 60-mL syringe and filter with 0.22-μm filter prior to use.

4. Kellenberger's UA (store in the dark): to 20 mL veronal-acetate stock, add 51 mL ddH_2O, 28 mL 0.1 *N* HCl, and 0.5 g UA. Sonicate to dissolve and store at room temperature in brown bottle.
5. Spurr resin (EMS, cat. no. 14300).
6. 1 *M* Tris stock, pH 7.5: dissolve 141 g Tris base in 800 mL H_2O; bring to pH 7.5 with 1 *N* HCl; add water to 1 L.
7. 1 *M* Dithiothreitol stock (Sigma, cat. no. D9779): dissolve 15.42 g in 100 mL; aliquot to 1 mL in 1.5 mL Eppendorf microfuge tubes; store at –20°C.
8. 0.5 *M* Ethylenediaminetetraacetic acid stock 0.5 *M* (Sigma, cat. no. E-5134): Dissolve 186.12 g in 500 mL ddH2O.
9. 2 *M* Sorbitol stock (Sigma, cat. no. S-1876): dissolve 364.34 g in 1000 mL.
10. β-Glucuronidase (Sigma, cat. no. G-0876). Use 100 μL/mL for spheroplasting.
11. Zymolyase (100T; Seikagaku cat. no. 120493): prepare 150 μL (1-mg/mL aliquots); store at –20°C. Use 50 μL/mL for spheroplasting.
12. 0.2 *M* phos-citrate (2X stock): weigh out 3.48 g K_2HPO_4 (or 4.56 g $K_2HPO_4 \cdot 3H_2O$) and 1.4 g citric acid; bring to 100 mL. The pH should be approx 5.9. This is mixed 1:1 with 2 *M* sorbitol for a working solution of phoscitrate-sorbitol buffer.
13. TDES (10 mL): mix 1 mL 1 *M* Tris-HCl, pH 7.5; 0.25 mL 1 *M* dithiothreitol; 0.1 mL 0.5 *M* ethylenediaminetetraacetic acid; 6 mL 2 *M* sorbitol; and 2.65 mL H_2O.
14. 0.2 *M* phos-citrate buffer (100 mL, 2X stock): prepare 0.2 *M* K_2HPO_4M containing 1.4 g citric acid. pH will be approx 5.9. This is mixed 1:1 with 2 *M* sorbitol for a working solution of phoscitrate-sorbitol buffer.

2.3. Immunoelectron Microscopy

1. Formaldehyde stock (EMS, cat. no. 19208), 16% in water (can be aliquoted and stored frozen at –20°C). Heat 16 g granular paraformaldehyde in 100 mL ddH_2O while stirring to 60°C; add two or three drops 1 *N* NaOH and continue stirring until dissolved. Dilute for use.
2. Glutaraldehyde stock: 10% ampules, EM grade (EMS, cat. no. 16120). Sealed ampules prepared in double-distilled water; stored under inert N_2 at 4°C.
3. Sodium m-periodate (Sigma, cat. no. S-1878). Use in periodate-lysine-paraformaldehyde fixative (*see* **step 4**).
4. PLP fixative: prepare 0.1 *M* lysine in 0.05 *M* phosphate buffer, pH 7.4, by dissolving 1.827 g lysine hydrochloride in 50 mL ddH_2O. Adjust to pH 7.4 with 0.1 *M* Na_2HPO_4 and bring volume to 100 mL with 0.1 *M* phosphate buffer, pH 7.4. Prepare 8% formaldehyde as described in **step 1**. Just prior to use, combine 75 mL 0.1 *M* lysine with 25 mL 8% formaldehyde and add 214 mg of $NaIO_4$ (sodium metaperiodate). The final composition will be 10 m*M* $NaIO_4$, 75 m*M* lysine, 37.5 m*M* phosphate, and 2% formaldehyde, pH 6.2.
5. Neutral UA, pH 7.4 (for adsorption staining) (EMS, cat. no. 22400): prepare a 4% solution of UA in double-distilled water and 0.3 *M* solution of oxalic acid in double-distilled water; mix equal parts (1:1) and adjust to pH 7.4 with 10% NH_4OH.
6. Sucrose (2.3 *M*) containing 20% polyvinylpyrrolidone (PVP; Sigma, cat. no. PVP10) in phosphate buffer: mix 20 mL 2.8 *M* sodium carbonate (0.59 g in 20 mL

ddH$_2$O) with 10 mL 100 m*M* phosphate buffer and 40 mg of sodium azide. Add 40 g PVP (MW 10,000) and dissolve by alternating between sonicating and warming in a 60°C water bath. Add 100 mL 2.3 *M* sucrose, aliquot, and store at 4°C.

7. Phosphate-buffered saline (PBS), 10X stock (Eisen's buffer): dissolve 25.62 g NaH_2PO_4 in 200 mL ddH$_2$O; dissolve 225 g $Na_2HPO_4 \cdot 7H_2O$ in 800 mL ddH$_2$O; and dissolve 876.6 g NaCl in 8 L ddH$_2$O. Mix the three solutions and bring the final volume to 10 L. Stock may be stored at room temperature. Dilute 1:10 for use. pH will be 7.4.
8. Postimmunolabeling embedding solution: 3.2% polyvinyl alcohol (PVA; Sigma, cat. no. P-8136), 0.2% methylcellulose (Sigma, cat. no. MD0430), 0.2% UA. Dissolve 1.6 g PVA in 25 mL ddH$_2$O; dissolve 0.1 g methylcellulose in 25 mL ddH$_2$O by heating to 60°C with stirring and quickly cooling on ice; mix the two solutions together and store at 4°C. For use, dispense 1 mL and add 0.2 mL 2% UA.

3. Methods

The most important considerations in any successful fixation are the choice of fixative and fixation conditions. (This is especially important for immunolocalization studies.) For conventional morphology, there are well-established, routine protocols variously using glutaraldehyde alone or in combination with formaldehyde, generally followed by osmium tetroxide or permanganate for fixing both yeast and mammalian cells.

For immunocytochemistry, however, standard fixation protocols do not exist as antigens differ significantly in their sensitivity to fixation, and one standard protocol cannot possibly satisfy all of the competing concerns. Therefore, optimal conditions must be determined empirically for each antigen, with the essential goal to immobilize antigens (hindering their movement by diffusion) while retaining their antigenicity and preserving acceptable ultrastructure. It is important to emphasize that this is always a compromise as a protocol that yields excellent ultrastructure will necessarily be poor at preserving antigenicity; the aim is to find an acceptable balance. The most commonly used fixatives for immunocytochemistry are organic solvents (used exclusively for light microscopy [LM]) and aldehydes (both LM and EM).

Formaldehyde is a monomeric aldehyde prepared fresh from paraformaldehyde and is the most widely used fixative for immunocytochemistry. It preserves cell structure by forming Schiff base intermediates with free amino groups of proteins, thereby stabilizing adjacent proteins. Importantly, formaldehyde fixes at equilibrium with itself and as such is considered a "reversible" fixative. That is, formaldehyde-fixed cells and tissue will begin to unfix in the absence of excess formaldehyde. Compared to organic solvents, formaldehyde is a considerably better fixative for preserving ultrastructure as well as antigenicity, although in some cases a reduction in antigenicity is observed.

We use formaldehyde routinely as a fixative for immunofluorescence microscopy and, in combination with glutaraldehyde, for IEM. When sensitivity to glutaraldehyde has been encountered, formaldehyde alone has been utilized for IEM (8% for 15 min, followed by 4% for variable amounts of time, 20 min to 24 h). Ultrastructural detail, as well as antigenicity, is often remarkably well preserved. Importantly, one should always prepare formaldehyde fresh from paraformaldehyde to ensure maximum purity. Formalin (37% formaldehyde) should never be used for EM or LM as it contains 10% methanol as a redox stabilizer and has an unpredictable influence on fixation.

Glutaraldehyde was originally introduced by Luft in 1959 and is markedly superior to formaldehyde as a fixative for preservation of cell structure because it is bifunctional and efficiently immobilizes proteins by covalently cross-linking primary amino groups. Whenever possible, glutaraldehyde in combination with formaldehyde is the fixative of choice for IEM applications. However, the use of glutaraldehyde in concentrations above 0.05% often results in a measurable, unacceptable loss of antigenicity. In some cases (e.g., that of the mannose 6-phosphate receptor), there is a complete loss of antigenicity in the presence of as little as 0.025% glutaraldehyde; in other cases, antigenicity can be preserved and satisfactory results obtained by controlling the time of fixation and concentration of glutaraldehyde.

Osmium tetroxide was originally introduced by George Palade in 1952 ***(21)***; and it has since become the most widely utilized fixative for conventional ultrastructural analyses. It functions not only as a fixative but also as an electron dense stain, stabilizing proteins and preserving fine detail. It is most important as a post-/secondary fixative, preserving lipids and imparting electron density to virtually all cell components. It is typically used in combination with glutaraldehyde. It is not at all useful for immunocytochemistry.

PLP was originally developed and introduced by Nakane and McLean in 1976 as an alternative fixative to glutaraldehyde for use in IEM and has gained popularity as a fixative for many immunocytochemical applications ***(22)***. Its advantage is that it has fewer deleterious effects on antigenicity and yet preserves an acceptable degree of structural integrity through stabilization of carbohydrate moieties (by periodate oxidation followed by lysine cross-linking of carbohydrate groups). We use it routinely when the sensitivity of a given antigen to glutaraldehyde is such that it cannot be used for fixation.

Permanganate was introduced by Luft in 1956 as an alternative to OsO_4 and prior to the introduction of glutaraldehyde ***(23)***. It gained popularity because of the resulting clarity of membranes and ease of fixation. The clarity of membranes results from the oxidation/extraction of cytoplasmic components (mainly proteins), and it penetrates deeper and faster than other fixatives. Although Luft originally utilized PO_4-buffered $KMnO_4$ at near-neutral pH, it is often utilized in unbuffered

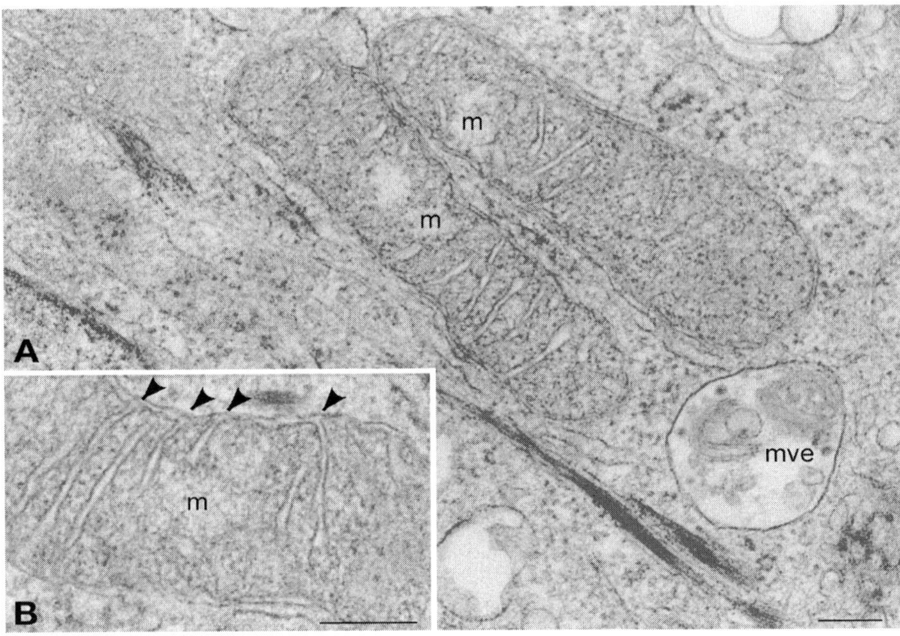

Fig. 1. Conventional electron microscopy of HeLa cells in culture (**A**), (**B**). Arrowheads indicate contact sites in mitochondrion (m). mve, multivesicular endosome. Bar: 0.25 μm.

aqueous solutions at concentrations ranging from 0.5 to 10% and times from 10 min to many hours.

3.1. Conventional EM of Mammalian Cells in Culture (Figs. 1 and 2)

1. A 100-mm dish of 70–80% confluent cells are fixed with 3.0% formaldehyde plus 1.5% glutaraldehyde in 100 m*M* Na cacodylate containing 5 m*M* Ca^{2+}, 2.5% sucrose, pH 7.4, for 1 h at room temperature.
2. Wash in 100 m*M* Na cacodylate containing 2.5% sucrose, pH 7.4, three times for 5 min each wash.
3. Cells are lifted and pelleted at progressively increasing *g* forces to minimize shear (1000*g* for 5 min, 3000*g* for 5 min, 6000*g* for 5 min, and finally 12,000*g* for 5 min; *see* **Note 1**.)
4. The pellet is dislodged from the tube bottom using an applicator stick sharpened to a fine tip and transferred to a scintillation vial (care should be taken to minimize disruption of the pellet).
5. Cells are postfixed with Palade's OsO_4 for 1 h on ice, protected from light, in a fume hood.
6. Rinse pellet once with Kellenberger's and incubate for 1–2 h at room temperature (or preferably overnight) in the dark.

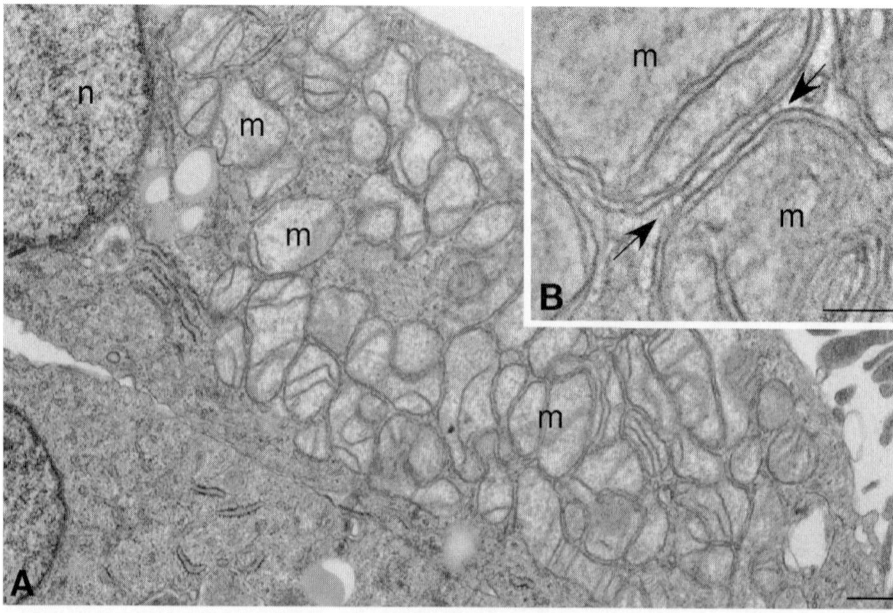

Fig. 2. Conventional EM reveals mitochondrial tethering by truncated Mfn1 (**A**), (**B**). Tannic acid treatment following osmium fixation allows for visualization of Mfn1 inter-mitochondrial tethers, indicated by arrows (**B**). n, nucleus; m, mitochondria. Bar: 0.25 μm.

7. Do one quick rinse in double-distilled water, then one quick rinse in cold (4°C) 50% ethanol.
8. Dehydrate with graded series of cold (4°C) ethanol (70, 95, 100%), then three 15-min washes in fresh 100% ethanol at room temperature, and finally two 5-min exchanges with propylene oxide (PO).
9. Place in 50% PO/50% EPON overnight, uncovered, under vacuum.
10. Replace with fresh 100% EPON and leave under vacuum for 4–6 h.
11. Embed in fresh 100% EPON using beem capsules or flat embedding molds (put typed or pencil-written label in dummy capsules with wooden stick, at least two capsules per sample). Place in 60°C oven for 24–48 h.
12. 80 nm sections are cut on a Leica UCT ultramicrotome, collected onto 400-mesh high-transmission grids, poststained with lead citrate and UA, and observed. (*see* **Note 2**).

3.2. Conventional EM of Yeast

There are essentially two well-established protocols for visualizing yeast mitochondria by conventional EM. The simplest method utilizes potassium permanganate and requires no spheroplasting ***(23)***; the second involves spheroplasting followed by the sequential application of reduced osmium and

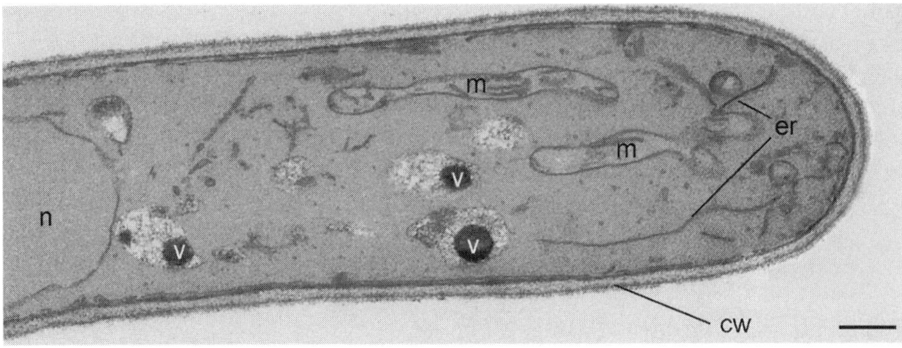

Fig. 3. Conventional EM of *Schizosaccharomyces pombe* fixed with potassium permanganate. cw, cell wall; er, endoplasmic reticulum; m, mitochondria; n, nucleus; v, vacuole. Bar: 1.0 μm.

thiocarbohydrazide ***(24)***. Both protocols are excellent for visualizing mitochondria as each results in a diminution of cytoplasmic contrast, thus greatly enhancing the visualization of endomembranes, especially those of mitochondria.

3.2.1. Permanganate Fixation of Yeast (see **Fig. 3**)

1. Cells are grown in synthetic minimal medium (or other medium) overnight to a density no greater than 0.6 OD_{600} units. In some experiments, temperature shifts are induced 1–3 h prior to fixation.
2. Cells are spun down at 300*g* for 10 min and resuspended in 1 mL 100 m*M* Na cacodylate containing 3% glutaraldehyde, 5 m*M* $CaCl_2$, and 5 m*M* $MgCl_2$, pH 7.4, for 1 h.
3. Cells are pelleted/washed through three changes of 100 m*M* Na cacodylate, pH 7.4; dispersed and subsequently embedded in ultralow-temperature gelling agarose (~1:1); cooled; and cut into small pieces (~1 mm^3).
4. Blocks are postfixed in 4% $KMNO_4$ prepared in double-distilled water for 1 h at room temperature (*see* **Note 3**).
5. Blocks are washed thoroughly (four times, 15 min total) in double-distilled water.
6. Blocks are then treated with 0.5% sodium meta-periodate for 15 min at room temperature to aid in infiltration (*see* **Note 4**).
7. Blocks are then washed three times in double-distilled water for 15 min.
8. Blocks are placed into filtered 2% UA (double-distilled water, ~ pH 4.5) overnight at room temperature (protected from light).
9. Blocks are dehydrated through a graded series of ethanol (50–100% at 4°C); left overnight in fresh 100% ethanol at room temperature; washed twice for 15 min (each wash in 100% ethanol), followed by two washes with propylene oxide; and placed into 50:50 mixture of PO and Spurr resin. Samples are rotated overnight at 1*g* (i.e., very slowly) (*see* **Note 5**).

10. The next day, the samples are transferred to 100% Spurr resin and left under vacuum for 24 h.
11. The following day, samples are given two changes of resin over 6–8 h, left overnight in a third change, and allowed to sit under vacuum through the day.
12. Subsequently, the samples are placed in beem capsules containing fresh Spurr resin and placed in the oven at 60–80°C for 24–48 h.
13. Sections are cut, placed onto 400-mesh high-transmission nickel grids, and post-stained with lead citrate (2–5 min). Do not stain with UA as the cytoplasmic staining intensity will be too dense (*see* **Note 6**).

3.2.2. OTO Fixation of Yeast (see ***Fig. 4****)*

1. Cells are grown in YPD or other media; inoculate early in the day prior to when you want them ready. Never let them grow over 0.6 OD_{600} U/mL. Harvest approx 100 mL (i.e., about 25–50 OD_{600} total).
2. Resuspend in 1 mL 100 m*M* Na cacodylate containing 3% glutaraldehyde, 5 m*M* $CaCl_2$, 5 m*M* $MgCl_2$, 2.5% sucrose, pH 7.4, and fix for 1 h at room temperature.
3. Cells are gently pelleted in Eppendorf tubes at 300*g* for 2 min.
4. Wash once with 1 mL TDES (*see* **Subheading 2.2., item 13**).
5. Resuspend in 1 mL TDES and incubate at room temperature for 10 min. This step softens the cell wall.
6. Spin cells at 500*g* for 2 min.
7. Wash in 1 mL 0.1 *M* phoscitrate/1 *M* sorbitol.
8. Resuspend in 0.5 mL phos-citrate/sorbitol. Add 50 μL β-glucuronidase, 25 μL 10 mg/mL zymolyase. Incubate at 30°C for 60 min with occasional agitation. (Increase the amount of zymolyase or incubation time for mutants with thick cell walls.)
9. Cells are gently pelleted at less than 2000*g* for 4 min.
10. Wash in 1 mL 0.1 *M* cacodylate/5 m*M* $CaCl_2$/1 *M* sorbitol.
11. Disperse cells and embed in 2% ultralow-temperature agarose (made in water). Cool. Cut into small pieces (~1 mm^3).
12. Postfix blocks in 1% OsO_4/1% potassium ferrocyanide in 0.1 *M* cacodylate/5 m*M* $CaCl_2$, pH 6.8. Fix at room temperature for 30 min.
13. Wash blocks thoroughly (four times in double-distilled water, 10 min total).
14. Transfer blocks to 1% thiocarbohydrazide at room temperature for 5 min (make 1% thiocarbohydrazide in water, stir for 1 h, let undissolved material settle out, and use the solution for the incubation with blocks).
15. Wash blocks in double-distilled water (four times, 1 min each).
16. Transfer blocks to 1% OsO_4/1% potassium ferrocyanide in cacodylate buffer, pH 6.8 (5 min at room temperature).
17. Wash cells four times with double-distilled water (15 min total) (*see* **Note 7**).
18. Optional step: *en bloc* stain in Kellenberger's UA for 2 h to overnight.
19. Dehydrate through a graded series of ethanol. Carry out 50 to 100% on ice (3–5 min/wash), then follow with four 15-min washes in 100% at room temperature.
20. Transfer blocks to 1:1 ethanol/propylene oxide (10 min).
21. Transfer blocks to 100% propylene oxide (two washes, 5 min each).

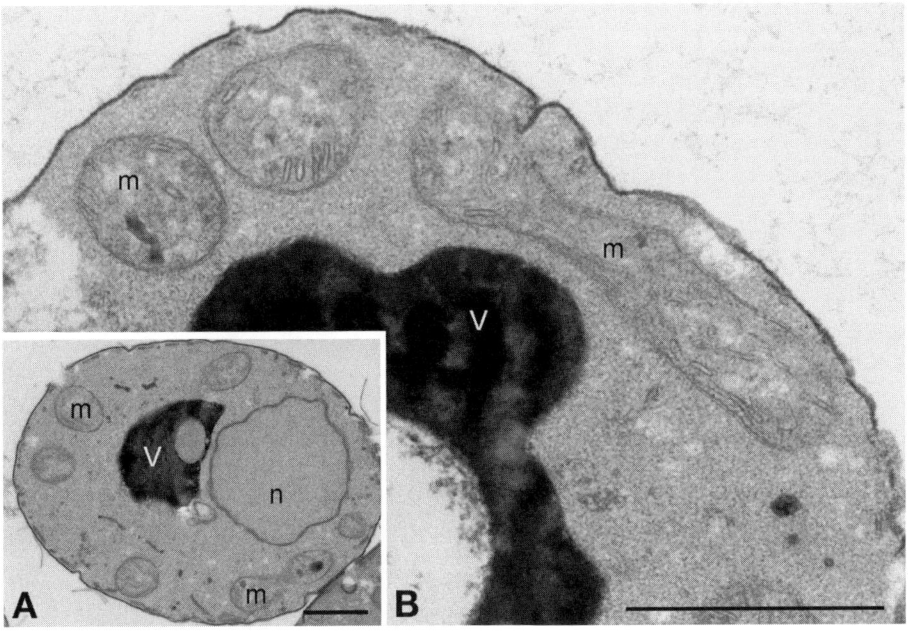

Fig. 4. Conventional EM of *S. cerevisiae* fixed by OTO fixation. m, mitochondria; n, nucleus; V, vacuole. Bar: 1.0 µm.

22. Transfer blocks to 1:1 PO/Spurr or EPON resin (hard formulation) for overnight under vacuum (*see* **Note 8**).
23. Transfer blocks to fresh Spurr or EPON resin; leave for 4–6 h.
24. Transfer to beem capsules and polymerize in fresh Spurr resin overnight (18–24 h). Section and poststain with lead and UA (20 min UA, 2 min lead citrate).

3.2.3. *In Vitro Mitochondria* (see **Fig. 5**)

1. Mitochondria are fixed in suspension by the addition of fixative (3% paraformaldehyde, 1.5% glutaraldehyde, and 2.5% sucrose contained in 100 m*M* cacodylate, pH 7.4, 4°C) for 1 h.
2. Mitochondria are then washed in 100 m*M* cacodylate and spun into a tight pellet (>13,000*g*).
3. Postfix in Palade's OsO_4 for 1 h, 4°C and subsequently *en bloc* stain in Kellenberger's UA overnight.
4. The pellets are dehydrated through a graded series of ethanol as described in **Subheading 3.3.2.**, **step 19**, infiltrated with EPON, and allowed to polymerize 24–48 h at 60°C.
5. 80-nm sections are cut on a Leica UCT ultramicrotome, collected onto 400-mesh high-transmission grids, poststained with lead citrate and UA, and observed (*see* **Notes 9–11**).

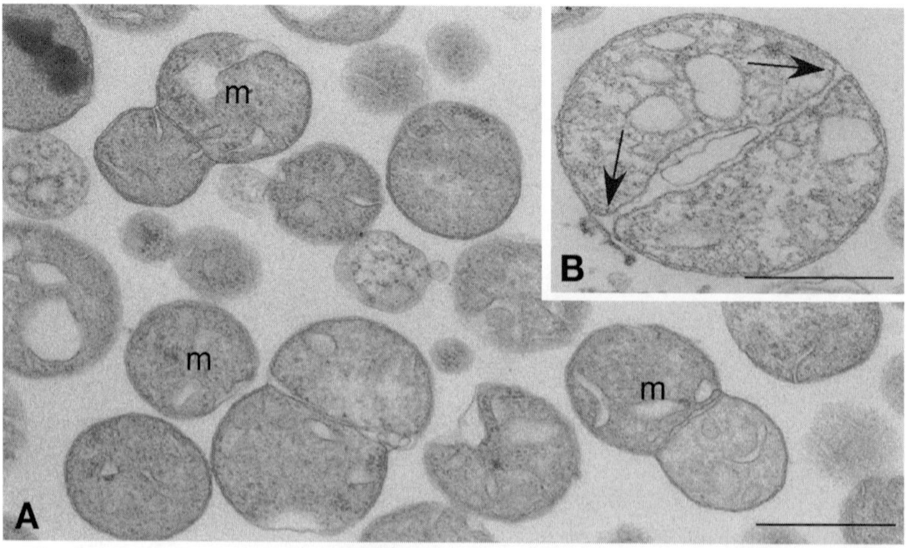

Fig. 5. Conventional EM on isolated *S. cerevisiae* mitochondria (**A**), (**B**). (**A**) Low magnification of mitochondrial pellet reveals numerous mitochondria in early stage of fusion. (**B**) High magnification clearly shows segregation of inner membranes (indicated by arrows) during mitochondrial fusion. Bar for (**A**) 0.5 μm; bar for (**B**), 0.25 μm.

3.3. Immunoelectron Microscopy

3.3.1. Immunogold Labeling of Ultrathin Cryosectioned Cells and Tissue (see *Fig. 6*)

1. Cells are fixed in suspension for 15 min by adding an equal volume of freshly prepared 8% formaldehyde contained in 100 m*M* PO_4 buffer, pH 7.4.
2. The cells are pelleted, resuspended in fresh fixative (8% formaldehyde, 100 m*M* $NaPO_4$, pH 7.4), and incubated for an additional 18–24 h at 4°C (*see* **Note 12**).
3. The cells are washed briefly in PBS and resuspended in 1% low gelling temperature agarose.
4. The agarose blocks are trimmed into 1-mm^3 pieces, cryoprotected by infiltration with 2.3 *M* sucrose/30% PVP (10,000 MW)/PBS, pH 7.4, for 2 h, mounted onto cryopins, and rapidly frozen in liquid nitrogen.
5. Ultrathin cryosections are cut on a Leica UCT ultramicrotome equipped with a fetal calf serum cryo-attachment and collected onto formvar/carbon-coated nickel grids.
6. The grids are washed through several drops of 1X PBS containing 2.5% FCS and 10 m*M* glycine, pH 7.4, then blocked in 10% FCS for 30 min and incubated overnight in primary antibody (*see* **Notes 13** and **14**).
7. After washing, the grids are incubated for 2 h in 5-nm gold conjugated to secondary antibodies against appropriate species (Jackson ImmunoResearch).

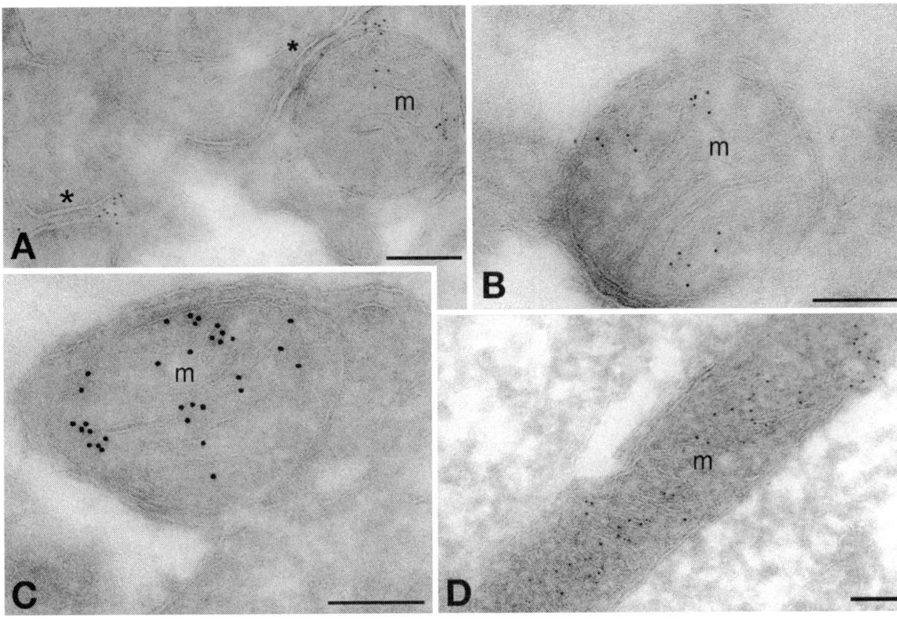

Fig. 6. Immunogold labeling of ultrathin cryosections (**A**)–(**D**). BCL-XL localizes to mitochondria and synaptic cleft and mitochondrial inner/outer membranes in mouse hippocampus (**A**), (**B**); adenosine triphosphate synthase localizes to mitochondrial cristae in isolated mouse hippocampal neurons (**C**); SIRT3, a human SIR2 homolog, localizes to mitochondria in HeLa cells (**D**). (**A**), (**B**), and (**D**) are 5 nm Au; (**C**) is 10 nm Au. m, mitochondria; asterisk, synapse. Bar: 0.5 μm.

8. The grids are washed through several drops of PBS followed by several drops of double-distilled water, floated on a 1-mL drop of neutral UA, pH 7.4, for 10 min, quickly washed through 5 drops ddH_2O, and floated onto an aqueous solution containing 3.0% PVA (MW 10,000)/0.2% methyl cellulose (400 centiposes)/0.1% UA.
9. Grids are then embedded by removing excess solution in **step 8** using no. 50 hardened Whatman filter paper and examined.

3.3.2. Immunogold Labeling of Isolated Mitochondria (see ***Fig. 7***)

1. Mitochondria are fixed in suspension by the 1:1 addition of 8% formaldehyde contained in PBS, pH 7.4, at room temperature for 15 min and subsequently pelleted.
2. They are resuspended in fresh 4% formaldehyde contained in PBS, pH 7.4, and allowed to fix an additional 12–15 h at 4°C.
3. Mitochondria are pelleted; washed briefly in PBS, resuspended in 2% low temperature gelling agarose, and allowed to cool.
4. The agarose blocks are then trimmed into 1-mm^3 pieces, cryoprotected in 2.3 *M* sucrose containing 20% PVP (MW 10,000) for 2 h, mounted onto cryopins, and frozen rapidly in liquid N_2.

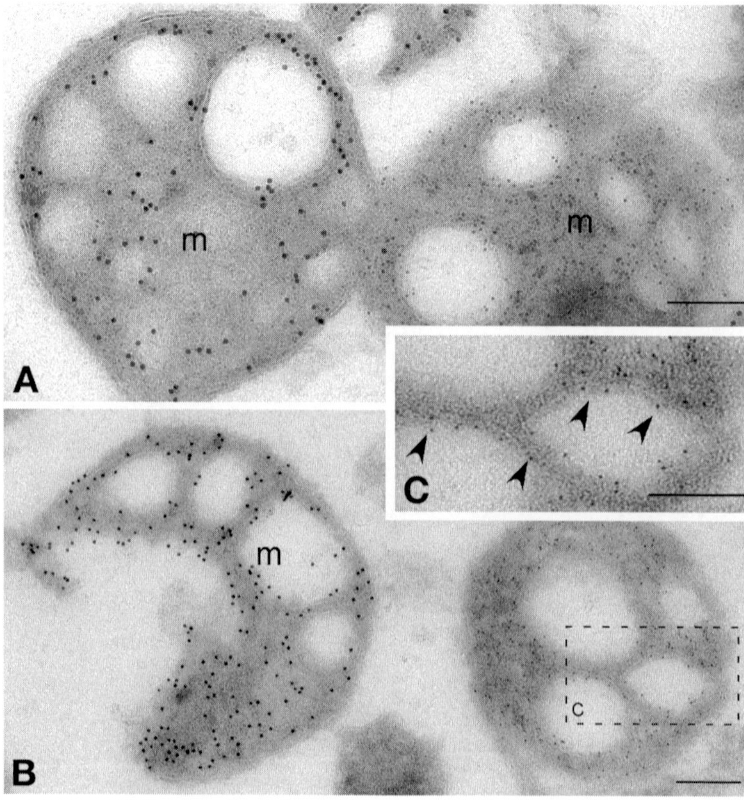

Fig. 7. Immunogold labeling of ultrathin cryosections from an isolated mitochondrial fusion assay (**A**)–(**C**). Preparations from two strains of *S. cerevisiae*, one expressing DsRed (10 nm Au) and the other GFP (5 nm Au) are mixed and monitored over time to assay for fusion. Panels (**A**)–(**C**) show segregation of label at time 0 min. (**C**) is high magnification of boxed area in (**B**). m, mitochondria. Bar: 0.1 µm.

5. Ultrathin cryosections are then cut on a Reichert UCT ultramicrotome equipped with a FCS cryo-stage at –100°C and collected onto formvar/carbon-coated nickel grids.
6. Grids are washed briefly through 8–10 drops 1X PBS containing 2.5% FCS and 0.01 *M* glycine, pH 7.4.
7. The grids are blocked in 10% FCS for a minimum of 15 min at room temperature and incubated overnight with primary antibody diluted to 10 µg/mL in 10% FCS (*see* **Notes 15–17**). In **ref.** ***20***, a mixture of mouse anti-GFP (Molecular Probes) and rabbit anti-DsRed (BD Biosciences) were both used at 10 µg/mL specific antibody (**Fig. 7**).
8. After washing as in **step 6**, grids are then incubated with donkey antimouse/rabbit/and so on 5-nm gold conjugate or donkey antirabbit/mouse/and so on

10-nm gold (Jackson Research Laboratories) and diluted 1:50 for 2 h (*see* **Notes 18** and **19**).

9. The grids are washed several times in PBS as in **step 6**, washed through several drops of double-distilled water, and subsequently floated on 1 mL of a mixture containing 3.0% PVA (MW 10,000), 0.2% methylcellulose (400 centiposes), and 0.2% UA.
10. Grids are then embedded by removing excess solution in **step 8** using no. 50 hardened Whatman filter paper and examined.

4. Notes

1. Great care should be taken when harvesting and spinning cells. Cultured cells are extremely sensitive to scraping and shear forces.
2. The pellet should always be sectioned vertically and examined top to bottom. This is because stratification of cells occurs during centrifugation. The healthier, more intact cells will settle near the bottom, whereas the more disrupted, unhealthy cells are found near the top of the pellet. Moreover, in the case of transient transfections, generally the transfected cells will be near the top of the pellet.
3. The fixation time along with concentration of permanganate can be varied from 2 to 8% to modulate contrast as desired.
4. The Na meta-periodate step in **step 6** must be closely adhered as greater time will result in excessive erosion of the cell wall; less time will result in inadequate permeablization.
5. Spurr resin should be used exclusively in lieu of EPON because of its considerably greater fluidity.
6. Poststain sections with lead citrate only to avoid excessive cytoplasmic contrast.
7. To further enhance contrast of coat proteins, insert a tannic acid incubation between **steps 17** and **18** (after the second osmication, before the UA). After the water washes, incubate blocks in 1–2% tannic acid/100 m*M* cacodylate, pH 7.4, for 30 min at room temperature. The standard is to use 1% tannic acid, but you can increase it to 2%. If 2% tannic acid is used, then be sure to check the pH of the solution; you will need to adjust it back to pH 7.4. Finally, again wash thoroughly with water prior to the UA incubation.
8. EPON is preferable to Spurr generally because it has better sublimination characteristics, thereby yielding better contrast. Because the cell walls have been removed, infiltration with EPON should not be a problem.
9. As always, pellets should be sectioned vertically and sampled top to bottom as the most intact/dense mitochondria will generally be near the bottom; the least intact/most swollen and disrupted mitochondria will be near the top.
10. Proper sampling is essential when assaying for unpredictable phenotypes.
11. When optimum visualization of contact sites/cristae is crucially important, thinner, 40- to 60-nm sections are desirable, recognizing that thinner sections will have a concomitant loss in contrast.
12. There is no single fixation regimen that is suitable for all applications. Generally, one should always titrate the antibody/antigen vs the various fixatives and time. In our center, we exclusively use variations on formaldehyde, glutaraldehyde, or PLP.

13. We prefer FCS as a blocking agent and diluent instead of bovine serum albumin, although many labs use FCS and bovine serum albumin interchangeably. FCS tends to yield lower background staining.
14. Incubation/labeling times can be varied considerably. Our standard is to label with the primary overnight (at 4°C) and 2–3 h at room temperature with the secondary. However, when antigens are abundant and the antibodies are of high affinity and titer, we have used incubation times as short as 10 min.
15. The blocking time can be highly variable, ranging from 15 min to 2 h and depends on the nonspecific staining/crossreactivity of the primary antibody.
16. One can vary the concentration of FCS in the block and the diluent in **step 7** from as little as 2.5% up to 15%. Concentrations of FCS above 15% will block binding of even the highest affinity antibodies.
17. Primary antibodies should always be purified, with a simple immunoglobulin G fraction derived from a 50% ammonium sulfate precipitation the most desirable. Affinity purification can be used when needed to clean a particularly sticky reagent but at a loss of reagent shelf life and the highest affinity antibodies.
18. Protein A gold conjugates can be substituted in **step 8** to prevent clustering of gold particles.
19. In double-labeling experiments, one should always pair the larger gold (10- to 12-nm) secondary conjugate with the most abundant antigen (or highest affinity antibody) and use the smaller secondary gold (5- to 6-nm) conjugate with the least abundant antigen (or lower titer antibody). It is not uncommon to experience a 10- to 50-fold drop in labeling intensity when comparing 10- vs 5-nm secondary antibodies.

References

1. Geuze, H. J. (1999) A future for EM in cell biology? *Trends Cell Biol.* **9,** 92–93.
2. Griffiths, G. (2001) Bringing electron microscopy back into focus for cell biology. *Trends Cell Biol.* **11,** 153–154.
3. Chalfie, M., Tu, Y., Euskirchen, G., Ward, W. W., and Prasher, D. C. (1994) Green fluorescent protein as a marker for gene expression. *Science* **263,** 802–805.
4. Wang, L., Jackson, W. C., Steinbach, P. A., and Tsien, R. Y. (2004) Evolution of new nonantibody proteins via iterative somatic hypermutation. *Proc. Natl. Acad. Sci. U. S. A.* **101,** 16,745–16,749.
5. Geuze, H. J., Slot, J. W., Strous, G. J., Lodish, H. F., and Schwartz, A. L. (1983) Intracellular site of asialoglycoprotein receptor-ligand uncoupling: double-label immunoelectron microscopy during receptor-mediated endocytosis. *Cell* **32,** 277–287.
6. Wall, D. A., Wilson, G., and Hubbard, A. L. (1980). The galactose-specific recognition system of mammalian liver: the route of ligand internalization in rat hepatocytes. *Cell* **21,** 79–93.
7. Willingham, M. C. and Pastan, I. (1980) The receptosome: an intermediate organelle of receptor mediated endocytosis in cultured fibroblasts. *Cell* **21,** 67–77.
8. Brown, W. J., Goodhouse, J., and Farqubar, M. G. (1986). Mannose-6-phosphate receptors for lysosomal enzymes cycle between the Golgi complex and endosomes. *J. Cell Biol.* **103,** 1235–1247.

9. Roth, J. and Berger, E. G. (1982). Immunocytochemical localization of galactosyltransferase in HeLa cells: codistribution with thiamine pyrophosphatase in trans-Golgi cisternae. *J. Cell Biol.* **92,** 223–229.
10. Roth, J., Taatjes, D. J., Lucocq, J. M., Weinstein, J., and Paulson, J. C. (1985). Demonstration of an extensive trans-tubular network continuous with the Golgi apparatus stack that may function in glycosylation. *Cell* **43,** 287–295.
11. Farquhar, M. G., Hendricks, L. H., Noda, T., and Velasco, A. (1992). in *Electron Microscopic Cytochemistry and Immunocytochemistry in Biomedicine* (Ogawa, K. and Barka, T., eds.), CRC Press, Boca Raton, FL, p. 441–479.
12. Stow, J. L., de Almeida, J. B., Narula, F. J., Holtzman, E. J., Ercolani, L., and Ausiello, D. A. (1991). A heterotrimeric G protein, G alpha i-3, on Golgi membranes regulates the secretion of a heparan sulfate proteoglycan in LLC-PK1 epithelial cells. *J. Cell Biol.* **114,** 1113–1124.
13. Nelson, J. (1992) Regulation of cell surface polarity from bacteria to mammals. *Science* **258,** 948–955.
14. Rodriguez-Boulan, E. and Nelson, W. J. (1989) Morphogenesis of the polarized epithelial cell phenotype. *Science* **245,** 718–725.
15. Claude, A. and Fullam, E. F. (1945) An electron microscope study of isolated mitochondria, method and preliminary results. *J. Exp. Med.* **81,** 51–62.
16. Palade, G. E. (1952) The fine structure of mitochondria. *Anat. Rec.* **114,** 427–451.
17. Koshiba, T., Detmer, S. A., Kaiser, J. T., Chen, H., McCaffery, J. M., and Chan, D. C. (2004) Structural basis of mitochondrial tethering by mitofusin complexes. *Science* **305,** 858–862.
18. Bleazard, W., McCaffery, J. M., King, E. J., et al. (1999) The dynamin-related GTPase Dnm1 regulates mitochondrial fission in yeast. *Nat. Cell Biol.* **1,** 298–304.
19. Mozdy, A. D., McCaffery, J. M., and Shaw, J. M. (2000) Dnm1p GTPase-mediated mitochondrial fission is a multi-step process requiring the novel integral membrane component Fis1p. *J. Cell Biol.* **151,** 367–380.
20. Meeusen, S., McCaffery, J. M., and Nunnari, J. (2004) Mitochondrial fusion intermediates revealed in vitro. *Science* **305,** 1747–1752.
21. Palade, G. E. (1952) A study of fixation for electron microscopy. *J. Exp. Med.* **95,** 285–298.
22. McLean, W. and Nakane, P. F. (1974) Periodate-lysine-paraformaldehyde fixative. A new fixation for immunoelectron microscopy. *J. Histochem. Cytochem.* **22,** 1077–1083.
23. Luft, J. H. (1956) Permanganate; a new fixative for electron microscopy. *J. Biophys. Biochem. Cytol.* **2,** 799–802.
24. Willingham, M. C. and Rutherford, A. V. (1984) The use of osmium-thiocarbohydrazide-osmium (OTO) and ferrocyanide-reduced osmium methods to enhance membrane contrast and preservation in cultured cells. *J. Histochem. Cytochem.* **32,** 455–460.

34

Mitochondrial Division in *Caenorhabditis elegans*

Shilpa Gandre and Alexander M. van der Bliek

Summary

The study of mitochondrial division proteins has largely focused on yeast and mammalian cells. We describe methods to use *Caenorhabditis elegans* as an alternative model for studying mitochondrial division, taking advantage of the many wonderful resources provided by the *C. elegans* community. Our methods are largely based on manipulation of gene expression using classic and molecular genetic techniques combined with fluorescence microscopy. Some biochemical methods are also included. As antibodies become available, these biochemical methods are likely to become more sophisticated.

Key Words: *C. elegans*; division; Drp1; fusion; nematode.

1. Introduction

Mitochondria are dynamic organelles that constantly move, divide, and fuse in living cells *(1)*. Mitochondrial fusion is important for mixing of mitochondrial deoxyribonucleic acids (DNAs) and thus for maintenance of functional mitochondria. Mitochondrial division is implicated in the process of programmed cell death. The recent discovery of proteins involved in these processes has opened new avenues toward understanding the dynamic nature of mitochondria. In our lab, we use *Caenorhabditis elegans* as a model organism to study processes that affect mitochondrial morphology *(2)*. *Caenorhabditis elegans* was chosen because it has the tractability of a simple genetic system with numerous resources while retaining most, if not all, the complex specializations that are inherent to multicellular eukaryotes.

Excellent descriptions of the biology of nematodes, topics that are studied with *C. elegans*, and most common techniques can be found in **refs.** ***3–6***. The worm community also shares much information through the World Wide Web. Key Web sites are http://elegans.swmed.edu/ and http://www.wormbase.org/.

From: *Methods in Molecular Biology, vol. 372: Mitochondria: Practical Protocols*
Edited by: D. Leister and J. M. Herrmann © Humana Press Inc., Totowa, NJ

Unfortunately, there is no room here to describe classic genetic techniques or more modern approaches to screening, such as screening for deletion mutants with polymerase chain reaction (PCR) or screening with the Ahringer ribonucleic acid interference (RNAi) library *(7)*. Instead, we describe techniques to study mitochondrial morphology *per se* in *C. elegans*.

2. Materials

2.1. Making Transgenic Animals by Microinjection

1. M9 buffer: 3 g KH_2PO_4, 6 g Na_2HPO_4, 5 g NaCl, and 1 mL 1 *M* $MgSO_4$ in a volume adjusted to 1 L with water.
2. Nematode growth medium (NGM) agar: Add 3 g NaCl, 17 g agar, 2.5 g peptone, and 1 mL cholesterol (5 mg/mL in ethanol) to 975 mL H_2O. Autoclave this solution and then, under sterile conditions, add 1 mL 1 *M* $CaCl_2$, 1 mL 1 *M* $MgSO_4$, and 25 mL 1 *M* potassium phosphate, pH 6.0. Mix after each addition.
3. Qiagen plasmid purification kit (Qiagen).
4. *C. elegans* mutants *glo-1, sid-1, rde-1, rrf-3*, the Bristol N2 strain that serves as wild type and the *Escherichia coli* strains HT115(DE3) and OP50 can be obtained from the *C. elegans* Genetics Center (http://biosci.umn.edu/CGC/CGChomepage.htm).
5. The 2% low melting point agarose solution for agarose pads is made by adding the agarose to distilled water, heating it to 65°C, and cooling to 42°C.
6. Borosilicate glass capillaries with an outer diameter of 1.2 mm (World Precision Instruments, Sarasota, FL, USA).
7. Microelectrode (injection needle) puller, manufactured (Sutter Scientific Instrument Co., Novato, CA, USA).
8. Kimax glass capillaries, 0.8–1.1 × 100 mm (Kimble Products, Vineland, NJ, USA).
9. Halocarbon oil series 700 (Halocarbon Products Corp., River Edge, NJ, USA).
10. 5% hydrofluoric acid solution is made by adding 1 mL hydrofluoric acid to 4 mL distilled water.

2.2. Knockdown of Gene Expression by RNAi

For knockdown of gene expression by RNAi, use an in vitro transcription reaction kit (Promega Corp., Madison, WI, USA).

2.3. Imaging Mitochondria in Caenorhabditis elegans

1. Aldicarb (Chem Service, West Chester, PA, USA).
2. Rhodamine 6G (Sigma-Aldrich, St. Louis, MO, USA).
3. MitoTracker (Invitrogen Inc., Carlsbad, CA, USA).

2.4. Isolation of Mitochondria From Caenorhabditis elegans

1. Mitochondrial isolation buffer (IB): 210 m*M* mannitol, 70 m*M* sucrose, 0.1 m*M* ethylenediaminetetraacetic acid (EDTA), pH 8.0, 5 m*M* Tris-HCl, pH 7.4, and 1 m*M* phenylmethylsulfonyl fluoride (PMSF).
2. 50 mL Potter-Elvehjem homogenizer (Kontes Glass Co., Vineland, NJ, USA).

2.5. Isolation of Nucleic Acids From Caenorhabditis elegans

1. Pestle and mortar.
2. Liquid nitrogen in an appropriate container suitable for pouring small amounts.
3. Protective eyeware and gloves.
4. 3 *M* NaAc, pH 4.5.
5. TE buffer: 1 m*M* Tris-HCl, pH 8.0, 1 m*M* EDTA, pH 8.0.
6. Solutions for sucrose flotation: 0.1 *M* NaCl, 60% (w/v) sucrose.
7. DNA lysis buffer: 0.1 *M* NaCl, 10 m*M* Tris-HCl, pH 8.0, 10 m*M* EDTA, pH 8.0, 1% sodium dodecyl sulfate (SDS), 1% β-mercaptoethanol, 100 μg/mL proteinase K (Sigma).
8. 5X Mg^{++}-free M9 buffer: add 3.4 g Na_2HPO_4, 1.5 g KH_2PO_4, 0.25 g NaCl, and 0.5 g NH_4Cl to water to make 100 mL.
9. For soaking buffer, make a 20-fold dilution of the 5X Mg^{++}-free M9 buffer with 3 m*M* Spermidine (Sigma) and a gelatin solution made by adding 0.05% (w/v) gelatin to distilled water, which is then autoclaved and filtered.
10. Trizol reagent (Invitrogen).
11. Bradford reagent for protein estimation (Bio-Rad).
12. To make 5X Laemmli sample buffer, mix 5 mL glycerol, 2.56 mL β-mercapto-ethanol, 2.13 mL Tris-HCL, pH 6.8, 1 g SDS, and trace amounts of bromophenol blue.

3. Methods

3.1. Making Transgenic Animals by Microinjection

3.1.1. General Considerations

1. There are several outstanding descriptions of microinjection procedures ***(5,6)***. We nevertheless include a description of our procedure because it is key to functional analysis of mitochondria in *C. elegans* and includes some pointers specific for our needs.
2. Microinjection takes practice, especially if this is the first time that one has handled worms. We therefore recommend first learning basic skills of maintaining and transferring worms.
3. An alternative to introducing DNA by microinjection is the use of biolistics. This procedure, however, is not as commonly used by the worm community.

3.1.2. Preparing the DNA for Injection

1. As marker for transformation, we often use a plasmid-encoding collagen with a dominant mutation derived from the strain *rol-6(su1006)*. This *rol-6* marker causes the worms to move in circles and rotate around their body axis, properties that are easy to spot in a field of nontransformed worms ***(8)***. Many other transformation markers exist. Some complement a recessive mutation and thus restore wild-type growth or behavior (*dpy-20*, *unc-119*, etc.); others rescue a lethal mutant (*pha-1*), so they can be used as a selectable marker ***(9)***. We prefer the *rol-6* marker because it causes the body to twist, so there are always some body wall muscles

aligned with the focal plane of the microscope. We usually add 50 ng/μL of the *rol-6 (su1006)*-encoding plasmid pRF4 to the injection mix (*see* **Note 1**).

2. RNAi-inducing or overexpression plasmids are added to manipulate gene expression in a particular target cell. We most often use the *myo-3* promoter to express these constructs in body wall muscles. The complementary DNA (cDNA) or genomic sequence could, however, also be expressed with its own promoter. Use of cDNA has the advantage of a shorter insert. Use of genomic DNA has the advantage of improved expression caused by the presence of introns. Depending on the promoter, expression constructs are generally injected at a concentration of 1–10 ng/μL.
3. Plasmids encoding fluorescent organelle markers, such as Pmyo-3::mito::GFP (green fluorescent protein), are added at final concentrations of 1–5 ng/μL, depending on their effectiveness.
4. Carrier DNA and water are added to bring the total DNA concentration to 100 ng/μL. We generally use a nonworm plasmid, such as pBluescript (Stratagene Inc., La Jolla, CA, USA), as carrier DNA. The use of lower concentrations of carrier DNA causes the formation of shorter extrachromosomal arrays. These are more frequently lost during cell division, which leads to a higher rate of mosaicism and low rates of transmission of the transgenic array in the germline (*see* **Note 2**).

3.1.3. Preparing Injection Pads

1. To make agarose pads on which to mount worms for injection, put a 50-μL drop of molten 2% low melting point agarose in the middle of a 48 × 65 mm coverslip and quickly put another coverslip on top, applying mild pressure to flatten the drop and make a thin film of agarose.
2. After the agarose has solidified, gently remove the coverslip by sliding it off and let the agarose dry by exposing it overnight to air. Pads can be stored indefinitely at room temperature. If necessary, they can be dried further in a vacuum oven.

3.1.4. Microinjection Needles

1. Pull needles for injection, starting with settings recommended by the manufacturer of the needle puller. Adjust those settings by trial and error to make needles with a tapered portion about 1 cm long. The exact degree of tapering determines the success of the injection. A needle that is too thin will bend when pushed up against the worm; a needle that is too thick will cause excessive damage to the worm cuticle. Injection needles are prepared in advance and stored in dust-free conditions. Freshly pulled needles have closed tips.
2. Mount the needle in a needle holder fitted with tubing connected to a valve that releases nitrogen gas from a pressurized gas tank. The valve is controlled by an electrical switch in a foot pedal. The valve releases pulses of pressurized nitrogen gas adjusted to 30 psi.
3. Open the tip of the needle for microinjection by etching with hydrofluoric acid. To do this, place the tip of the needle in 5% hydrofluoric acid and apply three quick pulses of pressure. Transfer the tip of the needle to distilled water and apply

pressure. Pass the needle back and forth between acid and water until a very fine stream of bubbles is visible. Larger bubbles indicate an opening that is too large. The larger amount of injection solution released by large needle openings can kill the worms.

4. Make a loading pipet by heating the center of a 100-μL glass capillary with a gas flame until it melts. Then, quickly remove the capillary from the flame and pull at both ends to make the middle portion thin. Break the middle, thus making two loading pipets.
5. To prevent contaminating particles from clogging the injection needle, centrifuge the injection solution for 10 min at 16,000*g* in a microcentrifuge and suck some of the top portion of the supernatant into a loading pipet.
6. Fill the injection needle by inserting the loading pipet into the back end of an injection needle and expel the solution with pressure from a mouth pipet. Once loaded, press the injection needles in molding clay and set upright for 5 min to let the injection solution move to the tip of the needle by capillary action.

3.1.5. Microinjection Procedure

1. At 1 d before injection, transfer between 20 and 40 healthy L4 stage worms to a fresh worm plate so that healthy, well-fed, young adults will be ready when needed.
2. At the beginning of an injection session, transfer these worms to an NGM plate without bacteria so that they lose some of the contaminating bacterial paste before mounting onto the injection pad.
3. Place a loaded needle in the needle holder of the micromanipulator. Attach the tubing to the control valve and pressurized nitrogen and test with a few quick pulses of pressure to ensure that the needle is open and injection mix is flowing freely.
4. Place a drop of Halocarbon oil in the center of the injection pad.
5. Transfer the coverslip with halocarbon oil to an inverted microscope. We use an old Olympus microscope with a gliding stage and 10× and 40× differential interference constant (DIC) objectives with long working distance for this purpose.
6. Bring the needle down into the oil drop and focus with the 10× objective. Apply pressure to see if the injection mixture is flowing freely. If the needle is clogged by an air bubble or dust particle, then it is best to try another needle because these are often difficult to dislodge. Center the tip of the needle under the 40× objective. From this point, we only move the needle along the *z*-axis. The worm is moved toward the needle with the gliding stage.
7. Place the coverslip with agarose pad onto the inverted lid of a 6-cm Petri plate and place this under a dissecting microscope with a transmitted light base. Placing the coverslip onto the lid brings it close to the height of worms on a growth plate, reducing the need to refocus when switching between worm plates and agarose pad.
8. Transfer a healthy young adult worm to the drop of hydrocarbon oil with a platinum wire worm pick. Push the worm gently onto the agarose pad until it becomes immobilized by sticking to the agarose. The worm gradually desiccates under the halocarbon oil and will die within 10 min. It is therefore important to acquire

enough skill with the different steps to manage the entire process, from mounting the worm to injection and retrieval, within a short period of time.

9. Put the coverslip with the mounted worm on the inverted microscope and bring the worm and needle tip in focus using the 10× objective. Position the worm at a 90° angle to the needle. Bring the worm close to the tip of the needle by moving the agarose pad with the gliding stage before switching to the 40× objective.
10. Move the stage such that the distal gonad cytoplasm is aligned with the tip of the needle. The *C. elegans* hermaphrodite gonad consists of two U-shaped arms, anterior and posterior to the vulva. Near the bend of the arm, plasma membranes begin to form around individual nuclei lining the walls of the gonad. Germline transformation is achieved by microinjection of DNA into the cytoplasm of the gonad syncytium.
11. Focus on the line of nuclei along the gonad and bring the tip of the needle into focus in that focal plane. Once this is achieved, move the stage until the tip of the needle makes a dent in the cuticle. Gently tap the back of the micromanipulator to make the needle penetrate the worm cuticle and enter the gonad.
12. Apply short pulses of pressure to the needle using the foot pedal until the solution of DNA is seen streaming into the gonad syncytium. The injection is deemed successful if the nuclei briefly swell from absorbing fluid.
13. Remove the needle by gently moving the stage. Check whether the needle is still freely flowing so it can be used again. Reposition the stage to inject the second gonad arm or move on to the next worm.
14. To retrieve injected worms from the agarose pad, move the pad to the dissecting microscope. Put a drop of M9 solution on top of the worm. The worm will be released from the pad and start thrashing. Suck the worm gently into a drawn-out glass capillary with a relatively wide opening and use this to transfer the worm to a fresh NGM plate. It often takes a few minutes for a worm to recover from the injection and desiccation under the halocarbon oil.
15. At the end of the injection session, transfer the injected worms with a sterile platinum pick to fresh plates to rid the worms of contaminating bacteria that were introduced during the microinjection procedure.
16. Over the next few days, check progeny of the injected worms for the appearance of transgenic animals. The number of transgenic animals per injected worm depends on the skill of the injector, quality of the DNA, genotype of the injected strain, and so on. From a typical session in which 20–40 worms were injected, we expect up to 40 transgenic progeny. Transfer the transgenic animals to fresh NGM plates.
17. Typically, 1 in 20 of the first-generation transgenic progeny has an extrachromosomal array that is stable enough to be transmitted to a reasonable fraction of subsequent progenies (20–60%). These are maintained as lines by periodically picking transgenic animals onto fresh NGM plates.

3.1.6. Integration by γ-Irradiation

1. To eliminate mosaicism and problems stemming from partial transmission, the transgenic arrays can be integrated into chromosomal DNA by γ-irradiation ***(10)***. For this, chose transgenic lines with a transmission frequency of 25–40%.

2. Subject well-fed plates with transgenic worms to 4000 rad from a ^{137}Cs source. Alternatively, X-rays can be used.
3. Pick 40 irradiated L4 worms that clearly have the transgenic array and transfer those to individual NGM agar plates with bacteria.
4. Allow the worms to grow for two generations so that potential integrations can become homozygous by self-fertilization and then clone 400 different F2 progeny by transferring them to individual plates.
5. Look for plates with 100% transmission of the transgene. Expect two to four integrated lines per 400 F2 animals. These lines are deemed independent if they were derived from different worms in **step 3**.
6. Stable transgenic lines should be back-crossed several times with wild-type animals to remove adventitious mutations that may have been introduced by the irradiation.

3.2. Knockdown of Gene Expression by RNAi

3.2.1. Transgenic Worms Expressing Snapback or Antisense Constructs

1. RNAi can be induced with a transgene that expresses antisense or snapback (hairpin) RNA ***(11)***. This method allows inactivation in a subset of cells as determined by the choice of tissue-specific promoters. It has the advantage that genes can be studied that are lethal if knocked down in a whole animal but allow survival when the knockdown is restricted to selected tissues. We typically use the *myo-3* promoter, which is a strong body wall muscle-specific promoter. Alternatively, conditional RNAi can be achieved with a heat shock promoter. Occasionally, sense constructs also induce RNAi, but this effect is not reliable enough for studying loss of function (*see* **Note 3**).
2. Clone a 400-bp or larger fragment of cDNA or of a gene of interest in the antisense orientation downstream of the chosen promoter (*see* **Note 4**). To make a snapback construct, first clone the antisense fragment. Then, clone the sense fragment downstream of the antisense fragment. Placing the antisense fragment upstream addresses concerns of unintentionally expressing a portion of the protein. A linker is needed to form a loop between the two sequences. This linker can be made by extending the antisense fragment to be 50 or 100 bp longer than the sense fragment.
3. If necessary, prevent spreading of RNAi to tissues that are not targeted by the transgene with a spreading defective mutant, such as *sid-1*, as background ***(12)***.
4. Inject the antisense or snapback construct at a concentration of 50–75 ng/μL. The injection mix should also contain a transformation marker (e.g., 50 ng/μL of the *rol-6* plasmid pRF4), mitochondrial markers (e.g., 1 ng/μL outer membrane yellow fluorescent protein [YFP] and 2 ng/μL mitochondrial matrix targeted cyan fluorescent protein [CFP]) and carrier DNA (e.g., 100 ng/μL pBluescript).

3.2.2. Injection or Soaking With In Vitro Synthesized Double-Stranded RNA

1. A portion of the gene of interest or a cDNA is cloned into the plasmid pBluescriptII (Stratagene), which has T7 and T3 RNA polymerase promoters

flanking the multicloning site. Typically, 400 bp of coding sequence is enough to elicit an RNAi response.

2. Plasmid DNA is purified with a Qiagen kit. Other methods for DNA isolation are fine. After this procedure, all solutions and manipulations should be ribonuclease (RNase) free.
3. Set up two restriction digests, each with 5 µg plasmid DNA in 100 µL reaction mix. One digest uniquely cleaves the plasmid on one side of the insert, the other on the other side.
4. Add 300 µL 1 *M* NH_4Ac, 10 m*M* EDTA, and 0.2% SDS to stop the reaction and 0.3 µL 20 mg/mL glycogen as carrier.
5. Add 400 µL phenol/chloroform (1:1), vortex the tube, and centrifuge for 1 min in a microfuge at 17,000*g*. Transfer the aqueous phase (upper layer) to a new tube.
6. Add 400 µL chloroform, vortex, and centrifuge for 1 min at 17,000*g*. Transfer the aqueous phase (upper layer) to a new tube.
7. Add 1 mL 100% ethanol, chill to –20°C, and centrifuge for 10 min at 17,000*g*. Carefully remove and discard the supernatant.
8. Add 1 mL ice-cold 70% ethanol to wash the pellet. Gently invert the tube, centrifuge for 1 min at 13,000 rpm, and discard the supernatant.
9. Dry the pellet and resuspend the DNA in 16 µL RNase-free TE.
10. Check the amount of linearized DNA by running 1 µL on an agarose gel.
11. Transfer 4 µL of the DNA solution (approx 1 µg) to a new tube.
12. Add 4 µL 5X transcription buffer, 4 µL 5X ribonucleoside triphosphate (rNTP) mix, 1 µL RNasin, 2 µL 100 m*M* dithiothreitol, 4 µL double-distilled water, and 1 µL T3 or T7 RNA polymerase at 20 U/µL (all from Promega).
13. Incubate for 90 min at 25°C when using T3 RNA polymerase or at 37°C when using T7 RNA polymerase.
14. Transfer 1 µL of both sense and antisense reactions to new tubes and store these at –20°C for later use as gel marker.
15. Combine the remainder of the sense and antisense reactions (19 µL each) in one tube.
16. Add 380 µL 1 *M* NH_4 acetate, 10 m*M* EDTA, and 0.2% SDS to stop the reaction and 0.3 µL 20 mg/mL glycogen as carrier.
17. Add 400 µL phenol/chloroform (1:1), vortex the tube, and centrifuge for 1 min in a microcentrifuge at 17,000*g*. Transfer the aqueous phase (upper layer) to a new tube.
18. Add 400 µL chloroform, vortex, and centrifuge for 1 min at 17,000*g*. Transfer the aqueous phase (upper layer) to a new tube.
19. Incubate for 10 min at 68°C, followed by 30 min at 37°C to anneal.
20. Add 1 mL 100% ethanol, chill to –20°C, and centrifuge for 10 min at 17,000*g*. Carefully remove and discard the supernatant.
21. Add 1 mL ice-cold 70% ethanol to wash the pellet. Gently invert the tube, centrifuge for 1 min at 17,000*g*, and discard the supernatant.
22. Dry the pellet and resuspend the RNA in 10 µL RNase-free TE.
23. Check the RNA concentration and duplex formation by running a 1-µL sample on a standard agarose gel, along with the samples of individual sense and antisense reactions (**step 14**).

24. Load the RNA into a needle and inject immediately or store the RNA at –70°C for later use.
25. Inject the double-stranded RNA (dsRNA) into the gonadal syncytia of young adult worms similar to the injection of DNA. Injections in other places of the body are also possible. We typically inject worms already expressing mitochondrial GFP to help ascertain the effect of a particular dsRNA on mitochondrial morphology (*see* **Note 5**).
26. As an alternative to injection, dsRNA can be introduced by soaking the worms in a solution with dsRNA ***(13)***. For this, collect L4 larvae in M9 buffer and wash by pelleting with a microcentrifuge followed by resuspending in M9 buffer.
27. Transfer the larvae to a fresh NGM agar plate without bacteria and let them move around for several minutes to remove more bacteria.
28. Resuspend the RNA from **step 21** in 10 μL soaking buffer.
29. Add 4 μL of this solution (4–20 μg) to a 200-μL PCR tube.
30. Add four to eight cleaned L4 larvae to the RNA solution and incubate for 24 h at 20°C.
31. Transfer the worms to fresh NGM plates with bacteria and look at progeny derived from the soaked worms.

3.2.3. Feeding RNAi

1. Timmons et al. demonstrated RNAi by feeding worms with bacteria expressing dsRNA for the gene of interest ***(14)***. Genomic DNA or cDNA is cloned into the vector pL4440, which is a modified version of pBluescript with a T7 promoter on each side of the multiple cloning site. This construct is transformed into HT115(DE3), an RNase III-deficient *E. coli* strain with isopropyl-β-D-thiogalactopyranoside (IPTG)-inducible T7 polymerase activity. Our feeding protocol is an adaptation of one from the Ahringer lab *(7)*.
2. Pick an isolated colony of HT115 bacteria, transformed either with empty vector or vector with gene of interest, and grow a small culture in LB media with 50 μg/mL ampicillin by incubating between 6 and 18 h at 37°C.
3. Place a drop of the bacterial suspension on NGM agar plates containing 25 μg/mL carbenicillin and 1 m*M* IPTG. Carbenicillin selects for pL4440. Incubate overnight at room temperature.
4. The following day transfer an L4 stage hermaphrodite (in our case, expressing the *rol-6* transgenic marker and mitochondrial GFP) onto a first plate, minimizing the amount of OP50 bacteria brought along from the growth plate.
5. Incubate the plate for several days to a week at 20°C to allow for RNAi to take effect.
6. Mount and observe the worms by fluorescence microscopy. We generally look at the effect on mitochondria in young adults of the F1 generation. A few of the F1 adults are transferred to a second set of plates so that mitochondrial morphology can be assessed in F2 adult worms as well.

***3.3. Imaging Mitochondria in* Caenorhabditis elegans**

3.3.1. Expression of Mitochondrial GFP Variants Using Tissue-Specific Promoters

1. In our lab, we used *ges-1*, *pes-10*, *col-12,* and *myo-3* promoters for intestinal cell, early embryonic, hypodermal cell, and body wall muscle expression, respectively,

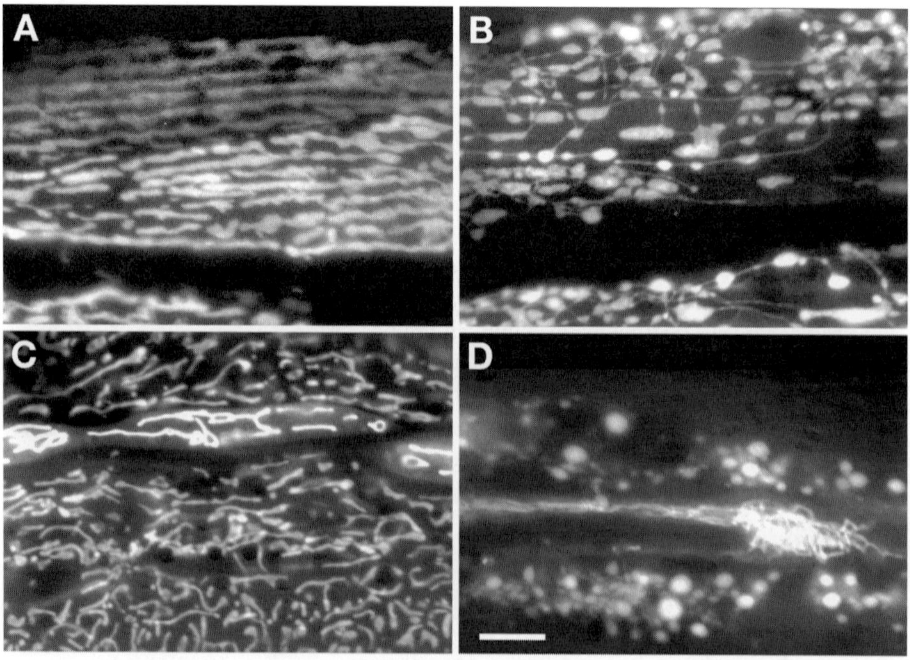

Fig. 1. Detection of *C. elegans* mitochondria with GFP or YFP. The transgene is present on an extrachromosomal array. (**A**) and (**B**) Expression using the *myo-3* gene promoter, which is specific for body wall muscles. The transgene encodes YFP with mitochondrial outer membrane targeting sequences. (**C**) and (**D**) Expression using the *col-12* promoter, which is specific for hypodermal cells. The transgene encodes GFP targeted to the mitochondrial matrix. **A** and **C** are from a wild-type worm; **B** and **D** are from worms treated with *drp-1* RNAi feeding bacteria. The DRP-1 protein was previously shown to be instrumental in mitochondrial outer membrane division *(2)*. The mitochondrial outer membrane often shows extensive connectivity when worms are treated with *drp-1* PNAi (**B**), while the matrix is disconnected and forms blebs (**D**). Scale bar: 10 μm.

as well as the endogenous promoters of the genes studied. Mitochondrial morphology is most clearly seen with *col-12* and *myo-3* promoters (**Fig. 1**). All our expression constructs were ultimately derived from vectors made in the laboratory of Andrew Fire (http://genome-www.stanford.edu/group/fire). The *myo-3* promoter was derived from the plasmid pPD96.52; other promoters consist of PCR fragments of genomic DNA (typically in the range of 0.5–1 kb) ligated into a Fire vector (**Fig. 1**).

2. For organelle targeting, we most often use the mitochondrial leader sequence from the Fire plasmid pPD96.32, which directs proteins to the mitochondrial matrix. We also use a mitochondrial outer membrane-targeting sequence derived from the yeast outer membrane protein TOM70. For this, the N-terminal 30 amino acids of TOM70 were amplified from yeast genomic DNA and fused to the N-terminus

of YFP *(2)*. YFP remains exposed to the cytosol, but unfortunately this can disrupt mitochondrial morphology when expressed at higher levels. As an alternative, we redirect YFP to the intermembrane space by placing the TOM70 sequence and YFP from the first construct behind the mitochondrial targeting sequence from pPD96.32. Localization of this new marker to the mitochondrial outer membrane has been verified with a *drp-1* mutant background where mitochondrial outer membrane and matrix compartments are clearly separated.

3. As fluorescent proteins, we use GFP, YFP, and CFP with appropriate filter sets from Chroma Technologies (Rockingham, VT, USA). YFP was made by introducing S65G, V68L, S72A, and T203Y mutations into GFP. CFP was derived from the plasmid pPD115.55, which contains Y66W, Y145F, M153T, and V163A mutations in GFP.
4. The choice of worm strains for injection depends on the experiment, but when background fluorescence is an issue, we use *glo-1* mutant animals. These worms have greatly reduced autofluorescence from gut granules, which helps to eliminate flare in the fluorescence microscope *(15)*.

3.3.2. Staining With Rhodamine 6G or MitoTracker

1. Pipet 0.5 mL of a 2.5 mg/mL stock of rhodamine 6G in ethanol onto 60-mm NGM agar plates seeded with bacteria. Spread the dye onto the agarose surrounding the bacterial lawn. The effective concentration will be approx 2.5 μg/mL.
2. Adult worms are transferred to these plates and allowed to feed for 2–3 h.
3. The worms are then transferred to a fresh NGM agar plate seeded with bacteria but without rhodamine G6 and allowed to feed for 1 h to decrease background fluorescence and gut staining.
4. Worms are then mounted on agarose pads and observed with a microscope (*see* next subheading).

3.3.3. Fluorescence Microscopy of Live Worms

1. Worms are mounted on slides with an agarose pad. We usually have a box of slides with agarose pads prepared in advance.
2. To make these slides, apply a 15-μL drop of melted 2% LMP agarose in M9 buffer to a standard microscope slide.
3. Quickly, before the agarose solidifies, spread the drop by pressing a coverslip onto the slide. The agarose should spread to a thin disk with a diameter of 1–2 cm. Allow a few minutes for the agarose to solidify at room temperature.
4. Gently remove the coverslip and let the agarose film dry at room temperature.
5. Place a 15-μL drop of M9 buffer with 1 m*M* Aldicarb on the agarose pad (*see* **Note 6**).
6. Using a platinum wire pick, transfer 15–25 worms from the NGM plate to the drop of buffer on the agarose pad (*see* **Note 7**).
7. Gently place a coverslip on the drop, taking care to crush the worms. The slide does not have to be sealed for quick observations of live worms, but be careful with focusing so worms are not squashed. Worms can be observed with a 100× oil immersion lens using filter sets for GFP, CFP, or YFP.

8. For longer exposures of live worms and for time-lapse photography, seal the slide with the nontoxic sealant VALAP. VALAP is 1:1:1 mixture of Vaseline, lanolin, and paraffin kept molten at 50°C. Spread VALAP along the edges of the coverslip as thinly as possible using a Q-tip. Excess VALAP can be removed with a scalpel.

3.4. Large-Scale Isolation of Mitochondria From *Caenorhabditis elegans*

3.4.1. Preparing Egg Plates

1. Mitochondria can be isolated from worms as a first step toward purifying mitochondrial proteins, for lipid analysis, to assess mitochondrial oxygen consumption rates, and so on. To obtain a good yield of mitochondria, many worms are needed. This can be achieved by growing worms on plates made with chicken eggs.
2. Make six hard-boiled eggs by boiling them for 12 min. Cracking of the eggs during boiling can be prevented by placing the raw eggs in water while it is still cold.
3. Chill the boiled eggs in cold water and peel the shell. Discard the yolks from three of the six eggs.
4. Add the remainder to 100 mL deionized water and mix with a Waring blender. Add more water if needed until the egg mixture is a thick but smooth paste.
5. Autoclave 500 mL NGM medium without Ca, Mg, or cholesterol.
6. Pour the NGM medium into an autoclaved 13×9 inch (33×23 cm) Pyrex baking pan.
7. Let the agar solidify by cooling, if necessary in a cold room.
8. Spread the egg mixture onto the NGM surface using a spatula.

3.4.2. Growing Worms on Egg Plates

1. Prepare ten 60-mm plates of worms grown on NGM agar seeded with OP50 bacteria. The plates should have just become full to have a large healthy starting population.
2. Add 1.5 mL M9 to each NGM worm plate and bring the worms into suspension by pipeting up and down with a glass Pasteur pipet.
3. Transfer the worm suspensions from all 10 plates onto the egg plate (*see* **Note 8**).
4. Grow for 3–4 d at room temperature (*see* **Note 9**).
5. Harvest worms from one egg plate by washing off the plate with 100 mL M9 buffer.
6. Pellet the worms by centrifugation for 5 min at 1000*g*.
7. Resuspend the pellet in 10 volumes ice-cold 0.1 *M* NaCl, add an equal volume of ice-cold 60% sucrose solution, and mix by inverting the tube.
8. Centrifuge for 2 min at 500*g* in a cooled benchtop centrifuge.
9. Transfer the uppermost layer, which mainly consists of floating worms, to a new tube using a glass Pasteur pipet.
10. Dilute the worm suspension with 10 volumes of ice-cold 0.1 *M* NaCl and centrifuge for 2 min at 500*g*.
11. Discard the supernatant and resuspend the pellet in 50 volumes of ice-cold 0.1 *M* NaCl.

3.4.3. Isolating Mitochondria

1. Pellet the worms by centrifuging for 5 min at 2000*g* ***(16)***.
2. Resuspend 5 g worms in 10 mL ice-cold IB with 1 m*M* PMSF.
3. Homogenize with 15 strokes of a chilled Potter-Elvehjem homogenizer.
4. Collect the homogenate in a 50-mL Falcon tube, increase the volume to 25 mL with IB with PMSF, and centrifuge for 10 min at 750*g*.
5. Transfer the supernatant to a fresh tube and resuspend the pellet in 10 mL IB with PMSF.
6. Homogenize the remaining pellet a second time with 15 more strokes of the homogenizer and repeat **steps 4** and **5**.
7. Combine the supernatants from **steps 5** and **6** and save an aliquot as total worm lysate.
8. Centrifuge the supernatants for 10 min at 12,000*g*. Save an aliquot of the resulting supernatant as postmitochondrial supernatant.
9. Resuspend the mitochondrial pellet in 12 mL IB and centrifuge for 10 min at 750*g*.
10. Transfer the supernatant to a new tube without disrupting the pellet and centrifuge for 10 min at 12,000*g*. Discard the supernatant.
11. Resuspend the mitochondrial pellets from **steps 9** and **10** in IB and combine in one tube.

***3.5. Isolation of Nucleic Acids and Proteins From* Caenorhabditis elegans**

3.5.1. Isolation of Caenorhabditis elegans *Genomic DNA*

1. Wash worms off plates with M9 buffer and collect in 15-mL conical glass tubes.
2. Pellet the worms by centrifugation for 5 min at 500-g in a tabletop centrifuge.
3. Resuspend the worms in 10 volumes of M9 and repellet by centrifugation.
4. Transfer the worm pellet to a pestle and freeze by adding liquid nitrogen.
5. Prechill the mortar and grind the worms, making sure that they stay frozen.
6. Quickly transfer the frozen powder into preheated lysis buffer. Avoid also transferring residual liquid nitrogen because that will boil on contact with lysis buffer. Always wear protective safety glasses.
7. Cap the tube and incubate for 1 h at 60°C and mix occasionally by inverting the tube.
8. Extract with 1 volume phenol, equilibrated to pH 7.5. Separate the phases by centrifugation for 5 min at 500*g* in a tabletop centrifuge. Transfer the (upper) aqueous phase to a new tube.
9. Extract the aqueous phase twice with phenol:chloroform (1:1) avoiding contaminants from the interface. Separate phases by centrifugation for 5 min at 500*g*.
10. Extract the aqueous phase once with chloroform.
11. Add 0.1 volume 3 *M* sodium acetate, pH 4.5, and 0.7 volumes isopropanol. Mix gently by inverting the tube and incubate for 3 min at room temperature.
12. Collect the DNA by centrifugation for 5 min at 1000*g*.
13. Discard the supernatant and wash the pellet with 75% ethanol chilled on ice.
14. Dry the pellet to air, add TE buffer, and dissolve the DNA by incubating for 1 h or more at room temperature (*see* **Note 10**).

3.5.2. Isolation of Total RNA

1. Collect the worms in M9 as described for DNA isolation.
2. Add 1 volume worms to 9 volumes of Trizol reagent and disrupt the worms and their DNA by passing 10 times through an 18- or 20-gage needle.
3. Incubate for 10 min at room temperature with periodic vortexing.
4. Centrifuge for 10 min at 500*g* to pellet insoluble material and transfer the supernatant to a new tube.
5. Extract with 1 volume phenol/chloroform and transfer the upper aqueous phase to a new tube.
6. Extract with 1 volume of chloroform and transfer the upper aqueous phase to a new tube.
7. Precipitate the RNA by adding 0.1 volume 3 *M* NaAc at pH 5.2 and 2.5 volumes ethanol, mixing and incubating for 30 min to –20°C.
8. Pellet the RNA by centrifugation for 10 min at 500g.
9. Decant the supernatant and wash the pellet with 70% ethanol.
10. Resuspend the pellet in RNase-free water.

3.5.3. Protein Extraction

1. Collect worms from two 6-cm NGM plates with 5 mL M9 buffer per plate and centrifuge for 5 min at 1000*g*.
2. Resuspend the pellet in 200 µL ice-cold Tris-buffered saline (TBS) (0.05 *M* Tris-HCl, pH 7.6, 0.15 *M* NaCl) with 1.5% *n*-octyl glucoside (Sigma-Aldrich) and protease inhibitor cocktail (Roche Molecular Biochemicals, Indianapolis, IN).
3. Homogenize with 15 strokes of a chilled, tight-fitting Potter-Elvehjem homogenizer.
4. To check whether cellular material has been released, take a small sample and look for empty cuticles under a microscope.
5. Incubate for 30 min on ice.
6. Centrifuge 30 min at 10,000*g* and 4°C and transfer the supernatant, which contains cellular proteins, to a new tube.
7. Estimate total protein with the Bradford method of protein estimation. Approximately 1 mg total protein can be obtained from worms collected from two fully grown 6-cm plates. Protein extracts can be stored at –70°C (*see* **Note 11**).

4. Notes

1. Nicking of the plasmid DNA reduces the chance of getting a stable extrachromosomal array. We therefore routinely make fresh plasmid DNA, dispense the DNA into smaller aliquots, and store those in a –20°C freezer without freeze-thaw function.
2. Genomic DNA isolated from *C. elegans*, instead of plasmids, as carrier for microinjection can improve germline expression of transgenic DNA ***(17)***.
3. Overexpression constructs sometimes induce RNAi, an effect that is usually undesirable because it is not always apparent whether the protein is over- or under-expressed. Such unintentional RNAi can be prevented by injecting overexpression constructs into an RNAi-defective mutant such as *rde-1* ***(18)***.

4. The size of the RNAi-inducing fragment influences the efficacy of gene knockdown. Generally, we use fragments 400 bp or larger. We find that for some genes even larger fragments are needed for optimal knockdown. In addition, some investigators use genetic backgrounds that are more sensitive to RNAi, such as the *rrf-3* mutant ***(19)***.
5. After recovery, the injected animals first lay some unaffected eggs, followed by a few partially affected eggs. RNAi is likely to be strongest in eggs laid 2–6 h after the injection. The effect decreases after 1 or 2 d. To distinguish these different levels of knockdown, transfer the injected animals to a fresh plate at 4–6 h after injection and then again at 16- to 24-h intervals. If the results of injections are variable, then it may help to have a single injected worm per plate.
6. We routinely use aldicarb, which is an inhibitor of acetylcholinesterase, to immobilize the worms for viewing by fluorescence microscopy. Levamisole, which is an inhibitor of the acetylcholine receptor, may also be useful, but the more widely used anesthetic azide is not recommended because it is an inhibitor of Ox-Phos and may therefore influence mitochondrial morphology.
7. To look at normal mitochondrial morphology, the worms that are picked should be healthy young adults. These worms should not come from a starved plate, and they should be picked from the bacterial lawn instead of away from the bacteria. Because growth temperature also affects mitochondrial morphology, it is preferred to keep this constant, for example, at 20°C.
8. Egg plates smell bad. Cover the plates with clear plastic wrap and store in a fume hood. When growing worms, leave a small opening in the plastic wrap to give the animals oxygen.
9. The large surface area of the plates and airflow in the fume hood can rapidly dry the plates. If the plates are hermetically sealed, then they stay moist. However, condensation on the plastic wrap will allow worms to crawl away from the food and onto the plastic. It is therefore better to leave an opening that allows evaporation and keeps worms from crawling away from the food. Drying of the plates can be prevented by adding 10–20 mL deionized water once a day. The agarose in the NGM plates also serves as a buffer for moisture.
10. If removal of RNA from the preparation is important, then deoxyribonuclease-free RNase A can be added at a final concentration of 10 μg/mL. Incubate for 1 h at 37°C and perform a second precipitation with sodium acetate and isopropanol, followed by a wash with 75% ethanol (**steps 11–14** of the DNA isolation protocol). If removal of contaminating bacterial DNA is important, then a more elaborate scheme to wash worms at the beginning of the protocol can be used ***(6)***.
11. Small preparations suffice for Western blot analysis. We typically load 100 adult worms boiled in 1X Laemmli sample buffer. It is also recommended to shear the DNA with several passes through a 20-gage hypodermic needle.

Acknowledgments

We thank other members of our lab for help in gathering these protocols and the worm community at large for providing such wonderful resources. Work in our lab is supported by grants from the American Cancer Society (RSG-01-147-01-CSM), the National Institutes of Health (GM051866), and the Jonsson Comprehensive Cancer Fund at the University of California at Los Angeles.

References

1. Rube, D. A. and van der Bliek, A. M. (2004) Mitochondrial morphology is dynamic and varied. *Mol. Cell Biochem.* **256–257,** 331–339.
2. Labrousse, A. M., Zapaterra, M., Rube, D. A., and van der Bliek, A. M. (1999) *C. elegans* dynamin-related protein *drp-1* controls severing of the mitochondrial outer membrane. *Mol. Cell* **4,** 815–826.
3. Wood, W. B. (ed.) (1988) *The Nematode* Caenorhabditis elegans, Cold Spring Harbor Laboratory Press, Cold Spring Harbor, NY.
4. Riddle, D. L., Blumenthal, T., Meyer, B. J., and Priess, J. R. (eds.) (1997) *C. elegans II,* Cold Spring Harbor Laboratory Press, Cold Spring Harbor, NY.
5. Epstein, H. F., and Shakes, D. C. (eds.) (1995) Caenorhabditis elegans*: Modern Biological Analysis of an Organism*, Academic Press, San Diego, CA.
6. Hope, I. A. (ed.) (2000) C. elegans: *A Practical Approach*, Oxford University Press, Oxford, UK.
7. Fraser, A. G., Kamath, R. S., Zipperlen, P., Martinez-Campos, M., Sohrmann, M., and Ahringer, J. (2000) Functional genomic analysis of *C. elegans* chromosome I by systematic RNA interference. *Nature* **408,** 325–530.
8. Fire, A. and Waterston, R. H. (1989) Proper expression of myosin genes in transgenic nematodes. *EMBO J.* **8,** 3419–3428.
9. Granato, M., Schnabel, H., and Schnabel, R. (1994) Pha-1, a selectable marker for gene transfer in *C. elegans*. *Nucl. Acids Res.* **22,** 1762–1763.
10. Wightman, B., Ha, I., and Ruvkun, G. (1993) Posttranscriptional regulation of the heterochronic gene lin-14 by lin-4 mediates temporal pattern formation in *C. elegans*. *Cell* **75,** 855–862.
11. Tavernarakis, N., Wang, S. L., Dorovkov, M., Ryazanov, A., and Driscoll, M. (2000) Heritable and inducible genetic interference by double-stranded RNA encoded by transgenes. *Nat. Genet.* **24,** 180–183.
12. Winston, W. M., Molodowitch, C., and Hunter, C. P. (2002) Systemic RNAi in *C. elegans* requires the putative transmembrane protein SID-1. *Science* **295,** 2456–2459.
13. Tabara, H., Grishok, A., and Mello, C. C. (1998) RNAi in *C. elegans*: soaking in the genome sequence. *Science* **282,** 430–431.
14. Timmons, L., Court, D. L., and Fire, A. (2001) Ingestion of bacterially expressed dsRNAs can produce specific and potent genetic interference in *Caenorhabditis elegans*. *Gene* **263,** 103–112.
15. Hermann, G. J., Schroeder, L. K., Hieb, C. A., et al. (2005) Genetic analysis of lysosomal trafficking in *Caenorhabditis elegans*. *Mol. Biol. Cell* **16,** 3273–3288.
16. Jonassen, T., Marbois, B. N., Faull, K. F., Clarke, C. F., and Larsen, P. L. (2002) Development and fertility in *Caenorhabditis elegans* clk-1 mutants depend upon transport of dietary coenzyme Q8 to mitochondria. *J. Biol. Chem.* **277,** 45,020–45,027.

17. Kelly, W. G., Xu, S., Montgomery, M. K., and Fire, A. (1997) Distinct requirements for somatic and germline expression of a generally expressed *Caenorhabditis elegans* gene. *Genetics* **146,** 227–238.
18. Tabara, H., Sarkissian, M., Kelly, W. G., et al. (1999) The *rde-1* gene, RNA interference, and transposon silencing in *C. elegans*. *Cell* **99,** 123–132.
19. Simmer, F., Tijsterman, M., Parrish, S., et al. (2002) Loss of the putative RNA-directed RNA polymerase RRF-3 makes *C. elegans* hypersensitive to RNAi. *Curr. Biol.* **12,** 1317–1319.

VI

Functional Genomics

35

Yeast Mitochondrial Transcriptomics

Mathilde Garcia, Xavier Darzacq, Frederic Devaux, Robert H. Singer, and Claude Jacq

Summary

Although 30 years ago it was strongly suggested that some cytoplasmic ribosomes are bound to the surface of yeast mitochondria, the mechanisms and the raison d'être of this process are not understood. For instance, it is not perfectly known which of the several hundred nuclearly encoded genes have to be translated to the mitochondrial vicinity to guide the import of the corresponding proteins. One can take advantage of several modern methods to address a number of aspects of the site-specific translation process of messenger ribonucleic acid (mRNA) coding for proteins imported into mitochondria. Three complementary approaches are presented to analyze the spatial distribution of mRNAs coding for proteins imported into mitochondria. Starting from biochemical purifications of mitochondria-bound polysomes, we describe a genomewide approach to classify all the cellular mRNAs according to their physical proximity with mitochondria; we also present real-time quantitative reverse transcription polymerase chain reaction monitoring of mRNA distribution to provide a quantified description of this localization. Finally, a fluorescence microscopy approach on a single living cell is described to visualize the in vivo localization of mRNAs involved in mitochondria biogenesis.

Key Words: DNA microarrays; mRNA localization; Q-RT-PCR; single-cell FISH.

1. Introduction

The complex cellular processes that supervise the building of mitochondria are not all identified. The protein import process is probably the best-clarified step. Three decades of smart biochemical experiments, mainly conducted in yeast, have led to a precise description of the machineries involved *(1,2)*. One of the next objectives will be to integrate these biochemical concepts into a living process in which each step will be regulated according to cellular requirements. In that respect, the question of the process by which cytoplasmically translated proteins are delivered to mitochondria is still a matter of speculation. It was

From: *Methods in Molecular Biology, vol. 372: Mitochondria: Practical Protocols*
Edited by: D. Leister and J. M. Herrmann © Humana Press Inc., Totowa, NJ

suggested *(3)* that proteins could be translated to the vicinity of mitochondria, and that cotranslational protein import into mitochondria could represent an alternative view *(4)* to an uncoupled process. One the other hand, the well-established evidence that some mitochondrial proteins can be imported in vitro has motivated the general belief that mitochondria-bound polysomes may not represent an obligatory process and, to some extent could be experimental artifacts *(5)*.

However, several experiments conducted with new methodologies have considerably revitalized this question. First, by means of a genomewide approach, it was shown *(6)* that a large proportion of messenger ribonucleic acids (mRNAs) for nuclear-encoded, mitochondrially localized proteins are translated to the vicinity of mitochondria. The fact that a strong correlation was found *(6)* between the genes with mRNA translated to the vicinity of mitochondria and their prokaryotic origin (as suggested in **ref.** *7*) gives credence to the results of microarray analyses. After these analyses, a mitochondrial localization of RNA (MLR) value ranging from 0 to 100 was given. Genes with a high MLR value (80–100) correspond to mRNAs likely to be found to the vicinity of mitochondria. In vivo fluorescent microscopy analyses have established that this corresponds to the actual cellular spatial distribution of these mRNAs *(6)*. Finally, genetic experiments conducted with the gene *ATP2* have shown that this specific mRNA spatial distribution is highly dependent on the integrity of its 3′-UTR sequence, which is also required for correct and functional mitochondrial activity *(6)*.

The putative role of restricted translation process in the biogenesis of mitochondria can be addressed through different strategies. We present three powerful experimental approaches that should shed new complementary light on this interesting question.

2. Materials

The strain CW252 *(8)* isogenic to W303 should be favored because of its intron-less mitochondrial genome, allowing easier detection of mitochondrial transcripts.

2.1. Mitochondria-Bound and Total RNA Isolation for Quantitative Polymerase Chain Reaction and Microarrays

1. YPGal: 1% (w/v) Bacto™ peptone, 1% (w/v) Bacto yeast extract, 2% (w/v) galactose. Autoclave 30 min at 110°C. Store at room temperature.
2. Gal-rich medium: 1% (w/v) Bacto peptone, 1% (w/v) Bacto yeast extract, 2% (w/v) galactose, 0.1% (w/v) KH_2PO_4, 0.12% (w/v) $(NH_4)_2SO_4$. 1 mL mix is made and dispensed in two 2-mL Erlenmeyer flasks and then autoclaved for 30 min at 110°C.
3. Preincubation buffer: 100 m*M* Tris-HCl, pH 9.3, 0.5 *M* β-mercaptoethanol. Do not store; should be prepared just before use.

4. Digestion buffer: 20 m*M* potassium phosphate buffer, pH 7.4, 1.35 *M* sorbitol. Do not store; should be prepared the day before the mitochondria isolation.
5. Zymolyase 100T (Seikagatu Corp., 120493).
6. Washing medium: 1% (w/v) Bacto peptone, 1% (w/v) Bacto yeast extract, 1 *M* sorbitol, 0.1% (w/v) KH_2PO_4, 0.12% (w/v) $(NH_4)_2SO_4$. Autoclave 30 min at 110°C.
7. Regeneration medium: 1% (w/v) Bacto peptone, 1% (w/v) Bacto yeast extract, 2% (w/v) galactose, 1 *M* sorbitol, 0.1% (w/v) KH_2PO_4, 0.12% (w/v) $(NH_4)_2SO_4$. Autoclave 30 min at 110°C.
8. Sorbitol-cycloheximide ice cube: 1 *M* sorbitol, 200 µg/mL cycloheximide. Should be prepared the day before the mitochondria isolation and stored at –20°C.
9. Cycloheximide solution: 100 mg/mL cycloheximide in ethanol solvent. Prepare 1 mL the day of mitochondria isolation; store at 4°C before use. Should not be conserved more than 1 d.
10. Sorbitol-cycloheximide buffer: prepare 1 *M* sorbitol stock solution. Autoclave 30 min at 110°C. Store at room temperature. Just before use, add 200 µg/mL cycloheximide to 100 mL 1 *M* sorbitol.
11. Mannitol buffer: 0.6 *M* mannitol, 30 m*M* Tris-HCl, pH 7.6, 5 m*M* MgAc, 100 m*M* KCl. Autoclave 30 min at 110°C. Store at room temperature. Just before use, complete with 5 m*M* β-mercaptoethanol, 200 µg/mL cycloheximide, 500 µg/mL heparin, and 1 L for 20 g of yeast dry weight of protease inhibitors (Sigma).
12. 30 mL Thomas Glass Potter with striated tip for more efficient cell breaking.

2.2. RNA Purification for Quantitative Polymerase Chain Reaction and Microarrays

1. TES buffer: 10 m*M* Tris-HCl, pH 7.5, 10 m*M* ethylenediaminetetraacetic acid, 0.5% (v/v) sodium dodecyl sulfate (SDS). Autoclave 30 min at 110°C. Store at room temperature.
2. Phenol-chloroform mix: phenol:chloroform 5:1. Store at 4°C.
3. Ready Red. Store at 4°C.
4. Sodium acetate: 3 *M* NaAc, pH 5.3. Autoclave 30 min at 110°C. Store at room temperature.
5. RNA purification kit: NucleoSpin RNA II kit from Macherey-Nagel.
6. Water: use molecular biology water for good RNA stability.

2.3. Quantitative Reverse Transcriptase Polymerase Chain Reaction

1. Polymerase chain reaction (PCR) machine (Mastercycler Eppendorf) is used for incubation steps.
2. Primers: random hexamers from Roche (1034731) and oligo dT from Invitrogen (yo1212). Store at –20°C.
3. Reverse transcriptase (RT) primer mix: 5 µg random hexamers (2.5 µL), 2 µg oligo dT (4 µL). Complete with ribonuclease (RNase)-free water to a final volume of 23 µL. Prepare on ice just before use.
4. BRL Superscript II kit (Gibco): this kit contains 5X SSII buffer, Superscript enzyme, and dithiothreitol (DTT) (0.1 *M*). Store at –20°C.

5. Deoxynucleotide 5′-triphosphate (dNTP) mix: 2.5 m*M* deoxyadenosine 5′-triphosphate, 2.5 m*M* deoxythymidine 5′-triphosphate, 2.5 m*M* deoxycytidine 5′-triphosphate, 2.5 m*M* deoxyguanosine 5′-triphosphate. Store at –20°C.
6. Reaction mix: 8 µL SSII buffer, 4 µL DTT, 2 µL dNTP. Prepare on ice just before use.
7. PCR extract kit: for retrotranscription product purification, use Nucleospin Extract Kit (Macherey-Nagel).
8. LightCycler instrument (Roche, 2011468).
9. LightCycler capillaries (20 µL) (Roche, 1909339).
10. Barrier tips (neptum) should be used to avoid deoxyribonucleic acid (DNA) contamination from pipetman.
11. QuantiTech Sybr Green PCR kit (Qiagen, 204143).
12. Primers: for every target transcript, a primer pair (for and rev primers) should be designed following kit instruction. Each primer is dissolved to a final concentration of 100 pmol/µL (10X primer). For and rev primers are then mixed to a final concentration of 10 pmol/µL (quantitative PCR [Q-PCR] primer mix). 10X primers and primer mix are stored at –20°C. Three types of primer pairs are used for precise quantification of RNA transcript localization: target transcript primers are used to quantify RNA of interest in each fraction, normalization primers are used to quantify mitochondrial RNA (e.g., *COX1* and *COX2*) and determine mitochondria purification yield, and contamination marker primers (e.g., *ACT1* and *RPL10*) are used to evaluate cytosolic RNA contamination.

2.4. Labeled Complementary DNA Synthesis for Microarray Analyses

1. Mastercycler personal (Eppendorf).
2. Random hexamers and oligo dT (12–18) (Invitrogen).
3. Molecular biology-grade water.
4. Superscript II RNase H reverse transcriptase, RT 5X first-strand buffer, and 0.1 *M* DTT (Invitrogen).
5. Cy3- and Cy5-linked deoxyuridine triphosphate (dUTP) from Amersham.
6. dNTP set, 100 m*M*.

2.5. RNA Hydrolysis Before Complementary DNA Purification

1. Pure NaOH.
2. 37% HCl.

2.6. Purification of Labeled Complementary DNA for Microarray Hybridization

1. 2K15 centrifuge (Sigma).
2. 5415 D centrifuge (Eppendorf).
3. 95% ethanol, spectrophotometry grade.
4. Qiaquick PCR purification kit.
5. 3 *M* Na acetate, pH 5.2.

2.7. Prehybridization of Microarrays

1. Yeast open reading frame microarray from the Service de Genomique du Departement de Biologie (www.transcriptome.ens.fr): about 6000 oligonucleotides representing the complete set of yeast open reading frames, deposited in duplicates on Ultragaps aminosilane slides (Corning).
2. 20X Standard Saline Citrate (SSC) from Qbiogen.
3. 20% (w/v) SDS.
4. 30% (w/v) bovine serum albumin (BSA).
5. Isopropanol (Merck).
6. 50-mL polypropylene tubes.

2.8. Microarray Hybridization

1. ArrayIT hybridization chamber (Telechem).
2. HS60 (60X, 22-mm) coverslips (Grace Biolabs).
3. 2X hybridization buffer: 50% (w/v) formamide, 10X SSC, 0.2% (w/v) SDS.
4. 70°C heating block.
5. 42°C water bath.

2.9. Microarray Washing

1. Washing buffer 1: 1X SSC, 0.2% (w/v) SDS.
2. Washing buffer 2: 0.1X SSC, 0.2% (w/v) SDS.
3. Washing buffers 3 and 4: 0.1X SSC.
4. CR412 centrifuge (Jouan) for 50-mL tube centrifugation.

2.10. Microarray Scanning and Image Analysis

1. Genepix 4000B scanner (Axon).
2. Genepix Pro 5.1 software.
3. PC Dell Dimension 8250, 42.4-GHz Pentium, 1-GB RAM, 75.50-GB hard drive; Windows 2000 or Windows XP.
4. Excel software.

2.11. Fixation of Cells and Spheroplasting for Fluorescent In Situ Hybridization

1. Buffer B: 1.2 *M* sorbitol (from 3 *M* autoclaved solution), 0.1 *M* potassium phosphate, pH 7.5 (from 1 *M* autoclaved solution); store at room temperature.
2. Formaldehyde, electron microscopy grade (Electron Microscopy Science, Fort Washington, PA).
3. Spheroplast buffer, 28.6 m*M* β-mercaptoethanol (Sigma, St. Louis, MO), 20 m*M* vanidyl ribonucleoside complex (New England Biolabs, Beverly, MA), 120 U/mL RNase inhibitor (Roche, Indianapolis, IN); in buffer B; prepare fresh.
4. Lyticase stock (Roche, Indianapolis, IN): 25,000 U/mL in water; store at –20°C.
5. 70% ethanol.
6. 22 × 22 mm type 1 coverslips (Fisher).

7. 0.1 *N* hydrochloric acid; dilute fresh.
8. 0.01% (w/v) poly-L-lysine (Sigma); dilute fresh.

2.12. Probe Synthesis Labeling and Purification for Fluorescent In Situ Hybridization

1. Amino-modifier C6 dT (Glen Research, Sterling, VA).
2. Synthesis is performed on an Applied Biosystems automated DNA/RNA synthesizer (model 392/394) using a 0.2-μm scale cartridge.
3. Fluorophores typically used for labeling are fluorescein isothiocyanate (FITC) (Molecular Probes, Eugene, OR), Cy3, Cy3.5, Cy5, Cy5.5 Fluorolink™ monofunctional dye (Amersham Biosciences, Piscataway, NJ).
4. Carbonate buffer: 0.1 *M* sodium carbonate, pH 9.0; store frozen at –20°C in 500-μL aliquots.
5. Sephadex G50 (Sigma), rehydrated and degassed in 10 m*M* TEAB (*see* **item 6**) and packed into a 25-mL plastic pipet by gravity flow.
6. 2 *M* Triethylamine bubbled (TEAB): weigh 101 g triethyl amine (Sigma) into a flask, add 200 mL water, and insert a Pasteur pipet in the solution connected to a dry ice chamber. Allow the Pasteur pipet to bubble overnight in the solution to verify that the pH is below 8.0. Adjust volume to 500 mL and store at 4°C. Triethyl amine is extremely corrosive and should not be exposed to plastic when pure; the bubbling operation should be conducted under a fume hood.
7. 10 m*M* TEAB: dilute from 2 *M* TEAB and store at 4°C.

2.13. Fluorescent In Situ Hybridization

1. 2X SSC: dilute from 20X SSC (Roche).
2. Phosphate-buffered saline (PBS): dilute from 10X PBS (Roche).
3. Competitor nucleic acids mix: 2.5 mg/mL sonicated salmon sperm DNA (Sigma), 2.5 mg/mL *Escherichia coli* transfer RNA (Sigma); store at –20°C in 100-μL aliquots.
4. Formamide solution: 40% (w/v) formamide (Sigma) in 2X SSC.
5. Solution F: 80% (w/v) formamide, 10 m*M* sodium phosphate, pH 7.0.
6. Solution H: 4X SSC, 4 μg/μL RNase-free BSA (Roche), 20 m*M* vanidyl ribonucleoside complex, 0.24 U/μL RNase inhibitor.
7. Triton wash solution: 0.1% (v/v) Triton X-100 in 2X SSC.
8. 4,6-Diamidino-2-phenylindole (DAPI) solution: 0.5 μg/mL DAPI (Molecular Probes) in PBS; store at 4°C.
9. Mounting medium: ProLong Antifade Kit (Molecular Probes).
10. Nail polish.

3. Methods

The following three methods aim at assessing the spatial distribution of nuclear-encoded mRNAs coding for mitochondrially localized products. The first two methods, quantitative reverse transcriptase polymerase chain reaction

(Q-RT-PCR) and microarray analyses, rely on the biochemical purification of the mitochondria-bound polysomes, whereas the third method allows observation of the in vivo localization of specific mRNAs. The aim of the real-time Q-RT-PCR analysis (**Subheadings 3.2.** to **3.3.**) is to provide a quantitative assessment of the spatial distribution of a specific mRNA based on a biochemical purification of mitochondria-bound polysomes (*see* **Note 1**, **Fig. 1**). The genomewide approaches using DNA microarrays (**Subheadings 3.4.–3.10.**), although less precise, have the clear advantage of allowing global analyses of RNA subpopulations. Therefore, microarray analyses identify subgroups of colocalized RNA and allow searching for correlations between mRNA location and protein properties or characterization of new mRNAs located to the mitochondria and likely to encode mitochondrial proteins ***(6)***. The microarray protocol used for mRNA mitochondrial location analyses is a standard one, identical to the protocols used for gene expression studies. However, the methods used for data analyses are different (*see* **Note 2**, **Fig. 2**).

Finally, single-cell fluorescent *in situ* hybridization (FISH) experiments (**Subheadings 3.11.–3.15.**) represent a necessary in vivo complement of the two preceding methods. Messenger ribonucleoproteins (mRNP) cytoplasmic localization can be directly observed by fluorescence microscopy, either in live cells using green fluorescent protein reporter proteins ***(9)*** or in fixed cells where endogenous mRNAs can be detected by FISH. New developments in probe design and in fluorophore chemistry allow routine detection of single molecules of mRNA within their cellular environment ***(10,11)***. FISH is particularly well suited or dissecting the different mechanisms governing mRNA localization in yeast ***(12)***.

The protocol we present, adapted from **refs. *13*** and ***14***, is designed to simultaneously compare the special distribution of different mRNAs relative to each other or to the mitochondria. FISH is particularly adapted in this case because mitochondria can be unambiguously detected using probes directed to the mitochondrial ribosomal RNAs (rRNAs), allowing for the simultaneous detection of mRNAs and mitochondrion in a single step. As an example, we show the simultaneous detection of the mitochondria-addressed *ATP2* mRNA compared to the *YRA1* mRNA, which is used as a nonlocalized control.

3.1. Mitochondria-Bound and Total RNA Isolation for Q-PCR and Microarrays

1. Preculture: CW252 yeast strain is grown in 20 mL YPGal at 30°C for 24 h with agitation (250 rpm).
2. Culture: an appropriate aliquot of preculture is transferred in 1 L Gal-rich medium and incubated for 1 night at 30°C with agitation; the next morning the OD_{600} should be between 0.8 and 3 (CW252 generation time in Gal-rich medium is around 2.5 h). The OD_{600} must be properly measured before mitochondria

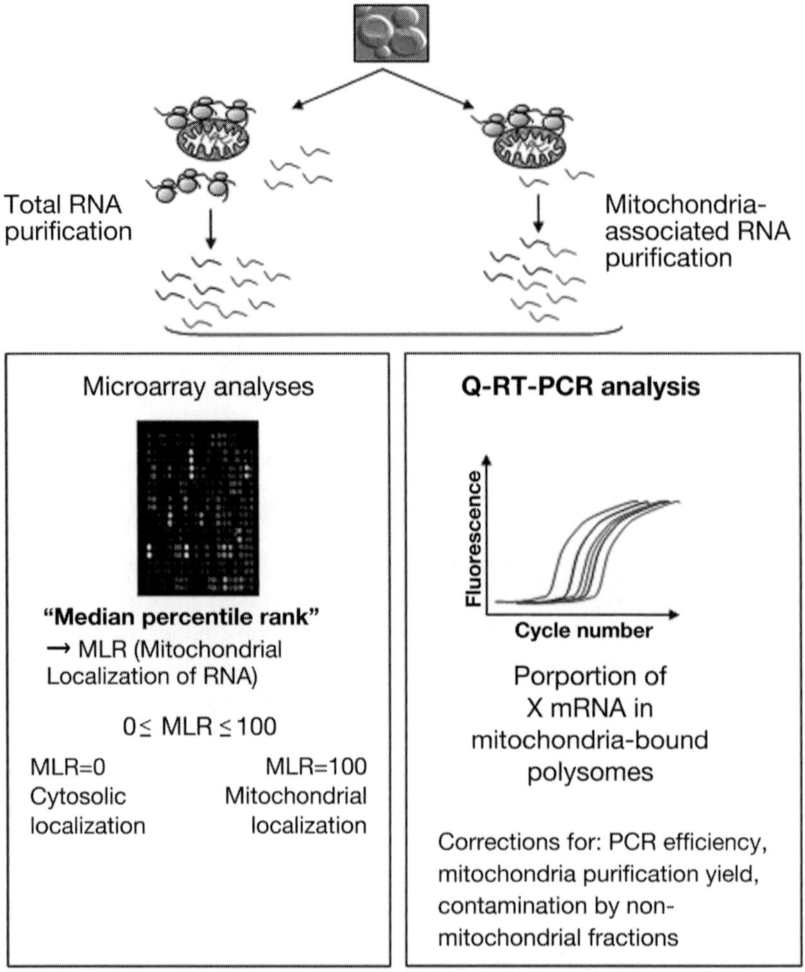

Gene	MLR	% mRNA in mitochondria-bound polysomes	
		Before Correction	After Correction
ACT1	24	17%	0%
ATP2	100	59%	50%
ATP3	90	49%	38%
COX4	27	7%	0%
COX6	30	11%	0%

Fig. 1. Biochemical methods to analyze the spatial distribution of mRNAs coding for proteins imported into mitochondria and example of results for *ACT1* (contamination marker), *ATP2*, *ATP3* (mitochondria-associated RNA), *COX 4*, and *COX6* (non-mitochondria-associated RNA).

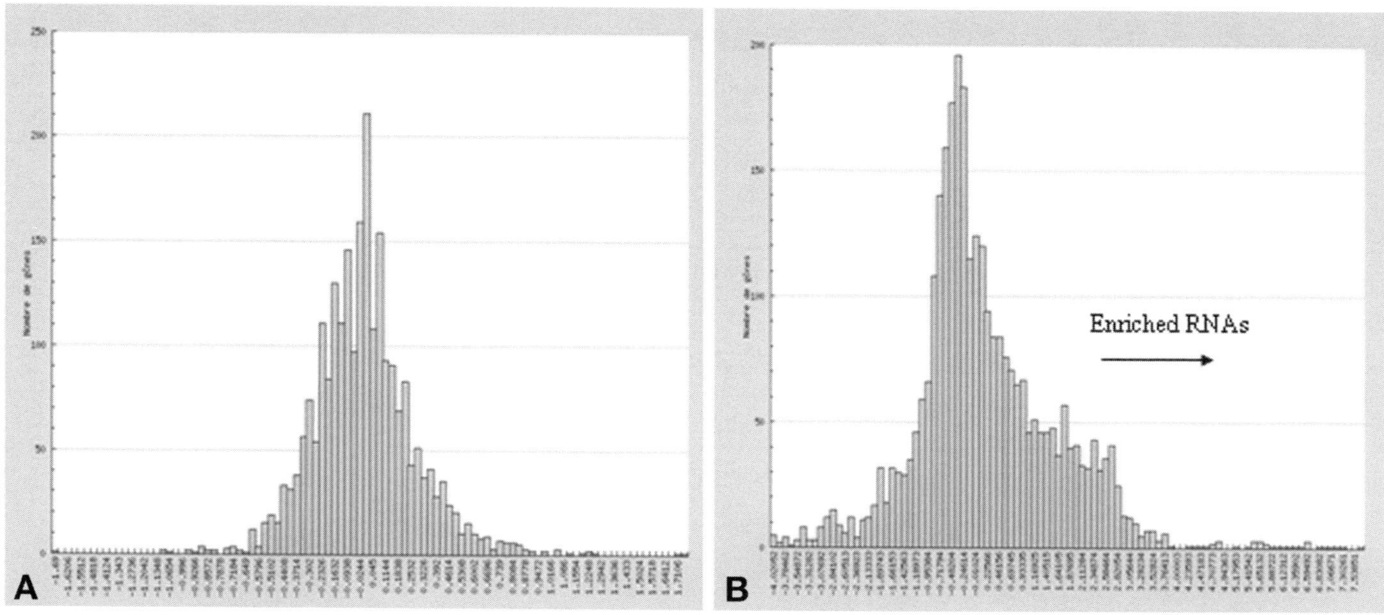

Fig. 2. The distribution of mitochondrial location analyses can be skewed. (**A**) Example of distribution of the Cy5/Cy3 fluorescence ratios (Rf) that can be obtained from a standard global gene expression experiment. Both repressed and activated genes are expected; the distribution is similar to a normal distribution. (**B**) Example of distribution of Rf that can be obtained from microarray experiments comparing total RNAs and mitochondria-associated RNAs. Only RNAs enriched in the Cy5 channel are expected: the distribution is skewed by enriched RNA.

isolation to evaluate dry weight (DW) using the formula DW = 0.28 × OD_{600} × V (V is the volume in liters).

3. Cells are collected by centrifugation for 15 min at 4300*g* at 4°C and washed with water stored at 4°C. Note that all centrifugation steps allowing washing and changing culture medium are performed for 15 min at 4300*g* at 4°C.
4. The cells are suspended in 20 mL per gram dry weight of preincubation buffer and then incubated in a 200-mL Erlenmeyer flask at 30°C with agitation for 10 min. Cells are then washed several times with 20 m*M* potassium phosphate buffer, pH 7.4; a total volume of 1 L buffer is used for this washing step.
5. Cells are suspended in 1-L Erlenmeyer flask to a final OD_{600} of 12 in digestion buffer. 5 mg zymolyase is added to the culture, and the mix is incubated for 10 min with agitation for enzymatic digestion of cell walls. After incubation, digestion should be performed at 80%. To verify digestion efficiency, the decrease of OD_{600} can be measured by mixing 50 µL culture in 1 mL water. Vigorously shake before measuring OD_{600} to perform cell lysis by osmotic shock in water. After digestion, wash cells with washing medium and then incubate in 200 mL regeneration medium in 1-L Erlenmeyer flask at 30°C with gentle agitation for 3 h.
6. During the incubation step, prepare cycloheximide solution and weigh 20 mg heparin, which will be added to 40 mL mannitol buffer just before use (*see* **step 9**).
7. Add 600 µL cycloheximide solution to the culture to block translation machinery and incubate at 30°C with agitation for 10 min additional. Set apart 8 mL culture, which will be used to prepare total RNA, dispense them in 2-mL Eppendorf tubes, and centrifuge at 18,000*g* at 4°C for 3 min. Wash pellet with 1 *M* sorbitol. Store the cell pellet at –80°C.
8. Perform a thermal shock to stop cell metabolism by dispensing culture in a beaker containing sorbitol-cycloheximide ice cubes. From that moment, all steps must be performed at 4°C in a cold room. Centrifuge culture at 4300*g* for 15 min at 4°C and wash with 100 mL sorbitol-cycloheximide buffer.
9. Suspend the pellet in 4 mL mannitol buffer. Cells are broken by 20 strokes in a glass potter, transferred into a 50-mL falcon tube, and centrifuged 8 min at 1700*g* to remove nucleus. Supernatant is taken off and transferred in a new Falcon. The pellet is suspended once more in 3 mL mannitol buffer, submitted to 20 strokes in the glass potter, and again centrifuged at 1700*g* for 8 min. The supernatant is added to the previous one, and a last step of centrifugation allows complete nuclear removal. Supernatant is taken off and transferred into a 15-mL tube that fits in high-speed centrifuge adapters. Centrifugation at 14,600*g* for 30 min leads to a red pellet of mitochondria, which is washed one time with mannitol buffer before storing at –80°C.

3.2. RNA Purification for Q-PCR and Microarrays

1. Suspend cells and mitochondrial pellets in 400 µL TES buffer and transfer in 1.5-mL Eppendorf tubes. Add 400 µL phenol-chloroform mix and incubate 15 min at 65°C. During the incubation step, vortex tubes 30 s every 5 min to homogenize. Centrifuge at room temperature for 15 min at 18,000*g*.

2. Transfer the aqueous phase in a new 1.5-mL Eppendorf tube and add 400 μL phenol-chloroform mix. Vortex three times for 30 s; centrifuge 15 min at 18,000*g*.
3. Transfer the aqueous phase in a new 1.5-mL Eppendorf tube and add 300 μL chloroform. Vortex three times for 20 s; centrifuge 2 min at 18,000*g*.
4. Transfer aqueous phase in tubes containing 30 μL sodium acetate, add 600 μL ethanol, and incubate 1 h at –20°C for RNA precipitation. Centrifuge 15 min at 18,000*g* at 4°C and resuspend the pellet in 20 μL RNase-free water.
5. Quantify RNA by measuring 260-nm absorbance (1 absorbance unit corresponds to 40 μg/μL RNA); about 250 μg total RNA purification and 50 μg from mitochondria purification are obtained in routine experiments.
6. Perform a purification step using the RNA purification kit. The RNA cleanup is an essential step for the quality of the reverse transcription. Follow the recommendations of the Macherey-Nagel kit (total RNA preparation from biological fluids section). Elute the column with 60 μL water first and then reelute with the same 60 μL.
7. Quantify purified RNA. For Q-RT-PCR, adjust concentration to 50 ng/μL. Store RNA at –20°C. When used, always keep RNA tubes in ice to avoid degradation.

3.3. Quantitative RT-PCR

1. Retrotransciption: for each sample, add 1 μL RNA to 23 μL RT primer mix. Incubate for 10 min at 70°C. Put on ice and add 14 μL reaction mix and 2 μL Superscript polymerase. Incubate 10 min at 23°C, then 1 h at 42°C. Store at –20°C. It is recommended to perform a control reaction without added RNA to verify the absence of contamination in the RT mix.
2. Purification: purification of RT product is an important step for Q-PCR efficiency. Just follow the recommendations of the Macherey-Nagel PCR extract kit (protocol for direct purification of PCR products). Elute with 50 μL elution buffer, then dilute to 1/10, 1/50, and 1/100 for Q-PCR.
3. Real-time Q-PCR: for each sample dilution (1/10, 1/50, 1/100), mix in a capillary tube 10 μL 2X Sibr mix and 8 μL H_2O (from QuantiTech Sybr Green PCR kit), 1 μL RT dilution, and 1 μL desired Q-PCR primer mix. Perform a LightCycler program as follows: a first step of initial polymerase activation of 15 min at 95°C, a second step of 55 cycles of amplification that consists of 15 s at 95°C (denaturation step), 30 s at 54°C (annealing step), and 20 s at 72°C (extension step). Transition rate is 20°C/s. Fluorescence acquisition is performed during the extension step. After amplification, the cycler performs a melting curve of product by increasing slowly (0.1°C/s) from a low temperature (65°C) to a high temperature (95°C) and measuring the decrease in the fluorescence. This allows verification of the amplification of a unique and specific product during Q-PCR. The program ends with a cooling step (20°C/s) to reach 40°C. Do not forget to carry out a Q-PCR with purified RNA diluted to 1/10 as a template to check the absence of DNA in RNA purifications.
4. Q-PCR data analysis is conducted as indicated in **Note 1**.

3.4. Labeled Complementary DNA Synthesis for Microarray Analyses

1. Prepare two 0.2-mL PCR tubes containing about 10 μg of either mitochondria-bound or total RNAs.
2. Add 5 μg random hexamers and 2 μg oligo dT in each tube. Add water to a final volume of 23 μL.
3. Incubate the tubes at 70°C for 10 min in a Mastercycler PCR machine.
4. While the tubes are incubating, prepare the following mixture: 16 μL RT 5X first-strand buffer plus 8 μL DTT 0.1 *M*, 4 μL of dNTP mix, and 4 μL Superscript II 200 U/μL enzyme.
5. Put the PCR tube on ice. Add 1 μL Cy3-dUTP 1 m*M* in one of the tubes. Add 1 μL Cy5-dUTP 1 m*M* in the other tube. When reproducing experiments, make a dye swap (inverse the labeling). Add 16 μL of mix to each tube.
6. Leave the tubes 5 min at room temperature.
7. Incubate for 2 h at 42°C in a Mastercycler PCR machine.

3.5. RNA Hydrolysis Before Complementary DNA Purification

1. Add 15 μL 0.1 *M* NaOH.
2. Incubate for 10 min at 70°C in the Mastercycler.
3. Neutralize the pH by adding 15 μL 0.1 *M* HCl.

3.6. Purification of Labeled Complementary DNA for Microarray Hybridization

1. Pool the two PCR tubes in 1 Eppendorf centrifuge tube. Add 1/10 volume 3 *M* Na acetate and 2.5 volumes ethanol.
2. Precipitate 30 min at –80°C. Centrifuge 30 min at 4°C at 18,000*g*. Discard supernatant. Resuspend pellet in 40 μL water and add 4 μL 3 *M* Na acetate.
3. Add 200 μL Qiagen PB buffer. Load sample to a Quiaquick PCR purification column and centrifuge at 13,500*g* for 1 min at room temperature in the 5415 D centrifuge.
4. Remove liquid, add 600 μL PE buffer, and centrifuge for 1 min. Remove liquid and centrifuge 1 min more to dry the column. Remove liquid. Place the column in a new centrifuge tube. Add 30 μL of water prewarmed at 42°C. Centrifuge 1 min. Keep the eluate for hybridization step.

3.7. Prehybridization of Microarrays

This treatment aims at inactivating the free aminosilane groups before hybridization to avoid nonspecific interaction of the labeled complementary DNA (cDNA) with the slide. It can be done during the 2-h incubation left during the labeling reaction.

1. Preheat 5X SSC, 0.1% (w/v) SDS, 1% (w/v) BSA mix in a Falcon tube at 42°C.
2. Place the slide in the Falcon tube and incubate for 45 min.
3. Rinse the slides five times (30 bottom-up moves) in five 50-mL Falcon tubes filled with water.

4. Put the slide in isopropanol and leave it to dry on a paper towel. Avoid any contact between the DNA spots (opposite to the serial number) and the paper. Stock the slide in a plastic box protected from light and moisture. Use prehybridized slides within 12 h.

3.8. Microarray Hybridization

1. Preheat the purified cDNA and hybridization buffer at 70°C for 3 min.
2. Prepare the hybridization chamber by putting 40 μL water in the holes and place the slide into the chamber (on yeast microarrays, the DNA spots are opposite the serial number label).
3. Add 35 μL hybridization buffer to the 30 μL purified cDNA. Put this mix onto the slide (avoiding bubbles as much as possible). Put the on coverslip carefully to make the liquid spread as homogeneously as possible (heterogeneity in the hybridization may lead to local heterogeneity in the final signal).
4. Close the hybridization chamber, drop it in a water bath at 42°C, and leave it to incubate overnight.

3.9. Microarray Washing

1. Get the chamber out of the water bath and dry it with a paper towel (*keep it horizontal*). Take the slide and move it up and down in a 50-mL Falcon tube filled with washing buffer 1 prewarmed at 42°C to make the coverslip fall down. Put the DNA spot side opposite the coverslip and drop it in the Falcon tube. Move the Falcon bottom-up about 30 times to rinse the slide.
2. Proceed the same in washing buffers 2–4 at room temperature.
3. Centrifuge the slides 5 min at 500*g* at room temperature in a new, clean, empty Falcon tube. The slide should not dry before the centrifugation step, so put the slide directly from washing buffer 4 into the centrifuge in which you have put the empty Falcon tube. Do not close the tube to avoid condensation. After centrifugation, slides are ready for scanning.

3.10. Microarray Scanning and Image Analysis

1. Scan the slide with the Genepix 4000B following the Axon's instructions. Use the "histogram" window to get the distribution of the signal in Cy3 (green) and Cy5 (red) channels. Look at the percentage of saturating pixels and dynamic range in each channel. Set up the Photomultiplicator voltage so that you have no saturating spots while using most of the dynamic range of fluorescence measurements (2^{16} values).
2. For image analyses, use the Genepix Pro software recommendations. The image analysis is rather automatic. Just control that the localization of the spots is correct and flag as "bad" the spots in which signal is obviously caused by artifacts (dusts, slide damage, etc.) rather than DNA hybridization. Get the "result" *.gpr file for further analysis of the data.
3. For microarray data, analysis *see* **Note 2**.

3.11. Fixation of Cells and Spheroplasting for FISH

1. Cells are grown in appropriate conditions (medium and temperature according to the strain used) to early log phase in 50 mL medium with shaking.
2. Cells are fixed 45 min at room temperature with occasional shaking by addition of the formaldehyde directly to the culture to a final concentration of 4%.
3. Cells are pelleted at 3000g and washed three times with 10 mL ice-cold buffer B.
4. Cell pellet is resuspended by gentle pipeting in 750 mL spheroplast buffer in a 1.5-mL tube; 24 μL lyticase stock solution are added, and cells are incubated at 30°C for 5–20 min with occasional inversion (the incubation time needs to be optimized according to the strain used).
5. Cells are pelleted 4 min at 3500*g* and 4°C, washed once with 1 mL buffer B, and resuspended in 750 μL to 1.4 mL buffer B depending on the desired concentration of cells for imaging.
6. Spread 100 μL cells with the tip of a pipet on a poly-L-lysine-treated cover slip (in the six-well tissue culture plate) and incubated at 4°C for 30 min (vibrations can prevent cells from adhering and should be avoided at this step).
7. Slowly add 3 mL buffer B to the well (avoiding direct flow).
8. Buffer B is replaced by 70% ethanol. Cells need to stay in 70% ethanol for at least 15 min at –20°C and can be kept for weeks under these conditions, sealing the plate with parafilm to avoid ethanol evaporation.

3.12. Probe Synthesis Labeling and Purification for FISH

1. Typically 4–6 antisense oligonucleotides of 50 nucleotides each are selected. Probes used in the same hybridization should have the same GC content (the protocol as described is optimized for 50% GC 50 nucleotides oligonucleotides). Probes with a low self-annealing potential are favored. Probe selection is facilitated by the Oligo6 software (Molecular Biology Insights, Cascade, CO). Five thymidines from the sequence of the oligonucleotide are replaced by aminoallyl thymidines, respecting spacing of at least eight nucleotides between each to avoid quenching. Probes used in **Fig. 3** are described in **Table 1**.
2. Probes are synthesized on an Applied Biosystems automated DNA/RNA synthesizer (model 392/394) according to manufacturer recommendations.
3. Probes are purified on oligonucleotide purification cartridge (OPC) columns (Applied Biosystems, Foster City, CA) according to manufacturer instructions.
4. Pure probes are dehydrated and resuspended in 100 μL water, and their concentration is determined by ultraviolet (UV) spectrophotometry. Dry 5–20 μg probe and use for labeling.
5. Cyanine probes (1 dry aliquot) are resuspended in 20 μL carbonate buffer. The total amount of resuspended dye is used to resuspend the dried probe by pipeting and vortexing. FITC probes need to be resuspended in dimethyl sulfoxide, and the labeling volume needs to be increased not to exceed 20% dimethyl sulfoxide. Labeling reaction is kept in the dark at room temperature for at least 12 h.

6. The probe is diluted with 50 µL 10 m*M* TEAB and loaded onto the Sephadex G50 column that is running 10 m*M* TEAB by gravity flow. At this point, the labeled probe runs faster than the free dye and should be visible as a faint band compared to the intensity of the free dye. (FITC dyes are best followed under short UV illumination.) The probe is collected in 1.5-mL tubes (typically 2.5–4 mL) and dehydrated.
7. Probe pellets are pooled and resuspended in 100 µL water; DNA concentration as well as labeling efficiency are determined by spectrophotometry (typically, 60–95% labeling efficiency is obtained). Probes are diluted to 40 ng/µL with TE and stored at –20°C.

3.13. Fluorescent In Situ Hybridization

1. Place 1 box of type 1 coverslips that are 22 × 22 mm in a beaker containing 250 mL 0.1 *N* hydrochloric acid and boil for 20 min.
2. Abundantly rinse coverslips with distilled water to remove any traces of acid; autoclave and store at 4°C immersed in water (dried coverslips stick to each other and are difficult to manipulate).
3. Individual coverslips are placed on Whatman paper and 100 µL poly-L-lysine solution is applied and spread with a pipet tip on each coverslip; after 2 min, the poly-L-lysine is removed, and the coverslips are air dried for 3 h.
4. Each coverslip is placed on a six-well plate (treated side up) and washed three times for 10 min with 3 mL water.
5. Water is removed, and coverslips are rested at a 45° angle on the wall of the wells (this will prevent them from sticking to the bottom of the plate) and allowed to dry to completion.
6. Dry coverslips are rested in the bottom of the wells and can be stored at room temperature.
7. Combine 2–10 ng of each individual probe (not exceeding 50 ng total) in a 1.5-mL tube with 4 µL competitor nucleic acids mix and dry under vacuum.
8. While the probes are drying, rehydrate cells twice for 5 min at room temperature in 2X SSC and 5 min in formamide solution.
9. Resuspend the probes in 12 µL solution F and heat at 100°C for 3 min.
10. Add 12 µL solution H and mix; 20 µL of the mix are dropped on the bottom of a Petri dish "hybridization chamber." Place 1 coverslip on the drop (cells face down). Fill the cap of a 50-mL tube with formamide solution and place in the dish to ensure humidification. Seal the Petri dish with parafilm and incubate for a minimum of 3 h (optimal after 10 h) at 37°C.
11. Remove coverslips, place back in six-well tissue plates, and wash twice for 15 min at 37°C with prewarmed formamide solution.
12. Wash coverslips for 15 min with 0.1% Triton X-100 solution with gentle shaking at room temperature.
13. Wash coverslips twice for 15 min with 1X SSC with gentle shaking at room temperature.
14. Incubate coverslips 5 min in 1 mL DAPI solution at room temperature.
15. Wash coverslips twice for 5 min with 2 mL PBS at room temperature.

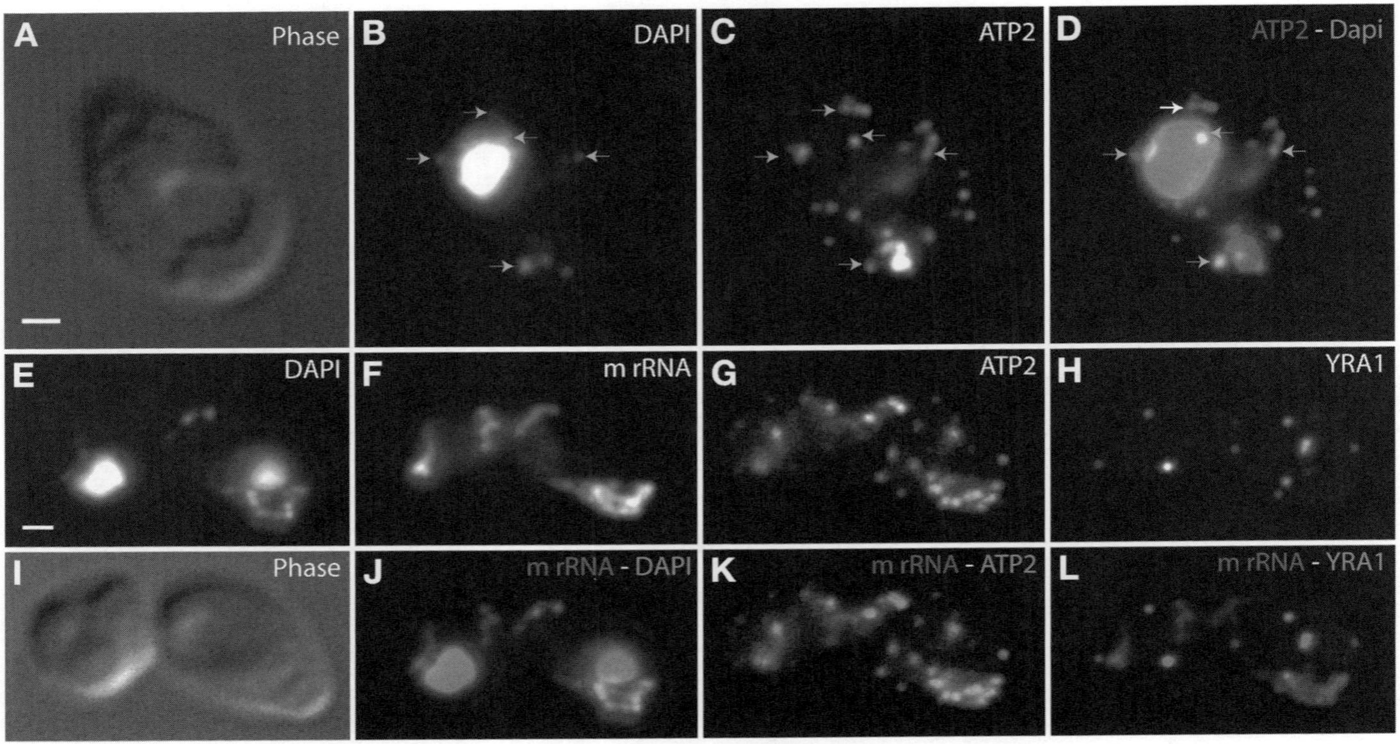
A
Phase
B
DAPI
C
ATP2
D
ATP2 - Dapi
E
DAPI
F
m rRNA
G
ATP2
H
YRA1
I
Phase
J
m rRNA - DAPI
K
m rRNA - ATP2
L
m rRNA - YRA1

Fig. 3.

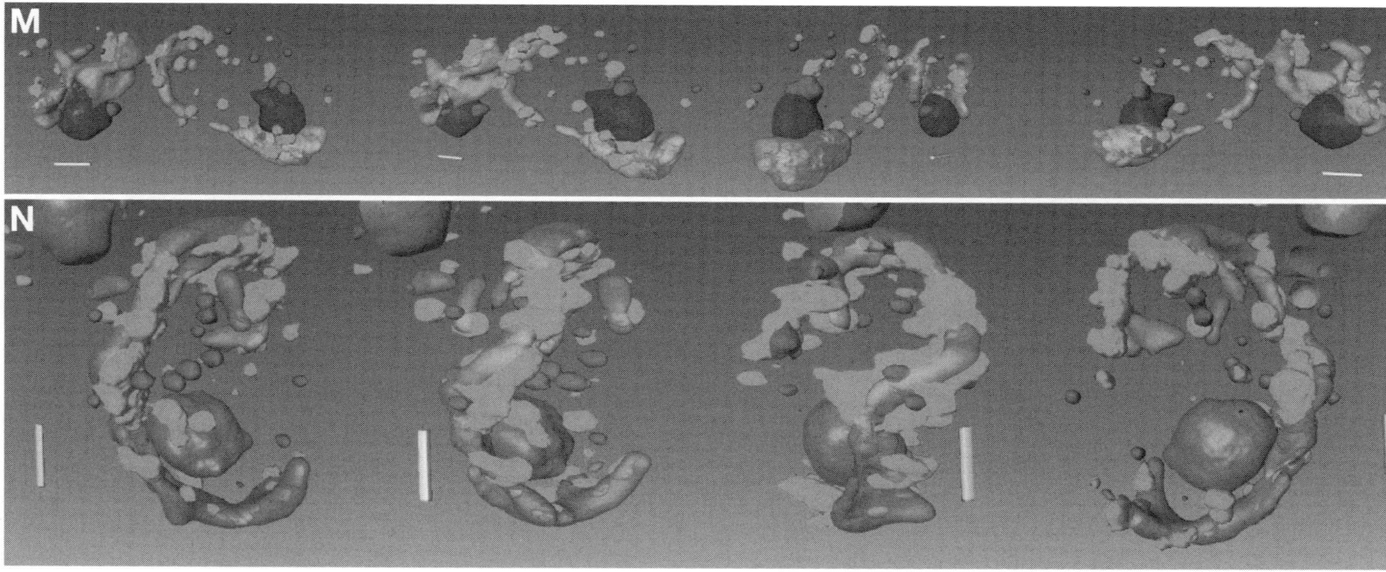

Fig. 3. Mitochondria and localization of mRNA molecules. (**A**)–(**D**) Single-plane distribution of *ATP2* mRNA: (**A**) Phase; (**B**) nucleus and mitochondria detected by DAPI staining; (**C**) *ATP2* mRNA detected by FISH using Cy3-labeled probes. Arrows show the position of DAPI-stained cytoplasmic structures that can be difficult to discriminate from the very bright nucleus. (**E**)–(**M**) simultaneous detection of (**E**) DAPI, (**F**) the mitochondrial rRNA (Cy3.5), (**G**) *ATP2* (Cy5), and (**H**) *YRA1* (Cy3). Three consecutive planes were merged. The simultaneous detection of three different fluorophores requires the use of narrow-band filters, resulting in a noticeable diminution of the signal-to-noise ratios (compare **G** and **C**) (**I**) phase; (**J**) merge, DAPI in light gray (green) from **E** and mitochondrial rRNA from **F**; (**K**) merge mitochondrial rRNA in dark grey (red) from **F** and *ATP2* mRNA in light gray (green) from **G**; (**L**) merge mitochondrial rRNA in dark gray (red) from **F** and *YRA1* mRNA in light gray (green) from **H**; (**M**) 3D representation of the cellular volume (same cells as in **E**–**L**) *ATP2* mRNA and mitochondrial rRNA are made green (Cy5) and purple (Cy3.5), respectively, and they appear as elongated intermingled gray structures. *YRA1* mRNA can be observed in red (Cy3); it appears as round, single, gray structures. (**N**) Same as **M** showing the variability of *ATP2* addressing from cell to cell (compare **M** and **N**). Scale bar: 1 μm.

Table 1
Probes Used

ATP2-1	**t**ACCAGTG**T**CAAGAACCTTT**T**CACCACGGACCAAACCT**T**CGGTACCA**T**CCA
ATP2-2	**t**ACCAACACCGG**T**GAAAACGGAAAAACCACCA**T**GGGCCTTGGCGATAT**T**GT
ATP2-3	A**T**ATTCAGCGA**T**CGTCAAACCAG**T**TAAAGCGACTC**T**GGCTCTGGCTCC**T**G
ATP2-4	**T**GGCAGGAGCAGGA**T**CTGTTAAATCA**T**CGGCTGGAACA**T**AAACGGCT**T**GC
ATP2-5	**t**GGCGACG**T**CATAATGTTCT**T**GACCGACAACGGCGGCA**T**CCAATAACCT**T**G
15S m rRNA	**t**AAACCATTA**T**GATTAACGC**T**CGCCCTCTTTG**T**GTTACCGCGACTGC**T**GGC
21S m rRNA-1	**t**GACCCGAAAGGGAACCGGAACCCCGAAGAGGGG**T**TCACACC**T**ATTAAAAA**t**a
21S m rRNA-2	AGC**T**GCATAGGGTC**T**TTCCGTCTTGC**T**GAAGGTACA**T**AGCATCTTCAC**T**ACGAT

Aminoallyl-modified thymidines are bold; sequence complementary to the target gene is capitalized. An additional aminoallyl T was added 5′ or 3′ of ATP2-1, ATP2-2, ATP2-5, 15S mitochondrial rRNA, and 21S mitochondrial rRNA-1 to increase the labeling of the probes in which the sequence did not offer five optimally spaced thymidines.

16. Mount coverslips on a drop of mounting solution; remove excess solution with a kimwipe and seal coverslips with nail polish.
17. Protect coverslips from light; these can be kept at –20°C for years.
18. Perform imaging on an Olympus BX61 upright microscope using a 100 × 1.35-numerical aperture (NA) objective; illumination is provided by a 100-W mercury lamp.
19. Choose filter cubes to resolve spectrally the used fluorophores from each other (Chroma, Rockingham, VT) and perform test experiments in which only one fluorochrome is used to address possible leaks of the staining in the other used channel.
20. Perform three-dimensional (3D) sampling, acquiring 40 images spaced by 200 nm in the *z*-axis.
21. Perform image acquisition using IPlab (Schanalytics) and image processing using Image J (W. S. Rasban, National Institutes of Health, Bethesda, MD, http://rsb.info.nih.gov/ij/).
22. Typically, the detection of a particular RNA using five different antisense oligonucleotides (a total of 25 dye molecules in average) provides sufficient signal-to-noise ratio to ensure detection of single molecules. The diameter of mRNPs is below the resolution of the light microscope; therefore, molecules of mRNAs present in a radius range inferior to 200 nm are detected as single objects. The number of molecules present in a specific structure is proportional to the total fluorescence.

The mRNAs that cluster to the vicinity of mitochondria often cannot be resolved as individual molecules.

23. The mitochondrial network is a 3D structure, and images observed on single planes are difficult to interpret because mitochondria and their associated mRNAs can be seen in different planes. The visual inspection of the successive planes of the 3D stack is usually sufficient to detect mitochondrial association. For display purposes, individual planes can be selected (**Fig. 3A–D**). Several stacks can be combined using a maximum projection algorithm (**Fig. 3E–L**). A global view of the 3D cytoplasmic volume can be restored using software solutions such as Imaris (Bitplane, Exton, PA) or Amira (Mercury Computer Systems, San Diego, CA) offering a virtual representation of the volumes (**Fig. 3M,N**)

We are currently developing software solutions that would facilitate the automatic scoring of mRNPs in the vicinity of the mitochondria to establish genetic screens that allow us to dissect the molecular mechanism of mitochondrial mRNA addressing.

4. Notes

1. Q-PCR data analysis:

Notations: In notations, *c* stands for cellular extract and *m* for mitochondrial extract.

Vm	Total volume of yeast culture used to mitochondrial isolation
Vc	Volume used for total RNA preparation
Qm	Quantity of RNA purified from mitochondrial isolation
Qc	Quantity of RNA purified from cells
$Qpcr$	Quantity of RNA used for Q-PCR experiment
$Q_c^{PCR}X$	PCR mix quantification of X RNA in total cellular RNA
$Q_m^{PCR}X$	PCR mix quantification of X RNA mitochondrial fraction
$Q_c^T X$	Total quantification of X RNA in total cellular RNA
$Q_m^T X$	Total quantification of X RNA in mitochondrial fraction
${}^cQ_m^T X$	Total quantification of X RNA in mitochondrial fraction after correction by contamination rate
$Q_{Cy}^T X$	Total quantification of X RNA in cytoplasm
$Rm(X)$	X RNA spatial distribution: percentage of RNA bound to mitochondria
Rc	Contamination rate
${}^cRm(X)$	X RNA spatial distribution after correction by contamination rate
E	Q-PCR efficiency (i.e., number of DNA targets obtained after a PCR cycle from one target template; $0 \leq E \leq 2$)
K	Threshold for amplified DNA during Q-PCR used to determine the initial template quantity

Cpm(*X*) Mitochondrial threshold cycle for X RNA, the Q-PCR cycle at which the amplification plots reach the threshold in mitochondrial RNA

Cpc(*X*) Cellular threshold cycle for X RNA, the Q-PCR cycle at which the amplification plots reach the threshold in total RNA

Q-PCR fluorescence curve analyses: The PCR reaction profile can be divided in three steps: an early background phase, an exponential phase, and a plateau. During the exponential phase, the amplification course is described by the equation $Q_n = Q_0 \times E^n$, where Q_n is the amount of target at cycle *n*, Q_0 is the initial amount of target, and *E* is the efficiency of amplification.

To compare target initial amount in different samples, a threshold *K* for amplification is set. *Cp* is the corresponding cycle number required to correlate real-time fluorescence curves to initial template concentration according to the equation $K = Q_0 \times E^{cp}$.

Different methods are used to determine *Cp*; for review, see Randy Rasmussen's paper at http://www.idahotec.com/lightcycler_u/lectures/quantification_on_lc.htm. In the laboratory, we use the second derivative maximum method, for which no human decision is required to help the software find the exponential portion of the amplification.

Determination of Q-PCR Efficiency, E, for a given primer mix. The equation $K = Q_0 \times E^{cp}$ can be linearized to $\log(K) = \log(Q_0) + C_p \times \log(E)$. So, arranging the form gives the following standard curve equation:

$$Cp = -\frac{1}{\log(E)} \times \log(Q_0) + \frac{\log(K)}{\log(E)}$$

Instead of realizing an external standard curve using genomic DNA, results from the different RNA dilutions can be exploited to determine *E*. Considering the initial quantity of template in nondiluted RNA $Q_{0(1/1)}$ for each dilution:

$$Q_{0(1/d)} = \frac{Q_{0(1/1)}}{d}$$

and the standard curve equation can be modified as follows:

$$Cp = -\frac{1}{\log(E)} \times \log\left(\frac{1}{d}\right) - \frac{1}{\log(E)} \times \log(Q_{0(1/1)}) + \frac{\log(K)}{\log(E)}$$

$$Cp = -\frac{1}{\log(E)} \times \log\left(\frac{1}{d}\right) + b$$

So, the slope of the curve

$$Cp = f\left[\log\left(\frac{1}{d}\right)\right]$$

gives a direct assessment of Q-PCR efficiency for the studied target.

Determination of mRNA spatial distribution. Mitochondrial RNAs (*COX1, COX2*) are used to normalize Q-PCR results and determine mitochondrial purification yield η. This yield takes into consideration efficiency of different steps from biochemical purification to RT-PCR.

$$\eta = \frac{Q_m^T COX1}{Q_c^T COX1}$$

Considering the following equations:

$$Q_m^T COX1 = Q_m^{PCR} COX1 \times \frac{Qm}{Qpcr} \text{ and } Q_m^{PCR} COX1 = \frac{K}{E^{Cpm(cox1)}}$$

$$Q_c^T COX1 = Q_C^{PCR} COX1 \times \frac{Qc}{Qpcr} \times \frac{Vm}{Vc} \text{ and } Q_c^{PCR} COX1 = \frac{K}{E^{Cpc(cox1)}}$$

The purification yield is

$$\eta = \frac{E^{Cpc(cox1)}}{E^{Cpm(cox1)}} \times \frac{Qm}{Qc} \times \frac{Vc}{Vm}$$

This yield is used to determine mitochondrial spatial distribution for each RNA:

$$Rm(X) = \frac{Q_m^T X}{Q_c^T X} \times \frac{1}{\eta} \times 100$$

$$Rm(X) = \frac{E^{Cpc(X)}}{E^{Cpm(X)}} \times \frac{Qm}{Qc} \times \frac{Vc}{Vm} \times \frac{1}{\eta} \times 100$$

$$Rm(X) = \frac{E^{Cpc(X)}}{E^{Cpm(X)}} \times \frac{E^{Cpm(COX1)}}{E^{Cpc(COX1)}} \times 100$$

Mitochondrial spatial distribution can be corrected considering contamination rate, which is the mitochondrial localization rate of RNA without connection to mitochondria (e.g., *ACT1*, *RPL10*).

Correction of contaminations with nonmitochondrial fractions. Mitochondrial fraction is always contaminated with other cellular fractions. One can take into account these contaminations if one assumes that some mRNAs are not connected with mitochondria biogenesis, like *ACT1* or *RPL10*. Note that other mRNAs may be considered more pertinent to assess contamination level.

Since cellular RNA arises from cytoplasm and mitochondria contribution and RNA purified in mitochondrial fraction involves specific RNA interaction and contamination from cytoplasm,

$$Q_c^T X = {}^cQ_m^T X + Q_{Cy}^T X$$

$$Q_c^T X = {}^cQ_m^T X + Rc \times Q_{Cy}^T X$$

This leads to

$${}^cQ_m^T X = \frac{1}{1 - Rc}\left(Q_m^T X - Rc \times Q_c^T X\right)$$

Finally, the corrected mitochondrial spatial distribution can be expressed as

$${}^cRm(X) = \frac{1}{1 - Rc}\left(Rm(X) - Rc\right)$$

Typical values for mitochondrial spatial distribution before and after correction are presented in **Fig. 1** for *ACT1*, *ATP2*, *ATP3*, *COX4,* and *COX6* transcripts (spatial distribution of *ACT1* is considered to represent the contamination rate).

2. *Microarray data analysis and normalization.* Basic analysis and normalization of data can be conducted using Excel software. A dedicated database can take in charge data management and integrates more sophisticated tools for data analysis (statistical analysis of microarray ***(15)*** or dedicated R packages [http://www.bioconductor.org/], for instance).

   The genomewide comparison of mitochondrially associated RNA vs total RNA requires analysis methods different from standard genomewide gene expression analyses. In this case, we expect a skewed distribution with a significant number of mitochondrially associated RNAs "getting out" of the distribution of total RNA on one side (**Fig. 2**). To solve this problem, several methods are available ***(16)***. The most widely used, which we describe here, is the median percentile rank method, which associates to each mRNA a value between 0 and 100, depending on its position in the ratio distribution, thus reflecting the reproducibility of its enrichment among replicate experiments.

   The first steps of data processing are similar to standard global gene expression analysis. For the spots that have been flagged as "good" during the image analysis, we keep from the result file (*.gpr) the median of the foreground (**F**) and of the local background (**B**) for channel Cy5 (635 nm) and Cy3 (532 nm). The Cy5/Cy3 fluorescence ratio (Rf) is then equal to (F635-B635)/(F532-B532). The mRNA are then sorted according to their Rf. The median percentile ranks (mpR) for each mRNA are calculated using the percentile rank function in Excel. The percentile rank of a mRNA with Rf = X is simply the percentage of mRNA with an Rf that is less than X. We multiply this value (ranging from 0 to 1 in Excel) by 100 to scale the value between 0 and 100 for better convenience. The spots that have been flagged as bad or absent at the image analysis step are given an mpR value of NA.

   What we called the MLR ***(6)*** is the median of the mpR of each mRNA among six microarray experiments. A high MLR thus reflects a reproducible enrichment of the corresponding RNA in the mitochondrially bound polysomes compared with its abundance in total RNAs.

Acknowledgments

We are grateful to present and past members of the R. H. Singer laboratory for developing and constantly improving the FISH protocol, Daniel Zenklusen for sharing the Yra1 probes, Melissa Lopez-Jones and Tatjana Trcek for the synthesis of the modified oligonucleotides, and Shailesh M. Shenoy for assistance with the microscopy. More general and updated applications of FISH can be found at www.singerlab.org. R. H. S. is funded by the National Institutes of Health (GM57071). We also thank all members of C. J.'s laboratory who contributed to the development of DNA microarray technology adapted to the study of mitochondria, especially Thierry Delaveau for his help in the Q-PCR technology and Sophie Lemoine for helpful discussions concerning statistical methods for microarray data analyses. Genevieve Dujardin is thanked for her advice and gift of strains. C. J. is funded by ARC 3310.

References

1. Schatz, G. and Dobberstein, B. (1996) Common principles of protein translocation across membranes. *Science* **271,** 1519–1526.
2. Neupert, W. (1997) Protein import into mitochondria. *Annu. Rev. Biochem.* **66,** 863–917.
3. Kellems, R. E., Allison, V. F., and Butow, R. A. (1975) Cytoplasmic type 80S ribosomes associated with yeast mitochondria. IV. Attachment of ribosomes to the outer membrane of isolated mitochondria. *J. Cell Biol.* **65,** 1–14.
4. Verner, K. (1993) Co-translational protein import into mitochondria: an alternative view. *Trends Biochem. Sci.* **18,** 366–371.
5. Suissa, M. and Schatz, G. (1982) Import of proteins into mitochondria. Translatable mRNAs for imported mitochondrial proteins are present in free as well as mitochondria-bound cytoplasmic polysomes. *J. Biol. Chem.* **257,** 13,048–13,055.
6. Marc, P., Margeot, A., Devaux, F., Blugeon, C., Corral-Debrinski, M., and Jacq, C. (2002) Genome-wide analysis of mRNAs targeted to yeast mitochondria. *EMBO Rep.* **3,** 159–164.
7. Karlberg, O., Canback, B., Kurland, C. G., and Andersson, S. G. (2000) The dual origin of the yeast mitochondrial proteome. *Yeast* **17,** 170–187.
8. Saint-Georges, Y., Bonnefoy, N., di Rago, J. P., Chiron, S., and Dujardin, G. (2002) A pathogenic cytochrome *b* mutation reveals new interactions between subunits of the mitochondrial bc1 complex. *J. Biol. Chem.* **277,** 49,397–49,402.
9. Bertrand, E., Chartrand, P., Schaefer, M., Shenoy, S. M., Singer, R. H., and Long, R. M. (1998) Localization of ASH1 mRNA particles in living yeast. *Mol. Cell* **2,** 437–445.
10. Femino, A. M., Fay, F. S., Fogarty, K., and Singer, R. H. (1998) Visualization of single RNA transcripts *in situ*. *Science* **280,** 585–590.

11. Femino, A. M., Fogarty, K., Lifshitz, L. M., Carrington, W., and Singer, R. H. (2003) Visualization of single molecules of mRNA *in situ*. *Methods Enzymol.* **361,** 245–304.
12. Long, R. M., Singer, R. H., Meng, X., Gonzalez, I., Nasmyth, K., and Jansen, R. P. (1997) Mating type switching in yeast controlled by asymmetric localization of ASH1 mRNA. *Science* **277,** 383–387.
13. Long, R. M., Elliott, D. J., Stutz, F., Rosbash, M., and Singer, R. H. (1995) Spatial consequences of defective processing of specific yeast mRNAs revealed by fluorescent *in situ* hybridization. *RNA* **1,** 1071–1078.
14. Chartrand, P., Bertrand, E., Singer, R. H., and Long, R. M. (2000) Sensitive and high-resolution detection of RNA *in situ*. *Methods Enzymol.* **318,** 493–506.
15. Tusher, V. G., Tibshirani, R., and Chu, G. (2001) Significance analysis of microarrays applied to the ionizing radiation response. *Proc. Natl. Acad. Sci. USA* **98,** 5116–5121.
16. Buck, M. J. and Lieb, J. D. (2004) ChIP-chip: considerations for the design, analysis, and application of genome-wide chromatin immunoprecipitation experiments. *Genomics* **83,** 349–360.

36

Plant Mitochondrial Transcriptomics by Quantitative RT-PCR

Rachel Clifton and James Whelan

Summary

Transcriptomic analysis using quantitative reverse transcriptase polymerase chain reaction (QRT-PCR) facilitates analysis of nuclear and mitochondrial-encoded mitochondrial genes, enabling mechanisms and regulation of signaling pathways to be explored. To illustrate this technique, we use genes of the mitochondrial respiratory chain. We show that several components of the mitochondrial respiratory chain respond to stress, in particular the alternative oxidase. This chapter describes a method involving total ribonucleic acid (RNA) isolation and QRT-PCR for the detection and analysis of transcriptional changes that accompany seven commonly used chemical stresses. This methodology describes an accurate technique to determine quantitatively absolute transcript levels and a platform to facilitate comparison between responses to other stress stimuli.

Key Words: Alternative respiratory pathway; gene expression; quantitative RT-PCR.

1. Introduction

Whole transcriptome profiling of budding yeast revealed that more than 14% of all genes are induced or repressed in response to a wide range of stresses. Genes involved in mitochondrial functions featured heavily among the induced genes *(1–3)*. More than 95% of mitochondrial proteins are encoded by the nuclear genome, and many of the large multisubunit complexes present in the mitochondria are comprise of proteins encoded in both the nuclear and mitochondrial genomes *(4)*. Thus, regulation of plant stress responses, including mitochondrial function, involves the coordination of nuclear and organelle gene expression.

Reliable and affordable expression analysis of many of nuclear-encoded and the mitochondrial-encoded mitochondrial genes is outside the technical limits of current commercially available whole-genome arrays, such as the Affymetrix

From: *Methods in Molecular Biology, vol. 372: Mitochondria: Practical Protocols*
Edited by: D. Leister and J. M. Herrmann © Humana Press Inc., Totowa, NJ

ATH1 GeneChips® *(5)*. Thus, quantitative reverse transcriptase polymerase chain reaction (QRT-PCR), an extremely sensitive and reliable technique ideal for the quantitation of low-abundance transcripts, is the method of choice for investigating mitochondrial transcriptomics.

As expression profiling has revealed dramatic variation in the level and types of genes expressed across tissue types and with development, care needs to be taken in designing the experimental setup and the model system to be used in the analysis of changes in gene expression caused by applied stresses. In Addition, many metabolic processes in plants have been demonstrated to have components influenced by or linked to circadian or light regulation *(6)*. The use of a suspension cell culture enables analysis of the responses of a single tissue type, and maintenance under a light-dark cycle enables the distinction between transcriptome variations caused by natural cycles and variations caused by addition of chemical.

Mitochondrial transcript analysis in response to applied stresses involves three steps: (1) development of an appropriate experimental setup; (2) development of an accurate QRT-PCR assay including design of gene-specific primers and optimization of reaction conditions for each transcript; and (3) informed analysis of the transcript data. We use components of the plant mitochondrial respiratory chain as a model to investigate this system, specifically genes encoding the alternative oxidase type II NAD(P)H dehydrogenases, as well as mitochondrial- and nuclear-encoded subunits of the adenosine triphosphate (ATP) synthase complex and a gene encoding a cytochrome-*c* and a complex I subunit *(7)*. Analysis of transcript expression of genes encoding mitochondrial respiratory chain proteins and energy-dissipating components using QRT-PCR in response to a range of chemical perturbations of an *Arabidopsis* suspension cell culture revealed gene-specific and unique regulation of mitochondrial- and nuclear-encoded respiratory chain subunits.

2. Materials

2.1. Arabidopsis *Suspension Cells and Treatments*

1. *Arabidopsis thaliana* cell culture: a heterotrophic *Arabidopsis* cell culture, established from callus of ecotype *Landsberg erecta* stem explants maintained by weekly subculture *(8)*. Cell culture was originally obtained from L. J. Sweetlove (Department of Plant Sciences, University of Oxford, Oxford, UK).
2. *Arabidopsis* suspension cell medium: 1 sachet/L Murashige and Skoog salt mixture (N1145; Invitrogen), 30 g/L sucrose, 500 μL 1 mg/mL naphthalene acetic acid (K salt), 50 μL 1 mg/mL kinetin (*see* **Note 1**). Adjust to pH 5.8 using 1 *M* KOH and autoclave in 100-mL aliquots in cell culture flasks sealed with aluminum foil. Store for up to 1 mo protected from light.
3. Chemical treatments were prepared fresh to the stock concentrations indicated (**Table 1**) and filtered sterilized. Each chemical was added to a conical flask containing 120 mL cell culture as indicated to obtain final working concentration:

Table 1
Chemicals Used to Treat *Arabidopsis* Suspension Cells and Their Stock and Working Concentrations

Treatment	Supplier	Stock concentration	Working concentration	Volume added to 120 mL culture
Chloramphenicol	Sigma	240 m*M* in ethanol	200 μ*M*	100 μL
Erythromycin	Sigma	240 m*M* in ethanol	200 μ*M*	100 μL
Rotenone	Sigma	10 m*M* in ethanol	40 μ*M*	480 μL
Salicylic acid	Sigma	100 m*M* in ethanol	100 μ*M*	120 μL
Oligomycin	Sigma	1 m*M* in ethanol	0.125 μ*M*	15 μL
Hydrogen peroxide	Sigma	30% (w/v)	10 m*M*	130 μL
Glucose	Sigma	2.17 *M* in *Arabidopsis* suspension cell medium	3% (w/v)	1 mL

4. Miracloth (Merck) is cut into squares with an approx length of 10 cm or size sufficient to cover the vacuum filter device used. Prior to filtering the cells, three layers of Miracloth are moistened in sterile water and layered over the filter device.

2.2. Total Ribonucleic Acid (RNA) Isolation and Complementary Deoxyribonucleic Acid (cDNA) Synthesis

2.2.1. Total RNA Isolation

1. It is important that all materials used in the RNA isolation procedure are sterile and free of ribonucleases (RNases). Use only sterile 1.5-mL microcentrifuge tubes and pipet tips and ensure all solutions have not been contaminated.
2. Mortar and pestles (60 × 30 mm). Soak mortar and pestles in a weak sodium hydroxide solution for at least 2 h, then rinse with sterile water and autoclave before use.
3. RNeasy Plant Mini Kit (Qiagen, Doncaster, Australia).
4. RNase-free deoxyribonuclease (DNase) I (Qiagen).
5. DNAfree™ (Ambion, Austin, TX, USA).
6. β-Mercaptoethanol, biotechnology grade.
7. Sterile water, molecular biology grade (*see* **Note 1**).

2.2.2. Total RNA Quantitation

1. 10X TAE buffer: 48.4 g/L Tris, 11.8 ml/L glacial acetic acid, 10 m*M* ethylenediaminetetraacetic acid. Dilute 1 in 10 in sterile water for use as 1X TAE.
2. Ethidium bromide.
3. 1% (w/v) Agarose gel in TAE buffer containing ethidium bromide to 0.1 μg/mL.
4. 5X Loading buffer: 50% (v/v) glycerol, 0.05% (v/v) bromophenol blue.
5. Sterile water, molecular biology grade (*see* **Note 1**).
6. Horizontal electrophoresis chamber and power supply.
7. Ultraviolet (UV) light source for visualizing ethidium bromide-stained agarose gels.
8. 120-μL Quartz cuvette.
9. Spectrophotometer.

2.2.3. cDNA Synthesis

1. Water baths at 65, 42, and 30°C.
2. Expand™ reverse transcriptase system (Roche).
3. Sterile 1.5-mL microcentrifuge tubes.
4. Random hexamer primer (Roche) diluted to 100 pmol/µL in water and stored in aliquots at –20°C.
5. RNasin 40 U/µL (Roche).
6. Deoxynucleotide 5′-triphosphates (dNTPs) (each at 10 m*M*) (Roche).
7. Sterile water, molecular biology grade.
8. QIAquick® PCR purification kit (Qiagen).
9. Bovine serum albumin (BSA) (Roche) is diluted to 0.08% (w/v) in water and stored in aliquots at –20°C (*see* **Note 2**).

2.3. Quantitative RT-PCR Standard Template Production

2.3.1. Cloning the Standard Template DNA

1. Oligonucleotide primers designed for each cloning and QRT-PCR reaction are obtained from appropriate suppliers in lyophilized form and resuspended in water to 200 pmol/µL with 0.008% (w/v) BSA. Aliquots of working dilutions of each primer at 20 pmol/µL in water with 0.008% (w/v) BSA are prepared and stored at –20°C.
2. Expand™ high-fidelity PCR system (Roche).
3. dNTPs (each at 10 m*M*) (Roche).
4. Sterile water, molecular biology grade.
5. 0.2-mL PCR tubes.
6. Thermocycler.
7. 1% (w/v) Agarose gel in TAE buffer (*see* **Subheading 2.2.2.**).
8. UV light source for visualizing ethidium bromide-stained agarose gels.
9. 96-Well microtiter plate (NUNC™).
10. QIAquick PCR purification kit (Qiagen).
11. TOPO TA cloning kit (Invitrogen, Sydney, Australia).

2.3.2. Quantitating the Standard Template DNA

1. 96-Well microtiter plate (NUNC).
2. PicoGreen® double-stranded DNA (dsDNA) quantitation kit (Molecular Probes).
3. FLUOstar OPTIMA (BMG Labtech).
4. Stocks and dilutions of template DNA standards: each template DNA standard is diluted in sterile water with a final BSA concentration of 0.008% (w/v) to generate a master stock of 1 fmol/µL. Aliquots of a working stock of 0.01 fmol/µL are made by adding 5 µL master stock (at 1 fmol/µL) to 50 µL 0.08% (w/v) BSA and 445 µL sterile water. Store all working and master stocks at –20°C and minimize freeze-thawing. Prepare 10-fold serial dilutions of the working stock with a final concentration of 0.008% (w/v) BSA, fresh as required.

2.4. Quantitative RT-PCR Optimization and Analysis

2.4.1. Primer Design and Optimization

1. TAIR (http://www.arabidopsis.org/).
2. ClustalW programs (http://clustalw.genome.jp/).

2.4.2. Optimizing QRT-PCR Conditions

1. PCR reaction plates (Bio-Rad).
2. High-Quality sealing tape (Bio-Rad).
3. iCycler™ (Bio-Rad).
4. iQ™ SYBR Supermix (Bio-Rad).
5. BSA (Roche).
6. iCycler iQ™ optical system software (v 3.0; Bio-Rad).

2.5. Bioinformatic Analysis of QRT-PCR-Derived Transcript Data

For bioinformatic analysis, use GeneCluster 2.0 (http://www-genome.wi.mit.edu/cancer/software/genecluster2/gc2.html; ***9,10***).

3. Methods

3.1. Arabidopsis *Suspension Cells and Treatments*

1. *Arabidopsis* suspension cells are maintained in 250-mL conical flasks under long-day conditions of 16 h of 100 μmol photons/m^2/s light followed by 8 h dark revolving at 150 rpm in an orbital shaker. After 7 d, cell growth is approx in the middle of the log phase, and subculturing of 20 mL of the culture into 100 mL *Arabidopsis* cell culture medium initiates the cycle again (*see* **Note 3**).
2. All the materials for the treatment protocol are made ready: chemical solutions of the treatments are prepared; flasks of cells are labeled to indicate the treatment they are receiving; a vacuum filtration device is prepared to filter cell samples during the time-course with Miracloth filters and collection vessels (*see* **Note 4**) prepared; and liquid nitrogen is ready to snap freeze and store the samples.
3. From each flask, the pretreatment sample is collected 2 h into the light phase of the light-dark cycle (*see* **Note 5**). The aluminum foil lids sealing the cell culture flasks are punctured, and from each flask 10 mL culture are removed with a sterile pipet and vacuum filtered; the filtered cells are scraped onto a collection vessel with a sterile spatula and snap frozen by liquid nitrogen.
4. To each flask, an appropriate volume of chemical is added directly into the cell suspension; the flask is resealed and returned to the orbital shaker, initiating the time course. Samples are then removed at 3, 12, and 24 h posttreatment. For each chemical treatment, three independent flasks are treated for replicate analysis. Samples from three untreated flasks are also collected as controls. Once snap frozen, it is important that samples remain frozen at all times and should be stored at −80°C until isolation of total RNA.

3.2. Total RNA Isolation and cDNA Synthesis

3.2.1. Total RNA Isolation

1. Total RNA is isolated from each sample separately. Using a precooled mortar and pestle, grind sample to a fine powder under liquid nitrogen. Transfer approx 100 mg (*see* **Note 6**) of powdered sample to a precooled 1.5-mL microcentrifuge tube—do not allow the sample to thaw.
2. Add 450 μL buffer RLT from the RNeasy Plant Mini Kit), premixed with 4.5 μL β-mercaptoethanol, onto the frozen powdered sample and vortex vigorously to ensure that the sample thaws in contact with buffer RLT (*see* **Note 7**).
3. Pipet lysate into a QIAshredder spin column in a 2-mL collection tube (Qiagen) and centrifuge for 2 min at 20,800*g*. Transfer the supernatant of the flowthrough fraction to a new 1.5-mL microcentrifuge tube without disturbing the cell pellet.
4. Add 0.5 volumes 100% ethanol to the cleared lysate, mix immediately by pipeting, transfer directly to an RNeasy minicolumn in a 2-mL collection tube from the RNeasy Plant Mini Kit, and centrifuge for 15 s at 10,000*g*. Discard the flowthrough.
5. Add 350 μL buffer RWI to the RNeasy minicolumn and centrifuge for 15 s at 10,000*g* to wash the column. Discard the flowthrough.
6. In a clean microcentrifuge tube, add 10 μL DNase I to 70 μL buffer RDD, both from the RNase-free DNase I kit (Qiagen), and mix gently by inversion (*see* **Note 8**). Transfer this DNase I mix directly onto the membrane of the RNeasy minicolumn and incubate at room temperature for 30 min.
7. Add 350 μL buffer RWI to the RNeasy minicolumn and centrifuge for 15 s at 10,000*g* to wash the column. Discard the flowthrough and collection tube.
8. Transfer RNeasy minicolumn to a new 2-mL collection tube, add 500 μL buffer RPE, and centrifuge for 15 s at 10,000*g* to wash the column. Discard the flowthrough.
9. Add an additional 500 μL buffer RPE to the RNeasy column and centrifuge for 2 min at 10,000*g* to dry the silica gel membrane.
10. Transfer RNeasy column to a new 1.5-mL microcentrifuge tube and add 40 μL sterile water directly onto the membrane. Incubate at room temperature for 1 min, then centrifuge for 1 min at 10,000*g* to elute RNA. To ensure maximal RNA yields, repeat this step by adding 40 μL sterile water directly onto the membrane, incubating for 1 min and centrifuging for 1 min at 10,000*g*.
11. Add 80 μL 10X DNase I buffer and 1.2 μL DNase I (2.4 U), both from the DNAfree kit (Ambion) to the total RNA. Mix gently by inversion and incubate for 30 min at 37°C.
12. Resuspend the DNase inactivation reagent, from the DNAfree kit (Ambion), by vortexing and add 8 μL of this slurry to the total RNA solution. Mix well and incubate at room temperature for 2 min. To pellet the inactivated DNase I, centrifuge for 1 min at 20,800*g*. Transfer supernatant containing total RNA to new 1.5-mL microcentrifuge tube and keep on ice.

3.2.2. Total RNA Quantitation

1. Prepare a 1% agarose gel by mixing 30 mL 1X TAE buffer with 3 g agarose in a 100-mL conical flask. Microwave to dissolve agarose and, when solution is just cool enough to touch, add ethidium bromide to 0.1 μg/mL and pour into a small gel tray with well comb. Once set, remove the well comb and complete the assembly of the gel unit, submerging the gel in 1X TAE buffer.
2. In a new 1.5-mL microcentrifuge tube, mix 2 μL of the total RNA with 2 μL 5X loading buffer and 6 μL sterile water. Load sample onto the agarose gel and apply a constant 70 V for about 40 min, then assess the quality of the total RNA isolation by examining the gel under UV light. Two clear bands representing the 28S and 18S ribosomal RNA should be apparent with minimal smearing if RNA integrity has been maintained.
3. Determine the quantity of total RNA isolated by spectrophotometric analysis. Prepare in triplicate microcentrifuge tubes containing 5 μL total RNA and 115 μL sterile water. Using a quartz cuvette, blank the spectrophotometer using water, then measure the absorbance of the RNA samples at 260 and 280 nm. Use the average of the three measurements to determine the quantity of RNA isolated according to the following calculation: Concentration of total RNA (μg/mL) = 40 (dilution factor) × A_{260} × 120 (total volume).

3.2.3. cDNA Synthesis

1. A cDNA synthesis reaction and a negative control reaction are performed for each total RNA sample. Label sterile microcentrifuge tubes for all the following steps for both synthesis and control reactions. The negative control reverse transcription reaction contains all the components in the reverse transcription reaction except for the reverse transcriptase enzyme. This negative control is used to determine if RNA isolations are free of genomic contamination.
2. To a sterile 1.5-mL microcentrifuge tube add 1 μg total RNA; the volume will depend on the concentration as determined by spectrophotometric analysis. Then add 1 μL 100 pmol/μL random hexamer primer (Roche) and sterile water to a volume of 10.5 μL. Incubate for 10 min at 65°C, then place on ice for 2 min.
3. Add in the following order: 4 μL Expand reverse transcriptase buffer, 2 μL 10 m*M* DTT, both from the Expand reverse transcriptase system (Roche); and 2 μL dNTPs (each at 10 m*M*) and 1 μL 40 U/μL RNasin, both from Roche (*see* **Note 9**). To the cDNA synthesis reactions, add 1 μL 50 U/μL Expand reverse transcriptase (Roche). Briefly mix reactions and centrifuge to collect solution, then incubate for 10 min at 30°C, then for 45 min at 42°C. Place reactions on ice.
4. Purify the cDNA using the QIAquick PCR purification kit (*see* **Note 10**). At the elution step, 30 μL sterile water is added to the center of the QIAquick membrane and incubated at room temperature for 2 min to ensure maximal cDNA elution, then centrifuged to 1 min at 18,000*g*.
5. The purified cDNA is then diluted 1:10 in sterile water with BSA to a final concentration of 0.008% (w/v) and stored in aliquots to minimize freeze-thawing at –20°C.

3.3. Quantitative RT-PCR Standard Template Production

3.3.1. Cloning the Standard Template DNA

1. Design cloning primers to amplify the product of your gene (*see* **Note 11**). Ensure primers lie within the transcribed product of the gene and will amplify a single product if the genome of the organism you are studying contains other genes with high sequence identity to your gene of interest, such as are found in multiple gene families.
2. Using the cloning primers and cDNA generated from total RNA as a template, the product that will be used as the standard template DNA is amplified using the Expand high-fidelity PCR system (Roche). To a 0.2-mL PCR tube, add 20 pmol of the forward and reverse cloning primers, 3 μL cDNA diluted 1:10, 1X Expand high-fidelity buffer, 3 m*M* $MgCl_2$, 1 m*M* dNTPs, and 1 U Expand high-fidelity enzyme, all from Roche. Make up to 25 μL with sterile water. Load the sealed PCR tube onto a thermocycler and apply cycling conditions consisting of an initial denaturation step at 94°C for 2 min, followed by 35 cycles of 94°C for 30 s, 55°C for 30 s, 72°C for 1 min per kilobase of expected amplicon, followed by a final extension step at 72°C for 5 min.
3. Once cycling is complete, transfer half the reaction volume to a new microcentrifuge tube, add 2.5 μL loading buffer, and load sample onto a 1% agarose gel. Prepare and run gel as for the RNA analysis. Visualize the gel under UV light; a single band of appropriate size should be apparent (*see* **Note 12**).
4. Purify the DNA fragment from the remaining product using the QIAquick PCR purification kit as per instructions, eluting product in 30 μL sterile water.
5. The purified product is then cloned using the TOPO TA cloning kit according to the manufacturer's instructions, and the identity of the insert is confirmed by restriction digest analysis and sequencing.
6. The plasmid containing the cloned template DNA is then used in the place of the cDNA in a PCR reaction as described in **step 2**. The purified product of this reaction is then quantitated; *see* **Subheading 3.3.2.**

3.3.2. Quantitating the Standard Template DNA

1. The amount of template DNA is quantitated using the PicoGreen dsDNA quantitation kit. Several dilutions of the λ DNA supplied in the PicoGreen dsDNA quantitation kit (0, 0.05, 0.1, 0.5, and 1 μg/mL) and two dilutions of the template DNA sample (1:100 and 1:200) are prepared in the supplied 1X TE buffer.
2. A 100-μL reaction consisting of equal volumes of the template DNA sample and a 1:100 dilution of the PicoGreen reagent is loaded into a 96-well microtiter plate in duplicate, mixed gently, and incubated for 5 min in the dark.
3. Sample fluorescence is then detected and analyzed using a FLUOstar OPTIMA. We measure the fluorescence from a fixed area for each well and use the data from the λ DNA dilution series to generate a standard curve. Using this standard curve, the average concentration of the diluted template DNA samples is obtained. The molar concentration of the standard DNA template solution is then calculated

using the following equation: DNA (fmol/μL) = (μg × 10^6 pg/1 μg) × (1 pmol/660) × (1/N), where N is the length of the standard DNA template in basepairs, and 660 is the average molecular weight of a dsDNA base.

4. Each template DNA standard is diluted in sterile water with a final BSA concentration of 0.008% (w/v) to generate a master stock of 1 fmol/μL from which working stocks of 0.01 fmol/μL with 0.008% (w/v) BSA are prepared (*see* **Subheading 2.3.2.**).

3.4. Quantitative RT-PCR Using the Bio-Rad iCycler

3.4.1. Primer Design and Optimization

Primers for QRT-PCR amplification are designed so it is possible to cross exon/intron boundaries, to produce an amplicon between 150 and 250 bp within the region amplified by the cloning primers, and to be gene specific, preventing amplification of other genes of high sequence identity (*see* **Note 13**). An example of primer design for a multigene family is shown in **Fig. 1**.

3.4.2. Optimizing QRT-PCR Conditions

1. Using serial dilutions of the standard DNA template, QRT-PCR conditions for maximum efficiency are optimized on the iCycler. Primer concentration and cycling conditions are varied to determine reaction conditions that yield optimal primer efficiency (100%) (*see* **Note 14**) and no primer-dimer or nonspecific amplification as determined by melt curve analysis and agarose gel electrophoresis of the PCR products.
2. We use 25-μL reactions containing 12.5 μL iQ SYBR Supermix, 2.5 μL 10^{-1} cDNA or standard DNA template, BSA to a final concentration of 0.008% (w/v), and 0.3–0.9 μ*M* forward and reverse QRT-PCR gene-specific primers (*see* **Note 15**). All reactions are performed in duplicate, including the dilution series of the template DNA standard, all cDNA samples for transcript analysis, and negative controls. Reactions are loaded onto PCR plates and sealed with high-quality sealing tape, then mixed gently, centrifuged at 1500*g* for 1 min, and loaded into the iCycler.
3. For analysis of members of multigene families, we use a "touchdown" PCR program to provide more specific primer hybridization consisting of the following cycling parameters: denaturation at 95°C for 3 min; amplification of 15 cycles at 95°C for 15 s, touchdown annealing from 85 to 55°C for 30 s and decreasing 2°C per cycle, 72°C for 30 s, followed by amplification without touchdown in the annealing phase: 30–40 cycles at 95°C for 15 s, 50°C for 30 s, 72°C for 30 s with data acquisition collected during the annealing and extension; melting curve analysis by 50 cycles of 15 s starting at 70°C with a transition rate of +0.5°C per cycle and continuous data acquisition; cooling to 4°C.
4. Results are analyzed using the iCycler iQ optical system software. Briefly, a threshold cycle for each sample is calculated, representing the cycle at which the fluorescence signal in that sample first increased significantly above the background level.

A

```
At3g22360      ------------------------------------------------------------
At3g27620      ------------------------------------------------------------
At3g22370      ----------ATGATGATAACTCGCGGTGGAGCCAAGGCGGCGAAATCGCTGTTAGTGGC 50
At1g32350      ------------------------------------------------------------
At5g64210      ATGAGTCAACTCATTACGAAAGCAGCTTTACGAGTTCTGCTTGTCTGCGGCAGAGGAAAC 60

At3g22360      ------ATGATGATGAGTCGTCGCTATGGAGCCAAGCTAATGGAA-ACTGCTGTGACTCA 53
At3g27620      ------ATGATCACTA--CATTACTCCGTCGCTCTCTCCTTGACGCATCAAAGCAAGCGA 52
At3g22370      GGCTGGACCACGTTTGTTCTCGACGGTCCGTACGGTTTCGTCTCACGAGGCTTTATCAGC 110
At1g32350      --------------------------------------------------ATGTCCTACA 10
At5g64210      TGTAACATGTTCGTGAGTTCTGTTTCCTCCACCTCTGTTATGAAAAGTCCGTATGAGATA 120

At3g22360      CAGCCATCTTTTGAATCCTAGGGTTCC----TCTCGTGACGGAAAATATTAGGGTTCCGG 109
At3g27620      CATCCATTAATGGAATTCTATTTCACCAACTTGCTCCAGCGAAATATTTAGGGTTCCGG 112
At3g22370      AAGCCATATTTTGAAGCCTGGTGTTACA----TCTGCTTGGATATGGACTAGAGCTCCGA 166
At1g32350      GATCGATTTACCGCACTCTTCGACCGGTACTATCATCTTCGGTACAATCCTCCGGTTTAG 70
At5g64210      ACGGCACCGATGCGAATTCATGACTGGTGC-GGCGGTTTTGG--CGATTTCAAGATCGGC 177
                    *        *                         *            * *

At3g22360      CGATGGGAGTCGTGAG-AGTTTTCAGCAAGATGACATTTGAGAAGAAGAAAACTACGGAG 168
At3g27620      CAGTGGGAGGTCTGAG-AGATTTCAGCAAGATGACATTTGAAAAGAAGAAAACATCGGAA 171
At3g22370      CGATTGGAGGTATGAG-ATTCGCTAGCACGATCACTCTGGGAGAGAAAACTCCGATGAAG 225
At1g32350      GAATTGGAGGATTCAG-GGGACATCTCATTA-GCCACTTGCCCAATGTTCGGCTATTGAG 128
At5g64210      TCTAAGCATGTGCAAG|GAAATTTTAACTTGAGGTGGATGGGTATGAGTTCTGCATCGGCG 237
                      * *      **          *   *      * *            *

At3g22360      GA---GAA-----------AGGATCTAGCGG----------TGGC---AAGGCT---GAT 198
At3g27620      GAAGAGGA-----------AGGATCCGGTGA----------CGGCGTCAAGGTTAATGAT 210
At3g22370      GAGGAGGACGCGAATCAGAAGAAAACAGAGAACGAATCCACCGGTGGAGACGCCGCCGGA 285
At1g32350      CTCTGACACGTCA-TCTCCGGTAAGTGGGAA------TAACCAGCCAGAAAATC--CTAT 179
At5g64210      ATGGAGAA-----------GAAAGATGAAAAT--------TTGACGGTAAAGAAG--GGT 276
                      *              *                          *

At3g22360      CAAGGTAACAAAGGGGAGCAATTAATCGTTAGCTACTGGGGAGTGAAGCCG------ATG 252
At3g27620      CAAGGTAACAAAGGGGAACAATTAATAGTTAGTTATTGGGGCGTGAAGCCA------ATG 264
At3g22370      GGTAATAACAAGGGAGATAAAGGAATCGCGAGCTATTGGGGTGTTGAACCT------AAT 339
At1g32350      CCGAACGGCCGATGGTA--AAGTTATATCCACTTATTGGGGTATACCTCCT------ACT 231
At5g64210      CAAAACGGCGGGGGGTCAGTAGCGGTTCCAAGTTATTGGGGAATAGAGACAGCAAAGATG 336
                       *    *      *    *    * ** *****  *     *       *

At3g22360      AAGATCACCAAAGAAGATGGAACTGAATGGAAATGGAGTTGCTTTAGG|CCATGGAGACA 312
At3g27620      AAAATCACCAAAGAAGATGGAACTGAATGGAAATGGAGTTGCTTTCGG|CCATGGGAGACT 324
At3g22370      AAGATTACTAAAGAAGATGGTTCTGAATGGAAGTGGAACTGTTTCAGG|CCATGGGAAACG 399
At1g32350      AAGATCACTAAACCGGACGGTTCAGCTTGGAAGTGGAATTGTTTTCAGCCATGGGATTCA 291
At5g64210      AAAATTACTCGGAAAGATGGATCGGATTGGCCTTGGAACTGTTTTATG|CCATGGGAGACT 396
               ** ** **       ** **  * *  ***   ****  ** **   *********  *
```

B

	Locus	Gene	Forward primer 5' – 3'	Reverse primer 5' – 3'	Product	Comments
Working primers	At3g22370	AOX1a	GACGGTCCGTACGGTTTCG	CTTCTGATTCGCGTCCTCCTCCT	133 bp	Single product
	At3g22360	AOX1b	CACAGCCATCTTTTGAATCCTAGGG	CATTCAGTTCCATCTTCTTTGG	183 bp	Single product
	At3g27620	AOX1c	GCATCAAAGCAAGCGACATCC	TGGCTTCACGCCCCAATAAC	184 bp	Single product
	At1g32350	AOX1d	CATCTTCGGTACAATCCTCC	CACTTCCAAGCTGAACCGTC	183 bp	Single product
	At5g64210	AOX2	GGCGATTTCAAGATCGGCTC	GTTCCAAGGCCAATCCGATC	176 bp	Single product
Trial primers - Unsuccessful	At3g22360a	AOX1b	CACAGCCATCTTTTGAATCCTAGGG	GGAACTGAATGGAAATGGAGTTG	194 bp	Primer dimer
	At3g22360b	AOX1b	GAGGAGAAAGGATCTAGCGG	CAGATCTGATTTATATGTCTCCC	119 bp	Non-specific amplification
	At3g22360c	AOX1b	GAGGAGAAAGGATCTAGCGG	GGAACTGAATGGAAATGGAGTTG	185 bp	Primer dimer
	At3g22360d	AOX1b	CATCTTTTGAATCCTAGGGTTC	CATTCAGTTCCATCTTCTTTGG	180 bp	Primer dimer
	At3g27620	AOX1c	CTATTTCACCAACTTGCTCCAGC	GGAACTGAATGGAAATGGAGTTG	190 bp	Unreliable standard curves
	At1g32350	AOX1d	CGACCGGTACTATCATCTTCG	GGCCTTTAGTGATCTTAGTAGG	175 bp	Primer dimer
	At5g64210	AOX2	GTGGATGGGTATGAGTTCTGC	CCAATCCGATCCATCTTTCCG	115 bp	Non-specific amplification

Fig. 1. Primer design for QRT-PCR analysis of the *Arabidopsis AOX* gene family. (**A**) Multiple sequence alignment of the coding sequences of the five *AOX* genes produced by ClustalW; only the least-conserved N-terminal regions are shown, numbered from the start ATG codon. Highlighted regions represent successful QRT-PCR primer-binding sites; forward primers are marked by black highlighting and reverse primers by gray highlighting. The symbol I marks exon-exon boundaries, and the asterisk marks conserved nucleotides. (**B**) Table illustrating the primer design process, with

A standard curve is generated from the threshold cycles of the dilution series of the template DNA standard (generally in the range of 10^{-3} to 10^{-7} fmol/μL), which is used to determine the starting concentration of transcript in each cDNA sample based on its threshold cycle. By using an external standard, absolute and relative levels of each transcript examined can be estimated.

5. Each cDNA preparation is analyzed for potential genomic contamination by performing QRT-PCR analysis on the no-reverse transcriptase negative control reaction using a non-intron-spanning primer pair that is generally expressed at low-to-moderate levels in the tissue under examination. Under these conditions, no product should be detected; the presence of product will indicate genomic contamination, and the sample should be discarded (*see* **Note 16**).
6. Under each treatment condition, transcript levels of the gene of interest at each point over the treatment time-course are collected for each flask (performed for each of the three biological replicates/flasks). For each biological replicate, the transcript data from each time-point are expressed as a ratio of the pretreated value (at time = 0) in that flask. For each time-point, separately, the average of the ratio to the pretreated value for the three flasks is determined, along with the standard error for this value (*see* **Note 17**). This process is repeated for the untreated samples, and the difference in transcript abundance between the treated samples and the untreated treated samples at each time-point is expressed as a ratio, with a standard error. This data are then used in bioinformatic or statistical analysis.

3.5. Bioinformatic Analysis of QRT-PCR-Derived Transcript Data

1. To look for genes that respond in a similar manner to certain treatments, a GeneCluster 2.0 matrix is constructed for each treatment analyzed. In this matrix, each row represents a separate gene, and the four columns house the fold change expression data derived by QRT-PCR over the four time-points: pretreatment (0 h) and 3, 12, and 24 h posttreatment. The "find classes" tool in GeneCluster 2.0 is used under default settings with the cluster range varied. An example result is given in **Fig. 1A**.
2. To look for classes of treatments inducing common response patterns over the complement of genes analyzed, a GeneCluster 2.0 matrix is constructed in which the matrix rows are comprised of expression data for all genes with four time-points, with the data from each treatment in a separate row. Again, we use the find classes function of GeneCluster 2.0 under default settings to generate unsupervised self-organizing maps to extract classes in the data set representing common response trends to different treatments. An example result is given in **Fig. 1B**.

---

all primers designed and tested for the gene family listed. For several genes, multiple primer sets were examined before a primer combination resulting in a reliable single product without any crossreactivity was found. Primer pairs that were found experimentally to be unsuitable for QRT-PCR analysis are marked as unsuccessful, and the comments column indicates the reason. Suitable primer pairs are listed as working primers.

4. Notes

1. All water-soluble solutions were prepared in double deionized water passed through a NANO-Pure water purification system (Barnstead, Dubuque, IA, USA). Water used in nucleic acid manipulations was obtained from the Milli-Q Plus Ultrapure system (Millipore, Sydney, Australia) and sterilized by autoclaving prior to use.
2. We have found addition of BSA to a final volume of 0.008% (w/v) of all solutions used in QRT-PCR reactions, including primers, cDNA samples, and DNA standards, reduces nonspecific binding of the nucleic acids to materials in the microcentrifuge tubes, allowing accurate quantitation and preparation of low-concentration samples and decreasing the susceptibility of samples to freeze-thawing degradation.
3. We have found under these conditions that cells will be light green, and their plastids contain thylakoid membranes with small grana stacks.
4. We have found aluminum foil folded into small pockets is an ideal storage device for the cell samples. The filtered cells are scraped off the filters onto the foil, which is then folded to seal the sample and dropped into liquid nitrogen. The foil pockets can be easily labeled with permanent markers and are able to withstand snap freezing and prolonged storage at –80°C.
5. It is important that the chemical treatment proceeds as rapidly as possible once the pretreatment sample has been taken. This ensures that the pretreatment sample is representative of the transcript profile just prior to the addition of the treatment.
6. Do not use more than 100 μg tissue as this may result in incomplete lysis, resulting in lower RNA yield and purity. In addition, the RNeasy columns have a maximum RNA-binding capacity; exceeding this capacity will result in inconsistent yields of total RNA.
7. Perform all steps using β-mercaptoethanol in a fume hood and dispose of pipet tips and waste solutions containing β-mercaptoethanol appropriately.
8. DNase I is extremely sensitive to physical denaturation; mixing should only be carried out by gently inverting the tube. DNase I should be stored in single-use aliquots at –20°C; thawed aliquots can be stored at –4°C for up to 6 wk and should not be refrozen. All work performed with DNase I is carried out in a laminar flow using filtered tips. This is to ensure no materials used for work with DNA come into contact with DNase I.
9. A master mix containing Expand reverse transcriptase buffer, DTT, dNTPs, and RNasin can be made, with volumes scaled according to the number of reactions to be performed. We find this is more efficient.
10. We have found purifying the cDNA using the QIAquick PCR purification kit and then diluting the purified cDNA 1 in 10 removes any contaminants that may inhibit subsequent QRT-PCR reactions.
11. Ideally, the cloning primers will amplify the bulk of the transcribed product of the gene in a PCR for which cDNA is used as a template. This will result in the standard template DNA resembling transcribed gene product as closely as possible; thus, the reaction conditions, such as primer efficiency, observed when using the standard as template should be equivalent to the conditions seen with sample cDNA.

12. If additional nonspecific PCR products are seen following separation by agarose gel electrophoresis, then the desired DNA fragment can be excised using a clean razor blade and placed in a preweighed 1.5-mL microcentrifuge tube, then purified using the QIAquick gel extraction kit according to manufacturer's instructions.
13. For analysis of genes belonging to multiple gene families, an alignment of the transcribed products of all members is required. Use coding sequence to ensure that primers only bind to transcribed regions; alternatively, 5′-UTR regions can be considered when coding sequences have very high identity across a gene family. Sequences can be downloaded from TAIR and aligned using ClustalW programs.
14. Primer efficiency is used to describe the rate of amplification of the DNA in a QRT-PCR reaction; an efficiency of 100% implies the amount of DNA doubles each cycle. Several factors can affect the primer efficiency of a QRT-PCR reaction, including the amount of primer and the amount of salt (Mg^{2+}) in the reaction. Primer concentrations above optimal can result in nonspecific binding or primer-dimer amplification. Mg^{2+} is required for polymerase function; however, salt in overabundance will decrease the annealing specificity of the primers and may result in nonspecific products. To determine the optimal reaction conditions for each primer pair, we perform optimizing iCycler reactions with primer concentrations between 0.3 and 0.9 μ*M* and Mg^{2+} concentrations between 1 and 5 m*M*. The conditions, which result in a primer efficiency closest to 100%, are then used for that primer pair. Primer design can also have an impact on efficiency, with factors such as tendency toward hairpin and other secondary structures affecting annealing ability.
15. When setting up an iCycler run, we recommend preparing a master mix containing the primers, iQ SYBR Supermix, BSA, and sterile water. Aliquot this master mix into the wells of the PCR plate and then add the sample, either template DNA standard or cDNA, directly into the master mix. For each run, a master mix sufficient for all reactions using that primer pair should be prepared. This will ensure reaction conditions are constant across all samples analyzed and enable confidant comparison of threshold cycles for all samples in that run. In addition, we ensure all samples derived from each treatment (i.e., the triplicate cDNA preps sampled in duplicate from each time-point) are analyzed by the same iCycler run, thus ensuring all samples under each treatment condition are analyzed under constant PCR conditions, minimizing the effects of minor variations in pipeting or reaction setup.
16. We have found using two DNase I treatment sets removes all potentially contaminating genomic DNA from the RNA isolation. Always run a positive control (e.g., a reaction using cDNA or the standard template DNA for that primer pair) to ensure the absence of product is caused by the absence of genomic DNA and not incorrect reaction setup.
17. We find most of the variability is associated with different RNA samples rather than the different reverse transcription reactions or duplicate iCycler runs. Overall, we find variation within and across replicates minimal, and we are consistently able to detect statistically significant changes in transcript levels in response to treatments.

References

1. Chen, D., Toone, W. M., Mata, J., et al. (2003) Global transcriptional responses of fission yeast to environmental stress. *Mol. Biol. Cell.* **14,** 214–229.
2. Gasch, A. P., Spellman, P. T., Kao, C. M., et al. (2000) Genomic expression programs in the response of yeast cells to environmental changes. *Mol. Biol. Cell* **11,** 4241–4257.
3. Causton, H. C., Ren, B., Koh, S. S., et al. (2001) Remodeling of yeast genome expression in response to environmental changes. *Mol. Biol. Cell* **12,** 323–337.
4. Gray, M. W., Burger, G., and Lang, B. F. (1999) Mitochondrial evolution. *Science* **283,** 1476–1481.
5. Czechowski, T., Bari, R. P., Stitt, M., Scheible, W.-R., and Udvardi, M. K. (2004) Real time RT-PCR profiling of over 1400 *Arabidopsis* transcription factors: unprecedented sensitivity reveals novel root- and shoot-specific genes. *Plant J.* **38,** 366–379.
6. Schaffer, R., Landgraf, J., Accerbi, M., Simon, V., Larson, M., and Wisman, E. (2001) Microarray analysis of diurnal and circadian-regulated genes in *Arabidopsis*. *Plant Cell* **13,** 113–123.
7. Clifton, R., Lister, R., Parker, K. L., et al. (2005) Stress induced co-expression of alternative respiratory chain components in *Arabidopsis thaliana*. *Plant Mol. Biol.* **58,** 193–212.
8. Sweetlove, L. J., Heazlewood, J. L., Herald, V., et al. (2002) The impact of oxidative stress on *Arabidopsis* mitochondria. *Plant J.* **32,** 891–904.
9. Golub, T. R., Slonim, D. K., Tamayo, P., et al. (1999) Molecular classification of cancer: class discovery and class prediction by gene expression monitoring. *Science* **286,** 531–537.
10. Tamayo, P., Slonim, D., Mesirov, J., et al. (1999) Interpreting patterns of gene expression with self-organizing maps: methods and application to hematopoietic differentiation. *Proc. Natl. Acad. Sci. U. S. A.* **96,** 2907–2912.

37

Proteomics of Yeast Mitochondria

Jörg Reinders and Albert Sickmann

Summary

Because virtually all cellular processes are based on proteins, detailed knowledge of the mitochondrial proteome represents an integral part of understanding mitochondrial function. The analysis of very complex protein mixtures such as entire cell organelles makes high demands on analysis techniques in order to ensure integrity of the obtained data set. The use of *Saccharomyces cerevisiae* as a model system allows the isolation of mitochondria of utmost purity in large amounts. Various approaches can be accomplished in the subsequent analysis to achieve the most complete overview possible. Combinations of orthogonal proteomics techniques include two-dimensional polyacrylamide gel electrophoresis (2D-PAGE), one-dimensional sodium dodecyl sulfate-polyacrylamide gel electrophoresis (1D-SDS-PAGE), and nano-LC-MS/MS (nano-liquid chromatography with tandem mass spectrometry) as well as multidimensional high-performance liquid chromatography with tandem mass spectrometry (HPLC-MS/MS). The inherent limitations of the individual methods can be countervailed by parallel application of these approaches.

Key Words: Database search; mass fingerprint; mitochondrial proteomics; multidimensional HPLC; nano-LC-MS/MS; 1D-SDS-PAGE; separation; subcellular fractionation; tandem mass spectrometry; 2D-PAGE.

1. Introduction

Making an inventory of the protein content is the first step in characterization of the mitochondrial proteome, which grants the basis for molecular understanding of various mitochondrial functions. Because only a few proteins are encoded in the mitochondrial deoxyribonucleic acid (DNA), the bigger part is imported, turning mitochondria into a system tailor-made for proteomic studies. Mitochondria are more than just the power plants of the cell. They are involved in various cellular functions, such as amino acid and lipid metabolism or apoptosis. However, proteins related to energy budget are the most abundant ones, although they only account

From: *Methods in Molecular Biology, vol. 372: Mitochondria: Practical Protocols*
Edited by: D. Leister and J. M. Herrmann © Humana Press Inc., Totowa, NJ

for about 15% of the estimated 800–1000 proteins present in yeast mitochondria ***(1)***. Because of the broad range of protein abundances, extreme requirements are set concerning separation power, sensitivity, and particularly the dynamic range of the applied methods. Because each procedure has inherent limitations, only a combination of orthologous methods can offer an unbiased overview.

The first analysis step is the isolation of mitochondria of the highest possible purity as a prerequisite of major importance for proteomic study ***(2)***. A combination of differential centrifugation and gradient centrifugation steps has proven useful for this purpose. Moreover, purity has to be documented by suitable and independent methods such as Western blots, with a sufficient number of antibodies against representative mitochondrial and nonmitochondrial marker proteins ***(3)***. Therefore, antibodies covering a large variety, preferably all cell compartments, should be used. After lysis, the reduction of sample complexity may be solved on the protein level (i.e., by electrophoretic means) or on the peptide level after proteolytic digest (e.g., by chromatographic methods).

The classical, but still state-of-the-art, approach for protein separation is two-dimensional polyacrylamide gel electrophoresis (2D-PAGE) ***(4,5)***. Separation of proteins according to isoelectric point in the first and molecular weight in the second dimension allows parallel display of up to 10,000 protein species on a single gel, including separation of isoforms. 2D-PAGE is an imaging procedure of high reproducibility and sensitivity that is utterly dependent on the staining method. Furthermore, it can be coupled to various subsequent separation and detection systems, most of all liquid chromatography and mass spectrometry. However, 2D-PAGE-based approaches are biased against certain groups of proteins, such as low abundant and very hydrophobic ones as well as proteins with extreme p*I* or size. Thus, especially proteins of high interest (e.g., membrane proteins) are often under-represented in such studies because of specific losses caused by precipitation after isoelectric focusing and during transfer to the second dimension.

To circumvent the intrinsic limitations of the classical 2D-PAGE, gel systems compatible with detergent-containing lysis buffers such as one-dimensional sodium dodecyl sulfate PAGE (1D-SDS-PAGE) can be used, thereby reducing solubility problems. However, separation range and resolution of 1D-SDS-PAGE are not sufficient to display all individual protein components of highly complex samples. Therefore, subsequent separation methods have to be applied (*see* **Fig. 1**), such as reversed-phase chromatography on the peptide level after proteolytic cleavage of gel-separated proteins ***(6)***. The lack of isoform separation is countervailed by the superior sensitivity concerning detection of membrane proteins.

Some proteins may not be identified by 2D-PAGE or the approach combining 1D-SDS-PAGE on the protein and nano-reversed-phase high-performance liquid chromatography (HPLC) on the peptide separation level. Thus, an additional procedure can be chosen to transfer the protein separation problem

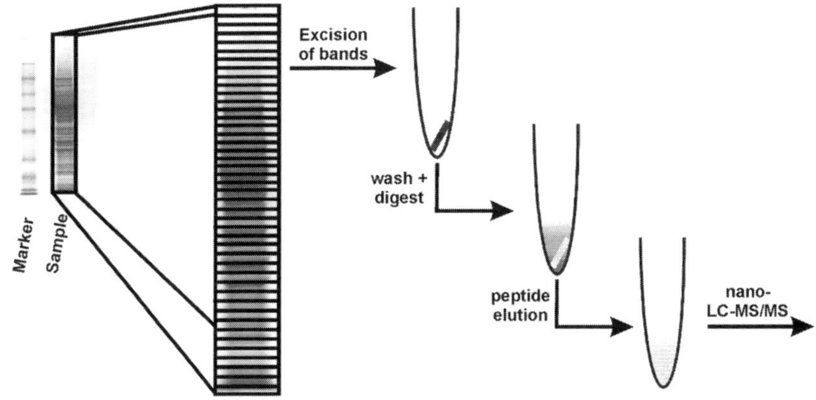

Fig. 1. Workflow for SDS-PAGE and subsequent nano-LC-MS/MS-analysis. The entire gel lane is divided into 1-mm slices that are excised and washed. After proteolytic in-gel digest, the obtained peptides are eluted from the gel piece and subjected to nano-LC-MS/MS.

entirely to the peptide level. By initial proteolytic cleavage of proteins, sample complexity is raised tremendously. However, protein properties differ significantly; peptides show similar physicochemical characteristics and are thus better accessible to distinct separation methods. Therefore, peptide separation capacities are much higher than capacities of protein separations, which are more or less suitable for certain classes of proteins only. Peptide separation is usually accomplished by multidimensional liquid chromatography *(7)* that is—at least theoretically—able to resolve tens of thousands of peptides. Furthermore, mass spectrometric (MS) coupling of this method can be automated but is afflicted with the need for high-throughput bioinformatics capable of interpreting the huge amount of obtained data.

A combination of complementary approaches on the protein and peptide levels (*see* **Fig. 2**) may facilitate an almost complete overview of the proteome of entire cell organelles by countervailing mutual drawbacks. Because every method is limited in sensitivity, dynamic range, and suitability for certain sample types, some proteins will escape detection. The combination of orthogonal approaches can diminish the number of unidentified proteins. However, guaranteed complete overviews of a (sub)proteome may not be created as it cannot be excluded that a protein escapes all applied methods.

2. Materials

2.1. Isolation of Mitochondria

1. YPG medium: 1% (w/v) yeast extract, 2% (w/v) Bacto™ peptone, 3% (v/v) glycerol, pH 5.0.

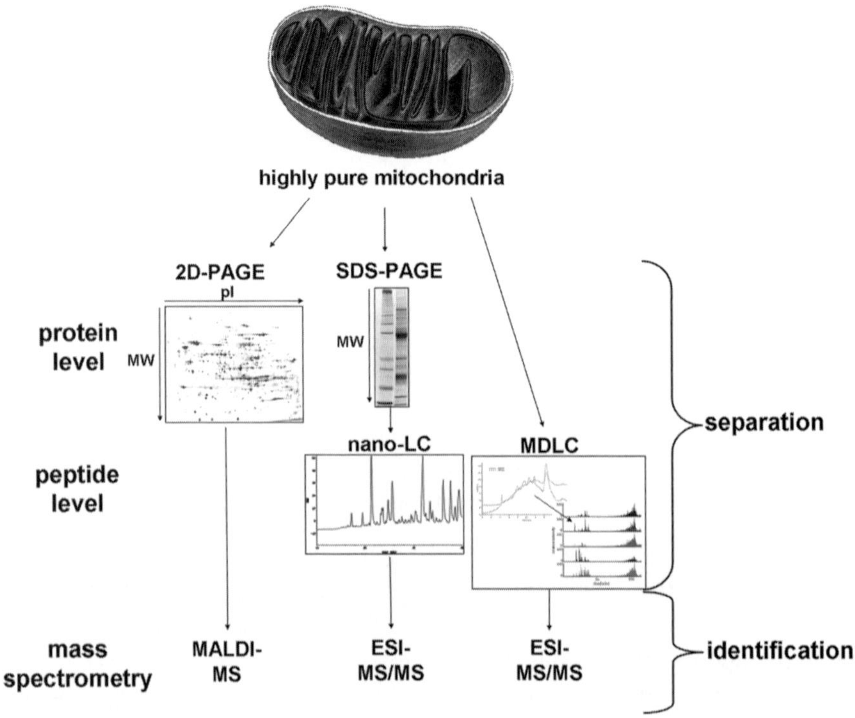

Fig. 2. Strategy for the analysis of entire cell organelles such as mitochondria. The isolated mitochondria are subjected to various separation and detection methods to rule out method-dependent biases. The most complete overview of the organelle proteome can be obtained from the combination of orthogonal techniques. Therefore, three different 2D separation approaches are accomplished. One method operates exclusively on the protein level (2D-PAGE); the second is carried out on both protein and peptide levels (1D-SDS-PAGE and nano-LC), and the last is exclusively applied on the peptide level (multidimensional LC).

2. Dithiothreitol (DTT) buffer: 100 m*M* Tris-H_2SO_4, pH 9.4, 10 m*M* DTT.
3. Zymolase buffer: 1.2 *M* sorbitol, 20 m*M* potassium phosphate, pH 7.4.
4. Homogenization buffer: 0.6 *M* sorbitol, 10 m*M* Tris-HCl, pH 7.4, 1 m*M* ethylenediaminetetraacetic acid (EDTA), 1 m*M* phenylmethylsulfonyl fluoride, 0.2% (w/v) bovine serum albumin.
5. SEM buffer: 250 m*M* sucrose, 1 m*M* EDTA, 10 m*M* 3-(*N*-morpholino)propanesulfonic acid (MOPS), pH 7.2.
6. EM buffer: 1 m*M* EDTA, 10 m*M* MOPS, pH 7.2.

2.2. Lysis of Mitochondria

1. Glass beads, 1-mm diameter.
2. Sonication bath.

2.3. Two-Dimensional Polyacrylamide Gel Electrophoresis

For 2D-PAGE, use GE Healthcare Ettan™ 2D-PAGE system or similar.

2.3.1. Isoelectric Focusing (First Dimension)

For isoelectric focusing (first dimension) rehydration buffer, use 7 *M* urea, 2 *M* thiourea, 2% (w/v) CHAPS (3-[(cholamidopropyl)dimethylammonio]-propanesulfonate), 0.002% (w/v) bromophenol blue, 50 m*M* DTT, 2% (v/v) immobilized pH gradient (IPG) buffer (filtered using a mixed bed ion exchanger, e.g., Serdolit® MB-1, Serva Electrophoresis GmbH, Heidelberg, Germany).

2.3.2. Casting of Polyacrylamide Gels

1. Gel solution: 12.5% (w/v) acrylamide, 375 m*M* Tris-HCl, pH 8.8, 0.1% (w/v) SDS, 0.1% ammonium persulfate (APS), 0.0138% (v/v) *N,N,N'*-tetramethyl-ethylenediamine.
2. Storage buffer: 375 m*M* Tris-HCl, pH 8.8, 0.1% (w/v) SDS.

2.3.3. SDS-PAGE (Second Dimension)

1. SDS equilibration buffer 1: 50 m*M* Tris-HCl, pH 8.8, 6 *M* urea, 30% (v/v) glycerol, 2% (w/v) SDS, 0.002% (w/v) bromophenol blue, 130 m*M* DTT.
2. SDS equilibration buffer 2: 50 m*M* Tris-HCl, pH 8.8, 6 *M* urea, 30% (v/v) glycerol, 2% (w/v) SDS, 0.002% (w/v) bromophenol blue, 280 m*M* iodoacetamide.
3. SDS running buffer: 25 m*M* Tris base, 192 m*M* glycine, 0.1% (w/v) SDS.
4. Agarose solution: 0.5% (w/v) agarose, 0.001% (w/v) bromophenol blue in SDS running buffer.

2.3.4. Silver Staining of Gels

1. Fixation buffer: 50% (v/v) ethanol, 10% (v/v) acetic acid.
2. Sensitization buffer: 0.5 *M* sodium acetate, 0.2% (w/v) sodium thiosulfate, 30% (v/v) ethanol.
3. Staining buffer: 0.1% (w/v) silver nitrate, 0.03% (v/v) formaldehyde.
4. Developing solution: 2.5% (w/v) sodium carbonate, pH 10.9, 0.03% (v/v) formaldehyde.
5. Stop solution: 50 m*M* EDTA.

2.3.5. Colloidal Coomassie Staining of Gels

Staining solution: 34% (v/v) methanol, 2% (v/v) orthophosphoric acid, 17% (w/v) ammonium sulfate, 0.066% (w/v) Coomassie G-250.

2.3.6. Washing and Tryptic Digestion of Gel Spots

1. Washing buffer A: 50 m*M* NH_4HCO_3, pH 7.8.
2. Washing buffer B: 50% (v/v) washing buffer A, 50% (v/v) acetonitrile.
3. Trypsin solution: 25 ng/µL sequencing grade trypsin in 50 m*M* NH_4HCO_3, pH 7.8.

2.3.7. Matrix-Assisted Laser Desorption/Ionization-Mass Spectrometry

1. Matrix-assisted laser desorption/ionization-time-of-flight(/time-of-flight) mass spectrometer [MALDI-TOF(/TOF)-mass spectrometer], such as Bruker Daltonics Ultraflex TOF/TOF or similar.
2. Matrix solution: saturated α-cyano-4-hydroxy-cinnamic acid solution in 0.05% (v/v) trifluoroacetic acid, 50% (v/v) acetonitrile.

2.4. One-Dimensional Polyacrylamide Gel Electrophoresis

1. Invitrogen NuPAGE™ system or similar.
2. Lithium dodecyl sulfate (LDS)-sample buffer (4X): 106 m*M* Tris-HCl, 141 m*M* Tris base, 2% (w/v) LDS, 10% (v/v) glycerol, 0.51 m*M* EDTA, 0.22 m*M* Serva® Blue G250, 0.175 m*M* phenol red, pH 8.5 (as purchased from Invitrogen, Karlsruhe, Germany).

2.5. Nano-Liquid Chromatography-Tandem Mass Spectrometry

1. Electrospray ionization-tandem mass spectrometer (MS/MS) with nanospray source (e.g., LCQDeca XP^{Plus}, ThermoElectron).
2. Bioinert nano-HPLC system such as Dionex nano-LC (liquid chromatographic) system (Famos™, Switchos™, Ultimate™) or similar.
3. Trapping solvent: 0.1% (v/v) trifluoroacetic acid.
4. Nano-flow solvent A: 0.1% (v/v) formic acid.
5. Nano-flow solvent B: 0.1% (v/v) formic acid, 84% (v/v) acetonitrile.

2.6. Multidimensional Liquid Chromatography

Preparative, bioinert HPLC system such as Dionex BioLC™ or similar.

2.6.1. First Dimension: Strong Cation Exchange Chromatography

1. Solvent A: 50 m*M* potassium phosphate, pH 3.0.
2. Solvent B: 50 m*M* potassium phosphate, pH 5.5, 0.25 *M* NaCl, 25% (v/v) acetonitrile.

2.6.2. First Dimension: Strong Anion Exchange Chromatography

1. Solvent A: 50 m*M* Tris-HCl, pH 8.0.
2. Solvent B: 50 m*M* Tris-HCl, pH 8.0, 0.25 *M* NaCl, 25% acetonitrile (v/v).

3. Methods

3.1. Isolation of Highly Pure Mitochondria (3)

1. *Saccharomyces cerevisiae* cells are grown in YPG medium to an OD of 2.0–2.5.
2. The cells are pelleted at 3000*g* for 5 min and washed with distilled water.
3. Afterward, they are resuspended in 2 mL/g (wet weight) of DTT buffer and shaken slowly for 20 min at 30°C.
4. The cells are washed with zymolase buffer and incubated with 5 mg/g (wet weight) zymolase-20T in 7 mL/g (wet weight) of zymolase buffer for 45 min at 30°C for conversion into spheroblasts.

5. Homogenization is accomplished by 15 strokes in a glass-Teflon™ potter in 6.5 mL/g (wet weight) of ice-cold homogenization buffer.
6. Cell debris and nuclei are removed after dilution with 1 volume ice-cold homogenization buffer by centrifugation at 1500*g* for 5 min at 4°C.
7. The supernatant is centrifuged at 3000*g* for 5 min (4°C), and mitochondria are pelleted from the supernatant at 12,000*g* for 15 min (4°C).
8. The pellet is washed with SEM buffer and again pelleted at 12,000*g* for 15 min.
9. The obtained crude mitochondrial fraction is adjusted to 5 mg/mL in SEM buffer. The mitochondria are stable and retain function within this buffer. The obtained purity is sufficient for most studies concerning mitochondrial function or import, but higher purity is required for accurate proteomic analyses.
10. This crude mitochondrial fraction is further processed by 10 strokes in a glass-Teflon potter to remove further contaminations (e.g., cytoskeletal proteins attached to the outer membrane of the mitochondria).
11. A three-step sucrose gradient centrifugation (1.5 mL 60%, 4 mL 32%, 1.5 mL 23%, 1.5 mL 15% (w/v) sucrose in EM buffer) is used for further purification. Centrifugation is carried out at 134,000*g* in a Beckmann SW41 Ti swinging-bucket rotor for 1 h at 2°C.
12. The purified mitochondria are recovered from the 60%/32% interface and diluted with 2 volumes SEM buffer.
13. The mitochondria are pelleted at 12,000*g* at 2°C.
14. Mitochondria are resuspended in SEM buffer to a final concentration of 10 mg/mL (*see* **Note 1**).

By these additional steps, residual proteins from other organelles can be depleted, leaving the mitochondria virtually devoid of contamination. Documenting the performance of the isolation procedure is mandatory to ensure reproducible purification results. Thus, purity of the mitochondria should be examined by suitable methods (e.g., Western blotting against various marker proteins) (*see* **Table 1** and **Fig. 3**). Thereby, both representativity of the chosen marker proteins and sensitivity of the used antibodies are crucial.

3.2. Lysis of Mitochondria

1. An appropriate amount of highly pure mitochondria is pelleted by centrifugation at 12,000*g* for 15 min.
2. The pellet is resuspended and lysed on addition of two or three glass beads and three alternate sonication steps and incubations on ice for 1 min each (*see* **Note 2**). To avoid contaminations from proteins attached to the glass beads, they should be cleaned with 1 *M* HCl prior to use.

3.3. Two-Dimensional Polyacrylamide Gel Electrophoresis

3.3.1. Isoelectric Focusing (First Dimension)

1. IPG strips (24 cm, pH 3.0–10.0, nonlinear) are rehydrated with 450 µL rehydration buffer overnight (*see* **Note 3**).

Table 1
Possible Marker Proteins for Testing Purity of Mitochondria

Marker protein	Localization
Porin	mitochondrion
mtHSP70	mitochondrion
Tom40	mitochondrion
Tom70	mitochondrion
Tim23	mitochondrion
Sec61	endoplasmic reticulum
Sss1	endoplasmic reticulum
ALP	vacuole
BMH1/2	cytosol
PGK	cytosol
Pex13	peroxisome
Nsp1	nucleus

2. The mitochondrial lysate is adjusted to 3 µg/µL in rehydration buffer, and 100 µL are applied via a sample cup at the acidic side of the gradient. In case larger volumes have to be used, the sample may be diluted in rehydration buffer and applied directly within the rehydration, whereas higher sample concentrations can lead to protein precipitation.
3. For isoelectric focusing, the voltage is raised from 500 to 4000 V in 1.5 h and kept at 4000 V for 13 h for a total of 55,400 Vh. The IPG strip is cooled by a Peltier element throughout the process to minimize protein degradation. The strip has to be covered with cover fluid (paraffin oil, etc.) to prevent drying, which would cause urea precipitation. Furthermore, aerial CO_2 is excluded as it could disturb the pH gradient on contact.
4. The strips are washed with distilled water and subjected to the second dimension (SDS-PAGE). Very hydrophobic proteins such as membrane proteins tend to precipitate during isoelectric focusing as the solubility of a protein is at its minimum at its isoelectric point. In addition, proteins with unusual properties such as very high or low molecular weight, as well as highly acidic or alkaline p*I* and the like, may be underrepresented in 2D-PAGE-based approaches. These protein classes should be addressed by other methods ***(7,8)*** rather than 2D-PAGE.

3.3.2. Casting of Polyacrylamide Gels

1. The freshly prepared gel solution is poured into the gel-pouring stand, carefully avoiding air bubbles, and is directly overlaid with 1 mL water-saturated butanol per gel.
2. Polymerization is accomplished at room temperature for at least 4 h because residual acrylamide monomers may otherwise react with the sample. Such unwanted protein modifications may both change the proteins' physicochemical characteristics, causing shifts on the 2D gel, and hamper mass spectrometric identification of the proteins because of altered peptide masses. In case of overnight polymerization

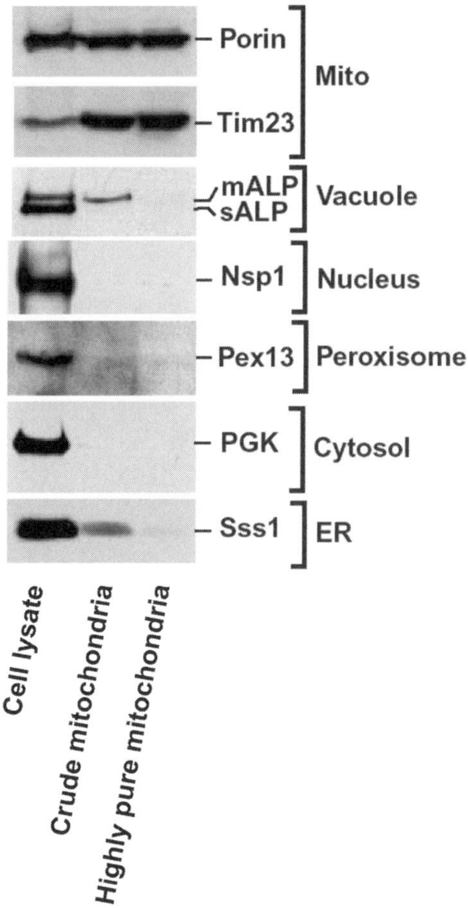

Fig. 3. Western blot against marker proteins of different localizations. Mitochondrial marker proteins are enriched throughout the isolation process, whereas nonmitochondrial proteins are depleted.

at 4°C, exchange the water-saturated butanol with storage buffer after 2 h (*see* **Note 4**) to avoid washout effects of SDS from the gel. Gel-to-gel variance can also be diminished by using gels cast in the same rather than different batches. Furthermore, freshly poured gels usually show better and more reliable performance than stored ones.

3.3.3. SDS-PAGE (Second Dimension)

1. The IPG strips are incubated with SDS equilibration buffer 1 for 15 min under gentle shaking for reduction of disulfide bonds.
2. The strips are rinsed with water and incubated with SDS equilibration buffer 2 for another 15 min for carbamidomethylation of the free cysteine residues (*see* **Note 5**).

3. Transfer the strips onto the SDS gels and overlay them with hot agarose solution. Any air bubbles between the IPG strips and the SDS gel can cause smearing effects and should therefore be avoided.
4. After congealing of the agarose, a marker is applied, and the SDS-PAGE is started using 5 W per gel for 30 min and continued at 17 W per gel. The gels should be cooled throughout the electrophoresis to diminish protein degradation. They are subjected to the respective staining procedure directly after electrophoresis.

3.3.4. Silver Staining (**9**)

1. Gels are incubated in fixation buffer overnight, avoiding diffusion of the gel spots.
2. Gels are incubated in the sensitization buffer (*see* **Note 6**) for 2 h prior to three 20-min watering steps.
3. The gels are gently shaken in staining buffer for 30 min and washed with 2.5% (w/v) sodium carbonate, pH 10.9, for 1 min.
4. The gels are kept in the developing solution for 2–3 min (*see* **Note 7**).
5. The development process is stopped by buffer exchange to stop solution (*see* **Fig. 4**).
6. Gels can be stored in water for up to 2 d. Longer storage times cause diffusion and washout effects.
7. Protein spots are excised and subjected to washing and digestion followed by MS detection of peptides.

3.3.5. Colloidal Coomassie Staining

1. The staining solution has to be freshly prepared prior to the staining procedure. Therefore, the ammonium sulfate should be added in small portions as the last component; otherwise, precipitation occurs. The colloidal Coomassie staining solution shows a metallic glance on the liquid surface.
2. The gels are directly transferred to the staining solution and incubated at room temperature from 2 h to 2 d (usually overnight) depending on the desired staining intensity.
3. Background staining is removed by washing the gels in distilled water. They should be stored in water for no longer than 2 d before washing and digestion (*see* **Subheading 3.3.6.**).

3.3.6. Washing and Tryptic Digestion of Gel Spots (**10**)

1. The gel pieces are washed three times alternately with 10 μL washing buffer A and 10 μL washing buffer B. By this washing-and-shrinking procedure, buffers, salts, detergents, and the like in the gel piece are removed. Although Coomassie staining is also removed by this method, silver stain is retained. Removal of Coomassie staining before MS is mandatory because residual Coomassie can suppress peptide ionization. As only 2–3% of the protein content of a 2D gel spot 1D-SDS band are stained by silver staining, removal of silver ions is not necessary for subsequent MS analysis.
2. The gel spots are dried, rehydrated with 4 μL trypsin solution, and incubated overnight at 37°C. In case of large gel pieces, 4 μL solvent may not be sufficient

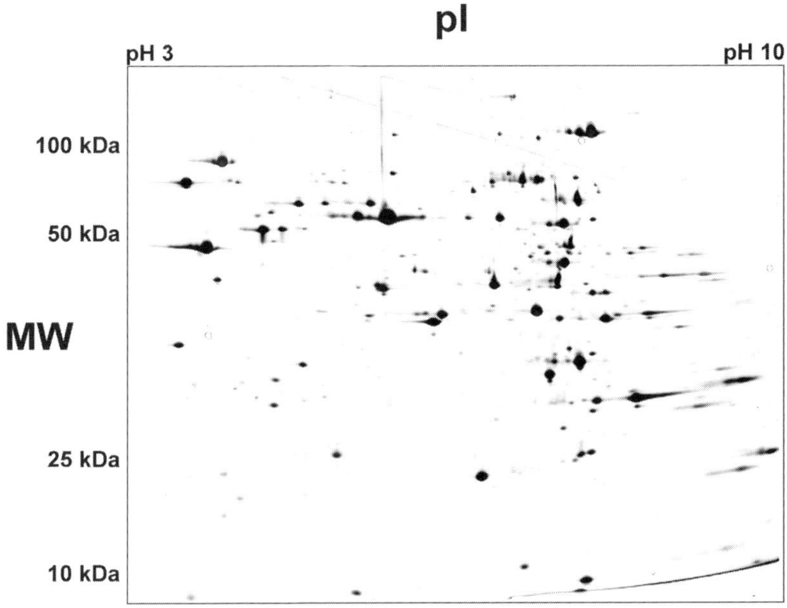

Fig. 4. 2D-PAGE silver-stained image of isolated mitochondria. The protein mixture is separated according to p*I* (first dimension) and molecular weight (second dimension). The protein spots are excised, washed, digested, and subjected to MALDI-mass fingerprinting.

to rehydrate the gel fully; thus, a few microliters of washing buffer A may be added.

3.3.7. Matrix-Assisted Laser Desorption/Ionization-Mass Spectrometry

1. The peptides are eluted from the gel pieces using 10 µL 0.1% (v/v) trifluoroacetic acid.
2. Equal volumes (typically 0.3–0.5 µL) of the eluate and matrix solution (*see* **Note 8**) are mixed directly on the MALDI target. After crystallization, MALDI-mass fingerprinting can be accomplished. In case of poor signals, the use of C18 tips (e.g., ZipTips™ C18, Millipore Corp., Bedford, MA, USA; Perfect Pure™ C18 Tips, Eppendorf AG, Hamburg, Germany; or OMIX™ C18 Tips, Palo Alto, CA, USA) may concentrate the peptides and thereby enhance signal-to-noise ratios.
3. The obtained spectra are searched against the *Saccharomyces* Genome Database (http://www.yeastgenome.org/) using the Mascot™ algorithm (http://www.matrixscience.com/) with trypsin as protease, carbamidomethylated cysteines as fixed modification (if carbamidomethylation was accomplished), and methionine oxidation as variable modification.

3.4. One-Dimensional SDS-PAGE

1. 50–100 µg mitochondria are lysed in 20 µL LDS-sample buffer.

2. The samples are heated to 90°C for 5 min to denaturate the proteins fully and are applied to a 10% Bis-Tris gel (MOPS-buffer system). Application of higher amounts of sample may result in smearing effects.
3. The gel is run at 50 V for approx 10 min until the proteins have entered the gel matrix. Then, the voltage is raised to 200 V, and the electrophoresis is stopped when the running front reaches the end of the gel. Throughout the entire process, the gel is cooled to 4°C to diminish protein degradation effects.
4. After removing the gel from the electrophoresis chamber, it is subjected to silver or colloidal Coomassie staining as prescribed in **Subheadings 3.3.4.** and **3.3.5**.
5. The entire lane is cut into approx 1-mm slices that are washed and digested tryptically (*see* **Subheading 3.3.6.**).
6. The peptides are eluted using 20 µL 5% (v/v) formic acid, and the eluate is used for nano-LC-MS/MS.

3.5. Nano-LC-MS/MS

1. The samples are adjusted to approx 20 µL using 5% (v/v) formic acid if necessary.
2. The sample is injected into the sample loop by the autosampler and flushed onto a trapping column (C18, 300-µm id, 1-mm length, LC Packings, Amsterdam, The Netherlands) at a flow rate of 30 µL/min. Thereby, the peptides are concentrated and desalted. Subsequently, the peptides are eluted onto the separation column (C18, 75-µm id, 15-cm length, LC Packings) by a second pump operating at approx 250 nL/min. Separation is accomplished using a 1-h binary gradient from 5 to 50% of solvent B. The LC system is directly coupled to an ion-trap mass spectrometer operating in triple-play mode via a nano-electrospray ionization ion source.
3. The obtained MS/MS spectra are searched against the *Saccharomyces* Genome Database using either the Sequest™ ***(11)*** or the Mascot algorithm.

3.6. Multidimensional Liquid Chromatography

1. Digest 2 mg purified mitochondria with trypsin, chymotrypsin, GluC, or subtilisin (protein:protease ratio 50:1 in 10 m*M* NH_4HCO_3; *see* **Note 9**). The digests are carried out at 37°C overnight or, in the case of subtilisin, for 2 h.
2. The samples are split in half and subjected to the first dimension, either strong anion exchange (SAX) or strong cation exchange (SCX) chromatography.

3.6.1. First Dimension: SCX Chromatography

1. The samples are adjusted to pH 3.0 using formic acid and applied to a 2.1-mm id SCX column 25 cm long (Polymer Laboratories GmbH, Darmstadt, Germany) at a flow rate of approx 150 µL/min.
2. Elution is accomplished using a 1-h binary gradient from 0 to 99% of solvent B *(7)*. Because the applied sample is far too complex to obtain distinct signals by chromatography, 1-min fractions are collected and stored immediately at –80°C.
3. Use 10-µL aliquots for subsequent nano-LC-MS/MS-analysis (*see* **Subheading 3.5.**; *see* **Note 10**).

3.6.2. First Dimension: SAX Chromatography

1. In case of SAX chromatography in the first dimension, the sample is directly applied to a 2.1-mm id SAX column 15 cm long (Polymer Laboratories GmbH) at a flow rate of approx 150 μL/min.
2. Elution is done using a 1-h binary gradient from 0 to 99% of solvent B *(7)*.
3. The 1-min fractions are collected and stored at –80°C.
4. Use 10-μL aliquots for nano-LC-MS/MS-analysis as the second dimension (*see* **Subheading 3.5.**; *see* **Note 10**).

4. Notes

1. The isolated mitochondria may be stored in SEM buffer at –80°C for several weeks.
2. Lysis can be performed in various buffer systems. Usually, the choice of buffer conditions is dependent on the subsequently applied methods. Solubilization of very hydrophobic proteins (i.e., membrane proteins) may be enhanced by using detergent-containing buffers, but compatibility to later separation procedures has to be kept in mind.
3. Rehydration buffer may be stored without DTT and IPG buffer (which would decompose) at –20°C for several weeks. Add the appropriate amount of DTT and IPG buffer directly before use.
4. Gel plates may be removed from the gel-pouring stand, and the gels may be stored at 4°C for a week if kept wet.
5. The carbamidomethylation step is strongly recommended. Otherwise, the free cysteines may be partly modified by residual acrylamide monomers in the fresh gel matrix. The gel resolution in the alkaline region (>pH 9.0) is usually inferior to the resolution in the lower pH region ("streaking"). This problem can partly be overcome by use of DeStreak™ *(12)* instead of the carbamidomethylation procedure. Note that the DTT in the first dimension has to be replaced by DeStreak using this technique.
6. The sensitivity of silver staining may be improved by addition of 0.5% (v/v) glutardialdehyde to the sensitization buffer. Unfortunately, MS analysis of protein spots is impossible using this additive.
7. The duration of the development step is strongly dependent on the sample amount. Application of the developing solution may be prolonged until spots arise or reduced if high background staining occurs.
8. Colder MALDI matrices such as 2,5-dihydroxy-benzoic acid may also be used to decrease spontaneous peptide fragmentation, but usually α-cyano-4-hydroxy-cinnamic acid yields the most homogeneous crystallization. Furthermore, hot matrices may aid in the acquisition of post-source decay (PSD) spectra in case of MS/MS-capable mass spectrometers.
9. Try to work at high protein concentrations as the solutions may have to be diluted prior to ion exchange chromatography (pH adjustment). Furthermore, the lower the applied sample volume for HPLC, the better is the resolution of the separation. However, too high concentrations (>10 mg/mL) may lead to precipitation effects.

Parallel application of different proteases enhances the number of identifiable proteins that could be not digestable by a certain protease. Thus, an aliquot of each digest should be checked for completeness of digestion by 1D-SDS-PAGE and by subsequent silver staining.

10. The peptide amount transferred to the second dimension may be individually lowered or raised. The fractions derived from the first dimension may also be dried down for volume reduction. Adjustment of the pH of the fractions before the second dimension is mandatory anyway.

Acknowledgments

We thank Dr. Christof Meisinger and Professor Nikolaus Pfanner for optimization of the purification protocol for yeast mitochondria. This work was supported by the DFG SI 835-2/1.

References

1. Sickmann, A., Reinders, J., Wagner, Y., et al. (2003) The proteome of *Saccharomyces cerevisiae* mitochondria. *Proc. Natl. Acad. Sci. U. S. A.* **100,** 13,207–13,212.
2. Stasyk, T. and Huber, L. A. (2004) Zooming in: fractionation strategies in proteomics. *Proteomics* **4,** 3704–3716.
3. Meisinger, C., Sommer, T., and Pfanner, N. (2000) Purification of *Saccharomcyes cerevisiae* mitochondria devoid of microsomal and cytosolic contaminations. *Anal. Biochem.* **287,** 339–342.
4. Klose, J. (1975) Protein mapping by combined isoelectric focusing and electrophoresis of mouse tissues. A novel approach to testing for induced point mutations in mammals. *Humangenetik* **26,** 231–243.
5. O'Farrell, P. H. (1975) High resolution two-dimensional electrophoresis of proteins. *J. Biol. Chem.* **250,** 4007–4021.
6. Simpson, R. J., Connolly, L. M., Eddes, J. S., Pereira, J. J., Moritz, R. L., and Reid, G. E. (2000) Proteomic analysis of the human colon carcinoma cell line (LIM 1215): development of a membrane protein database. *Electrophoresis* **21,** 1707–1732.
7. Wagner, Y., Sickmann, A., Meyer, H. E., and Daum, G. (2003) Multidimensional nano-HPLC for analysis of protein complexes. *J. Am. Soc. Mass. Spectrom.* **14,** 1003–1011.
8. Hartinger, J., Stenius, K., Hogemann, D., and Jahn, R. (1996) 16-BAC/SDS-PAGE: a two-dimensional gel electrophoresis system suitable for the separation of integral membrane proteins *Anal. Biochem.* **240,** 126–133.
9. Heukeshoven, J. and Dernick, R. (1988) Improved silver staining procedure for fast staining in PhastSystem Development Unit. I. Staining of sodium dodecyl sulfate gels *Electrophoresis* **9,** 28–32.
10. Sickmann, A., Dormeyer, W., Wortelkamp, S., Woitalla, D., Kuhn, W., and Meyer, H. E. (2000) Identification of proteins from human cerebrospinal fluid, separated by two-dimensional polyacrylamide gel electrophoresis. *Electrophoresis* **21,** 2721–2728.

11. Eng, J. K., McCormack, A. L., Yates, I., and John, R. (1994) An approach to correlate tandem mass spectral data of peptides with amino acid sequences in a protein database. *J. Am. Soc. Mass. Spectrom.* **5,** 976–989.
12. Olsson, I., Larsson, K., Palmgren, R., and Bjellqvist, B. (2002) Organic disulfides as a means to generate streak-free two-dimensional maps with narrow range basic immobilized pH gradient strips as first dimension. *Proteomics* **2,** 1630–1632.

38

Arabidopsis Mitochondrial Proteomics

Joshua L. Heazlewood and A. Harvey Millar

Summary

Significant efforts have sought to uncover the protein profile of *Arabidopsis* mitochondria to act as a model for the mitochondrial proteome from plants. A combination of techniques have been undertaken to achieve this goal. We outline a basic two-dimensional polyacrylamide gel electrophoresis (2D-PAGE) separation of mitochondrial proteins, in-gel trypsination techniques, complex protein lysate digestions, and the identification of proteins by matrix-assisted laser desorption/ionization (MALDI) and electrospray ionization (ESI) mass spectrometry.

Key Words: *Arabidopsis*; mass spectrometry; proteomics; 2D-PAGE.

1. Introduction

Recent advances in mass spectrometry (MS) and the completion and extensive annotation of the nuclear and organelle genome sequences have been instrumental in driving the current explosion in plant proteomics. Mitochondria present an excellent system in which to undertake these proteomic analyses. They are relatively discrete membrane-bound organelles found in abundant numbers in most eukaryotic cells, including plants. Procedures for undertaking mitochondrial isolations from plants are well established and are capable of producing excellent yields of relatively pure fractions. Significantly, their protein complement (approx 1500 proteins) is only just outside the dynamic range of current limitations in proteomics.

Two studies, both using two-dimensional polyacrylamide gel electrophoresis (2D-PAGE) gels, published in 2001 began the analysis of the *Arabidopsis* mitochondrial proteome ***(1,2)***. A variety of more targeted studies have since used blue native gels, one-dimensional sodium dodecyl sulfate (SDS)-PAGE gels and diagonal 2D-SDS-PAGE gels to subdivide the mitochondrial proteome

From: *Methods in Molecular Biology, vol. 372: Mitochondria: Practical Protocols*
Edited by: D. Leister and J. M. Herrmann © Humana Press Inc., Totowa, NJ

further *(3–6)*. Detailed insights into the protein components of complexes I–V of the respiratory chain have also been undertaken using a combination of blue native and SDS-PAGE gels *(7–9)*. The proteins from these gel-based studies have been identified using a mixture of peptide mass fingerprinting by matrix-assisted laser desorption/ionization-time-of-flight (MALDI-TOF), pattern matching of tandem mass spectrometry (MS/MS) spectra from collision-induced dissociation experiments on individual peptides, and Edman N-terminal sequencing of proteins of interest. A larger analysis using nongel proteomic approaches based on liquid chromatography and MS/MS has provided a set of over 400 nonredundant proteins from *Arabidopsis* mitochondrial samples ***(10)***.

The following techniques outline a basic procedure for the arraying of an *Arabidopsis* mitochondrial proteome using 2D-PAGE and the methods used for the identification of proteins from such a gel. It also provides basic insight into the more complicated procedures involved in the analysis and identification of mitochondrial proteins from complex mixtures using MS.

2. Materials

2.1. 2D-PAGE

1. Microfuge.
2. Sample buffer: 6 *M* urea (*see* **Note 1**), 2 *M* thiourea, 2% (w/v) 3-[(3-cholamido-propyl) dimethylammonio]-1-propanesulfonate, 2% (v/v) immobilized pH gradient (IPG) buffer 3-10NL (GE Healthcare), 0.3% (w/v) dithiothreitol, trace amounts of bromophenol blue. Add tributylphosphine to 2 m*M* just prior to use (*see* **Note 2**).
3. Colloidal Coomassie stain: 17% (v/v) ammonium sulfate, 34% (v/v) methanol, 3% (v/v) phosphoric acid, 0.1% (w/v) brilliant blue G-250 (*see* **Note 3**).
4. Destain buffer: 0.5% (v/v) phosphoric acid.
5. SDS transfer buffer: 4 *M* urea, 125 m*M* Tris-HCl, pH 6.8, 20% (v/v) glycerol, 2% (w/v) SDS, 5% (v/v) β-mercaptoethanol, trace amounts bromophenol blue.
6. 12% Separating gel: 375 m*M* Tris-HCl pH 8.8, 0.1% (w/v) SDS, 33.6:1 acrylamide: bisacrylamide, 0.1% (w/v) ammonium persulfate, 0.04% (v/v) *N,N,N′,N′*-tetra-methylethylenediamine.
7. 4% stacking gel: 125 m*M* Tris-HCl, pH 6.8, 0.1% (w/v) SDS, 33.6:1 acrylamide: bisacrylamide, 0.1% (w/v) ammonium persulfate, 0.04% (v/v) *N,N,N′,N′*-tetra-methylethylenediamine.
8. DryStrip sealing solution: 1% (w/v) low-melt agarose, 1X SDS running buffer.
9. 10X SDS running buffer: 0.248 *M* Tris base, 1.92 *M* glycine, 1% (w/v) SDS.
10. DryStrips 3-10NL (GE Healthcare).
11. Multiphor II electrophoresis system (GE Healthcare).
12. Multitemp III thermostatic circulator (GE Healthcare).
13. EPS 3501 XL power supply (GE Healthcare).
14. Immobiline DryStrip reswelling tray (GE Healthcare).
15. Immobiline DryStrip kit (GE Healthcare).

16. Ettan DALT*six* large vertical system (GE Healthcare).
17. Mineral oil (GE Healthcare).

2.2. Trypsin Digestion

2.2.1. In-Gel Digestion

1. Destain solution: 50% (v/v) acetonitrile, 25 m*M* ammonium hydrogen carbonate.
2. Trypsin stock: 100 mg/mL trypsin in 0.01% (v/v) trifluoroacetic acid (*see* **Note 4**).
3. Digestion solution: 25 m*M* ammonium hydrogen carbonate, 12.5 µg/mL trypsin.
4. Extraction solution: 50% (v/v) acetonitrile, 5% (v/v) formic acid.
5. 100% acetonitrile.
6. Resuspension solution: 5% (v/v) acetonitrile, 0.1% (v/v) formic acid.
7. Low-binding polypropylene 96-well microplates (Nalge Nunc International).
8. Yellow pipet tips with approx 15 mm cut from the end of each tip, leaving a diameter of approx 1.5 mm.
9. Ultra Micro Tips (0.5–10 µL) (Quality Scientific Plastics).
10. Amplification tape for 96-well microplates (Nalge Nunc International).
11. Vacuum concentrator system with rotor for microplates.

2.2.2. Protein Lysate Digestion

1. Digestion buffer: 100 m*M* ammonium hydrogen carbonate.
2. Trypsin stock: 1 mg/mL trypsin in 0.01% (v/v) trifluoroacetic acid.

2.3. Mass Spectrometry

2.3.1. Matrix-Assisted Laser Desorption/Ionization

1. α-Cyano-4-hydroxycinnamic acid.
2. 70% (v/v) Acetonitrile and 0.1% (v/v) formic acid.
3. C_{18} ZipTip pipet tips (Millipore).
4. Wetting solution: 50% (v/v) acetonitrile.
5. Equilibrating solution: 0.1% (v/v) formic acid.
6. Washing solution: 5% (v/v) acetonitrile and 0.1% (v/v) formic acid.
7. Elution solution: 50% (v/v) acetonitrile and 0.1% (v/v) formic acid.

2.3.2. Electrospray Ionization

1. A nano- or capillary-capable high-performance liquid chromatography (HPLC) system.
2. Reverse-phase C_{18} HPLC column or strong cation exchange (SCX) HPLC column (*see* **Note 5**).
3. 500 m*M* Ammonium acetate.
4. Vivapure C_{18} Microspin columns (Sartorius).

3. Methods

One of the most important parameters for the successful study of the *Arabidopsis* mitochondrial proteome is the quality and resultant purity of the mitochondrial isolation procedure. The utilization of two density gradients for mitochondrial isolations from *Arabidopsis* cell culture results in a preparation with more than 95% purity on a protein basis and is outlined in **ref. *11***.

3.1. 2D-PAGE

This procedure is based on the utilization of a Multiphor II electrophoresis unit and Immobiline DryStrips 3-10NL, although it could be adapted to suit most common setups (**Fig. 1**).

3.1.1. Isoelectric Focusing

1. Add approx 3 volumes of acetone to a mitochondrial sample (0.5–1 mg total protein) and precipitate overnight at –20°C. If mitochondrial sample is provided as a pellet, then resuspend in approx 100 μL H_2O before adding acetone.
2. Centrifuge the precipitated mitochondrial sample in a microfuge at 20,000*g* for 20 min at 4°C. Remove supernatant and allow the pellet to air dry for approx 20 min.
3. Initially resuspend pellet in approx 100 μL sample buffer, then add enough buffer to reach approx 350 μL total volume (*see* **Note 6**).
4. Add 2 μL of a 1/10 tributylphosphine solution diluted in sample buffer.
5. Add 350 μL reconstituted sample to the reswelling tray by pipeting the solution along the length of the groove.
6. Remove the plastic backing from the IPG DryStrip, holding the positive end of the strip, and gently lay the strip face down (the side covered with plastic) into the groove starting from the nonnumbered end. This will allow bubbles to be displaced to the other end of the reswelling tray, which contains a small well.
7. Slide the DryStrip back and forth to prevent sticking.
8. Overlay DryStrip with 2–3 mL of mineral oil.
9. Allow the DryStrip to hydrate overnight (~10 h) (*see* **Note 7**).
10. The following day, measure out 100 mL mineral oil and pour approx 5 mL onto the cooling plate, which has been attached to a level Multiphor apparatus.
11. Position the DryStrip tray onto the cooling tray; avoid creating large bubbles.
12. Pour 10 mL mineral oil onto the DryStrip tray and place the Immobiline Strip Aligner on the tray, again avoiding production of large bubbles.
13. Connect the cooling system to the cooling plate and set to 20°C.
14. Cut two 11-cm isoelectric focusing (IEF) electrode strips and wet with approx 1 mL H_2O.
15. Remove hydrated strip from reswelling tray using forceps.
16. Remove excess oil by placing the strip gel side facing up on some damp towels. Avoid touching the gel on anything as it will stick.
17. Lay hydrated strip gel side facing up into the groves of the Strip Aligner with the positive end facing the anode end of the Multiphor apparatus.
18. If using multiple DryStrips, then ensure that the positive ends are lined up.
19. Lightly blot the IEF electrode strips with a tissue to remove excess water and place across the ends of the DryStrip (perpendicular to the DryStrips at both the anode

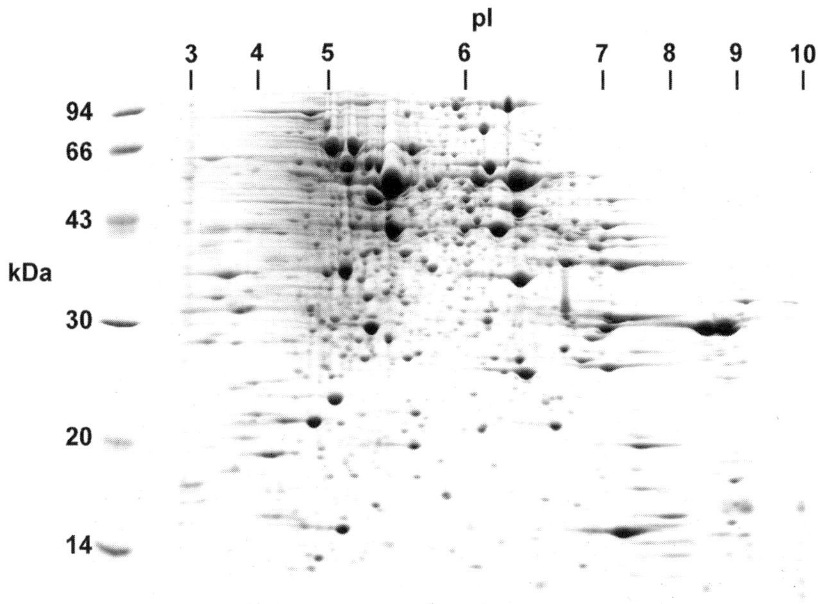

Fig. 1. The *Arabidopsis* mitochondrial proteome arrayed using a 3-10NL DryStrip. Mitochondria were isolated from 7-d dark-grown *Arabidopsis* cell culture using a two-step density gradient procedure. The 3-10NL IEF DryStrip was hydrated overnight with approx 1 mg acetone-precipitated protein. The second dimension was carried out using a 12% acrylamide gel. Horizontal numbers represent p*I* range on the 3-10NL DryStrip and demonstrate its nonlinear (NL) feature. Vertical numbers represent the mass of the molecular markers (kDa).

and cathode ends) on top of the gel. Ensure the IEF electrode strips are in contact with the gel by lightly pushing with a pair of forceps (*see* **Note 8**).

20. Position the electrodes over the IEF electrode strips and pour the rest of the mineral oil (~80 mL) over the DryStrip tray covering the DryStrips.
21. Place lid on unit and connect power supply.
22. Program power supply running conditions as in **Table 1** and initiate program (*see* **Note 9**).

3.1.2. Second-Dimension PAGE

1. When IEF has completed, remove DryStrips from the IEF apparatus, allow excess oil to drain, place into precut glass tubing (~25 cm) containing SDS transfer buffer, and stop each end. Ensure solution is covering DryStrip and place on horizontal rocking platform for 30 min (*see* **Note 10**).
2. Remove DryStrip from tube containing SDS transfer buffer and lay across the top of a 12% separating gel with a 4% stacker; seal in place with DryStrip sealing solution. Ensure that the positive/acidic end is marked to confirm orientation.

Table 1
Electrophoresis Conditions Used for the First-Dimensional Isoelectric Focusing of *Arabidopsis* Mitochondrial Proteins With a 3-10NL DryStrip

Phase	Voltage (V)	Current (mA)	Power (W)	Volt hours (Vh)
1/1	500	1	5	1
2/1	500	1	5	2500
3/1	3500	1	5	10.0 k
4/1	3500	1	5	39.0 k

3. Run electrophoresis at 25 mA per gel with 500 V maximum until bromophenol blue reaches the bottom of the gel (~5–6 h).
4. Dismantle apparatus when complete and place gels in colloidal Coomassie stain for approx 12 h (overnight) on a rocking platform.
5. Major proteins can be observed after a few hours; after staining, add destain buffer and place on a rocking platform for approx 3 h.
6. Scan gels using a flatbed scanner.

3.2. Trypsin Digestion

These protein digestion protocols utilize trypsin, although they have been successfully employed with other proteases (e.g., chymotrypsin) with only minor modifications. Typically, we use sequencing-grade trypsin (Roche) but have also employed Trypsin Gold (Promega).

3.2.1. In-Gel Digestion

1. Take up to two gel plugs containing your protein of interest from the destained gel.
2. Deposit the gel plugs into a polypropylene 96-well microplate using a new Ultra Micro Tip (*see* **Note 11**).
3. Add 50 µL destain solution to each sample; seal microplate with amplification tape and agitate on a shaking platform for 45 min at maximum.
4. Discard the destain solution using a pipet and repeat **step 3**.
5. Again, discard the destain solution. These steps remove most of the colloidal Coomassie stain from the gel plugs, ensuring more optimal digestion by trypsin (*see* **Note 12**).
6. Remove amplification tape and place 96-well microplate on a 50°C heat block for approx 30 min until the gel pieces are completely desiccated.
7. Add 10–15 µL digestion solution to each sample, seal with amplification tape, and place in a 37°C incubator overnight.
8. The following day, add 10–15 µL 100% acetonitrile and agitate on a shaking platform for 15 min (*see* **Note 13**).
9. Remove liquid from gel pieces and place in corresponding wells of a new 96-well microplate.
10. Add 10–15 µL extraction solution to the gel pieces and agitate on a shaking platform for 15 min. Remove the supernatants and pool in a new 96-well microplate.

11. Repeat **step 10**.
12. Place 96-well microplate containing pooled peptide extractions into a vacuum concentrator and evaporate aliquots for approx 30 min or until trace amounts of liquid remain. Seal microplate with amplification tape and store at –20°C until required.
13. When a sample is required for analysis, add 5–20 µL resuspension solution and agitate with a pipet or shaker to aid in resolubilizing peptides.

3.2.2. Protein Lysate Digestion

1. Resolubilize mitochondrial protein in ammonium hydrogen carbonate to a final concentration of 10 m*M*.
2. Add trypsin from freshly made stock solution to 1/10 (w/w) to mitochondrial protein.
3. Place the digest at 37°C overnight.

3.3. Mass Spectrometry

The method of sample delivery and analysis is clearly dictated by availability of hardware or costs involved with analysis. The two most common sample delivery methods in proteomics are MALDI and ESI; the type of data, quality, and analysis undertaken will depend on the hardware used.

3.3.1. Matrix-Assisted Laser Desorption/Ionization

The MALDI method of ionization is best suited for gel-separated proteins producing samples of relatively low complexity (three to four proteins). MALDI analysis has the advantage of being a relatively high-throughput technique, although it produces poorer quality data with respect to matching confidences. Generally, these data are comprised of intact peptide masses (peptide mass fingerprint), creating more ambiguity when interrogating for a match. More recently, MS/MS capabilities have been made available using this source to provide more confident levels of protein matching. The MALDI source is invariably attached to a TOF mass spectrometer for synergistic reasons that involve timed points of ionization. One of the advantages of the TOF is its mass accuracy, but this is counterbalanced by a poor duty cycle, which complicates its role when quantification of sample is required. Methods have been developed that take chromatographic separations of more complex samples and directly spot them onto MALDI plates for analysis (liquid chromatography [LC]/MALDI).

1. Resuspend extracted samples in 5–10 µL resuspension solution (*see* **Subheading 3.2.1.**).
2. Prime ZipTip with 10 µL wetting solution and aspirate to waste.
3. Equilibrate ZipTip by taking up and aspirating equilibration solution two times.
4. With the ZipTip attached, take up and aspirate the 5- to 10-µL sample in resuspension solution 10 times into the well to bind peptides. Finally, aspirate the depleted sample.
5. Take up 10 µL washing solution and aspirate; repeat twice.
6. Using a regular pipet tip, place 1–2 µL elution solution in a new microplate well.

7. With the ZipTip, take up and aspirate the elution solution 10 times into the new well of the microplate (*see* **Note 14**).
8. To make matrix solution, add approx 10 mg (the end of a small spatula) of α-cyano-4-hydroxycinnamic acid to a 1 mL solution of 70% (v/v) acetonitrile and 0.1% (v/v) formic acid (*see* **Note 15**).
9. Spot 0.5 μL matrix solution onto a MALDI plate and, before the matrix dries, spot 0.5 μL ZipTip-cleaned sample into this solution.
10. Place plate in MALDI source for analysis.

3.3.2. Electrospray Ionization

In proteomics, the ESI source is commonly used to analyze complex mixtures and is capable of interfacing with a wide range of mass spectrometers that usually provide MS/MS capabilities; these include the ion trap, the hybrid quadrupole-TOF, and the Fourier transform ion cyclotron resonance system. The analysis of a sample using ESI can often be achieved manually using syringes provided with the system, but by far the most convenient and sensitive procedures involve an HPLC apparatus connected to the source. Samples can then be loaded into an autosampler and peptides concentrated and separated on a reverse -phase C_{18} column prior to analysis by the mass spectrometer.

Most ESI-based analysis in proteomics produces MS/MS fragmentation data in which the parent ion (peptide) is fragmented (collision-induced dissociation), producing far greater matching confidence. The standard type of HPLC used for proteomic analyses has been the capillary flow HPLC, which enables a flow rate from 2 to 50 μL/min, although more recently the use of nanoflow has become popular, enabling flows down to 50 nL/min, thereby greatly improving sensitivity. Reductions in sample complexity have also been addressed through online chromatographic techniques employing multiple HPLC columns for fractionation.

3.3.2.1. LC/MS/MS

1. Typically, both gel-separated samples and protein lysates are analyzed using ESI.
2. For gel-separated samples, resuspend the extracted sample in 5–20 μL resuspension solution (*see* **Subheading 3.2.1.**); for protein lysates, the digested sample is adequate.
3. For gel-separated samples of low complexity, 10- to 30-min methods can be designed with eluting solvent gradients as short as 2–5 min. For more complex samples or if attempting to gain maximal coverage, analysis methods involving longer elution gradients comprising 1–3 h can be successfully used.
4. Load samples into the autosampler of the HPLC and queue for analysis.

3.3.2.2. LC/LC/MS/MS (Multidimensional Protein Identification Technology)

This technique was developed for the online fractionation and analysis of complex samples using multiple column separations ***(12)***. Two columns are required for this analysis with differing chemistries: a standard reverse-phase column (C_{18}) and an SCX column. This technique usually requires at least one switching valve if attempted online, although for simplicity the first separation stage can be undertaken offline and each fraction run as in **Subheading 3.3.2.1**.

1. Desalt sample using a C_{18} Microspin column (*see* **Note 16**).
2. Arrange the columns on the switching valve so they are in a series. The SCX column should be the first in the series. Ensure that each column can be switched to a waste line.
3. Load the digested sample (usually a complex lysate) onto the SCX column set in series with the C_{18} column.
4. Apply an acetonitrile gradient to approx 80%, analyzing any eluted peptides with the mass spectrometer.
5. Using the autosampler of the HPLC, sequentially load "shots" of increasing concentrations of ammonium acetate (from 0 to 500 m*M*) over the SCX column, eluting any peptides onto the C_{18} column. Use about 10–20 different concentration points, with an emphasis on 0–250 m*M*. After each salt shot, run an acetonitrile gradient to 80%, eluting peptides off the C_{18} column into the mass spectrometer. Ensure the C_{18} column is switched to a waste line to prevent excess salt from going to the mass spectrometer (*see* **Note 17**).

4. Notes

1. Although preparing stocks of sample buffer and freezing at –20°C is possible, issues can arise from multiple freeze-thawing events. Over time, urea will readily degrade to ammonium and cyanate in solution, and this decomposition will accelerate if the solution is heated. Thus, care should also be taken in maintaining a constant temperature (around 20°C) during the hydration of DryStrips and during IEF. Isocyanic acid can subsequently react with amide groups in proteins (N-terminus and the side chains of arginine and lysine) as well as cysteine side chains, resulting in protein carbamylation. This uncontrolled modification results in a protein population with varying degrees of modification and causes the appearance of horizontal reiterations of a protein on the 2D-PAGE, which are often interpreted as evidence of biological protein modifications such as phosphorylation. Consequently, sample buffer should be freshly prepared and stored as frozen aliquots. When thawed, any remaining buffer not used should be discarded. Some of the effects outlined can be resolved through reduction and alkylation steps undertaken prior to IEF by adding iodoacetamide to the sample. This addition results in the S-carboxymethylation of cysteine ***(13)***. This controlled alkylation step is more specifically used to prevent uncontrolled alkylation by reactive components such as unpolymerized acrylamide ***(14)***. We have found little advantage in using an alkylation step with *Arabidopsis* mitochondrial samples; however, if reduction and alkylation steps are performed, then they should be undertaken prior to the first dimension and not between the IEF and SDS-PAGE step, which is quite common in many older protocols ***(13)***.

2. Tributylphosphine has recently been included in most IEF sample buffers as it is a powerful uncharged reductant, features that provide excellent reducing conditions during IEF. Many protocols have completely replaced dithiothreitol or β-mercapto-ethanol in the sample buffer with this reducing agent as it can be used in lower concentrations and appears to increase protein solubility ***(13,15)***. Several surfactants have been employed in the sample buffer to improve the solubility of hydrophobic proteins in IEF. The addition of *N*-decyl-*N,N*-dimethyl-3-ammonio-1-propane sulfonate (SB 3-10) requires lowering of the urea concentration to 5 *M* in the sample buffer because of its poor solubility in high concentrations of urea ***(16)***; the addition of the more recently developed amidosulfobetaine 14 (ASB 14) is capable of withstanding urea concentrations at around 7–8 *M* ***(17,18)***. The use of either surfactant in the sample buffer is likely to improve the solubility of hydrophobic proteins and allow for their separation on the 2D-PAGE.
3. For optimal performance, the colloidal Coomassie stain should be made the day before it is required. When making this solution, a strict order of solute addition must be observed to avoid precipitation. To the ammonium sulfate, methanol is added while stirring, followed by the phosphoric acid and finally water to approx 80% of the final volume. In another beaker, the brilliant blue G-250 should be dissolved in water (use approx 20% of the final volume). These two solutions should be allowed to stir for several hours before combining, covered with plastic wrap, and allowed to stir overnight. The resulting solution should be deep blue and contain small blue-black particles.
4. One of the major problems when using a protease such as trypsin is the loss in enzyme activity because of autolysis. Although it is suggested that a solution in 0.01% (v/v) trifluoroacetic acid will last up to a week at 2–8°C or for months if stored at –80°C, we have observed some loss in activity from freezing-thawing or keeping a stock at 4°C. For these unmodified enzymes, it is probably safer to use a newly prepared aliquot each time a digestion is required. The advantage of using a protease that is autolytic rather than a more stable form is that known trypsin-derived tryptic fragments can be used as an internal control for mass accuracy, which can be critical when using MALDI-TOF-based peptide mass analysis. Furthermore, because each sample will contain identical amounts of protease, it provides a simple assessment of the digestion and extraction success of each sample as well as a means by which the mass spectrometer's performance can be evaluated. A disadvantage is that, once solubilized, it must be used immediately and any remaining enzyme discarded. To overcome perceived problems in autolysis, stabilized or modified forms of trypsin are available; two examples are Trypsin Gold (Promega) and trypsin modified sequencing grade (Roche). These enzymes work effectively, although we have found that a higher concentration of enzyme is required to achieve best results. Typically, this would be 200–400 μg per 10 μL gel digest compared with 125 μg outlined above.
5. The sensitivity gains when using electrospray at low flow rates are one of the principle reasons for the popularity of capillary and nano-based LC systems in proteomics. As a result, columns are now readily available with internal diameters

down to 75 μm. One of the major disadvantages of lower flow rates are the run time considerations, especially when analyzing multiple samples. The choice of which internal diameter to use will depend on the type of LC system available. Most capillary-based analysis run columns with internal diameters of 0.5 mm (flow ~10 μL/min) and 0.3 mm (flow ~4 μL/min), although nanosystems will run columns with internal diameters of 150 μm (flow ~1 μL/min) and 75 μm (flow ~250 nL/min). There can be considerable time gains on either setup if multiple pumps are used. Essentially, the capillary pump will be used to fast load the sample onto a trap column; the second pump (a nanopump) provides the gradient for peptide elution at the desired flow rate, which can be performed while the sample is loaded by the first pump.

6. This volume is recommended for the hydration of 18-cm Immobiline DryStrips; volumes will differ depending on strip length. The amount of protein that can be loaded to achieve successful focusing will also vary depending on the strip utilized ***(19)***. Although instructions for 3-10NL DryStrips indicate that up to 1.5 mg of protein can be successfully used on 18-cm DryStrips ***(20)***, in our hands a "well-focused" sample will be accomplished with up to 1 mg mitochondrial protein.
7. Because of the length of time hydrating the DryStrip, care should be taken to ensure that the hydration occurs at a constant temperature to prevent issues associated with urea precipitation at low temperatures or urea degradation at higher temperatures. If necessary, reswelling can be undertaken on the cooling plate with the cooling system set at 20°C.
8. To maximize the focusing region of the DryStrip, try to place the IEF electrode strip on the edge of the hydrated gel; only half the width of the IEF electrode strip needs to be in contact with the gel for efficient electrophoresis to occur.
9. The entire program should take approx 20 h. A means of assessing the success of IEF is by noting the current at the end of the run. Resistance should be minimal, with values around 100 μA, although this value will vary depending on the composition of the original sample. Ensure that transitions between phases are set as "step" and not "gradient."
10. Many older published protocols include iodoacetamide in the transfer buffer to provide a reduction and controlled alkylation step. Many of the problems associated with uncontrolled modifications of cysteine residues will already have occurred during hydration and IEF ***(13)***. As noted, the reduction and alkylation step should occur before IEF is carried out (*see* **Note 1**).
11. The use of cut yellow tips for harvesting gel spots provides a cheap and readily available means of ensuring no cross-contamination of samples occurs. In contrast, using a scalpel blade or the like will require some washing step to ensure little or no crossover occurs.
12. These two wash steps appear to remove enough of the Coomassie stain for a successful trypsin digest to proceed. In some instances where the spot is very intense, these two washes will not remove all of the Coomassie; this seems to have a minimal impact on the digestion of the sample.

13. After the overnight digest, the hydrated gel pieces should be surrounded by a pool of digestion solution. This attribute appears to be a good indication of a successful digestion. If this has not occurred, then modify the number of gel pieces used or increase the volume of digestion solution.
14. The utilization of a desalting step is vital for a successful signal using MALDI. The ZipTip provides an excellent means when dealing with small volumes associated with gel digests and the subsequent spotting on the MALDI plate for analysis. The presence of salts in a sample will cause significant ion suppression and thus must be removed. If using MALDI extensively, then it is possible to minimize salts by decreasing the concentration of ammonium hydrogen carbonate in the digestion solution to around 10 m*M*. This concentration is still capable of maintaining the solution at pH 8.0.
15. A freshly prepared solution appears to provide the best ionization conditions for MALDI analysis and provides a cheaper source of matrix compared to commercially available premixes. Most MALDI analyses of peptides are now undertaken with α-cyano-4-hydroxycinnamic acid as the matrix as it produces an even distribution of crystals compared to compounds like 2,5-dihydroxybenzoic acid.
16. The digested sample must be desalted before loading onto the SCX column, or peptides will pass through it to waste.
17. There are some reports that indicated it is cleaner to run increasing salt gradient over the SCX column as it results in less sample bleed when compared to the salt shots ***(21)***.

Acknowledgments

This work is supported through grants provided by the Australian Research Council Discovery Program, an ARC QEII Research Fellowship to A. H. M., and a UWA Postdoctoral Research Fellowship to J. L. H.

References

1. Kruft, V., Eubel, H., Jansch, L., Werhahn, W., and Braun, H. P. (2001) Proteomic approach to identify novel mitochondrial proteins in Arabidopsis. *Plant Physiol.* **127,** 1694–1710.
2. Millar, A. H., Sweetlove, L. J., Giege, P., and Leaver, C. J. (2001) Analysis of the *Arabidopsis* mitochondrial proteome. *Plant Physiol.* **127,** 1711–1727.
3. Werhahn, W. and Braun, H. P. (2002) Biochemical dissection of the mitochondrial proteome from *Arabidopsis thaliana* by three-dimensional gel electrophoresis. *Electrophoresis* **23,** 640–646.
4. Herald, V. L., Heazlewood, J. L., Day, D. A., and Millar, A. H. (2003) Proteomic identification of divalent metal cation binding proteins in plant mitochondria. *FEBS Lett.* **537,** 96–100.
5. Millar, A. H. and Heazlewood, J. L. (2003) Genomic and proteomic analysis of mitochondrial carrier proteins in Arabidopsis. *Plant Physiol.* **131,** 443–553.
6. Brugière, S., Kowalski, S., Ferro, M., et al. (2004) The hydrophobic proteome of mitochondrial membranes from *Arabidopsis* cell suspensions. *Phytochemistry* **65,** 1693–1707.

7. Eubel, H., Jansch, L., and Braun, H. P. (2003) New insights into the respiratory chain of plant mitochondria. Supercomplexes and a unique composition of Complex II. *Plant Physiol.* **133,** 274–286.
8. Heazlewood, J. L., Howell, K. A., and Millar, A. H. (2003) Mitochondrial complex I from *Arabidopsis* and rice: orthologs of mammalian and fungal components coupled with plant-specific subunits. *Biochim. Biophys. Acta* **1604,** 159–169.
9. Heazlewood, J. L., Whelan, J., and Millar, A. H. (2003) The products of the mitochondrial *orf25* and *orfB* genes are FO components in the plant F_1F_O ATP synthase. *FEBS Lett.* **540,** 201–205.
10. Heazlewood, J. L., Tonti-Filippini, J. S., Gout, A. M., Day, D. A., Whelan, J., and Millar, A. H. (2004) Experimental analysis of the *Arabidopsis* mitochondrial proteome highlights signaling and regulatory components, provides assessment of targeting prediction programs, and indicates plant-specific mitochondrial proteins. *Plant Cell* **16,** 241–256.
11. Millar, A. H., Liddell, A., and Leaver, C. J. (2001) Chapter 3 in *Mitochondria*, Vol. 65 (Pon, L. A., and Schon, E. A., eds.), Academic Press, San Diego, CA, pp. 53–74.
12. Washburn, M. P., Wolters, D., and Yates, J. R., 3rd (2001) Large-scale analysis of the yeast proteome by multidimensional protein identification technology. *Nat. Biotechnol.* **19,** 242–247.
13. Herbert, B., Galvani, M., Hamdan, M., et al. (2001) Reduction and alkylation of proteins in preparation of two-dimensional map analysis: why, when, and how? *Electrophoresis* **22,** 2046–2057.
14. Bordini, E., Hamdan, M., and Righetti, P. G. (1999) Matrix-assisted laser desorption/ionisation time-of-flight mass spectrometry for monitoring alkylation of β-lactoglobulin B exposed to a series of N-substituted acrylamide monomers. *Rapid Commun. Mass Spectrom.* **13,** 2209–2215.
15. Herbert, B. R., Molloy, M. P., Gooley, A. A., Walsh, B. J., Bryson, W. G., and Williams, K. L. (1998) Improved protein solubility in two-dimensional electrophoresis using tributyl phosphine as reducing agent. *Electrophoresis* **19,** 845–851.
16. Rabilloud, T., Adessi, C., Giraudel, A., and Lunardi, J. (1997) Improvement of the solubilization of proteins in two-dimensional electrophoresis with immobilized pH gradients. *Electrophoresis* **18,** 307–316.
17. Chevallet, M., Santoni, V., Poinas, A., et al. (1998) New zwitterionic detergents improve the analysis of membrane proteins by two-dimensional electrophoresis. *Electrophoresis* **19,** 1901–1909.
18. Herbert, B. (1999) Advances in protein solubilisation for two-dimensional electrophoresis. *Electrophoresis* **20,** 660–663.
19. Berkelman, T., and Stenstedt, T. (2002) *2-D Electrophoresis Using Immobilized pH Gradients: Principles and Methods*, GE Healthcare, 80-6429-60, Edition AC, Uppsala, Sweden.
20. GE Healthcare. (2003) *Instructions: Immobiline DryStrip*, GE Healthcare, 71-5024-30, Edition AC, Uppsala, Sweden.
21. Vollmer, M., Hörth, P., and Nägele, E. (2003) *Tools and Considerations to Increase Resolution of Complex Proteome Samples by Two-Dimensional Offline LC/MS*, Agilent Technologies Palo Alto, CA.

39

MitoP2, an Integrated Database for Mitochondrial Proteins

Holger Prokisch and Uwe Ahting

Summary

The impact of mitochondria on several fundamental cellular processes is reflected in their involvement in the pathophysiology of common diseases such as Parkinson's disease, diabetes, and obesity and a wide range of monogenic disorders primarily associated with energy impairment or metabolic diseases. The importance of mitochondria is also reflected by the steep increase of proteins, which has been localized to this organelle. In yeast, more than 500 of the expected 700–800 mitochondrial proteins are already annotated. In the mammalian species, the expected numbers are estimated to be in the range of 1500–2000 proteins, and the currently annotated entries reach almost 700. In addition to the studies dealing with single proteins, there are many high-throughput approaches that improve the description of the mitochondrial proteome. They include computational predictions of signaling sequences, proteome mapping, mutant screening, expression profiling, protein–protein interaction, and cellular sublocalization studies. The MitoP2 database (http://www.mitop2.de/) was established to structure, explore, and customize the available data on mitochondrial proteins, functions, and diseases. MitoP2 provides a comprehensive picture of the mitochondrial proteome by focusing on (1) the orthology between species, including *Saccharomyces cerevisiae*, mouse, humans, and *Arabidopsis thaliana*; (2) the definition of mitochondrial reference sets in these species; (3) the integration of data predictive for mitochondrial localization or function stemming from genomewide approaches; (4) the allocation of a gateway for functional data from model systems and genetics of mitochondriopathies; and (5) the calculation of a combined score for each protein summarizing the indirect evidence for a mitochondrial localization. All data are accessible via search tools and linked to the original data source. By providing an overview of functional annotations from different databases, the MitoP2 database lends itself to genetic investigations of human mitochondriopathies.

Key Words: Bioinformatics; *in silico* prediction; mitochondrial database; mitochondrial localization; mitochondrial proteome; MitoP2; proteomics.

From: *Methods in Molecular Biology, vol. 372: Mitochondria: Practical Protocols*
Edited by: D. Leister and J. M. Herrmann © Humana Press Inc., Totowa, NJ

1. Introduction

Although a complete list of the molecular components of mitochondria is not available at the moment, such a description would be a crucial prerequisite for system biology approaches. In mammals, only about half of the expected mitochondrial proteins are known. In yeast, with a well-characterized mitochondrial proteome, it is still impossible to acquire an accurate part list ***(1)***. The annotation of mitochondrial proteins in the generic databases is often incomplete and does not always distinguish between proteins that have a confirmed mitochondrial subcellular localization and those that are only candidates according to preliminary experimental results or *in silico* predictions.

Two comprehensive subcellular localization studies performed in yeast ***(2,3)*** have predicted that approx 12–13% of the coding capacity of the nuclear genome is devoted to mitochondrial proteins. This accounts for about 800 mitochondrial proteins, of which 525 are already certainly known, leaving about 250 proteins to be determined. Altogether, the high-throughput experiments deliver more than 4000 mitochondrial candidates, indicating a substantial number of false positives.

To collect and analyze this variety of information on mitochondrial proteins, the database MitoP2 was created ***(4)***. It integrates data sets from genomewide approaches in five species together with functional annotations and genomic information from diverse databases. A reference set of mitochondrial proteins in yeast, mouse, and humans is manually annotated, and evidence scores for potential new mitochondrial proteins are calculated. The database is widely used and has been successfully applied to identify a disease gene by a positional candidate approach ***(5)***. The following description is based on the April 2005 update. Five different species can be selected from the MitoP2 home page (http://www.mitop2.de/; *see* **Note 1**): yeast, neurospora, mouse, human, and *Arabidopsis thaliana*, the last as a test version.

2. Exploring the Mitochondrial Proteome of Yeast

Saccharomyces cerevisiae is the most intensively investigated eukaryotic model organism, with a wealth of annotations available for single proteins and accumulating data from genomewide experiments. The yeast section of the MitoP2 database contains all 6516 open reading frames (ORFs) of the yeast genome based on the most comprehensive and actual *Saccharomyces* Genome Database (SGD) ***(6)***. For each ORF, MitoP2 uses the description, the subcellular localization, and the functional annotation from SGD. The functional annotations are complemented by the functional catalog entries from the Munich Information Center for Protein Sequences (MIPS) ***(7)***, which also provides the physical features, including the deoxyribonucleic acid (DNA) and protein sequence. These features are regularly updated. In addition to the single protein

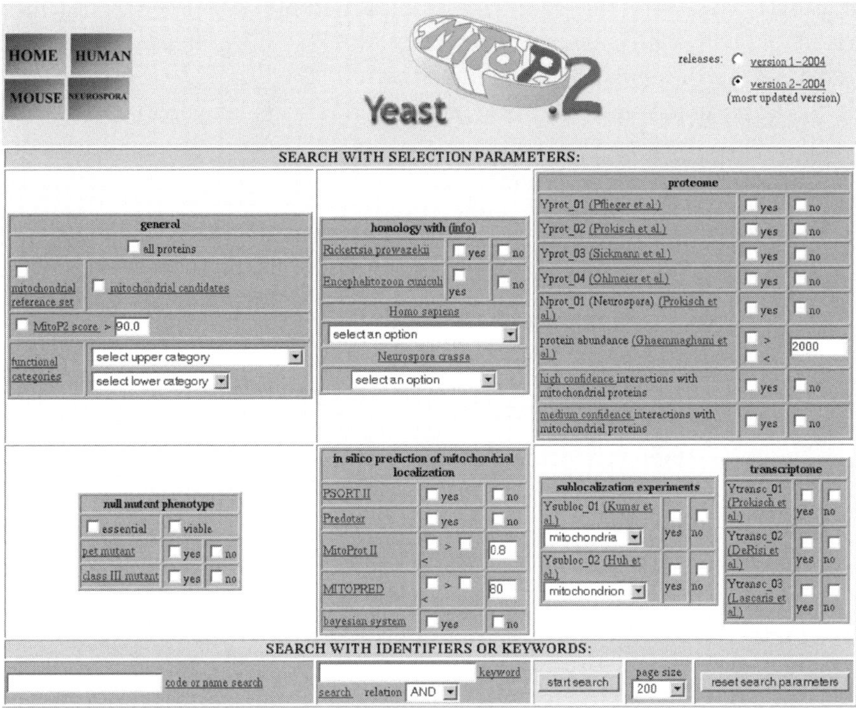

Fig. 1. Search mask of the yeast section of the MitoP2 database at http://www.mitop2.de/. The search mask is structured in several search boxes. All search options (underlined) are linked to the original literature or to an explanation for this selection.

entries, MitoP2 is collecting and processing data sets from genomewide approaches, such as proteome studies, *in silico* predictions, mutant phenotype collections, and physical/functional/genetic interaction studies.

Via the yeast search mask, the user has access to all the compiled data and, by means of direct links, to the data source. The search mask is structured in several search boxes (**Fig. 1**). All search options (underlined) are linked to the original literature or to an explanation for this selection. The general search box allows selecting the mitochondrial reference set, which currently contains 525 proteins with experimental evidence for mitochondrial localization, or a subset of proteins divided in major functional categories. Alternatively, it is possible to search for new candidates (which are all proteins apart from the reference set) by using the MitoP2 evidence score.

The indirect support for mitochondrial localization from high-throughput data sets can be screened by search boxes for null mutant phenotypes from genomewide experiments ***(8,9)***; systematic subcellular localization studies

(2,3); three transcriptome data sets of differentially expressed genes under fermentable and nonfermentable growth conditions, upregulated in response to the diauxic shift or regulated by the Hap4 transcription factor ***(1,10,11)***; proteome analyses of purified mitochondria ***(1,12,13)***; protein abundance in copy number per cell ***(14)***; as well as low- and high-throughput protein–protein interactions with known mitochondrial proteins ***(15)***. Apart from such experimental data, the affiliation to the subcellular compartment can be judged according to the presence or absence of mitochondrial targeting sequences by *in silico* predictions ***(16–19)*** or by the homology to a known mitochondrial protein from another species (defined as bidirectional best BLAST (Basic Local Alignment Search Tool) hit or best BLAST hit with a score < E^{-10}) ***(20)***.

Further ways to estimate mitchondrial status of a given protein is to compare homologous proteins from *Rickettsia prowazekii*, a prokaryote believed to be closest to the evolutionary origin of mitochondria, and to look for the absence of a homolog in *Encephallitozoon cuniculi*, an eukaryote lacking mitochondria. The search mask of MitoP2 allows the selection of single or several combined options. In addition, gene names and key words can be used. It is also possible to exclude certain categories from a search or to define threshold levels.

The result of a search is given in output lists, which may include annotated mitochondrial proteins from the reference set, labeled in green, as well as candidates. When selecting for example mitochondrial subcellular localization according to the **ref.** ***(2)*** and an *in silico* prediction tool such as Predotar ***(19)***, MitoP2 extracts a list of 233 proteins (**Fig. 2**). This list contains 181 proteins from the mitochondrial reference set and 52 candidate proteins fulfilling the two selection criteria. The specificity in detecting mitochondrial proteins of this combined query is estimated from the percentage of the listed proteins ($n = 233$) that are present in the mitochondrial reference set ($n = 181$), in this case 78% (*see* **Note 2**). The sensitivity of the combination of these two approaches in detecting the known mitochondrial proteins is 34% (181 of 525).

Each individual protein entry—in a single line—contains the ORF description from SGD, the subcellular localization according to SGD or MitoP2 annotations, and the detailed information from high-throughput experiments and *in silico* calculations. In addition to the matrix provided by the columns with systematic experimental results and *in silico* data, the user can access detailed information for each single protein by selecting the ORF button. An extra page linked to each entry provides (1) a description of the individual protein compiled from SGD plus additional functional, genetic, and biophysical properties from MIPS (CYGD, Comprehensive Yeast Genome Database); (2) the corresponding entry from the Gene Ontology (GO) database ***(21)***; (3) a list of homologs from other species generated by bidirectional best BLAST hits; and (4) a compilation of low-, medium-, or high-confidence protein–protein interactions weighted

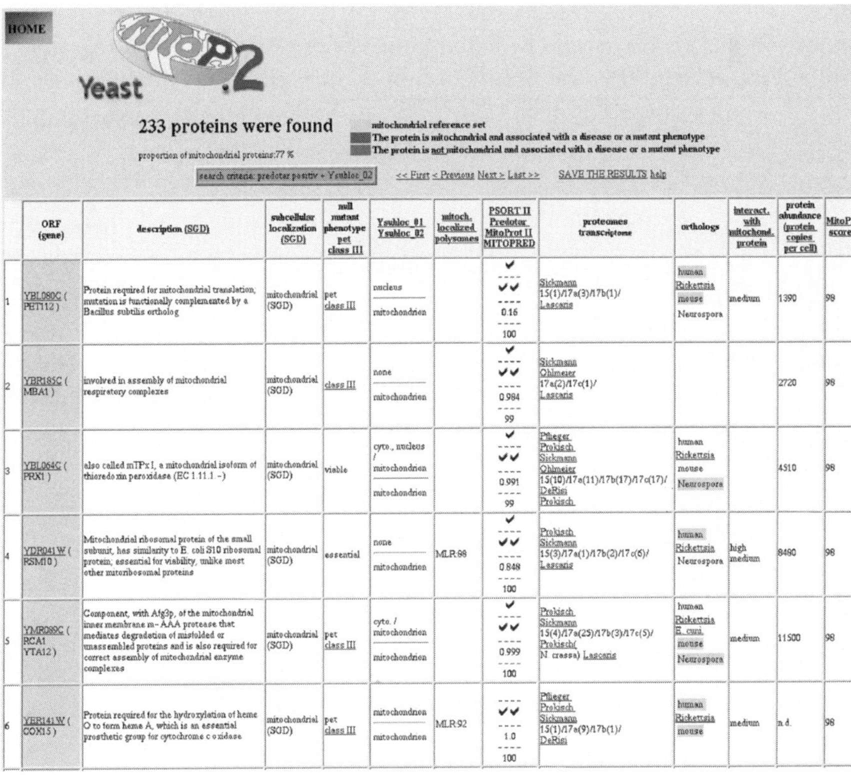

	ORF (gene)	description (SGD)	subcellular localization (SGD)	null mutant phenotype pet class III	Ysubloc_01 Ysubloc_02	mitoch. localized polysomes	PSORT II Predotar MitoProt II MITOPRED	proteomes transcriptome	orthologs	interact. with mitochond. protein	protein abundance (protein copies per cell)	MitoP2 score
1	YBL080C (PET112)	Protein required for mitochondrial translation; mutation is functionally complemented by a Bacillus subtilis ortholog	mitochondrial (SGD)	pet class III	nucleus / mitochondrion		✔ ---- ✔✔ ---- 0.16 ---- 100	Sickmann 15(1)/17a(3)/17b(1)/ Lascaris	human Rickettsia mouse Neurospora	medium	1390	98
2	YBR185C (MBA1)	involved in assembly of mitochondrial respiratory complexes	mitochondrial (SGD)	class III	none / mitochondrion		✔ ---- ✔✔ ---- 0.984 ---- 99	Sickmann Ohlmeier 17a(2)/17c(1)/ Lascaris			2720	98
3	YBL064C (PRX1)	also called mTPx1, a mitochondrial isoform of thioredoxin peroxidase (EC 1.11.1.-)	mitochondrial (SGD)	viable	cyto., nucleus / mitochondrion / mitochondrion		✔ ---- ✔✔ ---- 0.991 ---- 99	Pflieger Prokisch Sickmann Ohlmeier 15(10)/17a(11)/17b(17)/17c(17)/ DeRisi Prokisch	human Rickettsia mouse Neurospora		4510	98
4	YDR041W (RSM10)	Mitochondrial ribosomal protein of the small subunit, has similarity to E. coli S10 ribosomal protein; essential for viability, unlike most other mitoribosomal proteins	mitochondrial (SGD)	essential	none / mitochondrion	MLR:88	✔ ---- ✔✔ ---- 0.848 ---- 100	Prokisch Sickmann 15(3)/17a(1)/17b(2)/17c(6)/ Lascaris	human Rickettsia Neurospora	high medium	8480	98
5	YMR089C (RCA1 YTA12)	Component, with Afg3p, of the mitochondrial inner membrane m-AAA protease that mediates degradation of misfolded or unassembled proteins and is also required for correct assembly of mitochondrial enzyme complexes	mitochondrial (SGD)	pet class III	cyto. / mitochondrion / mitochondrion		✔ ---- ✔✔ ---- 0.999 ---- 100	Prokisch Sickmann 15(4)/17a(25)/17b(3)/17c(5)/ Prokisch(N. crassa) Lascaris	human Rickettsia E. cuni. mouse Neurospora	medium	11500	98
6	YER141W (COX15)	Protein required for the hydroxylation of heme O to form heme A, which is an essential prosthetic group for cytochrome c oxidase	mitochondrial (SGD)	pet class III	mitochondrion / mitochondrion	MLR:92	---- ✔✔ ---- 1.0 ---- 100	Pflieger Prokisch Sickmann 15(1)/17a(9)/17b(1)/ DeRisi	human Rickettsia mouse	medium	n.d.	98

Fig. 2. Section of an example result of a query in the yeast section of the MitoP2 database. MitoP2 extracted 233 of 6516 yeast proteins predicted to be in the mitochondria by the Predotar program *(19)* and localized to mitochondria by the high-throughput subcellular localization of Huh et al. *(2)*.

according to von Mering et al. ***(15)*** in addition to protein–protein interactions from single experiments. The high-throughput interaction data sets are provided with links to the original literature.

The annotated reference set for *S. cerevisiae* is based on single gene studies only and does not contain any information from any of the integrated high-throughput approaches. Thus, it allows a benchmarking of them because no bias toward either of the discussed high-throughput approaches was generated. Owing to the incompleteness of the reference set itself, the calculated specificities and sensitivities are conservative estimates (**Table 1**; *see* **Note 2**).

The power of identifying mitochondrial proteins can be increased by combing various genomewide data sets. A predictive score (MitoP2 score) based on the specificity of the best combination of approaches yields a sensitivity and specificity higher than those achieved by any single approach (*see* **Table 1**) ***(1)***.

Table 1
Comparison of Specificity and Sensitivity for Various Approaches Integrated in MitoP2 in Determining the Mitochondrial Localization of Proteins

Source	Total proteins	Specificity (*see* **Note 3**)	Sensitivity (*see* **Note 3**)
S. cerevisiae data sets			
MitoProt II score > 0.8 ***(16)***	790	35%	83%
MITOPRED score > 80 ***(17)***	1045	34%	68%
PSORT II ***(18)***	981	27%	51%
Predotar ***(19)***	832	36%	58%
Bayesian prediction ***(31)***	500	42%	40%
Deletion phenotype ***(8)***	381	50%	37%
Deletion phenotype ***(9)***	466	51%	45%
Mitopolysomes ***(32)***	303	23%	13%
Ysublocalisation_01 ***(3)***	364	64%	45%
Ysublocalisation_02 ***(2)***	527	68%	69%
Yproteome_01 ***(12)***	177	79%	27%
Yproteome_02 ***(1)***	1357	31%	83%
Yproteome_03 ***(13)***	749	51%	73%
Yproteome_04 ***(33)***	252	61%	29%
Ytranscriptome_01 ***(1)***	546	50%	52%
Ytranscriptome_02 ***(10)***	416	19%	15%
Ytranscriptome_03 ***(11)***	514	43%	43%
Human mitochondrial ortholog[a]	565	60%	65%
Mouse mitochondrial ortholog[a]	425	68%	55%
N. crassa mitochondrial ortholog[a]	337	84%	55%
MitoP2 score > 90	691	61%	81%
MitoP2 score > 95	395	78%	75%
HUMAN data sets			
MitoProt II score > 0.8 ***(16)***	2559	12%	43%
MITOPRED score > 80 ***(17)***	2892	15%	61%
PSORT II ***(18)***	6125	5%	45%
Predotar ***(19)***	2139	14%	44%
Hproteome_01 ***(27)***	736	37%	38%
Mproteome_01 ***(28)***	156	83%	10%
Mproteome_02 ***(29)***	478	60%	31%
MSublocalisation_01 ***(30)***	566	26%	80%
S. cerevisiae mitochondrial ortholog[a]	854	40%	47%
N. crassa mitochondrial ortholog[a]	523	48%	35%
No *E. cuniculi* ortholog[a]	38890	12%	14%
R. prowazeckii ortholog[a]	1426	14%	30%
MitoP2 score > 70	1002	52%	73%

[a]Defined as bidirectional best BLAST hit or best BLAST hit < E^{-10}.

A more detailed description for the calculation of the MitoP2 score is given in **Note 3** and in the database itself.

3. Exploring the Mammalian Mitochondrial Proteome

In contrast to the situation in yeast, the mammalian mitochondrial proteome is about twice as large but with much fewer data available from high-throughput experiments. The current list of human and mouse mitochondrial proteins in the MitoP2 database contains about 700 manually annotated entries that cover about half of the expected ones. More than 120 of these proteins are shown to be involved in human diseases; most of these proteins take part in the metabolism of amino acids, nucleic acid, lipids, heme, or coenzymes (MitoP2, OMIM [Online Mendelian Inheritance in Man]). In addition, defects in the mitochondrial respiratory chain/oxidative phosphorylation system are responsible for a panoply of human disorders, ranging from sporadic myopathies to fatal encephalomyopathies. Recent epidemiological studies showed that disorders of the mitochondrial respiratory chain, the classical mitochondrial diseases, affect at least 1 in 5000 of the population, making these disorders a common genetically determined disease entity *(22)*. About half of the patients carry mutations in the mitochondrial DNA (mtDNA); so far, more than 100 different pathogenic mtDNA point mutations and an even larger number of different mtDNA deletions have been found. The causes of a large proportion of the remaining 50% with nuclear mutations have yet to be determined.

The human and mouse sections of the MitoP2 data sets comprise 44,996 and 27,628 protein entries, respectively, extracted from the "nonredundant" Swiss-Prot data sets (last release no. 45 from 2004) *(23)*. Human-mouse orthologs were determined by a bidirectional best BLAST hit or best BLAST hit with a score $< E^{-10}$. The orthologs are used equally for both the annotation of the human and mouse mitochondrial reference sets and the integration of high-throughput data sets.

For each protein, the description, the chromosomal localization, the subcellular localization, and the cross references and literature were extracted from the Swiss-Prot database. Functional annotations were compiled from GO. The mitochondrial reference set proteins from mouse were also annotated according to the MIPS functional catalog *(7)* supplemented by the physical features, including the DNA and protein sequence. In these cases, an explanation or PubMed links for the mitochondrial annotation is given. Human proteins already associated with Mendelian disorders are extracted from OMIM and listed with the OMIM title *(24)*. They are marked in red if they are mitochondrial or in orange if not. In addition, the mouse entries are listed with phenotypes if a mouse model exists in the Mouse Genome Database (MGD) *(25)*. So far, more than 50 mouse models are investigated, having mutations in genes

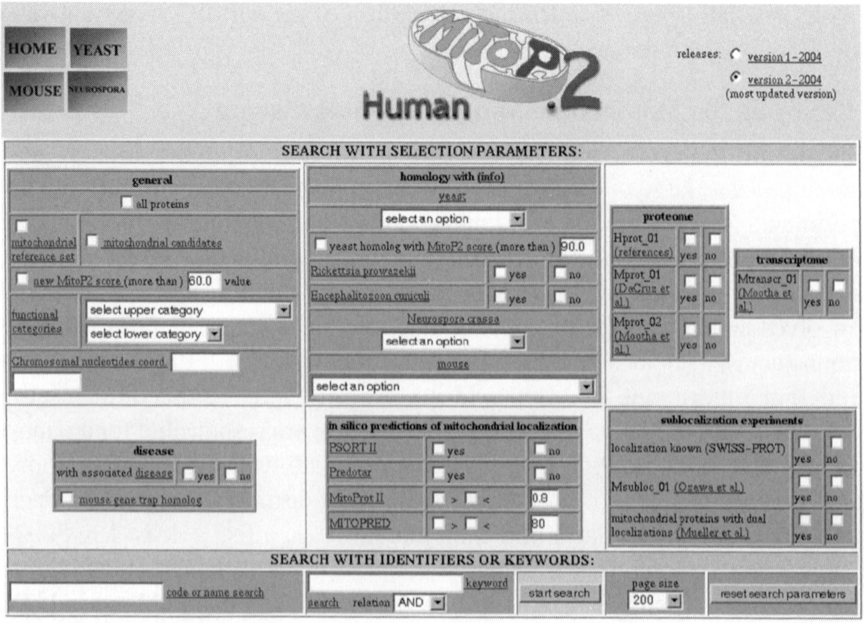

Fig. 3. Search mask of the human section of the MitoP2 database at http://www.mitop2.de/. The search mask is structured in several search boxes. All search options (underlined) are linked to the original literature or to an explanation for this selection.

coding mitochondrial proteins. For those users interested in creating mouse models, available mouse gene trap insertion cell lines from the International Gene Trap Consortium are listed as well (www.genetrap.org). All features are regularly updated.

Both the mouse and the human search masks contain the same search options (**Fig. 3**). The proteome analysis of mitochondria purified from heart tissue is so far the only high-throughput experiment available for humans and has been integrated under the category "proteome" ***(26,27)***. Three additional high-throughput analyses performed in the mouse have been added. They include two proteome experiments ***(28,29)*** and one green fluorescent protein subcellular localization study ***(30)***. Because not all mouse proteins have clear human orthologs and not all predicted proteins have entries in Swiss-Prot, the number of entries in human and mouse differ and are partly incomplete. The search mask includes homology searches between human, yeast, *E. cuniculi*, and *R. prowazekii*. In addition, it is possible to include searches with established algorithms [MitoProt II ***(16)***, PSORT II ***(18)***, Predotar ***(19)***, and MITOPRED ***(17)***] to predict the subcellular localization of a protein based on the amino acid sequence.

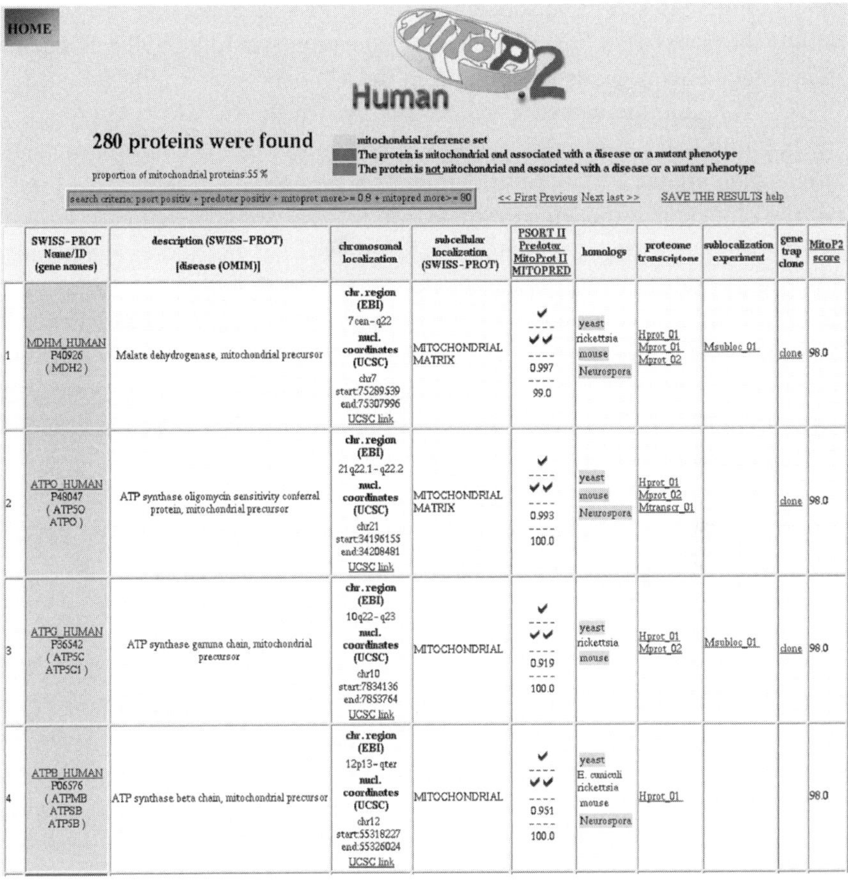

Fig. 4. Section of the result page of a query in the HUMAN section of the MitoP2 database. In this case, the search criteria were: PSORTII *(18)*, Predotar *(19)*, MitoProt II *(16)*, and MITOPRED *(17)*. MitoP2 extracts 280 human proteins, of which 55% are known to be mitochondrial ones.

Another category that can be included or excluded in a database search is gene neighborhood data based on similar expression profiles ("transcriptome") *(29)*. Again, the search mask allows single or combined searches for all these components.

By selecting, for instance, only PSORT II on the human search mask, MitoP2 records 6124 proteins, including 329 entries from the reference set (5%). By combining all four *in silico* predictions, MitoP2 lists 280 proteins, of which 55% are known mitochondrial ones (**Fig. 4**), demonstrating that the combination of evidence increases the specificity but reduces the sensitivity. Each entry in the resulting matrix is shown with its ID (Swiss-Prot primary accession number); protein description; subcellular localization as annotated in Swiss-Prot, including the Swiss-Prot link; chromosomal location according to the cytogenetic map and the sequence position linked to the University of California at Santa Cruz (UCSC) genome browser, and if it is known the OMIM title *(24)* of the corresponding disease. In addition, this page summarizes the information from *in silico* calculations, the high-throughput experiments, the availability of mouse gene trap clones, and the MitoP2 score.

A common question in human genetics that can be addressed by using MitoP2 is the identification of candidate genes. Elpeleg et al. *(5)*, for example, mapped an encephalomyophathy locus to a 21-Mb region on chromosome 13. The chromosome coordinates (i.e., 13:40878920 and 13:61359487) were used as selection criteria in MitoP2. In combination with a MitoP2 score greater than 60, MitoP2 lists 3 proteins of 113 ORFs in this region, two candidates and one well-characterized reference set protein identified in two proteome experiments (**Fig. 5**). The reference set entry, encoding the β-chain of the adenosine 5′-diphosphate (ADP)-forming succinyl coenzyme A (CoA) ligase, turned out to be mutated.

Figure 6 represents the entry for this succinyl CoA ligase as an example for a single protein entry in the human section of MitoP2, which provides in addition to the known matrix lane (1) a list of homologies to sequences from other species, including the BLAST E value and the percentage of the aligned protein length, linked to the corresponding MitoP2 pages; (2) if existent, a Swiss-Prot description of an associated disease; (3) the corresponding GO annotations for molecular protein function, biological processes in which the protein is involved, and cellular components; (4) the available literature about the protein and protein variants listed with authors and title; and (5) a table of cross references annotated in Swiss-Prot.

4. Notes

1. The following are the URLs for databases mentioned in the text: MitoP2, http://www.mitop2.de/; SGD, http://www.yeastgenome.org/; CYGD, http://mips.gsf.de/genre/proj/yeast/; Swiss-Prot, http://www.expasy.org/sprot/; UCSC genome browser, http://genome.ucsc.edu/; OMIM, http://www.ncbi.nlm. nih.gov/entrez/query.fcgi?db=OMIM; GO, http://www.geneontology.org/; MGD, http://www.informatics.jax.org/; Gene Trap, http://www.genetrap.org.
2. It seems intuitively evident that an approach to find mitochondrial-localized proteins, which finds all proteins present in a cell, has a high sensitivity but a poor specificity, and that another approach that finds only 20 proteins, which are all certainly mitochondrial located, has a high specificity but a low sensitivity. The specificity of an approach to identify mitochondrial-localized proteins is estimated by the proportion (in percentage) of the found proteins that are part of the mitochondrial reference set. The sensitivity of an approach to find

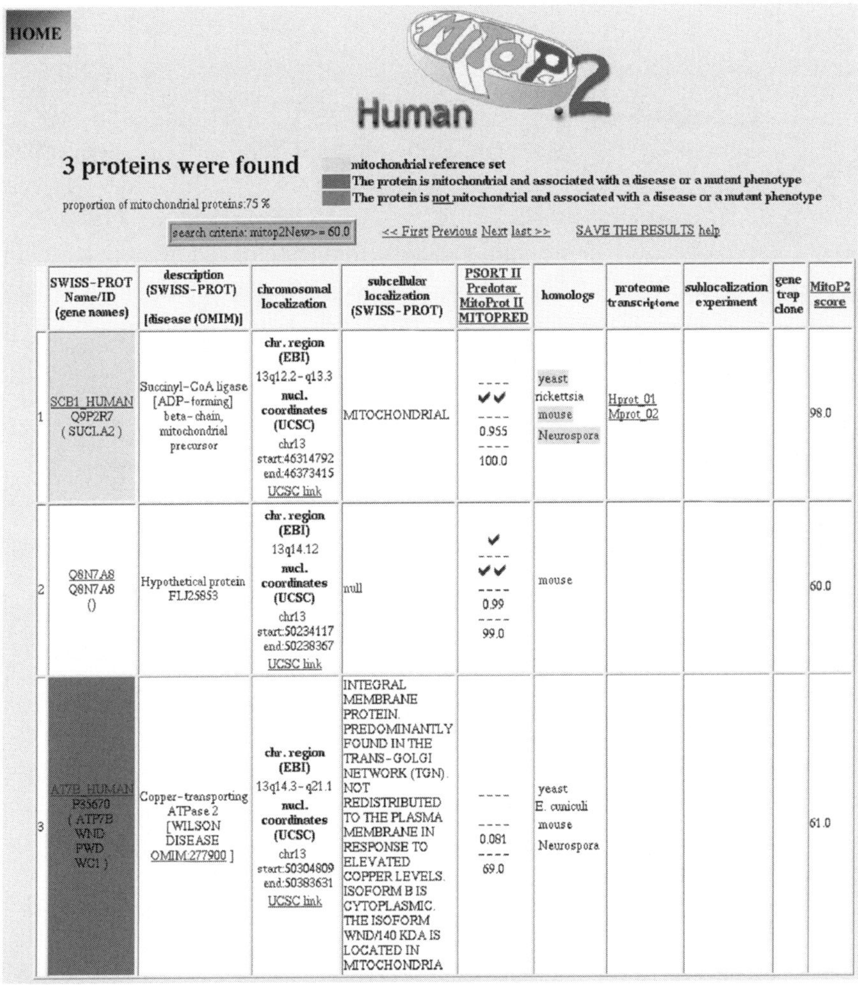

	SWISS-PROT Name/ID (gene names)	description (SWISS-PROT) [disease (OMIM)]	chromosomal localization	subcellular localization (SWISS-PROT)	PSORT II Predotar MitoProt II MITOPRED	homologs	proteome transcriptome	sublocalization experiment	gene trap clone	MitoP2 score
1	SCB1_HUMAN Q9P2R7 (SUCLA2)	Succinyl-CoA ligase [ADP-forming] beta-chain, mitochondrial precursor	**chr. region (EBI)** 13q12.2-q13.3 **nucl. coordinates (UCSC)** chr13 start:46314792 end:46373415 UCSC link	MITOCHONDRIAL	---- ✔✔ ---- 0.955 ---- 100.0	yeast rickettsia mouse Neurospora	Hprot_01 Mprot_02			98.0
2	Q8N7A8 Q8N7A8 ()	Hypothetical protein FLJ25853	**chr. region (EBI)** 13q14.12 **nucl. coordinates (UCSC)** chr13 start:50234117 end:50238367 UCSC link	null	✔ ---- ✔✔ ---- 0.99 ---- 99.0	mouse				60.0
3	AT7B_HUMAN P35670 (ATP7B WND PWD WC1)	Copper-transporting ATPase 2 [WILSON DISEASE OMIM:277900]	**chr. region (EBI)** 13q14.3-q21.1 **nucl. coordinates (UCSC)** chr13 start:50304809 end:50383631 UCSC link	INTEGRAL MEMBRANE PROTEIN. PREDOMINANTLY FOUND IN THE TRANS-GOLGI NETWORK (TGN). NOT REDISTRIBUTED TO THE PLASMA MEMBRANE IN RESPONSE TO ELEVATED COPPER LEVELS. ISOFORM B IS CYTOPLASMIC. THE ISOFORM WND/140 KDA IS LOCATED IN MITOCHONDRIA	---- ---- 0.081 ---- 69.0	yeast E. cuniculi mouse Neurospora				61.0

Fig. 5. MitoP2 can be used to identify mitochondrial proteins and candidates in a specific chromosomal region. In this case, the search criteria were the chromosomal region (13:40878920 and 13:61359487) and a MitoP2 score above 60 as performed by Elpeleg et al. *(5)*.

mitochondrial-localized proteins is estimated by the proportion (in percentage) of reference set proteins that is covered by the approach. Owing to the incompleteness of the reference sets, the specificities and sensitivities as defined here are conservative estimates.

3. The MitoP2 score is defined as follows: the percentage of known mitochondrial proteins from a set of proteins identified in a certain genomewide experiment (specificity) or in the overlap of several data sets (specificity of the combination

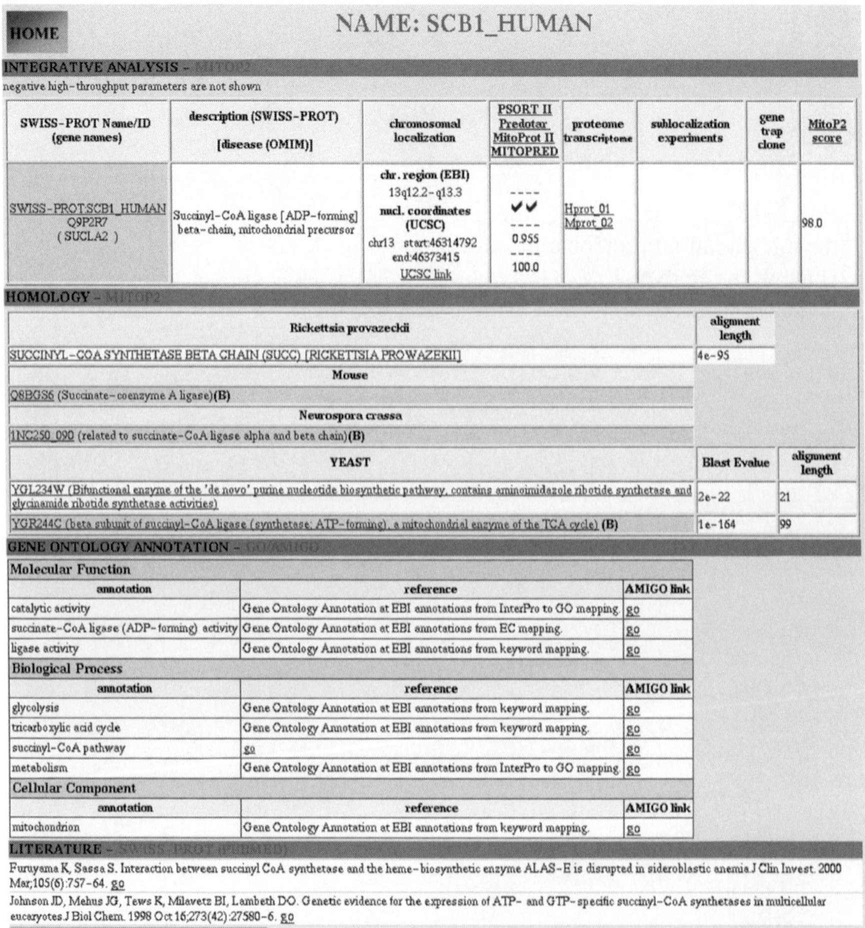

Fig. 6. Single protein entry for human succinyl CoA ligase in the MitoP2 database. MitoP2 presents for each protein entry the Swiss-Prot name and description, the chromosomal localization, the prediction results from mitochondrial prediction programs, homologous proteins in other organisms, gene ontology annotations, and relevant literature.

of several methods) was calculated. Most proteins belonged to more than one combination, and for those proteins multiple R values were calculated. For example, proteins identified by two approaches received three R values: the specificity of the first approach alone, the specificity of the second approach alone, and the specificity among the overlap of both approaches. The MitoP2 score was chosen to represent the highest R value calculated for a protein. The number gives a lower limit of the specificity of a defined combination as the mitochondrial reference data set is not complete.

Acknowledgments

The MitoP2 project is funded by the BMBF projects NGFN2 (Nationales Genomforschungsnetz 2) and BFAM (Bioinformatics for the Functional Analysis of Mammalian Genomes) and the MitEURO consortium.

References

1. Prokisch, H., Scharfe, C., Camp, D. G., 2nd, et al. (2004) Integrative analysis of the mitochondrial proteome in yeast. *PLoS Biol.* **2,** e160.
2. Huh, W. K., Falvo, J. V., Gerke, L. C., et al. (2003) Global analysis of protein localization in budding yeast. *Nature* **425,** 686–691.
3. Kumar, A., Cheung, K. H., Tosches, N., et al. (2002) The TRIPLES database: a community resource for yeast molecular biology. *Nucleic Acids Res.* **30,** 73–75.
4. Andreoli, C., Prokisch, H., Hortnagel, K., et al. (2004) MitoP2, an integrated database on mitochondrial proteins in yeast and man. *Nucleic Acids Res.* **32,** D459–D462.
5. Elpeleg, O., Miller, C., Hershkovitz, E., et al. (2005) Deficiency of the ADP-forming succinyl-CoA synthase activity is associated with encephalomyopathy and mitochondrial DNA depletion. *Am. J. Hum. Genet.* **76,** 1081–1086.
6. Cherry, J. M., Ball, C., Weng, S., et al. (1997) Genetic and physical maps of *Saccharomyces cerevisiae. Nature* **387,** 67–73.
7. Ruepp, A., Zollner, A., Maier, D., et al. (2004) The FunCat, a functional annotation scheme for systematic classification of proteins from whole genomes. *Nucleic Acids Res.* **32,** 5539–5545.
8. Dimmer, K. S., Fritz, S., Fuchs, F., et al. (2002) Genetic basis of mitochondrial function and morphology in *Saccharomyces cerevisiae. Mol. Biol. Cell.* **13,** 847–853.
9. Steinmetz, L. M., Scharfe, C., Deutschbauer, A. M., et al. (2002) Systematic screen for human disease genes in yeast. *Nat. Genet.* **31,** 400–404.
10. DeRisi, J. L., Iyer, V. R., and Brown, P. O. (1997) Exploring the metabolic and genetic control of gene expression on a genomic scale. *Science* **278,** 680–686.
11. Lascaris, R., Bussemaker, H. J., Boorsma, A., et al. (2003) Hap4p overexpression in glucose-grown *Saccharomyces cerevisiae* induces cells to enter a novel metabolic state. *Genome Biol.* **4,** R3.
12. Pflieger, D., Le Caer, J. P., Lemaire, C., Bernard, B. A., Dujardin, G., and Rossier, J. (2002) Systematic identification of mitochondrial proteins by LC-MS/MS. *Anal. Chem.* **74,** 2400–2406.
13. Sickmann, A., Reinders, J., Wagner, Y., et al. (2003) The proteome of *Saccharomyces cerevisiae* mitochondria. *Proc. Natl. Acad. Sci. USA* **100,** 13,207–13,212.
14. Ghaemmaghami, S., Huh, W. K., Bower, K., et al. (2003) Global analysis of protein expression in yeast. *Nature* **425,** 737–741.
15. von Mering, C., Krause, R., Snel, B., et al. (2002) Comparative assessment of large-scale data sets of protein-protein interactions. *Nature* **417,** 399–403.
16. Claros, M. G. (1995) MitoProt, a Macintosh application for studying mitochondrial proteins. *Comput. Appl. Biosci.* **11,** 441–447.
17. Guda, C., Fahy, E., and Subramaniam, S. (2004) MITOPRED: a genome-scale method for prediction of nucleus-encoded mitochondrial proteins. *Bioinformatics* **20,** 1785–1794.

18. Nakai, K. and Horton, P. (1999) PSORT: a program for detecting sorting signals in proteins and predicting their subcellular localization. *Trends Biochem. Sci.* **24,** 34–36.
19. Small, I., Peeters, N., Legeai, F., and Lurin, C. (2004) Predotar: A tool for rapidly screening proteomes for N-terminal targeting sequences. *Proteomics* **4,** 1581–1590.
20. Altschul, S. F., Madden, T. L., Schaffer, A. A., et al. (1997) Gapped BLAST and PSI-BLAST: a new generation of protein database search programs. *Nucleic Acids Res.* **25,** 3389–3402.
21. Ashburner, M., Ball, C. A., Blake, J. A., et al. (2000) Gene ontology: tool for the unification of biology. The Gene Ontology Consortium. *Nat Genet.* **25,** 25–29.
22. Schaefer, A. M., Taylor, R. W., Turnbull, D. M., and Chinnery, P. F. (2004) The epidemiology of mitochondrial disorders—past, present and future. *Biochim. Biophys. Acta.* **1659,** 115–120.
23. Boeckmann, B., Bairoch, A., Apweiler, R., et al. (2003) The SWISS-PROT protein knowledgebase and its supplement TrEMBL in 2003. *Nucleic Acids Res.* **31,** 365–370.
24. Bult, C. J., Blake, J. A., Richardson, J. E., et al. (2004) The Mouse Genome Database (MGD): integrating biology with the genome. *Nucleic Acids Res.* **32,** D476–D481.
25. Gaucher, S. P., Taylor, S. W., Fahy, E., et al. (2004) Expanded coverage of the human heart mitochondrial proteome using multidimensional liquid chromatography coupled with tandem mass spectrometry. *J. Proteome Res.* **3,** 495–505.
26. Taylor, S.W., Fahy, E., Zhang, B., et al. (2003) Characterization of the human heart mitochondrial proteome. *Nat Biotechnol.* **21,** 281–286.
27. Da Cruz, S., Xenarios, I., Langridge, J., Vilbois, F., Parone, P. A., and Martinou, J. C. (2003) Proteomic analysis of the mouse liver mitochondrial inner membrane. *J. Biol. Chem.* **278,** 41,566–41,571.
28. Mootha, V. K., Bunkenborg, J., Olsen, J. V., et al. (2003) Integrated analysis of protein composition, tissue diversity, and gene regulation in mouse mitochondria. *Cell* **115,** 629–640.
29. Ozawa, T., Sako, Y., Sato, M., Kitamura, T., and Umezawa, Y. (2003) A genetic approach to identifying mitochondrial proteins. *Nat. Biotechnol.* **21,** 287–293.
30. Hamosh, A., Scott, A. F., Amberger, J., Bocchini, C., Valle, D., and McKusick, V.A. (2002) Online Mendelian Inheritance in Man (OMIM), a knowledgebase of human genes and genetic disorders. *Nucleic Acids Res.* **30,** 52–55.
31. Drawid, A. and Gerstein, M. (2000) A Bayesian system integrating expression data with sequence patterns for localizing proteins: comprehensive application to the yeast genome. *J. Mol. Biol.* **301,** 1059–1075.
32. Marc, P., Margeot, A., Devaux, F., Blugeon, C., Corral-Debrinski, M., and Jacq, C. (2002) Genome-wide analysis of mRNAs targeted to yeast mitochondria. *EMBO Rep.* **3,** 159–164.
33. Ohlmeier, S., Kastaniotis, A. J., Hiltunen, J. K., and Bergmann, U. (2004) The yeast mitochondrial proteome, a study of fermentative and respiratory growth. *J. Biol. Chem.* **279,** 3956–3979.

Index